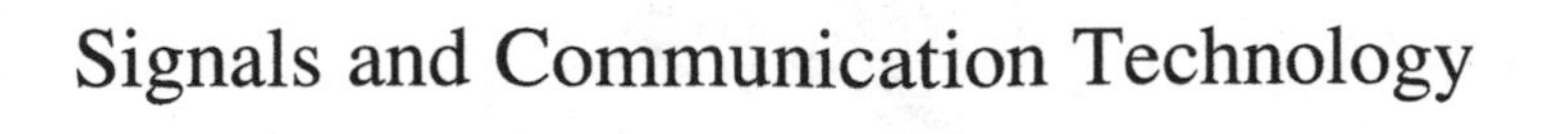
Signals and Communication Technology

For further volumes:
http://www.springer.com/series/4748

Aurelio Uncini

Fundamentals of Adaptive Signal Processing

Aurelio Uncini
DIET
Sapienza University of Rome
Rome
Italy

ISSN 1860-4862 ISSN 1860-4870 (electronic)
ISBN 978-3-319-35341-8 ISBN 978-3-319-02807-1 (eBook)
DOI 10.1007/978-3-319-02807-1
Springer Cham Heidelberg New York Dordrecht London

Printed on acid-free paper

Springer is part of Springer Science+Business Media (www.springer.com)

Preface

Since the ancient times of human history, there have been many attempts to define intelligence. Aristotle argued that all persons express similar intellectual faculties and that the differences were due to the teaching and example. In more recent times, intelligence has been defined as a set of innate cognitive functions, adaptive, imaginative, etc., arising from a human or animal biological brain. Among these, the capacity of adaptation is the main prerogative present in all definitions of intelligent behavior. From the biological point of view, adaptation is a property that all living organisms possess and that can be interpreted both as a propensity to the species improvement and as a conservative process tending to the species preservation over time of the life. From the psychological point of view, adaptation is synonymous with learning. In this sense, learning is a behavioral function, more or less conscious, of a subject that adapts her or his attitude as a result of experience: *learning is to adapt*.

In intelligent systems, whether biologically inspired, or entirely artificial, adaptation and methods to carry it out represent an essential prerogative. In this framework, adaptive filters are defined as information processing systems, analog or digital, capable of autonomously adjusting their parameters in response to external stimuli. In other words, the system learns independently and adapts its parameters to achieve a certain processing goal such as extracting the useful information from an acquired signal and the removal of disturbances due to noise or other sources interfering or, more generally, the adaptive filter provides the elimination of the redundant information. In fact, in support of this hypothesis, the British neuroscientist Horace B. Barlow in 1953 discovered that the frog brain has neurons which fire in response to specific visual stimuli and concluded that one of the main aims of visual processing is the reduction of redundancy. His works have been milestones in the study of the properties of the biological nervous system. Indeed, his researches demonstrate that the main function of a machine perception is to eliminate redundant information coming from receptors.

The usability of *adaptive signal processing* methods to the solution of real problems is extensive and represents a paradigm for many *strategic* applications. Adaptive signal processing methods are used in economic and financial sciences, in engineering and social sciences, in medicine, biology, and neuroscience, and in many other areas of high strategic interest. Adaptive signal processing is also a very active field of study and research that, for a thorough understanding, requires advanced interdisciplinary knowledge.

Objectives of the Text

The aim of this book is to provide advanced theoretical and practical tools for the study and determination of circuit structures and robust algorithms for the adaptive signals processing in different application scenarios. Example can be found in multimodal and multimedia communications, the biological and biomedical areas, economic model, environmental sciences, acoustics, telecommunications, remote sensing, monitoring, and, in general, modeling and prediction of complex physical phenomena.

In particular, in addition to presenting the fundamental theoretical base concepts, the most important adaptive algorithms are introduced, while also providing tools to evaluate the algorithms' performance. The reader, in addition to acquiring the basic theories, will be able to design and implement the algorithms and evaluate their performance for specific applications.

The idea of the text is based on years of teaching activities of the author, during the course *Algorithms for Adaptive Signal Processing* held at the Faculty of Information Engineering of "Sapienza" University of Rome. In preparing the book, particular attention was paid in the first chapters and in mathematics appendices, which make the text suitable to their readers without special prerequisites, other than those common to all first 3-year courses of the Information Engineering Faculty and other scientific faculties.

Adaptive filters are nonstationary nonlinear and time-varying dynamic systems, and, at times, to avoid a simplistic approach, the arguments may have some conceits that might be difficult to understand. For this reason, many of the subjects are introduced by considering different points of view and with multiple levels of analysis.

In the literature on this topic numerous and authoritative texts are available. The reasons that led to the writing of this work are linked to a philosophically different vision of intelligent signal processing. In fact, adaptive filtering methods can be introduced starting from different theories. In this work we wanted to avoid an "ideological" approach related to some specific discipline, but we wanted to put an emphasis on interdisciplinarity presenting the most important topics with different paradigms.

For example, a central importance argument, as the least mean squares (LMS) algorithm, is exposed in three distinct and, to some extent, original ways. In the first, following a more systemistic criteria, the LMS is presented by considering an energy approach through the Lyapunov attractor. In the second, with a "classic" statistical approach, is introduced the stochastic approximation of the gradient-descent optimization method. In the third mode, following a different approach, is presented considering the simple axiomatic properties of minimal perturbation. Moreover, it should be noted that this philosophy is not only a pedagogical exercise, but it is of fundamental importance in more advanced topics and theoretical demonstrations where, following a philosophy rather than the other, it happens very often to tread winding roads and dead end.

Organization and Structure of the Book

The sequence of arguments is presented in classical mode. The first part introduces the basic concepts of optimal linear filtering. Following the first and second order, online and batch processing techniques are presented. A particular effort has been made in trying to present arguments with a common formalism, while trying to remain faithful to the considered original references. The entire notation is defined in discrete time and the algorithms are presented in order to facilitate the reader to the writing of computer programs, for the practical realization of the applications described in the text.

The book consists of nine chapters, each one containing the references where the reader can independently deepen topics of particular interest, and three mathematics appendices.

Chapter 1 covers preliminary topics of the discrete-time signals and circuits and some basic methods of digital signal processing.

Chapter 2 introduces the basic definitions of adaptive filtering theory and the main filters topologies are discussed. In addition, the concept of cost function to be minimized and the main philosophies concerning adaptation methods are introduced. Finally, the main application fields of adaptive signal processing techniques are presented and discussed.

In Chap. 3, the Wiener optimal filtering theory is presented. In particular, the problems of the mean square error minimization and of its optimal value determination are addressed. The formulation of the normal equations and the optimal Wiener filter in discrete time is introduced. Moreover, the type 1, 2, and 3 multichannel notations and its multi-input-output optimal filter generalization are presented. Are also discussed corollaries, and presented some applications related to the random sequences prediction and estimation.

In Chap. 4, adaptation methods, in the case that the input signals are not statistically characterized, are addressed. The principle of least squares (LS), bringing back the estimation problem into an optimization algorithm, is introduced. The normal equations in the Yule–Walker formulation are introduced and the similarities and differences with Wiener optimal filtering theory are also discussed. Moreover, the minimum variance optimal estimators, the normal equations weighing techniques, the regularization LS approach, and the linearly constrained and the nonlinear LS techniques are introduced. The algebraic methods to matrix decomposition for solving the LS systems in the cases and of over/under-determined equations system are also introduced and discussed. The technique of singular value decomposition in the solution of the LS systems is discussed. The method of Lyapunov attractor for the iterative LS solution is presented, and the least mean squares and Kaczmarz algorithms, seen as an iterative LS solution, are introduced. Finally, the methodology of total least squares (TLS) and the matching pursuit algorithms for underdetermined sparse LS systems are presented and discussed.

Chapter 5 introduces the first-order adaptation algorithms for online adaptive filtering. The methods are presented with a classical statistical approach and the LMS algorithm with the stochastic gradient paradigm. In addition, methods for performance evaluation of adaptation algorithms, with particular reference to the convergence speed and tracking analysis, are presented and discussed. Some general axiomatic properties of the adaptive filters are introduced. Moreover, the methodology of stochastic difference equations, as a general method for evaluating the performance of online adaptation algorithms, is introduced. Finally, variants of the LMS algorithm, some multichannel algorithms applications, and delayed learning algorithms, such as the class filtered-x LMS in its various forms, the method filtered error LMS, and the method of the adjoint network, are presented and discussed.

In Chap. 6, the most important second-order algorithms for the solution of LS equations with recursive methods are introduced. In the first part of the chapter, the Newton's method and its version with time-average correlation estimation, defining the class of adaptive algorithms such as sequential regression, are briefly exposed. Subsequently, in the context of the second-order algorithms, a variant of the NLMS algorithm, called affine projection algorithm (APA), is presented. Thereafter the family of algorithms called recursive least squares (RLS) is presented, and their convergence characteristics are studied. In the following, some RLS variants and generalizations as the Kalman filter are presented. Moreover, some criteria to study the performance of adaptive algorithms operating in nonstationary environments are introduced. Finally, a more general adaptation law based on natural gradient approach, considering sparsity constraints, is briefly introduced.

In Chap. 7, structures and algorithms for the implementation of adaptive filters in batch and online mode, operating in transformed domain (typically the frequency domain), are introduced. In the first part of the chapter, the block LMS algorithm is introduced. Successively, two paragraphs about the frequency domain constrained algorithms known as frequency domain adaptive filters (FDAF), the unconstrained FDAF, and the partitioned FDAF are introduced. In the third paragraph, the transformed domain adaptive algorithms, referred to as transform-domain adaptive filters (TDAF), are presented. The chapter also introduces the multirate methods and the subband adaptive filters (SAFs).

In Chap. 8, the forward and backward linear prediction and the issue of the order recursive algorithms are considered. Both of these topics are related to implementative structures with particular robustness and efficiency properties. In connection with this last aspect, the subject of the filter circuit structure and the adaptation algorithm is introduced, in relation to the problems of noise control, scaling and efficient computation, and effects due to coefficients quantization.

Chapter 9 introduces the problem space-time domain adaptive filtering, in which the signals are acquired by homogeneous sensor arrays, arranged in different spatial positions. This issue, known in the literature as array processing (AP), is of fundamental interest in many application fields. In particular, the basic concepts of discrete space-time filtering are introduced. The first part of the chapter introduces the basics of the anechoic and echoic wave propagation model, the sensors directivity functions, the signal model, and steering vectors of some typical array geometries. The characteristics of noise field in various application contexts and the array quality indices are also discussed. In the second part of the chapter, methods for conventional beamforming are introduced, and the radiation characteristics are discussed, the main design criteria in relation to the optimization of quality indices. Moreover, the broadband beamformer with spectral decomposition and the methods of direct synthesis of the spatial response are introduced and discussed. In the third part of the chapter, the statistically optimal static beamforming is introduced. The LS methodology is extended in order to minimize the interference related to the noise field. In addition, the super-directive methods, the related regularized solution techniques, and the post-filtering method are discussed. The minimum variance broadband method (the Frost algorithm) is also presented. In the fourth part, the adaptive mode for the determination of the online beamforming operating nonstationary signal condition is presented. In the final part of the chapter, the issue of the time-delay estimation (TDE) and direction of arrival (DOA) estimation in the case of free-field narrow-band signals and in the case of broadband signals in reverberant environment is presented.

In addition, in order to have a possible self-contained text there are three appendices, with a common formalism to all the arguments, that recall to the reader some basic necessary prerequisites for a proper understanding of the topics covered in this book.

In Appendix A, some basic concepts and quick reference of linear algebra are recalled.

In Appendix B, the basic concepts of the nonlinear programming are briefly introduced. In particular, some fundamental concepts of the unconstrained and the constrained optimization methods are presented.

Finally, in Appendix C some basic concepts on random variables, stochastic processes, and estimation theory are recalled.

For editorial choice further study and insights, exercises, project proposals, the study of real application, and a library containing MATLAB (® registered trademark of The MathWorks, Inc.) codes for the calculation of main algorithms discussed in the this text are inserted into a second volume which is currently being written.

Additional materials to the text can be found at: http://www.uncini.com/FASP

Rome, Italy Aurelio Uncini

Acknowledgments

Many colleagues have contributed to the creation of this book giving useful tips, reading the drafts, or enduring my musings on the subject.

I wish to thank my collaborators, Raffaele Parisi and Michele Scarpiniti, of the Department of Information Engineering, Electronic and Telecommunication (DIET) of "Sapienza" University of Rome, and the colleagues from other universities: Stefano Squartini of the Polytechnic University of Marche—Italy; Alberto Carini of the University of Urbino—Italy; and Francesco Palmieri of the Second University of Naples—Italy; Gino Baldi of KPMG.

I would also like to thank all students and thesis students attending the research laboratory Intelligent Signal Processing & Multimedia Lab (ISPAMM LAB) at the DIET, where they have been implemented and compared many of the algorithms presented in the text. A special thanks goes to PhD students and Post Doc researchers, Danilo Comminiello and Simone Scardapane, who carried out an effective proofreading.

A special thanks to all the authors in the bibliography of each chapter. This book is formed by a mosaic of arguments, where each tile is made up of one atom of knowledge. My original contribution, if they are successful in my work, is only in the vision of the whole, i.e., in the picture that emerges from the mosaic of this knowledge.

Finally, a special thanks goes to my wife Silvia and my daughter Claudia to whom I subtracted a lot of my time and who have supported me during the writing of the work. The book is dedicated to them.

Abbreviations and Acronyms

$\varnothing$	Empty set
$\mathbb{Z}$	Integer number
$\mathbb{R}$	Real number
$\mathbb{C}$	Complex number
$(\mathbb{R},\mathbb{C})$	Real or complex number
acf	Autocorrelation function
AD-LMS	Adjoint LMS
AEC	Adaptive echo canceller
AF	Adaptive filter
AIC	Adaptive interference canceller
ALE	Adaptive line enhancement
AML	Approximate maximum likelihood
ANC	Active noise cancellation or control
ANN	Artificial neural network
AP	Array processing
APA	Affine projection algorithm
AR	Autoregressive
ARMA	Autoregressive moving average
ASO	Approximate stochastic optimization
ASR	Automatic speech recognition
AST	Affine scaling transformation
ATF	Acoustic transfer function
AWGN	Additive Gaussian white noise
BF	Beamforming
BFGS	Broyden–Fletcher–Goldfarb–Shanno
BI_ART	Block iterative algebraic reconstruction technique
BIBO	Bounded-input–bounded-output
BLMS	Block least mean squares
BLP	Backward linear prediction
BLUE	Best linear unbiased estimator
BSP	Blind signal processing

BSS	Blind signal separation
ccf	Crosscorrelation function
CC-FDAF	Circular convolution frequency domain adaptive filters
CF	Cost function
CFDAF	Constrained frequency domain adaptive filters
CGA	Conjugate gradient algorithms
CLS	Constrained least squares
CPSD	Cross power spectral density
CQF	Conjugate quadrature filters
CRB	Cramér–Rao bound
CRLS	Conventional RLS
CT	Continuous time
CTFS	Continuous time Fourier series
CTFT	Continuous time Fourier transform
DAM	Direct-averaging method
DCT	discrete cosine transform
DFS	Discrete Fourier series
DFT	Discrete Fourier transform
DHT	Discrete Hartley transform
DI	Directivity index
DLMS	Delayed LMS
DLS	Data least squares
DMA	Differential microphones array
DOA	Direction of arrivals
DOI	Direction of interest
DSFB	Delay and sum beamforming
DSP	Digital signal process/or/ing
DST	Discrete sine transform
DT	Discrete time
DTFT	Discrete time Fourier transform
DWSB	Delay and weighted sum beamforming
ECG	Electrocardiogram
EEG	Electroencephalogram
EGA	Exponentiated gradient algorithms
EMSE	Excess mean square error
ESPRIT	Estimation signal parameters rotational invariance technique
ESR	Error sequential regression
EWRLS	Exponentially weighted RLS
FAEST	Fast a posteriori error sequential technique
FB	Filter bank
FBLMS	Fast block least mean squares
FBLP	Forward–backward linear prediction
FDAF	Frequency domain adaptive filters
FDE	Finite difference equation

FFT	Fast Fourier transform
FIR	Finite impulse response
FKA	Fast Kalman algorithm
FLMS	Fast LMS
FLP	Forward linear prediction
FOCUSS	FOCal Underdetermined System Solver
FOV	Field of view
FRLS	Fast RLS
FSBF	Filter and sum beamforming
FTF	Fast transversal (RLS) filter
FX-LMS	Filtered-x LMS
GCC	Generalized cross-correlation
GP-LCLMS	Gradient projection LCLMS
GSC	Generalized sidelobe canceller
GTLS	Generalized total least squares
ICA	Independent component analysis
IC	Initial conditions
iid	Independent and identically distributed
IIR	Infinite impulse response
IPNLMS	Improved PNLMS
ISI	Inter-symbol interference
KF	Kalman Filter
KLD	Kullback–Leibler divergence
KLT	Karhunen–Loeve transform
LCLMS	Linearly constrained least mean squares
LCMV	Linearly constrained minimum variance
LD	Look directions
LDA	Levinson–Durbin algorithm
LHA	Linear harmonic array
LMF	Least mean fourth
LMS	Least mean squares
LORETA	LOw-Resolution Electromagnetic Tomography Algorithm
LPC	Linear prediction coding
LS	Least squares
LSE	Least square error
LSE	Least squares error
LSUE	Least squares unbiased estimator
MA	Moving average
MAC	Multiply and accumulate
MAF	Multi-delay adaptive filter
MCA	Minor component analysis
MEFEX	Multiple error filtered-x
MFB	Matched filter beamformer
MIL	Matrix inversion lemma

MIMO	Multiple-input multiple-output
MISO	Multiple-input single-output
ML	Maximum likelihood
MLDE	Maximum-likelihood distortionless estimator
MLP	Multilayer perceptron
MMSE	Minimum mean square error
MNS	Minimum norm solution
MPA	Matching pursuit algorithms
MRA	Main response axis
MSC	Magnitude square coherence
MSC	Multiple sidelobe canceller
MSD	Mean square deviation
MSE	Mean squares error
MUSIC	Multiple signal classification
MVDR	Minimum variance distortionless response
MVU	Minimum variance unbiased
NAPA	Natural APA
NGA	Natural gradient algorithm
NLMS	Normalized least mean squares
NLR	Nonlinear regression
OA-FDAF	Overlap-add frequency domain adaptive filters
ODE	Ordinary difference equation
OS-FDAF	Overlap-save frequency domain adaptive filters
PAPA	Proportional APA
PARCOR	Partial correlation
PBFDAF	Partitioned block frequency domain adaptive filters
PCA	Principal component analysis
PFDABF	Partitioned frequency domain adaptive beamformer
PFDAF	Partitioned frequency domain adaptive filters
PHAT	Phase transform
PNLMS	Proportionate NLMS
PRC	Perfect reconstruction conditions
PSD	Power spectral density
PSK	Phase shift keying
Q.E.D	Quod erat demonstrandum (this completes the proof)
QAM	Quadrature amplitude modulation
QMF	Quadrature mirror filters
RLS	Recursive least squares
RNN	Recurrent neural network
ROF	Recursive order filter
RTF	Room transfer functions
RV	Random variable
SAF	Subband adaptive filters
SBC	Subband coding

SBD	Subband decomposition
SCOT	Smoothed coherence transform
SDA	Steepest-descent algorithms
SDBF	Superdirective beamforming
SDE	Stochastic difference equation
SDS	Spatial directivity spectrum
SE-LMS	Signed-error LMS
SGA	Stochastic-gradient algorithms
SIMO	Single-input multiple-output
SISO	Single input single output
SNR	Signal-to-noise ratio
SOI	Source of interest
SP	Stochastic processes
SR-LMS	Signed-regressor LMS
SRP	Steered response power
SSE	Sum of squares errors
SS-LMS	Sign–sign LMS
STFT	Short-time transformation
SVD	Singular value decomposition
TBWP	Time-bandwidth-product
TDAF	Transform-domain adaptive filters
TDE	Time delay estimation
TF	Transfer function
TFR	Transfer function ratio
TLS	Total least squares
UCA	Uniform circular array
UFDAF	Unconstrained frequency domain adaptive filters
ULA	Uniform linear array
VLA	Very large array
VLSI	Very large-scale integration
WEV	Weights error vector
WGN	White Gaussian noise
WLS	Weighted total least
WMNS	Weighted minimum norm solution
WPO	Weighted projection operators
WSBF	Weighted sum beamforming

Contents

Chapter 1
Discrete-Time Signals and Circuits Fundamentals

1.1 Introduction

In all real physical situations, in the communication processes, and in the wider meaning terms, it is usual to think the signals as variable physical quantity or symbols, to which is associated a certain *information*. A signal that *carries information* is variable and, in general, we are interested in the time (or other)-domain variation: *signal* $\Leftrightarrow$ *function of time* or, more generally, *signal* $\Leftrightarrow$ *function of several variables*.

Examples of signals are continuous bounded functions of time as the human voice, a sound wave produced by a musical instrument, a signal from a transducer, an image, video, etc. In these cases we speak of signals defined in the time domain or of analog or continuous-time (CT) signals. An image is a continuous function of two spatial variables, while a video consists of a continuous bounded time-space function. Examples of one- and two-dimensional real signals are shown in Fig. 1.1.

In the case of one-dimensional signals, from the mathematical point of view, it is convenient to represent this variability with a time continuous function, denoted $x_a(t)$, where the subscript a stands for *analog*. A signal is defined as *analog* when it is in close analogy to a real-world physical quantity such as, for example, the voltage and current of an electrical circuit. The analog signals are then, by their nature, usually represented with real everywhere continuous functions. Sometimes, as in telecommunications modulation process case or in particular real physical situations, the signals can be defined in the complex domain. In Fig. 1.2 is reported an example of a complex domain signal written as $x(t) = x_R(t) + j \cdot x_I(t) = e^{-\alpha t}e^{j\omega t}$ where $x_R(t)$ and $x_I(t)$ are, respectively, the real and imaginary signal parts, ω is defined as *angular frequency* (or *radian frequency*, *pulsatance*, etc.), and α is defined as *damping coefficient*. Other times, as in the case of pulse signals, the boundedness constraint can be removed.

A. Uncini, *Fundamentals of Adaptive Signal Processing*, Signals and Communication Technology, DOI 10.1007/978-3-319-02807-1_1,

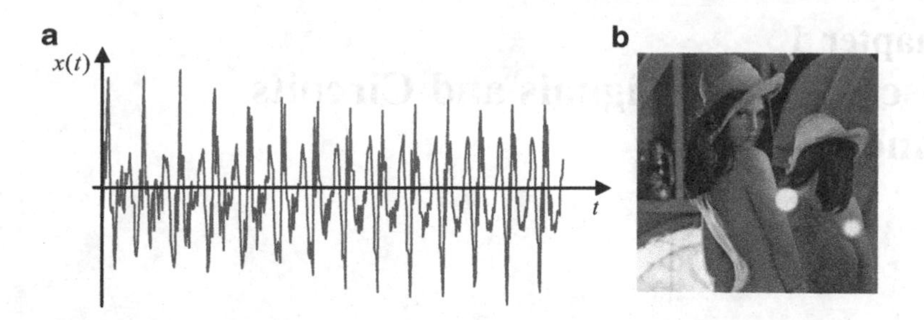

Fig. 1.1 Examples of real *analog* or *continuous-time* signals (**a**) human voice tract; (**b**) image of *Lena*

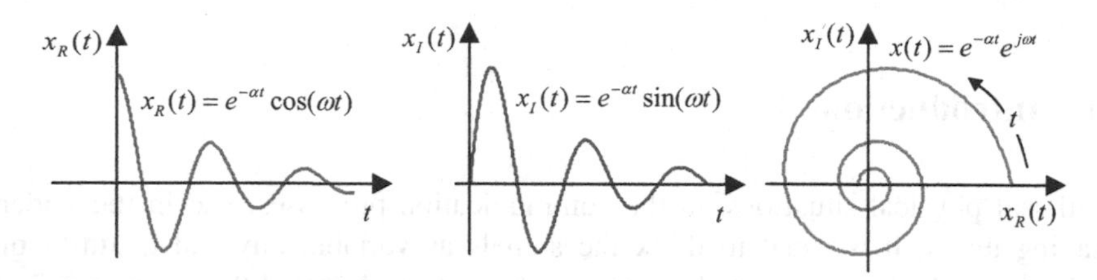

Fig. 1.2 Example of signal defined in the complex domain. Representation of a damped complex sinusoid

1.1.1 Discrete-Time Signals

In certain situations it is possible to define a signal, which contains a certain information, with a real or complex sequence of numbers. In this case, the signal is limited to a discrete set of values defined in a precise time instant. This signal is therefore defined as a *discrete-time signal* or *sequence* or *time series*.

For discrete-time (DT) signals description, it is usual to use the form $x[n]$, where the index $n \in \mathbb{Z}$ can be any physical variable (such as time, distance, etc.) but which frequently is a time index. In addition, the square brackets are used just to emphasize the discrete nature of the signal that represents the process.

Therefore, the DT signals are defined by a sequence that can be generated through an algorithm or, as often happens, by a *sampling process* that transforms, under appropriate assumptions, an analog signal into a sequence. Examples of such signals are audio *wave* files (with the extension .wav) commonly found in PCs. In fact, these files are DT signals stored on the hard drive (or memory) with a specific format. Previously acquired through your sound card or generated with appropriate algorithms, these signals can be listened, viewed, edited, processed, etc.

An example of a graphical representation of a sequence is shown in Fig. 1.3.

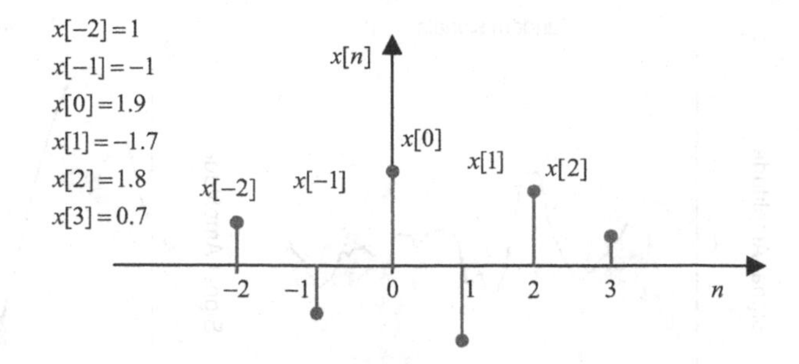

Fig. 1.3 Example of a discrete-time signal or sequence

1.1.2 Deterministic and Random Sequences

A sequence is said to be deterministic if it is fully *predictable* or if it is generated by an algorithm which *exactly* determines the value for each n. In this case the information content carried by the signal is null because it is entirely predictable.

A sequence is said to be *random* (or *aleatory* or *stochastic*) if it evolves over time (or in other domains) in unpredictable ways (or not entirely predictable). The characterization of a random sequence can be carried out by statistical quantities related to the signal which may present some regularity.

Even if not exactly predictable sample by sample, the random signals can be predicted in its average behavior. In other words, the sequence can be described, characterized, and processed, taking into consideration their statistical parameters rather than by an explicit equation (Fig. 1.4).

For more details and random signal characterization, see Appendix C on *stochastic processes*.

1.2 Basic Deterministic Sequences

In the study and DT signals applications, it is usual to encounter deterministic signals easily generated with simple algorithms. As we shall see in the following chapters, these sequences may be useful for DT systems characterization [1, 2].

1.2.1 Unitary Impulse

The *unitary impulse*, called also DT *delta function*, is a sequence, shown in Fig. 1.5a, defined as

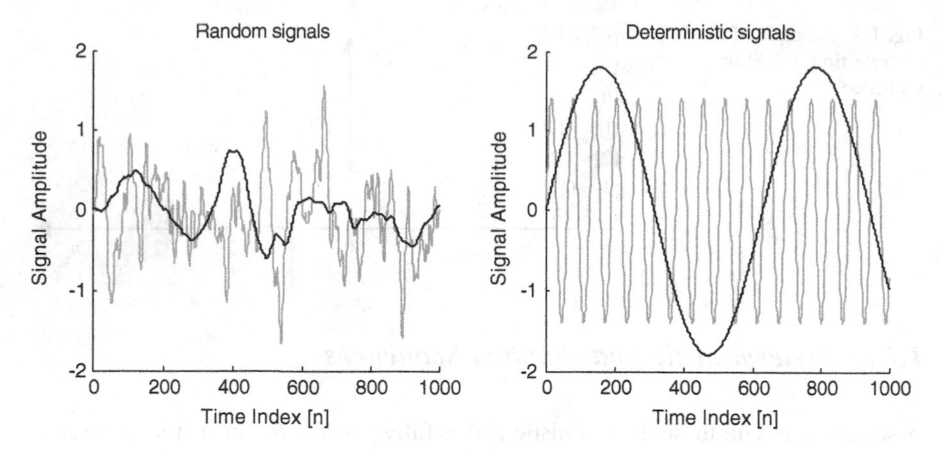

Fig. 1.4 An example of random and deterministic sequences

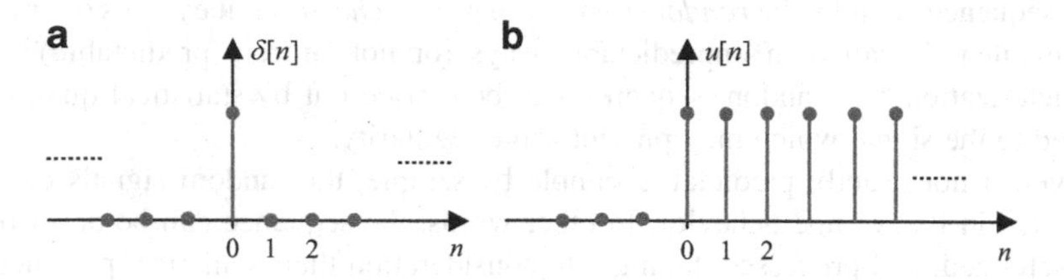

Fig. 1.5 Discrete-time signals (**a**) unitary impulse $\delta[n]$; (**b**) unit step $u[n]$

$$\delta[n] = \begin{cases} 1 & \text{for } n = 0 \\ 0 & \text{otherwise.} \end{cases} \tag{1.1}$$

Property An arbitrary sequence $x[n]$ can be represented as a sum of delayed and weighted impulses: sampling property. Therefore we can write

$$x[n] = \sum_{k=-\infty}^{\infty} x[k]\delta[n-k].$$

1.2.2 Unit Step

The *unit step* sequence is a sequence (see Fig. 1.5b) defined as

$$u[n] = \begin{cases} 1 & \text{for } n \geq 0 \\ 0 & \text{for } n < 0. \end{cases} \tag{1.2}$$

In addition, it is easy to show that the unit step sequence verifies the property

$$u[n] \therefore \begin{cases} u[n] = \sum_{k=0}^{\infty} \delta[n-k] \\ \delta[n] = u[n] - u[n-1]. \end{cases}$$

1.2.3 Real and Complex Exponential Sequences

The real and complex exponential sequence is defined as

$$x[n] = A\alpha^n \quad A,\ \alpha \in (\mathbb{R}, \mathbb{C}). \tag{1.3}$$

The exponential sequence can take various shapes depending on the actual values that can assume the α and A coefficients. Figure 1.6 shows the trends of real sequences for some values of α and A.

In the complex case we have that $A = |A|e^{j\phi}$ and $\alpha = |\alpha|e^{j\omega}$. Moreover, note that using Euler's formula, the sequence can be rewritten as

$$x[n] = |A||\alpha|^n e^{j(\omega n+\phi)} = |A||\alpha|^n \Big(\cos(\omega n + \phi) + j \sin(\omega n + \phi) \Big), \tag{1.4}$$

where the parameters A, α, ω, and ϕ are defined, respectively, as A amplitude, α damping coefficient, ω angular frequency (or radial frequency, pulsatance, ...), ϕ phase (Fig. 1.7).

From the above expression it can be seen that the sequence $x[n]$ has an envelope that is a function of the parameters α and its shape appears to be

$$\begin{aligned} |\alpha| &< 1 \quad \textit{decreasing with } n, \\ |\alpha| &= 1 \quad \textit{constant}, \\ |\alpha| &> 1 \quad \textit{increasing with } n. \end{aligned}$$

Special cases of the expression (1.4), for $\alpha = 1$, are shown below

$$\begin{aligned} &|A|e^{j(\omega n+\phi)} && \textit{complex sinusoid}, \\ &\cos(\omega n + \phi) = \frac{e^{j(\omega n+\phi)} + e^{-j(\omega n+\phi)}}{2} && \textit{real cosine}, \\ &\sin(\omega n + \phi) = \frac{e^{j(\omega n+\phi)} - e^{-j(\omega n+\phi)}}{j2} && \textit{real sinusoid}. \end{aligned}$$

1.3 Discrete-Time Signal Representation with Unitary Transformations

Let us consider real or complex domain finite duration sequences, indicated as

$$\mathbf{x} \in (\mathbb{R}, \mathbb{C})^{(N\times 1)} \triangleq \big[x[0] \quad x[1] \quad \cdots \quad x[N-1] \big]^T. \tag{1.5}$$

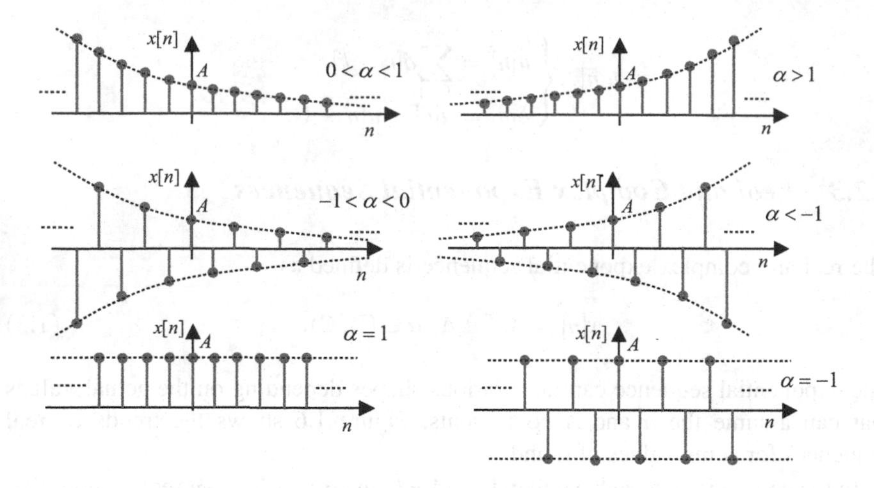

Fig. 1.6 Real exponential sequence trends for some of α and A values

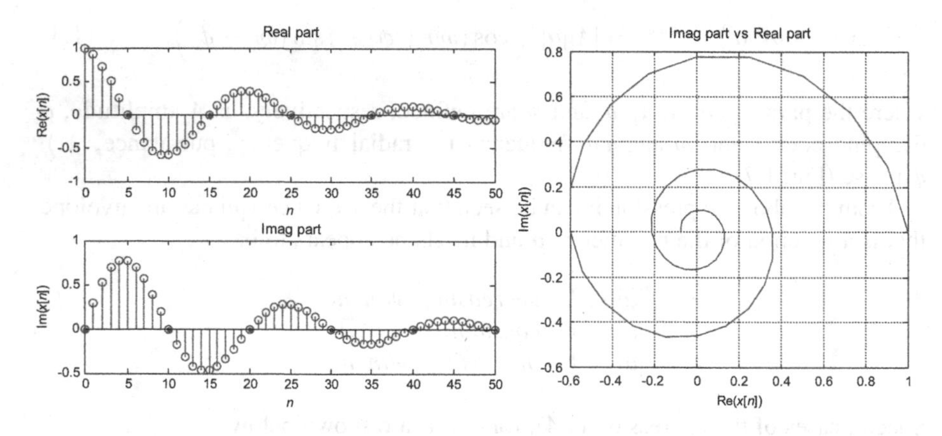

Fig. 1.7 Example of a damped complex sinusoid sequence generated by (1.4)

Indicating with $\mathbf{F} \in (\mathbb{R}, \mathbb{C})^{N \times N}$, a suitable invertible matrix, called *basis matrix* or *kernel matrix*, let us consider the linear transformation defined as

$$X = \mathbf{F} \cdot \mathbf{x}, \tag{1.6}$$

where $X \in (\mathbb{R}, \mathbb{C})^{(N \times 1)} \triangleq [X(0), X(1), \ldots, X(N-1)]^T$ is the vector containing the transformed values of $\mathbf{x}$. The vector X is a *representation* of the sequence (1.5) in the domain described by the basis matrix $\mathbf{F}$. Moreover, the inverse transform is

$$\mathbf{x} = \mathbf{F}^{-1} \cdot X. \tag{1.7}$$

The transformation (1.6) is called *unitary transform*, if for $\mathbf{F} \in \mathbb{R}^{N \times N}$ we have that

$$\mathbf{F}^{-1} = \mathbf{F}^T \quad \Leftrightarrow \quad \mathbf{F}\mathbf{F}^T = \mathbf{I}, \tag{1.8}$$

while for the complex case $\mathbf{F} \in \mathbb{C}^{N \times N}$, we have that

$$\mathbf{F}^{-1} = \mathbf{F}^H \quad \Leftrightarrow \quad \mathbf{F}\mathbf{F}^H = \mathbf{I}, \tag{1.9}$$

where the superscript (H) indicates the transposed complex conjugate matrix, also called the *Hermitian matrix* (see Sect. A.2.1).

Property The unitary transformation rotates the $\mathbf{x}$ vector without changing its length. Indicating with $\|\cdot\|$ the norm of a vector is $\|X\| = \|\mathbf{x}\|$. In fact, we have that

$$\|X\| = X^T X = [\mathbf{F} \cdot \mathbf{x}]^T \mathbf{F} \cdot \mathbf{x} = \mathbf{x}^T \cdot \mathbf{F}^T \mathbf{F} \cdot \mathbf{x} = \|\mathbf{x}\|. \tag{1.10}$$

Note that the matrix $\mathbf{F}$ can be expressed as

$$\mathbf{F} \triangleq K \begin{bmatrix} F_N^{00} & F_N^{01} & \cdots & F_N^{0(N-1)} \\ F_N^{10} & F_N^{11} & \cdots & F_N^{1(N-1)} \\ \vdots & \vdots & \ddots & \vdots \\ F_N^{(N-1)0} & F_N^{(N-1)1} & \cdots & F_N^{(N-1)(N-1)} \end{bmatrix}, \tag{1.11}$$

where F_N is real or complex number defined by the nature of the transformation, and K is a constant sometimes necessary to verify the unitary transformation property (1.8), and (1.9).

The basis matrix $\mathbf{F}$ can be *a priori* fixed and independent from the input data or *data independent*. For example, as we shall see later, this occurs in the discrete Fourier transform (DFT) and in other types of transformation presented and discussed below.

In the case the basis $\mathbf{F}$ is *data dependent*, as will be seen in Sect. 1.3.6, this can be calculated in an optimum way according to some criterion, although, in the case of nonstationary signals[1], with a significant computational cost increase.

[1] For *nonstationary signal*, for instance, may be considered a signal generated by a nonstationary system such as, for example, a sine wave oscillator that continuously varies the amplitude, phase, and frequency such that its statistical characteristics (mean value, rms value, etc.) are not constant.

1.3.1 The Discrete Fourier Transform

A DT sequence is said to be periodic, of period N, if

$$\tilde{x}[n] = \tilde{x}[n+N], \qquad -\infty < n < \infty.$$

Similarly to the Fourier series for analog signals, a periodic sequence can be represented as the sum of discrete sinusoids.

We define the discrete Fourier series (DFS) for a periodic sequence $\tilde{x}[n]$, as

$$\tilde{X}(k) = \sum_{n=0}^{N-1} \tilde{x}[n] e^{-j\frac{2\pi}{N}kn}, \tag{1.12}$$

$$\tilde{x}[n] = \frac{1}{N} \sum_{k=0}^{N-1} \tilde{X}(k) e^{j\frac{2\pi}{N}kn}. \tag{1.13}$$

The Fourier series is an exact representation of the periodic sequence.

A N length sequence can then be *exactly* represented with the couple of equations defined as direct discrete Fourier transform DFT and inverse DFT (IDFT)

$$X(k) = \sum_{n=0}^{N-1} x[n] e^{-j\frac{2\pi}{N}kn}, \qquad k = 0, 1, \ldots, N-1, \tag{1.14}$$

$$x[n] = \frac{1}{N} \sum_{k=0}^{N-1} X(k) e^{j\frac{2\pi}{N}nk}, \quad n = 0, 1, \ldots, N-1. \tag{1.15}$$

In the DFT use, we must always remember that we are representing a periodic sequence of period N. Table 1.1 shows some of DFT properties.

1.3.2 DFT as Unitary Transformation

The DFT can be interpreted as a unitary transformation if in (1.6) the matrix $\mathbf{F} \in \mathbb{C}^{N \times N}$ is evaluated considering the DFT definition (1.14). Indeed, if the matrix is formed by DFT components, the double summation (1.14) can be interpreted as a matrix–vector product. In order that Eq. (1.14) is identical to (1.6), the matrix $\mathbf{F}$ coefficients F_N are determined as

$$F_N = e^{-j2\pi/N} = \cos(2\pi/N) - j\sin(2\pi/N).$$

The DFT matrix is defined as $\mathbf{F} \triangleq \{f_{k,n}^{\text{DFT}} = F_N^{kn}\ k, n \in [0, N-1]\}$

Table 1.1 Main properties of the DFT

	Sequence	DFT
Linearity	$ax_1[n]+bx_2[n]$	$aX_1[k]+bX_2[k]$
Time shifting	$x([n-m])_N$	$e^{-j\frac{2\pi}{N}km}X(k)$
Frequency shifting	$e^{-j\frac{2\pi}{N}nm}x[n]$	$X(k-m)$
Temporal inversion	$x([-n])_N$	$X^*(k)$
Convolution	$\sum_{m=0}^{N-1}x[m]h([n-m])_N$	$X(k)H(k)$
Multiplication	$x[n]w[n]$	$\frac{1}{N}\sum_{r=0}^{N-1}X(r)H([k-r])_N$

$$f_{k,n}^{\mathrm{DFT}}=e^{-j\frac{2\pi}{N}kn}=\cos\frac{2\pi}{N}kn-j\sin\frac{2\pi}{N}kn,\quad \text{for}\quad k,n=0,1,\ldots,N-1 \tag{1.16}$$

in explicit terms

$$\mathbf{F}\triangleq K\begin{bmatrix}1 & 1 & 1 & \cdots & 1\\ 1 & e^{-j2\pi/N} & e^{-j4\pi/N} & \cdots & e^{-j2\pi(N-1)/N}\\ 1 & e^{-j4\pi/N} & e^{-j8\pi/N} & \cdots & e^{-j4\pi(N-1)/N}\\ \vdots & \vdots & \vdots & \ddots & \vdots\\ 1 & e^{-j2\pi(N-1)/N} & e^{-j4\pi(N-1)/N} & \cdots & e^{-j2\pi(N-1)^2/N}\end{bmatrix}. \tag{1.17}$$

The complex matrix $\mathbf{F}$ is also symmetric. By its definition, the reader can easily observe that (1.9) are satisfied $\left(\mathbf{F}^{-1}=\mathbf{F}^H \text{ and } \mathbf{F}\mathbf{F}^H=\mathbf{I}\right)$ provided that in (1.17), $K=1/\sqrt{N}$. The multiplication by the term $1/\sqrt{N}$ is inserted just to make the linear and unitary transformation. From the previous expressions, the DFT (1.14) and the IDFT (1.15) can be calculated, respectively, as

$$\begin{aligned}\boldsymbol{X}&=\mathbf{F}\cdot\mathbf{x},\\ \mathbf{x}&=\mathbf{F}^H\cdot\boldsymbol{X}.\end{aligned}$$

The DFT, formally, can be defined as an invertible linear transformation that *maps* real or complex sequence in another complex sequence. In formal terms it can be referred to as DFT $\Rightarrow f:(\mathbb{C},\mathbb{R})^N\to\mathbb{C}^N$.

1.3.3 Discrete Hartley Transform

In the case of real sequences it is possible, and often convenient, to use transformations defined in real domain $f:\mathbb{R}^N\to\mathbb{R}^N$. In fact, in the case of real signals having a complex arithmetic determines a computational load that is not always strictly necessary.

The discrete Hartley transform (DHT) is defined as

$$X(k) = \sum_{n=0}^{N-1} x[n]\left(\cos\frac{2\pi}{N}kn + \sin\frac{2\pi}{N}kn\right), \quad k = 0, 1, \ldots, N-1. \tag{1.18}$$

Whereby in (1.11) $F_N = \cos(2\pi/N) + \sin(2\pi/N)$, then the DHT matrix can be defined as $\mathbf{F} \triangleq \{f_{kn}^{\mathrm{DHT}} = F_N^{kn}\ k, n \in [0, N-1]\}$.

$$f_{k,n}^{\mathrm{DHT}} = \cos\frac{2\pi}{N}kn + \sin\frac{2\pi}{N}kn, \quad \text{for} \quad k, n = 0, 1, \ldots, N-1. \tag{1.19}$$

To verify the unitarity conditions (1.8), as for the DFT, $K = 1/\sqrt{N}$ applies. In practice, the DHT coincides with the DFT for real signals.

1.3.4 Discrete Sine and Cosine Transforms

In the discrete cosine transform (DCT) and in the discrete sine transform (DST) [3], the sequence can be only real $\mathbf{x} \in \mathbb{R}^N \triangleq [x(0), x(1), \ldots, x(N-1)]^T$ and represented in terms of real sine or cosine functions series. In formal terms $\mathrm{DCT}/\mathrm{DST} \Rightarrow f : \mathbb{R}^N \to \mathbb{R}^N$. In particular, the DCT/DST transformations are similar but not identical to the DFT and applicable only to real sequences. In the literature some variations are defined. Unlike the DFT, which is uniquely defined, the real DCT/DST transformations can be defined in different ways depending on the type of the periodicity definition imposed to the finite N-length sequence[2] $x[n]$ (see [1] for details). In the literature (at least) four variants are reported. That said *type II* or *DCT-II*, which is based on $2N$ periodicity, appears to be one of the most used.

1.3.4.1 Type II Discrete Cosine Transform

The cosine transform, DCT-II version, is defined as

$$X(k) = K_n \sum_{n=0}^{N-1} x[n] \cos\left[\frac{\pi}{N}\left(\frac{2n+1}{2}\right)k\right], \quad k = 0, 1, \ldots, N-1. \tag{1.20}$$

In unitary transformation terms, the matrix $\mathbf{F}$ coefficients in (1.6) (see [3, 4]) are defined as

---

[2] Given a $x[n]$ sequence, with $0 \leq n \leq N-1$, there are more ways to extend it as a periodic sequence depending on the aggregation of the segments and the type of, odd or even, chosen symmetry.

$$f_{k,n}^{\mathrm{DCT}} = K_k \cos \frac{\pi(2n+1)k}{2N}, \qquad \text{for} \qquad n,k = 0,1,\ldots,N-1, \tag{1.21}$$

where, in order to verify that $\mathbf{FF}^{-1} = \mathbf{I}$, we have that

$$K_k = \begin{cases} 1/\sqrt{N} & k=0 \\ \sqrt{2/N} & k>0. \end{cases} \tag{1.22}$$

The cosine transform of a N-length sequence can be calculated by reflecting the image on its edges, to obtain a $2N$-length sequence, taking the DFT and extracting the real part. There are also algorithms for the direct calculation with only real arithmetic operations.

1.3.4.2 Type II Discrete Sine Transform

The DST-II version is defined as

$$X(k) = K_n \sum_{n=0}^{N-1} x[n] \sin \frac{\pi(2n+1)(k+1)}{2N}, \qquad k = 0,1,\ldots,N-1. \tag{1.23}$$

Accordingly, the elements of the matrix $\mathbf{F}$ are defined as

$$f_{k,n}^{\mathrm{DST}} = K_n \sin \frac{\pi(2n+1)(k+1)}{2N}, \qquad \text{for} \qquad n,k = 0,1,\ldots,N-1 \tag{1.24}$$

with K_n defined as in (1.22). Note that the DCT, the DST, and other transformations can be computed with fast algorithms based on or similar to the DFT. For other transformations types, refer to the literature [3–10].

1.3.5 Haar Unitary Transform

Given a CT signal $x(t)$, $t \in [0,1)$, divided into $N = 2^b$ tracts, or sampled with sampling period equal to $t_s = 1/N$, the Haar transform can be defined as

$$X(k) = K \sum_{n=0}^{N-1} x(t_s \cdot n) \varphi_k(t), \quad k = 0,1,\ldots,N-1. \tag{1.25}$$

The CT Haar family functions $\varphi_k(t)$, $k = 0,1,\ldots,N-1$ are defined into the interval $t \in [0,1)$, and for the index k we have that

$$k = 2^p + q - 1, \qquad \text{for} \qquad p, q \in \mathbb{Z}, \tag{1.26}$$

where p is such that $2^p < k$, i.e., the largest power of two contained in k while $(q-1)$ is the remaining part, i.e., $q = k - 2^p + 1$.

For $k=0$, the Haar functions is defined as

$$\varphi_0(t) = 1/\sqrt{N} \tag{1.27}$$

while for $k>0$ we have

$$\varphi_k(t) = \frac{1}{\sqrt{N}} \begin{cases} 2^{p/2} & (q-1)/2^p \le t < \left(q-\frac{1}{2}\right)/2^p \\ -2^{p/2} & \left(q-\frac{1}{2}\right)/2^p \le t < q/2^p \\ 0 & \text{otherwise,} \end{cases} \qquad \text{for} \quad q = k - 2^p + 1. \tag{1.28}$$

From the above definition, one can show that p determines the amplitude and the width of the nonzero part of $\varphi_k(t)$ function, while q determines the position of the nonzero function tract. Figure 1.8 shows the plot of some Haar basis functions for $N=2^8$.

Remark The Haar basis functions can be constructed as *dilation* and *translation* of a certain elementary function indicated as *mother function*.

1.3.6 Data-Dependent Unitary Transformation

The data-dependent transformation matrix is a function of the input data which, since the input sequence can have time-varying statistical characteristic, is run-time computed. We define a N-length *sliding window* in order to select a input data tract, in which the statistic can be considered constant, defined as

$$\mathbf{x}_n \in (\mathbb{R}, \mathbb{C})^N \triangleq \left[x[n], x[n-1], \ldots, x[n-N+1]\right]^T. \tag{1.29}$$

One of the most common methods for the definition of data-dependent unitary transformation is based on the autocorrelation matrix of the input sequence x_n (see, for details, Appendix C) defined as

$$\mathbf{R} = E\left\{\mathbf{x}_n \mathbf{x}_n^H\right\} \tag{1.30}$$

or, practically, of its estimate. In fact, considering *ergodic processes*, the *ensemble average* $E\{\cdot\}$ may be replaced by a *time average* $\hat{E}\{\cdot\} = \frac{1}{N}\sum_{n=1:N}\{\cdot\}$, for which (1.30) can be estimated as

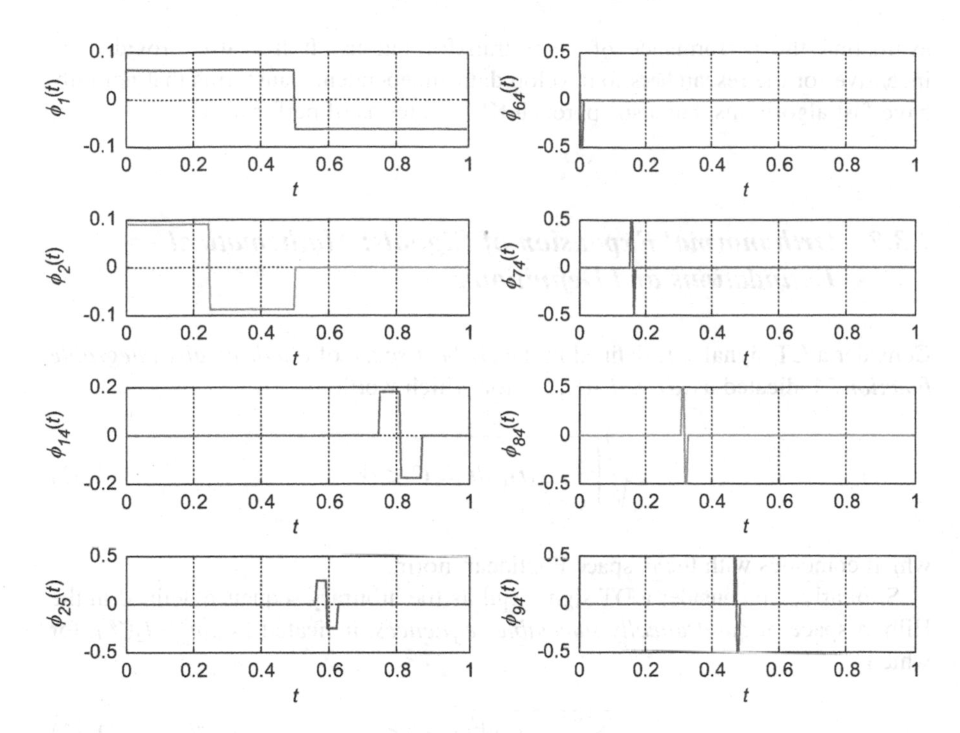

Fig. 1.8 Plot of some Haar's kernel functions calculated with (1.28)

$$\mathbf{R}_{xx} \in (\mathbb{R}, \mathbb{C})^{N\times N} = \frac{1}{N}\sum_{n=0}^{N-1} \mathbf{x}_n \mathbf{x}_n^H. \tag{1.31}$$

The correlation matrix $\mathbf{R}_{xx} \in (\mathbb{C}, \mathbb{R})^{N\times N}$ can always be diagonalized through the *unitary similarity transformation* (see Appendix A) defined by the relation $\mathbf{\Lambda} = \mathbf{Q}^H \mathbf{R}_{xx} \mathbf{Q}$, in which $\mathbf{\Lambda} \in (\mathbb{C}, \mathbb{R})^{N\times N} \triangleq \text{diag}(\lambda_0 \quad \lambda_1 \quad \cdots \quad \lambda_{N-1})$, where λ_k are the eigenvalues of the $\mathbf{R}_{xx}$ matrix. The unitary transformation $\mathbf{F} = \mathbf{Q}^H$, which diagonalizes the correlation, is the *optimal* data-dependent unitary transformation and is known as Karhunen–Loeve transform (KLT). The problem of choosing this optimal transformation is essentially related to the computational cost required for its determination. In general, the determination of data-dependent optimal transformation $\mathbf{F}$ has complexity order of $O(N^2)$.

Remark Choosing data-independent transformations, or signal representations related to a predetermined and *a priori* fixed of orthogonal vectors basis, such as DFT, DST, DCT, etc., the computational cost can be reduced to $O(N)$.

Moreover, transformations like DCT can represent a KLT approximation. In fact, it is known that DCT performance approaches that of KLT for a signal generated by a first-order Markov model with large adjacent correlation coefficient. In addition, another important aspect is that KLT has been used as a benchmark in evaluating the performance of other transformations. It has also provided an incentive for the researchers to develop data-independent transforms that not only have fast algorithms, but also approach KLT in terms of performance.

1.3.7 Orthonormal Expansion of Signals: Mathematical Foundations and Definitions

Consider a CT signal $x(t)$ defined in the *Hilbert space* of *quadratically integrable functions*, indicated as $x(t) \in L_2(\mathbb{R}, \mathbb{C})$, for which worth

$$\sqrt{\int_{t \in \mathbb{R}} |x(t)|^2 dt} = C < \infty \tag{1.32}$$

which coincides with the $\mathbb{R}$-space Euclidean norm.

Similarly, we consider a DT signal $x[n]$ as the arbitrary sequence defined in the Hilbert space of *quadratically summable sequences*, indicated as $x[n] \in l_2(\mathbb{Z})$, for which

$$\sqrt{\sum\nolimits_{n \in \mathbb{Z}} |x[n]|^2} = C < \infty. \tag{1.33}$$

Therefore, considering the finite duration sequence as a column vector $\mathbf{x}$, (1.33) coincides with the definition of L_2 vector norm

$$\sqrt{\mathbf{x}^T \mathbf{x}} = \|\mathbf{x}\|_2. \tag{1.34}$$

1.3.7.1 Inner Product

Given the CT signals $x(t) \in L_2(\mathbb{R}, \mathbb{C})$ and $h(t) \in L_2(\mathbb{R}, \mathbb{C})$, we define *inner product* in the context of Euclidean space, the relationship

$$\langle x(t), h(t) \rangle = \int_{-\infty}^{\infty} x(t) h^*(t) dt \tag{1.35}$$

while, for DT signals $x[n] \in l_2(\mathbb{Z})$ and $h[n] \in l_2(\mathbb{Z})$, the inner product can be defined as

$$\langle x[n], h[n] \rangle = \sum_{n \in \mathbb{Z}} h^*[n] x[n]. \tag{1.36}$$

Moreover, considering the finite duration sequences $\mathbf{x}$ and $\mathbf{h}$ as column vectors of the same length the inner product is defined as

$$\langle x[n], h[n] \rangle = \mathbf{x}^H \mathbf{h}. \tag{1.37}$$

Note that the previous definition coincides with the scalar vectors product (or dot product or inner product).

1.3.7.2 On the CT/DT Signals Expansion, in Continuous or Discrete Kernel Functions

As for the signals, which can be defined in CT or DT, also transformations can be defined in a continuous or discrete domain. In the case of the frequency domain, we have the continuous frequency transformations (FT) or the developments in frequency series (FS). Therefore, considering the possible combinations, there are four possibilities: continuous/discrete signals and integral/series transforms. In the classic case of time–frequency transformations, indicated generically as *Fourier transformations*, we have the following four possibilities.

(a) **Continuous-time-signal-integral-transformation (CTFT)** ***(Fourier transform)***

$$X(j\omega) = \int_{-\infty}^{\infty} x(t)e^{-j\omega t}dt, \qquad \text{or} \qquad X(j\omega) = \langle x(t), \varphi_\omega^*(t) \rangle \tag{1.38}$$

and

$$x(t) = \int_{-\infty}^{\infty} X(j\omega)e^{j\omega t}d\omega \qquad \text{or} \qquad x(t) = \langle X(j\omega), \varphi_\omega(t) \rangle. \tag{1.39}$$

Note that, in terms of inner-product, we have $\varphi_\omega(t) = e^{j\omega t}$.

(b) **Continuous-time-signal-series-expansion (CTFS)** ***(Fourier series)***

Let us $x(t)$ be a periodic signal of period T

$$X[k] = \frac{1}{T}\int_{-T/2}^{T/2} x(t)e^{-j2\pi kt/T}dt \tag{1.40}$$

and

$$x(t) = \sum\nolimits_k X[k]e^{j2\pi kt/T}. \tag{1.41}$$

(c) **Discrete-time-signal-integral-transformation**

$$X\left(e^{j\omega}\right) = \sum\nolimits_n x[n]e^{-j2\pi (f/f_s)n} \tag{1.42}$$

and

$$x[n] = \frac{1}{2\pi f_s} \int_{-\pi f_s}^{\pi f_s} X\left(e^{j\omega}\right) e^{j2\pi(f/f_s)n} d\omega. \tag{1.43}$$

Equations (1.42) and (1.43) coincide with the DT Fourier transform (DTFT) reintroduced below in Sect. 1.5.2.

(d) **Discrete-time-signal-series-expansion**

$$X[k] = \sum_{n=0}^{N-1} x[n] e^{-j2\pi nk/N} \tag{1.44}$$

and

$$x[n] = \frac{1}{N} \sum_{n=0}^{N-1} X[k] e^{j2\pi nk/N} \tag{1.45}$$

(1.44) and (1.45) meet DFT earlier introduced.

Note that, as introduced in the expressions (1.12) and (1.13), in the case of infinite length periodic sequences, it is possible to define the discrete DFS. In other words, the term discrete DFS is intended for use in lieu of DFT when the original function is periodic defined over an infinite interval.

1.3.7.3 Sequence Expansion into the Kernel Functions

Considering DT signals, indicating a certain function φ_k as *basis* or *kernel function*, the *expansion* of $x[n]$ in these kernel functions, in general terms, has the form:

$$\begin{aligned} x[n] &= \sum_{k \in \mathbb{Z}} \langle \varphi_k[l], x[l] \rangle \varphi_k[n] \\ &= \sum_{k \in \mathbb{Z}} X[k] \varphi_k[n], \end{aligned} \tag{1.46}$$

where the expression

$$X[k] = \langle \varphi_k[l], x[l] \rangle = \sum_l \varphi_k^*[l], x[l] \tag{1.47}$$

is the representation of the sequence $x[n]$ in the transformed domain $X[k]$ defined by the transformation $\langle \varphi_k[l], x[l] \rangle$.

The expansion (1.46) is called *orthonormal* if the basis function satisfies the orthonormality condition defined as

$$\langle \varphi_k[n], \varphi_l[n] \rangle = \delta[k-l] \tag{1.48}$$

and the set of basis functions is *complete*, i.e., *each* signal $x[n] \in l_2(\mathbb{Z})$ can be expressed with the expansion (1.46).

Property An important property of the orthonormal transformations is the *principle of energy conservation* (or *Parseval's Theorem*)

$$\|x\|^2 = \|X\|^2. \tag{1.49}$$

Property Indicating with $\widetilde{\varphi}_k$, a basis function such that

$$\langle \varphi_k[n], \widetilde{\varphi}_l[n] \rangle = \delta[k-l], \tag{1.50}$$

the expansion

$$\begin{aligned} x[n] &= \sum_{k \in \mathbb{Z}} \langle \varphi_k[l], x[l] \rangle \widetilde{\varphi}_k[n] \\ &= \sum_{k \in \mathbb{Z}} \widetilde{X}[k] \widetilde{\varphi}_k[n] \\ &= \sum_{k \in \mathbb{Z}} \langle \widetilde{\varphi}_k[l], x[l] \rangle \varphi_k[n] \\ &= \sum_{k \in \mathbb{Z}} X[k] \varphi_k[n], \end{aligned} \tag{1.51}$$

where

$$X[k] = \langle \widetilde{\varphi}_k[l], x[l] \rangle \quad \text{and} \quad \widetilde{X}[k] = \langle \varphi_k[l], x[l] \rangle \tag{1.52}$$

are indicated as biorthogonal expansion.

Note that in this case the energy conservation principle can be expressed as

$$\|x\|^2 = \left\langle X[k], \widetilde{X}[k] \right\rangle. \tag{1.53}$$

Examples of Expansion/Reconstruction of Haar, DTC, and DFT Representation

From the expression (1.46) the reconstruction of an expanded signal with a basis $\varphi_k(t)$ is performed as

$$x_n(t) = \langle \varphi_0, x \rangle \varphi_0(t) + \langle \varphi_1, x \rangle \varphi_1(t) + \cdots + \langle \varphi_n, x \rangle \varphi_n(t). \tag{1.54}$$

In practice, in discrete-time the signal between zero and one is divided (sampled) into $N = 2^b$ traits, for which we can write $\varphi_k(n)$, for $k, n = 0, 1, \ldots, 2^b - 1$.

The Haar expansion [5] for a window of $N = 2^b$ samples of signal is defined by the basis functions each of length equal to N, of the following type:

$$
\begin{aligned}
\varphi_0(n) &= \tfrac{1}{\sqrt{N}}\mathbf{1}_N \\
\varphi_1(n) &= \tfrac{1}{\sqrt{N}}\begin{bmatrix} \mathbf{1}_{N/2} & -\mathbf{1}_{N/2} \end{bmatrix} \\
\varphi_2(n) &= \tfrac{\sqrt{2}}{\sqrt{N}}\begin{bmatrix} \mathbf{1}_{N/4} & -\mathbf{1}_{N/4} & \mathbf{0}_{N/2} \end{bmatrix} \\
\varphi_3(n) &= \tfrac{\sqrt{2}}{\sqrt{N}}\begin{bmatrix} \mathbf{0}_{N/2} & -\mathbf{1}_{N/4} & \mathbf{1}_{N/4} \end{bmatrix} \\
\varphi_4(n) &= \tfrac{2}{\sqrt{N}}\begin{bmatrix} \mathbf{1}_{N/8} & -\mathbf{1}_{N/8} & \mathbf{0}_{3N/4} \end{bmatrix} \\
&\ \vdots \\
\varphi_i(n) &= \tfrac{2^{j/2}}{\sqrt{N}}\big[\mathbf{1}\big(k\cdot 2^{-j} \le t \le (k+1/2)\cdot 2^{-j}\big) \\
&\quad -\mathbf{1}\big((k+1/2)\cdot 2^{-j} \le t \le (k+1)\cdot 2^{-j}\big)\big] \quad \mathbf{0} \quad \text{otherwise} \\
&\ \vdots
\end{aligned}
\tag{1.55}
$$

where i is decomposed as $i = 2^j + k$, $j > 0$, $0 \le k \le 2^j - 1$, and $\mathbf{1}_N$ is defined as N "one" row vector $\mathbf{1}_N \in \mathbb{Z}^{1\times N} \triangleq [\,1 \ \ \cdots \ \ 1\,]$, and similarly $\mathbf{0}_N$ a vector of "zero" of equal length. In practice, one can easily verify that (1.55) coincides with the rows of the Haar matrix (1.28).

Remark The vector $\varphi_0[n] = \mathbf{1}_N$ corresponds to a *moving average filter* (discussed in Sect. 1.6.3), for which the average performance of $x[n]$ is described. In other words, for $k > 1$ it is responsible for the representation of the finer details.

In Fig. 1.9 is shown an example of a signal defined as

$$
x(t) = \begin{cases} \sin(2\pi t) + \cos(\pi t) - 1 & 0 \le t < \frac{1}{2} \\ e^{-(t-0.5)5} \cdot \cos(4\pi t) & \frac{1}{2} \le t < 1 \end{cases}
\tag{1.56}
$$

reconstructed with a different number of Haar basis. Note that in the experiment the signal is sampled with a sampling period equal to $t_s = 1/2^b$ with $b = 8$ and, therefore, *exactly reconstructed* with 256 basis functions.

Figure 1.10 shows the comparison of signal reconstruction with the basis of Haar, DCT-II, and DFT.

1.4 Discrete-Time Circuits

The processing of signals may occur in the CT domain with analog circuits or in DT domain with numerical circuits. In the case of analog signals is often used a *unifilar* systemic representation, as shown in Fig. 1.11a, in which the processing is defined by a mathematical operator $T\{\cdot\}$ such that $y(t) = T\{x(t)\}$. This schematization,

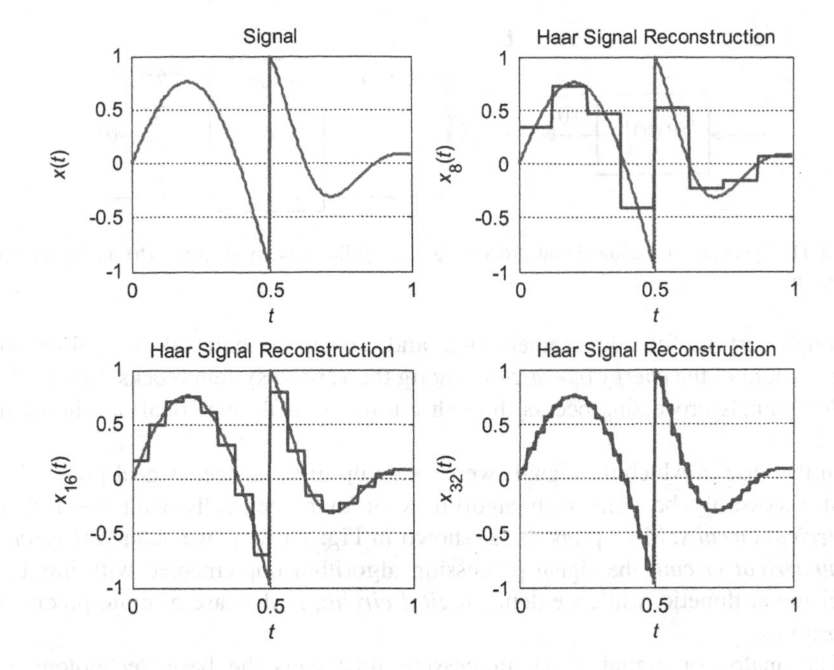

Fig. 1.9 Example of a signal represented with the Haar basic functions and reconstruction considering 8, 16, and 32 basis functions

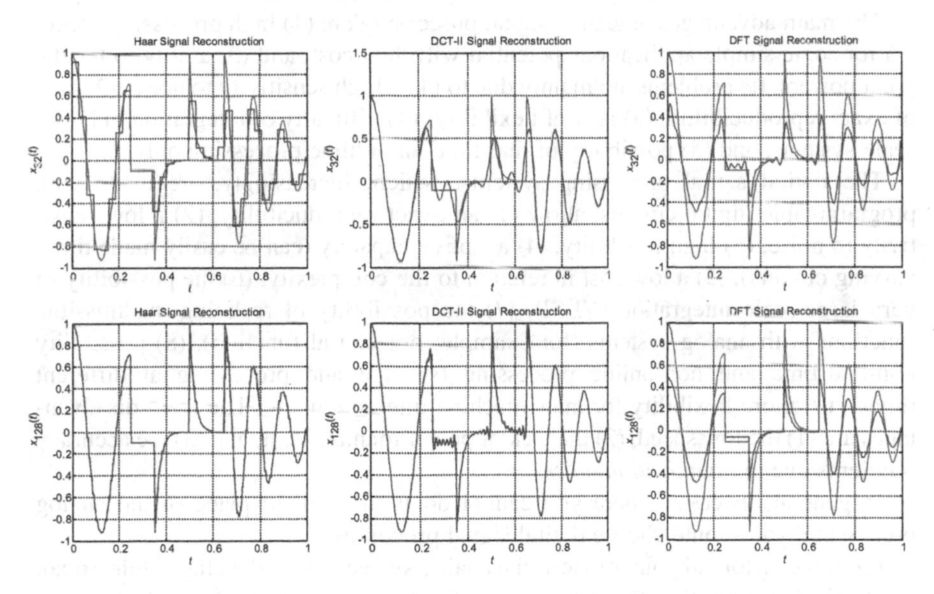

Fig. 1.10 Comparison of the signal reconstruction with the basic functions of Haar, DCT-II, and DFT, with 32 and 128 basis functions

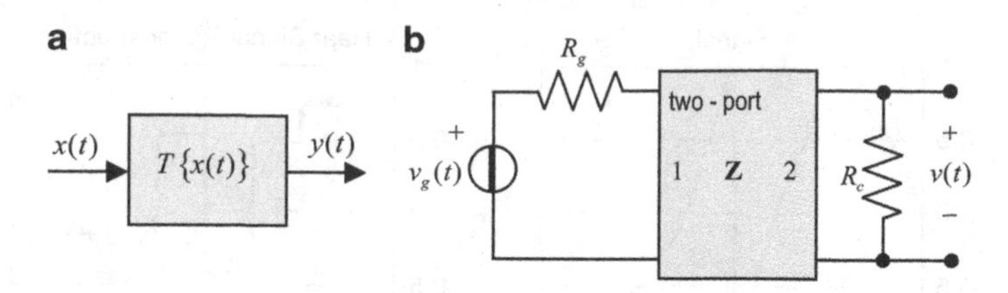

Fig. 1.11 Systems for analog signal processing (**a**) unifilar system diagram; (**b**) analog circuit approach

although very useful from mathematical and descriptive point of view, does not take account of the *energy interaction* among the various system blocks. In fact, the analog signals processing occurs through circuits (mostly electrical) as shown in Fig. 1.11b.

In the case in which the signals were made up of sequences, signal processing must necessarily be done with algorithms or more generally with the DT or *numerical circuits*. More properly, as shown in Fig. 1.12b, c, we define *DT circuit* or *numerical circuit*, the signal processing algorithm implemented with infinite-precision arithmetic, while we define *digital circuit*, in the case of finite-precision algorithms.

The analog or digital signal processing represents the basic technology of applications related to *Information and Communication Technology* (ICT) and, therefore, is a strategic discipline in virtually all of the so-called *high-tech* sectors.

The main advantages of analog signal processing are (1) high processing speed, (2) for some simple applications potential with low cost, and (3) ability to handle great powers. Its problems are mainly due to (1) a high sensitivity to noise, (2) lack of exact reproducibility, (3) lack of flexibility, (4) difficulty of integration in large-scale systems, and (5) possibility of real-time and online processing only.

The digital signal processing systems, which, increasingly, are made with programmable digital circuits, allow (1) an exact reproducibility, (2) a low sensitivity to noise, (3) high flexibility, (4) adaptive capacity (can be easily made time-varying circuits), (5) a low cost in relation to the complexity, (6) the possibility of very large-scale integration (VLSI), (7) the possibility of realizing inadmissible functions with analog systems (for example, not causal function), (8) possibility non-real-time and non-online processing (storage and processing at different times), (9) more flexibility for man–machine interaction, etc. The main disadvantages are (1) the low speed, (2) the lack of power management, and (3) the accuracy problems due to *quantization noise*.

In general, in cases where you cannot do without it, in place of the analog processing, you should choose digital signal processing.

From the historical point of view, the analog systems were developed (electrical, mechanical, pneumatic, etc.) in the first half of last century with the study of issues related to the synthesis of electrical circuits (dipoles, linear *RLCM* two-port, etc.) The development of theories concerning discrete systems has arisen, instead, with first digital computers' advent applied to the analog systems simulation. It can be

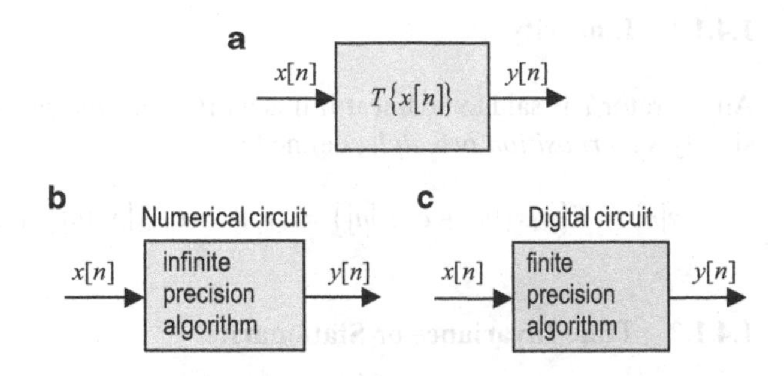

Fig. 1.12 A single-input–single-output (SISO) DT system maps the input $x[n]$ in the output $y[n]$, through the operator T (**a**) unifilar diagram; (**b**) infinite-precision algorithm or DT circuit; (**c**) finite-precision algorithm or digital circuit

said that the birth of digital signal processing (DSP) coincides with the formulation, in 1965, of the fast Fourier transform (FFT) algorithm due to Cooley and Tukey [11] that, through significant computational time reduction, has enabled the application of numerical systems also for real signals. During the same period, the DSP has become an autonomous discipline that is not necessarily linked to analog systems. In the following, and until now, there has been a huge development both theoretical and practical, which has led to new tools for processing and, more recently, also, the ability to interpret the signals.

The current use of analog circuits is essentially confined to very specific areas such as high frequency (VHF, UHF, microwave, etc.), in power applications (amplifiers, AC network filters, crossover networks, etc.), in the conditioning circuits, and for signals from special transducers (preamps, anti-aliasing filters, etc.).

Current applications of DSP techniques are innumerable. The development of non-real-time, where the signal is first stored and then processed without restricted time constraints, is the most diverse and implemented in virtually all fields: from weather (or generic time series) forecasts, bioengineering, the video and audio signals compression, VoIP technology, Internet media players, modeling and prediction of economic series, web network modeling, *big-data* analysis methods, etc.

At present, the real-time DT circuits are consistent with the adopted hardware speed, used in all technology fields. For example, in telecommunications: source coding, modulation, transmission, etc., in the processing of signals: voice, video, images, biological, seismic, radar, sonar, astronomy, music, multimedia, social information processing, array processing, etc.

1.4.1 General Properties of DT Circuits

As shown in Fig. 1.12a, we can assimilate a DT system to a mathematical operator, such that $y[n] = T\{x[n]\}$. Below, we outline some general properties for the operator T which also applies to the (hardware or software) DT circuits that implement it.

1.4.1.1 Linearity

An operator T is said to be linear if it is worth the *superposition effects principle*, or simply *superposition principle*, defined as

$$y[n] = T\{c_1x_1[n] + c_2x_2[n]\} \Rightarrow y[n] = c_1T\{x_1[n]\} + c_2T\{x_2[n]\}. \quad (1.57)$$

1.4.1.2 Time Invariance or Stationarity

In case the operator T was time invariant the *effects translation properties* are applied, defined as

$$y[n] = T\{x[n]\} \Rightarrow y[n - n_0] = T\{x[n - n_0]\}. \quad (1.58)$$

A DT circuit that satisfies the two previous properties is said to be *linear time invariant* (LTI).

1.4.1.3 Causality

The operator T is said to be *causal*, if its output at time index n_0 depends on input samples with time index $n \leq n_0$, or from the past or the present, but not from the future. For example, a circuit that realizes the *first-order backward difference*, i.e., characterized by the relation,

$$y[n] = x[n] - x[n - 1] \quad (1.59)$$

is causal. On the contrary, relation

$$y[n] = x[n + 1] - x[n], \quad (1.60)$$

the said *first-order forward difference*, is not causal. In this case, in fact, the output at time n depends on the future input at the time $n + 1$.

1.4.1.4 Bounded-Input–Bounded-Output Stability

An operator T is said to be *stable* if and only if for any bounded input there corresponds a bounded output.

In formal terms, if

$$y[n] = T\{x[n]\}$$

then

$$|x[n]| \leq c_1 < \infty \quad \Rightarrow \quad |y[n]| \leq c_2 < \infty, \qquad \forall n, \;\; x[n]. \tag{1.61}$$

This definition is also called DT bounded-input–bounded-output stability or BIBO stability.

Remark The DT bounded-input–bounded-output (DT-BIBO) stability definition, although formally very simple, is not, most of the times, useful for determining whether the operator T, or the circuit that realizes it, is or is not stable. Usually as best seen below, to verify that a circuit is stable are used simple criteria derived from the definition (1.61), taking into account the intrinsic structure of the circuit, or of some significant parameters that characterize it, and not of the input and output signals.

1.4.2 Impulse Response

The *impulse response*, as shown in Fig. 1.13, is defined as the circuit response when at its input is applied the unit impulse $\delta[n]$. This response is, in general, indicated as $h[n]$. So you can write.

$$h[n] = T\{\delta[n]\}. \tag{1.62}$$

1.4.3 Properties of DT LTI Circuits

A special class, as above indicated, often used in digital signal processing is that of the LTI circuits. These circuits are fully characterized by their impulse response $h[n]$.

Theorem If T is an LTI operator, this is *fully characterized* by its impulse response. For "fully characterized" means the property that, with the known input $x[n]$ and the impulse response $h[n]$, it is always possible to calculate the circuit output $y[n]$.

Proof From the time-invariant property if $h[n]=T\{\delta[n]\}$ then $h[n-n_0]=T\{\delta[n-n_0]\}$. It also appears that, from the sampling properties, the sequence $x[n]$ can be described as sum shifted impulses, i.e.,

$$x[n] = \sum_{k=-\infty}^{\infty} x[k]\delta[n-k] \tag{1.63}$$

so, it is

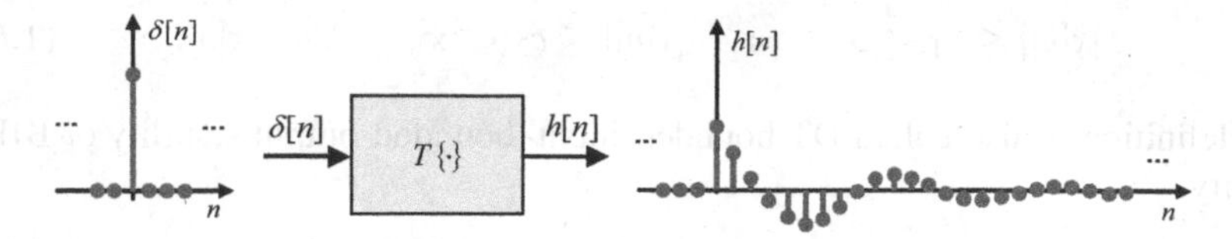

Fig. 1.13 Example of DT circuit response, to a unit impulse

$$y[n] = T\left\{ \sum_{k=-\infty}^{\infty} x[k]\delta[n-k] \right\}. \tag{1.64}$$

For linearity, it is possible to switch the T operator with the summation

$$y[n] = \sum_{k=-\infty}^{\infty} x[k]T\{\delta[n-k]\}$$

from which, by (1.62), is

$$y[n] = \sum_{k=-\infty}^{\infty} x[k]h[n-k], \qquad \text{for} \qquad -\infty < n < \infty. \tag{1.65}$$

1.4.3.1 Convolution Sum

The previous expression shows that, for a DT LTI circuit, known the impulse response and the input, the output computability is defined as the *convolution sum*. This operation, very important in DT circuits, also from software/hardware implementation point of view, is indicated as $y[n] = x[n] * h[n]$ or $y[n] = h[n] * x[n]$, where the symbol * denotes the DT convolution sum.

Remark The convolution can be seen as a superposition principle generalization. Indeed, from the previous development, it appears that the output can be interpreted as the sum of many shifted impulse responses.

Note, also, that (1.65), with simple variable substitution, can be rewritten as

$$y[n] = \sum_{k=-\infty}^{\infty} x[k]h[n-k] = \sum_{k=-\infty}^{\infty} h[k]x[n-k], \qquad \text{for} \qquad -\infty < n < \infty. \tag{1.66}$$

It is easy to show that the convolutional-type input–output link is a direct consequence of the superposition principle (1.57) and the translation property (1.58). In fact, the convolution defines a DT LTI system, or DT LTI system is completely defined by convolution.

1.4.3.2 Convolution Sum of Finite Duration Sequences

In the case of a finite duration sequences, indicated as

$$\mathbf{x}\in\mathbb{R}^{(N\times1)}\triangleq\begin{bmatrix}x[0] & x[1] & \cdots & x[N-1]\end{bmatrix}^T \tag{1.67}$$

for the $x[n]$ sequence, and

$$\mathbf{h}\in\mathbb{R}^{(M\times1)}\triangleq\begin{bmatrix}h[0] & h[1] & \cdots & h[M-1]\end{bmatrix}^T \tag{1.68}$$

for the impulse response, the summation extremes in the convolution sum assume finite values. Therefore (1.66) becomes

$$y[n]=\sum_{k=0}^{M-1}h[k]x[n-k],\qquad \text{for}\qquad 0\le n<(N-M-2). \tag{1.69}$$

Note that the output sequence duration is greater than that of the input. In case that one of the two sequences represents an impulse response of a physical system, the greater duration is interpreted with the presence of transient phenomena at the beginning and at the end of the convolution operation (Fig. 1.14).

1.4.4 Elements Definition in DT Circuits

Similarly as in CT, also in DT domain it is possible to define circuit elements through simple *constitutive relations*. In this case the nature of the signal is unique (only one quantity) and symbolic (a sequence of numbers). Then, in DT domain, the circuit element does not represent a physical low, but rather a causal relationship between its input–output quantities.

In DT circuits, being present only the "*through type*" quantities, it has only one *reactive element*[3]: the *delay unit* (indicated generally by the symbols D, z^{-1}, or q^{-1}). This allows the study of DT circuits through simple *unifilar diagrams*.

Figure 1.15 presents the definition of DT LTI circuit elements.

Example Consider the DT circuit in Fig. 1.16a. It is easy to determine the circuit input–output relationship by simple visual inspection. This is

[3] In electrical circuits a *reactive element* is defined by a constitutive relationship in which there is a time-dependence explicit by a differential of an electrical variable (e.g., current or voltage). For example, the constitutive relationship that defines the electrical element *capacitance* C [farad] is

$$i(t)=C\big(dv(t)/dt\big).$$

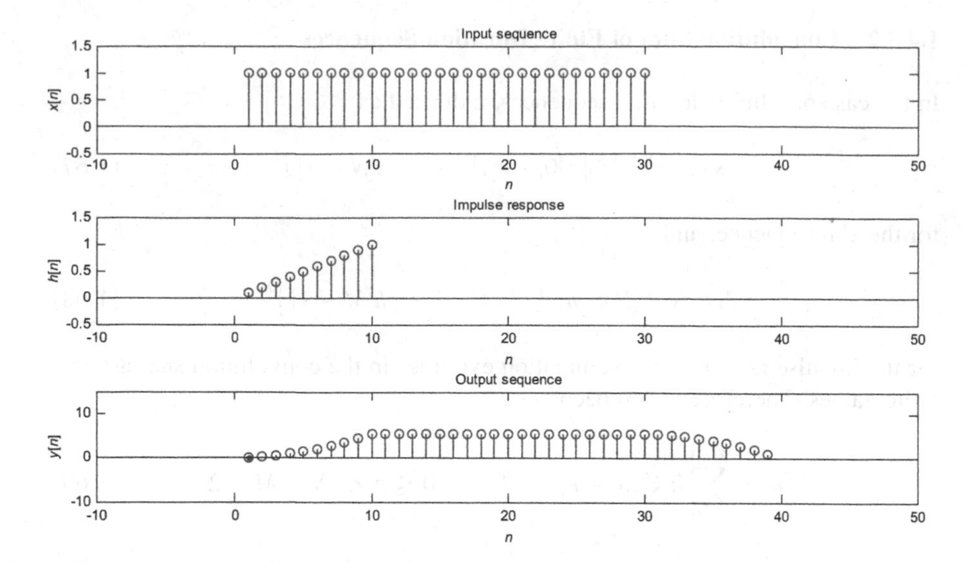

Fig. 1.14 Example of convolution sum between sequences of finite duration

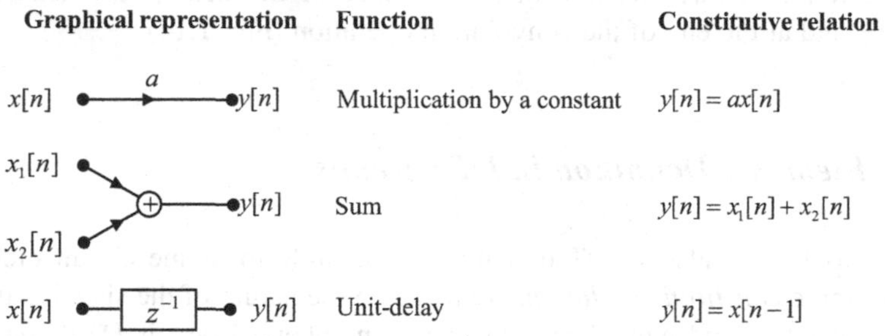

Graphical representation	Function	Constitutive relation
$x[n]$ —a→ $y[n]$	Multiplication by a constant	$y[n]=ax[n]$
$x_1[n]$, $x_2[n]$ → ⊕ → $y[n]$	Sum	$y[n]=x_1[n]+x_2[n]$
$x[n]$ — z^{-1} — $y[n]$	Unit-delay	$y[n]=x[n-1]$

Fig. 1.15 Definition and *constitutive relations* of the DT linear circuits

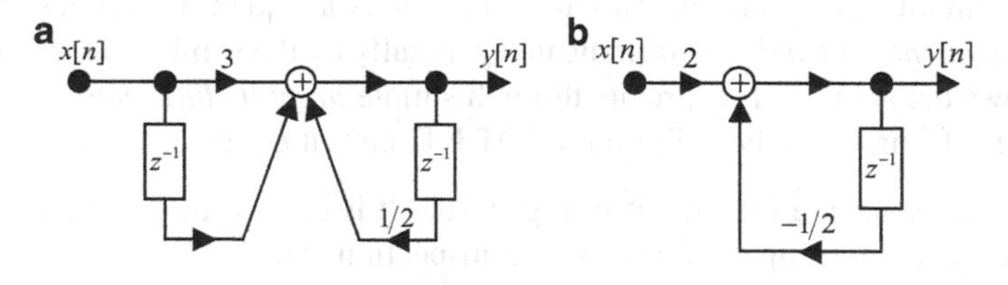

Fig. 1.16 Examples of DT circuits (**a**) with two delay elements; (**b**) with only one delay element

$$y[n] = 3x[n] + x[n-1] + \frac{1}{2}y[n-1].$$

The above expression is a *causal finite difference equation*. Therefore, it is evident that a DT LTI circuit, defined with the elements of Fig. 1.15, can always be related to an algorithm of this type. It follows that an algorithm, as formulated, is always attributed to a circuit. Consequently, we can assume the dualism: *algorithm* $\Leftrightarrow$ *circuit*.

Example Calculation of the impulse response Consider the circuit in Fig. 1.16b. By visual inspection we determine the difference equation that determines the causal input–output relationship

$$y[n] = 2x[n] - \frac{1}{2}y[n-1].$$

For the impulse response calculation we must assume zero initial conditions (IC) $(y[-1]=0)$. For an input $x[n] = \delta[n]$, we evaluate the specific output, getting

$$\begin{array}{ll} n=0 & y[0] = 2\cdot 1 + 0 = 2 \\ n=1 & y[1] = -\frac{1}{2}y[0] = -1 \\ n=2 & y[2] = -\frac{1}{2}y[1] = \frac{1}{2} \\ n=3 & y[3] = -\frac{1}{2}y[2] = -\frac{1}{4} \\ \vdots & \vdots \end{array}$$

Generalizing, for the kth sample, with simple consideration, the expression is obtained in closed form of the type

$$n=k \quad y[k] = (-1)^k 2/2^k$$

with plot in Fig. 1.17.

1.4.5 DT Circuits Representation in the Frequency Domain

The sinusoidal and exponential sequences as inputs for DT LTI circuits represent a set of eigenfunctions. In fact, the output sequence is exactly equal to the input sequence simply multiplied by a real or complex weight.

Suppose we want to measure experimentally the frequency response of a DT LTI circuit placing at its input a sinusoidal signal of unitary amplitude with variable frequency. Since the input is an eigenfunction, it is possible to evaluate the amplitude A_n and phase φ_n of the output sequence, for a set of frequencies that can be reported in a graphic form as represented by the diagram of Fig. 1.18, which is precisely the measured amplitude and phase responses.

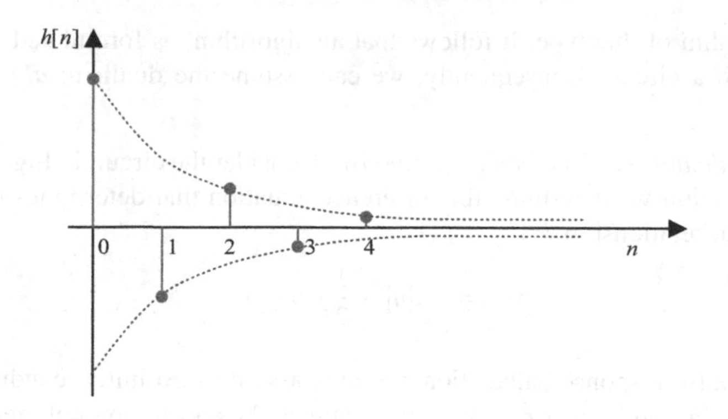

Fig. 1.17 Impulse response of the circuit in Fig. 1.16b

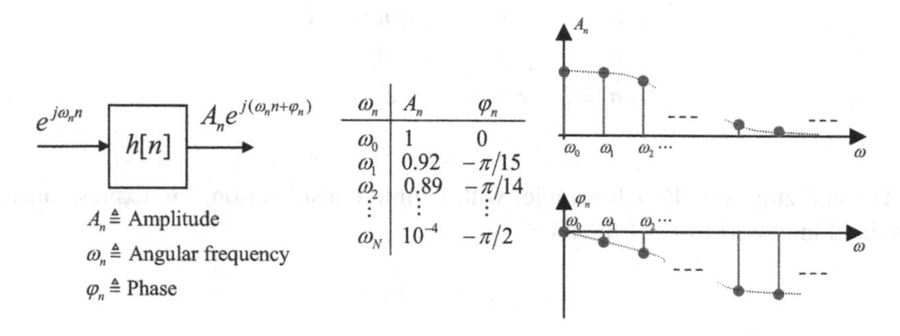

Fig. 1.18 Block diagram for the measurement of the frequency response (amplitude and phase) of a linear DT circuit

1.4.5.1 Frequency Response Computation

To perform the calculation in closed form of the frequency response, we proceed as in the empirical approach. The input is fed by a unitary-amplitude complex exponential of the type $x[n] = e^{j\omega n}$ and evaluating the output sequence.

The circuit's output can be calculated, known as its impulse response $h[n]$, through the convolution sum defined by (1.66). It has then

$$y[n] = \sum_{k=-\infty}^{\infty} h[k] e^{j\omega(n-k)} = \left(\sum_{k=-\infty}^{\infty} h[k] e^{-j\omega k} \right) e^{j\omega n}.$$

In the previous expression, it is observed that the output is calculated as the product between the input signal ($e^{j\omega n}$) and a quantity in brackets in the following indicated as

$$H(e^{j\omega}) = \sum_{k=-\infty}^{\infty} h[k]e^{-j\omega k}. \tag{1.70}$$

The complex function $H(e^{j\omega})$, defined as *frequency response*, shows that the steady-state response to a sinusoidal input is also sinusoid with the same frequency as the input, but with the amplitude and phase determined by the circuit characteristics represented by the function $H(e^{j\omega})$. For this reason, as we will see later in this chapter, under some conditions, this function is also called network function or transfer function (TF).

1.4.5.2 Frequency Response Periodicity

The frequency response is a periodic function with period 2π. In fact, if we write

$$H\left(e^{j(\omega+2\pi)}\right) = \sum_{n=0}^{\infty} h[n]e^{-j(\omega+2\pi)n}$$

noting that the term $e^{\pm j2\pi n} = 1$, it follows that $e^{-j(\omega+2\pi)n} = e^{-j\omega n}$. So it is true that

$$H\left(e^{j(\omega+2\pi n)}\right) = H(e^{j\omega}).$$

1.4.5.3 Frequency Response and Fourier Series

The $H(e^{j\omega})$ is a periodic function of ω; therefore, (1.70) can be interpreted as Fourier series with coefficients $h[n]$. From this observation we can derive the series coefficients from the well-known relationship

$$h[n] = \frac{1}{2\pi}\int_{-\pi}^{\pi} H(e^{j\omega})e^{j\omega n}d\omega. \tag{1.71}$$

Remark In practice, (1.70) allows us to evaluate the frequency domain circuits behavior, while the relation (1.71) allows us to determine the impulse response known the frequency response.

Equations (1.70) and (1.71) represent a *linear transformation*, which allow us to represent a circuit in the time or frequency domain. This transformation, valid not only for the impulse responses but also extendable to sequences, is exactly the DTFT previously defined by the expressions (1.42) and (1.43).

1.5 DT Circuits, Represented in the Transformed Domains

The signals analysis and DT circuits design can be facilitated if performed in the frequency domain. In fact, it is possible to represent signals and systems in various domains. Therefore, it is useful to see briefly the definitions and basic concepts of the *z*-transform and its relation with the Fourier transform.

1.5.1 The z-Transform

The *z*-transform of a sequence $x[n] \in l_2(\mathbb{Z})$, for $z \in \mathbb{C}$, is defined by the following equations pair:

$$X(z) = Z\{x[n]\} \triangleq \sum_{n=-\infty}^{\infty} x[n]z^{-n}, \qquad \textit{direct z-transform}, \tag{1.72}$$

$$x[n] = Z^{-1}\{X(Z)\} \triangleq \frac{1}{2\pi j}\oint_C X(z)z^{n-1}dz, \quad \textit{inverse z-transform}. \tag{1.73}$$

You can see that the $X(z)$ is an infinite power series in the z variable, where the sequence $x[n]$ plays the role of the series coefficients.

In general, this series converges to a finite value, only for certain values of z. A sufficient condition for convergence is given by

$$\sum_{n=-\infty}^{\infty} |x[n]||z^{-n}| < \infty. \tag{1.74}$$

The set of values for which the series converges defines a region in the complex z-plane, called region of convergence (ROC). This region has a shape delimited by two circles of radius R_1 and R_2 of the type

$$R_1 < |z| < R_2.$$

Example Let $x[n] = \delta[n - n_0]$, and the z-transform is

$$X(z) = z^{-n_0}.$$

Let $x[n] = u[n] - u[n - N]$, and it follows that $X(z)$ is

$$X(z) = \sum_{n=0}^{N-1} (1)z^{-n} = \frac{1 - z^{-N}}{1 - z^{-1}}.$$

In both examples we have seen, the sequence $x[n]$ has a finite duration. The $X(z)$ thus appears to be a polynomial in the z^{-1} variable, and the ROC is all the z-plane except the point at $z=0$. All finite length sequences have ROC of the type $0 < |z| < \infty$.

Example Let $x[n] = a^n u[n]$, and it follows that $X(z)$ is

$$X(z) = \sum_{n=0}^{\infty} a^n z^{-n} = \frac{1}{1 - az^{-1}}, \quad |a| < |z|.$$

In this case, the $X(z)$ turns out to be a geometric power series for which exists an expression in a closed form that defines the sum. This is a typical result for infinite length sequences defined for positive n. In this case the ROC is given by the form $|z| > R_1$.

Example Let $x[n] = -b^n u[-n-1]$, and it follows that $X(z)$ is

$$X(z) = \sum_{n=-\infty}^{-1} b^n z^{-n} = \frac{1}{1 - bz^{-1}}, \quad |z| < |b|.$$

The infinite length sequences $x[n]$ is defined for negative n. In this case the ROC has the form $|z| < R_2$.

The most general case, where $x[n]$ is defined for $-\infty < n < \infty$, can be seen as a combination of the previous cases. The ROC is thus $R_1 < |z| < R_2$.

There are theorems and important properties of the z-transform very useful for the study of linear systems. A non-exhaustive list of such properties is shown in Table 1.2.

1.5.2 Discrete-Time Fourier Transform

As introduced in Sect. 1.3.7.2, for signals which can be defined in CT or DT, also transformations can be defined in a continuous or discrete domain.

For a DT signal $x[n]$ it is possible to define a CT transform by the relations (1.70) and (1.71) that are not restricted only to circuit impulse response. In fact, this is possible by applying (1.70) and (1.71) to any sequence, provided the existence conditions.

A sequence $x[n]$ can be represented by the relations pair and (1.70) and (1.71), known as DTFT, rewritten as

Table 1.2 Main properties of the z-transform

	Sequence	z-transform
Linearity	$ax_2[n] + bx_2[n]$	$aX_1[z] + bX_2[z]$
Translation	$x[n-m]$	$z^{-m}X(z)$
Exponential weighing	$a^n x[n]$	$X(z/a)$
Linear weighing	$nx[n]$	$-z(dX(z)/dz)$
Temporal inversion	$x[-n]$	$X(z^{-1})$
Convolution	$x[n] * h[n]$	$X(z)H(z)$
	$x[n]w[n]$	$\frac{1}{2\pi j}\oint_C X(v)W(z/v)v^{-1}dv$

$$X(e^{j\omega}) = \sum_{n=-\infty}^{\infty} x[n]e^{-j\omega n}, \qquad \textit{direct DTFT transform}, \tag{1.75}$$

$$x[n] = \frac{1}{2\pi}\int_{-\pi}^{\pi} X(e^{j\omega})e^{j\omega n}d\omega, \qquad \textit{inverse DTFT transform}. \tag{1.76}$$

1.5.2.1 Existence Condition of the DTFT

The existence condition of the transform of a sequence $x[n]$ is simply its computability, namely:

(i) If $x[n]$ is *absolutely summable* then $X(e^{j\omega})$ exists and is a continuous function of ω (sufficient condition)

$$\sum_{n=-\infty}^{\infty} |x[n]| \leq c < \infty \quad \rightarrow \quad \textit{uniform convergence}.$$

(ii) If $x[n]$ is *quadratically summable*, then $X(e^{j\omega})$ exists and is a discontinuous function of ω (sufficient condition)

$$\sum_{n=-\infty}^{\infty} |x[n]|^2 \leq c < \infty \quad \rightarrow \quad \textit{not uniform convergence}.$$

(iii) If $x[n]$ is not absolutely or quadratically summable, then $X(e^{j\omega})$ can exist in special cases.

Example The DTFT of a complex exponential

$$x[n] = e^{-j\omega_0 n}, \quad -\infty < n < \infty$$

is equal to

$$X(e^{j\omega}) = \sum_{n=-\infty}^{\infty} 2\pi\underline{\delta}(\omega - \omega_0 + 2\pi n),$$

where $\underline{\delta}$ is the CT Dirac impulsive function.

Remark From the previous expressions one can simply deduce that:

- A stable circuit always has a frequency response,
- A circuit with bounded impulse response $(|h[n]| < \infty \ \forall n)$ and of finite time duration, called, as we shall see later, *Finite Impulse Response* (FIR), always has a frequency response and is therefore always stable.

1.5.2.2 DTFT and z-Transform Link

One can easily observe that (1.75) and (1.76) can be seen as a particular case of the z-transform [see (1.72) and (1.73)]. The Fourier representation is in fact achievable considering the z-transform only in the unit circle of the z-plane, as shown in Fig. 1.19.

As indicated in Fig. 1.19 the DTFT is generated by setting $z = e^{j\omega}$ in the z-transform. In the first of the two examples discussed above it is clear that, since the ROC of $X(z)$ includes the unit circle, also the DTFT converges. In other examples, the DTFT only exists if $|a| < 1$ and $|b| > 1$. Note that these conditions correspond to exponentially decreasing sequences and, therefore, to BIBO stable circuits.

1.5.2.3 Convolution Theorem

Table 1.2 shows the properties of the convolution for the z-transform. We show that this property (as well as others) is also valid for the DTFT.

A linear circuit with impulse response $h[n]$, in which input is present in a sequence $x[n]$, is subject to the relations

$$y[n] = h[n]*x[n] \Leftrightarrow Y(e^{j\omega}) = H(e^{j\omega})X(e^{j\omega})$$

and

$$y[n] = h[n]x[n] \Leftrightarrow Y(e^{j\omega}) = H(e^{j\omega})*X(e^{j\omega}).$$

That is, the convolution in the time domain is equivalent to multiplication in the frequency domain and vice versa.

Proof For (1.66) and (1.76), the output of the DT circuit can be written as

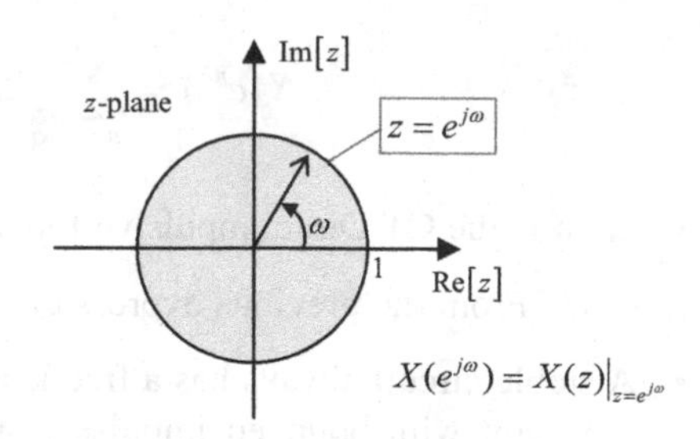

Fig. 1.19 DTFT and z-transform link

$$y[n] = \sum_{k=-\infty}^{\infty} h[k]x[n-k] = \sum_{k=-\infty}^{\infty} \left[h[k] \frac{1}{2\pi} \int_{-\pi}^{\pi} X(e^{j\omega}) e^{j\omega(n-k)} d\omega \right].$$

Separating the variables and, for property of linearity, switching the integration with the summation, we obtain

$$y[n] = \frac{1}{2\pi} \int_{-\pi}^{\pi} \left[\sum_{k=-\infty}^{\infty} h[k] e^{-j\omega k} \right] X(e^{j\omega}) e^{j\omega n} d\omega.$$

Form the transform definition (1.75), we can write

$$y[n] = \frac{1}{2\pi} \int_{-\pi}^{\pi} H(e^{j\omega}) X(e^{j\omega}) e^{j\omega n} d\omega.$$

Finally, for the definition (1.75), the output is equal to

$$Y(e^{j\omega}) = H(e^{j\omega}) X(e^{j\omega}).$$

With similar considerations, it is also easy to prove the inverse property.

1.5.2.4 Frequency and Phase Response

The frequency response $H(e^{j\omega})$ is a complex function of the complex variable depending on the angular frequency ω. It follows that $H(e^{j\omega})$ can be written, highlighting the real and imaginary part, as

$$H(e^{j\omega}) = H_{\mathrm{R}}(e^{j\omega}) + jH_{\mathrm{I}}(e^{j\omega}), \tag{1.77}$$

where $H_{\mathrm{R}}(e^{j\omega})$ and $H_{\mathrm{I}}(e^{j\omega})$ are two real functions representing, respectively, the real and imaginary part of the frequency response.

Moreover, the complex function $H(e^{j\omega})$ can be expressed in terms of modulus and phase:

$$H(e^{j\omega}) = |H(e^{j\omega})|e^{j\angle H(e^{j\omega})}, \tag{1.78}$$

where

$$\angle H(e^{j\omega}) = -\tan^{-1}\frac{H_I(e^{j\omega})}{H_R(e^{j\omega})}. \tag{1.79}$$

The expression (1.78) is sometimes written as $H(e^{j\omega}) = A(\omega)e^{j\phi(\omega)}$, where the real functions of real variable $A(\omega)$ and $\phi(\omega)$, respectively, represent the *amplitude response* and *phase response*.

Often, instead of the phase response, it is convenient to consider the *group delay* defined as

$$\tau(\omega) = -\frac{d(\angle H(e^{j\omega}))}{d\omega} \triangleq \textit{group delay}. \tag{1.80}$$

Example Calculate, as an example, the amplitude and phase response of a circuit characterized by the real exponential impulse response $h[n]$ of the type

$$h[n] = a^n u[n], \quad |a| < 1. \tag{1.81}$$

For (1.70) we have that

$$H(e^{j\omega}) = \sum_{n=0}^{\infty} h[n]e^{-j\omega n} = \sum_{n=0}^{\infty} (ae^{-j\omega})^n.$$

And for $|a| < 1$ the previous expression converges to

$$H(e^{j\omega}) = \frac{1}{1 - ae^{-j\omega}} = \frac{1}{1 - a(\cos\omega - j\sin\omega)} = \frac{1}{(1 - a\cos\omega) + ja\sin\omega}.$$

Calculating the module, we get

$$|H(e^{j\omega})| = \frac{1}{\sqrt{(1 - a\cos\omega)^2 + (a\sin\omega)^2}} = \frac{1}{\sqrt{1 - 2a\cos\omega + a^2}}.$$

Instead, for the phase we have that

$$\angle H(e^{j\omega}) = -\arctan\left(\frac{a\sin\omega}{1 - a\cos\omega}\right).$$

The amplitude and phase are plotted in Fig. 1.20.

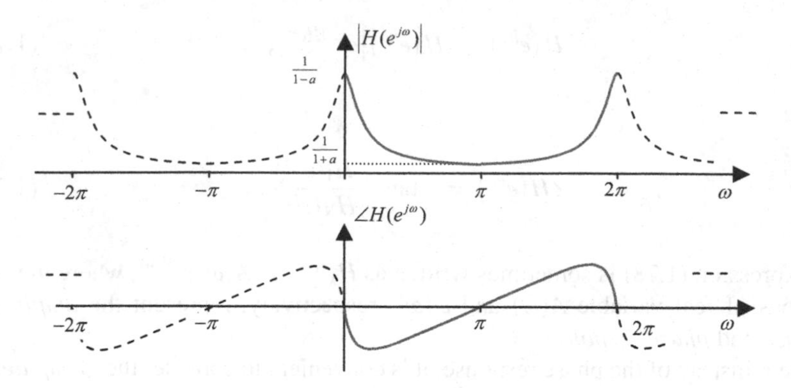

Fig. 1.20 Amplitude and phase response of the sequence $h[n] = a^n u[n]$ per $|a| < 1$

1.5.3 The z-Domain Transfer Function and Relationship with DTFT

For a DT circuit the transfer function (TF) $H(z)$ is defined as the z-domain ratio between the output and input, i.e.,

$$H(z) \triangleq \frac{Y(z)}{X(z)}. \tag{1.82}$$

As previously noted, for a DT LTI circuit, the frequency response is defined as

$$H(z)|_{z=e^{j\omega}} = H(e^{j\omega}).$$

For $X(z) = 1$, i.e., $x[n] = \delta[n]$, from the definition (1.82) and for the convolution theorem (see Table 1.2), it appears that the impulse response can be also defined as

$$h[n] = Z^{-1}\{H(z)\}, \tag{1.83}$$

where $Z^{-1}\{\cdot\}$ indicates the inverse z-transform. This expression generalizes the relationship with the Fourier series previously introduced [see (1.71)].

1.5.4 The DFT and z-Transform

As seen in Sect. 1.3, we can define the DFT as a sequence to series expansion transformation. Indeed, a DT sequence $\mathbf{x} \in (\mathbb{R}, \mathbb{C})^{N \times 1}$ can be represented in different domains using unitary transformations $\mathbf{Fx}$, where $\mathbf{F} \in (\mathbb{R}, \mathbb{C})^{N \times N}$ is proper unitary matrix called basis matrix.

In particular, considering the relationship (1.14) and (1.15), a N-length sequence can then be *exactly represented* with these couple of equations defined as DFT and IDFT.

Remark Note that the DTFT is a continuous function that represents the frequency content of a discrete nonperiodic signal. On the contrary, the DFT is a discrete function that represents periodic discrete periodic signal. Moreover, DFT has periodicity also in frequency domain, but, obviously, we calculate and represent just one period of DFT.

1.5.4.1 The FFT Algorithm

The N values of the DFT can be computed very efficiently by means of a family of algorithms called FFT [10–12]. The FFT algorithm is among the most widely used in many areas of digital signal processing, in the spectral estimation, in filtering, and in many other applications.

The fundamental paradigm for the development of the FFT is that of divide et impera. The calculation of the DFT is divided into most simple blocks and the entire DFT is calculated with the subsequent reaggregation of the various sub-blocks.

In the original algorithm of Cooley and Tukey[4] [11], the sequence length is a power of two $N = 2^B$ $(B = \log_2 N)$. The calculation of the entire transform is divided into two DFT long (N/2). Each of these is performed with a further subdivision in long sequences (N/4) and so on.

Remark Without entering on the merits of the algorithm, the details and variations may be found in [12] and [10]; we want to emphasize in this note that the computational cost of an FFT of a long sequence N is equal to $N\log_2 N$ while for a DFT it is equal to N^2.

1.6 DT Circuits Defined by Finite Difference Equations

In the class of causal DT-LTI circuits, the circuit that satisfies the finite difference equation (FDE) of order p, i.e., of the type, is of great practical importance [1–3].

$$\sum_{k=0}^{p} a_k y[n-k] = \sum_{k=0}^{q} b_k x[n-k], \qquad a_k, b_k \in \mathbb{R}, \qquad p \geq q. \tag{1.84}$$

For $a_0 = 1$, the above expression can be written in the *normalized form*

[4] Later, it was discovered that the two authors had independently reinvented an algorithm of Carl Friedrich Gauss in 1805 (and subsequently rediscovered in many other limited forms).

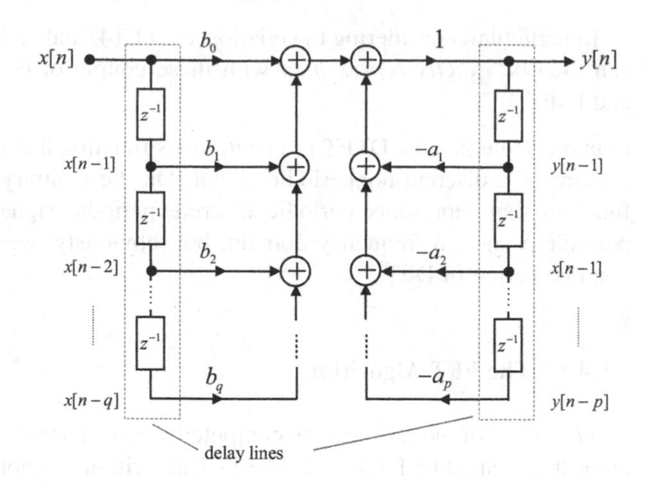

Fig. 1.21 Possible circuit representation of finite difference equation for $a_0 = 1$

$$y[n] = \sum_{k=0}^{q} b_k x[n-k] - \sum_{k=1}^{p} a_k y[n-k], \tag{1.85}$$

which appears to be characterized by a useful circuit representation shown in Fig. 1.21. From the definition (1.82), the TF of (1.84) appears to be a *rational function* of the type

$$H(z) = \frac{\sum_{k=0}^{q} b_k z^{-k}}{\sum_{k=0}^{p} a_k z^{-k}} = \left(\frac{b_0}{a_0}\right) \frac{\prod_{k=1}^{q} \left(1 - c_k z^{-1}\right)}{\prod_{k=1}^{p} \left(1 - d_k z^{-1}\right)}. \tag{1.86}$$

Therefore, the indices $(p,\ q)$ represent, respectively, the maximum degree of the polynomial in the TF's numerator and denominator. Please note that it can sometimes be convenient to indicate the summation of (1.86) such as $(M,\ N)$ representing the delay lines length (see Fig. 1.21) and it is obvious that in this case it is $M = q + 1$ and $N = p + 1$.

As for the physical realizability $p \geq q$, the FDE order is expressed by the degree of the denominator of (1.86).

1.6.1 Pole–Zero Plot and Stability Criterion

In (1.86) the polynomial roots of the numerator, indicated here as $z_1, z_2, \ldots, z_k, \ldots,$ are called *zeros*. The name "zero" is derived simply from the fact that $H(z) \rightarrow 0$ for $z \rightarrow z_k$. The polynomial roots of the denominator, herein referred to as $p_1, p_2, \ldots,$

$p_k, \ldots,$ are such that the TF $H(z) \to \infty$ for $z \to p_k$. These values are indicated as *poles*[5] of the TF.

As the frequency response (amplitude, phase, and group delay), a graphical representation of the $H(z)$ roots is largely used for circuits and systems characterization. The resulting graph, the said *pole–zero plot*, is very important in the design phase, for the evaluation of certain characteristics of the circuit TF.

As an example, consider a TF $H(z)$ defined as

$$H(z) = \frac{(1+0.75z^{-1})(1-0.5z^{-1})\left(1-0.9e^{j\frac{\pi}{2}}z^{-1}\right)\left(1-0.9e^{-j\frac{\pi}{2}}z^{-1}\right)}{\left(1-0.5e^{j\frac{\pi}{4}}z^{-1}\right)\left(1-0.5e^{-j\frac{\pi}{4}}z^{-1}\right)\left(1-0.75e^{j\frac{3\pi}{4}}z^{-1}\right)\left(1-0.75e^{-j\frac{3\pi}{4}}z^{-1}\right)}$$
$$= \frac{1.0+0.25z^{-1}+0.435z^{-2}+0.2025z^{-3}-0.30375z^{-4}}{1+0.35355z^{-1}+0.0625z^{-2}-0.13258z^{-3}+0.14062z^{-4}} \tag{1.87}$$

characterized by two pairs of complex conjugate poles and two real zeros and a pair of complex conjugate zeros.

Figure 1.22 shows the plot of the TF[6] characteristic curves. The poles position, indicated in the form $re^{\pm j\theta}$ (in our case $p_{1,2} = 0.5e^{\pm j\pi/4}$ and $p_{3,4} = 0.75e^{\pm j3\pi/4}$), determine two resonances at the respective pulsations visible in the figure. The zeros position on the real axis at π and 0 [rad] ($z_1 = 0.75e^{j\pi}$ and $z_2 = 0.5e^{j0}$), determines the attenuation of the magnitude response at the band extremities, while the pair of zeros ($z_{3,4} = 0.9e^{\pm j\pi/2}$) determines the anti-resonance at the band center. Note that the amplitudes of the resonance and anti-resonance are proportional, respectively, to the pole/zero radius.

1.6.1.1 BIBO Stability Criterion

Previously we have seen that a circuit is stable *if and only if* $\forall |x[n]| < \infty \Rightarrow |y[n]| < \infty$. For a circuit LTI is considered, moreover, the link between the impulse response and input, given by the convolution sum (1.65). If the input is limited, the condition for the limited output is then dependent on the characteristics of the impulse response.

A simple *necessary and sufficient* condition is the *absolute summability* of the impulse response $h[n]$ or

[5] It seems that the term pole is derived from the *pole* of the circus that underlies the tarp. The cusp shape assumed by the tensed canvas from the pole recalls the plot of the TF module $|H(z)|_{z \to p_k}$ for $z \to p_k$.

[6] The TF's plots were evaluated with the program MATLAB FDAtool.

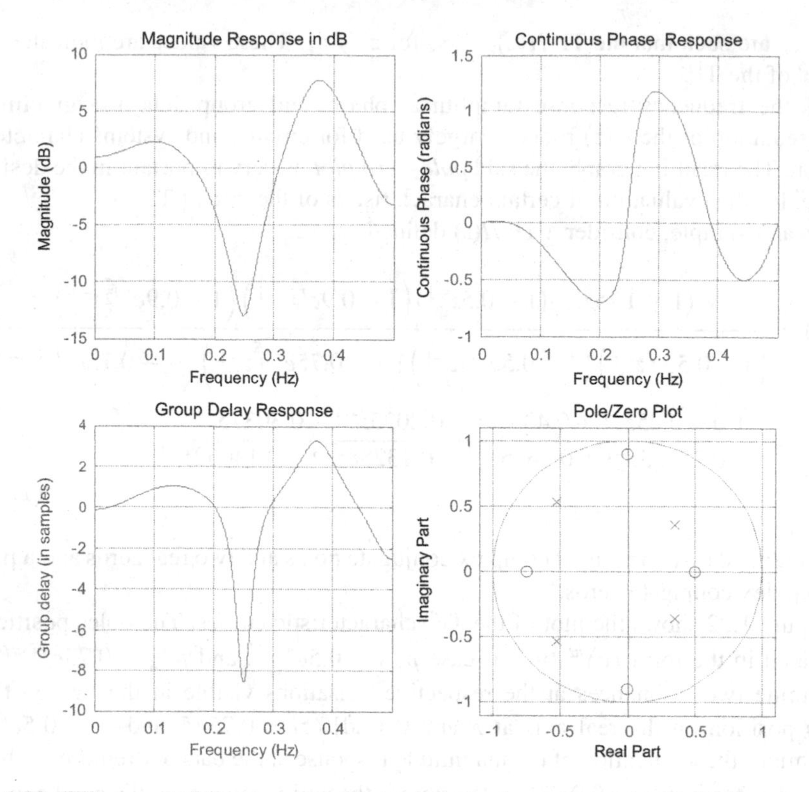

Fig. 1.22 TF's characteristic curves (amplitude response, phase, group delay, and pole–zero plot). Placing the poles and zeros in an appropriate manner it is possible to obtain, in an approximate way, a certain frequency response

$$\sum_{k=-\infty}^{\infty} |h[k]| \leq S < \infty. \tag{1.88}$$

In fact for $|x[n]| \leq C < \infty$ the sufficiency is easily proved considering that the output sequence is also bounded, i.e.,

$$|y[n]| = \left|\sum_{k=-\infty}^{\infty} h[k]x[n-k]\right| \leq C \cdot \sum_{k=-\infty}^{\infty} |h[k]| \leq C \cdot S < \infty.$$

For the necessity, the condition (1.88) is true if, for $C = \infty$, there exists (at least) an input for which the output is unbounded. For example, for a bounded input defined as $x[n] = |h[-n]|/h[-n]$, for the convolution sum (1.66), we have that the output is unbounded, i.e., $y[0] = \sum_{k=-\infty}^{\infty} |h[k]|^2/|h[k]| = C$, that proves the necessity condition.

Considering the z-transform of (1.88) [see (1.74)], an equivalent stability condition can be expressed as

$$\sum_{k=-\infty}^{\infty} \left| h[k]z^{-k} \right| < \infty. \tag{1.89}$$

For $|z| = 1$, this condition is equivalent to the ROC condition on the $H(z)$ that must include the unit circle.

The consequence of this observation for circuits modeled with causal FDE, for which the $H(z)$ is a rational function, is that all $H(z)$ TF's poles must be inside the unit circle. Thus, a causal circuit is stable *if and only if* it has all the poles within the unit circle.

1.6.2 *Circuits with the Impulse Response of Finite and Infinite Duration*

A DT LTI circuit may have an impulse response of finite or infinite duration. If the impulse response has a finite duration, the circuit is called FIR filter. If the impulse response has instead an unlimited duration the circuit is called infinite impulse response IIR filter.

In the FIR filter case the coefficients a_k, of the FDE (1.85), are all zero and, in general, the coefficients b_k are indicated in terms of the impulse response $h[k] = b_k$ for $k = 0, 1, \ldots, M - 1$. In this case the expression (1.85) becomes a simple finite convolution sum.

$$y[n] = \sum_{k=0}^{M-1} h[k]x[n-k]. \tag{1.90}$$

In this case, for (1.86) and (1.90) the $H(z)$ is of the type

$$H(z) = h[0] + h[1]z^{-1} + \cdots + h[M-1]z^{-M-1},$$

where all the poles are positioned in the origin, for which the circuit is always stable. Note that, by convention, the index M indicates the length of the impulse response and hence the maximum degree of the polynomial is equal to $(M - 1)$.

1.6.2.1 Convolution as a Product, Data-Matrix Impulse-Response Vector

In the case that the impulse response and the input signal are both sequences of finite duration the expression (1.90) can be interpreted as a matrix vector product. Let $x[n]$, $0 \leq n \leq N - 1$, and $h[n]$, $0 \leq n \leq M - 1$, with $M < N$, it follows that the output $y[n]$ has length equal to $L = N + M - 1$. Arranging the samples of the impulse response and the output in column vectors defined, respectively, as

$$\begin{aligned}\mathbf{h} &\triangleq \begin{bmatrix} h[0] & h[1] & \cdots & h[M-1] \end{bmatrix}^T, \\ \mathbf{y} &\triangleq \begin{bmatrix} y[0] & y[1] & \cdots & y[L-1] \end{bmatrix}^T. \end{aligned}$$

The system output, reinterpreting (1.90), can be written as

$$\begin{bmatrix} y[0] \\ \vdots \\ y[M-1] \\ \vdots \\ y[N-1] \\ \vdots \\ y[L-1] \end{bmatrix} = \begin{bmatrix} x[0] & 0 & \cdots & 0 \\ x[1] & x[0] & \cdots & 0 \\ \vdots & \vdots & \ddots & \vdots \\ x[M-1] & x[M-2] & \cdots & x[0] \\ x[M] & x[M-1] & \cdots & x[1] \\ \vdots & \vdots & \ddots & \\ x[N-1] & x[N-2] & \cdots & x[N-M] \\ 0 & x[N-1] & \cdots & x[N-M+1] \\ \vdots & \vdots & \ddots & \vdots \\ 0 & 0 & \cdots & x[N-1] \end{bmatrix} \begin{bmatrix} h[M-1] \\ h[M-0] \\ \vdots \\ h[0] \end{bmatrix} \tag{1.91}$$

or, in a more compact way, as

$$\mathbf{y} = \mathbf{X}\mathbf{h}, \tag{1.92}$$

where the $(L \times M)$ data matrix $\mathbf{X}$ contains the samples of the input signal arranged in columns, gradually shifted down one sample. The $\mathbf{X}$ matrix columns filling scheme is illustrated in Fig. 1.23.

Note that the first and the last $M-1$ rows of the matrix contain zeros due to the signal transient; by consequence, the first and the last $M-1$ output samples are characterized by a so-called *transient effect*.

1.6.2.2 Convolution as a Product, Convolution-Operator-Matrix Input Sequence

The relation (1.92) can also be written in the notation *convolution operator*, as

$$\mathbf{y} = \mathbf{H}\mathbf{x}. \tag{1.93}$$

Denoting by $\mathbf{h} \in \mathbb{R}^{M \times 1}$, the impulse response of the FIR filter, and with $\mathbf{x} \in \mathbb{R}^{N \times 1}$ the input sequence for $M < N$, is defined *convolution operator matrix* $\mathbf{H} \in \mathbb{R}^{(M+N-1 \times N)}$, the matrix containing the shifted impulse response $\mathbf{h}$ replicas, filled with zeros, as indicated in (1.94), such that the vectors $\mathbf{x}$ and $\mathbf{y}$ contain, respectively, the input and output window samples.

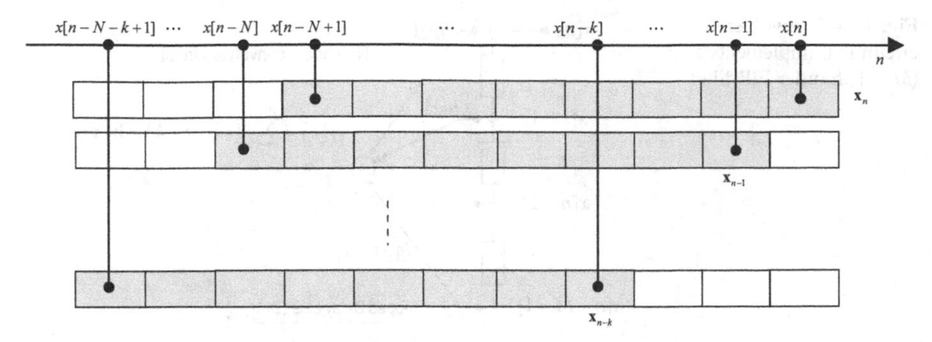

Fig. 1.23 Filling by columns, of the diagonal-constant data-matrix $\mathbf{X}^T$. The vector at the top corresponds to the first column and so on

$$\begin{bmatrix} y[0] \\ \vdots \\ \vdots \\ y[M-1] \\ \vdots \\ \vdots \\ y[N-1] \\ \vdots \\ \vdots \\ y[L-1] \end{bmatrix} = \begin{bmatrix} 0 & 0 & \cdots & \cdots & \cdots & 0 & h[M-1] \\ \vdots & \cdots & \cdots & \cdots & 0 & h[M-1] & h[M-2] \\ \vdots & \ddots & \ddots & \ddots & \ddots & \ddots & \ddots \\ \vdots & \cdots & \cdots & 0 & h[M-1] & \cdots & h[0] \\ \vdots & \cdots & 0 & h[M-1] & \cdots & h[0] & 0 \\ \vdots & 0 & \ddots & \ddots & \ddots & \ddots & \vdots \\ 0 & h[M-1] & \cdots & \cdots & \cdots & \cdots & 0 \\ h[M-1] & h[M-2] & \cdots & h[1] & h[0] & 0 & 0 \end{bmatrix} \begin{bmatrix} x[0] \\ x[1] \\ \vdots \\ x[N-1] \end{bmatrix} \tag{1.94}$$

Remarks Note that the matrices **X** and **H**, which appear in the convolution expressions (1.92) and (1.94), have identical elements on the diagonals, i.e., are *diagonal-constant matrices*. More formally, let $[a_{i,j}]$ the elements of the matrix **A** and we have that $[a_{i,j}] = [a_{i+1,j+1}]$; this particular matrix is defined as *Toeplitz matrix* and is very important for many applications described in the following chapters. More information and details on the properties of Toeplitz matrices will be provided later in the text (see also Sect. A.2.4).

1.6.2.3 Online Convolution as Inner Product Vectors

With similar reasoning we can express the nth output sample of a FIR filter as a vectors inner-product

$$y[n] = \mathbf{x}_n^T \mathbf{h} = \mathbf{h}^T \mathbf{x}_n, \qquad \text{for} \quad n = 0, 1, \ldots, L-1, \tag{1.95}$$

where $\mathbf{x}_n = \begin{bmatrix} x[n] & x[n-1] & \cdots & x[n-M+1] \end{bmatrix}^T$ indicates a M-length sliding window on the input sequence.

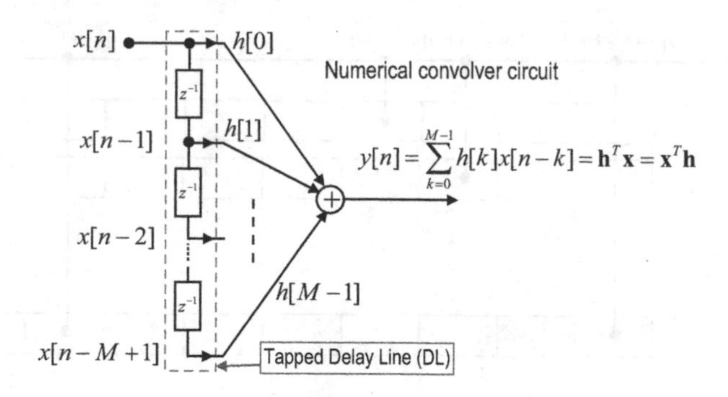

Fig. 1.24 Convolver circuit that implements a $(M-1)$th order FIR filter

Remark In general, given the FDE of the type (1.84), it describes a FIR filter if $a_0 = 1$, and $a_1 = a_2 = \cdots = a_N = 0$, i.e., the expression (1.90) where, in this case, it is usual to consider $h[k] = \{b_k\}$. The related circuit, the said *numerical time-domain convolver*, is illustrated in Fig. 1.24.

Remark A fundamental operation in digital filtering, which often determines the hardware signal processing architecture or digital signal processors (DSP), is the result of multiplying the filter coefficients by the signal samples and results in accumulation, i.e., multiply and accumulate (MAC). In this case, the convolution is implemented as the inner product (1.95), in which for each time instant the vector $\mathbf{x}_n$ is updated with the new input sequence sample as schematically shown in Fig. 1.25.

Almost all DSPs have a hardware multiplier and an instruction in *assembly* language, which directly implements the MAC, in a single machine cycle.

1.6.3 Example of FIR Filter—The Moving Average Filter

A filter that calculates the *moving average* is characterized by the following FDE:

$$y[n] = \sum_{k=0}^{M-1} x[n-k] = x[n] + x[n-1] + \cdots + x[n-M+1]. \tag{1.96}$$

The term M indicates the filter length. Its impulse response, shown in Fig. 1.26, has a finite duration.

The TF is

$$H(z) = \sum_{k=0}^{M-1} z^{-k} = \frac{1-z^{-M}}{1-z^{-1}} = \frac{z^M - 1}{z^{M-1}(z-1)}. \tag{1.97}$$

Developing the $H(z)$, we obtain

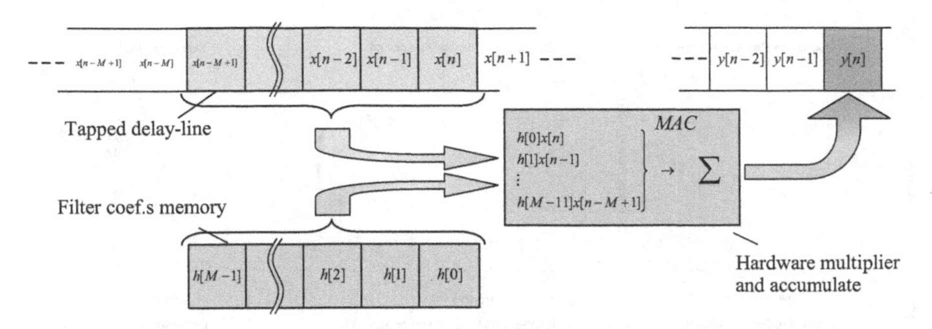

Fig. 1.25 FIR filtering as a linear combination and signal shift in the delay line

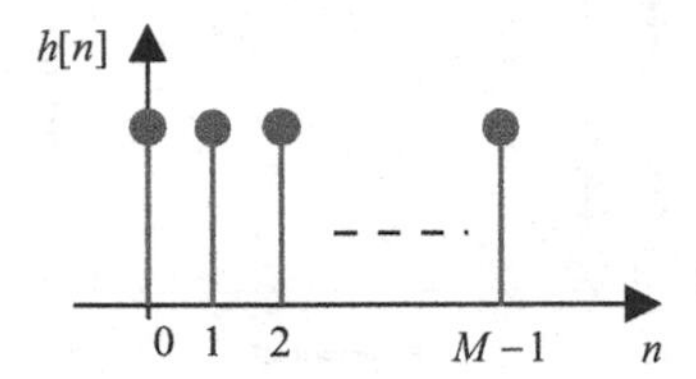

Fig. 1.26 Impulse response of a moving average of a $(M-1)$th order filter

$$H(z)=\frac{(z-1)\left(z-e^{j2\pi/M}\right)\left(z-e^{j4\pi/M}\right)\dots\left(z-e^{j2\pi(M-1)/M}\right)}{z^{M-1}(z-1)}.$$

The zero at $z=1$ is deleted from the respective pole. Dividing each term for z we get

$$H(z)=\prod_{k=1}^{M-1}\left(1-e^{j\frac{2\pi}{M}k}z^{-1}\right)$$

for which it has a pole of order M for $z=0$. The zeros are uniformly distributed in the unit circle, except for the deleted zero at $z=1$.

From (1.97), for $z=e^{j\omega}$, the DTFT of the moving average filter is equal to

$$H\left(e^{j\omega}\right)=\frac{1-e^{-j\omega M}}{1-e^{-j\omega}}=1-e^{-j\omega(M-1)/2}\frac{\text{sen}(\omega M/2)}{\text{sen}(\omega/2)}.$$

Figure 1.27 shows the frequency responses of the moving average filter, with coefficients, $h[k]=1/M$, $k=0, 1, \dots, M-1$, for $M=5$ and $M=10$. Note that the filter has, of course, low-pass-type characteristics.

The amplitude and phase response is then

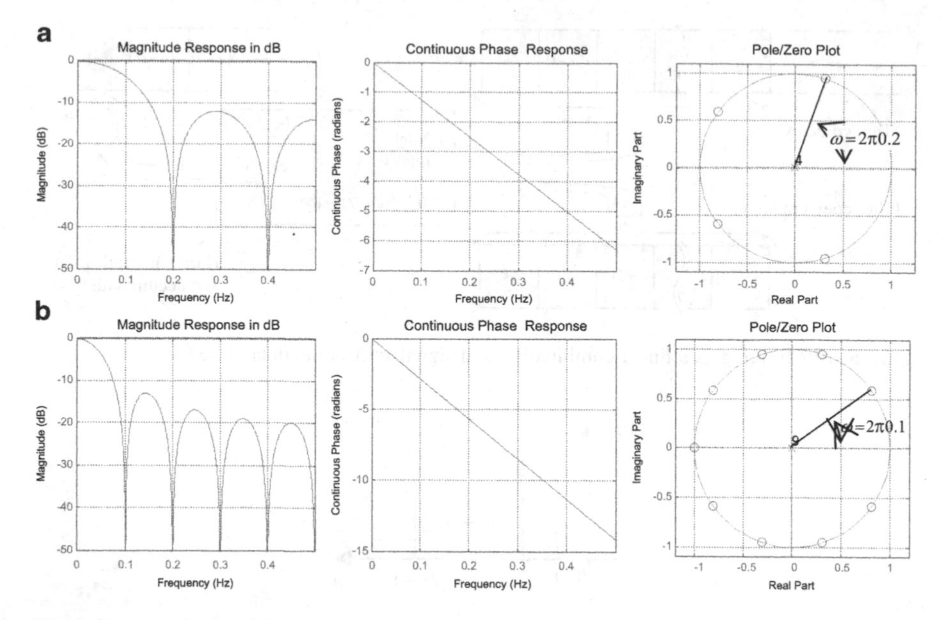

Fig. 1.27 Amplitude and phase response and pole–zero plot, in normalized frequency scale, of a moving average filter (**a**) of order 4 ($M = 5$) and (**b**) and of order 9 ($M = 10$)

$$\left|H\left(e^{j\omega}\right)\right| = \left|\frac{\operatorname{sen}(\pi f N)}{\operatorname{sen}(\pi f)}\right|, \quad -0.5 \le f \le 0.5, \tag{1.98}$$

$$\arg\left[H\left(e^{j\omega}\right)\right] = -\pi f(M-1). \tag{1.99}$$

1.6.4 Generalized Linear-Phase FIR Filters

A DT LTI circuit is called *linear phase* when its frequency response can be expressed in the form

$$H\left(e^{j\omega}\right) = \left|H\left(e^{j\omega}\right)\right| e^{-j\alpha\omega}, \tag{1.100}$$

where α is a real number [see, for example, expressions (1.98) and (1.99)]. In other words, for (1.80), the system with TF (1.100) presents a constant group delay

$$\tau(\omega) = \alpha. \tag{1.101}$$

In some applications, it may be necessary to define the phase behavior more generally. Such systems, called *generalized linear phase*, are characterized by a frequency response defined as

$$H\left(e^{j\omega}\right) = A\left(e^{j\omega}\right)e^{j\phi(\omega)} \tag{1.102}$$

with

$$\phi(\omega) = -\alpha\omega + \beta, \tag{1.103}$$

where A(ω) and $\phi(\omega)$ are real functions of ω, and α and β terms are constants. In the case of M-length FIR filters, the sufficient conditions to obtain a generalized linear phase are the symmetry or the anti-symmetry. That is,

$$h[n] = \pm h[N-1-n]. \tag{1.104}$$

Examples of impulse responses plot for generalized linear-phase FIR filters are shown in Fig. 1.28.

From the figure we observe that it is possible to define four types of filters of odd/even length and odd/even symmetry. In addition, we can observe that symmetry condition expressed in terms of the z-transform of (1.104) can be written as

$$H(z) = \pm z^{N-1}H\left(z^{-1}\right). \tag{1.105}$$

Note that this condition can be very useful in filter bank design [7].

Remark The linear-phase FIR filters are of central importance in many DSP practical applications. In addition, it can be proved that causal linear-phase IIR system, describable by generalized FDE, does not exist.

1.6.5 Example of IIR Filter

A filter with impulse response of infinite duration is characterized by a TF of the type (1.86) in which the term $a_0 = 1$. For example, in the case of second-order IIR filter, the FDE can be written as

$$y[n] = b_0x[n] + b_1x[n-1] + b_2x[n-2] - a_1y[n-1] - a_2y[n-2]$$

with TF

$$H(z) = \frac{b_0 + b_1z^{-1} + b_2z^{-2}}{1 + a_1z^{-1} + a_2z^{-2}}.$$

The II order form, called *II second-order cell*, is important since it is one of the fundamental filtering building blocks techniques, with which to make more complex circuit architecture.

Possible DT circuit schemes of the above equation are represented in Fig. 1.29.

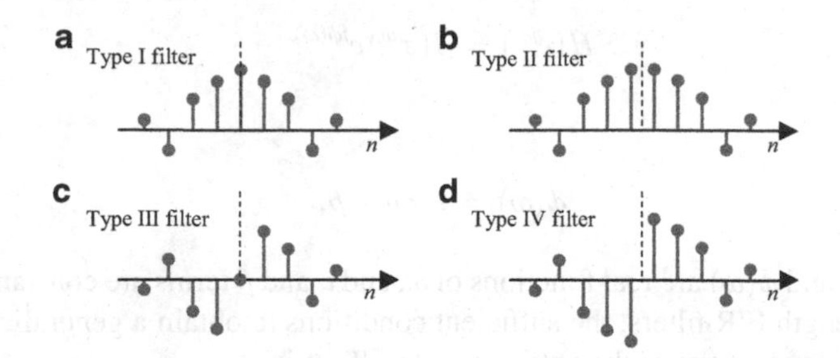

Fig. 1.28 The impulse response symmetry of generalized linear-phase FIR filters (**a**) Type I, even symmetry, odd M; (**b**) Type II, even symmetry, even M; (**c**) Type III, odd symmetry odd M; (**d**) Type IV, odd symmetry, even M

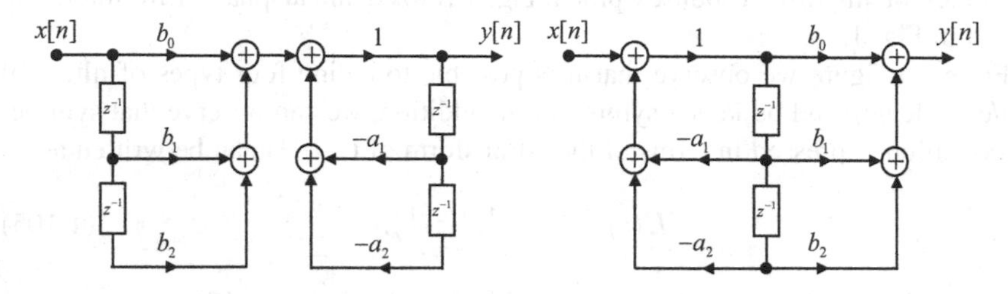

Fig. 1.29 Possible schemes for a second order IIR filter, also called second order cell

1.6.5.1 Digital Resonator

The *digital resonator* is a circuit characterized by a magnitude response peak around a certain frequency. In practice, the resonance is made with a pair of complex conjugate poles very close to the unit circle.

In the case the TF of second-order resonator can be written as

$$H(z) = \frac{1}{(1 - re^{j\theta}z^{-1})(1 - re^{-j\theta}z^{-1})},$$

where r is the radius of the pole whose value determines the resonance width, while the phase θ determines the frequency.

Figure 1.30 shows the characteristic plots of two digital resonators with the following network functions:

$$H(z) = \frac{1}{(1 - 0.95e^{j\frac{\pi}{2}}z^{-1})(1 - 0.95e^{-j\frac{\pi}{2}}z^{-1})} = \frac{1}{1 + 0.9025z^{-2}}$$

and

$$H(z) = \frac{1}{(1 - 0.707e^{j\frac{\pi}{2}}z^{-1})(1 - 0.707e^{-j\frac{\pi}{2}}z^{-1})} = \frac{1}{1 + 0.5z^{-2}}$$

for the Fig. 1.30a, b, respectively.

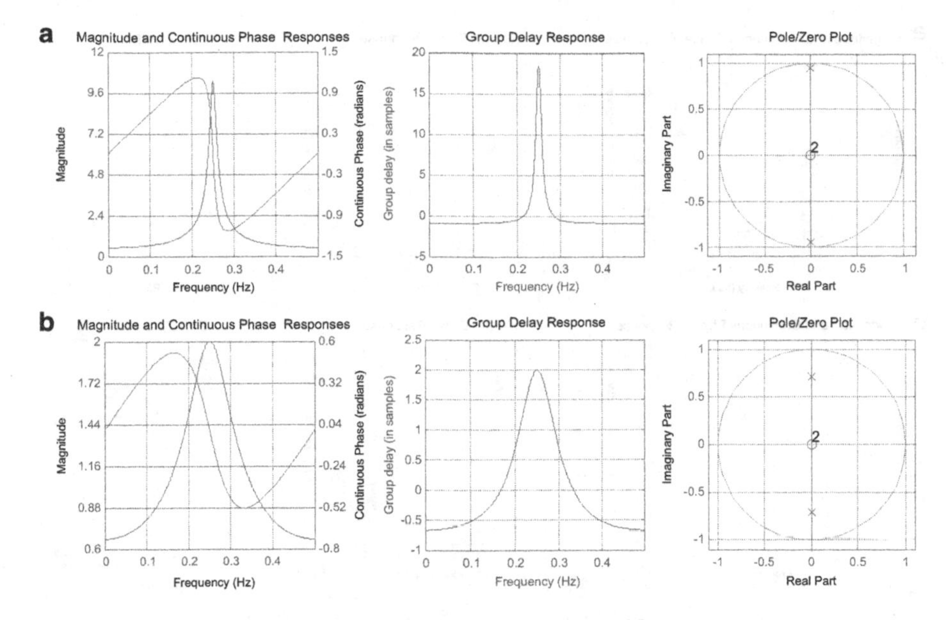

Fig. 1.30 Characteristic plots of IIR digital resonators (**a**) $r = 0.95$ and $\theta = \pi/2$; (**b**) $r = 0.707$ and $\theta = \pi/2$

1.6.5.2 Anti-resonant Circuits and Notch Filter

An anti-resonance can be easily obtained by placing a pairs of complex conjugate zeros on the unit circle in order to zero the TF at the location of the zero itself.

However, with this method any control over the bandwidth of anti-resonant filter is not possible. In practice, to achieve a good selectivity, or a narrow band anti-resonant filter, called *notch filter*, it is sufficient to place a pair of complex conjugate poles (with $r < 1$) to same frequencies, on the radius r of the zeros. The filter TF is then

$$H(z) = \frac{\left(1 - e^{j\theta}z^{-1}\right)\left(1 - e^{-j\theta}z^{-1}\right)}{\left(1 - re^{j\theta}z^{-1}\right)\left(1 - re^{-j\theta}z^{-1}\right)} \quad \text{with} \quad r < 1.$$

In this way the presence of the pole, although not completely canceling the effect of the zero in correspondence of the anti-resonance, makes the curve much narrower. Figure 1.31 shows the characteristic plots of two digital notch filters with the following TF:

$$H(z) = \frac{\left(1 - e^{j\frac{\pi}{4}}z^{-1}\right)\left(1 - e^{-j\frac{\pi}{4}}z^{-1}\right)}{\left(1 - 0.95e^{j\frac{\pi}{4}}z^{-1}\right)\left(1 - 0.95e^{-j\frac{\pi}{4}}z^{-1}\right)} = \frac{1 - 1.41421z^{-1} + z^{-2}}{1 - 1.34350z^{-1} + 0.90250z^{-2}}$$

and

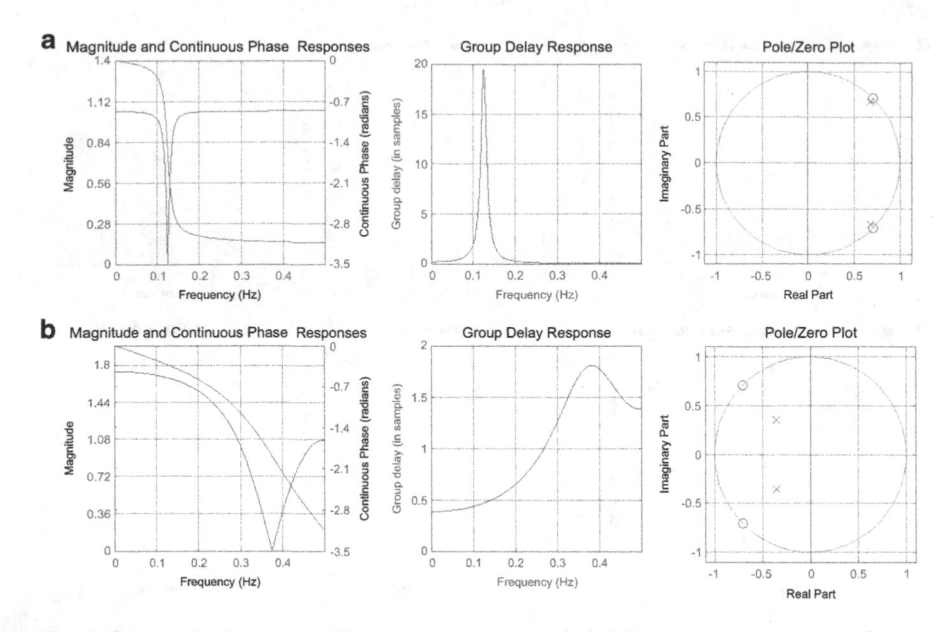

Fig. 1.31 Characteristic plots of IIR digital *notch* filter (**a**) $r = 0.95$ and $\theta = \pi/4$; (**b**) $r = 0.707$ and $\theta = (3\pi)/4$

$$H(z) = \frac{\left(1 - e^{j\frac{3\pi}{4}}z^{-1}\right)\left(1 - e^{-j\frac{3\pi}{4}}z^{-1}\right)}{\left(1 - 0.5e^{j\frac{3\pi}{4}}z^{-1}\right)\left(1 - 0.5e^{-j\frac{3\pi}{4}}z^{-1}\right)} = \frac{1 + 1.41421z^{-1} + z^{-2}}{1 + 0.707z^{-1} + 0.25z^{-2}}$$

for Fig. 1.31a, b, respectively.

1.6.5.3 All-Pass Filter

An all-pass filter has a TF in which the zeros are the reciprocals of the poles. It follows that the amplitude response is flat (the zero module cancels the pole module), while, being respectively the poles and zeros inside and outside the unit circle, the phases of the poles and the zeros have the same sign. The phase, therefore, can assume values even higher. In fact, the all-pass filter is not a *minimum-phase* filter.[7]

The N-order TF of an all-pass filter can be written as

[7] A stable circuit is said to be minimum phase if the zeros make a positive contribution to the phase or in the case of analog circuits are to the left of the imaginary axis or, in the case of DT circuits, are inside the unit circle.

$$H(z)=\prod_{k=1}^{N/2}\left[r_k^2\frac{\left(1-r_k^{-1}e^{j\theta_k}z^{-1}\right)\left(1-r_k^{-1}e^{-j\theta_k}z^{-1}\right)}{(1-r_ke^{j\theta_k}z^{-1})(1-r_ke^{-j\theta_k}z^{-1})}\right],$$

which with simple calculations can be shown to be equivalent to the TF written as

$$H(z)=\frac{z^{-N}D(z^{-1})}{D(z)}=\frac{a_N+a_{N-1}z^{-1}+\cdots+a_1z^{-(N-1)}+z^{-N}}{1+a_1z^{-1}+\cdots+a_Nz^{-N}}. \tag{1.106}$$

It may be noted that the numerator polynomial is the *mirror version* of the polynomial denominator

The input–output FDE relationship is

$$y[n]=a_Nx[n]+\cdots+x[n-N]-a_1y[n-1]-\cdots-a_Ny[n-N]. \tag{1.107}$$

For example, a first-order all-pass filter has a TF defined as

$$H(z)=\frac{z^{-1}+a_1}{1+a_1z^{-1}}=a_1^{-1}\frac{1+a_1^{-1}z^{-1}}{1+a_1z^{-1}} \tag{1.108}$$

from where we observe that zero is the reciprocal of the pole. Figure 1.32a is an example of characteristic plot for a first-order all-pass filter.

Figure 1.32b shows the characteristic curves of a second-order all-pass filter characterized by a TF corresponding to a pair of complex conjugate poles arranged on the axis $\pm\pi/4$ with $r=0.8$, i.e.,

$$H(z)=0.8^2\frac{\left(1-1.25e^{\pm j\pi/4}z^{-1}\right)}{\left(1-0.8e^{\pm j\pi/4}z^{-1}\right)}=\frac{0.64-1.13137z^{-1}+z^{-2}}{1-1.13137z^{-1}+0.64z^{-2}}. \tag{1.109}$$

An all-pass filter, widely used in audio applications, for echo effects, etc., artificial reverberators, is characterized by a TF of the type

$$H(z)=\frac{g+z^{-D}}{1+gz^{-D}} \tag{1.110}$$

for which the FDE that realizes is

$$y[n]=gx[n]+x[n-D]-gy[n-D]. \tag{1.111}$$

The circuit structure of this type of all-pass filter, also called generalized or *universal comb filter*, can be realized with a single delay line as shown in Fig. 1.34. Figure 1.33 shows the characteristic plots typical of a generalized comb filter.

Note the uniform distribution of the pole–zero around the unit circle and the group delay plot which has a comb shape (hence the name comb filter).

All methods described are reported in [1, 2, 13].

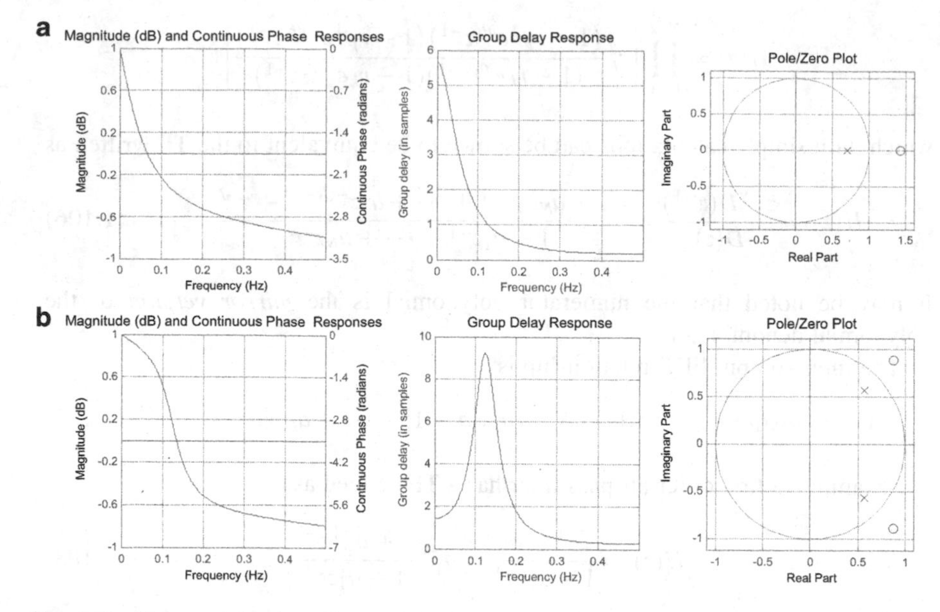

Fig. 1.32 Characteristic plots of typical all-pass filters (**a**) I order (1.108) with $a_1 = -0.7$; (**b**) II order (1.109)

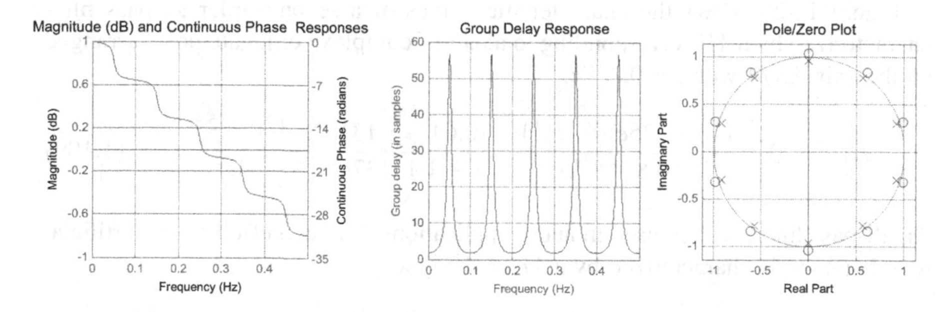

Fig. 1.33 Universal comb all-pass filter and its representation in direct form II with a single delay line

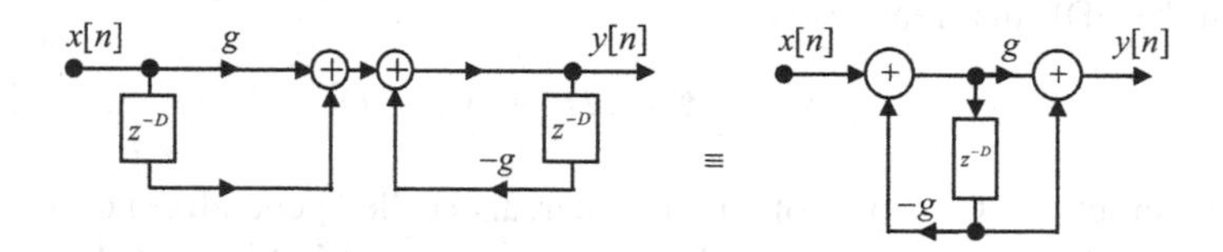

Fig. 1.34 Example of *universal comb filter* with Type I TF (1.10) with $D = 10$, $g = 0.7$

1.6.6 Inverse Filters

A circuit with TF $H_i(z)$ is called inverse of $H(z)$, if $H(z)H_i(z) = 1$, which implies that $H_i(z) = 1/H(z)$, i.e., in the time domain $h[n] * h_i[n] = \delta[n]$. In addition, the amplitude response $H_i(e^{j\omega})$ is a mirror version, with respect to the unitary amplitude, of $H(e^{j\omega})$.

Not for all circuits exists the inverse form. For example, an ideal low-pass filter (which completely eliminates all frequencies below a certain cutoff frequency) does not admit the inverse form. In fact, the eliminated frequencies may in no way be recovered.

A generic circuit with TF with zeros inside the unit circle

$$H(z) = \frac{b_0 \prod_{k=0}^{M} \left(1 - c_k z^{-1}\right)}{a_0 \prod_{k=N}^{N} \left(1 - d_k z^{-1}\right)}$$

admits a stable inverse circuit of the type

$$H(z) = \frac{a_0 \prod_{k=N}^{N} \left(1 - d_k z^{-1}\right)}{b_0 \prod_{k=0}^{M} \left(1 - c_k z^{-1}\right)}$$

i.e., the $H(z)$ zeros become the poles of the inverse filter $H_i(z)$.

The previous equations show that a causal stable and minimum-phase circuit is invertible and its inverse will be stable, causal, and minimum-phase.

Property Given a not minimum phase $H(z)$, this can always be expressed as the product of a minimum-phase function $H_{\text{mp}}(z)$ and an all-pass rational function $H_{\text{ap}}(z)$, i.e.,

$$H(z) = H_{\text{mp}}(z)H_{\text{ap}}(z). \tag{1.112}$$

For the proof of this property, suppose that $H(z)$ has a zero outside the unit circle, equal to $z = re^{\pm j\theta}$ (with $|r| > 1$), and all remaining poles/zeros are inside the unit circle. The $H(z)$, highlighting such zeros, can be rewritten as

$$H(z) = H_1(z)\left(1 - re^{j\theta}z^{-1}\right)\left(1 - re^{-j\theta}z^{-1}\right)$$

in which, by definition, the $H_1(z)$ is minimum phase. Also note that, by multiplying and dividing by the term $\left(1 - r^{-1}e^{\pm j\theta}z^{-1}\right)$, this expression can be rewritten as

$$H(z) = \underbrace{H_1(z)\left(1 - r^{-1}e^{j\theta}z^{-1}\right)\left(1 - r^{-1}e^{-j\theta}z^{-1}\right)}_{\text{minimum phase}} \underbrace{\frac{\left(1 - re^{j\theta}z^{-1}\right)\left(1 - re^{-j\theta}z^{-1}\right)}{\left(1 - r^{-1}e^{j\theta}z^{-1}\right)\left(1 - r^{-1}e^{-j\theta}z^{-1}\right)}}_{\text{all-pass}},$$

which demonstrates (1.112).

References

1. Oppenheim AV, Schafer RW, Buck JR (1999) Discrete-time signal processing, 2nd edn. Prentice Hall, Englewood Cliffs, NJ
2. Rabiner LR, Gold B (1975) Theory and application of digital signal processing. Prentice-Hall, Englewood Cliffs, NJ
3. Ahmed N, Natarajan T, Rao KR (1974) Discrete cosine transform. IEEE Trans Comput 23 (1):90–93
4. Martucci SA (1994) Symmetric convolution and the discrete sine and cosine transforms. IEEE Trans Signal Process SP-42(5):1038–1051
5. Haar A (1910) Zur Theorie der Orthogonalen Funktionensysteme. Math Ann 69:331–371
6. Mallat SG (1998) A wavelet tour of signal processing. Academic, San Diego, CA. ISBN 0-12-466605-1
7. Vetterli M, Kovačević J (2007) Wavelets and subband coding. Open-Access Edition, http://www.waveletsandsubbandcoding.org/
8. Vetterli M, Kovačević J, Goyal VK (2013) Foundations of signal processing. Free version, http://www.fourierandwavelets.org
9. Feig E, Winograd S (1992) Fast algorithms for the discrete cosine transform. IEEE Trans Signal Process 40(9):2174–2193
10. Frigo M, Johnson SG (2005) The design and implementation of FFTW3. Proc IEEE 93 (2):216–231
11. Cooley JW, Tukey J (1965) An algorithm for the machine calculation of complex Fourier Series. Math Comput 19:297–301
12. Brigam EO (1998) The fast fourier transform and its application. Prentice-Hall, Englewood Cliffs, NJ
13. Antoniou A (1979) Digital filters: analysis and design. MacGraw-Hill, New York

Chapter 2
Introduction to Adaptive Signal and Array Processing

2.1 Introduction

In the study of signal processing techniques, the term *adaptive* is used when a system (analogue or digital) is able to *adjust* their own parameters in response to external stimulations. In other words, an *adaptive system* autonomously changes its internal parameters for achieving a certain processing goal such as, for example, the minimization of the effect of noise overlying the signal of interest (SOI).

In the digital signal processing (DSP), a discrete-time (DT) circuit that accepts as input one or more signals, executes a prescribed processing, and produces one or more outputs, is defined as *numerical filter*. More specifically, the term filter refers to a device *hardware* or *software* able to process the input signals with the aim of extracting information on the basis of specific criteria. In this context, an adaptive filter (AF) can be defined as a *smart circuit* capable of adapting according to an established law. The adaptation law is defined as a function, stochastic, deterministic, or heuristic of external signals and of the AF's free parameters.

The usability of adaptive filtering methods to the solution of real problems is extensive as are the areas of practical interest. The AFs are widely used in many signal processing areas such as modeling, estimation, detection, sources separation, etc. For example, in order to create a model of as physical systems, AFs have potential applications in all high-tech areas (like biomedical, acoustical, telecommunications, mechanical, physic, economical, management, financial, etc.) [10–14, 17–19].

With the *neural networks* advent, which can be considered as a class of nonlinear AF, application field is further extended to the area of the *artificial intelligence* methodologies in order to provide consistent solutions also in the case of so-called *ill-posed* problems. More recently, such methods have been merged into a nascent discipline called *computational intelligence* [4–9, 22].

A. Uncini, *Fundamentals of Adaptive Signal Processing*, Signals and Communication Technology, DOI 10.1007/978-3-319-02807-1_2,

2.1.1 Linear Versus Nonlinear Numerical Filter

A numerical or digital filter is defined by a relationship between an input $\{x[n], x[n-1], \ldots\}$ and an output $\{y[n], y[n-1], \ldots\}$ [16]. In general, $x[n]$ and $y[n]$ can be stochastic or deterministic one or multidimensional sequences. In the case of *transversal* or FIR filters, the output depends only on the input with a relation of the type as follows:

$$y[n] = \Phi\{x[n], x[n-1], \ldots, x[n-M+1]\} \tag{2.1}$$

with M equal to the filter memory length. In the case of *recursive* or IIR filters, the output depends on the input signal and also on the past outputs as follows:

$$y[n] = \Psi\{x[n], x[n-1], \ldots, x[n-M+1], y[n-1], \ldots, y[n-N+1]\}, \tag{2.2}$$

where N is the length of the delay line on the output signal. The pair $(N-1, M-1)$ is defined as *filter order*.

In the case of linear FIR or IIR filters, the operators $\Phi(\cdot)$ and $\Psi(\cdot)$ take the form of a linear combination. In particular, the expressions (2.1) and (2.2) are written, in the case of transversal filters, as discrete-time convolution:

$$y[n] = \sum_{k=0}^{M-1} h_k x[n-k] \qquad \text{or} \qquad y[n] = \sum_{k=0}^{M-1} h[k]x[n-k] \tag{2.3}$$

while, in the case of recursive filters, as finite difference equation (FDE):

$$\begin{aligned} y[n] &= \sum_{k=0}^{M-1} b_k x[n-k] - \sum_{k=1}^{N-1} a_k y[n-k] \quad \text{or} \\ y[n] &= \sum_{k=0}^{M-1} b[k]x[n-k] - \sum_{k=1}^{N-1} a[k]y[n-k]. \end{aligned} \tag{2.4}$$

In these cases, the free parameters are, in the FIR case, the impulse response samples h_k, for $k=0, 1, \ldots, M-1$, while in the case IIR the FDE coefficients b_k, for $k=0, 1, \ldots, M-1$ and a_k for $k=1, 2, \ldots, N-1$. For filter design, we intend the determination of these free parameters and of the order, which in this case coincides with the degree of the numerator and denominator polynomials of the transfer function (TF) associated with (2.4).

The ability of a filter to perform a certain task is usually expressed through a criterion that minimizes a given cost function (CF), often referred as $J(\cdot)$, depending on the filter free parameters.

For example, said Ω the space of free parameters: $\mathbf{w} \in \mathbb{R}^{M\times 1} \subset \Omega$, i.e., $\mathbf{w} = \{h_k\}$ or $\mathbf{w} = \{a_k, b_k\}$, if you desire a certain frequency response, called $H_d(e^{j\omega})$ the *desired frequency response* and the $H(e^{j\omega})$ is the actual filter frequency response, the determination of **w** is performed by minimizing some distance between the two responses. It follows that the criterion for the determination of the filter parameters **w** can be reduced to an optimization problem that can be formalized as

$$\mathbf{w} \therefore \arg\min_{\mathbf{w}\in\Omega} \left\{J(\mathbf{w})\right\}, \quad \text{for} \quad J(\mathbf{w}) = \left\|H(e^{j\omega}) - H_d(e^{j\omega})\right\|^{L_p}, \tag{2.5}$$

where the CF $J(\mathbf{w})$ is defined by the distance, indicated with $\|\cdot\|^{L_p}$, between the two frequency responses and L_p denotes the norm of the distance (L_0, L_1, L_2, …, L_p, …, L_∞). Once chosen the order and determined the free parameters, the filter **w** is *frozen* on a circuit structure suitable to perform filtering process.

Remark We remind the reader that the vector norm $\|\mathbf{x}\|^{L_p}$, also referred to as $\|\mathbf{x}\|_p$, is defined as

$$\|\mathbf{x}\|_p \triangleq \left(\sum_{i=0}^{N-1} |x[i]|^p\right)^{1/p}, \quad \text{for} \quad p \geq 1. \tag{2.6}$$

For other definitions and details, see Appendix A (Sect. A.10).

2.2 Definitions and Basic Property of Adaptive Filtering

An AF is a circuit, generally defined in the DT domain, in which the free parameters **w** are continuously changed according to a priori defined criteria without an explicit user control. A principle diagram of a generic adaptive filter is shown in Fig. 2.1.

Since the parameters **w** are subject to change during the filtering process, the AF is *time-variant* (or *nonstationary*) system. So, to indicate the free parameters, the more adequate formalism is $\mathbf{w} \to \mathbf{w}[n]$ or $\mathbf{w} \to \mathbf{w}_n$. In convolution expressions or in the FDE, the parameters are no longer constant but a time function. For example, the (2.3) can be rewritten as

$$y[n] = \sum_{k=0}^{M-1} h_n(k)x[n-k] \qquad \text{or} \qquad y[n] = \sum_{k=0}^{M-1} h_k[n]x[n-k] \tag{2.7}$$

while the expression (2.4) can be rewritten as

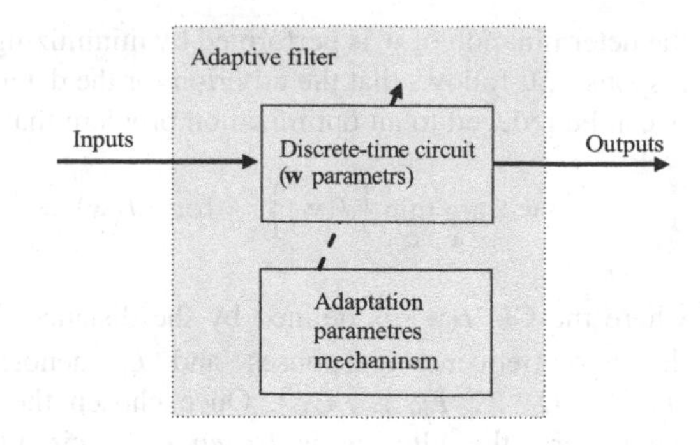

Fig. 2.1 Schematic diagram of an adaptive filter characterized by a certain relationship between the inputs and outputs, and the adaptation mechanism for changing the relationship itself

$$y[n]=\sum_{k=0}^{M-1}b_n(k)x[n-k]-\sum_{k=1}^{N-1}a_n(k)y[n-k] \quad \text{or as}$$
$$y[n]=\sum_{k=0}^{M-1}b_k[n]x[n-k]-\sum_{k=1}^{N-1}a_k[n]y[n-k]. \tag{2.8}$$

In more formal terms, an AF can be defined by the presence of two distinct parts:

1. An algorithm that processes the input signals and produces a certain output, such as in the linear case, that implements the relation (2.7) or (2.8).
2. An algorithm that computes the circuit parameters $\mathbf{w}[n]$ variation law of the type $\mathbf{w}[n+1]=\mathbf{w}[n]+\Delta\mathbf{w}[n]$ with $\Delta\mathbf{w}[n]$ determined by a certain predetermined criterion.

Therefore, an AF is characterized by the ability to change its parameters according to certain external signals. Although from the conceptual point of view, this is possible regardless of the continuous-time (CT) or DT circuit nature, it is obvious that the change of parameters in analog circuits appears to be, from the technological point of view, rather complex and in the following chapters we will refer to DT circuits only.

Figures 2.2 and 2.3 show an indicative scheme of adaptive filters.

The procedure that determines the variation law of the free parameters is also called *adaptation algorithm* or, depending on the context, *learning algorithm*. The learning with the aid of an external reference signal as shown in Fig. 2.2a, is defined as *supervised*. In the case where the learning is driven by a kind of self-organization without external references, we generally use the terms *unsupervised* or *blind* learning as shown in Fig. 2.2b. In particular, with reference to Fig. 2.3, we can define the following quantities:

- $x[n]$ is the input signal that can be considered of deterministic or stochastic nature. Considering the biological paradigm, $x[n]$ represents the data observation provided by the sensory organs.
- $d[n]$ is the external reference also said desired output. In other contexts, $d[n]$ indicates the *example* provided by the *supervisor* or *teacher*.

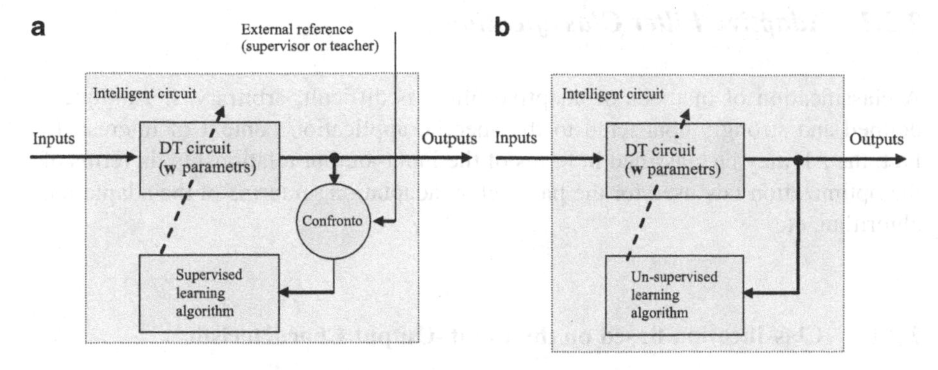

Fig. 2.2 Learning paradigms (**a**) with supervision; (**b**) without supervision

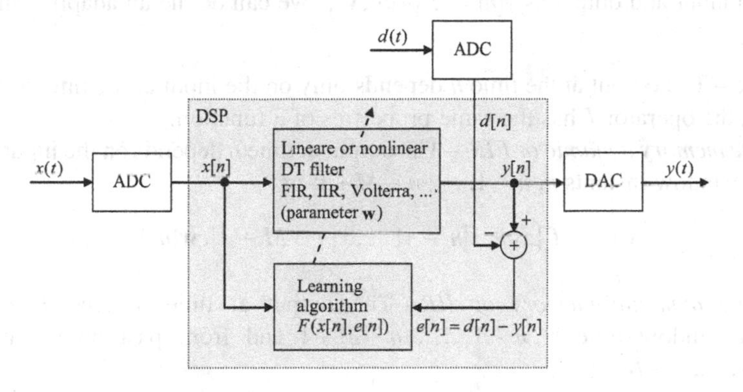

Fig. 2.3 Schematic diagram of a discrete-time *adaptive filter*

- $y[n]$ indicates the circuit output. This signal is in practice the task of the circuit.
- $e[n]$ represents the deviation between the reference provided by the supervisor and the output of the circuit: $e[n] = d[n] - y[n]$ and is defined as *error signal*.

The choice of the CF (or *optimization criterion*) in the supervised case is some function of the input and error, of the type $J(\mathbf{w}) = F(x[n], e[n])$, which, together with the learning algorithm, appear to be of central importance in applications.

In other words, the adaptation process can be defined as a dedicated procedure for determining, during the filtering operation itself, the optimum value of the set of free parameters **w** such that the filter is able to perform a certain predetermined task. This procedure consists in an optimization process that acts on the CF $J(\mathbf{w})$ minimization, through the acquired input signals and, if available, through a set of a priori known information, conjecture, etc.

2.2.1 *Adaptive Filter Classification*

A classification of methods of adaptive filters is difficult, arbitrary, not uniquely defined and strongly connected to the specific application context of interest. In fact, the AF may be classified in terms of the input–output relationship, in terms of the optimization law used for the parameters adaptation, in terms of the adaptation algorithm, etc.

2.2.1.1 Classification Based on the Input–Output Characteristic

Starting from the properties of an operator $T\{\cdot\}$ that defines the relationship between input and output as $y[n] = T\{x[n], \mathbf{w}\}$, we can define an adaptive filter as follows:

- *Static*—The output at the time n depends only on the input at the time n. In this case, the operator T has the same properties of a function.
- *Finite memory dynamic or FIR*—The output at time n depends on the input to the time window instants n, $n-1$, …, $n-M+1$, or

$$y[n] = T\{x[n], x[n-1], ..., x[n-M+1], \mathbf{w}[n]\}. \tag{2.9}$$

- *Infinite dynamic memory or IIR*—The output at time n depends on the input window time n, $n-1$, …, $n-M+1$ and from past output at time $n-1$, …, $n-N+1$

$$y[n] = T\{x[n], x[n-1], ..., x[n-M+1], y[n-1], ..., y[n-N+1], \mathbf{w}[n]\} \tag{2.10}$$

- *Linear*—For the operator $T\{\cdot\}$ the *superposition principle* is valid.
- *Nonlinear*: For the operator $T\{\cdot\}$ the *superposition principle* is not valid. In this case it is possible and usually necessary, according to specific application, to define further subclassifications due to the nature of the nonlinearity that can be monodrome, invertible, not invertible, static, dynamic, etc.

For example, as previously seen, a linear finite memory dynamic circuit is defined as a simply AF. In particular, in the case of infinite memory dynamic, is defined as *recursive*-AF or IIR-AF, while in the case of finite memory is referred to as *transversal*-AF or FIR-AF. Again, by way of example, an artificial neural network (ANN) belongs to the class of nonlinear static or dynamic circuits. In particular, ANNs with finite or infinite memory are often referred to as recurrent neural networks (RNN).

A possible AF classification on the basis of input–output characteristics, relative only to the dynamic case, is shown in Fig. 2.4.

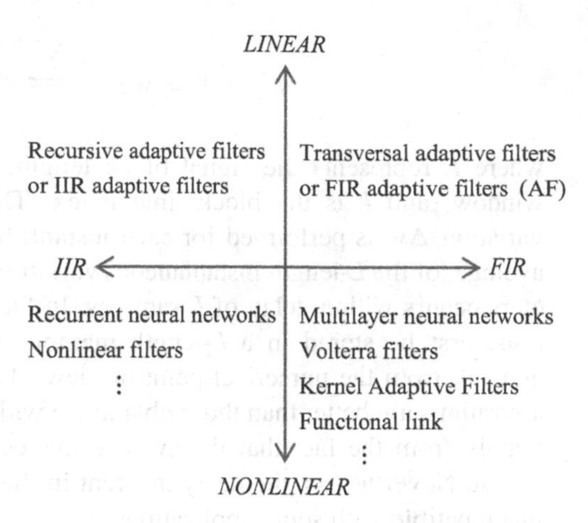

Fig. 2.4 Classification of AF on the basis of input–output characteristics

2.2.1.2 Classification Based on the Learning Algorithm

Another way to classify the AF is related to the adaptation capacity, namely the *learning algorithm* or, in other words, the way in which it is possible to calculate the free parameters $\mathbf{w}$ in function of the external signals.

The adaptation law may be defined by

$$\mathbf{w}[n+1] = \mathbf{w}[n] + \Delta\mathbf{w}[n] \qquad \text{or} \qquad \mathbf{w}[n] = \mathbf{w}[n-1] + \Delta\mathbf{w}[n], \tag{2.11}$$

where the index n may not be related to the time index of input–output signal.

- *Supervised algorithms*—In this case, the calculation of the parametric variation $\Delta\mathbf{w}[n]$ is a function of the input and reference signals (or of the error), for which we can write

$$\Delta\mathbf{w}[n] = F\{x[n], e[n]\}. \tag{2.12}$$

- *Unsupervised or blind algorithms*—The calculation of the parametric variation $\Delta\mathbf{w}[n]$ is not dependent on the reference signal but is a function, more or less explicit, of the circuit's input and output:

$$\Delta\mathbf{w}[n] = F\{x[n], y[n]\}. \tag{2.13}$$

- *Online algorithms*—With this class of learning algorithms, very important in some DSP application, the update of the parameters $\mathbf{w}$ is carried out whenever a new input sample is available. The parameters update is done continuously and for each new input, one output sample is produced with a *group-delay* of the corresponding TF. Therefore, the system could have (or it is desirable to have) some tracking capabilities.
- *Block algorithms*—With the so-called block algorithms, the calculation of parameters $\mathbf{w}$ is periodically done with a relation of the type

$$\mathbf{w}[k+1] = \mathbf{w}[k] + \frac{\sum_{i=0}^{L-1} \Delta\mathbf{w}[i]}{L}, \tag{2.14}$$

where L represents the signal block length, also referred to as the analysis window, and k is the block time index. The calculation of the parametric variation $\Delta\mathbf{w}$ is performed for each instant, but the update is made, taking an average, of the L-length instantaneous variations. The update of the filter parameters occurs with a delay of L samples. In fact, before the adaptation, samples must first be stored in a L-length memory buffer said analysis window. In general, from the numerical point of view, the results obtained with the block algorithms are better than those obtainable with online algorithms. This depends mainly from the fact that the average makes the parameters estimation more robust. Nevertheless, the delay inherent in the block algorithms update may be incompatible with some applications.

- *Batch algorithms*—The batch algorithms are defined as an extreme form of block algorithm where the calculation of the parameters $\mathbf{w}$ is performed prior knowledge of the entire input sequence. However, also in this case, the delay of the update may be incompatible with some applications. Note, also, that the difference between block algorithms and those in batch is only formal since, in practical cases, the two methods tend to be unified.

2.2.1.3 Classification Based on the Cost Function to be Optimized

Another important aspect in optimum filtering concerns the CF characteristics and by the way this is minimized during the adaptation process.

For CF minimization are available methodologies based on different paradigms, for example, statistical, deterministic, and heuristic. Therefore, it is possible, a classification on the basis of the learning paradigms and rules as follows:

- *Deterministic learning*—The filter input is supposed generated according to a precise *signal model* and the CF to be minimized or maximized is of *exact* type. An example of deterministic CF, very common for several learning algorithms, is simply the least squares error (LSE) measured instantaneously or over an N-length window, defined as

$$\hat{J}(\mathbf{w}) \triangleq \hat{E}\left\{ \left| e[n] \right|^2 \right\} = \frac{1}{N} \sum_{n=0}^{N-1} \left| e[n] \right|^2 \qquad \textit{least squares error}, \tag{2.15}$$

where, $\hat{E}\{\cdot\}$ represents the time-average functional and, in the case of supervised learning, the error is defined as $e[n] = d[n] - y[n]$.

In order to take into account of any uncertainties, we consider the measures affected by random errors. The system randomness is usually modeled as noise additively superimposed to the output signal and, sometime, to the desired output. Most of the theory, in these cases, is developed considering additive Gaussian white noise (AGWN).

Remark The *principle of least squares* (LS), introduced by Gauss in the late 1700s for planetary orbits determination, is the basis of a wide class of algorithms for estimation and consequently for *machine learning* and is of great importance in many real applications. With the LS methodology is not made any probabilistic assumption on the input process that is considered to be deterministic but is simply assumed a certain *signal model*. The main advantage consists in the great variety of possible practical applications at the expense, however, of the possibility of not obtaining the optimal solution (in a statistical sense) [15, 19].

- *Stochastic learning*—In the case where the CF is derived from an exact statistical approach, the input–output signals are considered stochastic processes (SP), characterized by a probability density function (pdf) and statistical averages such as the first- and second-order moments. What is minimized, in turn, is a certain statistical average [12, 13]. For example, a CF widely used for the development of many learning algorithms (supervised, batch, and online) is the mean square error (MSE) defined as

$$J(\mathbf{w}) = E\left\{\left|e[n]\right|^2\right\} \qquad \textit{mean square error (MSE)}, \tag{2.16}$$

where, $E\{\cdot\}$ represents the expectation functional (see Sect. C.2.1 for more details).

Example: Mean Value Estimation of a Sequence

To better understand the above concepts, we consider the example of the estimate of a constant parameter w starting from a series of noisy observations. The signal model of the LS method is of the type

$$x[n] = \overline{w} + \eta[n], \quad \text{for} \quad n = 0, 1, ..., N-1 \tag{2.17}$$

for which the estimate of the parameter w corresponds to the estimate of the mean value of the sequence $x[n]$, where $\eta[n] \sim N\left(0, \sigma_\eta^2\right)$ is, for hypothesis, Gaussian white noise with zero mean. So, the constant w is just the mean value to estimate (see Sect. C.3.2 for more details).

Batch algorithm

Suppose we want to estimate the time-average value of (2.17) with batch algorithm. By elementary and intuitive reasoning, a batch estimator of the mean value may be defined by the function as follows:

$$w = \frac{1}{N}\sum_{n=0}^{N-1} x[n]. \tag{2.18}$$

Note that (2.18) corresponds to the maximum likelihood (ML) estimator as defined in Sect. C.3.2.2. Furthermore, the LS estimator is *statistically optimal* since we have that

$$E\{w\} = \overline{w} \tag{2.19}$$

$$\mathrm{var}(w) = \frac{1}{N^2}\sum_{n=0}^{N-1} \mathrm{var}\big(x[n]\big) = \frac{1}{N^2}\Big[N\sigma_\eta^2\Big] = \frac{\sigma_\eta^2}{N} \tag{2.20}$$

indeed, the expectation of the estimate converges to the true value and the variance tends to zero with the increase of the sample length. The estimator is then said to minimum variance unbiased (MVU) (see Sect. C.3.1.6).

Online Algorithm

We consider now the estimation of the mean value, with online or recursive algorithm, always with LS criterion, also called *sequential LS*. With this type of implementation, the estimate of the mean value is updated whenever new sample is available. To start the recurrence, we consider that the first estimated mean value is identical to the first available value (initial condition or IC) and update this estimate whenever there is a new sample of the sequence. Intuitively, at the arrival of the second sample, the estimate is updated using the equation $w_1 = (x[0] + x[1])/2$ and so on. More formally, this procedure can be written as

$$\begin{array}{ll}
n = 0 & w_0 = x[0] \\
n = 1 & w_1 = \dfrac{x[1] + x[0]}{2} = \dfrac{x[1] + w_0}{2} \\
n = 2 & w_2 = \dfrac{x[2] + x[1] + x[0]}{3} = \dfrac{x[2] + 2 \cdot w_1}{3} \\
\vdots & \qquad \vdots \qquad\qquad\qquad \vdots
\end{array}$$

Generalizing the previous expressions, for the kth value of sequence, we have the following recursive formula:

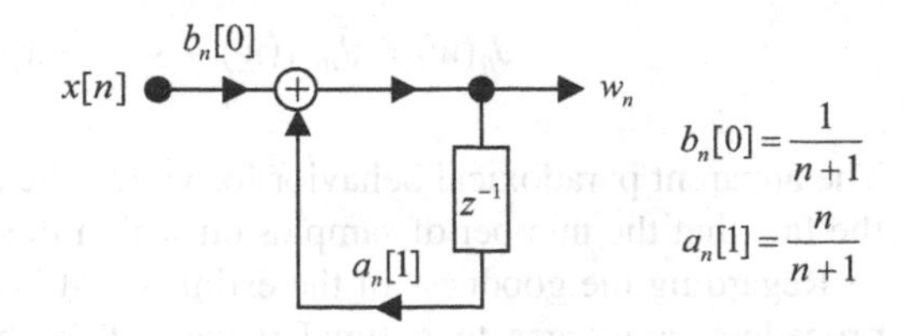

Fig. 2.5 Circuit diagram of an *online* mean value sequence estimator. Note that the *estimator filter* is time varying as they parameters depend on the time index n

$$\begin{aligned} w_k &= \frac{x[k] + k \cdot w_{k-1}}{k+1} \\ &= \frac{1}{k+1} x[k] + \frac{k}{k+1} w_{k-1}. \end{aligned} \tag{2.21}$$

Note that the expression (2.21) can be interpreted as a first-order IIR filter with time-variable coefficients as illustrated in Fig. 2.5.

In online estimating the partial data is immediately available, while in the case of batch algorithm, the mean value is calculated based on the knowledge of all the window signals, for which the data is available after a delay equal to the length of the analysis window itself.

Note that rearranging (2.21), this can be written as

$$w_n = w_{n-1} + \frac{1}{n+1}\left(x[n] - w_{n-1}\right), \tag{2.22}$$

where we see that the current estimate depends on the estimate at the previous time index plus a *correction factor*, the contribution of which decreases as n increases. The term on the right side of (2.22) can be considered as an *error* $\left(\varepsilon[n] = x[n] - w_{n-1}\right)$ in the prediction of the term w_n starting from the previous samples embedded in the term w_{n-1}. Regarding the minimum LSE defined in (2.15), this can be calculated recursively as

$$\hat{J}_{n-1}(w) = \sum_{k=0}^{n-1} \left(x[k] - w_{n-1}\right)^2 \tag{2.23}$$

and for (2.22) is

$$\hat{J}_{n}(w) = \sum_{k=0}^{n} \left(x[k] - w_{n}\right)^2. \tag{2.24}$$

Combining the above, after a few steps, it is shown that

$$\hat{J}_n(w) = \hat{J}_{n-1}(w) + \frac{n}{n+1}\left(x[n] - w_{n-1}\right)^2. \tag{2.25}$$

The apparent paradoxical behavior for which the error increases with n depends on the fact that the number of samples on which this is calculated increases.

Regarding the goodness of the estimates, it is easy to understand that the batch procedure converges to optimal value, while the online estimation can have a certain value of bias dependent to the choice of the initial conditions (for further details Appendix C and, for example, [1]).

2.3 Main Adaptive Filtering Applications

The use of adaptive circuits, linear and nonlinear, etc., is of central importance in various scientific and technological areas. Are shown below in general terms, and not related to a specific application domain, some signal processing situations (linear and nonlinear) typical of adaptive filters such as the identification, filtering, prediction, etc [10, 11, 17]. Regarding the application, we can identify the following four essential points that strongly characterize the structure: *the choice of model*, *the set of measures*, *the definition of a cost function*, and, finally, *the choice of the optimization algorithm.*

2.3.1 *Dynamic Physical System Identification Process*

The term *identification* means the determination of the mathematical relationship and its parameters that models and predicts the behavior of an unknown physical system. A general scheme of an identification system is shown in Fig. 2.6. In this case, without loss of generality, the physical system is defined in the CT domain while its mathematical model is defined in the DT domain.

2.3.1.1 Model Selection

In the case which the mathematical model is really general and not tracings the structure, in terms of physical laws, of the unknown system to identify, the predictor system behavior is characterized in terms of generic mathematical relationship between input and output sequences. The model behaves like a *black-box* which mimics the behavior of the physical system and is indicated as *behavioral model.* The modeling procedure is often referred to as *functional identification.*

In the case where there is a priori knowledge on the physical model, such as the nature of the mechanical laws of the system itself, the structure of the predictor's mathematical model is not generic and could realistically reproduce the nature of

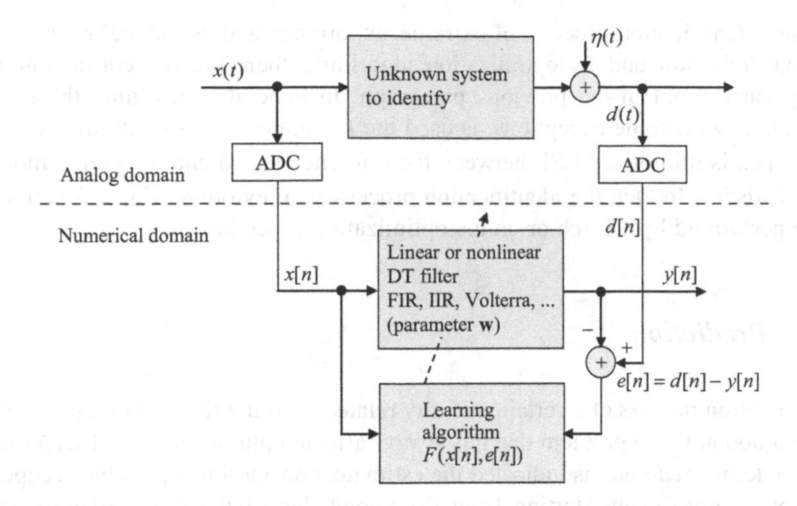

Fig. 2.6 AF scheme used for the identification of an unknown physical system

the physical system laws. For example, if the physical system was a mechanical device characterized by a second-order linear differential equation, the corresponding mathematical model could be achieved with a second-order FDE, i.e., an IIR filter. In this case, the identification procedure is called *structural identification* (or *white-box*) and would simply consist in the estimation of the coefficients of the IIR filter taken as a model. In these situations, if the system is *observable*, the estimated parameters of the model could be traced to the real parameters values of the physical dynamic system.

2.3.1.2 Set of Measures

In measurement process, it is very important the choice of the input signal which must be at *maximum information content*. In the case of identification of a linear system, it is appropriate the use of broadband spectrum signals. However, the amplitude of these signals must not be such as to bring the system in nonlinear operation regions.

For example, in the linear systems identification are often used binary signals, i.e., defined on only two values, in the case of voltage signal ($+V$, $-V$), and with random alternating variability, called pseudo random binary sequences (PRBS). The PRBS are, in fact, characterized by a white spectrum and able to excite all *natural modes* of the physical system and also have limited range that it can mitigate the effects of any parasitic or unwanted nonlinearity.

In the case of nonlinear systems identification, the choice of the measurement signals may be quite complex. A simple rule is to use signals with statistical characteristics similar to those typical in the normal operation of the physical system to be modeled.

In the identification process, of extreme importance and essential, are the cost function definition and its optimization algorithm, therefore the considerations already carried out in the previous paragraph. In general, concerning the norm to optimize, with some exceptions, is used the L_2 (indicated also with the symbol $||\cdot||_2$), i.e., is minimized LSE between the reference signal and the output model (or its statistic). In fact, the identification process, as previously widely described, can be performed by a batch or online optimization procedure.

2.3.2 *Prediction*

The estimation process of a certain quantity related to future time (or otherwise in a different domain) is a problem that has always affected philosophers and scientists. With the term *prediction* is indicated the estimate, considering a possible accepted error, of a future event starting from the knowledge of the time series of past observations.

More formally, we can think the prediction system as an operator, very similar to those defined in (2.9) and (2.10), with M past samples time window as argument, i.e., we can write as follows:

$$\hat{x}[n] = T\{x[n-1], ..., x[n-M], \mathbf{w}_f\} \tag{2.26}$$

in which $\hat{x}[n]$ denotes the predicted signal at time n. The expression (2.26) concerns the prediction of nth value of the sequence known the previous samples and, therefore, it is said *forward predictor*. It is possible, as will be seen better in the following, perform a backward prediction and, in this case, one would speak of *backward predictor* whose generic expression is given as follows:

$$\hat{x}[n-M] = T\{x[n], ..., x[n-M+1], \mathbf{w}_b\}. \tag{2.27}$$

The expression of prediction (2.26), as we can easily see, can be realized with the general scheme of Fig. 2.7 with $D = 1$ (one step forward prediction).

2.3.3 *Adaptive Inverse Modeling Estimation*

Another important application in many sectors is related to the estimation of the inverse of the physical model (a problem also known as *inverse filtering* or, for the linear case, *deconvolution*). Given a certain system, the estimation of the inverse model can be made by inserting the adaptive circuit upstream or downstream of the system itself.

As illustrated in Fig. 2.8, in the downstream case circuit that performs the estimate is said *equalizer* while, in the case in which the estimate is made with

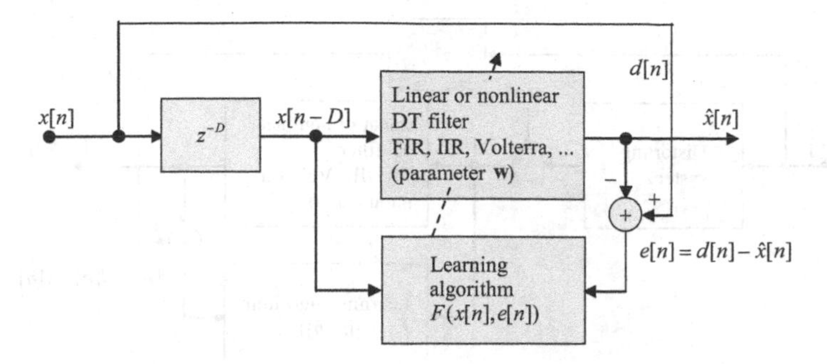

Fig. 2.7 Possible circuit scheme used as a adaptive *forward predictor*

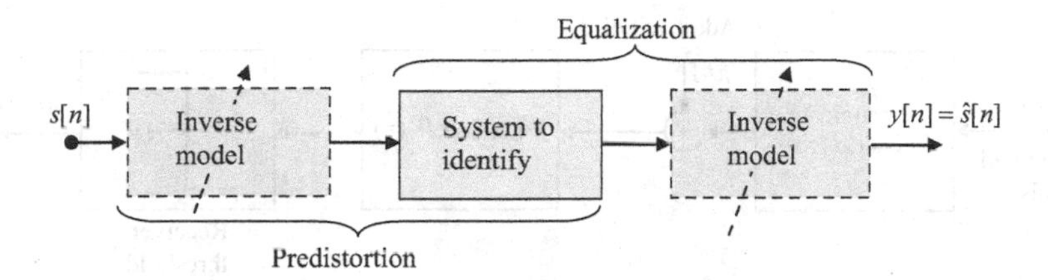

Fig. 2.8 The downstream or upstream estimation schemes for inverse modeling estimation

the circuit placed upstream of the system, the circuit is said *predistorter*. In the linear case, for the commutability property of linear operators, the downstream or upstream estimation schemes, in the case of convergence, lead to the same result. On the contrary, in the nonlinear case, the switching properties are no longer valid.

2.3.3.1 The Adaptive Channel Equalization

A common case of downstream linear estimation, typical in the field of data transmission is the so-called *adaptive channel equalization*.

A possible general scheme the adaptive learning of an equalizer circuit is shown in Fig. 2.9.

With reference to the principle diagram illustrated in Fig. 2.10, the TF $H(z)$ represents the combined response of the channel and the transmission filter. The additive disturbance depends on the thermal noise of electronic devices and other disturbances due, for example, to the interference with adjacent channels. The transmitted symbols $s[n]$, usually with the form of sinc(x) pulses, are distorted in various ways by the transmission channel.

Due to the nonideality of the channel impulse response, the distortion causes, generally, a *temporal spreading* of the transmitted pulse. This widening means that the transmitted pulse found to be different from zero for a time such as to interfere

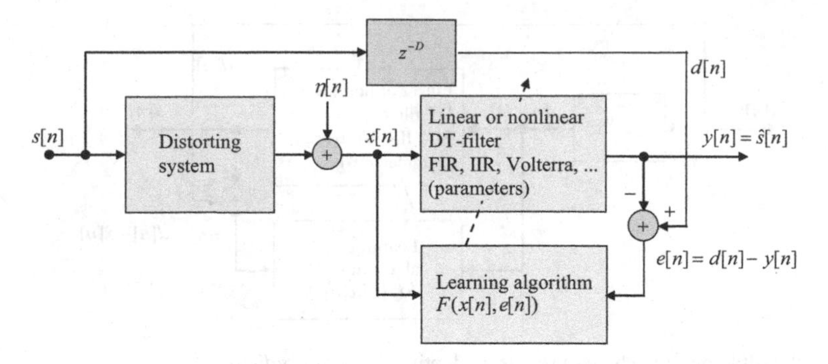

Fig. 2.9 Principle of a linear and nonlinear adaptive equalizer

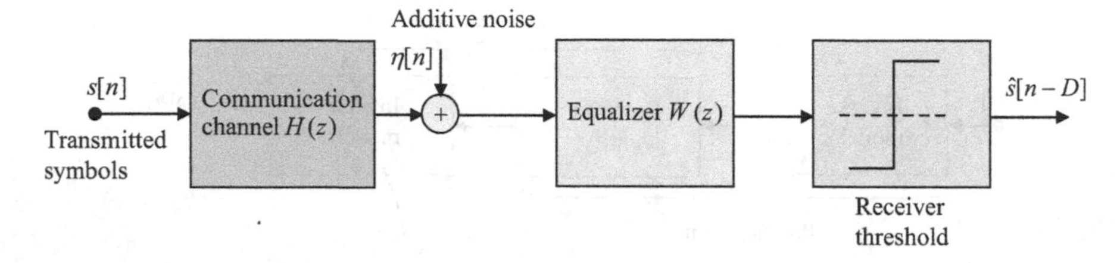

Fig. 2.10 Baseband transmission system representation with channel equalizer and receiver threshold

with the other adjacent transmitted pulses. For this reason, this phenomenon is referred to as intersymbol interference[1] (ISI).

Figure 2.11 reports a transmission system with two possible adaptation modes for the equalization filter.

In practice, for the initial adaptation period (or initial phase), it is used a reference signal already stored in the receiver. This sequence, said *preamble*, is the same that is sent to the receiver and which allows a first adaptation of the equalizer. With reference to Fig. 2.11, in the initial phase, the switch is in the *initial training* mode. After this first phase of adaptation, the receiver outputs are the same symbols of the input with probability tending to one. At this point, the switch can be moved to the position *decision directed* mode that allows to adapt the equalizer itself in a continuative way. Working with this technique, the equalizer is able to identify and track possible slow variation of the communication channel $H(z)$.

[1] Definition of ISI from USA Federal Standard 1037C, titled Glossary of Telecommunication Terms: *in a digital transmission system, distortion of the received signal, which distortion is manifested in the temporal spreading and consequent overlap of individual pulses to the degree that the receiver cannot reliably distinguish between changes of state, i.e., between individual signal elements.*

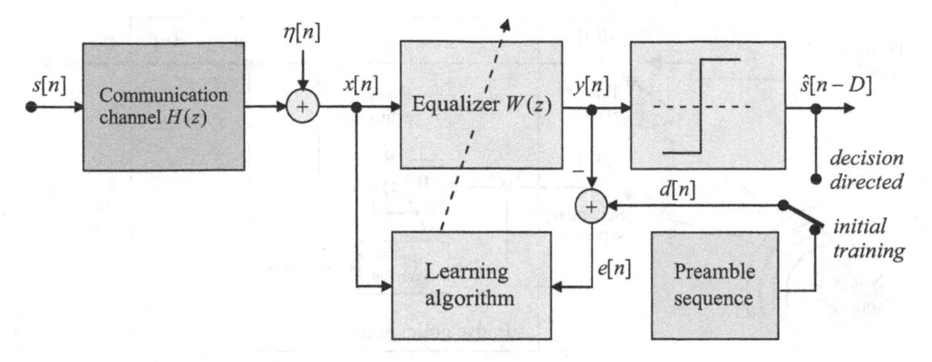

Fig. 2.11 Transmission system with equalizer adaptation scheme

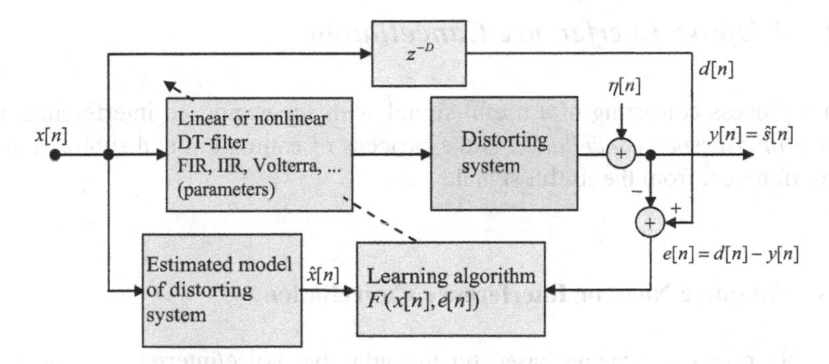

Fig. 2.12 Adaptive filter scheme used as an adaptive predistorter

2.3.3.2 Control and Predistortion

The term *predistortion* indicates the estimation of the inverse model performed upstream with respect to the distorting nonlinear physical system.

In the case of the linear AF and linear physical system, for which the blocks are switchable, it is obvious that the predistortion is equivalent to the problem of equalization. In this case the adaptation, where possible, is performed with the scheme of Fig. 2.9 (except then switch the two blocks).

In the nonlinear case, the problem of determining the predistortion network is much more complex. First of all, it is necessary to determine the conditions of existence and uniqueness of the solution. Note that not all of the nonlinearities are invertible and, moreover, in the case of not monodromicity of the distorting physical system, there would not be unique solution to the problem.

A possible principle scheme for the realization of the predistorter is shown in Fig. 2.12.

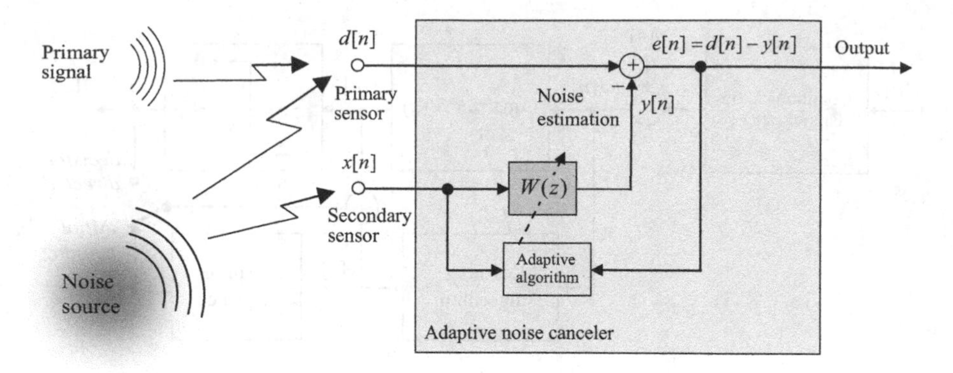

Fig. 2.13 Block diagram of an adaptive noise or interference canceller

2.3.4 Adaptive Interference Cancellation

Given a process consisting of a useful signal with superimposed interference, the *adaptive interference cancellation* is the process of estimation and subtraction of this interference, from the useful signal.

2.3.4.1 Adaptive Noise or Interference Cancellation

A possible principle scheme based on the adaptive noise/interference canceller (AIC) is illustrated in Fig. 2.13. The *primary sensor* receives mainly the SOI, called also primary signal, to which is superimposed an interfering noise source. The secondary (or reference) sensor captures the signal due to (mostly) the noise source. The adaptive filter is adjusted in such a way that a replica of the interfering signal (in practice, its estimate) is subtracted from the primary process [2, 10].

It can be observed that the adaptive noise canceller has different principle scheme than the general form of AF shown in Fig. 2.6. In this case the residual error signal, that in the other examples is used only for the filter parameters adaptation algorithm, represents the system's output. In AIC, the desired signal is represented by the signal acquired by the *primary sensor* and consists in SOI plus the unwanted noise. Note that the signal on the *secondary sensor*, in the context of adaptive noise/interference cancellation, is also said *reference input*.

2.3.4.2 Echo Cancellation

In general terms, in telephone communications, the possibility of having a return echo of their own voice is attributable to two separate cases. The echo signal can be generated by electrical circuits or by acoustic coupling between loudspeaker and microphone [20, 21].

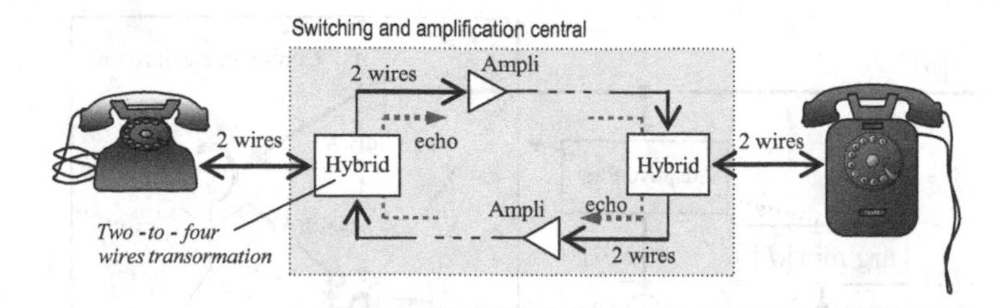

Fig. 2.14 Scheme of a two-wire telephone communication. The transformation from two-to-four wires, made by a hybrid circuit, is required for the insertion of amplification and signal switching stations

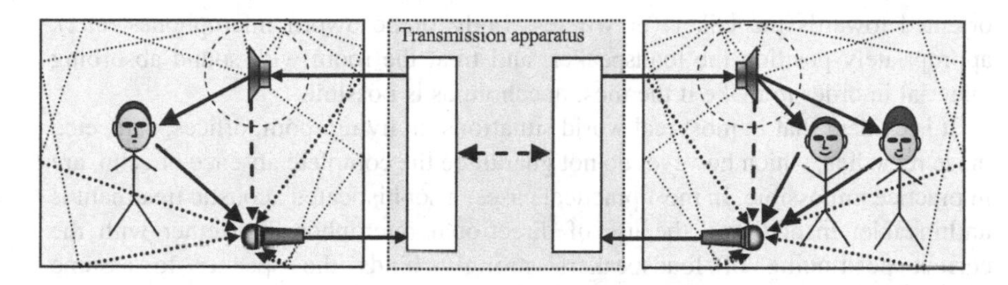

Fig. 2.15 Teleconference scenario. The microphone, in addition to capturing the voice of the subject, also acquires the signal coming from the loudspeaker and the various reflections due to the walls of the room (reverberation)

The echo generated by electrical circuits is due to unbalanced *two-to-four wire converter* also called *hybrid circuit*. As we know, at the telephone user terminal comes a cable with only two wires (the so-called *twisted pair*), for both directions of communication. In order to be able to switch and amplify the signal in both directions, it is necessary to insert, along the same line, a two-to-four wire converter. In this way, we have two twisted pairs, and each of them is related only to one direction of transmission. Therefore, the signal is switchable and amplifiable. A scheme of principle is illustrated in Fig. 2.14.

In the latest video conferencing or hands-free phone systems, echo can also be generated by the coupling noise between the loudspeaker and the microphone. The problem of acoustic echo can be easily explained by considering the typical teleconference scenario shown in Fig. 2.15. The microphone, as well as capturing the voice of the subject, acquires the signal from the loudspeaker which, together with reflections from the walls, is being returned to the sender (*far-end side*). So, at the sender side there will be a return echo that can seriously affect the intelligibility of communication.

The prevention of the echo return is therefore of central importance for the quality of the transmission itself and can be performed in various modes. In specific video conference rooms, as in the television studios where it is possible (or desirable) to intervene acoustically, we can use unidirectional microphones

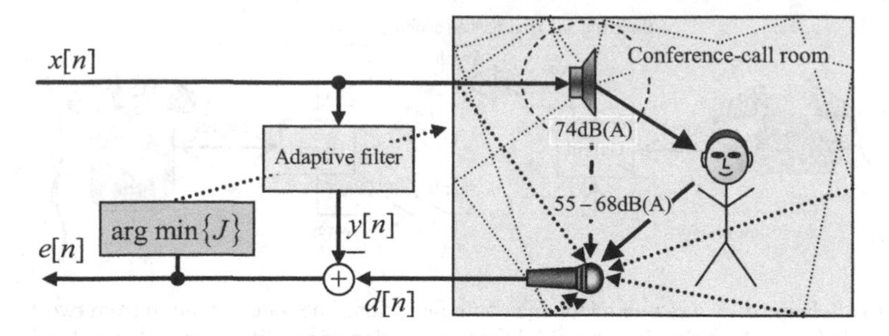

Fig. 2.16 Adaptive acoustic echo canceller scheme

oriented towards the talker (or wireless body or headworn microphones, etc.), appropriately position the loudspeaker, and treat the room with sound absorbing material in order to make it the most anechoic as is possible.

It is evident that in most real world situations, in living room, offices, cars, etc., those remedies, which however do not guarantee the complete absence of echo, are in practice impossible. In most practical cases, a sophisticated acoustic treatment is unthinkable; in addition, the use of directional microphones together with the correct positioning of loudspeakers strongly binds the speaker to assume predetermined fixed positions.

The echo cancellation can be performed with an adaptive filter, placed in parallel to the respective transmission sides, said adaptive echo cancellers (AEC). With reference to Fig. 2.16, the far-end signal $x[n]$ transmitted by the other side is filtered and subtracted from the microphone signal indicated as $d[n]$. The purpose of the adaptive filter is to model the acoustic path between the loudspeaker and the microphone in order to subtract from the signal to be transmitted $d[n]$ the far-end signal $x[n]$ together with all the reflections due to the walls of the room.

The reflected signal from the wall, or reverb, is simply the same signal attenuated and delayed (sign changed). The reverb can be modeled as a simple convolution operation. In practice, the echo canceller is a linear adaptive filter (usually FIR) that models the reverberation of the conference call room.

Although conceptually simple, acoustic echo cancellation reveals a problem of a certain complexity. A typical office room reverberation is of the order of hundreds of ms. For example, considering a sampling frequency of 16 kHz and a reverberation time of 100–200 ms, for the cancellation of the echo effect, the filter should have a length of not less than 1,600–3,200 coefficients (*taps*). The real-time adaptation of filters of this order is a problem that, in general, is faced with a dedicated processor said digital signal processor. Another important aspect, for which the research in the field of acoustic echo cancellation is still today strongly active, regards the convergence speed of the adaptation algorithms. If the speaker is not in a fixed position but moves relative to the microphone and the walls, the acoustic configuration changes continuously. So, the adaptive filter must perform a real-time tracking of acoustics variant of the system and, in these cases, the efficiency of the adaptation algorithm, in terms of convergence speed, plays a fundamental role.

Other aspects of current research on acoustic echo cancellation concern the extension to multichannel case, i.e., when there are multiple loudspeakers and/or more microphones. In this context, we think that the inclusion of the *positional audio* paradigm can be used in order to make video conference a more natural communication system (augmented reality). In this class of systems, at the position of the talker on the video, also corresponds a positional acoustic model. To have an adequate acoustic spatiality, as in the simple stereo case, at least two microphones and two loudspeakers appropriately driven are necessary.

Remark The acoustic transduction devices, such as microphones and especially loudspeakers, are by their nature (sometimes strongly) nonlinear and, for this reason, the adaptive acoustic systems that use such devices should take into account of such nonlinearity.

In the case of echo cancellation and even, as we shall see in the next section, in the active noise cancellation is almost always considered the hypothesis of linear acoustic transducer. The treatment of nonlinearity, in particular those of the loudspeaker, is a very promising active area of current research. Such nonlinearity, are dynamic, of difficult modeling and, in addition, negligible only in high cost devices.

2.3.4.3 Active Noise Control

The active noise cancellation or active noise control (ANC) consists in producing of an acoustic wave, said *antinoise*, in phase opposition with respect to the wave generated by the noise source. This wave has the objective of creating of a silence zone in a given region of space [2].

The schematic diagram of an ANC is illustrated in Fig. 2.17. The noise signal, acquired by a microphone placed near the noise source, is said primary source. The antinoise wave, generated from the loudspeaker, is known as the secondary source. To have a high degree of noise attenuation, the amplitude and phase of the secondary source must follow perfectly the primary source.

The ANC is a highly complex problem that requires precise control, temporal stability, and high computational resources. Typically, in practical situations, noise control is very effective in the low-medium audio frequencies range where, in addition to the active control, there is an adequate soundproofing acoustic treatment, which attenuates the wall reflections.

Figure 2.18 shows the principle diagram of an active noise canceller in a duct. This simple geometry, for wavelengths large relatively to the section of the duct, makes the acoustic wave one dimensional, and in this situation, the problem of noise control can be simplified.

In fact, an important aspect of ANC is that related to the geometry of the environment of intervention. In this regard, it is useful to classify four possible categories.

1. *One-dimensional tubes*—For example, in the silencing of the ducts and vents of air conditioning systems, in fume hoods, exhaust systems of vehicles, etc.

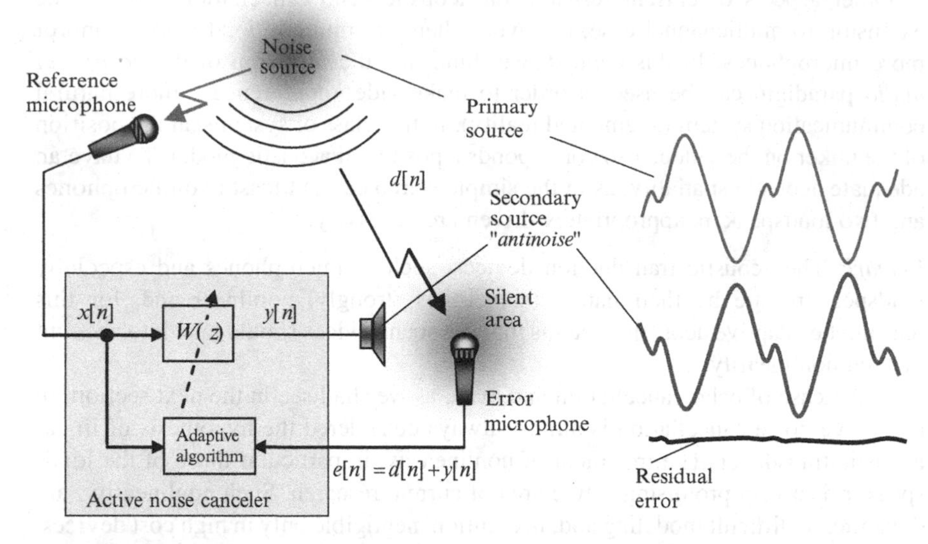

Fig. 2.17 Principle of operation of an active noise canceller (ANC). The loudspeaker makes a wave in phase opposition with respect to the noise at the point where the error microphone is located. The reference microphone should be placed as close as possible to the acoustic source of noise

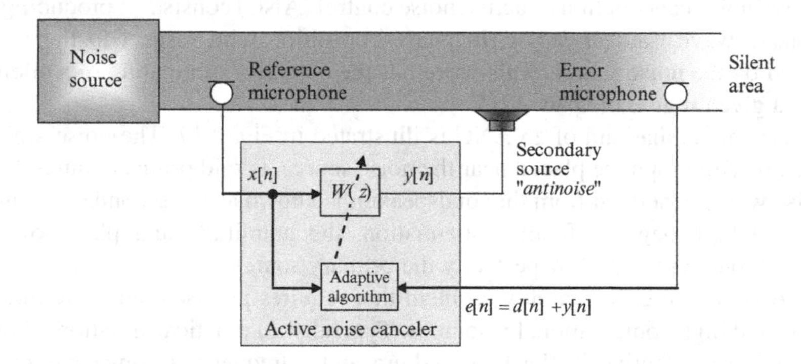

Fig. 2.18 The ANC in a narrow duct. The noise source may be, for example, a fan

2. *Confined spaces*—Operational situations where a certain reverb is present. For example, automotive interiors, rooms, etc.
3. *Free space*—Absence, or nearly so, of reflections.
4. *Personal protection*—For example, active headphones, or in situations where the size of the environment is very small compared to the concerned wavelengths.

Fig. 2.19 The Very Large Array (VLA) radio telescope in New Mexico (USA) uses 27 antennas, each with 25 m diameter, arranged with a "Y" shape. Each antenna can move along three rails, two of length equal to 21 km, and the other of length equal to 19 km (image courtesy of NRAO/AUI)

2.4 Array of Sensors and Array Processing

In many practical situations of adaptive filtering, regarding applications in acoustical, mechanical, and electromagnetic domains, the involved vibration modes are often very complex. In such circumstances, it is appropriate to use a multiplicity of sensors, in general, homogeneous.

The signals related to the same process are captured with a set of sensors or elements properly arranged in the space. The array is designed to capture processes related to the propagation of waves (acoustic or electromagnetic) resulting from one or more radiation sources. The energy field intercepted by the sensors' array is sampled in both the time and space domains. The processing of signals from sensors' arrays, homogeneous and spatially distributed, is referred to as array signal processing or simply array processing (AP) [3].

The application fields of the AP are manifold. Consider, for example, the acquisition of biomedical signals such as the electroencephalogram (EEG), the electrocardiogram (ECG), the tomography, or, in other fields, the antenna arrays, radar, the detection of seismic signals, the sonar, the microphone arrays for the acquisition of acoustic signals, etc.

As an example, Fig. 2.19 shows a picture of a famous radio telescope called the Very Large Array (VLA) in New Mexico—USA, consisting of an array of 27 parabolic, 25 m diameter antennas (see Fig. 2.20). The antennas are mounted on three rails arranged in a Y shape for which the array has a variable geometry.

The purpose of the AP is in principle the same as the classic DSP: the extraction of meaningful information from measured data.

Remark In adaptive filtering with only one sensor, the nature of the sampling is only temporal. In the case of arrays, we must also consider the geometry of the system. For which the filtering is performed, as well as in the time domain, also in

Fig. 2.20 Detail of the VLA parabolic antennas (image courtesy of NRAO/AUI)

the spatial domain, i.e., the nature of filtering in the time domain as well as in the spatial domain (i.e., discrete space–time filtering).

2.4.1 *Multichannel Noise Cancellation and Estimation of the Direction of Arrival*

Figure 2.21 shows the adaptive interference/noise cancellation microphone array. In this case, the capture of the noise sources is carried out with several microphones allowing higher performance in the case of complex acoustic of vibration modes. The system, in practice, can be considered a simple generalization of the single channel AIC illustrated in Sect. 2.3.4.

In many situations of practical interest, it is necessary or useful to identify the direction of arrival of an acoustic or electromagnetic radiation. In the case of narrowband signals, taking as reference the diagram of Fig. 2.22, the arrival of the wave (by hypothesis plane wave) is intercepted by the sensor closest to it. By measuring the delay in the arrival of the signal among the various sensors, with simple geometrical considerations, it is possible to estimate the radiation arrival angle.

2.4.2 *Beamforming*

Depending on the nature of the acquired field (electromagnetic, acoustic, and mechanical), the AP sensors can be antennas, microphones, vibrometric-mechanical transducers, accelerometers, etc. In any of these cases, sensors are

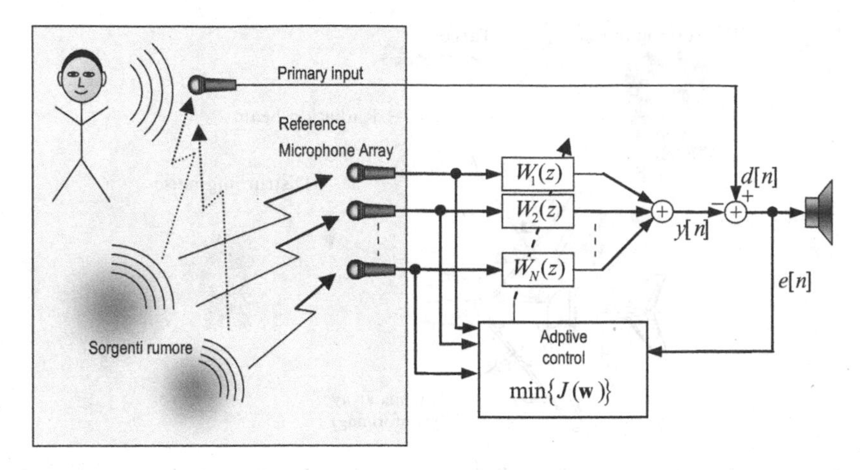

Fig. 2.21 Adaptive noise canceller with microphone array

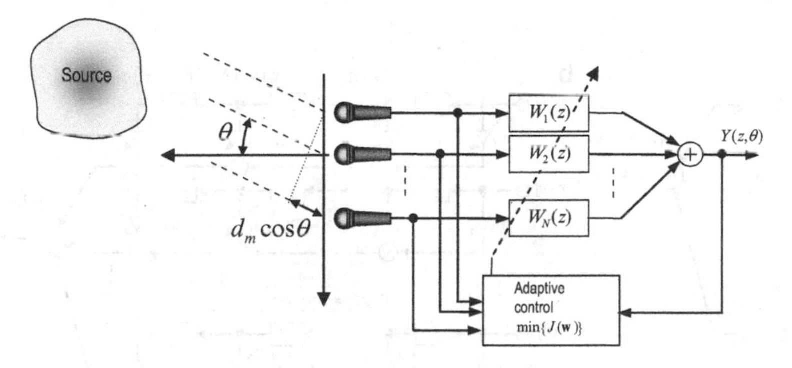

Fig. 2.22 Array of sensors for the detection of arrivals (DOA)

provided with a specific *radiation diagram* or *radiation beam* related to the characteristics of the transducer gain as a function of the arrival radiation angle.

For example, in the case of the directive antenna, with suitably positioned elements, the radiation pattern has the form of a narrow lobe, also said *beam*, whereby to increase the sensitivity on a certain predetermined direction, it must be physically turning and tilting the antenna along that direction.

The term *beamforming* indicates the possibility of synthesizing a certain radiation diagram, by means of an electronic control that provides an appropriate signal feeding to the elements of the array of sensors that are kept fixed.

For example, in the case of narrowband processes, typical in some TLC fields, it is sufficient to sum the signals of the individual elements with an appropriate phase. The beam angle is determined by the filtering, performed in the spatial domain, due to the position of the array of discrete elements.

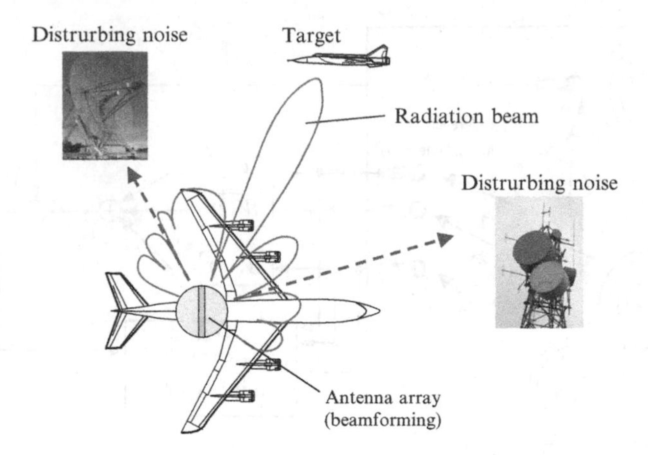

Fig. 2.23 Adaptive beamforming. The radiation beam is shaped in such a way so as to obtain the maximum gain in the direction of interest (DOI) and an attenuation to the disturbing signal's directions

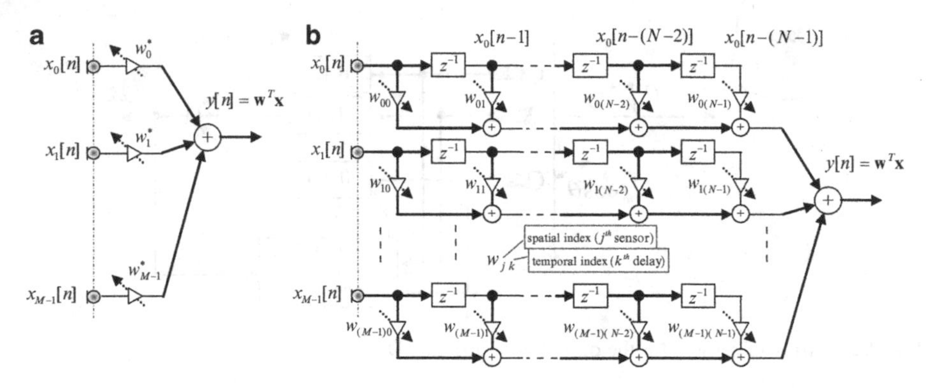

Fig. 2.24 The beamforming consists in a linear combination of the signals present on the receivers (**a**) in the case of narrowband sources (*phased array*), the outputs of the receivers are multiplied by a complex weight and then summed; (**b**) in the case of broadband sources, the signals on sensors are filtered with an FIR filter and then summed

Figure 2.23 shows an example of an adaptive beamformer, in particular a radar mounted on an aircraft able to generate simultaneously a high-gain beam towards a certain target and strong attenuation to the disturbing signals.

Figure 2.24a shows the principle diagram of *delay-and-sum* beamformer for narrowband signals. The output is a simple combination of the input signals carried out with complex coefficients. For this reason, this type of beamformer is also said *phased array*.

In broadband signals, beamforming, such as for example in the case of the acoustic speech signals, to obtain a certain radiation pattern, before the sum must

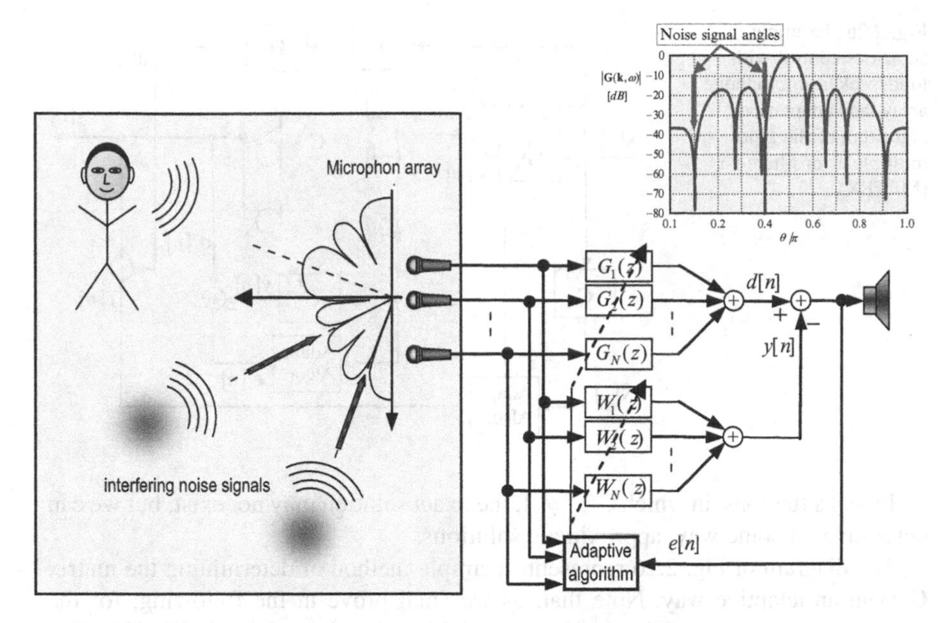

Fig. 2.25 Wideband audio beamforming with interference cancellation

also properly filter the signal. In these cases, in fact, a more sophisticated space–time processing is required as, for example, the one shown in Fig. 2.24b.

As a further example, Fig. 2.25 shows the principle diagram of a broadband microphone array for the capture of the voice signal, with adaptive cancellation of interference. In this case, the beamforming is more properly said generalized sidelobe canceller (GSC).

2.4.3 Room Acoustics Active Control

Another application example of the array processing techniques is illustrated in Fig. 2.26, where it is reported a possible scheme for the environmental-acoustics active control of the type *room-in-room*. The idea is to correct the response of a certain listening room, featuring by an array of network transfer functions $\mathbf{C}(z)$, called room transfer functions (RTF), with a target or reference RTF denoted by $\mathbf{H}(z)$ that is, for example, the acoustics of a certain auditorium.

The problem, similarly to the predistortion control techniques previously illustrated, is to determine a matrix of network functions $\mathbf{G}(z)$ such that $\mathbf{G}(z) \cdot \mathbf{H}(z) = \mathbf{C}(z)$. In practice, the matrix $\mathbf{G}(z)$ assumes the function of the room acoustics controller. If there is an acoustic treatment with sound absorbing panels, for which the room is devoid of echo (or *anechoic*), less than a delay between the source *i*th and the sensor *j*th, then we have $\mathbf{C}(z) = \mathbf{I}$ and by consequence $\mathbf{G}(z) = \mathbf{H}(z)$.

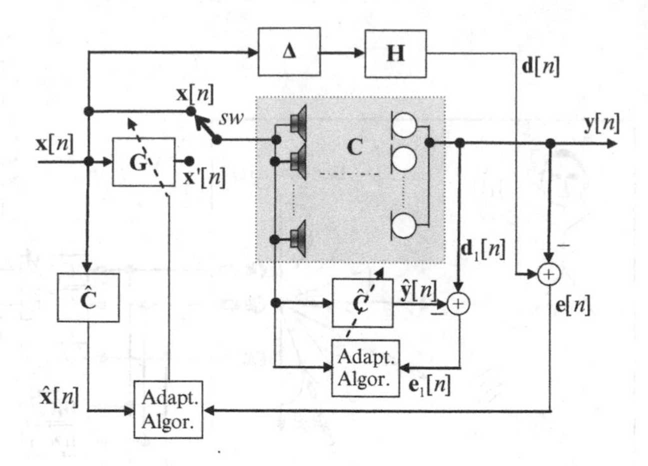

Fig. 2.26 Example room acoustics control with loudspeaker–microphone array, and adaptation algorithm of the type multiple error filtered-x (MEFEX)

In real situations, in which $\mathbf{C}(z) \neq \mathbf{I}$, the exact solution may not exist, but we can determine, in some way, approximate solutions.

The diagram of Fig. 2.26 represents a simple method of determining the matrix $\mathbf{G}(z)$ in an adaptive way. Note that, as we shall prove in the following, for the adaptive determination of $\mathbf{G}(z)$, it is required estimation of the RTF $\mathbf{C}(z)$. The system consists of two nested multichannel adaptive filters. The internal allows the RTF $\mathbf{C}(z)$ estimation, while the outer allows the determination of the controller $\mathbf{G}(z)$, filtering the input signals $\mathbf{x}[n]$ with the estimate RTF $\hat{\mathbf{C}}(z)$. For this reason, the method is called multiple error filtered-x (MEFEX) [2].

Remark The system architecture shown in Fig. 2.26 lends itself, with simple reasoning left to the reader, even for the implementation of a multichannel ANC that generalizes the scheme of Fig. 2.17.

2.5 Biological Inspired Intelligent Circuits

The theme in this text chiefly relates to the linear adaptive filtering and signal processing methods. The recent development of new nonlinear adaptive filters technique has opened new frontiers in both the disciplinary and applicative fields. For this reason, in this section, we want to introduce the fascinating topic of the *biological inspired intelligent circuits*, in the context of the adaptive filtering theory.

Among the various techniques for adaptive nonlinear filtering, the ANN are gaining increasing interest [4–8]. In fact, ANNs represent an emerging technology that finds its origins in many disciplines. In this context, however, ANNs are considered as a simple circuital paradigm, able to solve some specific adaptive signal processing problems. In very general terms, we can consider two classes of problems that can be solved by neural networks. The first is that of the *pattern*

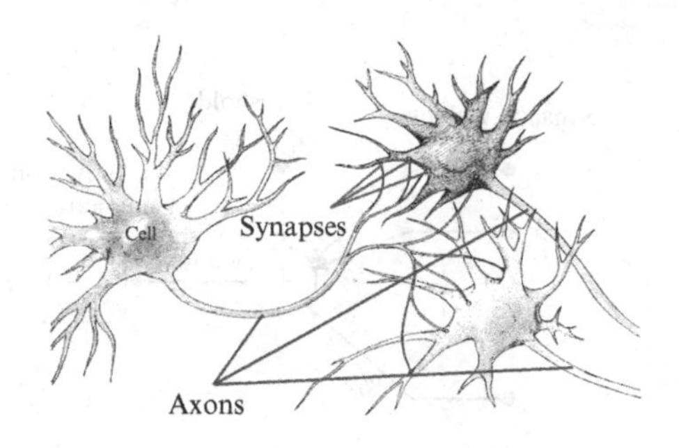

Fig. 2.27 The biological neurons

recognition, while the second class of applications, of greater interest in this volume, concerns the signal processing. The network input is fed with an orderly succession of data. When the output is unique, the ANN can be seen as a function of the general type (2.1) or (2.2); or a mathematical operator, in the case where the output is also a whole stretch of time function.

We can think of the ANN as circuits that attempt to emulate the "intelligent" behavior of a biological brain, which consists of an extensive network of elementary cells called *neurons*, which in very general terms, have the peculiarity of having innate abilities of learning and reasoning. More in particular, the main characteristics of the biological brain can be summarized as follows:

- *Local simplicity*: the biological neuron, shown schematically in Fig. 2.27, receives stimuli from other neurons to which it is connected and reproduces a pulse in the axon proportional to the weighted sum of its inputs.
- *Global Complexity*: the human brain has about 10^{12} to 10^{14} neurons each of them with about 10^4 connections.
- *Learning capabilities*: the connections strength varies when the network is exposed to external stimuli.
- *Fault tolerance*: in the case of brain damage, the performance degrades slowly with the increase of the damage.
- *Processing speed*: ability to solve very complex tasks in a short time (vision, memory, spatial, and temporal recognition in noisy and/or incomplete data).

Considering these biological assumptions, ANN can be defined as (Kohonen [8]): "*massively parallel circuit formed by the interconnection of simple adaptive elements that can interact with the objects of the real world in the same way of biological neural systems.*"

Over the past two decades, research on ANN involved and/or has intersected with other disciplines such as, for example, neurobiology, psychology, circuit theory, the statistical, estimation and information theories, etc.

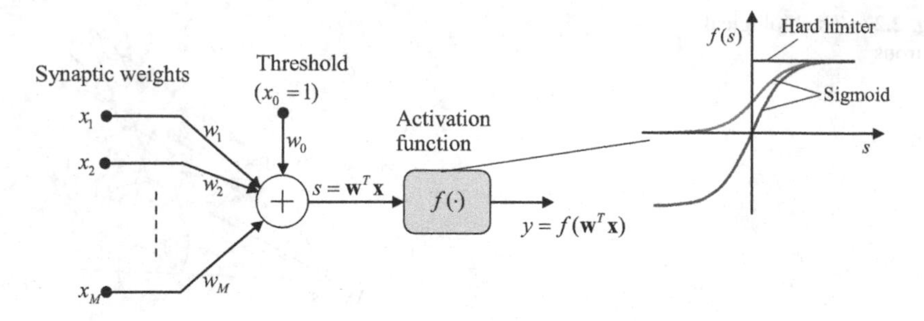

Fig. 2.28 Structure of the *formal neuron*

2.5.1 The Formal Neuron

By analogy to biological neural network, an ANN is also made of elementary processing circuits, defined as *formal neurons*, which have the structure of the type shown in schematic form in Fig. 2.28.

Observing the figure, it can be noted how each formal neuron, similar to the biological neuron, is composed of four characteristic elements: (1) input connections (*synapses*), (2) linear combiner, (3) activation function, and (4) connection output (*axon*).

In other words, the neuron is a circuit that performs a simple operation: the weighted sum of the inputs to which is added a constant value, called *threshold*, that produces a "high" output if the inputs' weighted sum exceeds the threshold value, on the contrary "low." With reference to Fig. 2.28, $\mathbf{x}$ is the input vector, $\mathbf{w}$ is the weight vector, and $w[0] = w_0$ is the threshold value, let $x[0] = 1$, the formal-neuron output can be expressed as $y = f(\mathbf{w}^T\mathbf{x})$ in which the function $f(\cdot)$ is, in general, a nonlinear function with sigmoid saturation characteristic.

2.5.2 ANN Topology

Many of the ANNs processing properties depend on the way in which individual neurons are connected.

Apart from the general problem of finding the optimal topological configuration, most of the studies on ANN, led to identify precise networks classes, each of them is particularly well suited for the solution of certain families of problems. Many neural networks are organized in layers and among these, one can identify an input layer, one output layer, and a number of intermediate layers, called hidden layers. Among the ANN models used in most applications, we remember the multilayer perceptron (MLP) [5] shown in Fig. 2.29.

When a neuron is connected to every other neuron in the network, regardless of the layer of belonging, one speaks then of *fully connected networks*. Instead, the

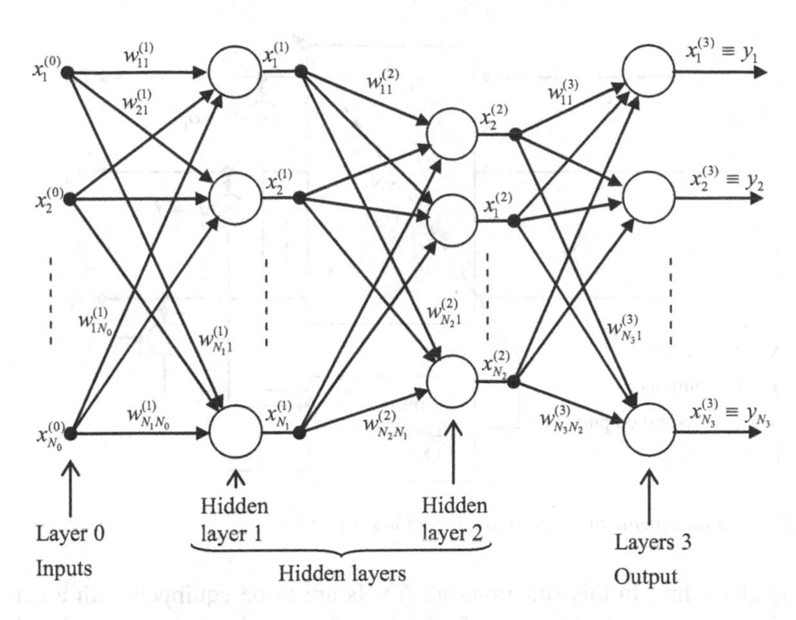

Fig. 2.29 Circuit diagram of the multilayer perceptron (MLP) network

networks are called RNN when the output of each neuron is connected to the input of the same neuron (for example, as in the Hopfield networks).

The MLP illustrated in Fig. 2.29 represents the configuration most widespread complex and distributed, and also better understood from a theoretical point of view. All neuron outputs of the each layer are connected directly (and solely) with the inputs of the subsequent layer, the absence of feedback allowing us to classify these networks with the term *feedforward*.

2.5.3 Learning Algorithms Paradigms

Similarly to what happens in the linear AF, the ANN learning can be performed considering very different philosophies. In particular for the ANN, the three main paradigms for learning are *supervised learning*, *unsupervised learning*, and *reinforcement learning*.

Figure 2.30, for example, shows a diagram of supervised learning in which the cost function to be minimized is of the type $J(\mathbf{w}) = F(\mathbf{e}[n], \mathbf{y}[n])$.

Said $\mathbf{w}$ the vector of the network's free parameters, learning with and without supervision has the same philosophy as already described in linear adaptive filtering [see (2.11), (2.12), and (2.13)].

In *reinforcement learning*, however, is the network itself that interacts with the environment. Each action of the network determines an environmental condition variation. Consequently, the environment produces a feedback that controls the

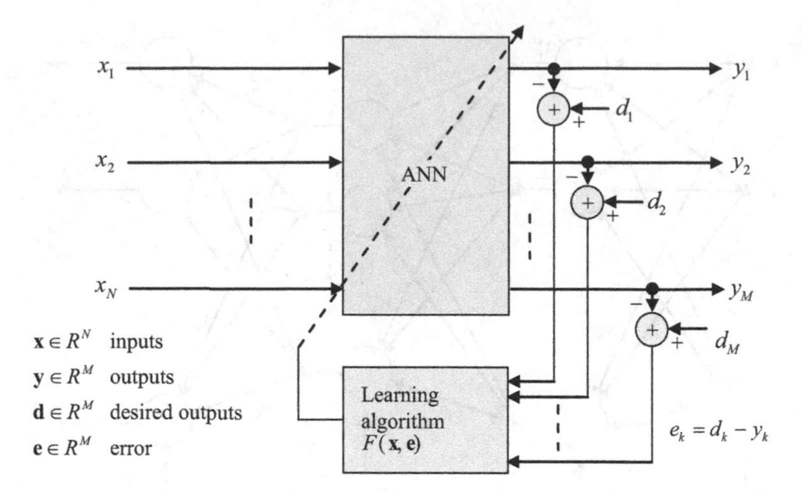

Fig. 2.30 General scheme of supervised learning algorithm

learning algorithm. In this situation, the ANNs are to be equipped with a certain perception capacity which allows the environment exploration and undertake a series of actions. In other words, in reinforcement learning real examples (input–output *training set*) are not present, but the solution space is explored and the minimization of the cost function is performed heuristically.

2.5.4 Blind Signal Processing and Signal Source Separation

The so-called blind signal processing (BSP) is one of the emerging areas of research in the context of adaptive signal processing.

The term *blind* is used when the adaptation does not require any reference signal. In other words, as shown in Fig. 2.31, the learning is performed without supervision. In practice, in the adaptation algorithm the minimization of a specific error is not provided, but the CF $J(\mathbf{w})$ is a function of only the input and output signals, namely, relatively to the diagram of the Fig. 2.31, $J(\mathbf{w}) = F(\mathbf{x}[n], \mathbf{y}[n])$.

The BSP methodologies have taken a leading role in strategic application areas such as, for example, digital communications, signal quality enhancement (images, video, audio, etc.), the equalization/reconstruction of signal, the technologies for medical diagnosis, multisensor systems, the geophysical, environmental, economic, data analysis, seismic exploration, remote sensing, data mining, nondestructive diagnostics systems, in the data fusion for monitoring and modeling of complex scenarios, etc.

In particular, in the BSP areas, the blind signal separation (BSS) consists in the information recovery from mixtures of statistically independent signals, acquired from sensor arrays. Each transducer receives a different combination of all sources,

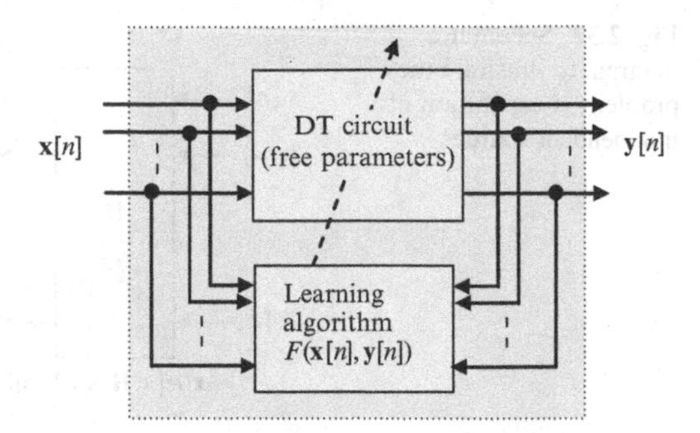

Fig. 2.31 Multichannel adaptive filter with blind learning scheme

and BSS methodologies separate the various information contents of the signals, even in the case of overlapping spectra. The blind approach is, in fact, radically different from the linear filtering in the time–frequency domain.

Another interesting aspect connected to the BSP is to intimate connection with the basic theories of *neuroscience* and *information theory* [9]. In parallel to the BSS, in fact, in recent years numerous studies have emerged on the unsupervised learning rules for ANN, based on those same paradigms and related to information theory and *independent component analysis* (ICA) [22].

2.5.4.1 Separation of Independent Sources

In the independent sources separation, the problem consists in the estimation of sources, indicated with the vector $\mathbf{s}[n]$, known as linear combination of these sources indicated with the vector $\mathbf{x}[n]$. In practice, the signal model can be written as

$$\begin{aligned} x_1[n] &= h_{11}s_1[n] + h_{12}s_2[n] + \cdots + h_{1N}s_N[n] + \eta_1[n] \\ x_2[n] &= h_{21}s_1[n] + h_{22}s_2[n] + \cdots + h_{2N}s_N[n] + \eta_2[n] \\ &\vdots \\ x_M[n] &= h_{M1}s_1[n] + h_{M2}s_2[n] + \cdots + h_{MN}s_N[n] + \eta_M[n] \end{aligned} \tag{2.28}$$

with $M \geq N$ and where h_{ij} represents the coefficients of the linear combination of the sources and the term $\eta_i[n]$ represents the measurement noise.

Said $\mathbf{H}$, the mixing coefficient matrix, the previous expression can be written as

$$\mathbf{x}[n] = \mathbf{s}[n]\mathbf{H} + \boldsymbol{\eta}[n]. \tag{2.29}$$

In relation to the formalism in the diagram illustrated in Fig. 2.32, the separation problem consists in the estimation of the demixing matrix $\mathbf{W}$ so that

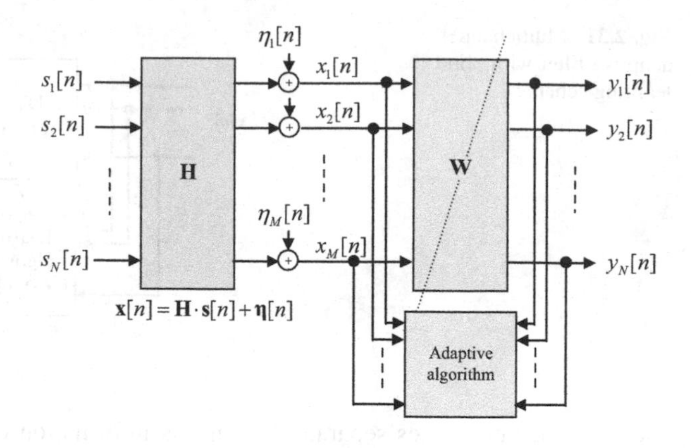

Fig. 2.32 Schematic diagram to illustrate the problem of separation of independent sources

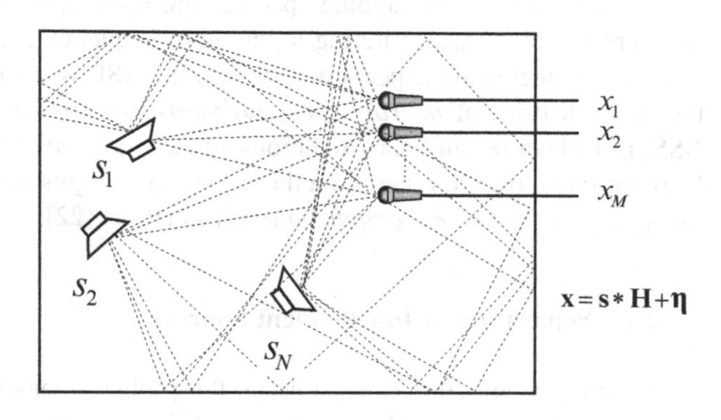

Fig. 2.33 Separation of sources with convolutional mixing

$$\mathbf{y}[n]=\mathbf{W}\mathbf{x}[n] \tag{2.30}$$

such that the vector $\mathbf{y}[n]$ represents an estimate of the source $\mathbf{s}[n]$, less than a gain and permutation factor (*trivial ambiguity*). In fact, in the sources separation, the useful information is contained in the signal waveform, rather than in amplitude or in the order in which the signals are presented. It follows that the problem of ambiguity of permutation and amplitudes poses no serious problems in the applications of the BSS techniques.

2.5.4.2 Deconvolution of Sources

In more general and physically realistic terms, the observations on the sensors can be linear combinations of reverberated versions of the input signal. In this case, the signal mixture is of convolutive type for which each coefficient of the matrix $\mathbf{H}$ is the impulse response of a dynamic system. In this case, the model of mixing, omitting the writing time index n, can be written as

$$
\begin{aligned}
x_1 &= s_1 * h_{11} + s_2 * h_{12} + \cdots + s_N * h_{1N} + \eta_1 \\
x_2 &= s_1 * h_{21} + s_2 * h_{22} + \cdots + s_N * h_{2N} + \eta_2 \\
&\vdots \\
x_M &= s_1 * h_{M1} + s_2 * h_{M2} + \cdots + s_N * h_{MN} + \eta_M.
\end{aligned}
$$

The sources estimate, even in this case, can be made using a network as in the model in Fig. 2.32. Consequently, however, each element w_{ij} of the matrix $\mathbf{W}$, defined in real or complex domain, is replaced by an FIR or IIR filter with TF equal to $W_{ij}(z)$ (Fig. 2.33).

The output is calculated as

$$\mathbf{y} = \mathbf{W} * \mathbf{x}$$

whereby

$$\mathbf{y} \approx \mathbf{s}$$

less than trivial ambiguity.

Remark For the separation, the basic assumption used is that the sources are statistically independent. This hypothesis is quite general and realistic when they are generated by different physically entities. This hypothesis about the sources has given rise to a new analysis tool precisely called ICA [9, 22].

Given the vastness of the *neural networks* topic, and more generally of the *machine learning for signal processing*, those arguments have not been included in this volume.

References

1. Kay SM (1993) Fundamental of statistical signal processing estimation theory. Prentice Hall, Englewood Cliffs, NJ
2. Kuo SM, Morgan DR (1999) Active noise control: a tutorial review. Proc IEEE 87(6):943–973
3. Van Trees HL (2002) Optimum array processing, Part IV: detection, estimation and modulation theory. Wiley, New York
4. R Lippmann (1987) An introduction to computing with neural nets. IEEE ASSP Magazine, vol 4, no 2, Aprile 1987
5. Rumelhart DE, McClelland JL, The PDP Research Group (1986) Parallel distributed processing: explorations in the microstructure of cognition, vol 1: foundations. MIT Press, Cambridge, MA
6. B Widrow, M Lehr (1990) 30 years of adaptive neural networks: perceptron, adaline and backpropagation. Proc IEEE 78(9), Settembre 1990
7. JJ Hopfield, DW Tank (1986) Computing with neural circuits: a model. Science 233: 625–633, August 1986
8. T Kohonen (2001) Self-organizing maps, 3rd extended edn. Springer series in information sciences, vol 30. Springer, Berlin

9. J Principe, A Chicocki, L Xu, E Oja, D Erdogmus (Guest eds) (2004) Special issue on—Information theoretic learning. IEEE Trans Neural Netw 15(4): 789–791, July 2004
10. Widrow B, Stearns SD (1985) Adaptive signal processing. Prentice Hall, Englewood Cliffs, NJ
11. Haykin S (1996) Adaptive filter theory, 3rd edn. Prentice Hall, Upper Saddle River, NJ
12. Wiener N (1949) Extrapolation, interpolation and smoothing of stationary time series, with engineering applications. Wiley, New York
13. Kailath T (1974) A view of three decades of linear filtering theory. IEEE Trans Inf Theory IT20(2):146–181
14. Box GEP, Jenkins GM (1970) Time series analysis: forecasting and control. Holden-Day, San Francisco
15. Bode HW, Shannon CE (1950) A simplified derivation of linear least squares smoothing and prediction theory. Proc IRE 38:417–425
16. Orfanidis SJ (1996) Introduction to signal processing. Prentice Hall, Upper Saddle River, NJ
17. Manolakis DG, Ingle VK, Kogon SM (2000) Statistical and adaptive signal processing. McGraw-Hill, New York
18. Burg JP (1968) A new analysis technique for time data series. NATO Advanced Study Institute on Signal Processing, Enschede, Netherlands
19. JA Cadzow (1990) Signal processing via least squares error modeling. IEEE ASSP Magazine, pp 12–31,October 1990
20. Benesty J, Gänsler T, Morgan DR, Sondhi MM, Gay SL (2001) Advances in network and acoustic echo cancellation. Springer, Berlin. ISBN 978-3-540-41721-7
21. Hänsler E, Schmidt G (eds) (2006) Topics in acoustic echo and noise control. Spinger, Berlin. ISBN 978-3-540-33212-1
22. E Oja, S Harmeling, L Almeida (Guest eds) (2004) Special issue on—Independent component analysis and beyond. Signal Process 84(2), February 2004

Chapter 3
Optimal Linear Filter Theory

3.1 Introduction

This chapter introduces the Wiener statistical theory of linear filtering that is a reference for the study and understanding of adaptive methods shown below in the text.

Although the original development of Wiener's theory was conducted in continuous time, for exposition consistency, it is preferred to introduce this topic directly in the discrete-time domain.

The issues of the mean squares error (MSE) minimization and the computation of its minimum value or minimum MSE (MMSE) are addressed. The normal equations and optimal solution computation using the discrete-time Wiener formulation are introduced and discussed. Attention has been directed, in particular, to the case of linear FIR filters, also known as adaptive transversal filters (AF). In addition, some multiple-input multiple-output (MIMO), algebraically equivalent, notations are presented.

Are also discussed corollaries as the geometric interpretation, the principle of orthogonality, and the principal component analysis (PCA) of the optimum filter.

Moreover, in the context of the Wiener theory, we present and discuss some classical AF application. The examples concern the computation of linear direct and inverse optimal filter models in some specific technological contexts.

3.2 Adaptive Filter Basic and Notations

Formally, the optimal Wiener filter is not a *true* AF [4]. The weights determination, in fact, is not a function of samples that instant by instant feed the filter itself, but is performed with an approach based on *a priori* knowledge of the second-order moments of the *stochastic processes* (SPs) of the input sequences [2–8]. In other words, the optimal filter is the same for all processes with the identical second-order

A. Uncini, *Fundamentals of Adaptive Signal Processing*, Signals and Communication Technology, DOI 10.1007/978-3-319-02807-1_3,

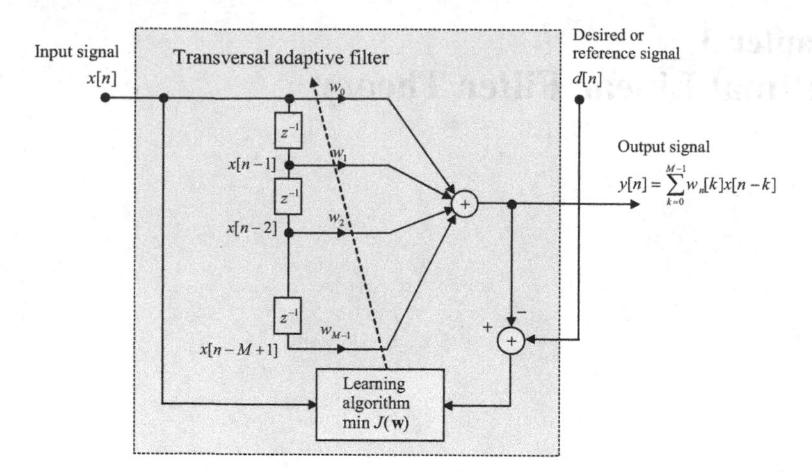

Fig. 3.1 Linear FIR adaptive filter also called *adaptive transversal filter*

statistics. However, most of the definitions and the formalism used in the text are common to this approach.

3.2.1 The Linear Adaptive Filter

To derive the single input single output (SISO) notation, consider the single channel FIR adaptive filter shown in Fig. 3.1.

Widely used in many applications, the adaptive transversal filter has an impulse response determined in order to estimate the reference signal, which in this context, is called the *desired output* and denoted as $d[n]$. The filter input–output relation is given by the discrete-time convolution between the input sequence $x[n]$ and the coefficients of the filter $w[n]$:

$$y[n]=\sum_{k=0}^{M-1} w_n[k]x[n-k]=w[n]*x[n], \tag{3.1}$$

where the index M represents the filter tapped-delay-line length.

3.2.1.1 Real and Complex Domain Vector Notation

For a compact representation is possible to use a vector notation. We define the weight vector $\mathbf{w}\in\mathbb{R}^{M\times 1}$ as

$$\mathbf{w} \in \mathbb{R}^{M \times 1} = \begin{bmatrix} w[0] & w[1] & \cdots & w[M-1] \end{bmatrix}^T, \tag{3.2}$$

containing the coefficients of the filter impulse response at time n. The vector of the input signal $\mathbf{x} \in \mathbb{R}^{M \times 1}$ is defined as

$$\mathbf{x} \in \mathbb{R}^{M \times 1} = \begin{bmatrix} x[n] & x[n-1] & \cdots & x[n-M+1] \end{bmatrix}^T, \tag{3.3}$$

which contains the window signal along the input delay line of the FIR filter, we can write the convolution (3.1) as the inner (or dot) product between input and weight vectors. That is,

$$y[n] = \mathbf{w}^T\mathbf{x} = \mathbf{x}^T\mathbf{w}. \tag{3.4}$$

For the definition of recursive algorithms using the vector representation, you may need to specify a time index n. You can then add to the above definitions a subscript n that is related to the updating step of the filter coefficients. Hence, indicating by $\Delta\mathbf{w}_n$ the coefficients' variation (calculated according to some law described below), we can write the adaptation rule as $\mathbf{w}_n = \mathbf{w}_{n-1} + \Delta\mathbf{w}_n$.

The vector $\mathbf{x}_n$ indicates the M-length time window input sequence at time n. That is defined as

$$\begin{aligned} \mathbf{x}_n &= \begin{bmatrix} x[n] & x[n-1] & \cdots & x[n-M+1] \end{bmatrix}^T \\ \mathbf{x}_{n-1} &= \begin{bmatrix} x[n-1] & x[n-2] & \cdots & x[n-M] \end{bmatrix}^T \\ \mathbf{x}_{n-2} &= \begin{bmatrix} x[n-2] & x[n-3] & \cdots & x[n-M-1] \end{bmatrix}^T, \\ &\vdots \end{aligned} \tag{3.5}$$

and in the absence of the 'n' index is worth $\mathbf{x} \to \mathbf{x}_n$ and $\mathbf{w} \to \mathbf{w}_n$. The vector $\mathbf{x}$ is sometimes referred to as *input vector regression*, or simply *input regression*.

In the case of complex domain signals, the sequence is defined as

$$x[n] = \mathrm{Re}\big(x[n]\big) + j\,\mathrm{Im}\big(x[n]\big) = x_{\mathrm{Re}}[n] + jx_{\mathrm{Im}}[n], \tag{3.6}$$

then we have that $\mathbf{x} \in \mathbb{C}^{M \times 1}$ and $\mathbf{x}_n = \mathbf{x}_{\mathrm{Re},n} + j\mathbf{x}_{\mathrm{Im},n}$, in particular we use the following convention:

$$\begin{aligned} \mathbf{x}_n &= \begin{bmatrix} x[n] & x[n-1] & \cdots & x[n-M+1] \end{bmatrix}^T \\ &= \begin{bmatrix} x^*[n] & x^*[n-1] & \cdots & x^*[n-M+1] \end{bmatrix}^H, \end{aligned} \tag{3.7}$$

while for the filter coefficients $\mathbf{w} \in \mathbb{C}^{M \times 1}$ we use the convention $\mathbf{w} = \mathbf{w}_{\mathrm{Re}} - j\mathbf{w}_{\mathrm{Im}}$:

$$\begin{aligned} \mathbf{w} &\triangleq \begin{bmatrix} w^*[0] & w^*[1] & \cdots & w^*[M-1] \end{bmatrix}^T \\ &= \begin{bmatrix} w[0] & w[1] & \cdots & w[M-1] \end{bmatrix}^H, \end{aligned} \tag{3.8}$$

note that also in this case we have $\mathbf{w} \to \mathbf{w}_n$.

With these definitions for the calculation of the output, the following notation is used[1]:

$$y[n] = \mathbf{w}^H\mathbf{x} = \left(\mathbf{x}^H\mathbf{w}\right)^*. \tag{3.9}$$

Note that, with this notation, the calculation of the filter's output in real and complex cases are formally similar. Moreover, defining vectors and matrix with the notation $\mathbf{w} \in (\mathbb{R},\mathbb{C})^{M\times 1}$, $\mathbf{X} \in (\mathbb{R},\mathbb{C})^{N\times M}$, …, with due attention to the conjugation operator, the extension in the complex domain of algorithms defined in the real domain can be made by simply replacing "H" → "T" and vice versa.

Remark As shown in Fig. 3.1, desired output is a sequence that, depending on the methodology used for the determination of the filter coefficients, can be defined as deterministic or random.

Without loss of generality, it is possible to think the output and reference sequences corrupted by noise indicated, respectively, as $\eta_y[n]$ and $\eta_d[n]$. This noise is often additive white with Gaussian distribution mentioned as WGN or AWGN with zero mean and uncorrelated with the input signal. In particular, it is characterized by its variance (or power) and indicated as $\eta[n] \triangleq N\left(\langle\eta[n]\rangle,\sigma_n^2\right)$. Later in this chapter, where not expressly stated, this noise is assumed zero.

3.2.2 *Composite Notations for Multiple-Input Multiple-Output Filter*

We extend the notation to the multiple-input multiple-output (MIMO) case, with P inputs and Q outputs as shown in Fig. 3.2. Indicating with

$$\mathbf{w}_{ij} \in (\mathbb{R},\mathbb{C})^{M\times 1} \triangleq \left[w_{ij}[0] \quad \cdots \quad w_{ij}[M-1]\right]^H, \quad i = 1,\ldots,Q,\ j = 1,\ldots,P, \tag{3.10}$$

the $P \cdot Q$ impulse responses, considered for simplicity, all of identical M length, between the jth input and the ith output. Indicating with

$$\mathbf{x}_j \in (\mathbb{R},\mathbb{C})^{M\times 1} \triangleq \left[x_j[n] \quad \cdots \quad x_j[n-M+1]\right]^T, \quad j = 1,2,\ldots,P, \tag{3.11}$$

the input signals present on the filters delay lines $\mathbf{w}_{ij}$ for $i = 1, 2, \ldots, P$. So, at the instant n, for the Q outputs, we can write

[1] It is noted that, in the complex case, for the filter's output calculation can be also used the following notation $y[n] = (\mathbf{w}^T\mathbf{x}^*)^* = \mathbf{x}^T\mathbf{w}^* = \mathbf{x}^T\,\hat{\mathbf{w}}$, for $\hat{\mathbf{w}} = \mathbf{w}^*$.

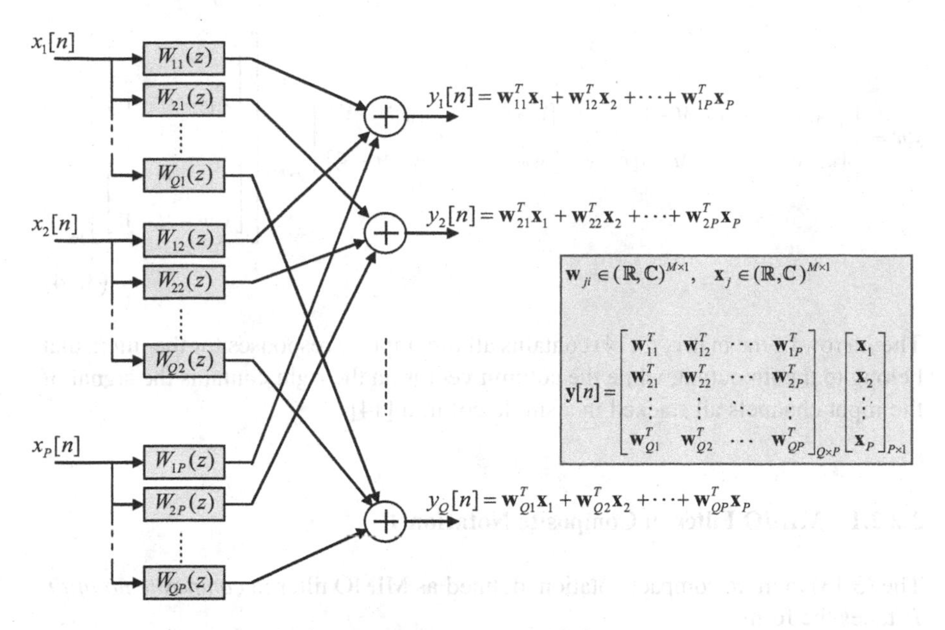

Fig. 3.2 Representation of P-inputs and Q-outputs MIMO filter

$$
\begin{aligned}
y_1[n] &= \mathbf{w}_{11}^T\mathbf{x}_1 + \mathbf{w}_{12}^T\mathbf{x}_2 + \cdots + \mathbf{w}_{1P}^T\mathbf{x}_P, \\
y_2[n] &= \mathbf{w}_{21}^T\mathbf{x}_1 + \mathbf{w}_{22}^T\mathbf{x}_2 + \cdots + \mathbf{w}_{2P}^T\mathbf{x}_P, \\
&\vdots \\
y_Q[n] &= \mathbf{w}_{Q1}^T\mathbf{x}_1 + \mathbf{w}_{Q2}^T\mathbf{x}_2 + \cdots + \mathbf{w}_{QP}^T\mathbf{x}_P.
\end{aligned}
\tag{3.12}
$$

The vector $\mathbf{x}_j$ is often referred to as the *data-record* relative to the jth input of the MIMO system.

Said $\mathbf{y}[n] \in (\mathbb{R},\mathbb{C})^{Q\times 1} = \begin{bmatrix} y_1[n] & y_2[n] & \cdots & y_Q[n] \end{bmatrix}^T$, the vector representing all the outputs of the MIMO filter at the time n (called output *snap-shot*), the output expression can be written as

$$
\mathbf{y}[n] \in (\mathbb{R},\mathbb{C})^{Q\times 1} = \begin{bmatrix} \mathbf{w}_{11}^T & \mathbf{w}_{12}^T & \cdots & \mathbf{w}_{1P}^T \\ \mathbf{w}_{21}^T & \mathbf{w}_{22}^T & \cdots & \mathbf{w}_{2P}^T \\ \vdots & \vdots & \ddots & \vdots \\ \mathbf{w}_{Q1}^T & \mathbf{w}_{Q2}^T & \cdots & \mathbf{w}_{QP}^T \end{bmatrix}_{Q\times P} \begin{bmatrix} \mathbf{x}_1 \\ \mathbf{x}_2 \\ \vdots \\ \mathbf{x}_P \end{bmatrix}_{P\times 1}, \tag{3.13}
$$

which written in extended mode, takes the form as

$$\mathbf{y}[n] = \begin{bmatrix} [w_{11}[0] & \cdots & w_{11}[M-1]] & \cdots & [w_{1P}[0] & \cdots & w_{1P}[M-1]] \\ & \vdots & & \ddots & & \vdots & \\ [w_{Q1}[0] & \cdots & w_{Q1}[M-1]] & \cdots & [w_{QP}[0] & \cdots & w_{QP}[M-1]] \end{bmatrix}_{Q\times PM} \begin{bmatrix} \begin{bmatrix} x_1[n] \\ \vdots \\ x_1[n-M+1] \end{bmatrix} \\ \vdots \\ \begin{bmatrix} x_P[n] \\ \vdots \\ x_P[n-M+1] \end{bmatrix} \end{bmatrix}_{PM\times 1}. \tag{3.14}$$

The *j*th row of the matrix (3.14) contains all the impulse responses for the filters that belong to the *j*th output, while the column vector on the right contains the signal of the input channels all stacked in a single column [14].

3.2.2.1 MIMO Filter in Composite Notation 1

The (3.13) in more compact notation, defined as MIMO filter in *composite notation 1*, takes the form

$$\mathbf{y}[n] = \mathbf{W}\mathbf{x}, \tag{3.15}$$

where, $\mathbf{W} \in (\mathbb{R},\mathbb{C})^{Q\times P(M)}$ is defined as

$$\mathbf{W} \in (\mathbb{R},\mathbb{C})^{Q\times P(M)} = \begin{bmatrix} \mathbf{w}_{11}^T & \mathbf{w}_{12}^T & \cdots & \mathbf{w}_{1P}^T \\ \mathbf{w}_{21}^T & \mathbf{w}_{22}^T & \cdots & \mathbf{w}_{2P}^T \\ \vdots & \vdots & \ddots & \vdots \\ \mathbf{w}_{Q1}^T & \mathbf{w}_{Q2}^T & \cdots & \mathbf{w}_{QP}^T \end{bmatrix}_{Q\times P}, \tag{3.16}$$

with the notation $Q \times P(M)$, we denote a partitioned $Q \times P$ matrix, where each element of the partition is a row vector equal to $\mathbf{w}_{ji}^T \in \mathbb{R}^{1\times M}$.

The vector $\mathbf{x}$, said *composite input*, is defined as

$$\mathbf{x} \in (\mathbb{R},\mathbb{C})^{P(M)\times 1} = \begin{bmatrix} \mathbf{x}_1 \\ \mathbf{x}_2 \\ \vdots \\ \mathbf{x}_P \end{bmatrix}_{P\times 1} = \begin{bmatrix} \mathbf{x}_1^T & \mathbf{x}_2^T & \cdots & \mathbf{x}_P^T \end{bmatrix}^T, \tag{3.17}$$

constructed as the vector of all stacked inputs at instant n ($\mathbf{x} \equiv \mathbf{x}_n$), i.e., $\mathbf{x}$ is formed by the input vectors $\mathbf{x}_{i,n}$ for $i = 1, \ldots, P$.

3.2.2.2 MIMO Filter in Composite Notation 2

Let us define the vector as

$$\mathbf{w}_{j:}^{H} \in (\mathbb{R},\mathbb{C})^{1\times P(M)} \triangleq \begin{bmatrix} \mathbf{w}_{j1}^{T} & \mathbf{w}_{j2}^{T} & \cdots & \mathbf{w}_{jP}^{T} \end{bmatrix}, \tag{3.18}$$

i.e., the *j*th row of the matrix **W**, and the *composite weights vector* **w**, built with vectors $\mathbf{w}_{j:}^{T}$ for all $j = 1, 2, \ldots, Q$, for which we can write

$$\mathbf{w} \in (\mathbb{R},\mathbb{C})^{(PM)Q\times 1} \triangleq \begin{bmatrix} \mathbf{w}_{1:} \\ \mathbf{w}_{2:} \\ \vdots \\ \mathbf{w}_{Q:} \end{bmatrix}_{Q\times 1}, \tag{3.19}$$

that is made with all matrix **W** rows, staked in a single column, i.e., $\mathbf{w} = \mathrm{vec}(\mathbf{W})$.

We define the *data composite matrix* $\mathbf{X} \in (\mathbb{R},\mathbb{C})^{Q(PM)\times Q}$ as

$$\mathbf{X} \in (\mathbb{R},\mathbb{C})^{(PM)Q\times Q} = \mathbf{I}_{Q\times Q} \otimes \mathbf{x} = \begin{bmatrix} \mathbf{x} & \mathbf{0} & \cdots & \mathbf{0} \\ \mathbf{0} & \mathbf{x} & \cdots & \mathbf{0} \\ \vdots & \vdots & \ddots & \vdots \\ \mathbf{0} & \mathbf{0} & \cdots & \mathbf{x} \end{bmatrix}_{Q\times Q}, \tag{3.20}$$

where the symbol $\otimes$ indicates the *Kronecker product* (Sect. A.13).

From the definitions (3.19) and (3.20) we can express the output as

$$\mathbf{y}[n] = (\mathbf{I} \otimes \mathbf{x})^{T}\mathrm{vec}(\mathbf{W}) = \begin{bmatrix} \mathbf{x}^{T} & \mathbf{0} & \cdots & \mathbf{0} \\ \mathbf{0} & \mathbf{x}^{T} & \cdots & \mathbf{0} \\ \vdots & \vdots & \ddots & \vdots \\ \mathbf{0} & \mathbf{0} & \cdots & \mathbf{x}^{T} \end{bmatrix}_{Q\times Q} \begin{bmatrix} \mathbf{w}_{1:} \\ \mathbf{w}_{2:} \\ \vdots \\ \mathbf{w}_{Q:} \end{bmatrix}_{Q\times 1} = \mathbf{X}^{T}\mathbf{w}, \tag{3.21}$$

i.e., the elements of the vector $\mathbf{y}[n]$ are defined as

$$y_{j}[n] = \mathbf{x}^{T}\mathbf{w}_{j}, \qquad \text{for} \quad j = 1, 2, \ldots, Q. \tag{3.22}$$

3.2.2.3 MIMO (*P*, *Q*) System as Parallel of *Q* Filters Banks

In some following developments, the matrix **W** is indicated as

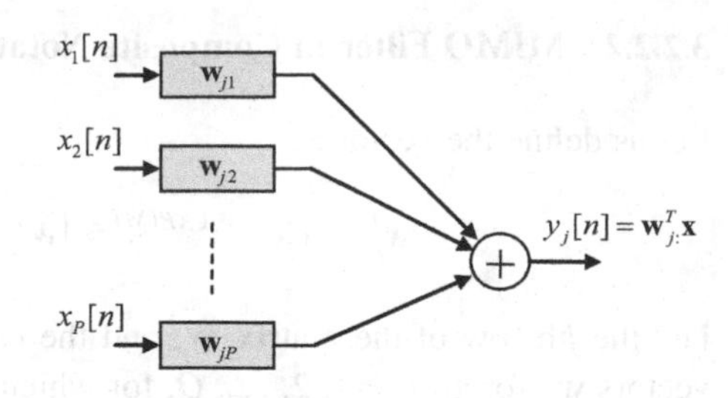

Fig. 3.3 Diagram of the *j*th MISO subsystem of the MIMO filter

$$\mathbf{W} \in (\mathbb{R}, \mathbb{C})^{Q \times P(M)} = \begin{bmatrix} \mathbf{w}_{1:}^T \\ \mathbf{w}_{2:}^T \\ \vdots \\ \mathbf{w}_{Q:}^T \end{bmatrix}_{Q \times 1}, \tag{3.23}$$

where with the vector $\mathbf{w}_{j:}^T \in \mathbb{R}^{1 \times P(M)}$, defined in (3.18), it indicates the *j*th row of the matrix $\mathbf{W}$ as shown in Fig. 3.3. This allows us to interpret (3.22) as a bank of P filters afferent to the *j*th system output. In fact, for each output channel is

$$\begin{aligned} y_j[n] &= \mathbf{w}_{j:}^T\mathbf{x} \\ &= \sum_{k=1}^{P} \mathbf{w}_{jk}^T\mathbf{x}_k. \end{aligned} \tag{3.24}$$

The MIMO AF can be interpreted as the parallel of Q multiple-input single-output (MISO) P channels filter banks each of which, as shown in Fig. 3.3, is characterized by the vector weights $\mathbf{w}_{j:}$. In other words, as best seen below, each of these banks can be independently adapted from each others.

3.2.2.4 MIMO Filter in Snap-Shot or Composite Notation 3

Considering the MISO system of Fig. 3.3, we define the vector

$$\mathbf{w}_j[k] \in (\mathbb{R}, \mathbb{C})^{P \times 1} = \begin{bmatrix} w_{j1}[k] & w_{j2}[k] & \cdots & w_{jP}[k] \end{bmatrix}_{P \times 1}^T, \tag{3.25}$$

containing the filter taps of the *j*th bank related to the delay k.

In a similar way, we define

$$\mathbf{x}[0] \in (\mathbb{R}, \mathbb{C})^{P \times 1} = \begin{bmatrix} x_1[0] & x_2[0] & \cdots & x_P[0] \end{bmatrix}^T, \tag{3.26}$$

the vector containing all inputs of the filter MISO at instant n, this vector is the input snap-shot. Furthermore, we define the vector $\mathbf{x}[k] \in \mathbb{R}^{P \times 1}$ as the signals present on the filter delay lines at the k-th delay.

With this formalism, the jth MISO channel output, combining (3.25) and (3.26), can be expressed in snap-shot notation as

$$y_j[n] = \sum_{k=0}^{M-1} \mathbf{w}_j^T[k]\mathbf{x}[k], \qquad \text{for} \quad j = 1, 2, \ldots, Q. \tag{3.27}$$

Note that defining the vectors $\mathbf{w}_j$ and $\mathbf{x}$, as

$$\mathbf{w}_j \in (\mathbb{R}, \mathbb{C})^{M(P)\times 1} = \begin{bmatrix} \mathbf{w}_j^H[0] & \mathbf{w}_j^H[1] & \cdots & \mathbf{w}_j^H[M-1] \end{bmatrix}_{M\times 1}^T, \tag{3.28}$$

$$\mathbf{x} \in (\mathbb{R}, \mathbb{C})^{M(P)\times 1} = \begin{bmatrix} \mathbf{x}^T[0] & \mathbf{x}^T[1] & \cdots & \mathbf{x}^T[M-1] \end{bmatrix}_{M\times 1}^T, \tag{3.29}$$

containing, respectively, all weights stacked of the jth bank and all inputs related to it, is $y_j[n] = \mathbf{w}_j^T\mathbf{x}$, analogous to (3.22) and (3.24).

Remark The MIMO composite notations 1, 2, and 3, defined by (3.15), (3.21), and (3.27), respectively from the algebraic point of view, are completely equivalent. However, note that for certain developments in the rest of the text, it is more convenient to use the one rather than the other notation.

3.2.3 Optimization Criterion and Cost Functions Definition

The free parameters $\mathbf{w}$ calculation of an AF is usually carried out according to some rule based on an optimization criterion that minimizes (or maximizes) a predefined cost function (CF). The criterion is usually chosen depending on the input signal characteristics.

If the nature of the input signal is stochastic, the CF is a function of some statistic of the error signal. In these cases, it is usual to consider the *statistical expectation* (or *expected value* or *ensemble-average value* or *mean value*), indicated by the operator $E\{\cdot\}$, of the square of error signal. Such quantity indicated as mean-square error (MSE) is defined as

$$J(\mathbf{w}) = E\Big\{ \big| e[n] \big|^2 \Big\}. \tag{3.30}$$

The minimization of (3.30) is performed by a stochastic optimization criterion called minimum mean square error (MMSE).

If the nature of $\mathbf{x}$ is deterministic, the CF is also a deterministic function of error signal expressed as certain *time-average value*. Usually, the CF is expressed as the *least squares error*. In this case, we more properly we use the sum of squared error (SSE) defined as

$$\hat{J}(\mathbf{w}) \triangleq \hat{E}\left\{|e[n]|^2\right\} = \frac{1}{N}\sum_n |e[n]|^2. \tag{3.31}$$

The filter output assumes the form of a linear combiner (3.4) and the error is defined as

$$e[n] = d[n] - y[n] = d[n] - \mathbf{w}^T\mathbf{x} = d[n] - \mathbf{x}^T\mathbf{w}. \tag{3.32}$$

The minimization of (3.31) is performed by an approximate stochastic criterion called least square error (LSE) and the class of algorithms derived from it is referred to as least squares (LS) algorithms.

The LSE criterion can be considered as an approximation of the stochastic MSE criterion where the notation (3.30), in practice, is replaced directly by time-average operator. Therefore it is

$$E\left\{|e[n]|^2\right\} \approx \hat{E}\left\{|e[n]|^2\right\}, \tag{3.33}$$

and, if this approximation is true, the SP is defined as an *ergodic process*.

Moreover, the expression (3.30) or (3.31) can be generalized as

$$J_p(\mathbf{w}) = E\left\{||e[n]||^p\right\}, \tag{3.34}$$

where p can assume values form 0, 1, 2, 3, ..., ∞.

3.2.4 Approximate Stochastic Optimization

The criterion that minimizes expression (3.30) is defined as MMSE. The optimum coefficients $\mathbf{w}_{\text{opt}}$ can be determined with MMSE by a rule of type

$$\mathbf{w}_{\text{opt}} = \min_{\mathbf{w}} \{J(\mathbf{w})\}. \tag{3.35}$$

By defining the *gradient* of CF as

$$\nabla J(\mathbf{w}) \triangleq \frac{\partial J(\mathbf{w})}{\partial \mathbf{w}}. \tag{3.36}$$

we can write

$$\mathbf{w}_{\text{opt}} \;\therefore\; \nabla J(\mathbf{w}) \to 0. \tag{3.37}$$

In the cases we have no *a priori* knowledge on the signal statistics, in particular the first- and second-order moments are unknown, for the determination of the filter parameters, we proceed by approximating the optimal statistical solution. Considering the available data, it is usual to refer to a new CF which can be formulated in more general form as

$$\hat{J}(\mathbf{w}) = E\{Q(\mathbf{w},\eta)\}, \tag{3.38}$$

where $Q(\mathbf{w},\eta)$ is an unknown distribution. The function $\hat{J}(\mathbf{w})$ represents an approximation of the stochastic CF (3.30). In other words, one minimizes not directly the gradient but rather its estimated value (stochastic gradient). For this reason, the learning paradigms arising from the minimization of a functional of the type (3.38) are also referred to as approximate stochastic optimization (ASO) methods.

Similar to the formalism of (3.37) we can write

$$\hat{\mathbf{w}}_{\text{opt}} \ \therefore \ \nabla\hat{J}(\mathbf{w}) \to 0. \tag{3.39}$$

The ASO algorithms can be derived in a recursive or nonrecursive (or *batch*) formulation. In batch formulation, the fundamental hypothesis is to know the entire signals (or a portion acquired by direct, usually noisy measures). In these cases, when it is possible to consider ergodic and stationary input processes, the expectation (3.38) can be replaced with its time average calculated over N signal (Fig. 3.4) samples as

$$\hat{J}(\mathbf{w}) = \frac{1}{N}\sum_{n=0}^{N-1} Q(\mathbf{w},\eta). \tag{3.40}$$

This chapter describes the batch methods for the cases of stochastic and deterministic CFs. The recursive or *online* techniques, in which the solution is updated when new input samples are available, are analyzed in Chaps. 4, 5 and 6.

3.3 Adaptation By Stochastic Optimization

In this section, the case in which the filter inputs are (real or complex) SPs described in terms of their *a priori* known second-order statistics is considered. The filter weights vector is considered as a deterministic unknown and the calculation of the its optimal value $\mathbf{w}_{\text{opt}}$ is made by directly minimization of the statistical CF MSE defined by (3.30).

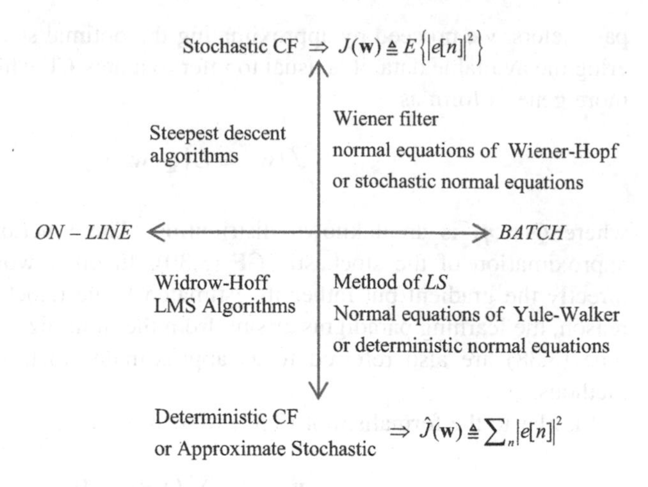

Fig. 3.4 Schematic representation of the learning algorithms

3.3.1 Normal Equations in Wiener–Hopf Notation

For the determination of the minimum of the function $J(\mathbf{w}) = E\{|e[n]|^2\}$, defined as MMSE, we proceed by calculating the gradient of $J(\mathbf{w})$ with the (3.36) and putting the result to zero as indicated by the (3.37). Let $\mathbf{w} \in (\mathbb{R},\mathbb{C})^{M\times 1}$ the vector of the unknown filter coefficients, for the derivative computation we consider the explicit error $e[n]$ representation. So for (3.4) we can write

$$e[n] = d[n] - \mathbf{w}^T\mathbf{x}. \tag{3.41}$$

The square error can be written as

$$e^2[n] = d^2[n] - \mathbf{w}^T\mathbf{x}d[n] - \mathbf{x}^T\mathbf{w}d[n] + \mathbf{w}^T\mathbf{x}\mathbf{x}^T\mathbf{w}. \tag{3.42}$$

The MSE (3.30) can be determined by taking the expectation of the previous expression:

$$J(\mathbf{w}) = E\{d^2[n]\} - E\{\mathbf{w}^T\mathbf{x}d[n]\} - E\{\mathbf{x}\mathbf{w}^Td[n]\} - E\{\mathbf{w}^T\mathbf{x}\mathbf{x}^T\mathbf{w}\}. \tag{3.43}$$

Recalling that the term $\sigma_d^2 = E\{d^2[n]\}$ is the *variance* of the signal $d[n]$, the term $\mathbf{g} = E\{\mathbf{x}d[n]\}$ represents the *cross-correlation vector* among the input $\mathbf{x}$ and the desired signal $d[n]$ and that $\mathbf{R} = E\{\mathbf{x}\mathbf{x}^T\}$ is the *autocorrelation matrix* of the input sequence, the expression (3.43) can be reduced to the following quadratic form

$$J(\mathbf{w}) = \sigma_d^2 - \mathbf{w}^T\mathbf{g} - \mathbf{g}^T\mathbf{w} + \mathbf{w}^T\mathbf{R}\mathbf{w}, \tag{3.44}$$

with gradient defined as (for vectors derivative rules see, for example [1]):

$$\begin{aligned}\nabla J(\mathbf{w}) &= \frac{\partial J(\mathbf{w})}{\partial \mathbf{w}} \\ &= \frac{\partial\left(\sigma_d^2 - \mathbf{w}^T\mathbf{g} - \mathbf{g}^T\mathbf{w} + \mathbf{w}^T\mathbf{R}\mathbf{w}\right)}{\partial \mathbf{w}} \\ &= 2(\mathbf{R}\mathbf{w} - \mathbf{g}).\end{aligned} \tag{3.45}$$

From the previous expression, you can write the following system of linear equations:

$$\mathbf{R}\mathbf{w} = \mathbf{g}, \tag{3.46}$$

known as *normal equations* in the Wiener–Hopf notation [2–5]. The solution of the system (3.46), also known as the Widrow–Hoff equations [2], can be written as

$$\mathbf{w}_{\text{opt}} = \mathbf{R}^{-1}\mathbf{g}. \tag{3.47}$$

Remark In the Wiener's optimal filtering theory, the filter's inputs are considered as SPs described in terms of their *a priori* known second-order statistics. The vector of the filter weights is considered as deterministic unknown and the calculation of optimal filter coefficients $\mathbf{w}_{\text{opt}}$ is made minimizing the statistical CF defined by the MSE (3.30). Note that many authors (e.g., [2–8]) define the *adaptive filter* (AF), the filter whose parameters are iteratively adjusted based on the new signal samples that gradually flows to its input. In this sense, the optimal filter with coefficients (3.47) is not formally defined as an AF, because it is *exactly* computed on the basis of their *a priori* known input statistics. In reality, the determination of the coefficients $\mathbf{w}_{\text{opt}}$, it is not a direct function of the signal flow input samples, but the filter is designed on the base on *a priori* knowledge of second-order moments of the input SPs. In other words, the optimal filter is the same for any input sequence with the same statistics. However, most of the definitions and the formalism used in the text are common to this theoretical approach. Nevertheless, as we shall see in following chapters, methods for adaptive filtering are derived from this theory, and the linear optimal Wiener estimator is a reference for the study and determination of the AFs properties.

3.3.1.1 Wiener–Hopf Normal Equations in Scalar Notation

The Wiener normal equations can be derived using scalar notation. Considering the CF (3.30) and the expression of the filter output in the real case we can write

$$e[n] = d[n] - \sum_{k=0}^{M-1} w[k]x[n-k]. \tag{3.48}$$

Refer to the *j*th element of the vector **w**, the derivative of (3.30) can be written as

$$\begin{aligned}\frac{\partial J(\mathbf{w})}{\partial w[i]} &= \frac{\partial E\{e^2[n]\}}{\partial w[i]} \\ &= 2E\left\{e[n]\frac{\partial e[n]}{\partial w[i]}\right\}, \qquad \text{for} \quad i = 0, 1, \ldots, M-1,\end{aligned} \tag{3.49}$$

where the error derivative is given by

$$\begin{aligned}\frac{\partial e[n]}{\partial w[j]} &= \frac{\partial}{\partial w[j]}\left[d[n] - \sum_{i=0}^{M-1} w[i]x[n-i]\right] \\ &= -x[n-j].\end{aligned}$$

From previous positions we can write

$$\begin{aligned}\frac{\partial J(\mathbf{w})}{\partial w[i]} &= -2E\left\{\left(d[n] - \sum_{i=0}^{M-1} w[i]x[n-i]\right)x[n-j]\right\} \\ &= -2E\{d[n]x[n-j]\} + 2\sum_{i=0}^{M-1} w[i]E\{x[n-i]x[n-j]\}.\end{aligned}$$

From a simple visual inspection of the above expression, the terms $E\{x[n-i]x[n-j]\}$ and $E\{d[n]x[n-j]\}$ represent the autocorrelation (acf) $r[n-i,\, n-j]$ and the cross-correlation (ccf) $g[n-j,\, n]$ sequences. Writing in a more compact mode, we have

$$\frac{\partial J(\mathbf{w})}{\partial w[i]} = 2\left(\sum_{i=0}^{M-1} r[n-i, n-j]w[i] - g[n-j, n]\right). \tag{3.50}$$

Equating to zero, we obtain the expression:

$$\sum_{i=0}^{M-1} r(n-i, n-j)w[i] = g(n-j, n), \tag{3.51}$$

which corresponds to a system of linear equations in the unknowns $w[i]$ known by the name of *normal equations*.

For $x[n]$ and $d[n]$ stationary SPs, the correlation functions no longer depend on the time index n but only to delays i and j, then we can write

$$r[n-i,n-j] \rightarrow r[j-i] \quad \text{and} \quad g[n-j,n] \rightarrow g[j].$$

It follows that the normal equations (3.51) can be rewritten as

$$\sum_{i=0}^{M-1} r[j-i]w[i] = g[j], \quad 0 \le j \le M-1. \tag{3.52}$$

Writing (3.52) vector form, we have $\mathbf{Rw} = \mathbf{g}$. So the previous coincides with (3.46).

Is interesting noted that the integral Wiener–Hopf equations (3.52) have been developed in the continuous-time domain in 1931 [4] and the first discrete-time formulation is due to Levinson and formulated in 1947 [5].

3.3.2 *On the Estimation of the Correlation Matrix*

For solution of (3.47), we observe that the autocorrelation matrix $\mathbf{R}$, in the case in that the sequence $\mathbf{x} \in (\mathbb{R},\mathbb{C})^{M\times 1}$, is defined as the *expectation of the outer product* of vector $\mathbf{x}$ (Sect. C.2.6). Formally,

$$\mathbf{R} \equiv E\{\mathbf{xx}^H\} = \begin{bmatrix} r[0] & r[1] & \cdots & r[M-1] \\ r^*[1] & r[0] & \cdots & r[M-2] \\ \vdots & \vdots & \ddots & \vdots \\ r^*[M-1] & r^*[M-2] & \cdots & r[0] \end{bmatrix}. \tag{3.53}$$

The term

$$r[k] \equiv E\{x[n]x^*[n-k]\} = E\{x[n+k]x^*[n]\}, \tag{3.54}$$

is, by definition, the acf of the sequence $x[n]$.

From previous definition, it is easy to show that the correlation matrix has the following property (Sect. C.1.8):

1. $\mathbf{R}$ is symmetric: in the real case is $\mathbf{R}^T = \mathbf{R}$ while in complex domain is $\mathbf{R}^H = \mathbf{R}$, and $r[-k] = r^*[k]$.
2. $\mathbf{R}$ is a Toeplitz matrix, i.e., has equal elements on the diagonals.
3. $\mathbf{R}$ is semidefinite positive for which $\mathbf{w}^H\mathbf{Rw} \ge 0$, $\forall\ \mathbf{w} \in (\mathbb{R},\mathbb{C})^{M\times 1}$. In practice, $\mathbf{R}$ is almost always positive definite $\mathbf{w}^H\mathbf{Rw} > 0$, or the matrix $\mathbf{R}$ is nonsingular and always invertible.

The vector $\mathbf{g} \in (\mathbb{R},\mathbb{C})^{M\times 1}$ is defined, as

$$\begin{aligned}\mathbf{g} &\equiv E\{\mathbf{x}d^*[n]\}\\ &= E\{x[n]d^*[n] \quad x[n-1]d^*[n] \quad \cdots \quad x[n-M+1]d^*[n]\}\\ &= \begin{bmatrix} g[0] & g[1] & \cdots & g[M-1]\end{bmatrix}^T.\end{aligned} \tag{3.55}$$

In (3.52) the terms $r[k]$ and $g[k]$ are defined, respectively, as the autocorrelation and cross-correlation coefficients.

3.3.2.1 Correlation Sequences Estimation

For the estimation of acf and ccf, the SP $x[n]$ is considered *ergodic* and the ensemble-average is computed as a simple time-average. Assuming N and M, respectively, the signal and the filter impulse-response lengths, the computation of the auto and cross-correlation sequences can be performed by a *biased estimator*. For $\mathbf{x} \in (\mathbb{R},\mathbb{C})^{N\times 1}$ we have

$$r[k] \triangleq \begin{cases} \dfrac{1}{N}\displaystyle\sum_{n=0}^{N-1-k} x[n+k]x^*[n] & 0 \le k \le M-1\\ r^*[-k] & -(M-1) \le k < 0, \end{cases} \tag{3.56}$$

or, equivalently, by the formula

$$r[k] \triangleq \begin{cases} \dfrac{1}{N}\displaystyle\sum_{n=k}^{N-1} x[n]x^*[n-k] & 0 \le k \le M-1\\ r^*[-k] & -(M-1) \le k < 0. \end{cases} \tag{3.57}$$

Assuming a finite sequence length, in the previous expression is implicitly used a rectangular window. In this case, it can be shown that the asymptotic behavior of the estimator is not optimal, but the estimate is biased. An alternative way to determine the autocorrelation sequence is to uses the formula of the *unbiased estimator* defined as

$$r_{np}[k] \triangleq \begin{cases} \dfrac{1}{N-k}\displaystyle\sum_{n=0}^{N-1-k} x[n+k]x^*[n] & 0 \le k \le M-1\\ r_{np}^*[-k] & -(M-1) \le k < 0. \end{cases} \tag{3.58}$$

From the expressions (3.56) and (3.57), let $r_v[n]$ be the *true* acf, and considering a white Gaussian input sequence, it is shown that for the unbiased estimator applies

$$E\{r_{np}[k]\} = r_v[k],$$

$$\lim_{N\to\infty}\left(\text{var}\{r_{np}[k]\}\right)=0,$$

while, for the biased estimator, we have that

$$E\{r[k]\}=\left(1-\frac{|k|}{N}\right)r_v[k]=r_v[k]-\frac{|k|}{N}r_v[k],$$

$$\text{var}\{r[k]\}=\left(\frac{N-|k|}{N}\right)^2\text{var}\{r_{np}[k]\}.$$

In the biased estimator, there is a systematic error (or *bias*), which tends to zero as $N\to\infty$, and a variance which tends to zero more slowly.

Remark Although the better asymptotic behavior of the unbiased estimator, the expression (3.58), due its definition, should be used with great caution because sometimes assume negative value and may produce numerical problems.

From similar considerations, the estimation of the ccf sequence $g[k]$ is obtained using the formula

$$g[k]=\frac{1}{N}\sum_{n=0}^{(N-1)-k}x[n+k]d^*[n],\qquad \text{for}\qquad k=0,1,\ldots,M-1. \tag{3.59}$$

Note that, for example, in MATLAB[2] there is a specific function for the estimation of biased and unbiased acf and ccf, `xcorr(x,y,MAXLAG,SCALEOPT)` (plus other options) through the expressions (3.56) and (3.57).

With regard to the **R** matrix inversion, given its symmetrical nature, different algorithms are available, particularly robust and with low computational cost, for example, there is the Cholesky factorization, the Levinson recursion, etc. [2, 3]. Some algorithms will be discussed later in the text.

3.3.2.2 Correlation Vectors Estimation

From definition (3.53), replacing the expectation with the time-average operator, such that $\hat{E}\{\cdot\}\sim E\{\cdot\}$, the *estimated time-average autocorrelation matrix*, indicated as $\mathbf{R}_{xx}\in\mathbb{R}^{M\times M}$, over N signal windows, is defined as

[2] ® MATLAB is a registered trademark of The MathWorks, Inc.

$$\mathbf{R}_{xx} = \frac{1}{N}\sum_{k=0}^{N-1}\mathbf{x}_{n-k}\mathbf{x}_{n-k}^T = \frac{1}{N}[\mathbf{x}_n \quad \mathbf{x}_{n-1} \quad \cdots \quad \mathbf{x}_{n-N+1}] \cdot \begin{bmatrix} \mathbf{x}_n^T \\ \mathbf{x}_{n-1}^T \\ \vdots \\ \mathbf{x}_{n-N+1}^T \end{bmatrix}. \tag{3.60}$$

Considering an N-length windows $[n - N + 1, n]$ and data matrix defined as $\mathbf{X} \in \mathbb{R}^{N \times M}$, the time-average autocorrelation matrix $\mathbf{R}_{xx} \in \mathbb{R}^{M \times M}$ can be written as

$$\begin{aligned}\mathbf{R}_{xx} &= \tfrac{1}{N}\underset{(M\times N)}{\mathbf{X}^T}\underset{(N\times M)}{\mathbf{X}} \\ &= \tfrac{1}{N}\begin{bmatrix} x[n] & \cdots & x[n-N+1] \\ \vdots & \ddots & \vdots \\ x[n-M+1] & \cdots & x[n-M-N+2] \end{bmatrix} \cdot \begin{bmatrix} x[n] & \cdots & x[n-M+1] \\ \vdots & \ddots & \vdots \\ x[n-N+1] & \cdots & x[n-M-N+2] \end{bmatrix}.\end{aligned} \tag{3.61}$$

With similar reasoning, it is possible to define the *estimated cross-correlation vector* over N windows $\mathbf{R}_{xd} \in \mathbb{R}^{M \times 1}$ as

$$\begin{aligned}\mathbf{R}_{xd} &= \frac{1}{N}\sum_{k=0}^{N-1}\mathbf{x}_{n-k}d[n-k] = \frac{1}{N}[\mathbf{x}_n \quad \mathbf{x}_{n-1} \quad \cdots \quad \mathbf{x}_{n-N+1}] \cdot \begin{bmatrix} d[n] \\ d[n-1] \\ \vdots \\ d[n-N+1] \end{bmatrix} \\ &= \tfrac{1}{N}\mathbf{X}^T\mathbf{d}.\end{aligned} \tag{3.62}$$

Remark If we consider the time-average operator instead of the expectation operator, the previous development shows that the LSE and MMSE formalisms are similar. It follows that for an ergodic process, the LSE solution tends to that of Wiener optimal solution for N sufficiently large.

3.3.3 Frequency Domain Interpretation and Coherence Function

An interesting interpretation of the Wiener filter in the frequency domain can be obtained by performing the DTFT of both sides of (3.52). We have that

$$R_{xx}\left(e^{j\omega}\right)W\left(e^{j\omega}\right) = R_{dx}\left(e^{j\omega}\right), \tag{3.63}$$

where the term $R_{xx}(e^{j\omega}) = \sum_{k=-\infty}^{k=\infty} r_{xx}[k]e^{-j\omega k}$ is defined as power spectral density (PSD) of the SP $x[n]$, $R_{dx}(e^{j\omega}) = \sum_{k=-\infty}^{k=\infty} g[k]e^{-j\omega k}$ is cross power spectral density

(CPSD) and $W(e^{j\omega})$ is the frequency response of the optimal filter. For which we have

$$W_{\text{opt}}\left(e^{j\omega}\right) = \frac{R_{dx}(e^{j\omega})}{R_{xx}(e^{j\omega})}. \tag{3.64}$$

The AF performances can be analyzed by frequency domain characterization of the error signal $e[n]$. In this case, you can use the *coherence function* between two stationary random processes $d[n]$ e $x[n]$, defined as

$$\gamma_{dx}\left(e^{j\omega}\right) \triangleq \frac{R_{dx}(e^{j\omega})}{\sqrt{R_{xx}(e^{j\omega})}\sqrt{R_{dd}(e^{j\omega})}}, \tag{3.65}$$

where $R_{dd}(e^{j\omega}) = \sum_{k=-\infty}^{k=\infty} r_{dd}[k]e^{-j\omega k}$ is the PSD of the process $d[n]$. Note that the PSD is a real and positive function that not preserves the phase information. Moreover, for the PSD and CPSD of linear SPs, are valid the following properties:

$$R_{dx}\left(e^{j\omega}\right) = R_{xd}^{*}\left(e^{j\omega}\right),\ R_{dx}\left(e^{j\omega}\right) = W\left(e^{j\omega}\right)R_{xx}\left(e^{j\omega}\right) \text{ and}$$
$$R_{yy}\left(e^{j\omega}\right) = \left|W\left(e^{j\omega}\right)\right|^2 R_{xx}\left(e^{j\omega}\right).$$

The coherence function is therefore a *normalized cross-spectrum* and the square of its amplitude

$$C_{dx}\left(e^{j\omega}\right) \equiv \left|\gamma_{dx}\left(e^{j\omega}\right)\right|^2 = \frac{\left|R_{dx}(e^{j\omega})\right|^2}{R_{xx}(e^{j\omega})R_{dd}(e^{j\omega})}, \tag{3.66}$$

is defined as magnitude square coherence (MSC). This function can be interpreted as a correlation in the frequency domain. In fact, if $x[n] = d[n]$, it follows that $\gamma_{dx}(e^{j\omega}) = 1$ (maximum correlation); conversely, if $x[n]$ is not correlated to $d[n]$ we have that $\gamma_{dx}(e^{j\omega}) = 0$. So, we have $0 \le \gamma_{dx}(e^{j\omega}) \le 1$ for each frequency.

To evaluate the maximum achievable performances of the optimal filter, the PSD of the error $R_{ee}(e^{j\omega})$ should be expressed as a function of MSC. The autocorrelation of the error $r_{ee}[k] = E\{e[n]e[n+k]\}$ is equal to that of the sum of two random processes, namely,

$$r_{ee}[k] = E\Big\{\left(d[n] - \mathbf{w}_n^T\mathbf{x}_n\right)\cdot\left(d[n+k], -\mathbf{w}_n^T\mathbf{x}_{n+k}\right\}.$$

From the above expression, with simple math not reported for brevity, the error PSD is

$$R_{ee}(e^{j\omega}) = R_{dd}(e^{j\omega}) - W^*(e^{j\omega})R_{dx}(e^{j\omega}) - WR_{xd}^*(e^{j\omega}) \\ + |W(e^{j\omega})|^2 R_{xx}(e^{j\omega}). \tag{3.67}$$

Combining this with (3.66) we obtain:

$$R_{ee}(e^{j\omega}) = [1 - C_{dx}(e^{j\omega})]R_{dd}(e^{j\omega}) + \left|W(e^{j\omega}) - \frac{R_{dx}(e^{j\omega})}{R_{xx}(e^{j\omega})}\right|^2 R_{xx}(e^{j\omega}), \tag{3.68}$$

where $C_{dx}(e^{j\omega})$ is defined from (3.66). The optimal filter $W_{\text{opt}}(e^{j\omega})$, which minimizes the previous expression turns out to be those calculated using the (3.64). With this optimal solution, the error PSD is defined as

$$R_{ee}(e^{j\omega}) = \left[1 - C_{dx}(e^{j\omega})\right]R_{dd}(e^{j\omega}). \tag{3.69}$$

Note that this expression indicates that the performance of the filter depends on the MSC function. In fact, the filter is optimal when $R_{ee}(e^{j\omega}) \to 0$. To achieve a good filter, we must have a high coherence $[C_{dx}(e^{j\omega}) \approx 1]$ at the frequencies of interest, for which $R_{ee}(e^{j\omega}) = 0$. Equivalently to have an adaptive filter with optimal performances, the reference signal $d[n]$ must be correlated to the input signal $x[n]$.

In other words, the MSC $C_{dx}(e^{j\omega})$ represents a noise measure, and a linearity measure of the relationship between the processes $d[n]$ and $x[n]$.

3.3.4 Adaptive Filter Performance Measurement

To evaluate the performance of the adaptive filter is usual to refer to a geometrical interpretation of the CF $J(\mathbf{w})$, also called *performance error surface*, and to a set of performance indices such as the *minimum error energy* and the *excess mean square error*, defined in the following.

3.3.4.1 Performance Surface

As previously stated in Sect. 3.3.1, the CF $J(\mathbf{w})$, is a quadratic function as given by

$$J(\mathbf{w}) = \sigma_d^2 - \mathbf{w}^T\mathbf{g} - \mathbf{g}^T\mathbf{w} + \mathbf{w}^T\mathbf{R}\mathbf{w}. \tag{3.70}$$

The CF defined by the MSE criterion (3.70), indicated as *performance surface* or *error surface*, represents an essential tool for defining the properties of the optimal filter and to analyze the properties of the adaptation algorithms discussed later in this and the next chapters.

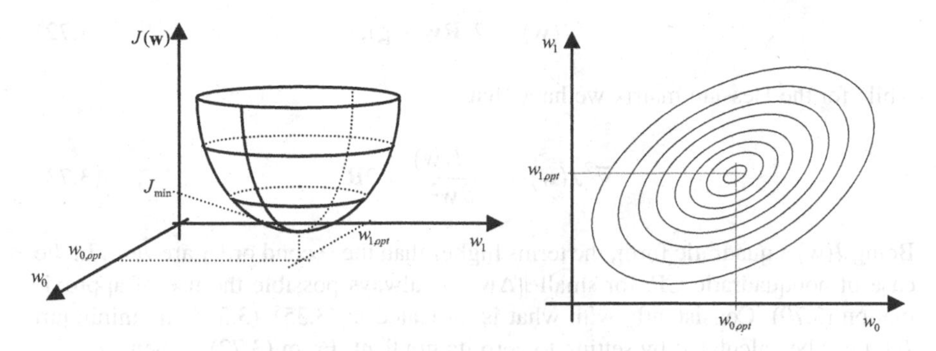

Fig. 3.5 Typical plot of a quadratic CF $J(\mathbf{w})$ for $M = 2$

Since the $J(\mathbf{w})$ is a quadratic form, its geometric characteristics are essential for both the determination of methodology for improvement, and for the determination of the theoretical limits of the algorithms used in the AF adaptation. In fact, these algorithms are realized for the, exact or approximate, estimation of its minimum value.

Property The function $J(\mathbf{w})$ is an hyperparaboloid with a minimum absolute and unique, indicated as MMSE. Figure 3.5 shows a typical trend of performance surface for $M = 2$. The function $J(\mathbf{w})$ has continuous derivatives and therefore is possible an approximation of a close point $\mathbf{w} + \Delta\mathbf{w}$, by using the Taylor expansion truncated at the second order:

$$J(\mathbf{w}+\Delta\mathbf{w}) = J(\mathbf{w}) + \sum_{i=1}^{M} \frac{\partial J(\mathbf{w})}{\partial w_i}\Delta w_i + \frac{1}{2}\sum_{i=1}^{M}\sum_{j=1}^{M} \frac{\partial^2 J(\mathbf{w})}{\partial w_i \partial w_j}\Delta w_i \Delta w_j,$$

or, in a more compact form

$$J(\mathbf{w}+\Delta\mathbf{w}) = J(\mathbf{w}) + (\Delta\mathbf{w})^T \nabla J(\mathbf{w}) + \tfrac{1}{2}(\Delta\mathbf{w})^T\left[\nabla^2 J(\mathbf{w})\right](\Delta\mathbf{w}), \qquad (3.71)$$

where the terms $\nabla J(\mathbf{w})$ and $\nabla^2 J(\mathbf{w})$, with elements $\partial J(\mathbf{w})/\partial \mathbf{w}$ and $\partial^2 J(\mathbf{w})/\partial \mathbf{w}^2$ are, respectively, the gradient vector and the Hessian matrix of the surface $J(\mathbf{w})$ (Sect. B.1.2).

To analyze the geometric properties of the performance surface, we have to study the gradient and the Hessian by deriving the expression[3] (3.70) respect to $\mathbf{w}$. For the gradient vector it is (3.45)

[3] We remind the reader that $(\partial \mathbf{x}^T\mathbf{a}/\partial \mathbf{x}) = (\partial \mathbf{a}^T\mathbf{x}/\partial \mathbf{x}) = \mathbf{a}$ and $(\partial \mathbf{x}^T\mathbf{B}\mathbf{x}/\partial \mathbf{x}) = (\mathbf{B} + \mathbf{B}^T)\mathbf{x}$. For vector and matrix derivative rules, see [1].

$$\nabla J(\mathbf{w}) = 2(\mathbf{R}\mathbf{w} - \mathbf{g}), \tag{3.72}$$

while for the Hessian matrix we have that

$$\nabla^2 J(\mathbf{w}) = \frac{\partial^2 J(\mathbf{w})}{\partial \mathbf{w}^2} = 2\mathbf{R}. \tag{3.73}$$

Being $J(\mathbf{w})$ a quadratic form, the terms higher than the second order are zero. In the case of nonquadratic CF, for small $\|\Delta\mathbf{w}\|$, is always possible the use of approximation (3.70). Consistently with what is indicated in (3.35)–(3.37), the minimum $J(\mathbf{w})$ can be calculated by setting to zero its gradient. From (3.72) is then

$$\nabla J(\mathbf{w}) \to 0 \quad \Rightarrow \quad \mathbf{R}\mathbf{w} - \mathbf{g} \to 0. \tag{3.74}$$

This result is, in fact, the *normal equations* in the notation of Wiener–Hopf $\mathbf{R}\mathbf{w} = \mathbf{g}$ already indicated in (3.46).

3.3.4.2 Minimum Error Energy

The minimum point of the error surface or MMSE, also called *minimum error energy* value, can be computed substituting in (3.70) the optimal vector $\mathbf{w}_{\text{opt}}$, calculated with (3.47), i.e., $J(\mathbf{w}_{\text{opt}}) \triangleq J(\mathbf{w})|_{\mathbf{w}=\mathbf{w}_{\text{opt}}} = J_{\min}$, so that it is

$$\begin{aligned} J_{\min} &= \sigma_d^2 - \mathbf{w}_{\text{opt}}^T \mathbf{g} \\ &= \sigma_d^2 - \mathbf{w}_{\text{opt}}^T \mathbf{R} \mathbf{w}_{\text{opt}} \\ &= \sigma_d^2 - \mathbf{g}^T \mathbf{R}^{-1} \mathbf{g}. \end{aligned} \tag{3.75}$$

3.3.4.3 Canonical Form of the Error Surface

It should be noted that the expression of the error surface (3.70) is a quadric form that can be expressed in vector notation as

$$J(\mathbf{w}) = \begin{bmatrix} 1 & \mathbf{w}^T \end{bmatrix} \begin{bmatrix} \sigma_d^2 & -\mathbf{g}^T \\ -\mathbf{g} & \mathbf{R} \end{bmatrix} \begin{bmatrix} 1 \\ \mathbf{w} \end{bmatrix}. \tag{3.76}$$

To derive the canonical form, the matrix in the middle is factored as the product of three matrices: lower-triangular, diagonal, and upper-triangular. For which the reader can easily verify that

$$\begin{bmatrix} \sigma_d^2 & -\mathbf{g}^T \\ -\mathbf{g} & \mathbf{R} \end{bmatrix} = \begin{bmatrix} 1 & -\mathbf{g}^T\mathbf{R}^{-1} \\ 0 & 1 \end{bmatrix} \begin{bmatrix} \sigma_d^2 - \mathbf{g}^T\mathbf{R}^{-1}\mathbf{g} & 0 \\ 0 & \mathbf{R} \end{bmatrix} \begin{bmatrix} 1 & 0 \\ -\mathbf{R}^{-1}\mathbf{g} & 1 \end{bmatrix}. \tag{3.77}$$

Substituting in (3.76) is

$$J(\mathbf{w}) = \sigma_d^2 - \mathbf{g}^T\mathbf{R}^{-1}\mathbf{g} + \left(\mathbf{w} - \mathbf{R}^{-1}\mathbf{g}\right)^T\mathbf{R}\left(\mathbf{w} - \mathbf{R}^{-1}\mathbf{g}\right) \qquad (3.78)$$

which is a canonical formulation alternative to (3.70).

3.3.4.4 Excess-Mean-Square Error

Note that for (3.75), for $\mathbf{w}_{\text{opt}} = \mathbf{R}^{-1}\mathbf{g}$ and omitting, for simplicity the writing of the argument $(\mathbf{w})$, by definition in (3.78) the error surface can be written as

$$J = J_{\min} + \left(\mathbf{w} - \mathbf{w}_{\text{opt}}\right)^T\mathbf{R}\left(\mathbf{w} - \mathbf{w}_{\text{opt}}\right). \qquad (3.79)$$

By defining $\mathbf{u}$, as *weights error vector* (WEV) such as

$$\mathbf{u} = \mathbf{w} - \mathbf{w}_{\text{opt}}, \qquad (3.80)$$

the MSE can be represented as a function of $\mathbf{u}$, as

$$J = J_{\min} + \mathbf{u}^T\mathbf{R}\mathbf{u}. \qquad (3.81)$$

The term

$$J_{\text{EMSE}} \triangleq J - J_{\min} = \mathbf{u}^T\mathbf{R}\mathbf{u}, \quad \textit{excess-mean-square error}, \qquad (3.82)$$

is defined as excess-mean-square error (EMSE). The correlation matrix is positive definite, it follows that it is also the excess of error, i.e., $\mathbf{u}^T\mathbf{R}\mathbf{u} \geq 0$. This shows that, in the case of the optimal solution, the error function is a unique and absolute minimum $J_{\min} = J(\mathbf{w}_{\text{opt}})$. It also defines the parameter *misadjustment* sometimes used in alternative to the EMSE, as

$$\mathcal{M} \triangleq \frac{J_{\text{EMSE}}}{J_{\min}} \quad \textit{mis adjustment}. \qquad (3.83)$$

3.3.5 Geometrical Interpretation and Orthogonality Principle

A geometric interpretation, very useful for a deeper understanding and for further theoretical developments presented below, is implicit in the calculation of the optimal solution $\mathbf{w}_{\text{opt}}$ of the Wiener filter. An important property, by solving the normal equation (3.46), is, in fact, the orthogonality between the vector of error $e[n]$ and the input signal $x[n]$. The orthogonality can be simply proved by multiplying both the sides of the expression of the error (3.41) by $\mathbf{x}$:

$$\mathbf{x}e[n] = \mathbf{x}d[n] - \mathbf{x}\mathbf{x}^T\mathbf{w}, \tag{3.84}$$

and taking the expectation of the above expression we have

$$E(\mathbf{x}e[n]) = \mathbf{g} - \mathbf{R}\mathbf{w}, \tag{3.85}$$

so, replacing the previous with the optimal value $\mathbf{w}_{\text{opt}} = \mathbf{R}^{-1}\mathbf{g}$, we have

$$E(\mathbf{x}e[n]) = 0, \tag{3.86}$$

which proves the orthogonality between the input signal and error. This result, the same well known in the Wiener theory [9–11], indicates that when the impulse response of the filter is the optimal, the error and input signals are uncorrelated.

Corollary Similar to (3.86), it is easy to prove that the principle of orthogonality is also valid for the output signal, i.e.,

$$E(\mathbf{y}e[n]) = 0. \tag{3.87}$$

The (3.87) is proved by writing the output of the filter explicitly as $E(\mathbf{w}_{\text{opt}}^T\mathbf{x}e[n])$, for the linearity of the expectation operator, we can write, in fact,

$$\mathbf{w}_{\text{opt}}^T E(\mathbf{x}e[n]) = 0,$$

so, for (3.86), the orthogonality of the error with the output sequence is also proved.

A graphical representation of the principle of orthogonality is illustrated in Fig. 3.6.

3.3.6 Principal Component Analysis of Optimal Filter

In order to evaluate some behaviors of the filter it is very useful to perform the eigenvalues and eigenvectors analysis of the autocorrelation matrix. From the geometry, it is shown that the correlation matrix $\mathbf{R} \in \mathbb{R}^{M \times M}$ can always be represented through the unitary similarity transformation [11–13] (Sect. A.9), defined by the relation

$$\mathbf{R} = \mathbf{Q}\mathbf{\Lambda}\mathbf{Q}^T = \sum_{k=0}^{M-1} \lambda_k \mathbf{q}_k \mathbf{q}_k^T, \tag{3.88}$$

or $\mathbf{\Lambda} = \mathbf{Q}^T\mathbf{R}\mathbf{Q}$, where $\mathbf{\Lambda} = \text{diag}(\lambda_0, \lambda_1, \ldots, \lambda_{M-1})$. The matrix $\mathbf{\Lambda}$, called *spectral matrix*, is a diagonal matrix formed with the eigenvalues λ_k of the matrix $\mathbf{R}$ (each autocorrelation matrix can be factorized in this way). The so-called *modal matrix*,

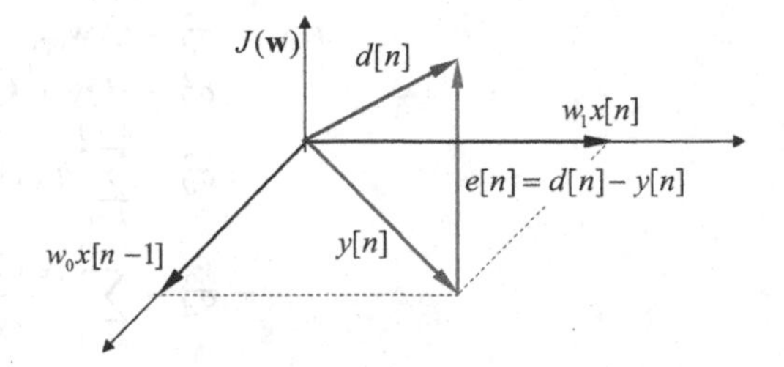

Fig. 3.6 Orthogonality of vectors of the input and output signals and error signal

defined as $\mathbf{Q} = [\,\mathbf{q}_0 \quad \mathbf{q}_1 \quad \cdots \quad \mathbf{q}_{M-1}\,]$, is orthonormal (such that $\mathbf{Q}^T\mathbf{Q} = \mathbf{I}$, namely $\mathbf{Q}^{-1} = \mathbf{Q}^T$). The vectors $\mathbf{q}_i$ (*eigenvectors* of the matrix $\mathbf{R}$) are orthogonal and with unitary length.

Suppose we apply the transformation defined by the modal matrix $\mathbf{Q}$, to the optimal solution of the Wiener filter, for which we can define a vector $\mathbf{v}$ such that $\mathbf{v} = \mathbf{Q}^T\mathbf{w}$ or $\mathbf{w} = \mathbf{Q}\mathbf{v}$. In addition, it should be noted that, given the nature of such transformation, the norms of $\mathbf{v}$ and $\mathbf{w}$ are identical. In fact, $\|\mathbf{w}\|^2 = \mathbf{w}^T\mathbf{w} = [\mathbf{Q}\mathbf{v}]^T\mathbf{Q}\mathbf{v} = \mathbf{v}^T\mathbf{Q}^T\mathbf{Q}\mathbf{v} = \|\mathbf{v}\|^2$, for which the transformation changes the direction but not the length of the vector. Substituting the notation (3.88) in the normal equation (3.46), at the optimal solution, we have

$$\mathbf{Q}\mathbf{\Lambda}\mathbf{Q}^T\mathbf{w}_{\text{opt}} = \mathbf{g} \quad \text{or} \quad \mathbf{\Lambda}\mathbf{Q}^T\mathbf{w}_{\text{opt}} = \mathbf{Q}^T\mathbf{g}, \tag{3.89}$$

let $\mathbf{g}' = \mathbf{Q}^T\mathbf{g}$ we can write

$$\mathbf{\Lambda}\mathbf{v}_{\text{opt}} = \mathbf{g}'. \tag{3.90}$$

The vector $\mathbf{g}'$ is defined as *decoupled cross-correlation*, as $\mathbf{\Lambda}$ is a diagonal matrix. Then, (3.90) is equivalent to a set of M distinct scalar equations of the type

$$\lambda_k v_{\text{opt}}(k) = g'(k), \quad k = 0, 1, \ldots, M-1, \tag{3.91}$$

with solution, for $\lambda_k \neq 0$, equal to

$$v_{\text{opt}}(k) = \frac{g'(k)}{\lambda_k}, \quad k = 0, 1, \ldots, M-1. \tag{3.92}$$

For (3.75) we have that

$$\begin{aligned} J_{\min} &= \sigma_d^2 - \mathbf{g}^T \mathbf{w}_{\text{opt}} \\ &= \sigma_d^2 - \left(\mathbf{Q}\mathbf{g}'\right)^T \mathbf{Q}\mathbf{v}_{\text{opt}} \\ &= \sigma_d^2 - \sum_{k=0}^{M-1} g'(k) v_{\text{opt}}(k) \\ &= \sigma_d^2 - \sum_{k=0}^{M-1} \frac{\left|g'(k)\right|^2}{\lambda_{\text{k}}}. \end{aligned} \tag{3.93}$$

The above equation shows that the eigenvalues and the decoupled cross-correlation influence the performance surface. The advantage of the decoupled representation (3.92) and (3.93) is that it is possible to study the effects of each parameter independently from the others.

To better appreciate the meaning of the above transformation, we consider the CF $J(\mathbf{w})$ as shown in Fig. 3.7. The MSE function $J(\mathbf{w})$ can be represented on the weights-plane of coordinates (w_0, w_1), with the isolevel curves that are of concentric ellipses with the center of coordinates $(w_{0,\text{opt}}, w_{1,\text{opt}})$ (optimal values), which corresponds to $J_{\min}(\mathbf{w})$.

Now suppose we want to define the new coordinates $\hat{\mathbf{u}}$, called *principal coordinates*, such that the axes are arranged in the center of the ellipsoid $J(\mathbf{w})$ and rotated along the maximum of the surface $J(\mathbf{w})$ as shown in Fig. 3.7. As said, the rotation–translation, for the calculation of $\hat{\mathbf{u}}$, is defined as

$$\mathbf{u} = \mathbf{w} - \mathbf{w}_{\text{opt}}, \quad \textit{WEV (see Sect. 3.3.4.4)}, \tag{3.94}$$

$$\hat{\mathbf{u}} = \mathbf{Q}^T \mathbf{u}, \quad \textit{rotation.} \tag{3.95}$$

With such a transformation the excess MSE, defined in (3.82), can be rewritten as

$$\begin{aligned} J_{\text{EMSE}} &= \hat{\mathbf{u}}\mathbf{R}\hat{\mathbf{u}} \\ &= \hat{\mathbf{u}}^T \boldsymbol{\Lambda} \hat{\mathbf{u}} \\ &= \sum_{k=0}^{M-1} \lambda_k |\hat{u}(k)|^2. \end{aligned} \tag{3.96}$$

The (3.96) shows that the penalty, paid for a deviation of a parameter from its optimal value, is proportional to the corresponding eigenvalue. In the case where the ith eigenvalue is equal to zero, would not be variations in (3.96).

The optimal solution (3.47), expressed in the principal coordinates $\hat{\mathbf{u}}$, appears to be

$$\mathbf{w}_{\text{opt}} = \mathbf{R}^{-1}\mathbf{g} = \mathbf{Q}\boldsymbol{\Lambda}\mathbf{Q}^T\mathbf{g} = \sum_{k=0}^{M-1} \frac{\mathbf{q}_k^T \mathbf{g}}{\lambda_k} \mathbf{q}_k = \sum_{k=0}^{M-1} \frac{g'(k)}{\lambda_k} \mathbf{q}_k. \tag{3.97}$$

The output of the optimum filter, expressed as principal component, is then

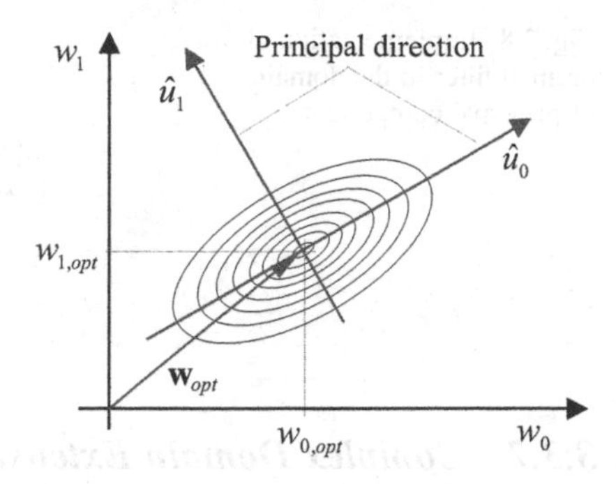

Fig. 3.7 Performance surface and principal component direction

$$y[n] = \mathbf{w}_{\text{opt}}^T \mathbf{x} = \sum_{k=0}^{M-1} \frac{g'(k)}{\lambda_k} \left(\mathbf{q}_k^T \mathbf{x} \right), \tag{3.98}$$

represented in the scheme of Fig. 3.8.

Remark The principal component analysis (PCA), as we shall see later in the text, is a tool of fundamental importance for the relevant theoretical and practical implications that this method entails. With this analysis, or more properly transformation, it is possible to represent a set of data according to their *natural coordinates*.

3.3.6.1 Condition Number of Correlation Matrix

In numerical analysis, the condition number $\chi(\cdot)$ associated with a problem represents the degree of its *numerical tractability*. In particular, in the calculation of the inverse of a matrix $\mathbf{R}$, in the case of L_2 norm, is shown that (Sect. A.12):

$$\chi(\mathbf{R}) = ||\mathbf{R}||_2 ||\mathbf{R}^{-1}||_2 = \frac{\lambda_{\max}}{\lambda_{\min}}, \tag{3.99}$$

with $\lambda_{\max}$ and $\lambda_{\min}$, respectively, the maximum and minimum eigenvalues of $\mathbf{R}$.

In the case of the Wiener filter, $\chi(\mathbf{R})$ provides indication on the shape of the error surface. For $\chi(\mathbf{R}) = 1$, the error surface is a regular paraboloid and its isolevel projections are perfect circles. It should be noted, as we shall see after, which $\chi(\mathbf{R})$ appears to be important for defining the *convergence performance* of an adaptive filter.

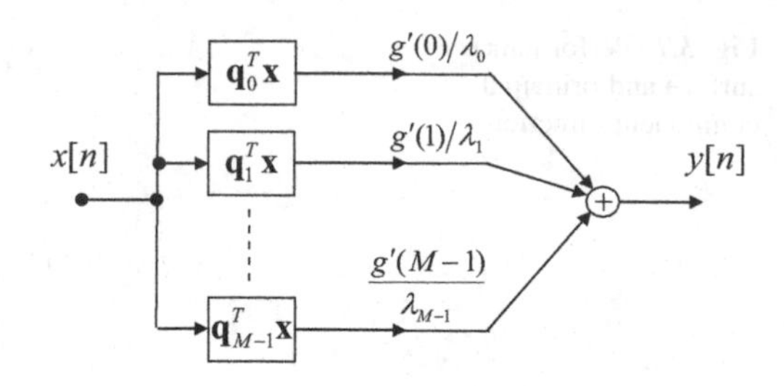

Fig. 3.8 Implementation of optimal filter in the domain of principal component

3.3.7 *Complex Domain Extension of the Wiener Filter*

In many practical situations, it is necessary to process sequences which by their nature are defined in the complex domain. For example, in the data transmission, it is usual to use the modulation process as phase shift keying (PSK) or quadrature amplitude modulation (QAM), in which the baseband signal is defined in the complex domain. Furthermore, the use of complex signals is essential in the implementation of the adaptive filtering in the frequency domain.

In this section, the results of the previous paragraphs are extended to the case where the signals $x[n]$, $d[n]$, and weights $w_i[n]$ have complex nature.

By definition the CF (3.30) in the complex domain becomes

$$J(\mathbf{w}) = E\left\{\left|e[n]\right|^2\right\} = E\{e[n]^* e[n]\}, \tag{3.100}$$

whereby $J(\mathbf{w})$ is real and, also in this case, is a quadratic form. In complex case, we have that $y[n] = \mathbf{w}^H\mathbf{x} = (\mathbf{x}^H\mathbf{w})^*$ and the complex error is $e[n] = d[n] - \mathbf{w}^H\mathbf{x}$ (or $e^*[n] = d^*[n] - \mathbf{x}^H\mathbf{w}$), the complex error surface is a simple extension of (3.70) and is defined as

$$\begin{aligned} J(\mathbf{w}) &= E\left\{\left(d[n] - \mathbf{w}^H, \mathbf{x}\right)\left(d^*[n] - \mathbf{x}^H\mathbf{w}\right)\right\} \\ &= E\{d^*[n]d[n]\} - \mathbf{w}^H E\{\mathbf{x}d^*[n]\} - E\{d[n]\mathbf{x}^H\}\mathbf{w} + \mathbf{w}^H E\{\mathbf{x}\mathbf{x}^H\}\mathbf{w} \\ &= \sigma_d^2 - \mathbf{w}^H\mathbf{g} - \mathbf{g}^H\mathbf{w} + \mathbf{w}^H\mathbf{R}\mathbf{w}. \end{aligned} \tag{3.101}$$

For the calculation of the optimum filter parameters it is necessary to perform the differentiation and solve the linear equations system such that $\nabla J(\mathbf{w}) \to 0$. In this case, the filter taps are complex and for the calculation of the gradient, must compute the partial derivative of (3.101) in an independent way with respect to the real and imaginary parts. In particular, in order to obtain the optimum filter coefficients, it should be solved simultaneously the following equations:

$$\frac{\partial J(\mathbf{w})}{\partial w_{j,\mathrm{Re}}} = 0 \quad \text{and} \quad \frac{\partial J(\mathbf{w})}{\partial w_{j,\mathrm{Im}}} = 0, \quad \text{for} \quad j = 0, 1, \ldots, M-1, \tag{3.102}$$

combined as

$$\frac{\partial J(\mathbf{w})}{\partial w_{j,\mathrm{Re}}} + j\frac{\partial J(\mathbf{w})}{\partial w_{j,\mathrm{Im}}} = 0, \qquad \text{for} \quad j = 0, 1, \ldots, M-1. \tag{3.103}$$

The above expression suggests the following definition of complex gradient:

$$\nabla J(\mathbf{w}) \triangleq \frac{\partial J(\mathbf{w})}{\partial w_{j,\mathrm{Re}}} + j\frac{\partial J(\mathbf{w})}{\partial w_{j,\mathrm{Im}}}, \tag{3.104}$$

and it is shown that the complex gradient of (3.101) is equal to

$$\nabla J(\mathbf{w}) = 2(\mathbf{R}\mathbf{w} - \mathbf{g}). \tag{3.105}$$

As for the real case, the optimal weight is for $\mathbf{Rw} - \mathbf{g} = \mathbf{0}$, where $\mathbf{R}$ is semipositive definite so that, even in the complex case, we have $\mathbf{w}_{\mathrm{opt}} = \mathbf{R}^{-1}\,\mathbf{g}$. This result is easily seen directly from (3.101) rewriting the canonical quadratic form as

$$J(\mathbf{w}) = \sigma_d^2 - \mathbf{g}^H\mathbf{R}^{-1}\mathbf{g} + \left(\mathbf{w} - \mathbf{R}^{-1}\mathbf{g}\right)^H\mathbf{R}\left(\mathbf{w} - \mathbf{R}^{-1}\mathbf{g}\right). \tag{3.106}$$

Being $\mathbf{R}$ positive defined, it appears that $\mathbf{g}^H\mathbf{R}^{-1}\mathbf{g} > 0$ and $(\mathbf{Rw} - \mathbf{g})^H\mathbf{R}^{-1}(\mathbf{Rw} - \mathbf{g}) > 0$. The minimum of (3.106) with respect to the variation of the parameters $\mathbf{w}$, is for $\mathbf{Rw} - \mathbf{g} = \mathbf{0}$.

Remark The previous development demonstrates the convention in Sect. 3.2.1.1, on the real-complex vector notation adopted in the text.

3.3.8 Multichannel Wiener's Normal Equations

Consider the MIMO adaptive filter with input–output relation (3.15), with reference to the formalism of Fig. 3.9, called $\mathbf{d}[n] \in (\mathbb{R},\mathbb{C})^{Q\times 1} = \begin{bmatrix} d_1[n] & d_2[n] & \cdots & d_Q[n] \end{bmatrix}^T$, the vector of desired outputs and $\mathbf{e}[n] \in (\mathbb{R},\mathbb{C})^{Q\times 1} = \begin{bmatrix} e_1[n] & e_2[n] & \cdots & e_Q[n] \end{bmatrix}^T$ the error vector, considering the composite-notation 1 (Sect. 3.2.2.1) for which the output snap-shot is $\mathbf{y}[n] = \mathbf{Wx}$, the error vector can be written as

$$\begin{aligned} \mathbf{e}[n] &= \mathbf{d}[n] - \mathbf{y}[n] \\ &= \mathbf{d}[n] - \mathbf{Wx}, \end{aligned} \tag{3.107}$$

i.e., explaining the individual error terms, it is

$$e_j[n] = d_j[n] - \mathbf{w}_{j:}^T\mathbf{x}, \quad j = 1, \ldots, Q. \tag{3.108}$$

The CF is defined as $J(\mathbf{W}) = E\{\mathbf{e}^T[n]\mathbf{e}[n]\}$, and for (3.107), we get

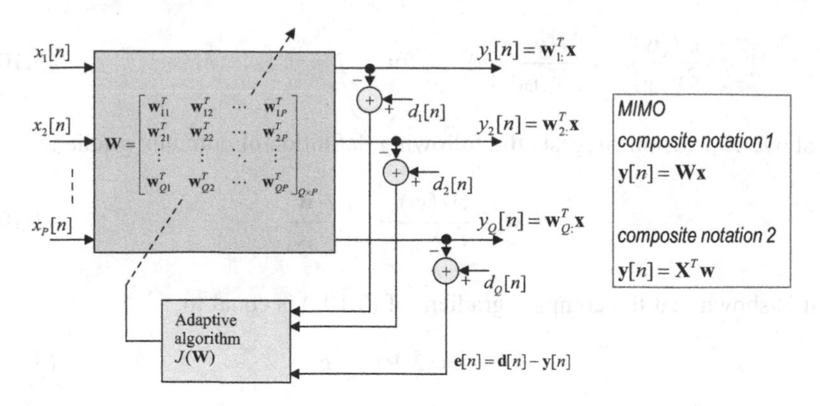

Fig. 3.9 Representation of MIMO adaptive filter

$$
\begin{aligned}
J(\mathbf{W}) &= E\left\{\mathbf{e}^T[n]\mathbf{e}[n]\right\} \\
&= \sum_{j=1}^{Q} E\{|e_j[n]|^2\} \\
&= \sum_{j=1}^{Q} J_j(\mathbf{w}_{j:}).
\end{aligned}
\tag{3.109}
$$

The above expression shows that the minimization of whole $J(\mathbf{W})$ or the minimization of independent terms $J_j(\mathbf{w}_{j\,:})$ produces the same result.

From the vector of all the inputs $\mathbf{x}$ definition (3.17), the multichannel correlation matrix can be defined as $\mathbf{R} = E\{\mathbf{x}\mathbf{x}^T\}$, for which it is given as

$$
\begin{aligned}
\mathbf{R} \in (\mathbb{R},\mathbb{C})^{P(M)\times P(M)} &= E\left\{\begin{bmatrix}\mathbf{x}_1 \\ \vdots \\ \mathbf{x}_P\end{bmatrix}\begin{bmatrix}\mathbf{x}_1^T & \cdots & \mathbf{x}_P^T\end{bmatrix}\right\} \\
&= \begin{bmatrix}
\mathbf{R}_{x_1x_1} & \mathbf{R}_{x_1x_2} & \cdots & \mathbf{R}_{x_1x_P} \\
\mathbf{R}_{x_2x_1} & \mathbf{R}_{x_2x_2} & \cdots & \mathbf{R}_{x_2x_P} \\
\vdots & \vdots & \ddots & \vdots \\
\mathbf{R}_{x_Px_1} & \mathbf{R}_{x_Px_2} & \cdots & \mathbf{R}_{x_Px_P}
\end{bmatrix}_{P\times P}.
\end{aligned}
\tag{3.110}
$$

This is a block Toeplitz structure, with $\mathbf{R}_{x_ix_j} = E\left\{\mathbf{x}_i\mathbf{x}_j^T\right\}$. Said $\mathbf{P}$ the cross-correlation matrix defined as

$$
\begin{aligned}
\mathbf{P} \in (\mathbb{R},\mathbb{C})^{P(M)\times Q} &= E\{\mathbf{x}\mathbf{d}^T\} \\
&= \begin{bmatrix}\mathbf{p}_{xd_1} & \mathbf{p}_{xd_2} & \cdots & \mathbf{p}_{xd_Q}\end{bmatrix}_{1\times Q},
\end{aligned}
$$

with $\mathbf{p}_{xd_j} = E\{\mathbf{x}d_j[n]\}$, the *MIMO Wiener's normal equations* are defined as

$$\mathbf{R}\mathbf{W} = \mathbf{P}, \tag{3.111}$$

where $\mathbf{R} \in (\mathbb{R},\mathbb{C})^{P(M)\times P(M)}$, $\mathbf{W} \in (\mathbb{R},\mathbb{C})^{P(M)\times Q}$, and $\mathbf{P} \in (\mathbb{R},\mathbb{C})^{P(M)\times Q}$, with solution

$$\mathbf{W}_{\text{opt}} = \mathbf{R}^{-1}\mathbf{P}. \tag{3.112}$$

Remark From the definition of the CF (0.109) (and from (3.23)) can be observed that the MIMO Wiener equations (3.111) can be decomposed in Q independent relationship of the type

$$\mathbf{R}\mathbf{w}_{j:} = \mathbf{p}_{xd_j}, \qquad \text{for} \quad j = 1,2,\ldots,Q. \tag{3.113}$$

The above expression enables to adapt the single subsystem, defined by the MISO bank filters $\mathbf{w}_{j:} \in (\mathbb{R},\mathbb{C})^{1\times P(M)} \triangleq \begin{bmatrix}\mathbf{w}_{j1}^H & \cdots & \mathbf{w}_{jP}^H\end{bmatrix}$, shown in Fig. 3.3, independently from the others using the same correlation matrix for all banks.

3.4 Examples of Applications

To explain from a more practical point of view, the method of Wiener for the estimation of the parameters of the optimum filter, below, is discussed with some applications. The first example consists in estimating the model of a linear dynamic system, the second in the *time delay estimation*, the third example discussed a problem of inverse (equalization type) model estimation, the fourth introduced the problem of *adaptive noise cancellation* with and without reference signal, and also, some typical application cases are discussed.

3.4.1 Dynamical System Modeling 1

Consider the problem of dynamic system identification as shown in Fig. 3.10. Suppose that the system to be modeled consists of a linear circuit with discrete-time transfer function (TF) equal to $H(z) = 1 - 2z^{-1}$ for which the model parameters to be estimated are $\mathbf{h} = \begin{bmatrix}1 & -2\end{bmatrix}^T$. Suppose also, that the TF, taken as a system model, is a two-tap linear FIR filter, such that $W(z) = w_0 + w_1 z^{-1}$.

For the optimum model parameter computation w_0 and w_1, suppose that the filter input sequence $x[n]$ is a zero-mean WGN with unitary variance $\sigma_x^2 = 1$. Moreover, suppose that the measure $d[n]$ is corrupted by an additive noise $\eta[n]$, also WGN zero-mean uncorrelated to $x[n]$ and with variance $\sigma_\eta^2 = 0.1$. For the determination of the optimum vector through the relation (3.47), we proceed with the determination of the correlation matrix and the cross-correlation vector. Since $x[n]$ is a white random process with unitary variance, by definition we have

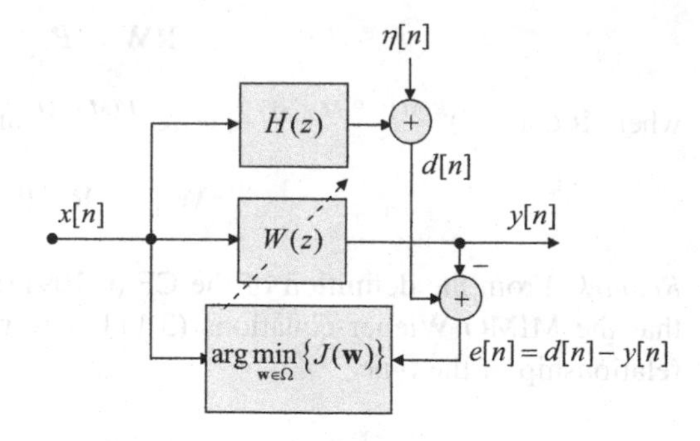

Fig. 3.10 Modeling of a linear dynamic system with Wiener filter

$E\{x[n]x[n-1]\} = E\{x[n-1]x[n]\} = 0$ and $E\{x^2[n]\} = \sigma_x^2 = 1$; for the matrix $\mathbf{R}$ is then

$$\mathbf{R} = E\{\mathbf{x}\mathbf{x}^T\} = \begin{bmatrix} E\{x^2[n]\} & E\{x[n], x[n-1]\} \\ E\{x[n-1], x[n]\} & E\{x^2[n-1]\} \end{bmatrix} = \begin{bmatrix} 1 & 0 \\ 0 & 1 \end{bmatrix}. \quad (3.114)$$

The $H(z)$ system output is defined by the following finite difference equation:

$$d[n] = x[n] - 2x[n-1] + \eta[n],$$

while, the cross-correlation vector $\mathbf{g}$ is

$$\begin{aligned} \mathbf{g} = E\{\mathbf{x}, d[n]\} &= \begin{bmatrix} E\{x[n], d[n]\} \\ E\{x[n-1], d[n]\} \end{bmatrix} \\ &= \begin{bmatrix} E\{x[n](x[n] - 2x[n-1] + \eta[n])\} \\ E\{x[n-1](x[n] - 2x[n-1] + \eta[n])\} \end{bmatrix}. \end{aligned} \quad (3.115)$$

Developing the terms of the previous expression, we have that $E\{x^2[n]\} = 1$ and $E\{x[n-1]x[n-1]\} = 1$; applies in addition, $E\{x[n]\eta[n]\} = E\{x[n-1]\eta[n]\} = 0$, so we obtain

$$\mathbf{g} = \begin{bmatrix} 1 \\ -2 \end{bmatrix}. \quad (3.116)$$

From the foregoing expressions, the Wiener solution turns out to be

$$\mathbf{w}_{\text{opt}} = \mathbf{R}^{-1}\mathbf{g} = \begin{bmatrix} 1 & 0 \\ 0 & 1 \end{bmatrix}^{-1} \begin{bmatrix} 1 \\ -2 \end{bmatrix} = \begin{bmatrix} 1 \\ -2 \end{bmatrix}, \quad (3.117)$$

in practice, for random inputs, the estimated parameters coincide with the parameters of the model: $\mathbf{w}_{\text{opt}} \equiv \mathbf{h}$.

3.4.1.1 Performance Surface and Minimum Energy Error Determination

The performance surface $J(\mathbf{w})$ (3.70) is

$$J(\mathbf{w}) = \sigma_d^2 - 2[w_0 \quad w_1]\begin{bmatrix} 1 \\ -2 \end{bmatrix} + [w_0 \quad w_1]\begin{bmatrix} 1 & 0 \\ 0 & 1 \end{bmatrix}\begin{bmatrix} w_0 \\ w_1 \end{bmatrix}. \tag{3.118}$$

Consider the variance σ_d^2

$$\begin{aligned} \sigma_d^2 = E\{d^2[n]\} &= E\left\{(x[n] - 2x[n-1] + \eta[n])^2\right\} \\ &= E\left\{x[n]^2\right\} + 4E\{x^2[n-1]\} + \sigma_\eta^2 = 0.1 + 1 + 4 = 5.1. \end{aligned} \tag{3.119}$$

Finally we have

$$J(\mathbf{w}) = 5.1 - 2w_0 + 4w_1 + w_0^2 + w_1^2, \tag{3.120}$$

whose performance graph[4] is reported in Fig. 3.11.

For the qualitative analysis of the shape of $J(\mathbf{w})$, observe that the expression (3.120) can be rewritten as

$$\begin{aligned} J(\mathbf{w}) &= \sigma_\eta^2 + 1 + 4 - 2w_0 + 4w_1 + w_0^2 + w_1^2 \\ &= \sigma_\eta^2 + (w_0 - 1)^2 + (w_1 + 2)^2. \end{aligned} \tag{3.121}$$

The latter shows that the minimum of the performance surface, i.e., the lowest error energy, coincides with the variance of the additive measurement noise. This is consistent, for this type of processes, as with previously developed in Sect. 3.3.4.2.

3.4.2 Dynamical System Modeling 2

Consider the problem of dynamical linear model identification as shown in Fig. 3.12, in the case that two noise sources are present.

The input of the filter $W(z)$ is

$$x[n] = u[n] + \eta_1[n], \tag{3.122}$$

while for the desired output we have that

[4] The graphs in the figure are drawn by means of the ® MATLAB `mesh` functions.

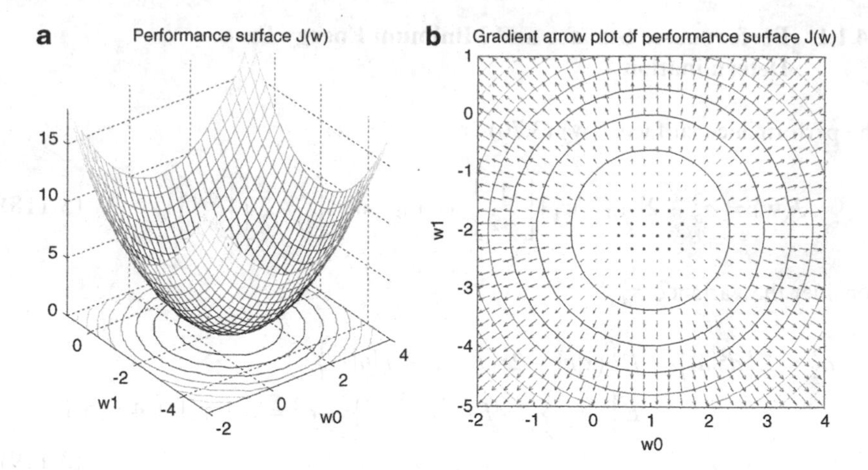

Fig. 3.11 Performance surface (3.120): (**a**) 3D plot; (**b**) isolevel projection and gradient trend (arrows)

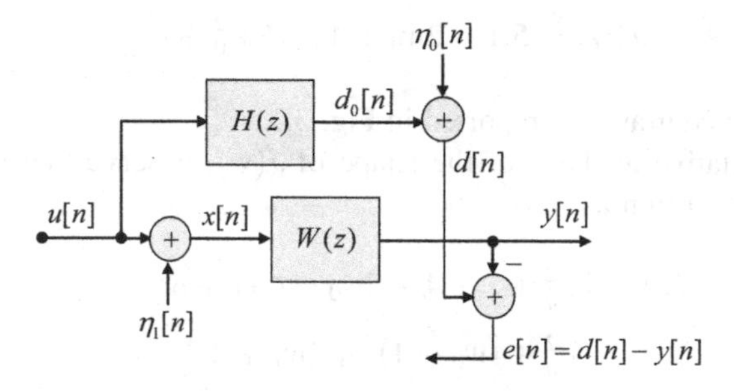

Fig. 3.12 Modeling of a linear dynamic system with two noise sources with Wiener filter

$$d[n] = \mathbf{h}^T\mathbf{u} + \eta_0[n]. \tag{3.123}$$

To determine the optimum Wiener filter, one proceeds computing the acf r_{xx} and r_{dx}

$$\begin{aligned} r_{xx}[k] &= E\{x[n]x[n-k]\} \\ &= E\Big\{\big(u[n] + \eta_1[n]\big)\big(u[n-k] + \eta_1[n-k]\big)\Big\} \\ &= E\{u[n]u[n-k]\} + E\{u[n]\eta_1[n-k]\} \\ &\quad + E\{\eta_1[n]u[n-k]\} + E\{\eta_1[n]\eta_1[n-k]\}. \end{aligned}$$

For uncorrelated $u[n]$ and $\eta_1[n]$, the terms $E\{u[n]\eta_1[n-k]\}$ and $E\{\eta_1[n]u[n-k]\}$ are zero, so we have that

$$r_{xx}[k] = r_{uu}[k] + r_{\eta_1\eta_1}[k], \tag{3.124}$$

or

$$R_{xx}(z) = R_{uu}(z) + R_{\eta_1\eta_1}(z). \tag{3.125}$$

To determine $r_{dx}[k]$ note that $u[n]$ is common for $x[n]$ and $d[n]$. Proceedings as in the previous case we have

$$\begin{aligned} r_{dx}[k] &= E\left[d[n]x[n-k]\right] \\ &= E\left[\left(d_0[n] + \eta_0[n]\right)\left(u[n-k] + \eta_1[n-k]\right)\right] \\ &= E\left(d_0[n]u[n-k]\right) + E\left(d_0[n]\eta_1[n-k]\right) \\ &\quad + E\left(\eta_0[n]u[n-k]\right) + E\left(\eta_0[n]\eta_1[n-k]\right), \end{aligned}$$

where $u[n]$, $\eta_0[n]$, and $\eta_1[n]$ are uncorrelated. Then we have that $r_{dx}[k] = r_{d_0u}[k]$ or

$$R_{dx}(z) = R_{d_0u}(z), \tag{3.126}$$

and then

$$R_{dx}(z) = H(z)R_{uu}(z). \tag{3.127}$$

For $|z| = 1$, with $z = e^{j\omega}$, the optimum Wiener filter from the above and for (3.64) is

$$W_{\text{opt}}\left(e^{j\omega}\right) = \frac{R_{dx}\left(e^{j\omega}\right)}{R_{xx}\left(e^{j\omega}\right)} = \frac{R_{uu}\left(e^{j\omega}\right)H\left(e^{j\omega}\right)}{R_{uu}\left(e^{j\omega}\right) + R_{\eta_1\eta_1}\left(e^{j\omega}\right)}. \tag{3.128}$$

In other words, the previous expression indicates that the optimum filter $W_{\text{opt}}(z)$ is equal to $H(z)$ when $R_{\eta_1\eta_1}(z) = 0$ or $\eta_1[n] = 0$ for each n. For further interpretation of (3.128), we define a parameter $K(e^{j\omega})$ as

$$K\left(e^{j\omega}\right) = \frac{R_{uu}\left(e^{j\omega}\right)}{R_{uu}\left(e^{j\omega}\right) + R_{\eta_1\eta_1}\left(e^{j\omega}\right)}. \tag{3.129}$$

Note that the terms $R_{uu}(e^{j\omega})$ and $R_{\eta_1\eta_1}(e^{j\omega})$ are PSD and, by definition, nonnegative real quantity. So we have that

$$0 \le K\left(e^{j\omega}\right) \le 1, \tag{3.130}$$

and

$$W_{\text{opt}}\left(e^{j\omega}\right) = K\left(e^{j\omega}\right)H\left(e^{j\omega}\right). \tag{3.131}$$

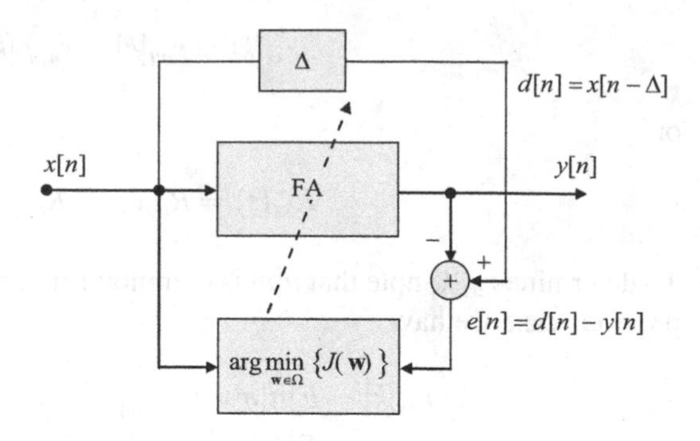

Fig. 3.13 Time delay estimation (TDE) scheme

3.4.3 Time Delay Estimation

Suppose as shown in Fig. 3.13, the delay to be estimated is equal to one sample $\Delta = 1$ and that the AF length is two. As in the previous example, also in this case, the problem can be interpreted as the identification of a TF that in this case is $H(z) = z^{-1}$ for which $\mathbf{h} = [0 \quad 1]^T$. Moreover, suppose that the AF input is a stochastic moving average (MA) process (Sect. C.3.3.3) defined as

$$x[n] = b_0\eta[n] + b_1\eta[n-1], \tag{3.132}$$

where $\eta[n] \triangleq N(0,\sigma_\eta^2)$ is a zero-mean WGN process, and there is no measure error.

For the determination of the matrix **R,** we note that

$$\begin{aligned} E\{x^2[n]\} &= E\{(b_0\eta[n] + b_1\eta[n-1])^2\} \\ &= E\{b_0^2\eta^2[n] + 2b_0\eta[n]b_1\eta[n-1] + b_1^2\eta^2[n-1]\} \\ &= b_0^2 + b_1^2 \end{aligned} \tag{3.133}$$

$$\begin{aligned} E\{x[n]x[n-1]\} &= E\Big\{(b_0\eta[n] + b_1\eta[n-1])(b_0\eta[n-1] + b_1\eta[n-2])\Big\} \\ &= b_0b_1. \end{aligned} \tag{3.134}$$

For the computation of the vector **g**, for $d[n] = x[n-1]$, note that

$$E\{d[n]x[n]\} = E\{x[n-1]x[n]\} = b_0b_1, \tag{3.135}$$

$$E\{d[n]x[n-1]\} = E\{x^2[n-1]\} = b_0^2 + b_1^2. \tag{3.136}$$

Remark In the experiments can be useful to have an SP $x[n]$ with unitary variance. In this case, from (3.133), this condition can be satisfied for $b_0^2 + b_1^2 = 1$ or, equivalently for $b_0 = \sqrt{1 - b_1^2}$.

For (3.133) and (3.134), we have that

$$\mathbf{R} = E\{\mathbf{x}\mathbf{x}^T\} = \begin{bmatrix} E\{x^2[n]\} & E\{x[n]x[n-1]\} \\ E\{x[n-1]x[n]\} & E\{x^2[n-1]\} \end{bmatrix} = \begin{bmatrix} b_0^2 + b_1^2 & b_0 b_1 \\ b_0 b & b_0^2 + b_1^2 \end{bmatrix}. \tag{3.137}$$

while for (3.135) and (3.136) we have that

$$\mathbf{g} = E\{\mathbf{x}d[n]\} = \begin{bmatrix} E\{x[n]d[n]\} \\ E\{x[n-1]d[n]\} \end{bmatrix} = \begin{bmatrix} b_0 b_1 \\ b_0^2 + b_1^2 \end{bmatrix}. \tag{3.138}$$

Let $a = b_0 b_1$ and $b = b_0^2 + b_1^2$, the normal equation is written as

$$\begin{bmatrix} b & a \\ a & b \end{bmatrix} \begin{bmatrix} w_0 \\ w_1 \end{bmatrix} = \begin{bmatrix} a \\ b \end{bmatrix}.$$

Let $\Delta = b^2 - a^2$, the Wiener solution $\mathbf{w}_{\text{opt}} = \mathbf{R}^{-1}\mathbf{g}$ is

$$\mathbf{w}_{\text{opt}} = \mathbf{R}^{-1}\mathbf{g} = \frac{1}{b^2 - a^2} \begin{bmatrix} b & -a \\ -a & b \end{bmatrix} \begin{bmatrix} a \\ b \end{bmatrix} = \frac{1}{b^2 - a^2} \begin{bmatrix} ba - ba \\ b^2 - a^2 \end{bmatrix} = \begin{bmatrix} 0 \\ 1 \end{bmatrix}.$$

Therefore, the Wiener solution is precisely a unit delay. Note that in this case the error is zero $e[n] = 0$.

3.4.3.1 Performance Surface and Minimum Energy Error Determination

The performance surface $J(\mathbf{w})$ (3.70) is

$$\begin{aligned} J(\mathbf{w}) &= \sigma_d^2 - 2\begin{bmatrix} w_0 & w_1 \end{bmatrix} \begin{bmatrix} a \\ b \end{bmatrix} + \begin{bmatrix} w_0 & w_1 \end{bmatrix} \begin{bmatrix} b & a \\ a & b \end{bmatrix} \begin{bmatrix} w_0 \\ w_1 \end{bmatrix} \\ &= \sigma_d^2 - 2(w_0 a + w_1 b) + 2a w_0 w_1 + b w_0^2 + b w_1^2. \end{aligned} \tag{3.139}$$

With minimum point at the optimum solution $\mathbf{w}_{\text{opt}} = [0\ 1]$. Figure 3.14 reports a typical plot of the performance surface $J(\mathbf{w})$.

3.4.4 Communication Channel Equalization

Let us consider the problem of communication channel equalization illustrated in Fig. 3.15, in which the channel is modeled as an L taps FIR filter $\mathbf{g} = [g[0], \ldots, g[L-1]]^T$. The channel TF $G(z)$ is defined as

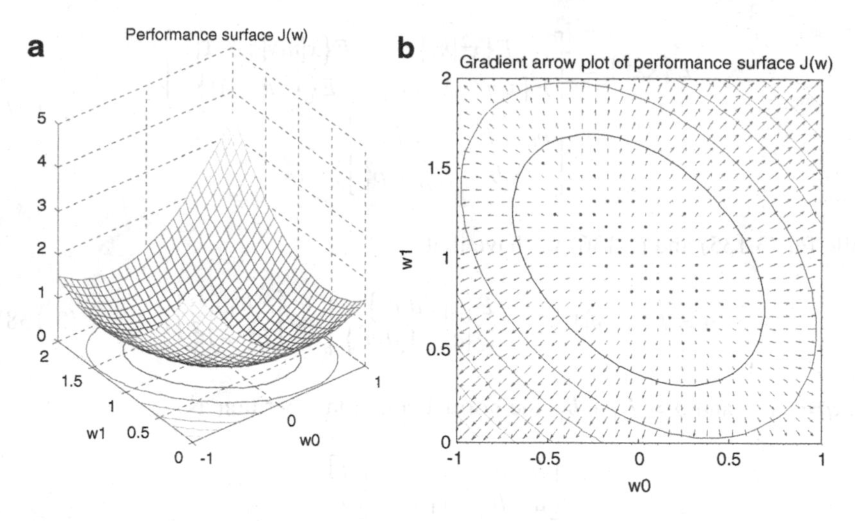

Fig. 3.14 Performance surface (3.139) for $b_1 = 0.707$ in (3.132): (**a**) 3D plot; (**b**) isolevel projection and gradient trend

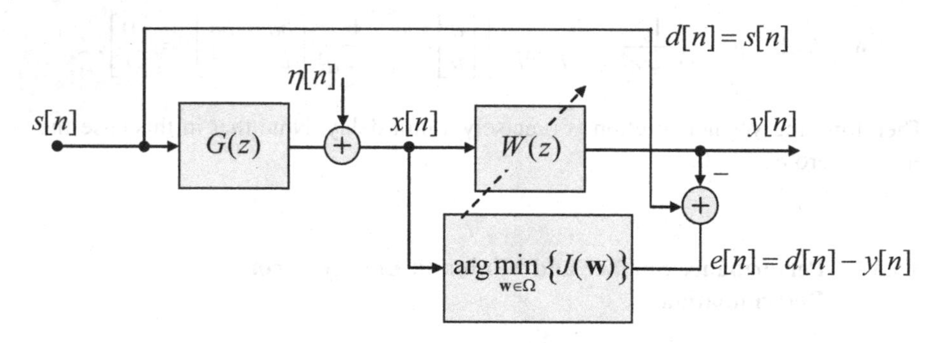

Fig. 3.15 Inverse model identification. Channel equalization example

$$G(z) = g[0] + g[1]z^{-1} + \cdots + g[L-1]z^{-L+1}. \tag{3.140}$$

The equalizer input $x[n]$ is then

$$x[n] = g[0]s[n] + \sum_{k=1}^{L-1} g[k]s[n-k] + \eta[n] = \mathbf{g}^T\mathbf{s} + \eta[n]. \tag{3.141}$$

The second term on the right side of the previous expression is the *intersymbol interference* (ISI), which describes interference superimposed to the symbols and that must be eliminated from the equalizer.

The equalizer task is thus to recover the symbols $s[n]$ corrupted by the channel's TF and by the superimposed noise. In the absence of noise, as already mentioned in the previous chapter (Sect. 2.3), the obvious solution is such that

$$W_{\text{opt}}(z) = 1/G(z), \tag{3.142}$$

whereby the causal solution exists only in the case where the $G(z)$ is minimum phase (i.e., its causal inverse corresponds to a stable circuit).

Considering the case with AGWN $\eta[n] \sim N(\sigma_\eta^2,0)$ whereby $s[n]$ and $\eta[n]$ are uncorrelated, for $|z|=1$ and $z=e^{j\omega}$ we have (for details Sect. C.2.7)

$$R_{xx}\left(e^{j\omega}\right) = R_{ss}\left(e^{j\omega}\right)\left|G\left(e^{j\omega}\right)\right|^2 + R_{\eta\eta}\left(e^{j\omega}\right). \tag{3.143}$$

From (3.141), $x[n]$ is the output of a linear system with impulse response $g[n]$ and input $s[n]$. It is also given as

$$R_{dx}\left(e^{j\omega}\right) \equiv R_{sx}\left(e^{j\omega}\right) = G^*\left(e^{j\omega}\right)R_{ss}\left(e^{j\omega}\right). \tag{3.144}$$

It should be noted that the previous result is independent of $\eta[n]$ that is uncorrelated.

From (3.143), (3.144), and for the (3.64), the optimum filter (Fig. 3.16) is

$$\begin{aligned} W_{\text{opt}}\left(e^{j\omega}\right) &= \frac{R_{dx}\left(e^{j\omega}\right)}{R_{xx}\left(e^{j\omega}\right)} \\ &= \frac{G^*\left(e^{j\omega}\right)R_{ss}\left(e^{j\omega}\right)}{R_{ss}\left(e^{j\omega}\right)\left|G\left(e^{j\omega}\right)\right|^2 + R_{\eta\eta}\left(e^{j\omega}\right)}. \end{aligned} \tag{3.145}$$

Equation (3.145) is the general solution of the problem without constraints on the length of the equalizer which could be also noncausal. Note that (3.145) includes the autocorrelation effects of the data $s[n]$ and of the noise $\eta[n]$.

To get a better interpretation, we divide the numerator and denominator of (3.145) with the first term of the denominator $R_{ss}(e^{j\omega})|G(e^{j\omega})|^2$. It is therefore

$$\begin{aligned} W_{\text{opt}}\left(e^{j\omega}\right) &= \frac{\dfrac{H^*\left(e^{j\omega}\right)R_{ss}\left(e^{j\omega}\right)}{R_{ss}\left(e^{j\omega}\right)\left|G\left(e^{j\omega}\right)\right|^2}}{\dfrac{R_{ss}\left(e^{j\omega}\right)\left|G\left(e^{j\omega}\right)\right|^2 + R_{\eta\eta}\left(e^{j\omega}\right)}{R_{ss}\left(e^{j\omega}\right)\left|G\left(e^{j\omega}\right)\right|^2}} \\ &= \frac{1}{1+\dfrac{R_{\eta\eta}\left(e^{j\omega}\right)}{R_{ss}\left(e^{j\omega}\right)\left|G\left(e^{j\omega}\right)\right|^2}} \cdot \frac{1}{G\left(e^{j\omega}\right)}. \end{aligned} \tag{3.146}$$

We define the parameter $\rho(e^{j\omega})$ as the ratio between the PSD of the signal and noise at the equalizer input:

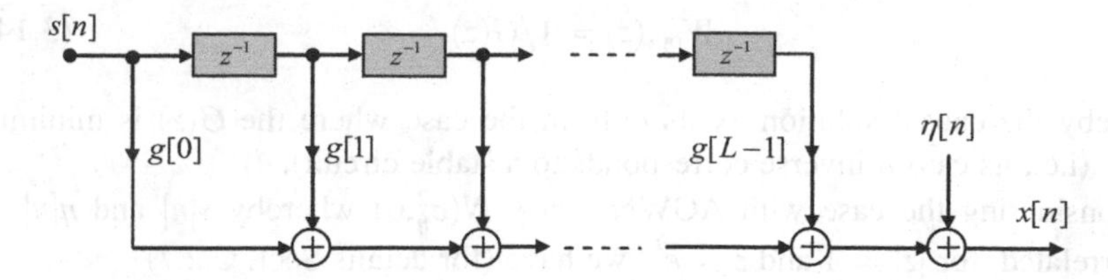

Fig. 3.16 Channel model

$$\rho\left(e^{j\omega}\right) \triangleq \frac{R_{ss}(e^{j\omega})|G(e^{j\omega})|^2}{R_{\eta\eta}\left(e^{j\omega}\right)}. \tag{3.147}$$

The terms $R_{ss}(e^{j\omega})|G(e^{j\omega})|^2$ and $R_{\eta\eta}(e^{j\omega})$ represent the signal and noise PSD at the channel output. Therefore, (3.146) can be rewritten as

$$W_{\text{opt}}\left(e^{j\omega}\right) = \frac{\rho(e^{j\omega})}{1+\rho(e^{j\omega})} \cdot \frac{1}{G(e^{j\omega})}. \tag{3.148}$$

Note that the frequency response of the optimum equalizer is inversely proportional to the channel's TF and that this proportionality depends on the frequency. Furthermore, the term $\rho(e^{j\omega})$ is, by definition, a nonnegative real quantity, for which

$$0 \leq \frac{\rho(e^{j\omega})}{1+\rho(e^{j\omega})} \leq 1. \tag{3.149}$$

The previous discussion shows that $W_{\text{opt}}(e^{j\omega})$ is proportional to the frequency response of the inverse of the communication channel with a proportionality parameter that is real and frequency dependent.

Example Consider a channel model with three real coefficients $\mathbf{g} = \left[-\frac{1}{3}, \frac{5}{6}, -\frac{1}{3}\right]$ and, for a preliminary analysis, without additive noise [7]. The receiver's input is

$$x[n] = -\frac{1}{3}s[n] + \frac{5}{6}s[n-1] - \frac{1}{3}s[n-2]. \tag{3.150}$$

From (3.142), the optimum equalizer is exactly the inverse of the channel's TF:

$$W_{\text{opt}}(z) = \frac{1}{G(z)} = \frac{-3}{1-\frac{5}{2}z^{-1}+z^{-2}}. \tag{3.151}$$

Developing into partial fractions, we get:

$$W_{\rm opt}(z) = \frac{1}{(1-\frac{1}{2}z^{-1})} - \frac{4}{(1-2z^{-1})}. \quad (3.152)$$

It should be noted that the previous TF has a pole outside the unit circle. This corresponds to a stable system only if the convergence region also includes the unit circle itself, i.e., only if one considers a noncausal equalizer. In this case, antitransforming (3.152) it follows that the impulse response of the optimum filter is a non-divergent (or stable) and noncausal, if it is defined as

$$w_{\rm opt}[n] = \begin{cases} 4(2)^n & n < 0 \\ \left(\frac{1}{2}\right)^n & n \geq 0, \end{cases} \quad (3.153)$$

shown in Fig. 3.17b. It should be noted that the convolution between the $h[n]$ and the equalizer response $\mathbf{w}_{\rm opt}[n]$ (3.153) produces just a unitary impulse (devoid of ISI), as shown in Fig. 3.17c.

Consider a binary input signal $s[n] \in (-1, +1)$, corrupted by AWGN with standard deviation evaluated for various levels of SNR as $\sigma_\eta = 10^{{\rm SNR}_{dB}/20}$ whose frequency trend is shown in Fig. 3.18. Note that when the input noise tends to zero, the equalizer approaches to the inverse of the channel response: $W_{\rm opt}(z) \approx 1/G(z)$.

3.4.5 *Adaptive Interference or Noise Cancellation*

Given a signal of interest (SOI) $s[n]$, the adaptive interference or noise cancellation[5] (AIC) consists in an attempt to subtract from $s[n]$ the uncorrelated additive noise or interference component [2].

$$R_{\eta_1\eta}(e^{j\omega}) = H(e^{j\omega})R_{\eta\eta}(e^{j\omega}). \quad (3.154)$$

As shown in Fig. 3.19, for the AIC systems two inputs are required. The first, called *primary reference*, is the SOI $s[n]$ corrupted by noise $\eta_1[n]$, while the other, called *secondary reference* or simple *reference*, presents a noise measure $\eta[n]$ that is correlated to the noise $\eta_1[n]$ added to the useful signal. Then we have

$$d[n] = s[n] + \eta_1[n], \quad (3.155)$$

$$x[n] = \eta[n]. \quad (3.156)$$

The AIC output is the signal error defined as

[5] To avoid possible ambiguity, we use the acronym AIC for adaptive noise/interference cancellation and ANC for active noise cancellation or control.

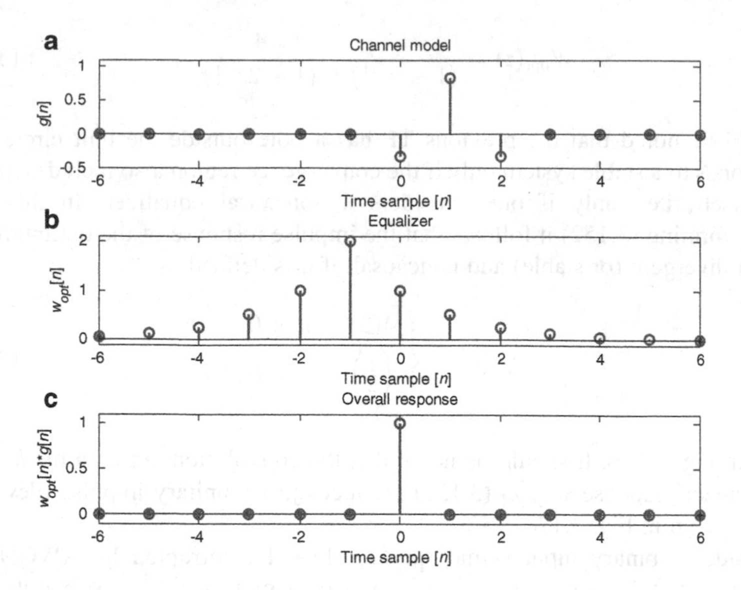

Fig. 3.17 Impulse response of (**a**) the communication channel (3.150); (**b**) its noncausal inverse is (3.153); (**c**) the convolution between the two previous

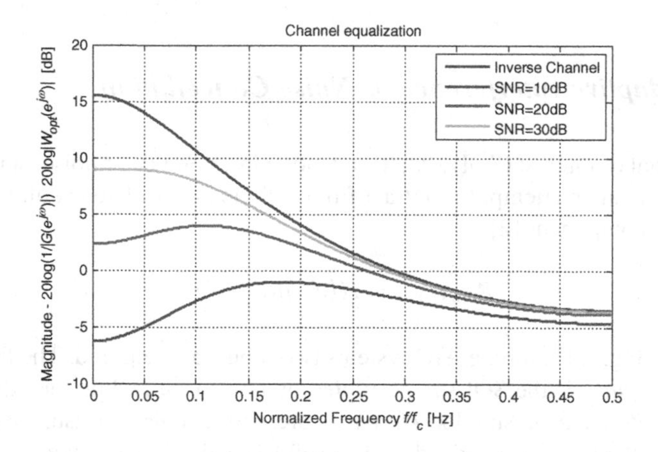

Fig. 3.18 The channel inverse $\left|G^{-1}(e^{j\omega})\right|_{\text{dB}}$ and the optimal filter $\left|W_{\text{opt}}(e^{j\omega})\right|_{\text{dB}}$ frequency responses in presence of AWGN $\sigma_\eta = 10^{\text{SNR}_{\text{dB}}/20}$ for $\text{SNR}_{\text{dB}} = 10, 20, 30$

$$e[n] = s[n] + \eta_1[n] - y[n], \tag{3.157}$$

with $y[n] = w[n]*\eta[n] = \mathbf{w}^T\boldsymbol{\eta}$.

By definition $s[n]$ is not correlated to the noise and, consequently, is not correlated to $\mathbf{w}^T\boldsymbol{\eta}$. So, by squaring (3.157) and taking the expectation, we get

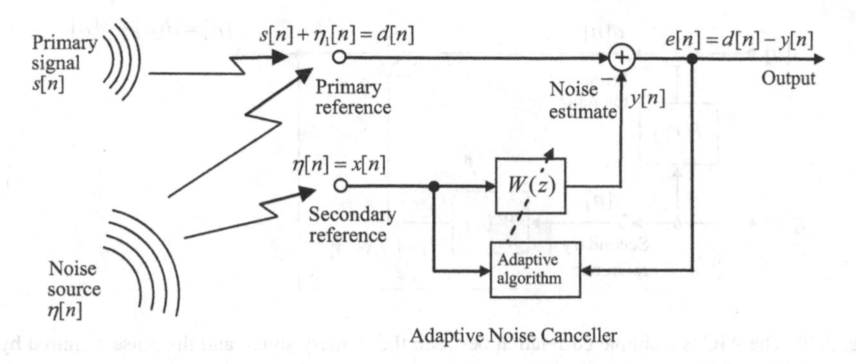

Fig. 3.19 Adaptive noise cancellation principle scheme

$$E\{e^2[n]\} = E\{s^2[n]\} + E\Big\{(\eta_1[n] - \mathbf{w}^T\boldsymbol{\eta})2\Big\} + 2E\Big\{s[n](\eta_1[n] - \mathbf{w}^T\boldsymbol{\eta})\Big\}. \quad (3.158)$$

Due to uncorrelation between $s[n]$ and $\eta[n]$, the last term of the previous expression is zero and the error minimization is for $y[n] = \eta_1[n]$. In this case, we have that $e[n] = s[n]$. Proceedings to minimization of the error (3.158) we have

$$\begin{aligned} J(\mathbf{w}) &= E\{s^2[n]\} + E\Big\{(\eta_1[n] - \mathbf{w}^T\boldsymbol{\eta})^2\Big\} \\ &= \sigma_s^2 + \sigma_{\eta_1}^2 - 2\mathbf{w}^T\mathbf{R}_{\eta_1\eta} + \mathbf{w}^T\mathbf{R}_{\eta\eta}\mathbf{w}, \end{aligned} \quad (3.159)$$

so for $\partial J(\mathbf{w})/\partial\mathbf{w} \rightarrow 0$ it follows $-2\mathbf{R}_{\eta_1\eta} + 2\mathbf{R}_{\eta\eta}\mathbf{w} = 0$; then we have

$$\mathbf{w}_{\text{opt}} = \mathbf{R}_{\eta\eta}^{-1}\mathbf{R}_{\eta_1\eta}. \quad (3.160)$$

Consider the scalar version of the previous expression (3.52)

$$r_{\eta\eta}[k] * w[k] = r_{\eta_1\eta}[j] \quad (3.161)$$

and taking the DTFT of both side, we have

$$W_{\text{opt}}(e^{j\omega}) = \frac{R_{\eta_1\eta}(e^{j\omega})}{R_{\eta\eta}(e^{j\omega})}. \quad (3.162)$$

As shown in Fig. 3.20, the correlation between primary and secondary noise sources can be modeled by an impulse response $h[n]$ such that $\eta_1[n] = \mathbf{h}^T\boldsymbol{\eta}$. Moreover, (by definition) $\eta[n]$ being WGN, in the stationary case for a linear system, is

$$R_{\eta_1\eta}(e^{j\omega}) = R_{\eta\eta}(e^{j\omega})H(e^{j\omega}).$$

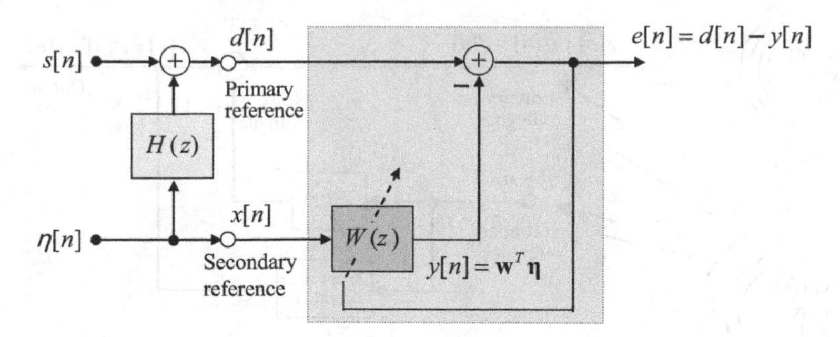

Fig. 3.20 The AIC is a simple correlation between the primary signal and the noise acquired by the secondary reference

In other words, the filter $h[n]$ represents the impulse response of the (acoustical, e.m., ...) path between primary and secondary sensors. For example, in the acoustic field, case $h[n]$ takes into account propagation delays and the wall reflections.

From the foregoing it appears that the AIC optimal filter $H(e^{j\omega})$ is the replica of the filter that models the path between the two references, i.e.,

$$W(e^{j\omega}) = H(e^{j\omega}). \tag{3.163}$$

This result is intuitive if you closely look at the AIC scheme. In fact, in this case

$$e[n] = s[n] + \mathbf{h}^T \boldsymbol{\eta} - \mathbf{w}^T \boldsymbol{\eta}, \tag{3.164}$$

and, in the case where the path between the reference was 'well-modeled', the error signal (AIC output) consists precisely in the useful signal and $e[n] = s[n]$.

Remark The expression (3.162) coincides with Wiener the formulation (3.64) in which the optimal filter is $W_{\text{opt}}(e^{j\omega}) = R_{dx}(e^{j\omega})/R_{xx}(e^{j\omega})$. Indeed, it should be noted that for (3.156), $R_{\eta\eta}(e^{j\omega}) \equiv R_{xx}(e^{j\omega})$ and, also, the PSD $R_{\eta_1\eta}(e^{j\omega})$ is equivalent to CPSD $R_{dx}(e^{j\omega})$ in the case of input signal $s[n]$ is zero or

$$\begin{aligned} R_{\eta_1\eta}(e^{j\omega}) &= R_{dx}(e^{j\omega})|_{s[n]=0} \\ &= H(e^{j\omega})R_{xx}(e^{j\omega}). \end{aligned} \tag{3.165}$$

In fact, from (3.64) we have that

$$\begin{aligned} W_{\text{opt}}(e^{j\omega}) &= \frac{R_{dx}(e^{j\omega})|_{s[n]=0}}{R_{xx}(e^{j\omega})} \\ &= \frac{R_{\eta_1\eta}(e^{j\omega})}{R_{\eta\eta}(e^{j\omega})} = H(e^{j\omega}). \end{aligned} \tag{3.166}$$

Remark One of the main applications of the AIC is to improve the quality of voice signal or speech enhancement. The aim is either the improvement of the perceived

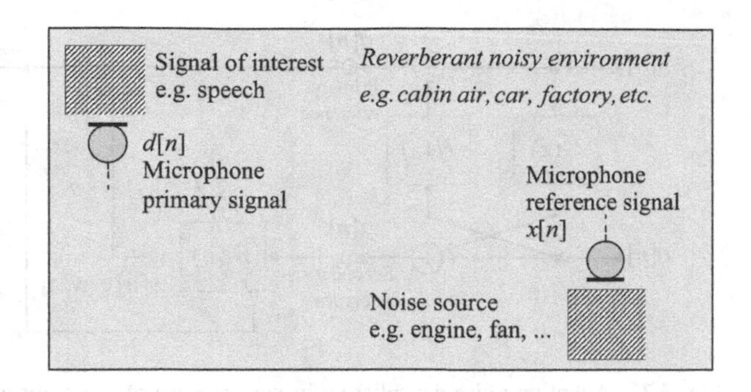

Fig. 3.21 Typical example of AIC systems application in reverberant noisy environment

acoustic quality in auditory communication systems, or the performance of automatic speech recognition (ASR) systems, which, in fact, strongly degrade their performance in the presence of noise.

In the case of reverberating environment, illustrated in Fig. 3.21, one of the main problems consists in the filter length L required for modeling the noise acoustic path. This path includes the delay between the noise source and the primary source and the reverberation effects.

3.4.5.1 Presence of the Useful Signal in the Secondary Reference

In real situations in the secondary reference in addition to the noise, a fraction of a signal correlated with $s[n]$ is present. This signal determines the partial cancellation of useful signal at the AIC output. Indicating the TF path between $s[n]$ and the secondary input with $G(z)$, an AIC more realistic diagram is shown in Fig. 3.22.

The presence of the $G(z)$ degrades the noise canceller performance and for the theoretical analysis should be included this effect. For a quantitative analysis we consider the expressions of the primary input and of the secondary reference:

$$d[n] = s[n] + \mathbf{h}^T \boldsymbol{\eta}, \tag{3.167}$$

and

$$x[n] = \eta[n] + \mathbf{g}^T \mathbf{s}. \tag{3.168}$$

Proceeding as in the previous case, the optimum Wiener filter can be directly determined by the expression $W_{\text{opt}}(e^{j\omega}) = R_{dx}(e^{j\omega})/R_{xx}(e^{j\omega})$. From the diagram of Fig. 3.22, being $s[n]$ and η[n] uncorrelated we can write

$$R_{xx}\left(e^{j\omega}\right) = R_{ss}\left(e^{j\omega}\right)\left|G\left(e^{j\omega}\right)\right|^2 + R_{\eta\eta}\left(e^{j\omega}\right). \tag{3.169}$$

The sequences $d[n]$ and $x[n]$ are related to $s[n]$ and $\eta[n]$ and to the respective impulse responses $g[n]$ and $h[n]$. To determine $R_{dx}(e^{j\omega})$, being $s[n]$ and $\eta[n]$ uncorrelated with

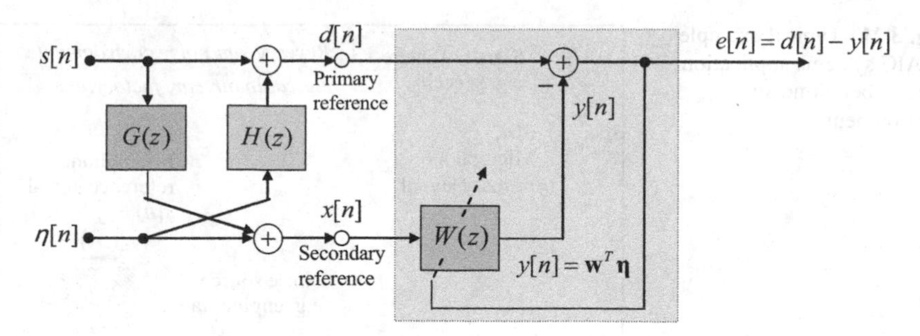

Fig. 3.22 Adaptive noise cancellation in the presence of cross-correlation between the primary input and the secondary reference

each other, the correlations with the reference $x[n]$ can be considered separately (Sect. C.2.7.6), we can therefore write

$$R_{dx}(e^{j\omega}) = R_{dx}(e^{j\omega})\big|_{\eta[n]=0} + R_{dx}(e^{j\omega})\big|_{s[n]=0}, \tag{3.170}$$

where the individual contributions are evaluated as

$$R_{dx}(e^{j\omega})\big|_{\eta[n]=0} = G^*(e^{j\omega})R_{ss}(e^{j\omega}), \tag{3.171}$$

and

$$R_{dx}(e^{j\omega})\big|_{s[n]=0} = H(e^{j\omega})R_{\eta\eta}(e^{j\omega}). \tag{3.172}$$

Substituting the above expressions in (3.170) we can obtain

$$R_{dx}(e^{j\omega}) = G^*(e^{j\omega})R_{ss}(e^{j\omega}) + H(z)R_{\eta\eta}(e^{j\omega}). \tag{3.173}$$

From the previous expressions, the optimal Wiener filter can be written as

$$\begin{aligned} W_{\text{opt}}(e^{j\omega}) &= \frac{R_{dx}(e^{j\omega})}{R_{xx}(e^{j\omega})} \\ &= \frac{G^*(e^{j\omega})R_{ss}(e^{j\omega}) + H(e^{j\omega})R_{\eta\eta}(e^{j\omega})}{R_{ss}(e^{j\omega})|G(e^{j\omega})|^2 + R_{\eta\eta}(e^{j\omega})}. \end{aligned} \tag{3.174}$$

Comparing the latter with (3.128) and (3.145), we note that (3.174) can be seen as a generalization of the previous results obtained by direct and inverse linear systems modeling. This can easily be verified by visual inspection of Fig. 3.22 where it can be seen that in the AIC scenario, either direct or inverse system modeling is present [7].

Therefore, the AIC output minimization consists in a trade-off between the cancelation of the SOI $s[n]$ and the noise cancellation. The noise is canceled when $W(z) \to H(z)$, while the condition for the cancelation of the signal $s[n]$ is $W(z) \to 1/G(z)$. In other words, the AIC considers the signals $\eta[n]$ and $s[n]$ in the same way.

3.4.5.2 AIC Performances Analysis

The performance measurements of the noise canceller can be made, as shown in [2], considering the improvement of the signal-to-noise ratio (SNR) between the primary input and the AIC output, in terms of PSD. To this end, we define the quantities $\rho_{\text{pri}}(e^{j\omega})$, $\rho_{\text{sec}}(e^{j\omega})$, and $\rho_{\text{out}}(e^{j\omega})$, as the SNR, at the primary input, at the secondary reference, and at the AIC output, respectively. So, for visual inspection of Fig. 3.22, we can directly write

$$\rho_{\text{pri}}\left(e^{j\omega}\right) = \frac{R_{ss}(e^{j\omega})}{|H(e^{j\omega})|^2 R_{\eta\eta}(e^{j\omega})}, \tag{3.175}$$

and

$$\rho_{\text{sec}}\left(e^{j\omega}\right) = \frac{|G(e^{j\omega})|^2 R_{ss}(e^{j\omega})}{R_{\eta\eta}(e^{j\omega})}. \tag{3.176}$$

The determination of $\rho_{\text{out}}(e^{j\omega})$ is more complex and can be done by evaluating the output PSD $R_{ee}(e^{j\omega})$, whereas the superposition of the effects of the individual contributions is due to the signals $s[n]$ and $\eta[n]$. From the AIC scheme, it is evidenced that the signal $s[n]$ reaches the output along two separate paths: the first directly, while the second through the TFs $G(z)$ and $W(z)$. So, we can write

$$R_{ee}\left(e^{j\omega}\right)\big|_{\eta[n]=0} = \left|1 - G\left(e^{j\omega}\right)W\left(e^{j\omega}\right)\right|^2 R_{ss}\left(e^{j\omega}\right). \tag{3.177}$$

Similarly, for the contribution due to the noise component, we have that

$$R_{ee}\left(e^{j\omega}\right)\big|_{s[n]=0} = \left|H\left(e^{j\omega}\right) - W\left(e^{j\omega}\right)\right|^2 R_{\eta\eta}\left(e^{j\omega}\right). \tag{3.178}$$

In the previous two expressions, substituting the optimal solution $W(e^{j\omega}) \to W_{\text{opt}}(e^{j\omega})$ calculated with the (3.174), one obtains, respectively,

$$R_{ee}\left(e^{j\omega}\right)\big|_{\eta[n]=0} = \left|\frac{(1 - G(e^{j\omega})H(e^{j\omega}))R_{\eta\eta}(e^{j\omega})}{R_{ss}(e^{j\omega})|G(e^{j\omega})|^2 + R_{\eta\eta}(e^{j\omega})}\right|^2 R_{ss}\left(e^{j\omega}\right)$$

and

$$R_{ee}\left(e^{j\omega}\right)\big|_{s[n]=0} = \left|\frac{(H(e^{j\omega})G(e^{j\omega}) - 1)G^*(e^{j\omega})R_{ss}(e^{j\omega})}{R_{ss}(e^{j\omega})|G(e^{j\omega})|^2 + R_{\eta\eta}(e^{j\omega})}\right|^2 R_{\eta\eta}\left(e^{j\omega}\right). \tag{3.179}$$

It follows that $\rho_{\text{out}}(e^{j\omega})$ can be computed as

$$\begin{aligned}\rho_{\text{out}}\left(e^{j\omega}\right) &= \frac{R_{ee}\left(e^{j\omega}\right)\big|_{\eta[n]=0}}{R_{ee}\left(e^{j\omega}\right)\big|_{s[n]=0}}\\ &= \frac{\left|(1 - G(e^{j\omega})H(e^{j\omega}))R_{\eta\eta}(e^{j\omega})\right|^2 R_{ss}(e^{j\omega})}{\left|(H(e^{j\omega})G(e^{j\omega}) - 1)G^*(e^{j\omega})R_{ss}(e^{j\omega})\right|^2 R_{\eta\eta}(e^{j\omega})},\end{aligned}$$

that, after some simplification, is

$$\rho_{\text{out}}\left(e^{j\omega}\right) = \frac{1}{|G(e^{j\omega})|^2}\frac{R_{\eta\eta}(e^{j\omega})}{R_{ss}(e^{j\omega})}. \tag{3.180}$$

Comparing the latter with (3.176), we can write

$$\rho_{\text{out}}\left(e^{j\omega}\right) = \frac{1}{\rho_{\text{sec}}\left(e^{j\omega}\right)}. \tag{3.181}$$

Equation (3.181), known as the *power inversion* (Widrow and Stearns [2]), indicates that the SNR at the AIC output is at most equal to the inverse of the SNR at the secondary reference. This result indicates that if we want to cancel the interfering noise, we need to reduce as much as possible the presence of the SOI $s[n]$ at the input of the secondary reference or, equivalently, the secondary reference must acquire only the noise. This suggests that the effective noise cancellation can be achieved by an appropriate physical isolation of the primary and secondary sensors.

3.4.6 AIC in Acoustic Underwater Exploration

In undersea seismic–acoustic exploration, signals from the hydrophones array are appropriately combined in a single signal (*beamformer*). Considering Fig. 3.23, the reference hydrophone is placed in proximity of the ship hull so as to acquire, mainly, the engine noise of the ship. The noise added to the useful signal is due to the direct noise contribution added to their reflections at various delays. A FIR filter with impulse response $h[n]$ can model these effects.

Denoting by $v[n]$ the noise produced by the hull, the model for noise subtraction is the classic AIC described above. Therefore, we can write

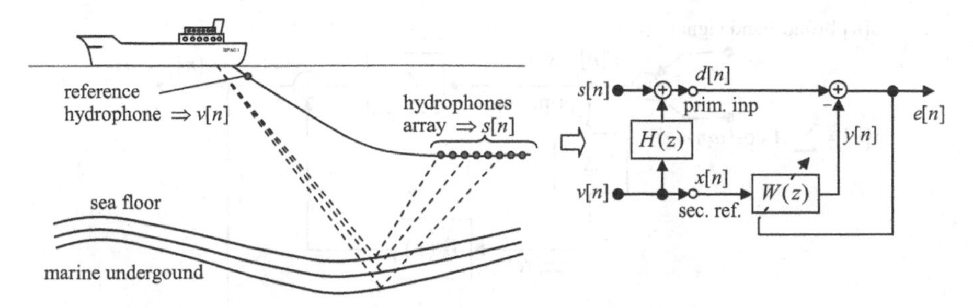

Fig. 3.23 Boat that carries an array of hydrophones for underwater exploration. The hydrophone placed close to the propeller capture mainly the hull noise $v[n]$

$$\begin{aligned} d[n] &= s[n] + \mathbf{h}^T \mathbf{v} \\ x[n] &= v[n]. \end{aligned} \tag{3.182}$$

In this type of problem, said *own-ship noise*, the primary signal is broadband, while the noise, mainly due to the boat's engine and propeller, is of a harmonic type, i.e., narrow band, and can be modeled as a sum of sinusoids of the type

$$v[n] = \sum_{i=0}^{M-1} A_i \cos\left(\omega_i nT + \phi_i\right). \tag{3.183}$$

In other words, the useful signal is broadband and the noise is a narrowband high correlated process.

3.4.7 AIC Without Secondary Reference Signal

Thus in the absence of reference signal, AIC systems may be defined in cases where the primary signal is constituted by the sum of two uncorrelated processes. The first is correlated narrow band, for example constituted by a sum of sinusoids, while the second is a broadband uncorrelated process. In the absence of secondary reference, the technique consists in the identification of the two parties and the subtraction of the noise from the useful part of the signal. Can be identified two distinct cases. Case 1: the useful signal is broadband and the noise is narrowband; case 2: the useful signal is narrowband while the noise is broadband.

3.4.7.1 Case 1: Broadband Signal and Narrowband Noise

In the case that the primary signal $s[n]$ is broadband and the superimposed noise is a narrowband process (for example due to a rotating machine). The signal of the primary reference can be defined as

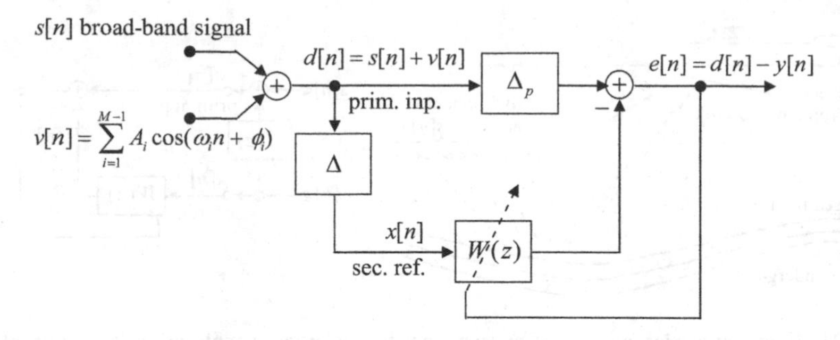

Fig. 3.24 Schematic of the AIC without primary reference for broadband useful signal and narrowband noise

$$d[n] = \underbrace{s[n]}_{\text{signals}} + \underbrace{\sum_{i=1}^{M-1} A_i \cos(\omega_i n + \phi_i)}_{\text{noise}}. \tag{3.184}$$

In this situation, the broadband part of the primary signal can be separated from the most correlated narrowband component which can be subtracted from the primary source in order to reduce the noise component. This is possible with the cancellation scheme illustrated in Fig. 3.24.

The determination of the delay Δ, necessary to decorrelate the useful signal, is calculated in such a way that it is

$$\Delta \therefore E\{s[n]s[n-\Delta]\} = 0. \tag{3.185}$$

Since the noise is a high correlated process, at the adaptive filter output is present only the noise signal, namely, the output of the filter is equal to $y[n] \sim \sum_{i=1}^{M-1} A_i \cos(\omega_i n + \phi_i)$. At convergence, this contribution is subtracted from the signal $d[n]$. To align the signals may be necessary to insert an appropriate delay Δ_p. So the AIC output tends to assume the value of the SOI $e[n] \sim s[n]$.

3.4.7.2 Case 2: Narrowband Signal and Broadband Noise: Adaptive Line Enhancement

In the case that the useful signal is narrowband and the noise broadband, the AIC system is defined as adaptive line enhancement (ALE). In this case, the correlated part of the primary signal can be separated from the noncorrelated noise component which can, therefore, be subtracted from the primary source. This situation is typical in cases where the SOI $s[n]$ is composed of one or more narrowband components as in the case of a sum of sinusoidal signals with unknown amplitude and frequency, immersed in broadband noise.

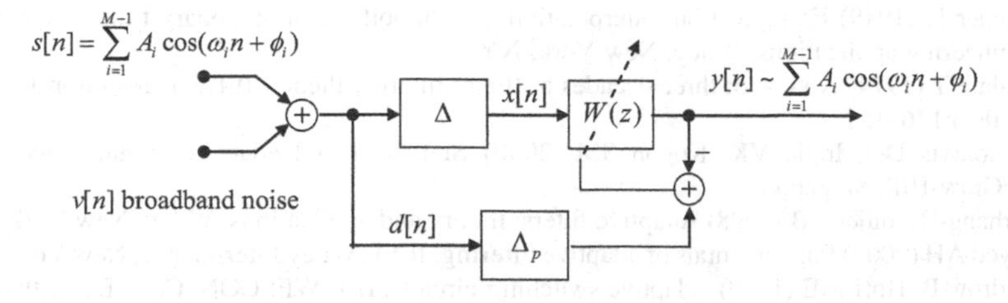

Fig. 3.25 Schematic representation of the adaptive line enhancer

An example of this scenario occurs in passive sonar used for the remote identification of ships and submarines. The noise from the boat is mainly due to the propulsion system, the auxiliary machinery, and the hydrodynamic effects. The *acoustic signature* of the vessel is made, typically, by a narrowband processes (spectral line), generated by the propulsion system, and superimposed on broadband noise due to hydrodynamics. The spectral lines level of the narrowband process part increases with the speed of the vessel and, in the case of submarine, also with the depth. The part of the signal due to the auxiliary machinery remains largely stable. In the source model, as well as fluctuations due to the variable nature of the source, you must also consider the changes introduced by the propagation and the frequency change (of the various spectral lines) due to the Doppler effect. For the instrumental measurement error, due to the hydrodynamic noise components, the estimation of the vessel acoustic signature appears to be a fairly complex. The amplitude of the narrowband components, related to the noise signal, is very low and also SNR can be lower than 0 dB.

In this scenario, the process at the receiver can be modeled as

$$d[n] = \underbrace{s[n]}_{\text{signal}} + \underbrace{\sum_{i=1}^{M-1} A_i \cos(\omega_i n + \phi_i)}_{\text{noise}} \tag{3.186}$$

where $v[n]$ represents the sources noise.

The AIC scheme is shown in Fig. 3.25. In this case, differently from Fig. 3.24, the useful signal is the output of the adaptive filter $y[n]$ and not the error signal, as in previous cases. In this case, in fact, the error is used only for the filter adaptation.

References

1. Petersen K.B., Pedersen M. S. (2012) The matrix cookbook Tech. Univ. Denmark, Kongens Lyngby, Denmark, Tech. Rep., Ver. November 15, 2012
2. Widrow B, Stearns SD (1985) Adaptive signal processing. Prentice-Hall, Englewood Cliffs, NJ
3. Haykin S (1996) Adaptive filter theory, 3rd edn. Prentice-Hall, Englewood Cliffs, NJ

4. Wiener N (1949) Extrapolation, interpolation and smoothing of stationary time series, with engineering applications. Wiley, New York, NY
5. Kailath T (1974) A view of three decades of linear filtering theory. IEEE Trans Inform Theor IT20(2):146–181
6. Manolakis DG, Ingle VK, Kogon SM (2000) Statistical and adaptive signal processing. McGraw-Hill, Singapore
7. Farhang-Boroujeny B (1998) Adaptive filters: theory and applications. Wiley, New York, NY
8. Sayed AH (2003) Fundamentals of adaptive filtering. IEEE Wiley Interscience, New York, NY
9. Widrow B, Hoff ME (1960) Adaptive switching circuits. IRE WESCON, Conv Rec 4:96–104
10. Orfanidis SJ (1996–2009) Introduction to signal processing, Prentice Hall, Englewood Cliffs, NJ. ISBN 0-13-209172-0
11. Golub GH, Van Loan CF (1989) Matrix computation. John Hopkins University press, London. ISBN ISBN 0-80183772-3
12. Strang G (1988) Linear algebra and its applications. Third Ed. ISBN: 0-15-551005-3, Thomas Learning ed
13. Noble B, Daniel JW (1988) Applied linear algebra. Prentice-Hall, Englewood Cliffs, NJ
14. Huang Y, Benesty J, Chen J (2006) Acoustic MIMO signal processing. Springer, series: signals and communication technology. ISBN: 978-3-540-37630-9

Chapter 4
Least Squares Method

4.1 Introduction

This chapter introduces the deterministic counterpart of the statistical Wiener filter theory. The problems of adaptation are addressed in the case where the filter input signals are sequences generated by linear deterministic models without any assumption on their statistics.

The basic idea is the principle of least squares (LS) introduced by Gauss that first formulated an estimation problem assimilating it to a simple optimization algorithm [2, 7, 19, 20, 32].[1]

In particular, we introduce the LS method and the *normal equations* in the original Yule–Walker notation. Moreover, some LS variants and the performance analysis of the solutions are presented and discussed. Methods for solving over/under-determined linear systems are also discussed. In addition, robust numerical methods based on matrix factorization, the paradigm of regularization, and general theory of total least squares (TLS) are considered. Finally, we present some methods to solve underdetermined sparse systems, minimizing an L_p-norm (with $0 \leq p < 1$).

4.1.1 The Basic Principle of Least Squares Method

The least squares principle can be illustrated with reference to Fig. 4.1. Consider a set of N real measurements, indicated as $y_0, y_1, \ldots, y_{N-1}$, relating to points (or time instants) $x_0, x_1, \ldots, x_{N-1}$, which represents the set of experimental data sometime indicated as the pair $[\mathbf{x},\mathbf{y}]$. Suppose we want to determine the parameters of a curve

[1] The method of least squares was introduced by the German mathematician Carl Friedrich Gauss in 1795, at age 18, and was used by him for the calculation of the orbit of the asteroid Ceres in 1821.

A. Uncini, *Fundamentals of Adaptive Signal Processing*, Signals and Communication Technology, DOI 10.1007/978-3-319-02807-1_4,

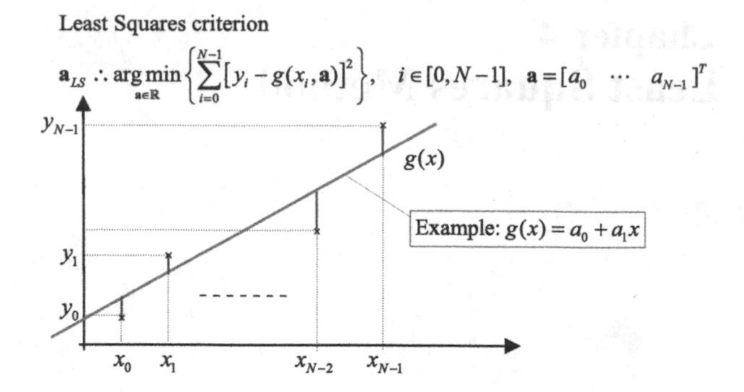

Fig. 4.1 The least squares method. The optimum solution $g(x_i)$ is determined by the minimization of the mean Euclidean distance between $g(x_i)$ and the available experimental measurements

$g(x_i, \mathbf{a})$ that best approximates (*best fits*) the experimental data according to a certain criterion. The name *least squares* derives from the optimum criterion used for the determination of the curve that minimizes the sum of squares of differences between $g(x_i, \mathbf{a})$ and y_i, for $i \in [0, N-1]$.

Unlike the Wiener theory, derived from an exact approach based on the *a priori* knowledge of the second-order ensemble averages of the processes, LS filtering methodology is defined considering a deterministic approach where the estimation problem is transformed to an optimization algorithm. In particular, instead of the ensemble averages, the time averages are used for which the optimum estimation also depends on the number of the available samples. In other words, the LS adaptation is based on the *Estimation Theory* where a deterministic optimization criterion is assumed [1].

Therefore, the LS approach is a paradigm able to define a large family of algorithms. In fact, the deterministic nature of the method allows numerous variations and specializations that may be derived by the vastness of the usable optimization methods, from the algebraic nature of the algorithms, from specific *a priori* knowledge, and by the use of L_p metric (that may not be the simple L_2). The choice of optimization criterion is governed by the nature of the cost function (CF) which, in turn, can be formulated on the specific problem. The algebraic nature derives, instead, from the determination of the solution, which generally implies the inversion of matrices which can be very large and *ill-conditioned*.

4.2 Least Squares Methods as Approximate Stochastic Optimization

The choice of the CF represents a central point for both the theoretical study and the adaptation algorithms development. In stochastic optimization, the CF is a function of the error statistic. In particular, as defined in (3.30), the expectation of the square error or MSE is defined as $J(\mathbf{w}) = E\{|e[n]|^2\}$.

A deterministic CF widely used in adaptive filtering, already introduced in the expression (3.31), is the sum of squared errors (SSE), defined as

$J(\mathbf{w}) \triangleq \hat{E}\left\{|e[n]|^2\right\} = \frac{1}{N}\sum_n |e[n]|^2$. The SSE can be considered as an approximation of stochastic MSE. In this sense, as already indicated in Sect. 3.2.4, the adaptation techniques deriving from the choice of deterministic CF are mentioned as approximate stochastic optimization (ASO) methods. It appears, in fact, that in the case of ergodic process $\hat{E}\left\{|e[n]|^2\right\} \approx E\left\{|e[n]|^2\right\}$.

4.2.1 Derivation of LS Method

The LS method is the basis of a wide class of non-recursive and recursive ASO algorithms. In the LS method, there are no assumptions on the statistic of the input sequences; the input is characterized by a stochastic generation model. For example, in the linear case we consider the AR, MA, or ARMA models (see Sect. C.3.3).

Therefore, for the development of learning algorithms, the explicit knowledge of the statistical functions of the involved processes that, therefore, are estimated is not necessary. The main advantage of the LS algorithms family consists in the large variety of possible practical applications at the expense, however, of the possibility of not being able to obtain the statistically optimal solution.

The LS method, as shown in Fig. 4.2, is widely used in the parameters estimation of the signal generation model. The deterministic signal $d[n]$ is generated using a stochastic model dependent on the unknown parameters $\mathbf{w}_d$ that, therefore, must be estimated. In these cases, the uncertainty sources are considered both the errors due to measurement noise and the inaccuracy of the assumed model. In general, these uncertainty sources are treated as additive noise with a certain distribution (typically Gaussian) superimposed to the available measure. What is observed is not the signal $d[n]$ but, because of the above uncertainty sources, his perturbed version $y[n]$.

The least squares estimator (LSE) is one that minimizes the least squares distance between the data $d[n]$ and the observed data $y[n]$ in the measurement interval $n \in [0, N-1]$. Therefore, the CF can be defined as

$$J(\mathbf{w}) = \sum_{n=0}^{N-1} |d[n] - y[n]|^2. \tag{4.1}$$

Unlike Wiener theory, where you know the statistics of not only the noise but also the signal, in LS method, the signal is considered deterministic even if generated by a stochastic model. The uncertainty is due to the noise usually supposed to be AWGN and zero mean.

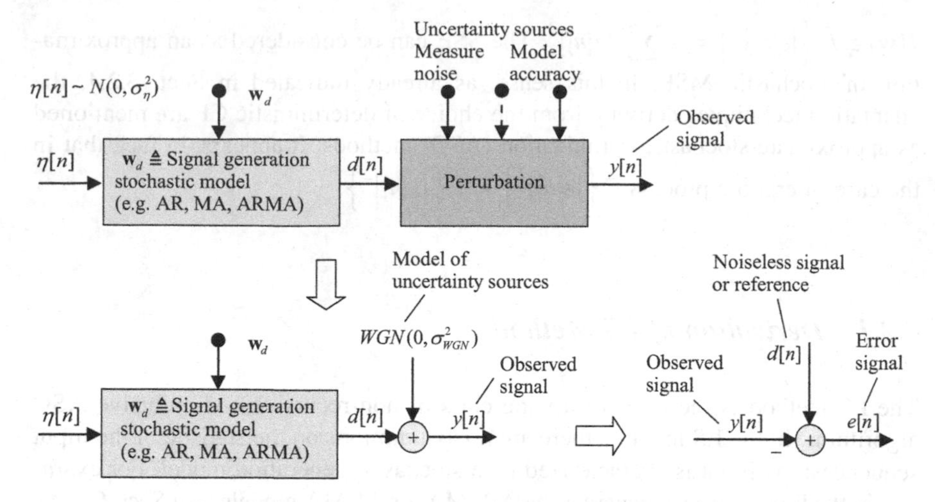

Fig. 4.2 Stochastic signal generation models for the derivation of the LS methods (modified from [1])

4.2.2 *Adaptive Filtering Formulation with LS Method*

In the LS method for adaptive filtering, shown schematically in Fig. 4.3, the signal $d[n]$, taken as a reference, is defined as the *desired output*. The output of the adaptive filter (AF) $y[n]$ represents an estimated value, and the AF parameters $\mathbf{w}$ represent the estimator. In other words, the N-length sequences $x[n]$ and $d[n]$ represent a single realization of *stochastic processes* (SP) that are unknown. Measurement errors and other sources of uncertainty are embedded in the noise superimposed on the observation. The model is valid in the case where the superimposed noise is Gaussian or not, even if the performance of the estimator depends on the statistic of the superimposed noise and on the choice of the model.

The filter weights vector $\mathbf{w}$ is a *random variable* (RV) array representing a linear MMSE estimator (see Sect. C.3.2.8). The determination of the estimator $\mathbf{w}$ is performed by processing the available data samples. For a general notation, we consider an N-length observation interval with time index limited by lower and upper bounds $n \in [n_1, n_2]$. As said, the estimate of the filter weights, or better the estimate of the estimator $\mathbf{w}$, is calculated by minimizing the sum of squared errors and the problem can be formulated as

$$\mathbf{w}_{\text{LS}} = \underset{\mathbf{w} \in (\mathbb{R},\mathbb{C})}{\arg\min} J(\mathbf{w}) \equiv \underset{\mathbf{w} \in (\mathbb{R},\mathbb{C})}{\arg\min} \sum_{n=n_1}^{n_2} |e[n]|^2. \tag{4.2}$$

In other words, the sum of squares errors (SSE) is minimized, corresponding to the energy of the error sequence, calculated in a deterministic way from the available data, rather than the squares error expectation. For hypothesis of ergodicity of the

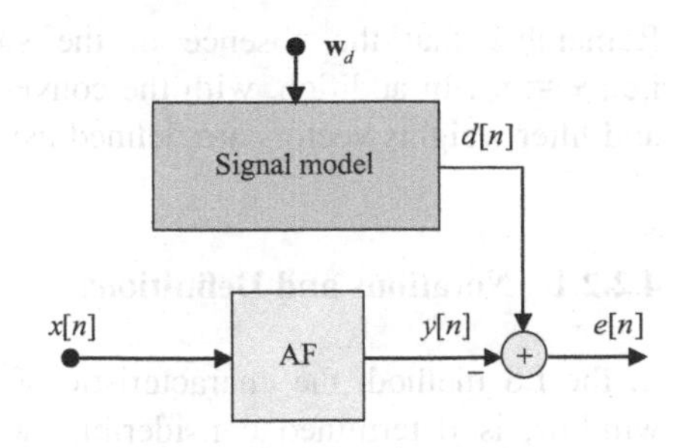

Fig. 4.3 Adaptive filtering: scheme for the derivation of the LS method

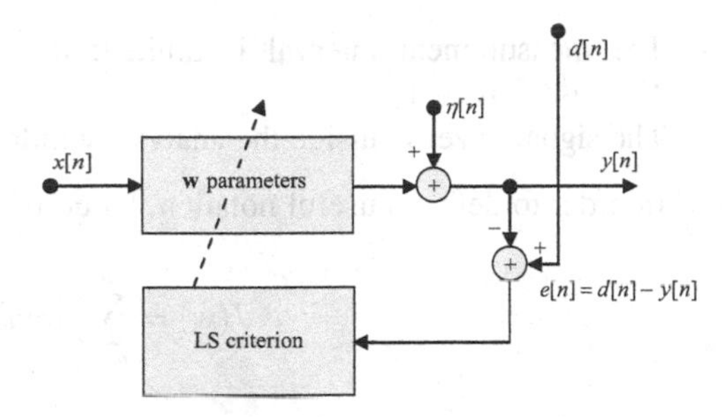

Fig. 4.4 Interpretation of the LS method in the context of adaptive filtering

process we consider the SSE, over a certain time window (4.2), as an MSE estimation.

For the LS methods development in the context of adaptive filtering, consider the diagram in Fig. 4.4 in which the noise is considered additively superimposed on the output sequence. For simplicity, let us consider the additive noise zero.

The output $y[n]$ is calculated as

$$y[n] = \sum_{i=0}^{M-1} \hat{w}^*[i]x[n-i] = \hat{\mathbf{w}}^H \mathbf{x}_n = \mathbf{x}_n^T \mathbf{w}, \tag{4.3}$$

where the vector $\mathbf{x}_n$ represents the signal on the filter delay line and, in order to simplify the notation, we pose $\mathbf{w} = \hat{\mathbf{w}}^*$. Therefore, the error with respect to the desired output $d[n]$, for the real and complex domain cases, is equal to

$$e[n] = d[n] - \mathbf{x}_n^T \mathbf{w}. \tag{4.4}$$

The filter $\mathbf{w}$ and regressor vector $\mathbf{x}_n$ are defined as

$$\mathbf{x}_n \in (\mathbb{R}, \mathbb{C})^{M\times 1} \triangleq \begin{bmatrix} x[n] & x[n-1] & \cdots & x[n-M+1] \end{bmatrix}^T, \tag{4.5}$$

$$\mathbf{w} \in (\mathbb{R}, \mathbb{C})^{M\times 1} \triangleq \begin{bmatrix} w^*[0] & w^*[1] & \cdots & w^*[M-1] \end{bmatrix}^T. \tag{4.6}$$

Remember that the absence of the subscript n indicates "at the instant n," i.e., $\mathbf{x} \equiv \mathbf{x}_n$. In addition, with the convention in Sect. 3.2.1.1 the input sequences and filter weights vectors are defined using the notation $\mathbf{x}, \mathbf{w} \in (\mathbb{R},\mathbb{C})^{M\times 1}$.

4.2.2.1 Notations and Definitions

In the LS method, the characteristic of the data block, defined by the N-length window, is determined considering the nature of the problem. The following conventions are assumed:

- The measurement interval is limited: $n \in [n_1,n_2]$ and has length equal to $N = n_2 - n_1 + 1$
- The signal is zero outside the analysis window

In order to define a useful notation, let consider the CF as

$$J(\mathbf{w}) = \sum_{n=n_1}^{n_2} |e[n]|^2. \tag{4.7}$$

For (4.4), explicitly writing the error at time instants: n_2, $n_2 - 1$, ..., n_1, we have that

$$\begin{aligned} e[n_2] &= d[n_2] - \mathbf{x}_{n_2}^T\mathbf{w} \\ e[n_2-1] &= d[n_2-1] - \mathbf{x}_{n_2-1}^T\mathbf{w} \\ &\vdots \\ e[n_1] &= d[n_1] - \mathbf{x}_{n_1}^T\mathbf{w}. \end{aligned} \tag{4.8}$$

Defining the vectors

$$\mathbf{e}_{n_2} = \begin{bmatrix} e[n_2] & e[n_2-1] & \cdots & e[n_1] \end{bmatrix}^T, \quad \textit{error vector}, \tag{4.9}$$

$$\mathbf{d}_{n_2} = \begin{bmatrix} d[n_2] & d[n_2-1] & \cdots & d[n_1] \end{bmatrix}^T, \quad \textit{desired output vector}, \tag{4.10}$$

$$\mathbf{x}_{n_2} = \begin{bmatrix} x[n_2] & x[n_2-1] & \cdots & x[n_2-M+1] \end{bmatrix}^T, \quad \textit{filters input}, \tag{4.11}$$

$$\mathbf{x}[n_2] = \begin{bmatrix} x[n_2] & x[n_2-1] & \cdots & x[n_1] \end{bmatrix}^T, \quad \textit{measurement interval}. \tag{4.12}$$

for simplicity, we omit writing the subscript n_2; the expression of the error can be defined in vector form as

$$\mathbf{e} = \mathbf{d} - \mathbf{X}\mathbf{w} \tag{4.13}$$

or, in explicit vector notation, we can write

$$\begin{bmatrix} e[n_2] \\ e[n_2-1] \\ \vdots \\ e[n_1] \end{bmatrix} = \begin{bmatrix} d[n_2] \\ d[n_2-1] \\ \vdots \\ d[n_1] \end{bmatrix} - \begin{bmatrix} x[n_2] & x[n_2-1] & \cdots & x[n_2-M+1] \\ x[n_2-1] & x[n_2-2] & \cdots & x[n_2-M] \\ \vdots & \vdots & \ddots & \vdots \\ x[n_1] & 0 & \cdots & 0 \end{bmatrix} \begin{bmatrix} w[0] \\ w[1] \\ \vdots \\ w[M-1] \end{bmatrix}. \tag{4.14}$$

Indicating the N row vectors of matrix $\mathbf{X} \in (\mathbb{R},\mathbb{C})^{N \times M}$, as $\mathbf{x}_k^T \in (\mathbb{R},\mathbb{C})^{1 \times M}$, $n_2 \leq k \leq n_1$, and the M column vectors as $\mathbf{x}[k] \in (\mathbb{R},\mathbb{C})^{1 \times N}$, $n_2 \leq k \leq (n_2 - M + 1)$, the data matrix can be defined as

$$\mathbf{X} = \begin{bmatrix} \mathbf{x}_{n_2} & \mathbf{x}_{n_2-1} & \cdots & \mathbf{x}_{n_1} \end{bmatrix}^T \quad \textit{row vector notation,} \tag{4.15}$$

$$\mathbf{X} = \begin{bmatrix} \mathbf{x}[n_2] & \mathbf{x}[n_2-1] & \cdots & \mathbf{x}[n_2-M+1] \end{bmatrix} \quad \textit{column vector notation.} \tag{4.16}$$

Note that the row vectors $\mathbf{x}_k^T$ of the matrix $\mathbf{X} \in (\mathbb{R},\mathbb{C})^{N \times M}$ contain the filter delay line samples (4.12), while column vectors $\mathbf{x}[k]$ contain the N-length analysis window sequence. The data matrix is defined such that convolution can be written as $\mathbf{y} = \mathbf{X}\mathbf{w}$ (see Sect. 1.6.2.1). Moreover, the matrix $\mathbf{X}$ can be filled in several ways discussed below in Sect. 4.2.3.

The SSE minimization is determined on an average of the N samples and the CF can be expressed as

$$\begin{aligned} E_e \equiv J(\mathbf{w}) &= \sum_{n=n_1}^{n_2} \left| e[n] \right|^2 \\ &= \mathbf{e}^H \mathbf{e} \\ &= \left\| \mathbf{d} - \mathbf{X}\mathbf{w} \right\|_2^2. \end{aligned} \tag{4.17}$$

As noted by the symbol E_e, the CF represents the energy of the error sequence.

4.2.2.2 Normal Equation in the Yule–Walker Form

Proceeding as for the Wiener filter, writing the CF (4.17) and considering the vector form (4.13), the following quadratic form can be defined:

$$\begin{aligned} J(\mathbf{w}) &= \left\| \mathbf{e} \right\|_2^2 \\ &= \left(\mathbf{d}^H - \mathbf{w}^H \mathbf{X}^H \right) \left(\mathbf{d} - \mathbf{X}\mathbf{w} \right) \\ &= E_d - \mathbf{w}^H \mathbf{X}^H \mathbf{d} - \mathbf{d}^H \mathbf{X}\mathbf{w} + \mathbf{w}^H \mathbf{X}^H \mathbf{X}\mathbf{w}, \end{aligned} \tag{4.18}$$

where $E_d = \mathbf{d}^H \mathbf{d}$. In terms of optimization theory, for the filter's weights calculation we have

$$\mathbf{w}_{\text{LS}} = \arg\min_{\mathbf{w}\in(\mathbb{R},\mathbb{C})} J(\mathbf{w}) = \arg\min_{\mathbf{w}\in(\mathbb{R},\mathbb{C})} \left\|\mathbf{d} - \mathbf{X}\mathbf{w}\right\|_2^2. \quad (4.19)$$

Differentiating (4.18) $\nabla J(\mathbf{w}) \triangleq \frac{\partial \hat{J}(\mathbf{w})}{\partial \mathbf{w}} = 2\mathbf{X}^H\mathbf{X}\mathbf{w} - 2\mathbf{X}^H\mathbf{d}$ and equating to zero $\nabla J(\mathbf{w}) = \mathbf{0}$, we obtain a system of N linear equations (built on window signal samples) with M unknowns, of the type

$$\mathbf{X}^H\mathbf{X}\mathbf{w} = \mathbf{X}^H\mathbf{d}, \quad (4.20)$$

with solution, for $N > M$,

$$\mathbf{w}_{\text{LS}} = \left(\mathbf{X}^H\mathbf{X}\right)^{-1}\mathbf{X}^H\mathbf{d}. \quad (4.21)$$

The expression (4.20) defines the LS system known with the name of the *Yule–Walker normal equations*, formulated for the first time in 1927 for the analysis of time series data (see, for example [2, 3]).

Remark By definition (see Sect. 3.3.2), we remind the reader that the time-average estimates of the correlations are evaluated as

$$E_d \triangleq \mathbf{d}^H\mathbf{d} = \sum_{j=n_1}^{n_2} \left|d[j]\right|^2, \quad \textit{desired output } d[n] \textit{ energy}, \quad (4.22)$$

$$\mathbf{R}_{xx} \triangleq \sum_{j=n_1}^{n_2} \mathbf{x}_j\mathbf{x}_j^H = \mathbf{X}^H\mathbf{X}, \quad \textit{time-average autocorrelation matrix}, \quad (4.23)$$

$$\mathbf{R}_{xd} \triangleq \sum_{j=n_1}^{n_2} \mathbf{x}_j d^*[n] = \mathbf{X}^H\mathbf{d}, \quad \textit{time-average crosscorrelation vector}, \quad (4.24)$$

where the factor $1/N$ is removed for simplicity. With these simplifications, in the case of ergodic process for $N \gg M$, we have that

$$\mathbf{R} \approx \frac{1}{N}\mathbf{R}_{xx} \quad \text{and} \quad \mathbf{g} \approx \frac{1}{N}\mathbf{R}_{xd}. \quad (4.25)$$

Equation (4.18) can then be written with formalism similar to (3.44) as

$$J(\mathbf{w}) = E_d - \mathbf{w}^H\mathbf{R}_{xd} - \mathbf{R}_{xd}^H\mathbf{w} + \mathbf{w}^H\mathbf{R}_{xx}\mathbf{w}, \quad (4.26)$$

and (4.20) as $\mathbf{R}_{xx}\mathbf{w} = \mathbf{R}_{xd}$. Note that (4.20), derived with algebraic criterion, has the same form of the Wiener equations $\mathbf{R}\mathbf{w} = \mathbf{g}$, derived with statistical methods. The solution of the system (4.20) in terms of (4.26) is

$$\mathbf{w}_{\text{LS}} = \mathbf{R}_{xx}^{-1}\mathbf{R}_{xd}, \quad (4.27)$$

where the *true* correlations are replaced by their estimates calculated on time averages.

Remark The matrix $\mathbf{X}^H\mathbf{X} \in (\mathbb{R},\mathbb{C})^{M\times M}$ that appears in the previous expressions is the correlation matrix defined in (4.23) and is square, nonsingular, and semi-positive definite. Therefore, even if the system of N equations in M unknowns admits no unique solution, it is possible to identify a single solution corresponding to that optimal in the LS sense. Using other criteria it is possible to find other solutions. Furthermore, observe that the matrix $\mathbf{X}^{\#} = (\mathbf{X}^H\mathbf{X})^{-1}\mathbf{X}^H$ appearing in (4.21) is defined as the *Moore–Penrose pseudoinverse* matrix for the case of overdetermined system that will be better defined later in Appendix A (see Sect. A.3.2) [8–11].

The previous development, if we consider the time-average operator instead of the expectation operator, shows that the LSE and MMSE formalisms are similar

$$\hat{E}\{\cdot\} = \sum_{n=n_1}^{n_2}(\cdot), \quad \textit{time-average operator} \rightarrow \text{LSE},$$

$$E\{\cdot\} = \int(\cdot), \quad \textit{expectation operator} \rightarrow \text{MMSE}.$$

It follows that for an ergodic process, the LSE solution tends to that of Wiener optimal solution for N sufficiently large.

4.2.2.3 Minimum Error Energy

The minimum error energy $E_{\text{LS}} \equiv J_{\min}(\mathbf{w}) \triangleq J(\mathbf{w})\big|_{\mathbf{w}=\mathbf{w}_{\text{LS}}}$ can be obtained by substituting the optimal LS solution $\mathbf{w}_{\text{LS}} = (\mathbf{X}^H\mathbf{X})^{-1}\mathbf{X}^H\mathbf{d}$ in (4.18). Therefore, we have

$$\begin{aligned} J_{\min}(\mathbf{w}) &= \mathbf{d}^H\mathbf{d} - \mathbf{d}^H\mathbf{X}\left(\mathbf{X}^H\mathbf{X}\right)^{-1}\mathbf{X}^H\mathbf{d} \\ &= \mathbf{d}^T\left[\mathbf{I} - \mathbf{X}\left(\mathbf{X}^H\mathbf{X}\right)^{-1}\mathbf{X}^H\right]\mathbf{d} \\ &= E_d - \mathbf{X}^H\mathbf{d}\mathbf{w}_{\text{LS}}. \end{aligned} \tag{4.28}$$

In terms of estimated correlations, the above can be written as

$$\begin{aligned} J_{\min}(\mathbf{w}) &= E_d - \mathbf{R}_{xd}^H\mathbf{R}_{xx}^{-1}\mathbf{R}_{xd} \\ &= E_d - \mathbf{R}_{xd}^H\mathbf{w}_{\text{LS}}. \end{aligned} \tag{4.29}$$

4.2.3 Implementing Notes and Time Indices

In real applications based on the LS methodology, the observed data samples block can be defined by a window (said *sliding window*) of appropriate length N, which flows on the signal. Its length is determined on the basis of the nature of the problem. Moreover, in the data matrix **X** definition for development of both theoretical analysis and calculation codes, it is necessary to determine accurately the various temporal indices.

In practical cases, the CF to be minimized, defined in (4.17), should be considered as causal and then with a time window of N samples back from the current time indicated with n. In other words, we consider the upper bound equal to the last time index $n_2 = n$, and a lower bound equal to $n_1 = n - N + 1$. The expression (4.17) is then rewritten as

$$J(\mathbf{w}) = \sum_{k=n-N+1}^{n} |e[k]|^2 = \|\mathbf{d} - \mathbf{X}\mathbf{w}\|_2^2. \tag{4.30}$$

With this convention, the vectors (4.9)–(4.12) are redefined as

$$\mathbf{e} = \begin{bmatrix} e[n] & e[n-1] & \cdots & e[n-N+1] \end{bmatrix}^T, \quad \textit{error vector}, \tag{4.31}$$

$$\mathbf{d} = \begin{bmatrix} d[n] & d[n-1] & \cdots & d[n-N+1] \end{bmatrix}^T, \quad \textit{desired output vector}, \tag{4.32}$$

$$\mathbf{x} = \begin{bmatrix} x[n] & x[n-1] & \cdots & x[n-M+1] \end{bmatrix}^T, \quad \textit{filter input}, \tag{4.33}$$

$$\mathbf{x}[n] = \begin{bmatrix} x[n] & x[n-1] & \cdots x[n-N+1] \end{bmatrix}^T, \quad \textit{measurement interval}. \tag{4.34}$$

End of the expression $\mathbf{e} = \mathbf{d} - \mathbf{X}\mathbf{w}$ is rewritten in extended mode as

$$\begin{bmatrix} e[n] \\ e[n-1] \\ \vdots \\ e[n-N+1] \end{bmatrix} = \begin{bmatrix} d[n] \\ d[n-1] \\ \vdots \\ d[n-N+1] \end{bmatrix} - \begin{bmatrix} x[n] & x[n-1] & \cdots & x[n-M+1] \\ x[n-1] & x[n-2] & \cdots & x[n-M] \\ \vdots & \vdots & \ddots & \vdots \\ x[n-N+1] & 0 & \cdots & 0 \end{bmatrix} \begin{bmatrix} w[0] \\ w[1] \\ \vdots \\ w[M-1] \end{bmatrix}. \tag{4.35}$$

Usually, the theoretical development refers to temporal indices k defined as $k \in [0, N-1]$. In practice, for the causality of the entire system, it is necessary to consider the relation

$$J(\mathbf{w}) = \sum_{m=n-N+1}^{n} |e[m]|^2 \equiv \sum_{k=0}^{N-1} |e[k]|^2, \tag{4.36}$$

i.e., subtracting the term $(n - N + 1)$ at the two extremes of the first summation. That is the conventional relationship between the indexes

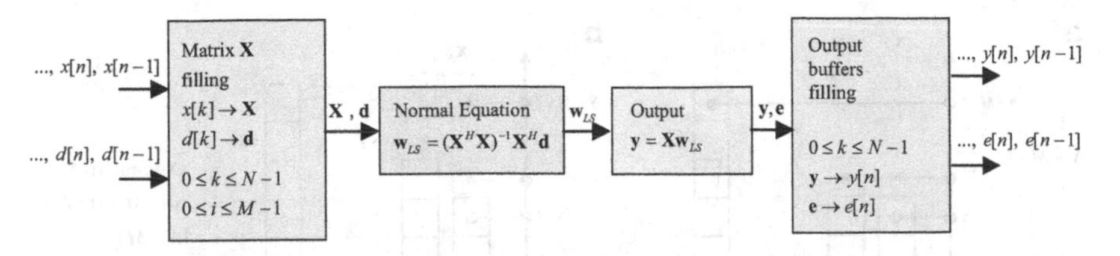

Fig. 4.5 Block-wise implementation scheme of the LS method

$$k = m - (n - N + 1). \tag{4.37}$$

It follows that the index k is in the range $0 \le k \le N - 1$. Note that, with the convention (4.37), the expression (4.35) is rewritten with an equivalent representation as

$$\begin{bmatrix} e[N-1] \\ \vdots \\ e[1] \\ e[0] \end{bmatrix} = \begin{bmatrix} d[N-1] \\ \vdots \\ d[1] \\ d[0] \end{bmatrix} - \begin{bmatrix} x[N-1] & \cdots & x[N-M+1] & x[N-M] \\ \vdots & \vdots & x[N-M] & x[N-M-1] \\ x[1] & x[0] & \ddots & \vdots \\ x[0] & 0 & \cdots & 0 \end{bmatrix} \begin{bmatrix} w[0] \\ w[1] \\ \vdots \\ w[M-1] \end{bmatrix}. \tag{4.38}$$

With reference to the scheme of Fig. 4.5, an important aspect in the LS method regards the choice of the data matrix **X** which can be made in different ways.

4.2.3.1 Data Matrix X from Single Sensor

From the expression (4.35), the data matrix **X** is defined considering that each column vector or each row vector corresponds to the same signal shifted by one sample. According to the windowing performed on the input data, there are various methods, illustrated schematically in Fig. 4.6a, for the choice of the data matrix **X**.

Post-windowing Method

The method known as post-windowing, shown in box (1) of Fig. 4.6a, already implicitly described by the expression (4.35), is one in which the data matrix is defined as $\mathbf{X} \in (\mathbb{R},\mathbb{C})^{N \times M}$, i.e.,

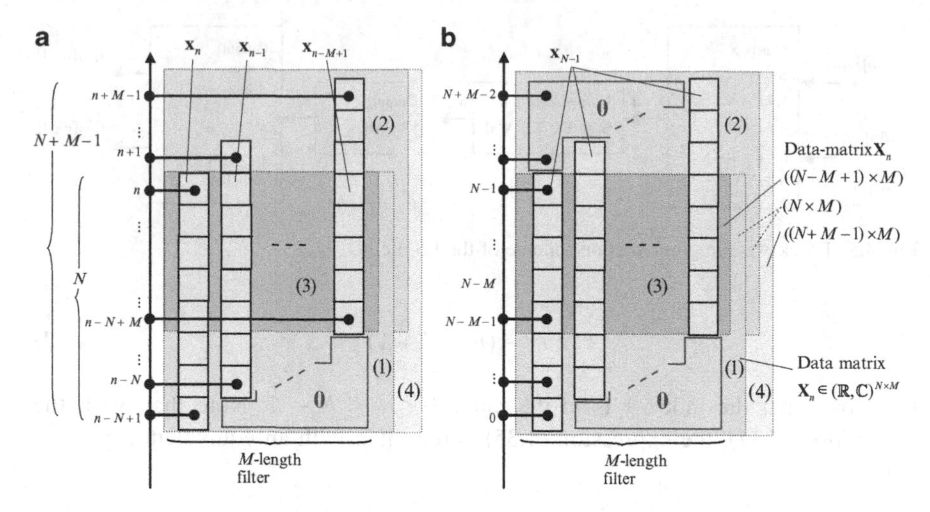

Fig. 4.6 Data matrix **X** filling in the case of *sliding window* for a single sensor signal: (**a**) time index n; (**b**) conventional time index $0 \le k \le N-1$

$$\mathbf{X} \triangleq \begin{bmatrix} \mathbf{x}_n^T \\ \mathbf{x}_{n-1}^T \\ \vdots \\ \mathbf{x}_{n-N+1}^T \end{bmatrix} = \begin{bmatrix} \mathbf{x}^T[n] \\ \mathbf{x}^T[n-1] \\ \vdots \\ \mathbf{x}^T[n-M+1] \end{bmatrix}^T$$
$$= \begin{bmatrix} x[n] & x[n-1] & \cdots & x[n-M+1] \\ x[n-1] & x[n-2] & \cdots & \vdots \\ \vdots & \vdots & \ddots & x[n-N-M+1] \\ x[n-N+1] & x[n-N] & \cdots & x[n-N-M+2] \end{bmatrix}, \tag{4.39}$$

where the row and column vectors are defined in (4.33) and (4.34).

Remark For (4.39), the filter output $\mathbf{y} = \mathbf{X}\mathbf{w}$ can be expressed as

$$y[k] = \mathbf{w}^H \mathbf{x}_k, \quad k = n, n-1, \ldots, n-N+1, \tag{4.40}$$

or as $\mathbf{y} = \mathbf{w}^H \mathbf{X}$, so we can write

$$\mathbf{y} = \begin{bmatrix} w[0] & \cdots & w[M-1] \end{bmatrix} \begin{bmatrix} \mathbf{x}[0] & \cdots & \mathbf{x}[M-1] \end{bmatrix}^T = \sum_{k=0}^{M-1} w[k]\mathbf{x}[k]. \tag{4.41}$$

In (4.40) the row vectors $\mathbf{x}_k^T$ are used, while in (4.41), the output is interpreted as a linear combination of the column vectors $\mathbf{x}[k]$ of the data matrix **X**.

Covariance Method

In the covariance method, no assumptions are made on the data outside of the N-length window analysis. The data matrix $\mathbf{X} \in (\mathbb{R},\mathbb{C})^{(N-M+1)\times M}$ is determined by the filling schema of the box (3) of Fig. 4.6a. It is then

$$\mathbf{X} \triangleq \begin{bmatrix} x[n] & x[n-1] & \cdots & x[n-M+2] & x[n-M+1] \\ x[n-1] & x[n-2] & \cdots & x[n-M+1] & x[n-M] \\ \vdots & \vdots & \ddots & \vdots & \vdots \\ x[n-N+M] & x[n-N+M-1] & \cdots & x[n-N+2] & x[n-N+1] \end{bmatrix}. \tag{4.42}$$

Pre- and Post-windowing or Autocorrelation Method

As shown in Fig. 4.6a, in the case that both windowing sides are considered, the data matrix $\mathbf{X}$ has dimension equal to $((N+M-1)\times M)$. Assuming zero by definition the data outside the range of measurement, $\mathbf{X} \in (\mathbb{R},\mathbb{C})^{(N+M-1)\times M}$ is defined as

$$\mathbf{X} = \begin{bmatrix} x[n+M-1] & x[n+M-2] & \dots & x[n] \\ x[n+M-2] & x[n+M-3] & \dots & x[n-1] \\ \vdots & \vdots & \therefore & \\ x[n+1] & x[n] & \cdots & x[n-M+2] \\ \hline x[n] & x[n-1] & \dots & x[n-M+1] \\ \vdots & \vdots & \therefore & \vdots \\ x[n-M+1] & x[n-M] & \cdots & x[n-2M] \\ x[n-M] & x[n-M-1] & \cdots & x[n-2M+1] \\ \hline \vdots & \vdots & \therefore & \vdots \\ x[n-N+2] & x[n-N+1] & \cdots & x[n-N-M+3] \\ x[n-N+1] & x[n-N] & \cdots & x[n-N-M+2] \end{bmatrix} \begin{matrix} \text{pre-wind. (2)} \\ \text{covar. (3)} \\ \text{post-wind. (1)} \\ \text{auto-corr. (4)} \end{matrix} \tag{4.43}$$

As shown in Fig. 4.6, in the previous expression, all the possible ways of choosing the type of windowing have been explicitly shown. The elements relating to data outside the range of measurement, i.e., with index $k > n$ and $k \leq (n-N)$, are zero.

Remark In the case that $N \gg M$, the covariance and autocorrelation techniques are coincident.

4.2.3.2 Data Matrix X from Sensors Array

In the case of array of sensors the data matrix $\mathbf{X}_n \in (\mathbb{R},\mathbb{C})^{N\times M}$ consists of columns that contain the signals from the individual sensors called *data records*. Each line contains the samples of all the sensors for each time instant that are called *snapshots*. That is then the following convention:

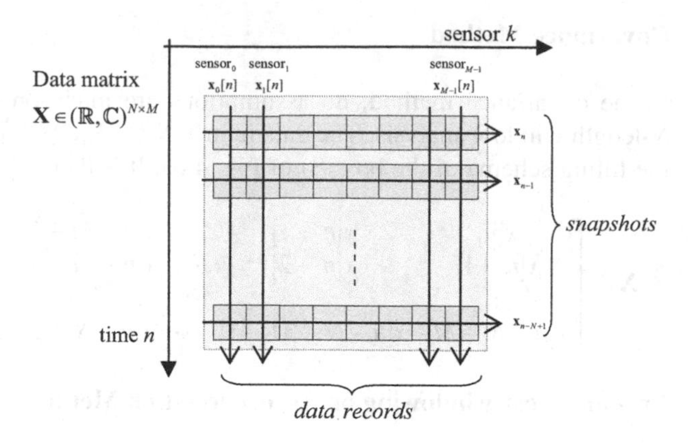

Fig. 4.7 Data matrix **X** definition in the case of a sensor' array. The columns of **X** contain the regressions samples from various sensors (*data records*). The rows contain the data of all the sensors in a certain time instant (*snapshots*)

$$\mathbf{X} \triangleq \begin{bmatrix} x_0[n] & x_1[n] & \cdots & x_{M-1}[n] \\ x_0[n-1] & x_1[n-1] & \cdots & x_{M-1}[n-1] \\ \vdots & \vdots & \cdots & \vdots \\ x_0[n-N+1] & x_1[n-N+1] & \cdots & x_{M-1}[n-N+1] \end{bmatrix}. \tag{4.44}$$

With reference to Fig. 4.7, we define *data record* vector

$$\mathbf{x}_k[n] \triangleq \begin{bmatrix} x_k[n] & x_k[n-1] & \cdots & x_k[n-N+1] \end{bmatrix}^T, \quad kth\ data\ record \tag{4.45}$$

corresponding to the kth column of the matrix, containing N signal samples between the extremes $[n, n-N+1]$ coming from the kth sensor. We define *snapshot*, the vector of the kth row of the matrix **X**, i.e.,

$$\mathbf{x}_k \triangleq \begin{bmatrix} x_0[k] & x_1[k] & \cdots & x_{M-1}[k] \end{bmatrix}^T, \quad snapshot\ kth\ time\ instant \tag{4.46}$$

containing samples at the kth instant from the M sensors. It follows that the matrix data (4.44) can be defined directly by the *snapshot* or *data record* vectors as

$$\mathbf{X} \triangleq \begin{bmatrix} \mathbf{x}_0[n] & \mathbf{x}_1[n] & \cdots & \mathbf{x}_{M-1}[n] \end{bmatrix} = \begin{bmatrix} \mathbf{x}_n & \mathbf{x}_{n-1} & \cdots & \mathbf{x}_{n-N+1} \end{bmatrix}^T. \tag{4.47}$$

4.2.4 Geometric Interpretation and Orthogonality Principle

An interesting geometric interpretation of the LS criterion can be made by considering the desired output vector **d** and the column vectors $\mathbf{x}[k] \in \mathbf{X}$, $0 \le k \le M-1$, as vectors of an N-dimensional space with inner product and lengths defined, respectively, as

$$\langle \mathbf{x}[i], \mathbf{x}[j] \rangle \triangleq \mathbf{x}^H[i]\mathbf{x}[j], \tag{4.48}$$

$$\left\| \mathbf{x}[i] \right\|_2^2 \triangleq \langle \mathbf{x}^H[i], \mathbf{x}[i] \rangle = E_x. \tag{4.49}$$

As indicated by (4.41) the output vector of the filter appears to be a linear combination of linearly independent column vectors of $\mathbf{X}$, or

$$\mathbf{y} = \sum_{k=0}^{M-1} w[k]\mathbf{x}[k], \tag{4.50}$$

so the M linearly independent vectors $\mathbf{x}[k]$ form an M-dimensional subspace, which in *linear algebra* is defined as the *column space* or *image* or *range* (see Sect. A.6.1). The dimension of the column space is called the *rank* of the matrix. This space indicated as $\mathcal{R}(\mathbf{X})$ is defined as the set of all possible linear combinations of linearly independent column vectors of the matrix $\mathbf{X}$, for which the filter output $\mathbf{y}$ lies in that space.

Note that in the context of the *estimation theory* $\mathcal{R}(\mathbf{X})$ is referred to as *estimation space.*

4.2.4.1 Orthogonality Principle

The desired output vector $\mathbf{d}$ lies outside of the estimation space. The error vector $\mathbf{e}$, given by the distance between the vectors $\mathbf{d}$ and $\mathbf{y}$, is minimal when it is perpendicular to the estimation space itself, i.e., $\min\{\mathbf{e}\} \therefore \mathbf{e} \perp \mathcal{R}(\mathbf{X})$. With this assumption, defined as the *orthogonality principle*, it appears that

$$\langle \mathbf{x}[k], \mathbf{e} \rangle = \mathbf{x}^H[k]\mathbf{e} = 0, \quad \text{for} \quad 0 \le k \le M-1, \tag{4.51}$$

that is, considering all the columns, we can write

$$\mathbf{X}^H(\mathbf{d} - \mathbf{X}\mathbf{w}_{\text{LS}}) = \mathbf{0}. \tag{4.52}$$

Rearranging the previous expression is then

$$\mathbf{X}^H\mathbf{X}\mathbf{w}_{\text{LS}} = \mathbf{X}^H\mathbf{d}. \tag{4.53}$$

The latter is precisely the Yule–Walker normal equations derived through the orthogonality principle. As in the case of MMSE presented in Chap. 3, the geometric interpretation and the imposition of orthogonality between the vectors represent a very powerful tool for the optimal solutions calculation of and for the determination of important properties.

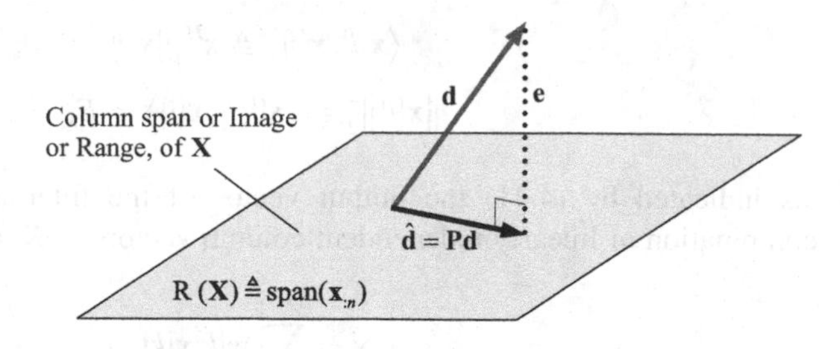

Fig. 4.8 Interpretation of LS solution as projection operator

4.2.4.2 Projection Operator and Column Space of X

An alternative interpretation of the LS solution is that obtained from the definition of the projection operator $\mathbf{P}$ of the matrix $\mathbf{X}$. Consider the vector $\hat{\mathbf{d}}$ as a projection of the vector $\mathbf{d}$ on the column space $\mathcal{R}(\mathbf{X})$ as shown in Fig. 4.8.

Indicating the set of all linearly independent column vectors of the data matrix $\mathbf{X}$ as $\mathbf{x}_{:n} \triangleq \begin{bmatrix} \mathbf{x}[n] & \cdots & \mathbf{x}[n-M'+1] \end{bmatrix}$, then $\mathcal{R}(\mathbf{X}) \triangleq \text{span}(\mathbf{x}_{:n})$. The vector $\hat{\mathbf{d}} = \mathbf{P}\mathbf{d}$ is then, by definition, characterized by the following properties:

- $\hat{\mathbf{d}}$ is obtained from the linear combination of column vectors $\mathbf{x}_{:n}$
- Among all vectors $\text{span}(\mathbf{x}_{:n})$, $\hat{\mathbf{d}}$ is that at minimum Euclidean distance from $\mathbf{d}$
- The difference $\mathbf{e} = \mathbf{d} - \hat{\mathbf{d}}$ is orthogonal to the space $\mathcal{R}(\mathbf{X})$

Note that the previous three properties correspond with the properties of orthogonality, as described in Sect. 3.3.5 satisfied by the vectors $\mathbf{y}$ and $\mathbf{e}$ of the LS filter. In fact, $\mathbf{y}$ is a linear combination of the vectors $\mathbf{x}[n]$. Moreover, $\mathbf{y}$ is obtained by the minimization of $\mathbf{e}^H\mathbf{e}$, where $\mathbf{e} = \mathbf{d} - \mathbf{y}$, so it is equivalent to the Euclidean distance minimization between $\mathbf{d}$ and $\mathbf{y}$. Finally, for the orthogonality principle, $\mathbf{e}$ is orthogonal to the space described by $\mathbf{x}_{:n}$, for which $\mathbf{y}$ represents the projection of $\mathbf{d}$ in the space described by the vectors $\mathbf{x}_{:n}$, or

$$\mathbf{y} = \mathbf{P}\mathbf{d}. \tag{4.54}$$

Since by definition $\mathbf{y} = \mathbf{X}\mathbf{w}_{\text{LS}}$, remembering that $\mathbf{w}_{\text{LS}} = (\mathbf{X}^H\mathbf{X})^{-1}\mathbf{X}^H\mathbf{d}$ it follows that the projection operator $\mathbf{P}$, related to the input data matrix $\mathbf{X}$ (see Sect. A6.5), is defined as

$$\mathbf{P} \in (\mathbb{R}, \mathbb{C})^{N\times N} \triangleq \mathbf{X}\left(\mathbf{X}^H\mathbf{X}\right)^{-1}\mathbf{X}^H, \quad \textit{projection operator}. \tag{4.55}$$

It is easy to show that the following properties are valid:

$$\begin{aligned} \mathbf{P} &= \mathbf{P}^H \\ \mathbf{P}^2 &= \mathbf{P}\mathbf{P}^H = \mathbf{P}, \end{aligned} \tag{4.56}$$

and also,

$$\mathbf{e} = (\mathbf{I} - \mathbf{P})\mathbf{y}, \tag{4.57}$$

wherein the matrix $(\mathbf{I} - \mathbf{P})$ is defined as a *orthogonal projection complement operator* (see Sect. A.6.5 for more details).

4.2.4.3 LS Solution Property

The LS system solution (4.20) has the following properties:

- The $\mathbf{w}_{\text{LS}}$ solution is unique if the matrix $\mathbf{X}$ is full rank, or $M' = M$ (all of its columns are linearly independent and, necessarily, $N \geq M$). In this case the solution is equal to $\mathbf{w}_{\text{LS}} = (\mathbf{X}^H\mathbf{X})^{-1}\mathbf{X}^H\mathbf{d}$
- If the solution is not unique, as in the underdetermined system case, it is possible to identify an appropriate solution (among the infinite). Solution, with minimum Euclidean norm $\|\mathbf{w}\|_2^2$, is obtainable considering a constrained optimization criterion defined as

$$\mathbf{w}_* \therefore \min_{\mathbf{w}} \|\mathbf{w}\|_2^2 \quad \text{subject to (s.t.)} \quad \mathbf{X}\mathbf{w} = \mathbf{d}. \tag{4.58}$$

The solution of the *constrained optimization problem*[2] (4.58) is equal to

$$\mathbf{w}_* = \mathbf{X}^H \left(\mathbf{X}\mathbf{X}^H\right)^{-1} \mathbf{d}. \tag{4.59}$$

Equation (4.59) can be demonstrated in various ways including the method of *Lagrange multipliers* discussed in Sect. 4.3.1.2 (see also Sect. B.3.2), or through matrix decomposition methods such as singular value decomposition (SVD) which will be discussed in Sect. 4.4.3.1.

4.2.5 LS Variants

The LS methodology is represented by a broad class of algorithms that includes several variants. In this section, we present some of them used to define more accurate solutions in the case that certain information is *a priori* known. The variants discussed are related to the CF definition that, in addition to the normal equations, may contain other constraints. Typically, such constraints are defined on the basis of knowledge about the nature of the measurement noise and/or based on *a priori* known assumptions about the optimal solution.

[2] Note that the *constrained optimization* is a methodology used very often for the determination of the optimal solution in particular adaptive filtering problems. See Appendix B for deepening.

Other variants, defined by the algebraic nature of the LS, will be discussed later in this chapter (see Sect. 4.4) and allow to define more robust and efficient computing structures [7, 12, 19, 20, 32].

4.2.5.1 Weighted Least Squares

A first variant of the LS method, which allows the use of any known information about the nature of the measurement noise, and allows a more accurate estimate of **w**, is the one called weighted least squares (WLS).

The hypothesis is to weight less the errors in the instants where the noise contribution is high. By defining $g_n \geq 0$, the weighing coefficient of the nth instantaneous error, we can write

$$\begin{aligned} J(\mathbf{w}) &= \sum_{n=0}^{N-1} g_n \left| e[n] \right|^2 \\ &= \mathbf{e}^H \mathbf{G} \mathbf{e}, \end{aligned} \tag{4.60}$$

where $\mathbf{G} \in \mathbb{R}^{N \times N}$ is a positive-definite diagonal matrix, called *weighing matrix*,

$$\mathbf{G} \triangleq \operatorname{diag}[g_k \geq 0, \quad k = 0, 1, \ldots, N-1] \tag{4.61}$$

of elements chosen with a value inversely proportional to the level of the noise. For compactness of notation, the weighted norm is often indicated in the form $\left\|\mathbf{e}\right\|_{\mathbf{G}}^2 \triangleq \left\|\mathbf{e}^H \mathbf{G} \mathbf{e}\right\|_2$. The CF to be minimized is equal to $\left\|(\mathbf{d} - \mathbf{X}\mathbf{w})\right\|_{\mathbf{G}}^2$ or, in other words,

$$\begin{aligned} J(\mathbf{w}) &= (\mathbf{d} - \mathbf{X}\mathbf{w})^H \mathbf{G} (\mathbf{d} - \mathbf{X}\mathbf{w}) = \mathbf{G} \left\|(\mathbf{d} - \mathbf{X}\mathbf{w})\right\|_2^2 \\ &= \left\|\mathbf{G}^{1/2} \mathbf{e}\right\|_2^2. \end{aligned} \tag{4.62}$$

This function corresponds to the *negative likelihood* when the noise is Gaussian and characterized by a covariance matrix equal to $\mathbf{G}^{-1}$.

From (4.62), differentiating and setting to zero $\nabla J(\mathbf{w}) = \mathbf{0}$, it is immediate to derive the linear system of equations for which the normal equation, in the over-determined case, takes the simple form

$$\mathbf{X}^H \mathbf{G} \mathbf{X} \mathbf{w} = \mathbf{X}^H \mathbf{G} \mathbf{d}, \tag{4.63}$$

with solution

$$\mathbf{w}_{\text{WLS}} = \left(\mathbf{X}^H \mathbf{G} \mathbf{X}\right)^{-1} \mathbf{X}^H \mathbf{G} \mathbf{d}. \tag{4.64}$$

In case $\mathbf{G} = \mathbf{I}$, the previous expression is, in fact, identical to (4.21). It is easily shown, moreover, that the minimum error energy is equal to

$$J_{\min}(\mathbf{w}) = \mathbf{d}^H(\mathbf{I} - \mathbf{P}_G)\mathbf{d}, \tag{4.65}$$

where $\mathbf{P}_G$ is the *weighed projection operator* (WPO) defined as

$$\mathbf{P}_G \triangleq \mathbf{X}\left(\mathbf{X}^H\mathbf{G}\mathbf{X}\right)^{-1}\mathbf{X}^H\mathbf{G}. \tag{4.66}$$

Remark The kth parameter of the $\mathbf{G}$ matrix can be interpreted as *weighing factor* of the kth equation of the LS system: if $g_k = 0$, for $N > M$, the kth equation is not taken into consideration. For example, in the case of spectral estimation the coefficients g_k of the weighing matrix may be determined on the basis of the presence of noise on the kth signal window corresponding to the kth LS equation (measurement noise): weighing less equations most noise contaminated would be a more robust spectral estimation.

Gauss–Markov Best Linear Unbiased Estimator

In the case of Gaussian noise is easy to see that the best choice of the weighing matrix is the inverse of the noise covariance matrix (indicated as $\mathbf{R}_{ee}^{-1}$). In this case, assuming zero-mean Gaussian measure noise the optimal weighing matrix is equal to

$$\mathbf{G} \triangleq \operatorname{diag}\left[\left(1/\sigma_k^2\right) \geq 0, \quad k = 0, 1, \ldots, N-1\right], \tag{4.67}$$

where with σ_k^2 is indicated the noise power relative to the kth equation. Therefore, we have that $\mathbf{G}^{-1} = \mathbf{R}_{ee} = E\left\{\mathbf{e}\mathbf{e}^H\right\}$ and the LS solution is

$$\mathbf{w}_{\text{BLUE}} = \left(\mathbf{X}^H\mathbf{R}_{ee}^{-1}\mathbf{X}\right)^{-1}\mathbf{X}^H\mathbf{R}_{ee}^{-1}\mathbf{d}, \tag{4.68}$$

and the more noisy equations would weigh less in the estimation of the parameters $\mathbf{w}$.

Remark With this choice (4.67) of the weighting matrix, the estimator, the *best* achievable, is called best linear unbiased estimator (BLUE).

4.2.5.2 Regularized LS

A second variant of the LS method is one that incorporates a certain additive term, called *regularization term*, on CF, so as to optimization is formulated a

$$J(\mathbf{w}) = \delta J_s(\mathbf{w}) + \hat{J}(\mathbf{w}), \tag{4.69}$$

where $\delta > 0$ is a suitable constant, $\hat{J}(\mathbf{w})$ is the noise energy (the usual CF), and the term $\delta J_s(\mathbf{w})$ is the *smoothness constraint* (also called energy stabilizer), which is usually some weights $\mathbf{w}$ function. A general and typical choice in LS problems is to define the CF as

$$J(\mathbf{w}) = \left\|\mathbf{w} - \overline{\mathbf{w}}\right\|_{\mathbf{\Pi}}^2 + \left\|\mathbf{d} - \mathbf{X}\mathbf{w}\right\|_2^2, \tag{4.70}$$

for which the term $\delta J_s(\mathbf{w})$ is defined as

$$\left\|\mathbf{w} - \overline{\mathbf{w}}\right\|_{\mathbf{\Pi}}^2 \triangleq (\mathbf{w} - \overline{\mathbf{w}})^H \mathbf{\Pi} (\mathbf{w} - \overline{\mathbf{w}}),$$

where $\mathbf{\Pi} \in \mathbb{R}^{M \times M}$ represents a weighing matrix that, in general, takes the form

$$\mathbf{\Pi} = \delta \mathbf{I}.$$

In practice, the optimization problem is formulated as

$$\mathbf{w}_* = \arg\min_{\mathbf{w}} \left[(\mathbf{w} - \overline{\mathbf{w}})^H \mathbf{\Pi} (\mathbf{w} - \overline{\mathbf{w}}) + \left\|\mathbf{d} - \mathbf{X}\mathbf{w}\right\|_2^2 \right]. \tag{4.71}$$

Unlike the SSE (4.17), the expression (4.70) contains the term $(\mathbf{w} - \overline{\mathbf{w}})^H \mathbf{\Pi} (\mathbf{w} - \overline{\mathbf{w}})$, where $\mathbf{\Pi}$ is positive definite and generally chosen as a multiple of the identity matrix, and $\overline{\mathbf{w}}$ is *a priori* known column vector.

Equation (4.70) allows to incorporate *a priori* knowledge on the solution $\mathbf{w}$. Suppose that $\mathbf{\Pi} = \delta \mathbf{I}$ and that δ is a large positive number. In this situation the first term of CF in (4.71) will assume a dominant value, i.e., the CF is "more minimized" when the distance between the vectors $\mathbf{w}$ and $\overline{\mathbf{w}}$ tends to a minimum. A large $\mathbf{\Pi}$ value assumes the significance of a high degree of confidence that the vector $\overline{\mathbf{w}}$ is near the optimum. In other words, for large $\mathbf{\Pi}$ it follows $\mathbf{w} \rightarrow \overline{\mathbf{w}}$. On the contrary, a small $\mathbf{\Pi}$ value implies a high degree of uncertainty on the initial hypothesis $\overline{\mathbf{w}}$.

The solution of (4.71) can be determined in various ways. In order to fulfill the direct differentiation with respect to $\mathbf{w}$, as done in the general description in Sect. 4.2.2.2, the change of variable $\mathbf{z} = \mathbf{w} - \overline{\mathbf{w}}$ and $\mathbf{b} = \mathbf{d} - \mathbf{X}\overline{\mathbf{w}}$ is introduced. For which the CF (4.71) becomes

$$J(\mathbf{z}) = \mathbf{z}^H \mathbf{\Pi} \mathbf{z} + \left\|\mathbf{b} - \mathbf{X}\mathbf{z}\right\|_2^2. \tag{4.72}$$

Differentiating and setting to zero we have that

$$\nabla J(\mathbf{z}) = \mathbf{\Pi} \mathbf{z} - \mathbf{X}^H (\mathbf{b} - \mathbf{X}\mathbf{z}) \equiv \mathbf{0},$$

for which the normal equations take the form

$$(\mathbf{\Pi} + \mathbf{X}^H \mathbf{X})(\mathbf{w} - \overline{\mathbf{w}}) = \mathbf{X}^H (\mathbf{d} - \mathbf{X}\overline{\mathbf{w}}), \tag{4.73}$$

with solution

$$\mathbf{w}_* = \overline{\mathbf{w}} + \left(\mathbf{\Pi} + \mathbf{X}^H\mathbf{X}\right)^{-1}\mathbf{X}^H(\mathbf{d} - \mathbf{X}\overline{\mathbf{w}}). \tag{4.74}$$

Finally, you can easily demonstrate that the minimum energy of error is

$$J_{\min}(\mathbf{w}) = (\mathbf{d} - \mathbf{X}\overline{\mathbf{w}})^H\left[\mathbf{I} + \mathbf{X}\mathbf{\Pi}^{-1}\mathbf{X}^H\right]^{-1}(\mathbf{d} - \mathbf{X}\overline{\mathbf{w}}). \tag{4.75}$$

4.2.5.3 Regularization and Ill-Conditioning of the R_{xx} Matrix

Another reason to introduce the regularization term is due to the fact that the measurement data noise, in combination with the likely ill-conditioning of the $\mathbf{X}^H\mathbf{X}$ matrix, can determine a high deviation from the correct solution.

The Russian mathematician Tikhonov was perhaps the first to study the problem of deviation from the true solution in terms of *regularization*. The problem is posed in the definition of a criterion for the selection of an approximate solution among a set of feasible solutions.

The basic idea of the *Tikhonov's regularization theory* consists in the determination of a compromise between a solution faithful to the noisy data and a solution based on *a priori* information available about the nature of the data (for example, knowledge of the model, the order of generation of the data, the statistics of the noise, etc.). In other words, the regularization imposes a smoothness constraint on the set of possible solutions.

In case there is no initial hypothesis on the solution, and there is only the problem of ill-conditioning of the matrix $\mathbf{X}^H\mathbf{X}$, in (4.70) arises $\overline{\mathbf{w}} = \mathbf{0}$ and $\mathbf{\Pi} = \delta\mathbf{I}$. In this case, the CF assumes the form

$$J(\mathbf{w}) = \delta\|\mathbf{w}\|_2^2 + \|\mathbf{d} - \mathbf{X}\mathbf{w}\|_2^2 \qquad \text{with} \qquad \delta > 0. \tag{4.76}$$

Some properties of the smoothness constraint may be determined by considering the gradient of CF (4.76), for which we can write

$$\nabla J(\mathbf{w}) = \mathbf{X}^H(\mathbf{d} - \mathbf{X}\mathbf{w}) - \delta\mathbf{w} \equiv \mathbf{0}. \tag{4.77}$$

From the above may be derived the normal equations in the form[3]

$$\left(\mathbf{X}^H\mathbf{X} + \delta\mathbf{I}\right)\mathbf{w} = \mathbf{X}^H\mathbf{d}, \tag{4.78}$$

with solution

$$\mathbf{w}_* = \left(\mathbf{X}^H\mathbf{X} + \delta\mathbf{I}\right)^{-1}\mathbf{X}^H\mathbf{d}. \tag{4.79}$$

[3] Note that this solution is equivalent to the *δ-solution* described in Sect. 4.3.1.2.

Note that the condition number of the matrix $(\mathbf{X}^H\mathbf{X} + \delta\mathbf{I})$ is given by

$$\chi(\mathbf{X}^H\mathbf{X} + \delta\mathbf{I}) = \frac{\lambda_{\max} + \delta}{\lambda_{\min} + \delta}, \tag{4.80}$$

with $\lambda_{\max}$ and $\lambda_{\min}$, respectively, the maximum and minimum eigenvalues of $\mathbf{X}^H\mathbf{X}$. It follows that the number

$$\chi(\mathbf{X}^H\mathbf{X} + \delta\mathbf{I}) < \chi(\mathbf{X}^H\mathbf{X}), \tag{4.81}$$

so, if for example $\lambda_{\max} = 1$ and $\lambda_{\min} = 0.01$ by choosing a value of $\delta = 0.1$, the condition number improves by a factor of 10 (from 100 to 10).

In other words, as asserted, the term $\delta\|\mathbf{w}\|^2$ acts as a stabilizer and prevents too deviated solutions.

4.2.5.4 Weighed and Regularized LS

In the case both *a priori* knowledge on the noise and assumptions on the solution are available, a CF that takes into account the knowledge can be defined as

$$\begin{aligned} \mathbf{w}_* &= \arg\min_{\mathbf{w}} \left[\|\mathbf{w} - \overline{\mathbf{w}}\|_{\boldsymbol{\Pi}}^2 + \|\mathbf{d} - \mathbf{X}\mathbf{w}\|_{\mathbf{G}}^2 \right] \\ &= \arg\min_{\mathbf{w}} \left[(\mathbf{w} - \overline{\mathbf{w}})^H \boldsymbol{\Pi} (\mathbf{w} - \overline{\mathbf{w}}) + (\mathbf{d} - \mathbf{X}\mathbf{w})^H \mathbf{G} (\mathbf{d} - \mathbf{X}\mathbf{w}) \right], \end{aligned} \tag{4.82}$$

so, by differentiating and setting to zero, the normal equations are defined as

$$(\boldsymbol{\Pi} + \mathbf{X}^H\mathbf{G}\mathbf{X})(\mathbf{w} - \overline{\mathbf{w}}) = \mathbf{X}^H\mathbf{G}(\mathbf{d} - \mathbf{X}\mathbf{w}), \tag{4.83}$$

with solution

$$\mathbf{w}_* = \overline{\mathbf{w}} + \left(\boldsymbol{\Pi} + \mathbf{X}^H\mathbf{G}\mathbf{X}\right)^{-1}\mathbf{X}^H\mathbf{G}(\mathbf{d} - \mathbf{X}\overline{\mathbf{w}}), \tag{4.84}$$

with minimum of CF

$$J_{\min}(\mathbf{w}) = (\mathbf{d} - \mathbf{X}\overline{\mathbf{w}})^H \left[\mathbf{G}^{-1} + \mathbf{X}\boldsymbol{\Pi}^{-1}\mathbf{X}^H\right]^{-1}(\mathbf{d} - \mathbf{X}\overline{\mathbf{w}}). \tag{4.85}$$

4.2.5.5 Linearly Constrained LS

The formulation of the LS method may be subject to constraints due to the specific needs of the problem. For example, constraints can be used to avoid trivial solutions or in order to formalize some knowledge *a priori* available. If the constraints are expressed with a linear relationship of the type $\mathbf{C}^H\mathbf{w} = \mathbf{b}$, $\mathbf{C}^H \in (\mathbb{R}, \mathbb{C})^{N_c \times M}$ with $M > N_c$ and $\mathbf{b} \in \mathbb{R}^{N_c \times 1}$, which define a linear system of N_c (number of constraints) equations, is defined on the specific application; then the problem can be formulated with the following CF:

$$\mathbf{w}_* \therefore \min_{\mathbf{w}} J(\mathbf{w}) \qquad \text{s.t.} \qquad \mathbf{C}^H\mathbf{w} = \mathbf{b}. \tag{4.86}$$

To determine the constrained LS (CLS) solution we may use the method of *Lagrange multipliers* (see Appendix B, Sect. B.3 for details) where the optimization problem (4.86) is expressed as a new CF defined as linear combination of the standard LS CF (4.18) and the homogeneous constraint equations. This new CF, called *Lagrange function* or *Lagrangian*, is indicated as $L(\mathbf{w},\boldsymbol{\lambda})$, and in our case can be written as

$$L(\mathbf{w}, \boldsymbol{\lambda}) = \left(\mathbf{d}^H - \mathbf{w}^H\mathbf{X}^H\right)(\mathbf{d} - \mathbf{X}\mathbf{w}) + \boldsymbol{\lambda}^H\left(\mathbf{C}^H\mathbf{w} - \mathbf{b}\right), \tag{4.87}$$

where $\boldsymbol{\lambda} \in (\mathbb{R}, \mathbb{C})^{N_c \times 1} = [\,\lambda_0 \quad \cdots \quad \lambda_{N_c-1}\,]^T$ is the vector of *Lagrange multipliers*. Therefore, the optimum (see Sect. B.3.2) can be determined by the solutions of a system of equations of the type

$$\nabla_{\mathbf{w}} L(\mathbf{w}, \boldsymbol{\lambda}) = \mathbf{0}, \quad \nabla_{\boldsymbol{\lambda}} L(\mathbf{w}, \boldsymbol{\lambda}) = \mathbf{0}, \tag{4.88}$$

which are tailored over the specific problem. The necessary condition so that $\mathbf{w}_*$ represents an optimal solution is that there exists $\boldsymbol{\lambda}_*$ such that the pair $(\mathbf{w}_*, \boldsymbol{\lambda}_*)$ satisfies the expressions (4.88). It follows that to determine the solution of (4.86), it is necessary to determine both the parameters $\mathbf{w}$ and the Lagrange multipliers $\boldsymbol{\lambda}$, through the minimization of (4.87) with respect to $\mathbf{w}$ and $\boldsymbol{\lambda}$.

Expanding (4.87) and taking the gradient respect to $\mathbf{w}$, we get

$$\frac{\partial L(\mathbf{w}, \boldsymbol{\lambda})}{\partial \mathbf{w}} = -2\mathbf{X}^H\mathbf{d} + 2\mathbf{X}^H\mathbf{X}\mathbf{w} + \mathbf{C}\boldsymbol{\lambda}, \tag{4.89}$$

and setting it equal to zero we have that

$$\begin{aligned}\mathbf{w}_c &= \left(\mathbf{X}^H\mathbf{X}\right)^{-1}\mathbf{X}^H\mathbf{d} - \tfrac{1}{2}(\mathbf{X}^H\mathbf{X})^{-1}\mathbf{C}\boldsymbol{\lambda} \\ &= \mathbf{w}_{LS} - \tfrac{1}{2}\left(\mathbf{X}^H\mathbf{X}\right)^{-1}\mathbf{C}\boldsymbol{\lambda}.\end{aligned} \tag{4.90}$$

To find $\boldsymbol{\lambda}$ we impose the constraint $\mathbf{C}^H\mathbf{w}_c = \mathbf{b}$, so that

$$\mathbf{C}^H\mathbf{w}_{\text{LS}} - \tfrac{1}{2}\mathbf{C}^H\left(\mathbf{X}^H\mathbf{X}\right)^{-1}\mathbf{C}\boldsymbol{\lambda} = \mathbf{b}$$

and hence, solving for $\boldsymbol{\lambda}$, we get

$$\boldsymbol{\lambda} = 2\left[\mathbf{C}^H\left(\mathbf{X}^H\mathbf{X}\right)^{-1}\mathbf{C}\right]^{-1}\left(\mathbf{C}^H\mathbf{w}_{\text{LS}} - \mathbf{b}\right).$$

Substituting the last in (4.90), we pose $\mathbf{R}_{xx} = \mathbf{X}^H\mathbf{X}$; the solution is

$$\begin{aligned}\mathbf{w}_{\text{c}} &= \mathbf{w}_{\text{LS}} - \mathbf{C}\left[\mathbf{C}^H\mathbf{R}_{xx}^{-1}\mathbf{C}\right]^{-1}\mathbf{R}_{xx}^{-1}\left(\mathbf{C}^H\mathbf{w}_{\text{LS}} - \mathbf{b}\right)\\ &= \mathbf{w}_{\text{LS}} - \mathbf{C}\left[\mathbf{C}^H\mathbf{R}_{xx}^{-1}\mathbf{C}\right]^{-1}\mathbf{C}^H\mathbf{R}_{xx}^{-1}\mathbf{w}_{\text{LS}} + \mathbf{C}\left[\mathbf{C}^H\mathbf{R}_{xx}^{-1}\mathbf{C}\right]^{-1}\mathbf{R}_{xx}^{-1}\mathbf{b}.\end{aligned} \tag{4.91}$$

Let $\mathbf{F} = \mathbf{C}\left[\mathbf{C}^H\mathbf{R}_{xx}^{-1}\mathbf{C}\right]^{-1}\mathbf{R}_{xx}^{-1}\mathbf{b}$, and considering the weighted projection operators (WPO) defined as

$$\widetilde{\mathbf{P}} = \mathbf{C}\left[\mathbf{C}^H\mathbf{R}_{xx}^{-1}\mathbf{C}\right]^{-1}\mathbf{C}^H\mathbf{R}_{xx}^{-1}, \quad \textit{WPO} \tag{4.92}$$

$$\mathbf{P} = \mathbf{I}_{M\times M} - \widetilde{\mathbf{P}}, \quad \textit{orthogonal complement WPO} \tag{4.93}$$

the expression (4.91) can be rewritten as

$$\mathbf{w}_{\text{c}} = \mathbf{P}\mathbf{w}_{\text{LS}} + \mathbf{F}. \tag{4.94}$$

From the previous equation, we note that the CLS represents a sort of corrected version of the unconstrained LS solutions.

For a better understanding consider a simple LS problem where you are seeking an optimal solution such that the $\mathbf{w}$ parameters are all identical to each other, i.e., $w[0] = w[1] = \cdots = w[M-1]$. For $M = 2$ and $N_{\text{c}} = 1$ a simple choice of the constraint that meets this criterion can be expressed as

$$\mathbf{C}^H\mathbf{w} = \mathbf{b} \Rightarrow \begin{bmatrix}1 & -1\end{bmatrix}\begin{bmatrix}w[0]\\ w[1]\end{bmatrix} = 0, \tag{4.95}$$

considering a simple geometric interpretation as described in Fig. 4.9. The constrained optimal solution $\mathbf{w}_{\text{c}}$ is lying on the so-called *constraint plane* defined as $\Lambda = \left\{\mathbf{w} : \mathbf{C}^H\mathbf{w} = \mathbf{b}\right\}$ that in our case is a simple line through the origin: $w[0] + w[1] = 0$. So the solution $\mathbf{w}_{\text{c}}$ is at minimum distance to the standard LS solution, i.e., corresponds to the tangent point between the isolevel curve of the CF $J(\mathbf{w})$ and the plane Λ.

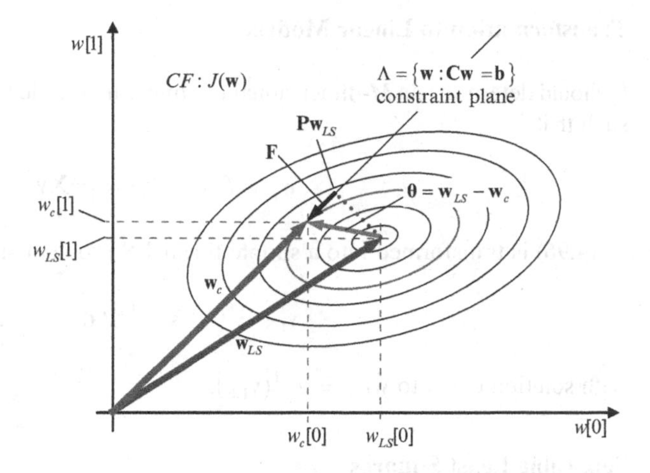

Fig. 4.9 Geometrical interpretation of linearly constrained LS

4.2.5.6 Nonlinear LS

In the case in which the relationship between the **w** parameters and the input **x** is nonlinear, the expression of the error (4.13) can be written as $\mathbf{e} = \mathbf{d} - f(\mathbf{x},\mathbf{w})$, where $f(\cdot)$ is suitable nonlinear function. The CF is written as

$$J(\mathbf{w}) = \left[\mathbf{d} - f(\mathbf{x},\mathbf{w})\right]^T \left[\mathbf{d} - f(\mathbf{x},\mathbf{w})\right] \tag{4.96}$$

and can describe the problem called in statistics nonlinear regression (NLR) [4]. The determination of the solution of (4.96), which depends on the nature of the nonlinear function, can be very difficult and may not exist.

For example, one of the most common NLR models is the exponential decay or exponential growth model defined as $w[0]e^{-w[1]x[n]}$ or other *s*-shaped functions that can be defined, for example, as

$$w[0]e^{-\left(w[1]+w[2]x[n]\right)}, \quad \frac{w[0]}{1+e^{-\left(w[1]+w[2]x[n]\right)}}, \quad \frac{w[0]}{\left[1+e^{-\left(w[1]+w[2]x[n]\right)}\right]^{w[4]}}, \ldots .$$

Another common form of NLR models is the rational function model defined as

$$\frac{\sum_{j=1}^{K} w[j]x^{j-1}}{1+\sum_{j=1}^{M} w[K+j]x^j}.$$

The solution of nonlinear LS is usually based on iterative approach and may suffer from the limitation of the numerical methods. However, for some types of nonlinearity, it is possible to determine simplified solutions through (1) transformation of the parameters to a linear model or (2) separability of the nonlinear function [4, 31].

Transformation to Linear Models

It should determine an M-dimensional nonlinear invertible transformation $\mathbf{v} = g(\mathbf{w})$ such that

$$f(\mathbf{x},\mathbf{w}) = f\left[\mathbf{x}, g^{-1}(\mathbf{v})\right] = \mathbf{X}\mathbf{v} \tag{4.97}$$

so (4.96) is transformed into a simple linear LS problem such that

$$\mathbf{v}_{\mathrm{LS}} = \left(\mathbf{X}^T\mathbf{X}\right)^{-1}\mathbf{X}^T\mathbf{d} \tag{4.98}$$

with solution equal to $\mathbf{w}_{\mathrm{LS}} = g^{-1}(\mathbf{v}_{\mathrm{LS}})$.

Separable Least Squares

By partitioning the vector of unknown parameters as

$$\mathbf{w} = \left[\mathbf{v}\ \ \mathbf{z}\right]^T, \qquad \mathbf{v} \in \mathbb{R}^{R\times 1},\ \mathbf{z} \in \mathbb{R}^{(M-R)\times 1} \tag{4.99}$$

it should determine a relationship such that the nonlinear function $f(\mathbf{x},\mathbf{w})$ can be written as the product

$$f(\mathbf{x},\mathbf{w}) = \mathbf{X}(\mathbf{v})\mathbf{z}. \tag{4.100}$$

Substituting in (4.96) we get

$$J(\mathbf{w}) = \left[\mathbf{d} - \mathbf{X}(\mathbf{v})\mathbf{z}\right]^T\left[\mathbf{d} - \mathbf{X}(\mathbf{v})\mathbf{z}\right]. \tag{4.101}$$

This model is linear in $\mathbf{z}$ while it is not in the remaining parameters $\mathbf{v}$. So for the unknown $\mathbf{z}$ we can write

$$\mathbf{z}_{\mathrm{LS}} = \left[\mathbf{X}^T(\mathbf{v})\mathbf{X}(\mathbf{v})\right]^{-1}\mathbf{X}(\mathbf{v})^T\mathbf{d}. \tag{4.102}$$

Then, from the expression of $\hat{J}_{\min}$ (4.28), the resulting error is

$$J(\mathbf{v},\mathbf{z}_{\mathrm{LS}}) = \mathbf{d}^T\mathbf{d} - \mathbf{d}^T\mathbf{X}(\mathbf{v})\left(\mathbf{X}(\mathbf{v})^T\mathbf{X}(\mathbf{v})\right)^{-1}\mathbf{X}(\mathbf{v})^T\mathbf{d} \tag{4.103}$$

and, in order to find the remaining part $\mathbf{v}_{\mathrm{LS}}$, the problem is reduced to the maximization of the function

$$\mathbf{d}^T\mathbf{X}(\mathbf{v})\left(\mathbf{X}(\mathbf{v})^T\mathbf{X}(\mathbf{v})\right)^{-1}\mathbf{X}(\mathbf{v})^T\mathbf{d} \tag{4.104}$$

with respect to $\mathbf{v}$ parameters, for example, by using a numerical iterative method described later in the text.

4.3 On the Solution of Linear Systems with LS Method

The study of the principles of the LS method is fundamental to many adaptive signal processing problems such as Fourier analysis, the optimal estimation parameters, the prediction, the deconvolution, etc. In this context, an aspect of particular importance is the solving method of linear equations system, related to the general problem formulation, described by the expression (4.13) [8–11].

4.3.1 About the Over and Underdetermined Linear Equations Systems

In general, in (4.13), the linear system matrix $\mathbf{X} \in (\mathbb{R},\mathbb{C})^{N\times M}$

$$\mathbf{X}\mathbf{w} = \mathbf{d} \tag{4.105}$$

is rectangular, and we can identify three distinct situations:

$$\begin{aligned} N &= M, \quad \textit{consistent system,} \\ N &> M, \quad \textit{overdetermined system,} \\ N &< M, \quad \textit{underdetermined system.} \end{aligned}$$

For $N = M$, it is $\text{rank}(\mathbf{X}) = N = M$; the exact solution is unique and is

$$\mathbf{X}\mathbf{w} = \mathbf{d} \quad \Rightarrow \quad \mathbf{w}_* = \mathbf{X}^{-1}\mathbf{d}. \tag{4.106}$$

4.3.1.1 Overdetermined Systems

In the case of overdetermined systems is $N > M$ and $\text{rank}(\mathbf{X}) = M$. By multiplying by $\mathbf{X}^H$ both sides of the linear system, we get the expression $\mathbf{X}^H\mathbf{X}\mathbf{w} = \mathbf{X}^H\mathbf{d}$, where $(\mathbf{X}^H\mathbf{X}) \in (\mathbb{R},\mathbb{C})^{M\times M}$ has $\text{rank}(\mathbf{X}^H\mathbf{X}) = M$ and is invertible. Note, also, that this result coincides exactly with the Yule–Walker normal equations (4.20) derived by minimizing (4.18). It follows, then, that the solution of the system (4.106) is expressed as

$$\mathbf{w}_* \equiv \mathbf{w}_{\text{LS}} = \left(\mathbf{X}^H\mathbf{X}\right)^{-1}\mathbf{X}^H\mathbf{d} = \mathbf{X}^{\#}\mathbf{d}. \tag{4.107}$$

The above expression coincides with the minimum energy error solution (4.21) or minimum L_2-norm. This energy is equal to

$$J_{\min}(\mathbf{w}) \equiv J(\mathbf{w}_*) = \mathbf{d}^H\left(\mathbf{I} - \mathbf{X}\mathbf{X}^{\#}\right)\mathbf{d} \geq 0, \tag{4.108}$$

where $\mathbf{X}^{\#} \triangleq \left(\mathbf{X}^H\mathbf{X}\right)^{-1}\mathbf{X}^H$ $(M \times N)$ is the Moore–Penrose pseudoinverse of $\mathbf{X}$ $(N \times M)$ for the overdetermined case.

4.3.1.2 Underdetermined Systems

In the underdetermined case we have that $N < M$ and $\text{rank}(\mathbf{X}) = N$. As already seen in (4.58), the solution is not unique. Among the infinite, we can find a solution such that the norm $J(\mathbf{w}) = \|\mathbf{w}\|_2^2$ is minimum

$$\mathbf{w}_* \therefore \arg\min_{\mathbf{w}} \|\mathbf{w}\|_2^2 \quad \text{s.t.} \quad \begin{cases} \mathbf{d} - \mathbf{X}\mathbf{w} = \mathbf{0} & \text{deterministic} \\ \|\mathbf{d} - \mathbf{X}\mathbf{w}\|_2^2 \leq \varepsilon & \text{stochastic.} \end{cases} \tag{4.109}$$

Proceeding as in Sect. 4.2.5.5, see (4.88), the optimal solution is obtained by trying to point out the relationships that allow the explicit computation of the vectors $\mathbf{w}_*$ and $\boldsymbol{\lambda}_*$.

For the problem (4.109) the Lagrangian takes the form

$$L(\mathbf{w}, \boldsymbol{\lambda}) = \mathbf{w}^H\mathbf{w} + 2\boldsymbol{\lambda}^H(\mathbf{d} - \mathbf{X}\mathbf{w}). \tag{4.110}$$

Therefore, conditions to meet (4.88) are

$$\nabla_{\mathbf{w}} L(\mathbf{w}, \boldsymbol{\lambda}) = 2\mathbf{w}_* - 2\mathbf{X}^H\boldsymbol{\lambda}_* = \mathbf{0}, \tag{4.111}$$

$$\nabla_{\boldsymbol{\lambda}} L(\mathbf{w}, \boldsymbol{\lambda}) = \mathbf{d} - \mathbf{X}\mathbf{w}_* = \mathbf{0}. \tag{4.112}$$

In this case the optimal solution can be obtained by observing that from (4.111) is $\mathbf{w}_* = \mathbf{X}^H\boldsymbol{\lambda}_*$ for which, pre-multiplying both members by $\mathbf{X}$, $\boldsymbol{\lambda}_* = \left(\mathbf{X}\mathbf{X}^H\right)^{-1}\mathbf{X}\mathbf{w}_*$. For the constraint expressed by (4.112) ($\mathbf{d} = \mathbf{X}\mathbf{w}_*$) we can write $\boldsymbol{\lambda}_* = \left(\mathbf{X}\mathbf{X}^H\right)^{-1}\mathbf{d}$ and substituting $\boldsymbol{\lambda}_*$ value in (4.111) we finally have

$$\mathbf{w}_* = \mathbf{X}^H\left(\mathbf{X}\mathbf{X}^H\right)^{-1}\mathbf{d} = \mathbf{X}^\#\mathbf{d}. \tag{4.113}$$

Note that $\mathbf{X}\mathbf{X}^H \in (\mathbb{R},\mathbb{C})^{N\times N}$ is invertible and that $\mathbf{X}^\# = \mathbf{X}^H\left(\mathbf{X}\mathbf{X}^H\right)^{-1}$ is the Moore–Penrose pseudoinverse in the case of underdetermined system. It is also

$$\begin{aligned} \|\mathbf{w}_*\|_2^2 &= \left(\mathbf{X}^H\left(\mathbf{X}\mathbf{X}^H\right)^{-1}\mathbf{d}\right)^H\left(\mathbf{X}^H(\mathbf{X}\mathbf{X}^H,)^{-1},\mathbf{d}\right) \\ &= \mathbf{d}^H\left(\mathbf{X}^H\mathbf{X}\right)^{-1}\mathbf{d}. \end{aligned} \tag{4.114}$$

Substituting (4.113) in (4.18), the minimum error energy is

$$\hat{J}_{\min}(\mathbf{w}) = E_d - \mathbf{w}_*^H\mathbf{X}^H\mathbf{d}. \tag{4.115}$$

Remark Unlike the overdetermined case, the proof of (4.113) is not immediate. To define the pseudoinverse for the case of underdetermined system you can consider the singular value decomposition (see Sect. A.11) of the matrix $\mathbf{X}$. This topic is introduced in Appendix A (see Sect. A.11.2), where the expression of the pseudoinverse in the cases $N > M$ and $N < M$ is demonstrated.

4.3.1.3 The δ-Solution Algorithm: Levenberg–Marquardt Variant

In expressions (4.107) and (4.113), the terms $\mathbf{X}^H\mathbf{X} \in (\mathbb{R},\mathbb{C})^{M\times M}$ or $\mathbf{X}\mathbf{X}^H \in \mathbb{R}^{N\times N}$ may be ill-conditioned and their inversion may cause numerical instability. In these cases, it is necessary to identify robust methods for the determination of the solution. This issue is still a topic of active research in this field. A simple mode, indicated as a *δ-solution*, also called Levenberg–Marquardt variant (see Sect. B.2.5), consists in adding to $\mathbf{X}^H\mathbf{X}$ or $\mathbf{X}\mathbf{X}^H$, a diagonal matrix $\delta\mathbf{I}$, in which the term $\delta > 0$ represents a minimum amplitude constant, such that the matrix is always invertible. In this case, the pseudoinverse is redefined as

$$\mathbf{X}^{\#} = \left(\mathbf{X}^H\mathbf{X} + \delta\mathbf{I}\right)^{-1}\mathbf{X}^H \qquad \text{or} \qquad \mathbf{X}^{\#} = \mathbf{X}^H\left(\mathbf{X}\mathbf{X}^H + \delta\mathbf{I}\right)^{-1}. \tag{4.116}$$

Remark The Levenberg–Marquardt variant is identical to the regularized LS solution already introduced earlier in Sect. 4.2.5.3.

Moreover the matrix equality $\left[\delta\mathbf{I} + \mathbf{X}^H\mathbf{X}\right]^{-1}\mathbf{X}^H = \mathbf{X}^H\left[\delta\mathbf{I} + \mathbf{X}\mathbf{X}^H\right]^{-1}$ is algebraically provable with the matrix inversion lemma (see Sect. A.3.4).

4.3.2 *Iterative LS System Solution with Lyapunov Attractor*

The algorithms with iterative solution, based on the gradient descent of the CF, can be derived through a general methodology starting from the previously described batch LS methods. The LS CF (4.18), $J(\mathbf{w}) = \left\|\mathbf{e}\right\|_2^2$, allows an interpretation in the context of the dynamical systems theory. In fact, the iterative solution algorithm can be assimilated to a *continuous nonlinear time-invariant dynamic system* described by differential equations system defined as

$$\dot{\mathbf{w}} = f\left(\mathbf{w}(t), \mathbf{x}(t)\right), \quad \mathbf{w}(0) = \mathbf{w}_0, \tag{4.117}$$

where $f(\cdot) : \mathbb{R}^M \to \mathbb{R}^M$, $\mathbf{w}$ is the state variable, $\dot{\mathbf{w}} = d\mathbf{w}/dt$, $\mathbf{x}$ the input, and $\mathbf{w}_0$ the IC. In the absence of external excitations, $\mathbf{w}_\text{e}$ is an *equilibrium point* if $f(\mathbf{w}_\text{e}) = 0$. The system is *globally asymptotically stable* if $\forall$ $\mathbf{w}(0)$, for every trajectory $\mathbf{w}(t)$, we have $\mathbf{w}(t) \to \mathbf{w}_\text{e}$ as $t \to \infty$ (implies $\mathbf{w}_\text{e}$ is the *unique* equilibrium point). While system is *locally asymptotically stable* near $\mathbf{w}_\text{e}$ if exists a radius $R > 0$ such that $\left|\mathbf{w}(0) - \mathbf{w}_\text{e}\right| \le R \Rightarrow \mathbf{w}(t) \to \mathbf{w}_\text{e}$ as $t \to \infty$. In any case, considering the energy of physical system, if the system loses energy over time, it must stop at a specific final *equilibrium point* state $\mathbf{w}_\text{e}$. This final state is defined as *attractor*.

In particular, the recursive algorithm can be viewed as a dynamic system of the type (4.117) of which (4.18) represents its energy. In such conditions, the system is subject to the stability constraint, indicated by the Lyapunov theorem.

Lyapunov Theorem If for dynamic system of the type (4.117), it is possible to define a *generalized energy function* $J(\cdot) : \mathbb{R}^M \rightarrow \mathbb{R}$ in the state variables, such that

$$\begin{aligned} J(\mathbf{w}) &> 0, \quad \forall \mathbf{w} \neq \mathbf{w}_e \\ J(\mathbf{w}) &= 0, \quad \mathbf{w} = \mathbf{w}_e, \end{aligned} \tag{4.118}$$

where $\mathbf{w}_e$ is a *locally asymptotically stable point*, i.e., $\forall\ \varepsilon > 0$; for $t \rightarrow \infty$, it follows that $\|\mathbf{w}(t) - \mathbf{w}_e(t)\| \leq \varepsilon$, such that

$$\frac{\partial J(\mathbf{w})}{\partial t} < 0, \quad \forall \mathbf{w} \neq \mathbf{w}_e \qquad \text{and} \qquad \left.\frac{\partial J(\mathbf{w})}{\partial t}\right|_{\mathbf{w}=\mathbf{w}_e} = 0. \tag{4.119}$$

Often, for simplicity we consider $\mathbf{w}_e = \mathbf{0}$ (or changing the coordinates so that $\mathbf{w}_e = \mathbf{0}$, i.e., use $\tilde{\mathbf{w}} = \mathbf{w} - \mathbf{w}_e$). Then, if the state trajectory converges to $\mathbf{w}_e$ as $t \rightarrow \infty$ (i.e., the system is globally asymptotically stable), then $J(\mathbf{w})$ is the so-called *Lyapunov function*.

Equation (4.119) indicates that the system stability can be tested without requiring the explicit knowledge of its actual physical energy, provided that it is possible to find a Lyapunov function that satisfies the constraints (4.118), (4.119). These constraints, in the case LS system, are obvious as it is a quadratic function. Then, for (4.118)–(4.119), we can write

$$\dot{J}(\mathbf{w}) = \frac{\partial J(\mathbf{w})}{\partial \mathbf{w}} \frac{\mathrm{d}\mathbf{w}}{\mathrm{d}t}. \tag{4.120}$$

Considering the approximations $\dot{J}(\mathbf{w}_n) \approx \Delta J(\mathbf{w}_n) = \|\mathbf{e}_n\|^2 - \|\mathbf{e}_{n-1}\|^2$ and $(\mathrm{d}\mathbf{w}/\mathrm{d}t) \approx \Delta\mathbf{w}_n = (\mathbf{w}_n - \mathbf{w}_{n-1})$, for a more constructive formulation, (4.120) can be rewritten as

$$\|\mathbf{e}_n\|^2 - \|\mathbf{e}_{n-1}\|^2 = \nabla^T J(\mathbf{w}) \cdot (\mathbf{w}_n - \mathbf{w}_{n-1}),$$

where the CF gradient is $\nabla J(\mathbf{w}) = 2\mathbf{X}^T\mathbf{X}\mathbf{w}_{n-1} - 2\mathbf{X}^T\mathbf{d} = 2\mathbf{X}^T(\mathbf{y} - \mathbf{d}) = -2\mathbf{X}^T\mathbf{e}_{n-1}$. Moreover, for (4.119) $\Delta J(\mathbf{w}_n) < 0$, so we can define a scalar parameter as $\alpha = \|\mathbf{e}_n\|^2 / \|\mathbf{e}_{n-1}\|^2 < 1$, such that we can write

$$\begin{aligned} \mathbf{w}_n - \mathbf{w}_{n-1} &= \frac{(\alpha - 1)\|\mathbf{e}_{n-1}\|^2}{\nabla^T J(\mathbf{w}) \nabla J(\mathbf{w})} \nabla J(\mathbf{w}) \\ &= \tfrac{1-\alpha}{2} \mathbf{X}^T \left[\mathbf{X}\mathbf{X}^T\right]^{-1} \mathbf{e}_{n-1}. \end{aligned} \tag{4.121}$$

4.3.2.1 Iterative LS

The recursive algorithm is determined incorporating all the scalars in the parameter μ_n and for $\delta > 0$, without loss of generality, considering the matrix equality (4.116).

Therefore, the expression (4.121) can be rewritten in the following equivalent forms of finite-difference equations (FDE) as

$$\begin{aligned}\mathbf{w}_n &= \mathbf{w}_{n-1} + \mu_n\mathbf{X}^H\left[\delta\mathbf{I} + \mathbf{X}\mathbf{X}^H\right]^{-1}(\mathbf{X}\mathbf{w}_{n-1} - \mathbf{d})\\ &= \mathbf{w}_{n-1} + \mu_n\left[\delta\mathbf{I} + \mathbf{X}^H\mathbf{X}\right]^{-1}\mathbf{X}^H(\mathbf{X}\mathbf{w}_{n-1} - \mathbf{d}).\end{aligned} \tag{4.122}$$

In addition, note that the term $\delta\mathbf{I}$ ($\delta \ll 1$) avoids division by zero and allows a more *regular* adaptation (see Sect. 4.2.5.3).

To ensure the algorithm stability, the parameter μ_n should be upper bounded. In fact, note that the algorithm coincides with that of Landweber [5], which converges to the LS solution $\mathbf{X}\mathbf{w} = \mathbf{d}$, when the parameters μ_n, here interpreted as *learning rates*, are such that $0 < \left(\mathbf{I} - \mu_n\mathbf{X}^H\mathbf{X}\right) < 1$. In other words, the learning rates are such that $0 < \mu_n < 1/\lambda_{\max}$ (where $\lambda_{\max}$ is the maximum eigenvalue of $\mathbf{X}^H\mathbf{X}$). The algorithm converges quickly in case that μ_n is close to its upper limit.

It is noted that for $N = 1$ the matrix $\mathbf{X}$ is a vector containing the sequence of the filter input $\mathbf{x}_n^H \in (\mathbb{R},\mathbb{C})^{1\times M}$, and (4.122) becomes

$$\mathbf{w}_n = \mathbf{w}_{n-1} + \frac{\mu_n}{\delta + \mathbf{x}_n^H\mathbf{x}_n}\mathbf{x}_n e[n]. \tag{4.123}$$

The quantity $e[n] = d[n] - \mathbf{w}_{n-1}^H\mathbf{x}$ is defined as *a priori* error or simply error. The expression (4.123) represents the *online adaptation algorithms* called normalized least mean squares (NLMS). The term "*normalized*" is related to the fact that the learning rate μ_n is divided by the norm of the input vector $\mathbf{x}_n^H\mathbf{x}_n$ (i.e., the energy of the input sequence). The algorithm (4.123) without normalization is denoted as least mean squares (LMS) and is one of the most popular online adaptive algorithms. Introduced by Widrow in 1959, the LMS and NLMS are reintroduced starting from different points and widely discussed below in Chap. 5.

A more efficient iterative block solution can be made considering the *order recursive* technique, partitioning the system into sets $\{i = 0, 1, \ldots, m\}$ not necessarily disjoint. For example, the method called block iterative algebraic reconstruction technique (BI-ART) can be written as

$$\mathbf{w}_n = \mathbf{w}_{n-1} + \mu_n\sum_{i=0}^{m}\frac{d_i - \mathbf{w}_{n-1}^H\mathbf{x}_i}{\mathbf{x}_i^H\mathbf{x}_i}\mathbf{x}_i, \tag{4.124}$$

where $\mathbf{x}_i$ is the ith row of $\mathbf{X}$, and the sum is carried out only in the subset $\{i = 0, 1, \ldots, m\}$. In the extreme case in which $m = 1$, the algorithm is the Kaczmarz method [6] also called *row-action-projection* method which can be written as

$$\mathbf{w}_n = \mathbf{w}_{n-1} + \mu_n\frac{d_i - \mathbf{w}_{n-1}^H\mathbf{x}_i}{\mathbf{x}_i^H\mathbf{x}_i}\mathbf{x}_i, \qquad \text{for} \qquad i = n \bmod (m+1), \tag{4.125}$$

where, for each iteration, $0 < \mu_n < 2$. Note that in this case the Kaczmarz algorithm is identical to the normalized NLMS (4.123). Furthermore, for $m > 1$ the algorithm described by (4.124) is, in the context of adaptive filtering, often referred to as affine projection algorithm (APA) also reintroduced and widely discussed in Chap. 6.

Remark The order recursive methods may result in very interesting variations of LS techniques both in robustness and for computational efficiency. This will be discussed specifically in Chap. 8.

4.3.2.2 Iterative Weighed LS

In the case of weighed LS (see Sect. 4.2.5.1) the CF is defined as $\hat{J}(\mathbf{w}) = \mathbf{e}^H\mathbf{G}\mathbf{e}$ and the expression of the estimate of the gradient is $\nabla\hat{J}(\mathbf{w}) = -2\mathbf{G}\mathbf{X}^H\mathbf{e}$. Then, the iterative update expression can be written as

$$\mathbf{w}_n = \mathbf{w}_{n-1} + \mu\left[\delta\mathbf{I} + \mathbf{X}^H\mathbf{G}\mathbf{X}\right]^{-1}\mathbf{X}^H\mathbf{G}\mathbf{e}_{n-1}. \quad (4.126)$$

Note, as will be seen later in Chap. 6, that a possible choice of the weighing matrix that cancels the eigenvalues spread of the matrix $\mathbf{X}^H\mathbf{X}$ is for $\mathbf{G} = (\mathbf{X}^H\mathbf{X})^{-1}$. It follows

$$\mathbf{w}_n = \mathbf{w}_{n-1} + \mu\mathbf{X}^H\mathbf{G}\mathbf{e}_{n-1}. \quad (4.127)$$

The weighing coincides with the inverse of the estimated autocorrelation matrix $\mathbf{G} = \mathbf{R}_{xx}^{-1}$ and (4.127) can be written as $\mathbf{w}_n = \mathbf{w}_{n-1} + \mu\mathbf{R}_{xx}^{-1}\mathbf{X}_n^H\mathbf{e}_{n-1}$. It is noted that for $N = 1$, the adaptation algorithm takes the form $\mathbf{w}_n = \mathbf{w}_{n-1} + \mu\mathbf{R}_{xx}^{-1}\mathbf{x}_n e^*[n]$, that is, the so-called LMS Newton algorithm also reintroduced in Chap. 6.

Remark The adaptive filtering is by definition based on the online recursive calculation of the coefficients $\mathbf{w}_n$, which are thus updated in the presence of new information available to the filter input itself. In later chapters, especially in Chaps. 5 and 6, these methodologies will be reintroduced in a more general way considering several different assumptions.

4.4 LS Methods Using Matrix Factorization

The methods derived from the LS formulation allow a formalization of the LS solution's estimate problem as an *algebraic problem*, defined by the solution of a linear over/under-determined equation system, directly built on blocks of signal data stored on the data matrix $\mathbf{X} \in (\mathbb{R},\mathbb{C})^{N\times M}$.

The algebraic nature of the approach to the solution estimation allows us to define several methodology variants. Above (see Sect. 4.2.5), some variations in the definition of CF requiring additional constraints able to formalize, in the CF itself, *a priori* knowledge about the nature of the noise and/or the optimal solution have been proposed.

In this section some LS variants, derived from the LS algebraic nature, based on either data matrix $\mathbf{X}$ or estimated correlation $\mathbf{X}^H\mathbf{X}$ matrix decomposition, are presented and discussed. This problem has been extensively studied in the literature and there are numerous techniques, usually based on algebraically equivalent matrix decompositions, with different robustness properties and/or computational cost. In fact as previously noted, the matrix $\mathbf{X}$ is constructed by inserting, for columns or rows, the filter input sequence shifted by one sample for which each column/row contains rather similar processes. In general, even in the case of array processing the columns are related to the same process sampled at different spatial points. Therefore, the $\mathbf{X}^H\mathbf{X}$ matrix is very often ill-conditioned and in many situations the robustness of the algorithm represents a very important aspect.

In Fig. 4.10 is shown a general scheme for the classification of estimation algorithms based on the LS class. The LS problem formulation derived from direct measurement of data blocks is usually called *amplitude domain* formulation, while that calculated by the correlation is also indicated as *power-domain* formulation.

4.4.1 LS Solution by Cholesky Decomposition

The Cholesky decomposition consists in the factorization of a symmetric or Hermitian positive-definite matrix $\mathbf{R} \in (\mathbb{R},\mathbb{C})^{M\times M}$ into the product of a lower triangular matrix and its transpose/Hermitian $\mathbf{R} = \tilde{\mathbf{L}}\tilde{\mathbf{L}}^H$.

A more general version of the previous factorization is defined as *upper-diagonal-upper* or LDL decomposition [8, 9]. The correlation matrix $\mathbf{R}$ or its time-average estimation $\mathbf{R}_{xx}$ (4.23) is decomposed into the product of three matrices:

$$\mathbf{R}_{xx} = \mathbf{L}\mathbf{D}\mathbf{L}^H, \tag{4.128}$$

where $\mathbf{L}$ is lower *unitriangular* matrix defined as

$$\mathbf{L} \triangleq \begin{bmatrix} 1 & 0 & \cdots & 0 \\ l_{10} & 1 & \cdots & 0 \\ \vdots & \vdots & \ddots & \vdots \\ l_{M-1,0} & l_{M-1,1} & \cdots & 1 \end{bmatrix} \tag{4.129}$$

while $\mathbf{D}$ is a diagonal matrix defined as

$$\mathbf{D} \triangleq \operatorname{diag}\left[\xi_0, \xi_1, \ldots, \xi_{M-1}\right]. \tag{4.130}$$

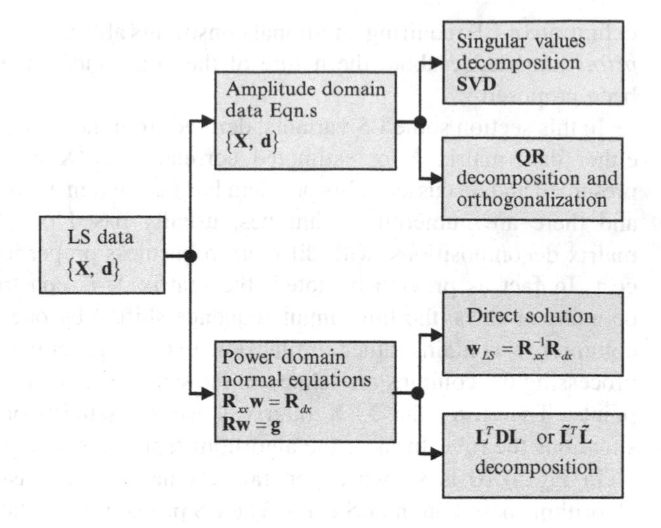

Fig. 4.10 A possible algorithms classification for the solution of the LS problem (modified from [7])

With the decomposition (4.128), the normal equation can be written as

$$\mathbf{LDL}^H\mathbf{w} = \mathbf{R}_{xd}. \tag{4.131}$$

By posing $\mathbf{L}^H\mathbf{w} = \mathbf{k}$, (4.131) can be solved for $\mathbf{k}$, using the lower triangular system as

$$\mathbf{k} = [\mathbf{LD}]^{-1}\mathbf{R}_{xd} \tag{4.132}$$

and for $\mathbf{w}$ by solving the upper triangular system. The estimate of the LS optimal solution is then

$$\mathbf{w}_{\text{LS}} = [\mathbf{L}^H]^{-1}\mathbf{k}. \tag{4.133}$$

Note the so-called *LDL* decomposition, as a form that is closely related to the eigen decomposition of real symmetric matrices, $\mathbf{R}_{xx} = \mathbf{Q}\mathbf{\Lambda}\mathbf{Q}^H$.

It is easily shown that the decomposition (4.128) allows the direct calculation of the minimum of the LS error (or in general MMSE) without the calculation of $\mathbf{w}_{\text{LS}}$, as

$$E_{\text{LS}} = E_d - \mathbf{k}^H\mathbf{D}\mathbf{k}. \tag{4.134}$$

Since $\mathbf{R}_{xx}$ is usually positive definite, the elements ξ_k in (4.130) are positive. We can then define a matrix $\widetilde{\mathbf{L}} = \mathbf{L}\mathbf{D}^{1/2}$, for which we can write the Cholesky decomposition of $\mathbf{R}_{xx}$ [8], as

$$\mathbf{R}_{xx} = \widetilde{\mathbf{L}}\widetilde{\mathbf{L}}^H. \tag{4.135}$$

In special cases, $\mathbf{R}$ is Toeplitz matrix, the *LDL* decomposition can be computed in $O(M^2)$ operations.

Remark In the solution of the normal equations with matrix transformations, in the case where certain numerical stability and estimation's robustness are required, it is commonly preferred to apply these transformations directly on the data [7, 8]. In previous section (see Sect. 4.2.5.3), it has been shown that the sensitivity of the solution $\mathbf{w}_{\mathrm{LS}}$, with respect to the data matrix $\mathbf{X}$ perturbations depends on the $\mathbf{R}_{xx}$'s condition number (ratio between the largest and the smallest eigenvalue), rather than the used algorithm. Note that the numerical accuracy required for the $\tilde{\mathbf{L}}$ matrix calculation directly from the data $\mathbf{X}$ is equal to half of that required for the $\tilde{\mathbf{L}}$ calculation from the correlation matrix $\mathbf{R}_{xx}$. Furthermore, the calculation of the product $\mathbf{X}^H\mathbf{X}$, needed to estimate of $\mathbf{R}_{xx}$, produces a certain loss of information and should be avoided in the case of low-precision arithmetic. As already indicated in the introduction of the paragraph, the algorithms for the $\tilde{\mathbf{L}}$ calculation from $\mathbf{X}$ are indicated as *square root methods* or techniques in the amplitude domain, while the methods that determine $\tilde{\mathbf{L}}$ from $\mathbf{R}_{xx}$ are known as power-domain techniques.

Moreover, note that the LS solution with Cholesky decomposition is strictly related to the recursive order methods with lattice-ladder structure (introduced in Sect. 8.3.5) that directly determine the decomposition (4.128).

4.4.2 LS Solution Methods with Orthogonalization

An orthogonal transformation is a linear transformation such that applied to a vector preserves its length. Given $\mathbf{Q}$ orthonormal (i.e., such that $\mathbf{Q}^{-1} = \mathbf{Q}^H$), the $\mathbf{y} = \mathbf{Q}^H\mathbf{x}$ transformation does not change the length of the vector to which it is applied; indeed we have that $\|\mathbf{y}\|_2^2 = \mathbf{y}^H\mathbf{y} = [\mathbf{Q}^H\mathbf{x}]^H\mathbf{Q}^H\mathbf{x} = \|\mathbf{x}\|_2^2$.

Note that $\mathbf{Q}$ is simply any orthogonal matrix and is not necessarily the *modal matrix* built with the eigenvectors of $\mathbf{R}$ as previously defined (see Sect. 3.3.6).

The procedures for the solution of the normal equations built directly on the measured data, although algebraically equivalent, may have different robustness. In this regard, the $\mathbf{Q}$ orthonormal transformation applied to the normal equations does not determine an increase of the error due to the numerical approximations (*round-off error*) but can lead to a greater estimate robustness and, if properly chosen, even a decrease in the computational cost. In general, we can determine two modes of use of orthogonal transformations for the solution of equations LS.

A first method consists in the transformation of the data matrix $\mathbf{X}$ in $\mathbf{Q}^H\mathbf{X}$, without affecting the estimation of the correlation $\mathbf{X}^H\mathbf{X}$. In fact, for any orthogonal matrix $\mathbf{Q}$ is

$$\begin{aligned}\mathbf{R}_{xx} &= \left(\mathbf{Q}^H\mathbf{X}\right)^H\mathbf{Q}^H\mathbf{X} \\ &= \mathbf{X}^H\mathbf{X}.\end{aligned} \tag{4.136}$$

In this situation the problem is to determine a certain transformation $\mathbf{Q}$ for which the LS system is redefined in a simpler form.

A second method consists of applying the orthogonalization matrix $\mathbf{Q}$ directly to the LS error defined as $\mathbf{e} = \mathbf{d} - \mathbf{X}\mathbf{w}$ [see (4.13)]. Since $\mathbf{Q}$ does not change the length of the vector to which it is applied, we have that

$$\arg\min_{\mathbf{w}} \left\|(\mathbf{d} - \mathbf{X}\mathbf{w})\right\|_2^2 = \arg\min_{\mathbf{w}} \left\|\mathbf{Q}^H(\mathbf{d} - \mathbf{X}\mathbf{w})\right\|_2^2. \tag{4.137}$$

Even in this case the problem is to find a matrix $\mathbf{Q}$ such that (4.137) results in a simplified form with respect to (4.19).

4.4.2.1 LS Solution with QR Factorization of X Data Matrix

Given an orthogonal matrix $\mathbf{Q} \in (\mathbb{R},\mathbb{C})^{N\times N}$ such that is

$$\mathbf{X} = \mathbf{Q}\begin{bmatrix} \mathcal{R} \\ \mathbf{0} \end{bmatrix}, \tag{4.138}$$

where $\mathbf{Q}$ is an orthogonal matrix such that $\mathcal{R} \in (\mathbb{R},\mathbb{C})^{M\times M}$ is an upper triangular matrix.

We remind the reader that the QR matrix factorization with coefficient $\mathbf{X} \in (\mathbb{R},\mathbb{C})^{N\times M}$ is defined as a decomposition of the type (4.138) (see [8–11]).

In the case in which $N > M$, it can be demonstrated that for a full rank data matrix ($\text{rank}(\mathbf{X}) = M$), the first M columns of $\mathbf{Q}$ form an orthonormal basis of $\mathbf{X}$; it follows that the QR calculation represents a way to determine an orthonormal basis of $\mathbf{X}$. This calculation can be made by considering various types of linear transformations including Householder, block Householder, Givens, Fast Givens, Gram–Schmidt, etc.

If we consider the expression of the error (4.137) we can write

$$\begin{aligned} \|\mathbf{e}\|_2^2 &= \left\|\mathbf{Q}^H\mathbf{e}\right\|_2^2 \\ &= \left\|\mathbf{Q}^H\mathbf{d} - \mathbf{Q}^H\mathbf{X}\mathbf{w}\right\|_2^2. \end{aligned} \tag{4.139}$$

Using a partition for the matrix $\mathbf{Q}$ defined as

$$\mathbf{Q} \triangleq [\mathbf{Q}_1 \;\; \mathbf{Q}_2], \tag{4.140}$$

where $\mathbf{Q}_1 \in (\mathbb{R},\mathbb{C})^{N\times M}$ and $\mathbf{Q}_2 \in (\mathbb{R},\mathbb{C})^{N\times(N-M)}$, we obtain the so-called *thin*-QR (see Fig. 4.11) and we can write

$$\mathbf{X} = \mathbf{Q}_1\mathcal{R}. \tag{4.141}$$

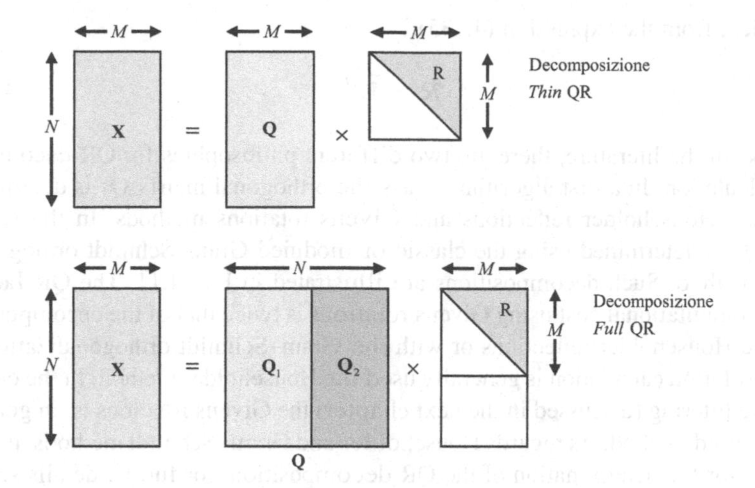

Fig. 4.11 Outline of the QR decomposition

By (4.139) and (4.140) we get

$$\mathbf{Q}^H\mathbf{d} = \begin{bmatrix} \mathbf{Q}_1^H\mathbf{d} \\ \mathbf{Q}_2^H\mathbf{d} \end{bmatrix}. \tag{4.142}$$

Substituting (4.142) and (4.141) in (4.139) we have that

$$\begin{aligned} \|\mathbf{e}\|_2 &= \left\| \begin{bmatrix} \mathcal{R}\mathbf{w} \\ \mathbf{0} \end{bmatrix} - \begin{bmatrix} \mathbf{Q}_1^H\mathbf{d} \\ \mathbf{Q}_2^H\mathbf{d} \end{bmatrix} \right\|_2 \\ &= \left\| \begin{bmatrix} \mathcal{R}\mathbf{w} - \mathbf{Q}_1^H\mathbf{d} \\ \mathbf{Q}_2^H\mathbf{d} \end{bmatrix} \right\|_2. \end{aligned} \tag{4.143}$$

A part of the previous system depends explicitly on the filter coefficients:

$$\mathbf{w}_{\mathrm{LS}} = \mathcal{R}^{-1}\mathbf{Q}_1^H\mathbf{d} \tag{4.144}$$

and also

$$J(\mathbf{w}_{\mathrm{LS}}) = \left\|\mathbf{Q}_2^H\mathbf{d}\right\|_2^2. \tag{4.145}$$

Note that $\mathcal{R} \in (\mathbb{R},\mathbb{C})^{M\times M}$ being triangular, the system (4.144) can be resolved with a simple backward substitution. Furthermore,

$$\begin{aligned} \mathbf{R}_{xx} &= \mathbf{X}^H\mathbf{X} \\ &= \mathcal{R}^H\mathcal{R} \end{aligned} \tag{4.146}$$

for which, from the expression (4.135),

$$\mathcal{R} = \tilde{\mathbf{L}}^H. \tag{4.147}$$

Remark In the literature, there are two different philosophies for QR decomposition calculation. In a first algorithms class the orthogonal matrix $\mathbf{Q}_1$ is determined using the Householder reflections and Givens rotations methods. In the second class, $\mathbf{Q}_1$ is determined using the classic or modified Gram–Schmidt orthogonalization method. Such decompositions are illustrated in Fig. 4.11. The QR factorization computational cost using Givens rotations is twice that of the decomposition with the Householder reflections or with the Gram–Schmidt orthogonalization. In the LS solution calculation is generally used the Householder method. In the case of adaptive filtering (discussed in the next chapter) the Givens rotations is, in general, the preferred method. As regards Householder and Gram–Schmidt methods, used in practice for the determination of the QR decomposition, for further details see the algebra texts as, for example, Golub–Van Loan [8].

4.4.3 LS Solution with the Singular Value Decomposition Method

Among the matrix methods, the singular value decomposition (SVD) (see Sect. A.11) is one of the most important and elegant algebraic techniques for real–complex rectangular matrices factorization. Moreover, in LS systems it plays a role of primary importance for both the theoretical analysis and the practical implications. Indeed the SVD makes possible a unified approach to the definition of the pseudoinverse matrix and to overdetermined and underdetermined LS solution. The subspaces associated with the SVD are related to the properties of subspaces of the processes involved in the LS system. Finally, as regards the computational aspects it is one of the most robust numerical methods for solving linear ill-conditioned systems [8, 9, 12–18].

4.4.3.1 Singular Value Decomposition Theorem

Given a data matrix $\mathbf{X} \in (\mathbb{R},\mathbb{C})^{N\times M}$ of any rank r such that $r \le K$ where $K = \min(N,M)$, there are two orthogonal unitary matrices $\mathbf{U} \in (\mathbb{R},\mathbb{C})^{N\times N}$ and $\mathbf{V} \in (\mathbb{R},\mathbb{C})^{M\times M}$ such that the columns of $\mathbf{U}$ contain the $\mathbf{X}\mathbf{X}^H$ eigenvectors, while the columns of $\mathbf{V}$ contain the $\mathbf{X}\mathbf{X}^H$ eigenvectors. Formally

$$\mathbf{U} \in (\mathbb{R},\mathbb{C})^{N\times N} = [\,\mathbf{u}_0 \quad \mathbf{u}_1 \quad \cdots \quad \mathbf{u}_{N-1}\,] = \text{eigenvect}\big(\mathbf{X}\mathbf{X}^H\big) \tag{4.148}$$

$$\mathbf{V} \in (\mathbb{R},\mathbb{C})^{M\times M} = [\,\mathbf{v}_0 \quad \mathbf{v}_1 \quad \cdots \quad \mathbf{v}_{M-1}\,] = \text{eigenvect}\big(\mathbf{X}^H\mathbf{X}\big) \tag{4.149}$$

such as to make valid the following equality (shown in Fig. A.3):

$$\mathbf{U}^H\mathbf{X}\mathbf{V} = \boldsymbol{\Sigma} \tag{4.150}$$

or, equivalently,

$$\mathbf{X} = \mathbf{U}\boldsymbol{\Sigma}\mathbf{V}^H \quad \text{or} \quad \mathbf{X}^H = \mathbf{V}\boldsymbol{\Sigma}\mathbf{U}^H. \tag{4.151}$$

The matrix $\boldsymbol{\Sigma} \in \mathbb{R}^{N\times M}$ has the following structure:

$$\begin{aligned} K &= \min(M,N) & \boldsymbol{\Sigma} &= \begin{bmatrix} \boldsymbol{\Sigma}_K & \mathbf{0} \\ \mathbf{0} & \mathbf{0} \end{bmatrix} \\ K &= N = M & \boldsymbol{\Sigma} &= \boldsymbol{\Sigma}_K \end{aligned} \tag{4.152}$$

where the diagonal matrix $\boldsymbol{\Sigma}_K \in \mathbb{R}^{K\times K}$ contains the ordered positive square root of eigenvalues of the matrix $\mathbf{X}^H\mathbf{X}$ (o $\mathbf{X}\mathbf{X}^H$), defined as *singular values*. In formal terms

$$\boldsymbol{\Sigma}_K = \text{diag}(\sigma_0, \sigma_1, ..., \sigma_{K-1}) \tag{4.153}$$

which are ordered in descending order

$$\sigma_0 \geq \sigma_1 \geq ... \geq \sigma_{K-1} > 0 \tag{4.154}$$

and are zero for index $i > \text{rank}(\mathbf{X})$, that is,

$$\sigma_K = \cdots = \sigma_{N-1} = 0. \tag{4.155}$$

Remark The singular values σ_i of $\mathbf{X}$ are in descending order. The column vectors $\mathbf{u}_i$ and $\mathbf{v}_i$, respectively, are defined as *left singular vectors* and *right singular vectors* of $\mathbf{X}$. Since $\mathbf{U}$ and $\mathbf{V}$ are orthogonal, it is easy to see that the matrix $\mathbf{X}$ can be written as the following product:

$$\begin{aligned} \mathbf{X} &= \mathbf{U}\boldsymbol{\Sigma}\mathbf{V}^H \\ &= \sum_{i=0}^{K-1} \sigma_i \mathbf{u}_i \mathbf{v}_i^H. \end{aligned} \tag{4.156}$$

For more properties, please refer to Appendix A (see Sect. A.11).

4.4.3.2 LS and SVD

An important use of the SVD is that related to the solution of the over/under-determined LS systems equations of the type

$$\mathbf{X}\mathbf{w} \approx \mathbf{d} \tag{4.157}$$

for which, for (4.151), we can factorize the data matrix $\mathbf{X}$ as

$$\mathbf{U}\mathbf{\Sigma}\,\mathbf{V}^H\mathbf{w} \approx \mathbf{d}. \tag{4.158}$$

For $r \leq K$, considering (4.156),

$$\begin{aligned}\mathbf{w}_{\mathrm{LS}} &= \mathbf{V}_1\mathbf{\Sigma}_r^{-1}\mathbf{U}_1^H\mathbf{d}\\ &= \sum_{i=0}^{r-1}\frac{\mathbf{u}_i^H\mathbf{d}}{\sigma_i}\mathbf{v}_i\end{aligned} \tag{4.159}$$

which shows that LS system solution can be performed at reduced rank r without explicit matrix inversion. The solution (4.159) is exactly as described by (4.19) in accordance with minimum quadratic norm:

$$\mathbf{w}_{\mathrm{LS}} = \arg\min_{\mathbf{w}} \left\|\mathbf{d} - \mathbf{X}\mathbf{w}\right\|_2^2$$

or, equivalently,

$$\mathbf{w}_{\mathrm{LS}}(i) = \begin{cases}\dfrac{d(i)}{\sigma_i} = \dfrac{\mathbf{u}_i^H\mathbf{d}}{\sigma_i}\mathbf{v} & \text{for} \quad i = 0, 1, ..., r-1\\ 0 & \text{for} \quad i = r, r+1, ..., K-1\end{cases} \tag{4.160}$$

and for the minimum error energy

$$J_{\min}(\mathbf{w}) = \sum_{i=r}^{N-1}\left|\mathbf{u}_i^H\mathbf{d}\right|^2. \tag{4.161}$$

Below a brief note for the SVD factorization, for the LS systems solution, is reported. Table 4.1 shows the computational cost for some methods of calculation.

4.4.3.3 SVD-LS Algorithm

- Computation of SVD $\mathbf{X} = \mathbf{U}\mathbf{\Sigma}\mathbf{V}^H$.
- Evaluation of the rank($\mathbf{X}$).
- Computation of $\widetilde{d}_i = \mathbf{u}_i^H\mathbf{d}$ per $i = 0, 1, ..., N-1$.
- Optimal LS solution computation $\mathbf{w}_{\mathrm{LS}} = \sum_{i=0}^{r-1}\sigma_i^{-1}\widetilde{d}_i\mathbf{v}_i$.
- Error computation $J(\mathbf{w}_{\mathrm{LS}}) = \sum_{i=r}^{N-1}\left|\widetilde{d}_i\right|^2$.

Table 4.1 Computational cost of some LS estimation algorithms for $N > M$ [7, 19, 20]

LS algorithm	Floating Points Operation (FLOPS)
Normal equation	$NM^2 + M^3/3$
Householder orthogonalization	$2NM^2 - 2M^3/3$
Givens orthogonalization	$3NM^2 - M^3$
Modified Gram–Schmidt	$2NM^2$
Golub–Reinsch SVD	$4NM^2 + 8M^3$
R-SVD	$2NM^2 + 11M^3$

4.4.3.4 SVD and Tikhonov Regularization Theory

The calculation of the rank of **X** can present some problems in the presence of noise superimposed on the signal or in the case where the data matrix is nearly singular. The SVD allows, in these cases, to estimate the actual **X** rank relative to the *signal subspace* only. In the presence of noise, in fact, it is unlikely the existence of an index r such that for $i > r$ is $\sigma_i = 0$. Then, it is appropriate to establish a threshold below which you force singular value to assume a null value. For this purpose, we define *numerical rank*, the index r value such that, set a certain threshold value ε, the following relation holds:

$$\sigma_r^2 + \sigma_{r+1}^2 + \cdots + \sigma_{K-1}^2 < \varepsilon^2.$$

Moreover, these singular values are forced to zero. In this case the Frobenius norm is

$$\left\|\mathbf{X} - \mathbf{X}_r\right\|_F = \sqrt{\sigma_r^2 + \sigma_{r+1}^2 + \cdots + \sigma_{K-1}^2} < \varepsilon \tag{4.162}$$

and the **X** matrix is said *rank deficient* or with *numerical rank r*.

Note that this result has important implications in signal modeling and in techniques for signal compression. The LS solution calculation for rank-deficient matrices, in fact, requires extreme care. When a singular value is very small, its reciprocal, which is a singular value of the pseudoinverse $\mathbf{X}^{\#}$, becomes a very large number and, for numerical reasons, it was found that LS solution deviates from the real one. This problem can be mitigated by forcing to zero singular values below a certain threshold. The threshold level is generally determined based on the machine numerical precision and on the basis of the measurements accuracy stored in the data matrix **X**. For example, Golub and Van Loan, in [8], suggest a threshold of $10^{-6}\sigma_0$. In fact, by choosing a numerical rank such that $\sigma_{\min} > \sigma_{K-1}$, the condition number (see Sect. A.12), $\chi(\mathbf{X}^H\mathbf{X}) = \sigma_0/\sigma_{\min}$, decreases.

Another way to determine the minimum threshold is based on the Tikhonov regularization theory. Already discussed in Sect. 4.2.5.2, we have seen that the sum of the term $\delta\|\mathbf{w}\|^2$ to CF acts as a stabilizer and prevents too large solutions. In fact, using the Lagrange multipliers method, it can be shown that the regularized solution, indicated as $\mathbf{w}_{\mathrm{LS},\delta}$, takes the form

$$\mathbf{w}_{\mathrm{LS},\delta} = \sum_{i=0}^{r-1} \frac{\sigma_i}{\sigma_i^2 + \delta} \left(\mathbf{u}_i^H \mathbf{d}\right) \mathbf{v}_i \tag{4.163}$$

also known as *regularized LS solution*. Note that for $\delta = 0$ $\mathbf{w}_{\mathrm{LS},\delta} = \mathbf{w}_{\mathrm{LS}}$. In the case that $\delta > 0$ and $\sigma_i \to 0$, also the term $\sigma_i/(\sigma_i^2 + \delta) \to 0$ while $(1/\sigma_i) \to \infty$. In addition, it can be shown that $\|\mathbf{w}_{\mathrm{LS},\delta}\|_2^2 \leq \|\mathbf{d}\|_2^2/\sigma_r$ and that $\|\mathbf{w}_{\mathrm{LS},\delta}\|_2^2 \leq \|\mathbf{d}\|_2^2/\sqrt{\delta}$.

Remark The SVD decomposition of the data matrix $\mathbf{X}$, represents one of the most important methods for the discrimination of the *signal and noise subspaces* [14]. In fact, let $r = \mathrm{rank}(\mathbf{X})$, the first r columns of $\mathbf{U}$ form an orthonormal basis of the column space, i.e., $\mathcal{R}(\mathbf{X}) = \mathrm{span}(\mathbf{u}_0, \mathbf{u}_1, \ldots, \mathbf{u}_{r-1})$, while the first r columns of $\mathbf{V}$ form an orthonormal basis for the nullspace (or kernel) $\mathcal{N}\left(\mathbf{X}^H\right)$ of $\mathbf{X}$, i.e., $\mathcal{N}\left(\mathbf{X}^H\right) = \mathrm{span}(\mathbf{v}_r, \mathbf{v}_{r+1}, \ldots, \mathbf{v}_{N-1})$.

From the previous development, it is possible to define the following expansion:

$$\mathbf{X} = \begin{bmatrix} \mathbf{U}_1 & \mathbf{U}_2 \end{bmatrix} \begin{bmatrix} \mathbf{\Sigma}_r & \mathbf{0} \\ \mathbf{0} & \mathbf{0} \end{bmatrix} \begin{bmatrix} \mathbf{V}_1^H \\ \mathbf{V}_2^H \end{bmatrix} = \mathbf{U}_1 \mathbf{\Sigma}_r \mathbf{V}_1^H = \sum_{i=0}^{r-1} \sigma_i \mathbf{u}_i \mathbf{v}_i^H,$$

where $\mathbf{V}_1$, $\mathbf{V}_2$, $\mathbf{U}_1$, and $\mathbf{U}_2$ are orthonormal matrices defined as

$$\begin{aligned} \mathbf{V} &= \begin{bmatrix} \mathbf{V}_1 & \mathbf{V}_2 \end{bmatrix} \quad \text{with} \quad \mathbf{V}_1 \in \mathbb{C}^{M \times r} \quad \text{and} \quad \mathbf{V}_2 \in \mathbb{C}^{M \times M-r} \\ \mathbf{U} &= \begin{bmatrix} \mathbf{U}_1 & \mathbf{U}_2 \end{bmatrix} \quad \text{with} \quad \mathbf{U}_1 \in \mathbb{C}^{N \times r} \quad \text{and} \quad \mathbf{U}_2 \in \mathbb{C}^{N \times N-r} \end{aligned}$$

in fact, being $\mathcal{R}(\mathbf{X}) \perp \mathcal{N}\left(\mathbf{X}^H\right)$, we have that $\mathbf{V}_1^H \mathbf{V}_2 = \mathbf{0}$ and $\mathbf{U}_1^H \mathbf{U}_2 = \mathbf{0}$.

4.5 Total Least Squares

Consider the problem of determining the solution of an overdetermined linear equations system

$$\mathbf{X}\mathbf{w} \approx \mathbf{d} \tag{4.164}$$

with the data matrix and the known term defined, respectively, as $\mathbf{X} \in (\mathbb{R},\mathbb{C})^{N \times M}$ and $\mathbf{d} \in (\mathbb{R},\mathbb{C})^{N \times 1}$. In the previous paragraphs we have seen that with the LS method, the solution can be determined by minimizing the L_2-norm of the error. However, note that in the definition of the LS method it has been implicitly assumed that the error affects only the known term $\mathbf{d}$ while it is assumed a *noise-free* data matrix $\mathbf{X}$. Moreover, for notation simplicity, indicating the error as $\Delta\mathbf{d} \equiv \mathbf{e}$, the LS method can be reinterpreted as a constrained optimization problem described by the following expression:

$$\mathbf{w}_{\text{LS}} \therefore \arg\min_{\mathbf{w}} \left\|\Delta\mathbf{d}\right\|^2 \qquad \text{s.t.} \qquad \mathbf{X}\mathbf{w} = \overline{\mathbf{d}} + \Delta\mathbf{d}, \tag{4.165}$$

where $\overline{\mathbf{d}}$, such that $\mathbf{d} = \overline{\mathbf{d}} + \Delta\mathbf{d}$, indicates the *true* value of the unperturbed known term.

The total least squares (TLS) method [17, 18, 21] represents a natural way for the *best* solution of the system (4.164). Referred to as in statistic, as *errors in variables model*, the development of the TLS method is motivated by the writing of the linear system in which the measurement error affects both the known term and the data matrix. Defining $\Delta\mathbf{X}$ the perturbation of the data matrix, such that $\mathbf{X} = \overline{\mathbf{X}} + \Delta\mathbf{X}$, where $\overline{\mathbf{X}}$ is the noiseless data matrix, the TLS method can be formalized by the expression

$$\mathbf{w}_{\text{TLS}} \therefore \arg\min_{\mathbf{w},\,\Delta\mathbf{X},\,\Delta\mathbf{d}} \left\|\Delta\mathbf{X}\ \Delta\mathbf{d}\right\|_{\text{F}}^2 \qquad \text{s.t.} \qquad \left(\overline{\mathbf{X}} + \Delta\mathbf{X}\right)\mathbf{w} = \overline{\mathbf{d}} + \Delta\mathbf{d}, \tag{4.166}$$

where $\|\cdot\|_{\text{F}}^2$ indicates the quadratic Frobenius norm (see Sect. A.10).

Denoting by $\Delta X_{i,j}$ and Δd_i, respectively, the elements of the matrix $\Delta\mathbf{X}$ and vector $\Delta\mathbf{d}$, this norm can be defined as

$$\left\|\Delta\mathbf{X}\ \Delta\mathbf{d}\right\|_{\text{F}}^2 = \sum_{i=0}^{N-1} \left|\Delta d\right|_i^2 + \sum_{i=0}^{N-1}\sum_{j=0}^{M-1} \left|\Delta X\right|_{i,j}^2. \tag{4.167}$$

Remark From the above discussion, the general form of the LS paradigm can be defined by considering the following three cases:

- *Least squares* (LS): $\Delta\mathbf{X} = 0,\ \Delta\mathbf{d} \neq 0$
- *Data least squares* (DLS): $\Delta\mathbf{X} \neq 0,\ \Delta\mathbf{d} = 0$
- *Total least squares* (TLS): $\Delta\mathbf{X} \neq 0,\ \Delta\mathbf{d} \neq 0$

where the perturbations $\Delta\mathbf{X}$ and $\Delta\mathbf{d}$ are generally considered zero-mean Gaussian stochastic processes.

For a better understanding of the three methodologies consider the case, illustrated in Fig. 4.12 (which generalizes the approach described in Fig. 4.1), in which the problem is to determine the straight line approximating a known set of experimentally measured data $[\mathbf{x},\mathbf{y}]$.

In the TLS methodology, it is supposed that the error is present on both $[\mathbf{x},\mathbf{y}]$ measures. By simple reasoning, observing Fig. 4.12, it follows that for a better estimate of the approximating straight line, you should minimize the sum of the perpendicular distances between the measures and the straight line itself.

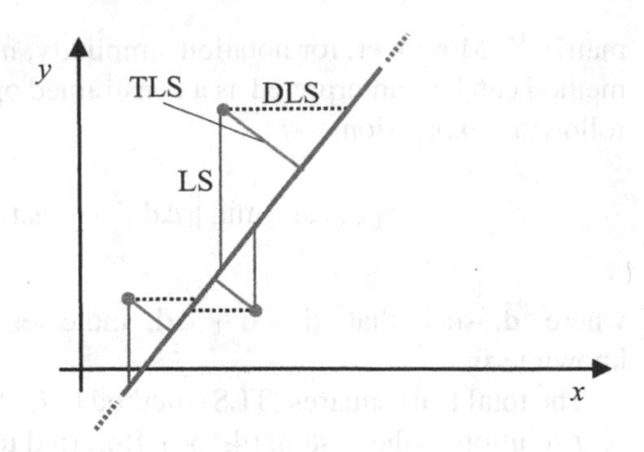

Fig. 4.12 Representation of the LS, TLS, and DLS optimization criteria. Choice of the distance to be minimized such that the *straight line* (optimally) approximates the available measures

In the case of LS the variable $\mathbf{x}$ is considered noiseless and the uncertainty is associated with the measure only of the quantity $\mathbf{y}$; it appears then that the error to be minimized is the sum of the distance parallel to the y-axis. Finally, the DLS technique is characterized by uncertainty only in the variable $\mathbf{x}$; in this case the quantity to minimize is equal to the sum of the distances parallel to the x-axis.

4.5.1 TLS Solution

Given the matrices $(\mathbf{X}, \mathbf{d})$ the TLS solution consists in the estimation of the $\overline{\mathbf{X}}$ matrix and $\overline{\mathbf{d}}$ vector, such that $\mathbf{w}$ satisfies the LS system $\overline{\mathbf{X}}\mathbf{w}_{\text{TLS}} = \overline{\mathbf{d}}$, or

$$(\mathbf{X} - \Delta\mathbf{X})\mathbf{w}_{\text{TLS}} = (\mathbf{d} - \Delta\mathbf{d}). \tag{4.168}$$

For the solution, note that the above expression can be written as

$$\left[\begin{bmatrix}\mathbf{d} & \mathbf{X}\end{bmatrix} - \begin{bmatrix}\Delta\mathbf{d} & \Delta\mathbf{X}\end{bmatrix}\right]\begin{bmatrix}-1 \\ \mathbf{w}_{\text{TLS}}\end{bmatrix} = \mathbf{0}. \tag{4.169}$$

By defining $\mathbf{S} \equiv \begin{bmatrix}\mathbf{d} & \mathbf{X}\end{bmatrix} \in (\mathbb{R}, \mathbb{C})^{N \times M+1}$ and $\Delta\mathbf{S} \equiv \begin{bmatrix}\Delta\mathbf{d} & \Delta\mathbf{X}\end{bmatrix} \in (\mathbb{R}, \mathbb{C})^{N \times M+1}$, respectively, as *augmented input matrix* and *augmented error matrix*, we have that

$$(\mathbf{S} - \Delta\mathbf{S})\begin{bmatrix}-1 \\ \mathbf{w}_{\text{TLS}}\end{bmatrix} = \mathbf{0}, \tag{4.170}$$

where the $\mathbf{S}$ matrix, for the presence of the noise, has full rank. If we assume that $N > M + 1$ it follows that $\text{rank}(\mathbf{S}) = M + 1$. The problem of the determination of the matrix $\Delta\mathbf{S}$, then, can be recast as the determination of the smallest perturbation of the augmented input matrix.

By expanding $\mathbf{S}$ with the SVD decomposition, we have that

$$\mathbf{S} = \sum_{i=0}^{M} \sigma_i \mathbf{u}_i \mathbf{v}_i^H, \tag{4.171}$$

where the terms σ_i represent the singular values of $\mathbf{S}$ in decreasing order and the vectors $\mathbf{u}_i$ and $\mathbf{v}_i$ are the left and right singular vectors, such that $\mathbf{u}_i^H \mathbf{u}_j = 0$ and $\mathbf{v}_i^T \mathbf{v}_j = 0$ for $i \neq j$. Since σ_M is the smallest singular value relative to the smallest perturbation, for the augmented error matrix $\Delta\mathbf{S}$, necessarily applies the expression

$$\Delta\mathbf{S} = \sigma_M \mathbf{u}_M \mathbf{v}_M^H. \tag{4.172}$$

By substituting the expressions (4.171) and (4.172) into (4.170) we can write

$$\left(\sum_{i=0}^{M} \sigma_i \mathbf{u}_i \mathbf{v}_i^H - \sigma_M \mathbf{u}_M \mathbf{v}_M^H \right) \begin{bmatrix} -1 \\ \mathbf{w}_{\text{TLS}} \end{bmatrix} = \left(\sum_{i=0}^{M-1} \sigma_i \mathbf{u}_i \mathbf{v}_i^H \right) \begin{bmatrix} -1 \\ \mathbf{w}_{\text{TLS}} \end{bmatrix} = \mathbf{0}. \tag{4.173}$$

Since $\mathbf{v}_M$ is orthogonal to the rest of the vectors $\mathbf{v}_0, \mathbf{v}_1, \ldots, \mathbf{v}_{M-1}$, the TLS solution for the coefficients filter vector can be written as

$$\begin{bmatrix} -1 \\ \mathbf{w}_{\text{TLS}} \end{bmatrix} = -\frac{\mathbf{v}_M}{v_{M,0}}, \tag{4.174}$$

where $v_{M,0}$ is the first nonzero element of the right singular vector $\mathbf{v}_M$ that satisfies (4.173). In other words, the TLS solution is described by right singular vectors corresponding to the smaller singular values of the augmented matrix $\mathbf{S}$.

An efficient approach for the singular vectors calculation consists in determining an optimal vector $\mathbf{v}$ such that following CF is minimum:

$$J(\mathbf{v}) = \frac{\mathbf{v}^H \mathbf{S}^H \mathbf{S} \mathbf{v}}{\|\mathbf{v}\|_2^2} \tag{4.175}$$

and the result is normalized such that we can write

$$\begin{bmatrix} -1 \\ \mathbf{w}_{\text{TLS}} \end{bmatrix} = -\frac{\mathbf{v}_{\text{opt}}}{v_{\text{opt},0}}, \tag{4.176}$$

where $\mathbf{v}_{\text{opt}}$ denotes the solution of the CF (4.175) minimization, and $v_{\text{opt},0}$ indicates the first element of the vector $\mathbf{v}_{\text{opt}}$. With simple calculations it appears that the optimal choice for $\mathbf{v}$ corresponds to the smallest eigenvalue of $\mathbf{S}^H\mathbf{S}$. So, for the TLS estimation, the described SVD procedure, also known (in other contexts) minor component analysis (MCA), can be used.

Remark The previous solution implies that the smallest singular value of the $\mathbf{S}$ matrix has unique value. If this hypothesis is not verified, the TLS problem would have infinite solutions. In this case, among the infinite solutions, is chosen that with a minimum norm $\|\mathbf{w}_{\mathrm{TLS}}\|_2^2$.

In the case of zero-mean, iid, Gaussian perturbations $\Delta\mathbf{S}$, it is possible to demonstrate that the TLS solution, which corresponds to that minimizes the CF (4.175), is an unbiased maximum-likelihood estimate. In other words, you can have a maximum-likelihood unbiased estimate, in the case of identical noise variances $\sigma_{\Delta X}^2 = \sigma_{\Delta d}^2$, on the data and on the known term.

4.5.2 Generalized TLS

The TLS method provides an unbiased estimate when the noise on the matrix $\mathbf{X}$ and the one on the vector $\mathbf{d}$ are iid with similar variances. However $\mathbf{X}$ and $\mathbf{d}$ may represent different physical quantities and these assumptions may therefore not be true.

In cases where $\sigma_{\Delta X}^2 \neq \sigma_{\Delta d}^2$ with $\Delta\mathbf{X}$ and $\Delta\mathbf{d}$ iid, we define the generalized TLS (GTLS), the algorithm that allows the determination of the optimal vector $\mathbf{w}$ by minimizing the following CF:

$$\begin{aligned} \mathbf{w}_{\mathrm{TLS}} &= \arg\min_{\mathbf{w}} J(\mathbf{w}) \\ &= \arg\min_{\mathbf{w}} \left(\gamma\|\Delta\mathbf{X}\|_F^2 + (1-\gamma)\|\Delta\mathbf{d}\|_F^2\right). \end{aligned} \tag{4.177}$$

The coefficient γ is defined by the expression

$$\frac{1-\gamma}{\gamma} = \beta = \frac{\sigma_{\Delta d}^2}{\sigma_{\Delta X}^2} \tag{4.178}$$

such that for $\gamma = 0$, (4.177) coincides with the standard LS, for $\gamma = 1$ with the DLS, and for $\gamma = 0.5$ it has just the TLS. In this case, considering (4.18) and (4.177), the CF to be minimized is the following:

$$J(\mathbf{w}) = E\{\mathbf{e}^H\mathbf{e}\}, \tag{4.179}$$

where the TLS error $\mathbf{e}$ is defined in the usual way as $\mathbf{e} = \mathbf{d} - \mathbf{X}\mathbf{w}$. So, remembering that $\mathbf{X} = \overline{\mathbf{X}} + \Delta\mathbf{X}$ and $\mathbf{d} = \overline{\mathbf{d}} + \Delta\mathbf{d}$ and defining $\overline{\mathbf{e}} = \overline{\mathbf{d}} - \overline{\mathbf{X}}\mathbf{w}$ we can write

$$\begin{aligned} \mathbf{e} &= \left(\overline{\mathbf{d}} + \Delta\mathbf{d}\right) - \left(\overline{\mathbf{X}} + \Delta\mathbf{X}\right)\mathbf{w} \\ &= \overline{\mathbf{e}} + (\Delta\mathbf{X} + \Delta\mathbf{d}). \end{aligned} \tag{4.180}$$

It is then

$$\begin{aligned} J(\mathbf{w}) &= E\Big\{\left[\bar{\mathbf{e}} + (\mathbf{\Delta X} + \mathbf{\Delta d})\right]^H \left[\bar{\mathbf{e}} + (\mathbf{\Delta X} + \mathbf{\Delta d})\right]\Big\} \\ &= E\{\bar{\mathbf{e}}^T\bar{\mathbf{e}}\} + \sigma_{\Delta d}^2 + \mathbf{w}^H\mathbf{R}_{\Delta X}\mathbf{w} \\ &= E\{\bar{\mathbf{e}}^T\bar{\mathbf{e}}\} + \sigma_{\Delta X}^2(\beta + \mathbf{w}^H\mathbf{w}), \end{aligned} \tag{4.181}$$

where, in the above expression, the noise component $\mathbf{\Delta X}$ is assumed uncorrelated and iid for which it is $\mathbf{R}_{\Delta X} = \sigma_{\Delta X}^2\mathbf{I}$ and parameter $\beta = \sigma_{\Delta d}^2/\sigma_{\Delta X}^2$represents the ratio between the noise powers. Moreover, in the minimization of the above expression, to eliminate the inherent estimate bias, due to the dependence of the noise from $\mathbf{w}$, it is convenient to redefine the CF (4.179) as

$$\begin{aligned} J(\mathbf{w}) &= \frac{1}{2}\frac{E\{\bar{\mathbf{e}}^H\bar{\mathbf{e}}\}}{\beta + \mathbf{w}^H\mathbf{w}} \\ &= \frac{1}{2}\frac{E\{\bar{\mathbf{e}}^H\bar{\mathbf{e}}\}}{\beta + \mathbf{w}^H\mathbf{w}} + \sigma_{\Delta X}^2. \end{aligned} \tag{4.182}$$

The previous CF removes, in fact, the effect of noise but implies that the ratio between the noise powers β must be *a priori* known. To derive an iterative algorithm, the CF can be rewritten with the expression

$$J(\mathbf{w}) = \sum_{i=0}^{M-1} J_i(\mathbf{w}), \tag{4.183}$$

where

$$\begin{aligned} J_i(\mathbf{w}) &= E\{\varepsilon_i^2[k]\} \\ &= \frac{1}{2}\frac{E\{e_i^2[k]\}}{(\beta + \mathbf{w}^H\mathbf{w})} \end{aligned} \tag{4.184}$$

with $e_i = w_i - \mathbf{x}_i^H\mathbf{w}$ end of the vector $\mathbf{x}_i^H$ is defined as ith column of $\mathbf{X}$. The estimate of the instantaneous gradient can be evaluated with the derivative $d\varepsilon_i/d\mathbf{w}$, for which

$$\frac{d\varepsilon_i}{d\mathbf{w}} = \frac{e_i[k]\mathbf{x}_i}{\beta + \mathbf{w}^H\mathbf{w}} - \frac{e_i^2[k]\mathbf{w}}{(\beta + \mathbf{w}^H\mathbf{w})^2}. \tag{4.185}$$

The iterative algorithm is therefore

$$\mathbf{w}_{k+1} = \mathbf{w}_k + \tilde{\eta}_k\tilde{e}_i[k](\mathbf{x}_i + \tilde{e}_i[k]\mathbf{w}_k), \tag{4.186}$$

where

$$\tilde{e}_i[k] = \frac{e_i[k]}{\beta + \mathbf{w}_k^H \mathbf{w}_k} - \frac{w_i - \mathbf{x}_i^H \mathbf{w}_k}{\beta + \mathbf{w}_k^H \mathbf{w}_k}.$$

Absorbing the term $\beta + \mathbf{w}_k^H \mathbf{w}_k$ (always positive) into the learning rate, so that

$$\eta_k = \frac{\tilde{\eta}_k}{\beta + \mathbf{w}_k^H \mathbf{w}_k}$$

the expression (4.186) can be rewritten as

$$\mathbf{w}_{k+1} = \mathbf{w}_k + \eta_k e_i[k] \left(\mathbf{x}_i + \tilde{e}_i[k] \mathbf{w}_k \right), \quad i = k \mod (M+1). \tag{4.187}$$

Remark The index i in the above expression is taken in cyclic mode $(M + 1)$ module, i.e., the columns of the matrix $\mathbf{X}$ and the elements w_i of the vector $\mathbf{w}$ are selected and processed in cyclic order.

4.6 Underdetermined Linear Systems with Sparse Solution

This section dealt with the problem of determining the solution for underdetermined LS systems, i.e., $\mathbf{Xw} \approx \mathbf{d}$, with $\mathbf{X} \in (\mathbb{R},\mathbb{C})^{N \times M}$ and $(N < M)$. The solution in this case is not unique, and among the infinite possibilities we can identify some that meet specific properties [22–29].

In the underdetermined case the LS system is said *over-complete* and the determination of solutions of interest can be formulated as a constrained optimization problem of the type already studied in Sect. 4.3.1.2, wherein the CF is defined as a function of the L_p-norm of the filter coefficients $\mathbf{w}$, i.e., $J_p(\mathbf{w}) = f(\|\mathbf{w}\|_p)$. In more formal terms, we can write

$$\mathbf{w}_* \therefore \arg\min_{\mathbf{w}} J_p(\mathbf{w}) \qquad \text{s.t.} \qquad \begin{cases} \mathbf{d} - \mathbf{Xw} = \mathbf{0} & \text{deterministic} \\ \left\| \mathbf{d} - \mathbf{Xw} \right\|_2 \le \mathbf{e} & \text{stochastic}, \end{cases} \tag{4.188}$$

where $f(\|\mathbf{w}\|_p)$ is an appropriate norm of the vector $\mathbf{w}$ of the type

$$J_p(\mathbf{w}) \equiv f\left(\left\| \mathbf{w} \right\|_p \right) = \sum_{i=0}^{M-1} |w_i|^p, \qquad \text{with} \qquad 0 \le p \le \infty. \tag{4.189}$$

The LS solution, as discussed in Sect. 4.3.1, among the infinite solutions determines the one with a minimum error energy, or minimum quadratic error norm $\|\mathbf{e}\|_2^2$. It is possible to find a solution depending of the norm order and, it is well known that some orders take a specific physical meaning. For example, in case of *infinity norm* $p = \infty$ the solution is indicated as the *minimum amplitude solution*. Moreover, for

$p = 1$ the problem can be formulated with the classical methods of linear programming, and there are many algorithms to determine the solution.

An interesting situation is when $0 \le p < 1$ wherein the vector solution, indicated as $\mathbf{w}_*$, contains elements equal to zero and the system is called *sparse*. The solution of a sparse system is often referred to as *minimum fuel solution*. In more formal terms, an underdetermined linear system has a *sparse solution* if the solution vector $\mathbf{w}_* \in (\mathbb{R},\mathbb{C})^{M\times 1}$ with $M > N$ has at most N nonzero elements. For example, in the case where $p = 0$ the solution represents a measure of the system sparseness, also called *numerosity*, as it defines the solution to a minimum number of non-null values. In formal terms

$$J_{p=0}(\mathbf{w}) = \text{num}\{i \therefore w_i \neq 0\}. \tag{4.190}$$

In general, there are numerous optimization algorithms such as those at minimum L_p-norm, with $0 \le p \le 1$, able to determine some solutions with precise mathematical properties, distinct from the remaining possible solutions.

Note that the (4.188) formulation is common in many real applications such as in the time–frequency representations, in the magnetic inverse problems, in the speech coding, in the spectral estimation, in the band-limited extrapolation, in the direction of arrival estimate, in the function approximation, in the fault diagnosis, and so on.

4.6.1 The Matching Pursuit Algorithms

Given the *over-complete* nature of the linear system (4.188), the number of basis in $\mathbf{X}$ is greater than the dimension of the desired signal. It follows that the sparse solution can represent a basis, i.e., the lowest representation, for the signal $\mathbf{d}$ itself.

In these cases, the problem consists in the selection of the *best basis* for the representation of the signal. This problem is known as *matching pursuit*.[4]

The *matching pursuit* consists then in determining the smallest subset of vectors, chosen on a redundant array, able to better represent the available data $\mathbf{d}$. For its determination, the signal is decomposed into a number of optimal bases, selected from a larger dictionary of bases, by means of optimization algorithms (called matching pursuit algorithms (MPA) or basis pursuit algorithms). In other words, in matching pursuit is necessary to identify a number of columns $\mathbf{x}_i$ of the matrix $\mathbf{X}$ that best represent the signal contained in the vector $\mathbf{d}$ (typically coming from sensors). This corresponds to the determination of a sparse solution of (4.188) for $p \le 1$.

The *minimum-numerosity* optimal base selection (for $p = 0$) can be made with, computationally very complex, enumerative methods of exhaustive search.

[4] The term *matching pursuit* indicates, in general, a numerical method for selecting the best projection (also known as *best matching*) of multidimensional data in a *over-complete* basis.

If you are interested in the selection of N vectors $\mathbf{x}_i$ that best represent $\mathbf{d}$, there are $M!/(M-N)!\,N!$ possible choice. By using exhaustive search, in fact, subsets of N equations can be obtained by removing, for each iteration j, $(N-M)$ columns of $\mathbf{X}$ and evaluating the L_p-norm of the optimal vector $\mathbf{w}_{*j} = \mathbf{X}_r^{-1}\mathbf{d}$, for each subset of these equations.

For high dimensionality problems, such methods are particularly inefficient. In fact, the determination of the smallest optimal base presents a complexity of order $O(NP)$ (called *NP-hard*). For large M the computational cost can be prohibitive and the "brute force" combinatorial approach cannot be made. Then the problem can be addressed in an alternative way with much faster and general sub-optimal search methods, able to find robust solutions, especially in the case where the data are corrupted by noise.

Property For a linear underdetermined system the optimal solution $\mathbf{w}_*$, which minimizes the L_p-norm, with the CF (4.188) with $p = 1$, contains at least N non-null elements. Also, if the column vectors $\mathbf{s}_i$ of the augmented matrix $\mathbf{S} \equiv [\mathbf{d}\ \mathbf{X}] \in (\mathbb{R},\mathbb{C})^{N\times M+1}$ satisfy the Haar condition,[5] then there is always a optimal vector $\mathbf{w}_*$ that has exactly N non-null components.

4.6.1.1 Best Basis Selection Problem Definition

The problem of the basis selection can be formulated in the following way. Let $D = \left[\mathbf{x}[k]\right]_{k=n}^{n-M+1}$ be a set of M vectors of length N, i.e., $\mathbf{x}[k] \in (\mathbb{R},\mathbb{C})^{N\times 1}$, such that $N \ll M$, and without loss of generality, have unit norm. Given a signal $\mathbf{d} \in (\mathbb{R},\mathbb{C})^N$, typically derived from measurement of a physical phenomenon, available with or without measurement error, the problem is to determine the most compact representation of the data $\mathbf{d}$, together with its tolerance, using a subset of basis vectors available in the dictionary D. In other words, we must determine the *sparsity index* r such that $\left[\mathbf{x}[k]\right]_{k=0}^{r-1}$ represents the "best" model for $\mathbf{d}$.

Because you are *pursuing* the goal of determining the smallest vectors set belonging to the dictionary D that best represent $\mathbf{d}$, these methodologies, as previously indicated, are called MPA. More precisely, considering a data matrix $\mathbf{X} \in (\mathbb{R},\mathbb{C})^{N\times M}$ formed with the dictionary vectors, defined as the set of column vectors $\mathbf{X} = \left[\mathbf{x}[n] \quad \cdots \quad \mathbf{x}[n-M+1]\right]$, the problem can be formulated as the determination of a solution $\mathbf{w}_* \in (\mathbb{R},\mathbb{C})^{M\times 1}$, with the minimum number (maximum N) of nonzero values such that $\left\|\mathbf{X}\mathbf{w} - \mathbf{d}\right\|_2 \leq \mathbf{e}$ or, in the deterministic case where $\mathbf{e} = \mathbf{0}$, such that $\mathbf{X}\mathbf{w} = \mathbf{d}$. Since the size of the null space of $\mathbf{X}$ is greater than zero $\left(\mathcal{N}(\mathbf{X}^H) > 0\right)$, the problem of minimization admits infinite solutions.

[5] A set of vectors $\mathbf{x} \in (\mathbb{R},\mathbb{C})^N$ satisfies the Haar condition if every set of N vectors is linearly independent. In other words, each subset selection of N vectors, from a base for the space $(\mathbb{R},\mathbb{C})^N$. A system of equations that satisfies the Haar condition is sometimes referred to as Tchebycheff system [21, 30].

4.6.2 Approximate Minimum $\mathbf{L_p}$-Norm LS Iterative Solution

According to (4.188), the determination of the sparse solution can be made by considering the LS (a minimum L_2-norm) as a weak approximation of the minimum L_p-norm solution. In fact, it is well known that the minimum energy solution, by definition, is never sparse by having typically all nonzero terms since, instead of concentrating the energy in a few points, it tends to *smear the solution* over a large number of values.

Formally, the problem can be defined by (4.188) where $f(\|\mathbf{w}\|_p) = \|\mathbf{w}\|_2$. In Sect. 4.3.1.2, we have seen that in the case of underdetermined LS system the solution is defined as

$$\mathbf{w}_{\mathrm{LS}} = \mathbf{X}^{\#}\mathbf{d} \tag{4.191}$$

with $\mathbf{X}^{\#} = \mathbf{X}^H(\mathbf{X}\mathbf{X}^H)^{-1}$ Moore–Penrose pseudoinverse matrix that, in general, produces a solution in which no elements of the $\mathbf{w}_{\mathrm{LS}}$ vector are zero. In other words, for $0 \le p \le 1$, you must select a few *best* columns of the $\mathbf{X}$ matrix. By applying an empirical approach, you can make the selection by imposing a sort of competition among the $\mathbf{X}$ columns vectors, which emphasized some of the columns and inhibits the other. At the end of this process (which can be iterated several times), only N columns survive while the others $(M - N)$ are forced to zero. The L_2 solution, together with the $\mathbf{X}$ columns selection criterion, represents a robust and computational efficient paradigm that represents a consistent approximation of the minimum L_p-norm (or sparse) solution.

4.6.2.1 Minimum Quadratic Norm Sparse Solution

A first approximate approach, called minimum norm solution (MNS), consists in an iterative procedure that selectively forces to zero a subset of the minimum energy solution. We proceed in the following modality.

Step 1 Estimate of the minimum L_2-norm solution, $\mathbf{w}_{\mathrm{LS}} = \mathbf{X}^H(\mathbf{X}\mathbf{X}^H)^{-1}$.

Step 2 On the basis of the obtained solution, remove some of the columns (at least one) corresponding to the $\mathbf{w}_{\mathrm{LS}}$ components with a minimum module (or other criteria) and force to zero such components.

Step 3 Calling $\mathbf{X}_r \in (\mathbb{R},\mathbb{C})^{N\times r}$, with $r \ge N$, the reduced data matrix (obtained by removing the columns of $\mathbf{X}$ with the procedure in step 2), estimate the remaining components of $\mathbf{w}_{1r} \in (\mathbb{R},\mathbb{C})^{r\times 1}$ vector as $\mathbf{w}_{1r} = \mathbf{X}_r^H(\mathbf{X}_r\mathbf{X}_r^H)^{-1}\mathbf{d} = \mathbf{X}_r^{\#}\mathbf{d}$.

Step 4 Repeat the procedure in steps 1–3, until the $(M - N)$, or as otherwise specified, the remaining columns of $\mathbf{X}$ are removed.

At the end of the procedure only N coefficients of $\mathbf{w}_*$, contained in the vector $\mathbf{w}_{Nr}$, are different from zero.

For a better understanding of the method, consider the following *minimum fuel problem* example (modified from [15]). Minimization of the $\|\mathbf{w}\|_1$ norm is subject to the constraint $\mathbf{X}\mathbf{w} = \mathbf{d}$, where the matrix $\mathbf{X}$ and the vector $\mathbf{d}$ are real and defined as

$$\mathbf{X} = \begin{bmatrix} 2 & -1 & 20 & -1 & -1 & 11 & 1 & 34 & -1 \\ -1 & 2 & 18 & 1 & 1 & 15 & -2 & 25 & 1 \\ 1 & 1 & 6 & 1 & -1 & 16 & 1 & 30 & -2 \end{bmatrix} \quad \mathbf{d} = \begin{bmatrix} 104 \\ 87 \\ 116 \end{bmatrix}. \quad (4.192)$$

The first step for the minimum L_1-norm solution consists in determining minimum energy (L_2) by means of (4.191)

$$\begin{aligned} \mathbf{w}_{\mathrm{LS}} &= \mathbf{X}^{\#}\mathbf{d} \\ &= [0.0917\ 0.2210\ -0.8692\ 0.2546\ -0.1684\ 2.0366\ 0.2019\ 2.8978\ -0.3798]^T. \end{aligned}$$

The second step is to select the three values of maximum modulus $w[2]$, $w[5]$, and $w[7]$. The others are set to zero

$$\mathbf{w}_{\mathrm{LS}}' = \begin{bmatrix} 0 & 0 & w[2] & 0 & 0 & w[5] & 0 & w[7] & 0 \end{bmatrix}^T$$

while the corresponding columns of $\mathbf{X}$ are eliminated and the new data matrix $\mathbf{X}_r$ reduces to

$$\mathbf{X}_r = \begin{bmatrix} 20 & 11 & 34 \\ 18 & 15 & 25 \\ 6 & 16 & 30 \end{bmatrix}. \quad (4.193)$$

In the third (and final step), the nonzero solutions of $\mathbf{w}_0'$ are determined as

$$\begin{aligned} \mathbf{w}_{1r} &= \mathbf{X}_r^{\#}\mathbf{d} \\ &= [-1 \quad 2 \quad 3]^T. \end{aligned}$$

The minimum L_1-norm solution is then

$$\mathbf{w}_{*L1} = [0 \quad 0 \quad -1 \quad 0 \quad 0 \quad 2 \quad 0 \quad 3 \quad 0]^T.$$

To ensure optimal performance, it is necessary to iterate the procedure several times by removing, at each iteration, only some columns of $\mathbf{X}$.

An alternative way for the removal of the $\mathbf{X}$ column consists in selecting the element of $\mathbf{w}_{\mathrm{LS}}$ such that, removed, the larger decrease of the norm $\|\mathbf{w}\|_1$ is determined.

Multichannel Extension

In many real-world signal processing, the observation vector $\mathbf{d}$ is available in multiple distinct time instants. In these cases it is possible to write more equations of the $\mathbf{X}\mathbf{w}_k = \mathbf{d}_k$, for $k = 0, 1, ..., K - 1$, which in compact form can be written as

$$\mathbf{X}\mathbf{W} = \mathbf{D}, \tag{4.194}$$

where $\mathbf{W} \in (\mathbb{R}, \mathbb{C})^{N \times K} = [\mathbf{w}_0 \quad \cdots \quad \mathbf{w}_{K-1}]$ and $\mathbf{D} \in (\mathbb{R}, \mathbb{C})^{N \times K} = [\mathbf{d}_0 \ldots \mathbf{d}_{K-1}]$. The goal of the optimization process is to find a sparse representation of the matrix $\mathbf{W}$ and it is therefore necessary that all the columns of $\mathbf{W}$ have the same sparse structure. The procedure for the determination of the solution is a simple extension of the one presented in the previous paragraph.

Step 1 Estimate of the LS solution (4.191), $\mathbf{W}_{\text{LS}} = \mathbf{X}^{\#}\mathbf{D}$.

Step 2 On the basis of the step 1 solution, identify and force to zero few rows (at least one) of $\mathbf{W}_{\text{LS}}$ and remove the corresponding columns of $\mathbf{X}$.

Step 3 Calling $\mathbf{X}_r \in (\mathbb{R},\mathbb{C})^{N \times r}$ with $r \geq N$ the reduced data matrix (obtained by removing the columns of $\mathbf{X}$ with the procedure in step 2), estimate the remaining components of $\mathbf{W}_{1r} \in (\mathbb{R},\mathbb{C})^{r \times M}$ as $\mathbf{W}_{1r} = \mathbf{X}_r^{\#}\mathbf{D}$.

Step 4 Repeat the procedure in steps 1–3 until $(M - N)$, or as otherwise specified, the remaining columns of $\mathbf{X}$ are removed.

4.6.2.2 Uniqueness of Solution

Consider the underdetermined system $\mathbf{X}\mathbf{w} = \mathbf{d}$, with $(N < M)$, and define the $\mathbf{X}_r \in (\mathbb{R}, \mathbb{C})^{N \times N_0}$ matrix constructed using the N_0 columns of $\mathbf{X}$ associated with the $N_0 \leq N$ desired null elements of the $\mathbf{w}_*$ vector. Moreover, let $\mathbf{X}_2 \in (\mathbb{R}, \mathbb{C})^{N \times (M - N_0)}$ be the matrix with $M - N_0$ columns of $\mathbf{X}$ associated with the zero entries of $\mathbf{w}_*$. If the reduced matrix $\mathbf{X}_r$ has full rank columns, $\mathbf{w}_*$ is the unique minimum L_1-norm solution s.t. $\mathbf{X}\mathbf{w} = \mathbf{d}$, if and only if

$$\|\mathbf{g}\|_\infty < 1, \qquad \text{with} \qquad \mathbf{g} = \mathbf{X}_2^T \left[\mathbf{X}_r^{\#}\right]^H \text{sign}\left[\mathbf{X}_r^{\#}\mathbf{d}\right]. \tag{4.195}$$

In the case that is also true the equality $\|\mathbf{g}\|_\infty \leq 1$ the solution is optimal but not unique.

Moreover, note that the presented iterative algorithms, while not guaranteeing the convergence to the optimal solution, are able to determine one of its good approximations.

4.6.2.3 Sparse Minimum Weighted L_2-Norm Solution

An MNS method variant consists in considering, inside the recurrence, a weighted quadratic norm minimization. Considering the expression of the CF (4.188) is then

$$f\left(\|\mathbf{w}\|_p\right) = \|\mathbf{G}^{-1}\mathbf{w}\|_2, \tag{4.196}$$

where $\mathbf{G}^{-1} \in (\mathbb{R},\mathbb{C})^{M\times M}$ is defined as a weighing matrix. The method is often referred to as weighted minimum norm solution (WMNS). In this case the solution is

$$\mathbf{w}_* = \mathbf{G}[\mathbf{X}\mathbf{G}]^{\#}\mathbf{d}. \tag{4.197}$$

In order to consider the cases of singular $\mathbf{G}$ matrix, in the definition of WMNS solution, the CF can be extended as $\|\mathbf{G}^{\#}\mathbf{w}\|$, so any solution can be generated with constraint $\mathbf{X}\mathbf{w} \approx \mathbf{d}$. In particular, for $\mathbf{G}$ diagonal, the CF is

$$\left\|\mathbf{G}^{\#}\mathbf{w}\right\|_2 = \sum_{i=0,\, g_i\neq 0}^{M-1}\left(\frac{w_i}{g_i}\right)^2, \quad \mathbf{G} = \mathrm{diag}(g_0, g_1, \ldots, g_{M-1}). \tag{4.198}$$

4.6.2.4 Low-Resolution Electromagnetic Tomography Algorithm

The $\mathbf{G}$ matrix is usually heuristically determined, and/or based on *a priori* knowledge in order to force the solution sparseness. For example, in the specific application problem of electromagnetic sensors, for the method referred to as LOw-Resolution Electromagnetic Tomography Algorithm (LORETA) [22], in (4.188) the WMNS is expressed as

$$f\left(\|\mathbf{w}\|_p\right) = \|\mathbf{w}^H\mathbf{G}^{-1}\mathbf{w}\|_2 \tag{4.199}$$

with solution

$$\mathbf{w}_* = \mathbf{G}\mathbf{X}^H\left[\mathbf{X}\mathbf{G}\mathbf{X}^H\right]^{\#}\mathbf{d}. \tag{4.200}$$

In particular, in the LORETA algorithm, the weighing matrix is defined as

$$\mathbf{G}^{-1} = \mathbf{B}\otimes \mathrm{diag}\left(\|\mathbf{x}_0\|_2, \|\mathbf{x}_1\|_2, \ldots, \|\mathbf{x}_{M-1}\|_2\right), \tag{4.201}$$

where $\otimes$ indicates the Kronecker product (see Sect. A.13), with $\mathbf{B}$ indicated the spatial discrete Laplacian operator which depends on the spatial location of the sensors, and $\|\mathbf{x}_i\|$ is shown with the L_2-norm of the ith column vector of $\mathbf{X}$.

4.6.2.5 Focal Underdetermined System Solver Algorithm

Proposed by Gorosnitsky and Rao in [23] and generalized and extended in [24, 25, 29], an alternative algorithm that generalizes previous approaches is called FOCal Underdetermined System Solver (FOCUSS). The system solution is strongly influenced by the initial condition that, depending on the application area, in turn, depends on the sensors characteristics (spatial distribution, noise, etc.) that can be determined by the procedure WMNS or LORETA. The FOCUSS algorithm consists in the repetition of the procedure WMNS adjusting, each iteration, the weighing matrix $\mathbf{G}$ until a large number of solution elements become close to zero in order to obtain a sparse solution.

For simplicity, consider the noiseless case so that $\mathbf{d}$ can be exactly represented by some dictionary columns. Again for simplicity, in the development define the vector $\mathbf{q}$ such that

$$\left\|\mathbf{q}\right\|_2 = \left\|\mathbf{G}^{\#}\mathbf{w}\right\|_2 \tag{4.202}$$

so, the optimization problem defined by WMNS (4.188) can be reformulated as

$$\mathbf{w}_* = \mathbf{G}\mathbf{q} \qquad \text{where} \qquad \mathbf{q} \therefore \arg\min_{\mathbf{q}} \left\|\mathbf{q}\right\|_2^2 \qquad \text{s.t.} \qquad \mathbf{X}\mathbf{G}\mathbf{q} = \mathbf{d}. \tag{4.203}$$

Starting from an initial solution $\mathbf{w}_0$ calculated, for example, with (4.197) or with (4.200), the algorithm FOCUSS in its basic form (see for [22] details) can be formalized by the following recursive expression:

$$\begin{aligned}
&\text{Step 1}: \quad \mathbf{G}_{Pk} = \left(\mathrm{diag}(\mathbf{w}_{k-1})\right) = \mathrm{diag}(w_{0,k-1}, w_{1,k-1}, \ldots, w_{M-1,k-1}),\\
&\text{Step 2}: \quad \mathbf{q}_k = (\mathbf{X}\mathbf{G}_{Pk})^{\#}\mathbf{d},\\
&\text{Step 3}: \quad \mathbf{w}_k = \mathbf{G}_{Pk}\mathbf{q}_k.
\end{aligned} \tag{4.204}$$

where $\mathbf{G}_{Pk}$ denotes *a posteriori* weighing matrix. In other words, at the kth iteration, $\mathbf{G}_{Pk}$ is a diagonal matrix that is *a priori* determined by $\mathbf{w}_{k-1}$ solution. Without loss of generality, to avoid biased zero solution, the initial value $\mathbf{w}_0$ of the WMNS solution is considered all nonzero elements. Note, also, that steps 2 and 3 of (4.204) represent a WMNS solution and that in the implementation, the algorithm can be written in a single step.

From vector (4.202) definition, the sparse solution determination is performed by forcing to zero the solutions w_i such that the ratio $(w_i/g_i) \rightarrow 0$ [see (4.198)]. So that the procedure produces (1) a partial reinforcement of some prominent indices of the current solution $\mathbf{w}_k$ and, (2) the suppression of the remaining (up to the limits) due to the achievement of the machine precision. Finally, the algorithm is stopped once the minimum number of desired solutions is reached.

Note that the algorithm does not simply increment the solutions that already at the beginning are large. During the procedure, these often become null while others,

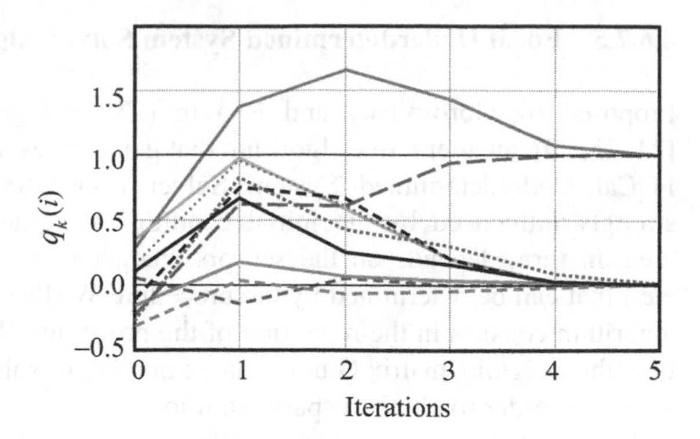

Fig. 4.13 FOCUSS algorithm. Trend of the elements $q_k(i)$ during the algorithm iterations for a (10×4) matrix **X** example (modified from [23])

small at the beginning, can emerge. Note also that CF (4.198) is never explicitly evaluated. The weights $w_i = 0$ and the corresponding subspaces are in fact implicitly deleted in (4.204) from the calculation of the product $(\mathbf{X}\mathbf{G}_{Pk})$. At the procedure end the vector elements will tend to assume values $q_k(i) \to 0$ for $w_k(i) \to 0$ and $q_k(i) \to 1$ for $w_k(i) \neq 0$.

Figure 4.13 shows the typical $q(i)$ elements convergence trend and it can be observed that after a small number of iterations converge to the value zero or one.

4.6.2.6 General FOCUSS Algorithm

The FOCUSS algorithm can be extended by introducing two variants. The first is to consider the term $\mathbf{w}_{k-1}^{l}$ in the recurrence (instead of $\mathbf{w}_{k-1}$) with $l \in \mathbb{N}^{+}$, and the second is to consider a pre-calculated additional matrix $\mathbf{G}_{Ak}$ at the beginning of the procedure, constant for all iterations and independent of the *a posteriori* constraint. This extension makes the algorithm more flexible and suitable for many different applications and provides a general method for the insertion of *a priori* information.

The form of the algorithm is then

$$\begin{aligned} &\text{Step 1}: \quad \mathbf{G}_{Pk} = \operatorname{diag}\left(\mathbf{w}_{k-1}^{l}\right), \\ &\text{Step 2}: \quad \mathbf{q}_k = (\mathbf{X}\mathbf{G}_{Ak}\mathbf{G}_{Pk})^{\#}\mathbf{d}, \\ &\text{Step 3}: \quad \mathbf{w}_k = \mathbf{G}_{Ak}\mathbf{G}_{Pk}\mathbf{q}_k. \end{aligned} \tag{4.205}$$

In case that a positivity constraint is imposed on the solution (i.e., $w_i > 0$), it is possible to extend the l exponent value to the real field for $l > 0.5$. This lower limit depends on the convergence algorithm properties not reported for brevity (for details, refer to [23]). The positivity constraint can be reinforced by incorporating in the algorithm a vector defined as $\mathbf{p}_k = \mathbf{w}_k - \mathbf{w}_{k-1}$. The iterative solution then becomes of the type $\hat{\mathbf{w}}_k = \mathbf{w}_{k-1} + \alpha\mathbf{p}_{k-1}$ where α represents the *adaptation step* chosen in order to have $\hat{\mathbf{w}}_k > 0$. More generally, it is possible to define other nondecreasing $\mathbf{w}_{k-1}$ functions, to be included into (4.205).

Implementation Notes

It is noted that calling $\mathbf{G}_k = \mathbf{G}_{Ak}\mathbf{G}_{Pk}$, for each iteration the FOCUSS algorithm requires the evaluation of $(\mathbf{XG}_k)^{\#}$ which corresponds to the $\mathbf{X}$ data matrix weighing at kth step. In the case when the term $(\mathbf{XG}_k)^{\#}$ was ill-conditioned the inverse calculation must be regularized in order to prevent too large $\mathbf{w}$ changes. For example, using the Tikhonov theory, the CF shall include an additive regularizing. For which the new CF becomes

$$\arg\min_{\mathbf{w}} \left[\left\| \mathbf{d} - \mathbf{Xw} \right\|_2^2 + \delta^2 \left\| \mathbf{G}_k \mathbf{w} \right\|_2^2 \right]. \tag{4.206}$$

When the condition number of $\mathbf{X}_G = \mathbf{XG}_k$ matrix is not very high, the solution (4.206) can be determined by solving the following normal equations:

$$\left(\mathbf{X}_G^H \mathbf{X}_G + \delta^2 \mathbf{I} \right) \mathbf{w}_{k+1} = \mathbf{X}_G^H \mathbf{d}. \tag{4.207}$$

4.6.2.7 FOCUSS Algorithm Reformulation by Affine Scaling Transformation

In this section we see how the optimal basis selection can be done through a *diversity measure*. The algorithm is derived by an L_p-norm $(p \le 1)$ diversity measure minimization that is, in turn, determined according to the entropy (defined in different modes) [28]. As we shall see the algorithm, which is closely related to the affine scaling transformation (AST), is equivalent to the previously described FOCUSS. The more general nature of the formulation allows for a new interpretation and extension of this class of algorithms. It also allows a more appropriate study of the convergence properties.

The optimization problem is formulated as in (4.188) where the CF $J_\rho(\mathbf{w})$, in this context called *diversity measure*, is a measure of the signal sparsity for which the function $J_\rho(\mathbf{w})$ can take various forms.

Diversity Measure $J_\rho(\mathbf{w})$

The most common form of the diversity measure, known in the literature for the linear inverse problems solution, is precisely that defined by (4.189). This measure was extended in [28], by introducing negative p values. Here are a few paradigms for the diversity measurement.

Diversity measure $\mathrm{L}_{(p\le 1)}$ *or generalized* $\mathrm{L_p}$*-norm* Such a diversity measure is defined as

$$J_\rho(\mathbf{w}) = \text{sign}(p)\sum_{i=0}^{M-1}|w_i|^p = \begin{cases} \sum_{i=0}^{M-1}|w_i|^p & 0 \leq p \leq 1 \\ -\sum_{i=0,\, w_i \neq 0}^{M-1}|w_i|^p & p < 0. \end{cases} \tag{4.208}$$

Note that the above expression, for $0 \leq p \leq 1$, represents a general form of entropy. The close connection with the expression a vector L_p-norm is such that this type of formulation indicated as $L_{(p \leq 1)}$ represents a p-norm-like *diversity measures* that, in fact, for negative p is not a true norm.

Diversity measurement with Gaussian entropy In this case the CF expression is

$$\begin{aligned} J_{\mathrm{G}}(\mathbf{w}) &= H_{\mathrm{G}}(\mathbf{w}) \\ &= \sum_{i=0}^{M-1} \ln|w_i|^2. \end{aligned} \tag{4.209}$$

Diversity measurement with Shannon entropy The CF expression is

$$\begin{aligned} J_{\mathrm{S}}(\mathbf{w}) &= H_{\mathrm{S}}(\mathbf{w}) \\ &= -\sum_{i=0}^{M-1} \widetilde{w}_i \log|\widetilde{w}_i|, \end{aligned} \tag{4.210}$$

where the $\widetilde{w}_i$ element can take different forms $\widetilde{w}_i = |w_i|$, $\widetilde{w}_i = |w_i| / \|w_i\|_1$, $\widetilde{w}_i = |w_i| / \|w_i\|_2$, or $\widetilde{w}_i = w_i$ per $w_i \geq 0$.

Diversity measurement with Renyi entropy The CF expression is

$$\begin{aligned} J_{\mathrm{R}}(\mathbf{w}) &= H_{\mathrm{R}}(\mathbf{w}) \\ &= \frac{1}{1-p} \log \sum_{i=1}^{M} (\widetilde{w}_i)^p, \end{aligned} \tag{4.211}$$

where $\widetilde{w}_i = |w_i| / \|w_i\|_1$ and $p \neq 1$.

Algorithm Derivation

Unlike the previous approach, considering the deterministic case, the algorithm derivation is made using the Lagrange multipliers method. Defining the Lagrangian $L(\mathbf{w},\boldsymbol{\lambda})$ such that

$$L(\mathbf{w},\boldsymbol{\lambda}) = J_{\rho}(\mathbf{w}) + \boldsymbol{\lambda}^{H}(\mathbf{d} - \mathbf{X}\mathbf{w}), \tag{4.212}$$

where $\boldsymbol{\lambda} \in (\mathbb{R},\mathbb{C})^{N\times 1}$ is the Lagrange multipliers vector, the necessary condition, so that $\mathbf{w}_*$ represents an optimal solution is that the vectors pair $(\mathbf{w}_*,\boldsymbol{\lambda}_*)$ satisfies the following expressions:

$$\begin{aligned} \nabla_{\mathbf{w}}L(\mathbf{w}_*,\boldsymbol{\lambda}_*) &= \nabla_{\mathbf{w}}J_{\rho}(\mathbf{w}) + \mathbf{X}^{H}\boldsymbol{\lambda}_* = 0 \\ \nabla_{\boldsymbol{\lambda}}L(\mathbf{w}_*,\boldsymbol{\lambda}_*) &= \mathbf{d} - \mathbf{X}\mathbf{w}_* = \mathbf{0}, \end{aligned} \tag{4.213}$$

where $\nabla_{\mathbf{w}}J_{\rho}(\mathbf{w})$ is the gradient of the diversity measure respect to the w_i elements. In the case of sparsity measurement, as defined by generalized L_p-norm (4.208), the expression of the gradient is equal to

$$\nabla_{w_i}J_{\rho}(\mathbf{w}) = |p| \cdot |w_i|^{p-2}w_i. \tag{4.214}$$

So substituting this into (4.213) yields a nonlinear equation in the variable $\mathbf{w}$ with solution not easy to calculate. To remedy this situation, the sparsity measure gradient can be represented in the following factorized form:

$$\nabla_{\mathbf{w}}J_{\rho}(\mathbf{w}) = \alpha(\mathbf{w})\boldsymbol{\Pi}(\mathbf{w})\mathbf{w}, \tag{4.215}$$

where $\alpha(\mathbf{w})$ and $\boldsymbol{\Pi}(\mathbf{w})$ are explicit functions of $\mathbf{w}$. For example, in the case of generalized L_p-norm (4.208) $\alpha(\mathbf{w}) = |p|$ and $\boldsymbol{\Pi}(\mathbf{w}) = \text{diag}\left(|w_i|^{p-2}\right)$. For (4.213) and (4.215), it follows that the solution (stationary point) satisfies the relations

$$\begin{aligned} \alpha(\mathbf{w}_*)\boldsymbol{\Pi}(\mathbf{w}_*)\mathbf{w}_* + \mathbf{X}^{H}\boldsymbol{\lambda}_* &= 0 \\ \mathbf{d} - \mathbf{X}\mathbf{w}_* &= \mathbf{0}. \end{aligned} \tag{4.216}$$

It is noted that for $p \leq 1$ the inverse matrix $\boldsymbol{\Pi}^{-1}(\mathbf{w}_*) = \text{diag}\left(|w_i|^{2-p}\right)$ exists for each $\mathbf{w}$. So solving (4.216) we obtain

$$\mathbf{w}_* = -\frac{1}{\alpha(\mathbf{w}_*)}\boldsymbol{\Pi}^{-1}(\mathbf{w}_*)\mathbf{X}^{H}\boldsymbol{\lambda}_*. \tag{4.217}$$

By substituting $\mathbf{w}^*$ in the second equation of (4.216) and solving for $\boldsymbol{\lambda}_*$, we get

$$\boldsymbol{\lambda}_* = -\alpha(\mathbf{w}_*)\left(\mathbf{X}\boldsymbol{\Pi}^{-1}(\mathbf{w}_*)\mathbf{X}^{H}\right)^{-1}\mathbf{d}. \tag{4.218}$$

Finally, replacing the latter in (4.217) we have

$$\mathbf{w}_* = \boldsymbol{\Pi}^{-1}(\mathbf{w}_*)\mathbf{X}^{H}\left(\mathbf{X}\boldsymbol{\Pi}^{-1}(\mathbf{w}_*)\mathbf{X}^{H}\right)^{-1}\mathbf{d}. \tag{4.219}$$

The latter is not useful to determine the solution since the optimal vector $\mathbf{w}_*$ appears both in the left and in the right sides. The expression, in fact, represents only a condition that must be satisfied by the solution. However, (4.219) suggests the following iterative procedure:

$$\mathbf{w}_{k+1} = \mathbf{\Pi}^{-1}(\mathbf{w}_k)\mathbf{X}^H\left(\mathbf{X}\mathbf{\Pi}^{-1}(\mathbf{w}_k)\mathbf{X}^H\right)^{-1}\mathbf{d} \tag{4.220}$$

that, being $\mathbf{\Pi}^{-1}(\mathbf{w}_k) = \text{diag}\left(|w_{k,i}|^{2-p}\right)$ for $p \leq 1$, does not pose particular implementative problems also in the case of sparse solution (which converges to zero for many elements w_i). It is known, in fact, that for $w_i = 0$, the corresponding diagonal element of $\mathbf{\Pi}^{-1}$ is zero.

Defining the matrix $\widetilde{\mathbf{\Pi}}^{-1}(\mathbf{w}_k) = \mathbf{\Pi}^{-\frac{1}{2}}(\mathbf{w}_k) = \text{diag}\left(|w_{k,i}|^{1-\frac{p}{2}}\right)$, a more compact form for (4.220) is the following:

$$\mathbf{w}_{k+1} = \widetilde{\mathbf{\Pi}}^{-1}(\mathbf{w}_k)\left(\mathbf{X}\widetilde{\mathbf{\Pi}}^{-1}(\mathbf{w}_k)\right)^{\#}\mathbf{d}. \tag{4.221}$$

It is noted that for $p = 2$, $\widetilde{\mathbf{\Pi}}^{-1}(\mathbf{w}_k) = \mathbf{I}$, for which the algorithm coincides with the standard LS formulation $\mathbf{w}_* = \mathbf{X}^{\#}\mathbf{d}$. Another interesting situation, in which $p = 0$, is that where the diagonal matrix is equal to $\widetilde{\mathbf{\Pi}}^{-1}(\mathbf{w}_k) = \text{diag}\left(|w_{k,i}|\right)$. To derive more rigorously the solution for $p = 0$, instead of using the generalized L_p-norm, you can use the Gaussian norm (4.209) for which the gradient in (4.213) can be expressed as

$$\nabla_{\mathbf{w}} J_\rho(\mathbf{w}) = 2\mathbf{\Pi}_G(\mathbf{w})\mathbf{w}, \tag{4.222}$$

where $\mathbf{\Pi}_G(\mathbf{w}) = \text{diag}\left(|w_i|^2\right)$.

Remark In the case of particularly noisy data, the expression (4.220) can be generalized by a regularization parameter, for which it is

$$\mathbf{w}_{k+1} = \mathbf{\Pi}^{-1}(\mathbf{w}_k)\mathbf{X}^H\left(\mathbf{X}\mathbf{\Pi}^{-1}(\mathbf{w}_k)\mathbf{X}^H + \delta_k\mathbf{I}\right)^{-1}\mathbf{d}, \tag{4.223}$$

where the term $\delta_k > 0$ represents the Tikhonov regularization parameter that can be chosen as a noise level function.

Multichannel Extension

In the multichannel case in which $\mathbf{XW} = \mathbf{D}$, the generalized norm may take the form

$$J_\rho(\mathbf{W}) = \text{sign}(p)\sum_{j=0}^{M-1}\|\mathbf{w}_j\|_2^p \quad 0 \leq p \leq 1 \qquad \text{s.t.} \qquad \mathbf{D} - \mathbf{XW} = \mathbf{0}. \tag{4.224}$$

The general FOCUSS expression is

$$\mathbf{W}_{k+1} = \mathbf{\Pi}^{-1}(\mathbf{W}_k)\mathbf{X}^H\left(\mathbf{X}\mathbf{\Pi}^{-1}(\mathbf{W}_k)\mathbf{X}^H\right)^{-1}\mathbf{W}, \tag{4.225}$$

where the matrix $\mathbf{\Pi}^{-1}(\mathbf{W}_k) = \operatorname{diag}\left(\|\mathbf{w}_{k,j}\|_2^{2-p}\right)$.

Remark The problem of finding sparse solutions to underdetermined linear problems from limited data arises in many real-world applications, as for example: spectral estimation and signal reconstruction, direction of arrival (DOA), compressed sensing, biomagnetic imaging problem, etc.

More details may be found in the literature. See for example [23–29].

References

1. Kay SM (1993) Fundamental of statistical signal processing estimation theory. Prentice Hall, Englewood Cliffs, NJ
2. Kailath T (1974) A view of three decades of linear filtering theory. IEEE Trans Inform Theor IT20(2):146–181
3. Box GEP, Jenkins GM (1970) Time series analysis: forecasting and control. Holden-Day, San Francisco, CA
4. Bates DM, Watts DG (1988) Nonlinear regression analysis and its applications. Wiley, New York
5. Landweber L (1951) An iteration formula for Fredholm integral equations of the first kind. Am J Math 73:615–624
6. Kaczmarz S (1937) Angenäherte Auflösung von Systemen linearer Gleichungen. Bulletin International de l'Académie Polonaise des Sciences et des Lettres Classe des Sciences Mathématiques et Naturelles Série A, Sciences Mathématiques 35:355–357
7. Manolakis DG, Ingle VK, Kogon SM (2000) Statistical and adaptive signal processing. McGraw-Hill, New York
8. Golub GH, Van Loan CF (1989) Matrix computation. John Hopkins University Press, Baltimore, MD. ISBN 0-80183772-3
9. Strang G (1988) Linear algebra and its applications, 3rd edn. Thomas Learning, Cambridge. ISBN:0-15-551005-3
10. Petersen KB, Pedersen MS, The matrix cookbook. http://matrixcookbook.com, Ver. February 16, 2008
11. Noble B, Daniel JW (1988) Applied linear algebra. Prentice-Hall, Englewood Cliffs, NJ
12. Haykin S (1996) Adaptive filter theory, 3rd edn. Prentice Hall, Englewood Cliffs, NJ
13. Cadzow JA, Baseghi B, Hsu T (1983) Singular-value decomposition approach to time series modeling. IEEE Proc Commun Radar Signal Process 130(3):202–210
14. van der Veen AJ, Deprettere EF, Swindlehurst AL (1993) Subspace-based signal analysis using singular value decomposition. Proc IEEE 81(9):1277–1308
15. Cichocki A, Amari SI (2002) Adaptive blind signal and image processing. Wiley, New York. ISBN 0-471-60791-6
16. Cichocki A, Unbehauen R (1994) Neural networks for optimization and signal processing. Wiley, New York
17. Van Huffel S, Vandewalle J (1991) The total least squares problems: computational aspects and analysis, vol 9, Frontiers in applied mathematics. SIAM, Philadelphia, PA

18. Golub GH, Van Loan CF (1980) An analysis of the total least squares problem. SIAM J Matrix Anal Appl 17:883–893
19. Farhang-Boroujeny B (1998) Adaptive filters: theory and applications. Wiley, New York
20. Sayed AH (2003) Fundamentals of adaptive filtering. Wiley, Hoboken, NJ. ISBN 0-471-46126-1
21. Golub GH, Hansen PC, O'Leary DP (1999) Tikhonov regularization and total least squares. SIAM J Matrix Anal Appl 21:185–194
22. Pascual-Marquia RD, Michel CM, Lehmannb D (1994) Low resolution electromagnetic tomography: a new method for localizing electrical activity in the brain. Int J Psychophysiol 18(1):49–65
23. Gorodnitsky IF, Rao BD (1997) Sparse signal reconstruction from limited data using FOCUSS: a re-weighted minimum norm algorithm. IEEE Trans Signal Process 45(3):600–616
24. Rao BD, Engan K, Cotter SF, Palmer J, Kreutz-Delgado K (2003) Subset selection in noise based on diversity measure minimization. IEEE Trans Signal Process 51(3):760–770
25. Wipf DP, Rao BD (2007) An empirical bayesian strategy for solving the simultaneous sparse approximation problem. IEEE Trans Signal Process 55(7):3704–3716
26. Zdunek R, Cichocki A (2008) Improved M-FOCUSS algorithm with overlapping blocks for locally smooth sparse signals. IEEE Trans Signal Process 56(10):4752–4761
27. He Z, Cichocki A, Zdunek R, Xie S (2009) Improved FOCUSS method with conjugate gradient iterations. IEEE Trans Signal Process 57(1):399–404
28. Xu P, Tian Y, Chen H, Yao D (2007) L_p norm iterative sparse solution for EEG source localization. IEEE Trans Biomed Eng 54(3):400–409
29. Rao BD, Kreutz-Delgado K (1999) An affine scaling methodology for best basis selection. IEEE Trans Signal Process 47(1):187–200
30. Cheney EW (1999) Introduction to approximation theory, 2nd edn. American Mathematical Society, Providence, RI
31. Golub G, Pereyra V (2003) Separable nonlinear least squares: the variable projection method and its applications. Inverse Probl 19 R1. doi:10.1088/0266-5611/19/2/201
32. Cadzow JA (1990) Signal processing via least squares error modeling. IEEE ASSP Magazine, pp 12–31, October 1990

Chapter 5
First-Order Adaptive Algorithms

5.1 Introduction

In the two previous chapters, attention was paid on the algorithms for the determination or estimation of filters parameters with a methodology that provides knowledge of the processes statistics or their *a priori* calculated estimation on an appropriate window signal length. In particular, with regard to the choice of the cost function (CF) to be minimized $J(\mathbf{w})$, the attention has been paid both to the solution methods of the Wiener–Hopf normal equations, which provide a *stochastic optimization* MMSE solution, and to the form of Yule–Walker that assumed a deterministic (or stochastic approximated) approach, by a *least squares error* (LSE) solution.

The approach based on the solution of the normal equations, which requires the knowledge or the estimation of certain quantities, is, by definition, of batch type and determines a systematic delay between the acquisition of the input signal and the availability of the solution to the filter output. This delay is at least equal to the analysis window length duration and, as already noted in Chap. 2, might not be compatible with the type of application. In these cases, in order to minimize this delay, an online approach is preferred. Note, also, that many authors consider that *adaptive filter* only whose parameters are updated with online approach.

In online adaptive filtering (or simply *adaptive filtering*) the optimal solution, which is the CF minimum, is estimated only after a certain number of iterations or *adaptation steps*. The problem becomes recursive and the optimal solution is reached after a certain number of steps, at limit infinite. For this reason, the algorithm is defined *online adaptation* and, at times, is referred to as *learning algorithm* [10, 11, 35].

A. Uncini, *Fundamentals of Adaptive Signal Processing*, Signals and Communication Technology, DOI 10.1007/978-3-319-02807-1_5,

Steepest-Descent and Stochastic-Gradient Adaptation Algorithms In the case where the CF was of statistical type, and predetermined together with the value of its gradient, i.e., the CF and its gradient are *a priori* known, the online adaptation procedures are called *search methods*. Belonging to that class are the so-called steepest-descent algorithms (SDA). In this case, the algorithms are derived from the recursive solution of the Wiener–Hopf stochastic normal equations.

Otherwise, if only you know a local estimate of the CF, and of its gradient, related to the value of the weights $\mathbf{w}_n$ at the nth adaptation step, and indicate respectively as $J(\mathbf{w}_n)$ and $\nabla J(\mathbf{w}_n)$, learning algorithms are called stochastic-gradient algorithms (SGA). In such cases, the methods of adaptation are derived from the recursive solution of the deterministic normal equations, i.e., the Yule–Walker form.

Algorithms Memoryless and with Memory In the case in which the adaptation rule depends only on the last sample present at the input, the class of algorithms is called *without memory* or *memoryless*.

On the contrary, in the algorithms with memory the gradient estimate depends, with a certain temporal depth defined by a certain *forgetting factor*, even on the estimates of the previous iterations. In general, in the case of stationary environment, the presence of a memory defines the fastest and most robust adaptation processes.

Order of the Adaptation Algorithm In the iterative optimization procedures, an important aspect concerns the order of the algorithm. In the *first-order algorithms*, the adaptation proceeds with only the knowledge of the CF first derivative with respect to the filter-free parameters. In the *second-order algorithm*, to decrease the number of iterations necessary for convergence to the optimum value, is also used information related to the CF second-order derivative (i.e., the Hessian function of $J(\mathbf{w})$) [5, 9, 42, 43].

In this chapter, the main first-order online algorithms for the recursive solution of the stochastic and deterministic normal equations are introduced. With reference to the generic diagram of transversal AF, illustrated in Fig. 5.1, the most common first-order SDA and SGA algorithms, with memory and memoryless, are presented. The second-order algorithms are presented in the next chapter.

5.1.1 *On the Recursive Formulation of the Adaptive Algorithms*

In the AF recursive formulation, the CF minimization is determined using an iterative procedure with a solution that evolves along the direction of the negative

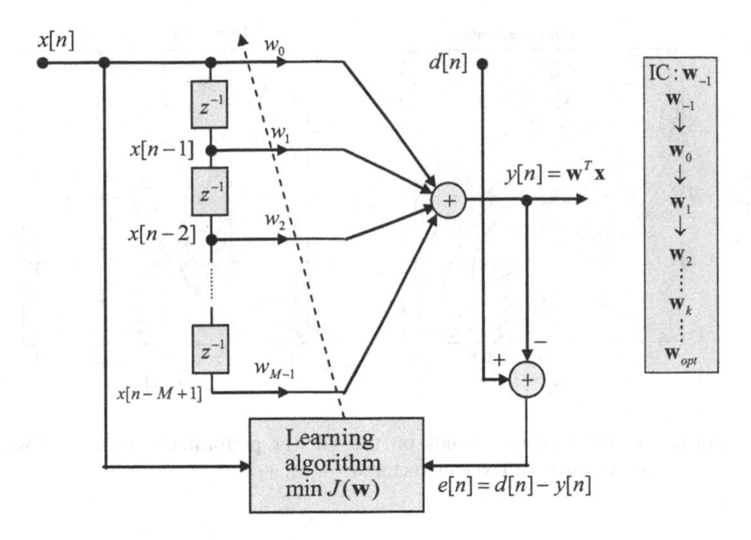

Fig. 5.1 Transversal adaptive filter

gradient of the CF itself. Starting from a certain AF's weights initial condition (IC) $\mathbf{w}_{-1} = \mathbf{0}$, or chosen randomly or on the basis of *a priori* known information, the estimation of the optimal solution occurs after a certain number (limit to infinity) of iterations

$$\mathbf{w}_{-1} \rightarrow \mathbf{w}_0 \rightarrow \mathbf{w}_1 \rightarrow \mathbf{w}_2 \rightarrow \ldots \mathbf{w}_k \ldots \rightarrow \mathbf{w}_{\mathrm{opt}}. \tag{5.1}$$

Referring to nonlinear programming methods (Sect. B.1) and, in particular, to unconstrained minimization methods (Sect. B.2), the estimator has a recursive form of the type

$$\mathbf{w}_k = \mathbf{w}_{k-1} + \mu \mathbf{v}_k, \tag{5.2}$$

or of the type

$$\mathbf{w}_{k+1} = \mathbf{w}_k + \mu \mathbf{v}_k, \tag{5.3}$$

where k is iteration index of the algorithm, also called adaptation index that, in some methods, may not represent the temporal index (in the case indicated with n) relative to the input signal. The vector $\mathbf{v}_k$ represents the direction of adaptation. The parameter μ represents the length of adaptation step also known as *learning rate* or *step size*. This parameter indicates how much to move down along the direction $\mathbf{v}_k$.

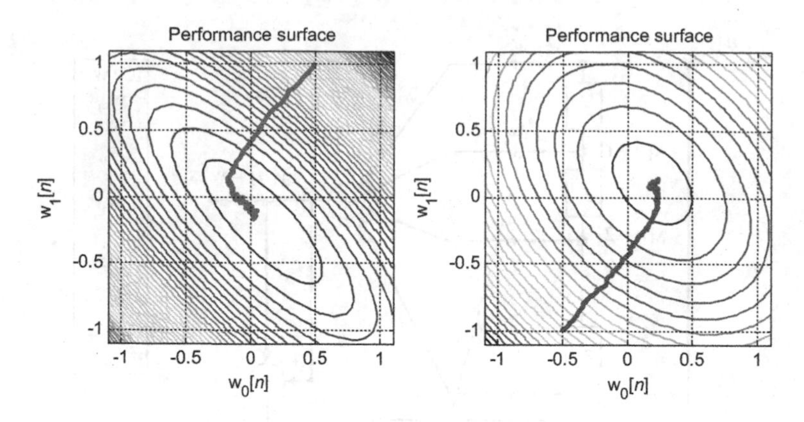

Fig. 5.2 Typical weights trajectory trends on the CF (or performance surface) $J(\mathbf{w})$, of a two-dimensional adaptive filter with weight vector $\mathbf{w} = [w_0\ w_1]$

Figure 5.2 shows two typical weights trajectories, superimposed on the isolevel curves of the surface error $J(w_0, w_1)$, relating to two AFs, adapted with the least mean squares (LMS) algorithm starting with random IC [the LMS, introduced in Sect. 4.3. 2.1, will be presented and discussed in detail later in this chapter (Sect. 5.3)].

A more general adaptation paradigm, which will be discussed in the next chapter, provides an update on the type

$$\mathbf{w}_{k+1} = \mathbf{M}(\mathbf{w}_k) + \mathbf{H}(\mathbf{v}_k), \tag{5.4}$$

where the operators $\mathbf{M}(\cdot)$ and $\mathbf{H}(\cdot)$, linear or nonlinear, may be defined by any *a priori* knowledge on the desired solution, and/or determined according to certain computing paradigms.

5.1.1.1 First-Order SDA and SGA Algorithms

For the recursive algorithms definition, suitable for AF implementation, you can trace back to the same philosophy followed for the determination of optimization methods developed in the previous chapter. In fact, even in this case, it is possible to develop statistical adaptation methods starting from the knowledge of the input processes, or working directly with deterministic functions of the signal samples. In the first case, the online algorithms are derived from the recursive solution of the normal equations in the Wiener–Hopf form, while, in the second case, they are derived from the recursive solution of the normal equations in the form of Yule–Walker [35, 37, 38].

In the literature, there are many recursive solution variations of the normal equations. These variants are generally derived from various definition and estimation ways of the CF, of its gradient, and, at times, of its Hessian. In the stochastic case, the CF is defined as $J(\mathbf{w}) = E\{|e[n]|^2\}$ (Sect. 3.2.3), while in the deterministic case applies, generally, the term $J(\mathbf{w}) = \sum_n |e[n]|^2$. As regards, for the gradient in the stochastic case we have that

$$\nabla J(\mathbf{w}) = \partial E\{|e[n]|^2\}/\partial \mathbf{w}, \qquad \textit{gradient}, \tag{5.5}$$

while for deterministic case we have

$$\nabla \hat{J}(\mathbf{w}) = \partial \sum\nolimits_n |e[n]|^2/\partial \mathbf{w}, \qquad \textit{stochastic gradient}. \tag{5.6}$$

In the case where the CF gradient at $k-1$ step is known (referred to as $\nabla J(\mathbf{w}_{k-1})$ or $\nabla J_{k-1}(\mathbf{w})$ or simply ∇J_{k-1}) it is possible to define some recursive techniques family for the solution, based on iterative unconstrained optimization algorithms (Sect. B.2). In the scientific literature, this class is referred to as *search methods* or *searching the performance surface* and the best known algorithm of the class is the so-called SDA. The SDA, in practice, allows the iterative solution of the Wiener–Hopf equations. Note that, given the popularity of the SDA, the *search methods* class is often simply indicated as SDA.

In adaptive filtering the gradient function is, in general, not known and for the optimization we refer to an estimate indicated as $\nabla \hat{J}(\mathbf{w}_{k-1})$. In this case it is usual to consider methods based on stochastic *search methods* approximations. This class is referred to as SGA and the most widespread family derived from that class is known as LMS algorithm.

From the general adaptation formulas (5.2) or (5.3) and from (5.5) and (5.6), the vector $\mathbf{v}_k$ is defined as follows:

$$\mathbf{v}_k = -\nabla J(\mathbf{w}_{k-1}), \qquad \textit{gradient vector, SDA algorithm}, \tag{5.7}$$

$$\mathbf{v}_k = -\nabla \hat{J}(\mathbf{w}_{k-1}), \qquad \textit{stochastic gradient vector, LMS (and variants)}. \tag{5.8}$$

The SDA and the LMS are first-order algorithms, because the adaptation is determined by the knowledge or estimate of the gradient, i.e., the CF first derivative made with respect to the filter parameters. Starting from a certain IC $\mathbf{w}_{-1}$, by (5.1), we proceed by solution updating along the direction, (5.7) or (5.8), opposite to the CF gradient with μ step size.

5.1.1.2 *A Priori* and *A Posteriori* Errors

Considering the model in Fig. 5.1, the error calculation is performed as the difference between the desired output and the actual filter output, i.e., $e[n] = d[n] - y[n]$. In the case implementation of (5.2), the calculation can be performed in two distinct modes. If the output $y[n]$ is calculated *before* the filters parameters update, the error is defined as *a priori error* or simply *error*

$$e[n] = d[n] - \mathbf{w}_{n-1}^T\mathbf{x}, \qquad \textit{a priori error}. \tag{5.9}$$

Otherwise, in the case that the error estimate was calculated *after* the filter update, the error is defined as *a posteriori error*

$$\varepsilon[n] = d[n] - \mathbf{w}_n^T\mathbf{x}, \qquad \textit{a posteriori error}. \tag{5.10}$$

As we will see later in this chapter, the two methods used to calculate the error are useful for both the definition of some properties and because, in some adaptation paradigms, in order to increase the robustness, the two modes can coexist within the same algorithm.

A desirable and usefully property for *all* adaptation algorithms is that the quadratic *a posteriori* error is always lower than the quadratic *a priori* error. That is,

$$\left|\varepsilon[n]\right|^2 < \left|e[n]\right|^2 \Leftrightarrow \sum_{k=n-N+1}^{n} \left|\varepsilon[k]\right|^2 < \sum_{k=n-N+1}^{n} \left|e[k]\right|^2, \qquad \forall n, N. \tag{5.11}$$

This condition is very important as it provides an energy constraint between *a priori* and *a posteriori* errors that can be exploited for the definition, as we shall see later, of many significant adaptation algorithms properties.

Note that considering the CF $J(\mathbf{w})$ as a certain dynamic system's *energy function*, the property (5.11) can be derived by considering the Lyapunov's theorem presented in Sect. 4.3.2 (4.119)–(4.222).

5.1.1.3 Second-Order SDA and SGA Algorithms

The adaptation filter performance can be improved by using a second-order update formula of the type

$$\mathbf{w}_{k+1} = \mathbf{w}_k - \mu_k \cdot \left[\nabla^2 J(\mathbf{w}_k)\right]^{-1} \nabla J(\mathbf{w}_k), \qquad \text{and} \qquad \nabla^2 J(\mathbf{w}_k) \neq \mathbf{0}, \tag{5.12}$$

where $\mu_k > 0$ is the opportune step size. Equation (5.12) is the standard form of the *discrete Newton's method* (Sect. B.2.4). Note that in (5.12) the terms

$$\nabla J(\mathbf{w}_k) = \partial J(\mathbf{w})/\partial \mathbf{w}, \tag{5.13}$$

and

$$\nabla^2 J(\mathbf{w}_k) = \partial^2 J(\mathbf{w})/\partial \mathbf{w}^2, \tag{5.14}$$

represent, respectively, the gradient and the Hessian matrix of the CF. In other words, in (5.12) the term $\nabla J(\mathbf{w}_k)$ determines the direction of the local gradient at the point $\mathbf{w}_k$, while considering the second derivative $\nabla^2 J(\mathbf{w}_k)$, the adaptation step length and the optimal direction towards the CF minimum are determined.

With reference to (5.4), the expression (5.12) can be considered a special case of a more general formulation of the type

$$\mathbf{w}_k = \mathbf{w}_{k-1} + \mu_k \mathbf{H}_k \mathbf{v}_k, \tag{5.15}$$

where $\mathbf{H}_k$ is a *weighing matrix* determinable in various modes. The product $\mu_k \mathbf{H}_k$ can be interpreted as a linear transformation to determine an optimum adaptation step (direction and length), such that the descent along the CF can be performed in very few steps.

In the unconstrained optimization literature, numerous techniques for the determination of the matrix $\mathbf{H}_k$ are available. The *Newton's algorithm* is simplest form. In fact, as indicated in (5.12), the weighing of equations (5.15) is made with the inverse Hessian matrix or by its estimate. That is,

$$\mathbf{H}_k = \left[\nabla^2 \hat{J}(\mathbf{w})\right]^{-1}. \tag{5.16}$$

More commonly in adaptive filtering, only a gradient estimate is known, and therefore, it is possible to determine only an estimate of the Hessian matrix (for example, by analyzing successive gradient vectors). In this case, the weighing matrix $\mathbf{H}_k$ takes the form

$$\mathbf{H}_k = \left[\nabla^2 \hat{J}(\mathbf{w}_{k-1})\right]^{-1}. \tag{5.17}$$

The learning rate can be constant μ_k or also determined with an appropriate optimization procedure.

5.1.1.4 Variants of Second-Order Methods

In the literature, there are numerous variations and specializations of the method (5.15). Some of the most common are below indicated.

The Levenberg–Marquardt Variants

In the Levenberg–Marquardt variant [1, 2], (5.15) is rewritten as

$$\mathbf{w}_k = \mathbf{w}_{k-1} - \mu_k \left[\delta \mathbf{I} + \nabla^2 \hat{J}(\mathbf{w}_{k-1})\right]^{-1} \nabla \hat{J}(\mathbf{w}_{k-1}), \tag{5.18}$$

in which the constant $\delta > 0$ (Sect. 4.3.1.3) should be chosen considering two opposing requirements: possibly small to increase the convergence speed and biased solution and sufficiently large such that the Hessian is always a positive definite matrix (Sect. B.2.5).

The Quasi-Newton Method

In many adaptation problems the Hessian matrix is not explicitly available. In the so-called *quasi-Newton* or *variable metric* methods (Sect. B.2.6), the inverse Hessian matrix is determined iteratively and in an approximate way. For example, in *sequential quasi-Newton* methods, the estimated inverse Hessian is evaluated considering two successive values of the CF gradient. In particular, in the method of Broyden–Fletcher–Goldfarb–Shanno (BFGS) [3], the adaptation takes the form

$$\begin{aligned}
\mathbf{w}_k &= \mathbf{w}_{k-1} + \mu_k \mathbf{d}_k \\
\mathbf{d}_k &\simeq \mathbf{w}_k - \mathbf{w}_{k-1} = -\mathbf{H}_{k-1}\nabla J(\mathbf{w}_{k-1}) \\
\mathbf{u}_k &\triangleq \nabla J(\mathbf{w}_k) - \nabla J(\mathbf{w}_{k-1}) \\
\mathbf{H}_k &= \left[\mathbf{I} - \frac{\mathbf{d}_k\mathbf{u}_k^T}{\mathbf{d}_k^T\mathbf{u}_k}\right]\mathbf{H}_{k-1}\left[\mathbf{I} - \frac{\mathbf{u}_k\mathbf{d}_k^T}{\mathbf{d}_k^T\mathbf{u}_k}\right] + \frac{\mathbf{d}_k\mathbf{d}_k^T}{\mathbf{d}_k^T\mathbf{u}_k},
\end{aligned} \tag{5.19}$$

where $\mathbf{H}_k$ denotes the current approximation of $\left[\nabla^2 J(\mathbf{w}_k)\right]^{-1}$. The step of adaptation μ_k is optimized with a procedure one-dimensional line search, the type described in (13) of Appendix B (Sect. B.2.3), which takes the form

$$\mu_k \therefore \arg\min_{\mu \geq 0} J\left[\mathbf{w}_{k-1} - \mu\mathbf{H}_{k-1}\nabla J(\mathbf{w}_{k-1})\right]. \tag{5.20}$$

The procedure is initialized with arbitrary IC $\mathbf{w}_{-1}$ and with the matrix $\mathbf{H}_{-1} = \mathbf{I}$. Alternatively, in the last of (5.19) the $\mathbf{H}_k$ can be calculated with the expression

$$\mathbf{H}_k = \mathbf{H}_{k-1} + \frac{(\mathbf{d}_k - \mathbf{H}_{k-1}\mathbf{u}_k)(\mathbf{d}_k - \mathbf{H}_{k-1}\mathbf{u}_k)^T}{(\mathbf{d}_k - \mathbf{H}_{k-1}\mathbf{u}_k)^T\mathbf{u}_k}. \tag{5.21}$$

The variable metric method is very advantageous from the computational point of view compared to that of Newton.

Methods of Conjugate Gradient of Fletcher–Reevs

The conjugate gradient algorithms (CGA) algorithms class is a simple modification compared to SDA and quasi-Newton methods, but with the advantage of a considerable increase of the convergence speed and the robustness and the decrease of internal memory required (the matrix $\mathbf{H}_k$ is not explicitly calculated). The standard form of the method is defined by the following recurrence:

$$\begin{aligned} \mathbf{w}_k &= \mathbf{w}_{k-1} + \mu_k \mathbf{d}_k \\ \mathbf{d}_k &\triangleq \beta_k \mathbf{d}_{k-1} - \nabla J(\mathbf{w}_{k-1}), \end{aligned} \tag{5.22}$$

where the parameter β_k, which affects the algorithm performance, can be evaluated according to different criteria (Sect. B.2.7). In general terms, it can be estimated with the following ratio:

$$\beta_k = \frac{\left\| \nabla J(\mathbf{w}_k) \right\|_2^2}{\left\| \nabla J(\mathbf{w}_{k-1}) \right\|_2^2}.$$

The parameter μ_k can be optimized with a one-dimensional line search procedure of the type

$$\mu_k \therefore \arg\min_{\mu \geq 0} J(\mathbf{w}_{k-1} + \mu \mathbf{d}_k). \tag{5.23}$$

Note that the increase of the convergence speed derives from the fact that the information of the search direction depends on the previous iteration $\mathbf{d}_{k-1}$ and that for a quadratic CF, it is conjugate with respect to the gradient direction. Theoretically the algorithm, for $\mathbf{w} \in \mathbb{R}^{M \times 1}$, converges in M, or less, iterations.

From the implementation point of view, to avoid numerical inaccuracy in the search direction calculation, or for the non-quadratic nature of the problem, the method requires a periodic reinitialization.

The CGA can be considered as an intermediate view between the SDA and the quasi-Newton method. Unlike the other procedures, the main CGA advantage derives from not the need to explicitly estimate the Hessian matrix which is, in practice, replaced by the parameter β_k. For further information, Sect. B.2.7.

5.1.1.5 Summary of the Second-Order SGA and SDA Methods

In general, with the recursive approach to optimal filtering, the adaptation has the form

$$\mathbf{w}_k = \mathbf{w}_{k-1} + \mu_k \mathbf{H}_k \mathbf{v}_k, \tag{5.24}$$

where, in the case of stochastic gradient, $\mathbf{v}_k$ and $\mathbf{H}_k$ are estimates of quantity

$$\begin{aligned} \mathbf{v}_k &\equiv -\nabla \hat{J}\left(\mathbf{w}_{k-1}\right) = -\nabla_{\mathbf{w}_{k-1}}\left(\hat{E}\left\{e^2[n]\right\}\right) \\ \mathbf{H}_k &\equiv \left[\nabla^2 \hat{J}\left(\mathbf{w}_{k-1}\right)\right]^{-1}. \end{aligned} \tag{5.25}$$

As we know, in fact, the expectation $E\{\cdot\}$ is replaced with the temporal operator denoted as $\hat{E}\{\cdot\}$ (or $< \cdot >$) that performs an estimate, whereas ergodic processes,

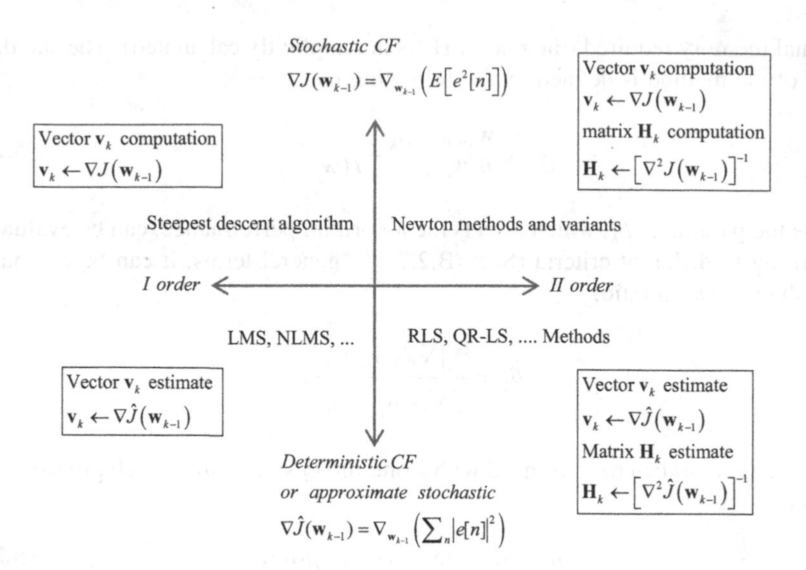

Fig. 5.3 Schematic representation of on-line learning algorithms for adaptive filtering

the first- and second-order ensemble averages, are replaced with time averages (5.6). The matrix $\mathbf{H}_k$ is the estimated inverse Hessian or, as in the simple LMS case, discussed in Sect. 5.3, there is $\mathbf{H}_k = \mathbf{I}$. These estimates can be made in various ways, more or less efficient ways, and it is therefore necessary to consider also the convergence properties.

Similarly to what was presented in Sect. 3.2.4 (Fig. 3.4), in Fig. 5.3 is shown a schematic representation of the first- and second-order stochastic and approximate stochastic online learning algorithms.

Remark In the case of batch algorithms stochastic and deterministic methods were presented in two different chapters. In the case of recursive algorithms such differentiation is less significant so we wish to present together the two paradigms. Given the vastness of the subject, the first-order methods are presented in this chapter, while in the next, those of the second order.

A schematic for the definition of recursive algorithms, described in this and the following chapter, is shown in Table 5.1

5.1.2 Performance of Adaptive Algorithms

An important aspect in adaptive filtering concerns with the performance measure. In order to characterize the quality of the performance, the adaptation process can be considered as a dynamic system described by the *transient* and *steady-state* response and according to the *stability criteria*, *convergence speed*, and *steady-state error*.

Table 5.1 Recursive solution of the normal equations: Stochastic and approximate stochastic approaches

Wiener–Hopf equations	Yule–Walker equations
$\mathbf{Rw} = \mathbf{g}$	$\mathbf{X}^T\mathbf{Xw} = \mathbf{X}^T\mathbf{d}$
Stochastic MSE criterion	Deterministic LS criterion
$J(\mathbf{w}) = E\{\lvert e[n]\rvert^2\}$	$\hat{J}(\mathbf{w}) = \mathbf{e}^T\mathbf{e}$
Exact gradient	Stochastic gradient
$\nabla J(\mathbf{w}) = \nabla_{\mathbf{w}}\left(E\{\lvert e[n]\rvert^2\}\right)$	$\nabla\hat{J}(\mathbf{w}) = \nabla_{\mathbf{w}}(\mathbf{e}^T\mathbf{e})$
exact knowledge of the gradient	noisy gradient estimate
Performance	Performance
$\lim_{k\to\infty} \mathbf{w}_k = \mathbf{w}_{\text{opt}}$	$\lim_{n\to\infty} E\{\mathbf{w}_n\} = \mathbf{w}_{\text{opt}}$
$\mathbf{w}_k$ deterministic unknown vector	$\mathbf{w}_n$ random variables vector
Search methods	*Stochastic-gradient alg.s (SGAn)*
Steepest-descent algorithm	Least mean squares (LMS)
Newton methods	Recursive least squares (RLS)
Quasi-Newton methods	Kalman filter
Other variants	Other variants

Furthermore, an important feature of adaptive algorithms regards the *tracking properties*. Given the specificity of adaptation algorithm, that property will be discussed in the next chapter (Sect. 6.6).

5.1.2.1 Adaptation Algorithm as Nonlinear Dynamic System

Considering Fig. 5.1, it is possible to observe that the adaptive algorithms are regulated by the error signal and, consequently, assimilated to nonlinear dynamical systems, generally stochastic and with feedback error control. Therefore, for the performance analysis it is necessary to refer to the *dynamical systems theory* and take into account the stability, the transient and steady-state behavior, etc. In practice, one can think of the adaptation algorithm, as a discrete-time dynamic system, governed by the finite difference equation that, in general, takes the form (5.15) rewritten as

$$\mathbf{w}_k = \mathbf{w}_{k-1} + \mu_k \mathbf{H}_k \mathbf{v}_k. \tag{5.26}$$

Depending on the quantities involved that can be deterministic or random variables, equation (5.26) is a deterministic or stochastic difference equation. The nonlinear nature of the system (5.26) is due to the presence of the product $\mathbf{H}_k\mathbf{v}_k$ which involves products between the process sequences.

In the case of stochastic CF, $\mathbf{w}$ represent a simple deterministic unknown vector and the optimal solution is the one provided by the Wiener filter $\mathbf{w}_{\text{opt}} = \mathbf{R}^{-1}\mathbf{g}$. Given the exact deterministic result, this can also be expressed in the frequency domain. In this case, the optimal filter can be defined as

$$W_{\text{opt}}\left(e^{j\omega}\right) = \frac{G(e^{j\omega})}{R(e^{j\omega})}. \tag{5.27}$$

This statistically optimal solution represents the performance upper limit we can expect from a linear adaptive filter with online algorithm.

5.1.2.2 Stability Analysis: Mean and Mean Square Convergence

Since the adaptive algorithm, are feedback error dynamic systems, it is necessary and important to the study of stability defined as bounded-input–bounded-output (BIBO). However, this analysis is difficult because of the nonlinear and nonstationary dynamical system nature, implicit in the actual algorithms formulation. From the statistical point, in SGA cases, the stochastic convergence is ensured everywhere if

$$\lim_{n\to\infty} \mathbf{w}_n = \mathbf{w}_{\text{opt}}, \tag{5.28}$$

and *almost everywhere* if, the said probability function $P\{\cdot\}$, that is,

$$P\left\{\lim_{n\to\infty} \left|w_n[i] - w_{\text{opt}}[i]\right| = 0\right\} = 1, \quad i = 0, 1, \ldots, M-1, \tag{5.29}$$

which defines the statistical *mean convergence*. In other words, (5.29) implies that some coefficient of the filter $\mathbf{w}_n$ does not converge with zero probability.

Another analysis type of the *mean square convergence* is defined as

$$\lim_{n\to\infty} E\left\{\left|w_n[i] - w_{\text{opt}}[i]\right|^2\right\} = c_i, \quad i = 0, 1, \ldots, M-1, \tag{5.30}$$

where c_i represents a small value (at the limit null). In fact, the use of the second-order moment allows to take into account, on average, of all samples in the sequence and provides an interpretation in terms of *error energy*.

5.1.2.3 Weights Error Vector

With reference to Table 5.1, in the case of adaptive algorithm convergence, the algorithm converges to the optimal value, i.e., the exact Wiener solution $\mathbf{w}_{\text{opt}} = \mathbf{R}^{-1}\mathbf{g}$. The vector $\mathbf{w}$ is, in this case, a simple algebraic unknown. So, whether you have the CF $J(\mathbf{w})$, and its gradient $\nabla J(\mathbf{w})$, we have the solution (5.28).

In cases where the exact CF is unknown and only a noisy estimate $\hat{J}(\mathbf{w}_n)$ is available, along with that of its gradient $\nabla\hat{J}(\mathbf{w}_n)$, then $\mathbf{w}_n$ is a RV. So, the performance measure can be characterized by considering a statistical function of its deviation from the optimal solution. In this case, for AF performance measuring, we should refer to a weights error vector (WEV) $\mathbf{u}_n$, defined as

$$\mathbf{u}_n = \mathbf{w}_n - \mathbf{w}_{\text{opt}} \tag{5.31}$$

and it is generally convenient to study its statistics considering the expected WEV, defined as

$$E\{\mathbf{u}_n\} = E\{\mathbf{w}_n\} - \mathbf{w}_{\text{opt}}. \tag{5.32}$$

5.1.2.4 Correlation Matrix of the Weights Error Vector

For the definition of adaptive algorithms transient and steady-state properties, a useful quantity is the WEV's correlation matrix, defined as

$$\mathbf{K}_n \triangleq E\left\{\mathbf{u}_n\mathbf{u}_n^T\right\}. \tag{5.33}$$

5.1.2.5 Mean Square Deviation of the Weights Error Vector

Another interesting quantity for the performance second-order statistical analysis is the scalar quantity $\mathcal{D}_n$ defined as

$$\mathcal{D}_n \triangleq E\left\{\left\|\mathbf{u}_n\right\|_2^2\right\}, \tag{5.34}$$

referred to as the weights error vector's mean square deviation (MSD). The MSD, although not a directly measurable quantity, represents a very important paradigm for the theoretical analysis of the statistical learning algorithms.

5.1.2.6 Steady-State Performance: Excess of Error

An AF is in steady state (*steady-state filter*) when, on average, its weights do not change during the process of adaptation. So, in formal terms we can write

$$E\{\mathbf{u}_n\} = E\{\mathbf{u}_{n-1}\} = s, \qquad \text{for} \quad n \to \infty \quad (\text{usually}\, s = 0), \tag{5.35}$$

$$\mathbf{K}_n = \mathbf{K}_{n-1} = \mathbf{C}, \qquad \text{for} \quad n \to \infty, \tag{5.36}$$

namely, the average and the WEV's correlation matrix tend to a constant value. In particular, it is also

$$E\left\{\left\|\mathbf{u}_n\right\|_2^2\right\} = E\left\{\left\|\mathbf{u}_{n-1}\right\|_2^2\right\} = k < \infty, \quad \text{for} \quad n \to \infty, \tag{5.37}$$

where k represents the *trace* of the matrix $\mathbf{K}_n$, shown as $k = \text{tr}(\mathbf{K}_n)$. Note from (5.34) and for the MSD, at steady state we have

$$\mathcal{D}_n = \text{tr}[\mathbf{K}_n] \quad \text{MSD}, \quad \text{for} \quad n \to \infty. \tag{5.38}$$

Of course, not all AFs reach the steady-state operating status. If the learning of rate is not small enough, the solution may diverge and the WEV $\mathbf{u}_n$ can grow without limit.

To monitor the steady-state performance, it is often useful to consider the value of the excess of the mean squares error (EMSE), which, as already introduced in Sect. 3.3.4.4, represents the deviation from the CF theoretical minimum (value that can take the CF), for which

$$J_n = J_{\min} + J_{\text{EMSE}}, \tag{5.39}$$

at steady state, i.e., for $n \to \infty$, we get

$$J_{\text{EMSE\_}\infty} \triangleq J_\infty - J_{\min}. \tag{5.40}$$

In other words, the steady-state error is evaluated by estimating the variation of the solution around the optimal solution. Furthermore, it is useful to define *misadjustment* parameter, sometimes used as an alternative to EMSE, as

$$\mathcal{M} \triangleq J_{\text{EMSE}} / J(\mathbf{w}_{\text{opt}}). \tag{5.41}$$

5.1.2.7 Convergence Speed and Learning Curve

To monitor the adaptation process, it is often useful to consider the CF value changes over the algorithm iterations. The graph trend of the CF is defined as *learning curve*. However, the CF $J(\mathbf{w})$ can take values in a range of several orders of magnitude and, for this reason, the learning curve is typically displayed with a logarithmic scale and often measured in decibels as $\text{MSE}_{\text{dB}} = 10 \log_{10} |J(\mathbf{w})|$ or $\text{MSE}_{\text{dB}} = 10 \log_{10} |e[n]|^2$.

In Fig. 5.4, for example, it shows the typical behavior of the learning curves of MSE_{dB} during a two-tap AF adaptation process (adapted with the LMS algorithm).

Remark In the SGA, is reported the trend of the *estimated* CF and, because of the estimation error, $\hat{J}_n$ is a very noisy quantity. Because of the stochastic nature of the SGA for a proper analysis of the learning curve it is necessary to refer to the *ensemble averages*, i.e., $J_n = E\{\hat{J}_n\}$, and not to the single trial. In practice, for a more accurate analysis, it is possible to realize more trials and, for the ergodicity,

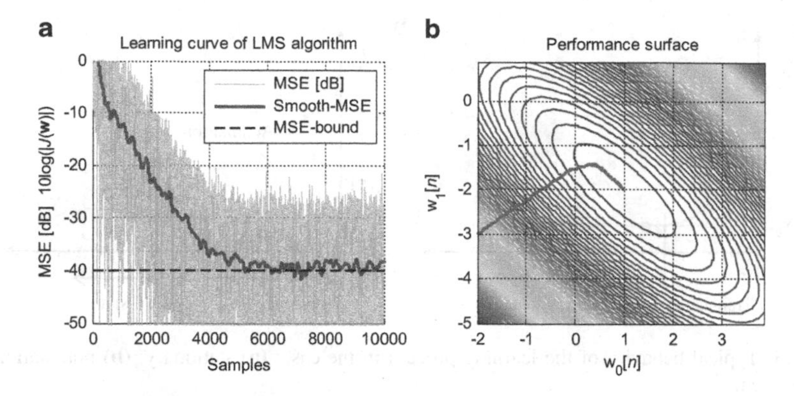

Fig. 5.4 Typical behavior of the learning curve for an adaptation algorithm that minimizes the locally estimated LS error for each iteration: (**a**) MSE_{dB} trend and its average value (smoothed curve obtained by a lowpass zero-phase IIR filtering); (**b**) weights w_0 and w_1 trajectory for the same experiment in curve (**a**)

make a simple *time average* or, for the single trial, smooth the noisy learning curve with a low-pass filter (optimal estimator) as was done in the dark curve of Fig. 5.4.

5.1.2.8 Tracking Properties

The AF task consists in the determination of the optimum vector $\mathbf{w}_{\text{opt}}$ after a certain number of learning steps, regardless of the IC $\mathbf{w}_{-1}$ and the input signals statistics. The adaptation process performances are assimilated to those of a dynamical system where the learning curve, which indicates the MSE trend as adaptation steps function, describes the transient properties, while the excess of the mean squares error $J_{\text{EMSE}}(\mathbf{w})$ describes the steady-state properties. In other words, in a stationary environment case, when the MSE reaches its minimum value, the adaptation process may stop.

In the case of nonstationary operating environment, it is important to consider the performance also in terms of tracking properties. As illustrated in Fig. 5.5, the optimum value $\mathbf{w}_{\text{opt}}$ is no longer static but is also time variant and indicated as $\mathbf{w}_{\text{opt}}[n]$ or $\mathbf{w}_{\text{opt},n}$. The subdivision of the learning algorithm in transient and steady-state responses is more complex and less significant. In these cases, in fact, in the learning curve transient phase must engage the $\mathbf{w}_{\text{opt}}[n]$ variation and is more properly referred to as *acquisition phase*. At the end of the acquisition phase the algorithm, which now is in *continuous adaptation*, is steady state, and it is more appropriate to measure the performance in terms of *tracking* property.

The non-stationarity may concern the input process $\mathbf{x}_n$, the desired output $d[n]$, or both. The adaptation algorithm requires, in general, the invertibility of the correlation $\mathbf{R}_n$, which means that the most critical problems are in the case wherein the non-stationarity relates to the input signal.

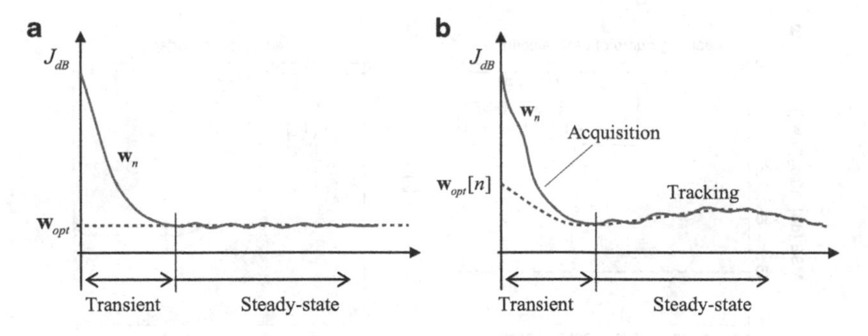

Fig. 5.5 Typical behavior of the learning process in the case: (**a**) stationary; (**b**) non-stationary environment

Remark The learning algorithm response is divided into the transitory (or acquisition) and steady-state phases. Therefore, the adaptation speed is a characteristic of the transitory phase while the tracking properties is a steady-state characteristic. These two properties are different and are characterized with different performance indices. The tracking, in fact, is possible only if the non-stationarity degree is slower than the AF acquisition speed. The general characterization of the tracking properties is dependent on the algorithm type and is treated more specifically in the next chapter.

5.1.3 General Properties of the Adaptation Algorithms

The determination of a general strategy for the performance measurement of an AF is a rather complex aim. In fact, learning algorithms are treated as discrete-time stochastic time-varying nonlinear dynamical systems. For the study of common properties, it is useful to refer in a general procedure that describes the most adaptation modes [6–8]. A form that generalizes (5.26) and represents a broad class of algorithms is described by the following:

$$\mathbf{w}_n = \mathbf{w}_{n-1} + \mu \mathbf{x}_n g\big(e[n]\big) \tag{5.42}$$

called *nonlinearity error adaptation*, where $g(\cdot)$ denotes an *a priori* determined nonlinear error function. Many adaptation algorithms, such as LMS, the NLMS etc., and other of the second-order described in the next chapter, can be viewed as a special case of (5.42). In terms of the WEV defined in (5.31), (5.42) is equivalent to the form

$$\mathbf{u}_n = \mathbf{u}_{n-1} + \mu \mathbf{x}_n g\big(e[n]\big). \tag{5.43}$$

In other words, the filter parameters update depends, as a nonlinear and stochastic function, on the output desired $d[n]$ and on the input regression $\mathbf{x}_n$. For this reason, the study of the unified measure of the adaptation algorithms performance represents a formidable challenge.

In this paragraph, general properties and rules for the definition and measurement of the AF performance are discussed.

5.1.3.1 The SGA Analysis Using Stochastic Difference Equation

The study of transient and steady-state properties of the adaptive algorithms can be derived from the solution to the stochastic difference equation (SDE) that describes it. In general terms, this analysis can be traced back to the following steps:

1. Definition of the adaptation nonlinear stochastic difference equation, for example, in the form (5.42) or in the form with transformed variables (5.43)
2. Solution to the equation defined in step 1 considering the expectation and/or the mean square of both sides
3. Study of the convergence, and/or other features, with calculation of the limit $n \rightarrow \infty$ for the solution to the step 2

As we will see in the specific cases, the presence of nonlinearity may create certain difficulties of a theoretical nature. In general, for the convergence analysis, simplificative assumptions, as the statistical independence of the processes, may be taken.

5.1.3.2 Minimal Perturbation Properties

The adaptive algorithms are presented as *approximate iterative solution of a global optimization problem.* Starting from a Lyapunov energy point of view (Sect. 4.3.2), and defining some general properties, this class of algorithms can be seen as *exact iterative solution of a local optimization problem.*

Property Any adaptive algorithm can be derived and characterized by considering the following simple and intuitive three axiomatic properties:

(i) The *a posteriori* error is always lower than the *a priori* error, i.e., $|\varepsilon[n]| < |e[n]|$
(ii) At the convergence for $n \rightarrow \infty$, the weights do not change during the adaptation, *minimum perturbation* properties
(iii) For $n \rightarrow \infty$, both the *a priori* and the *a posteriori* errors that tend to zero

Similarly to the development in Sect. 4.3.2 (4.119)–(4.222), in order to simplify the development, the property (i) is given as follows:

$$\varepsilon[n] = (1 - \alpha)e[n]. \tag{5.44}$$

Then, when $|1 - \alpha| < 1$, we have that $|\varepsilon[n]| < |e[n]|$.

In other words, as explained in [4, 5], an "optimal" adaptive algorithm must find a good balance between the *conservative* (keep the information acquired during the previous iterations) and *corrective* (sureness that the new available information increases the result accuracy) needs.

Remark The (5.44), quadratically averaged on multiple samples, expresses an *energetic constraint* between *a priori* and *a posteriori* errors involving the *passivity* of the adaptation circuit.

The minimal perturbation properties (ii) can be expressed by defining the quantity

$$\delta\mathbf{w} = \mathbf{w}_n - \mathbf{w}_{n-1}. \tag{5.45}$$

This allows the definition of a new CF that is $J(\mathbf{w}) = \|\delta\mathbf{w}\|_2^2$. It follows that, for (5.44) and (5.45), any adaptive algorithm that minimizes $J(\mathbf{w})$ can also be expressed as an *exact method of local minimization* that, in general terms, can be formulated as a constrained optimization problem of the type

$$\mathbf{w}_* \therefore \arg\min_{\mathbf{w}} \|\delta\mathbf{w}\|_2^2 \qquad \text{s.t.} \qquad \varepsilon[n] = (1-\alpha)e[n]. \tag{5.46}$$

The previous formalization has merely theoretical significance, since it is based on the *a priori* and *a posteriori* errors knowledge. For a more constructive use of the properties (i)–(iii), it is necessary to define the energetic constraint as a function of *a priori* error only. This can be done by multiplying both members of the (5.45) for the vector $\mathbf{x}^T$. So, it is possible to express the energy constraint $\varepsilon[n] = (1-\alpha)e[n]$ as a function of the *a priori* error only. Proceeding we have that

$$\begin{aligned} \mathbf{x}_n^T\delta\mathbf{w} &= \mathbf{x}_n^T\mathbf{w}_n - \mathbf{x}_n^T\mathbf{w}_{n-1} \\ &= -\varepsilon[n] + e[n] \\ &= -(1-\alpha)e[n] + e[n] \\ &= \alpha e[n]. \end{aligned} \tag{5.47}$$

Property (5.46) and (5.47) show that a generic adaptation algorithm can be defined as an optimization problem of the type $\mathbf{w}_* \therefore \arg\min \|\delta\mathbf{w}\|_2^2$. This is equivalent to the determination of a minimum Euclidean quadratic norm vector $\delta\mathbf{w}$, where a energetic constraint between the errors is imposed. The constraint, expressed as a function of only *a priori error*, has the form $\mathbf{x}_n^T\delta\mathbf{w} = \alpha e[n]$. Finally, we can then write

$$\mathbf{w}_* \therefore \arg\min_{\delta\mathbf{w}} \|\delta\mathbf{w}\|_2^2 \qquad \text{s.t.} \qquad \mathbf{x}_n^T\delta\mathbf{w} = \alpha e[n], \tag{5.48}$$

where, in particular, the parameter α is related to the specific adaptation algorithm.

Note that the expression $\mathbf{x}_n^T \delta \mathbf{w} = \alpha e[n]$ represents an underdetermined linear equations system in $\delta \mathbf{w}$, which admits infinite solutions. For $\|\mathbf{x}_n\|_2^2 = 0$ we have the trivial solution $\delta \mathbf{w} = 0$, while, for $\|\mathbf{x}_n\|_2^2 \neq 0$ applies

$$\delta \mathbf{w}_* = \mathbf{x}_n \left(\mathbf{x}_n^T \mathbf{x}_n\right)^{-1} \alpha e[n]. \tag{5.49}$$

From the expressions (5.45) and (5.49), the adaptation formula is expressible as

$$\mathbf{w}_n = \mathbf{w}_{n-1} + \frac{\alpha}{\|\mathbf{x}_n\|_2^2} \mathbf{x}_n e[n]. \tag{5.50}$$

Note that the parameter α, as will be later in this chapter, is specific to the adaptation algorithm. For example, as discussed in Sect. 5.3, in the case of LMS adaptation you have $\alpha = \mu \|\mathbf{x}_n\|_2^2$, while for the *normalized LMS* (NMLS) algorithm described in (Sect. 4.3.2.1), and revisited starting from different assumptions in Sect. 5.5, we have that $\alpha = \mu \|\mathbf{x}_n\|_2^2 / \left(\delta + \|\mathbf{x}_n\|_2^2\right)$.

Remark Note that the expression (5.48) does not define any form of adaptation algorithm. In fact, the parameter α can only be determined after the definition of the adaptation rule. In this sense, the previous development has not constructive characteristics, but implies important properties such as, for example, assimilation of the adaptation problem to an exact local minimization method, the passivity, etc., useful for the study of unified algorithms classes and/or to the definition of other classes. For more details, the readers can refer to [4–8].

5.1.3.3 Adaptive Algorithms Definition by Energetic Approach: The Principle of Energy Conservation

A unified approach to the study of adaptive algorithms, alternative to the stochastic difference equation, is based on the principle of energy conservation. Generalizing the properties (i)–(iii), as described in Sect. 5.1.3.2, the method is based on considering the energy balance between the *a priori* and *a posteriori* error for each time instant.

By definition, the desired output of an AF is $d[n] = \mathbf{x}_n^T \mathbf{w}_{\text{opt}} + v[n]$ ($v[n]$ is the measure noise); this allows us to express the *a priori* and error (5.9) as $e[n] = \mathbf{x}_n^T \mathbf{w}_{\text{opt}} - \mathbf{x}_n^T \mathbf{w}_{n-1} = -\mathbf{x}_n^T \mathbf{u}_{n-1}$ and such that we can write

$$e[n] \triangleq -\mathbf{x}_n^T \mathbf{u}_{n-1} \quad \text{and} \quad \varepsilon[n] \triangleq -\mathbf{x}_n^T \mathbf{u}_n. \tag{5.51}$$

So, multiplying by $\mathbf{x}_n^T$ both members of (5.43), the following relationship holds:

$$\varepsilon[n] = e[n] - \mu\|\mathbf{x}_n\|^2 g(e[n]). \tag{5.52}$$

Remark The expressions (5.43) and (5.52) represent an alternative way for the description of adaptation equation (5.42) in terms of *error's quantity* $\varepsilon[n]$, $e[n]$, $\mathbf{u}_n$, and $\mathbf{u}_{n-1}$. This type of formalism is useful since for the analysis of the adaptation characteristics, it is necessary to precisely define the trend of these quantities with respect to the time index n. It appears that to characterize the steady-state behavior it is necessary to determine the following quantity $E\left\{\|\mathbf{u}_n\|_2^2\right\}$, $E\left\{|\varepsilon[n]|^2\right\}$, and $E\left\{|e[n]|^2\right\}$ for $n \to \infty$. For stability, we are interested in the determination of the adaptation step μ values, such that the variances $E\left\{|\varepsilon[n]|^2\right\}$ and $E\left\{\|\mathbf{u}_n\|_2^2\right\}$ are minimal. For the analysis of transient behavior or, equivalently, the analysis of the convergence characteristics, it is necessary to study the trend of $E\left\{|\varepsilon[n]|^2\right\}$, of $E\{\mathbf{u}_n\}$, and of $E\left\{\|\mathbf{u}_n\|_2^2\right\}$.

Therefore, in general, we can affirm that for the learning algorithms performance analysis, it is necessary to determine the trend of the variances (or energies) of some quantities of error.

5.1.3.4 The Principle of Energy Conservation

Solving (5.52) for $g(\cdot)$ and substituting in (5.43) it is possible to eliminate the function. The elimination of the nonlinear function $g(\cdot)$, which determines the adaptation rule, makes the method general and independent from the specific algorithm. You can define two separate cases:

1. $\mathbf{x}_n = \mathbf{0}$. Is a degenerate condition in which

$$\|\mathbf{u}_n\|^2 = \|\mathbf{u}_{n-1}\|^2 \quad \text{and} \quad |e[n]|^2 = |\varepsilon[n]|^2. \tag{5.53}$$

2. $\mathbf{x}_n \neq \mathbf{0}$. Solving the (5.52) for $g(\cdot)$, we get

$$g(e[n]) = \frac{1}{\mu\|\mathbf{u}_n\|^2}(e[n] - \varepsilon[n]). \tag{5.54}$$

Replacing in the (5.43) we obtain the expression

$$\mathbf{u}_n = \mathbf{u}_{n-1} - \frac{\mathbf{u}_n}{\left\|\mathbf{u}_n\right\|^2}\left(\varepsilon[n] - e[n]\right), \tag{5.55}$$

that links the four error's quantity $e[n]$, $\varepsilon[n]$, $\mathbf{u}_n$, and $\mathbf{u}_{n-1}$, and it is not dependent on the learning rate μ. The previous expression can also be rewritten as

$$\mathbf{u}_n + \frac{\mathbf{u}_n}{\left\|\mathbf{u}_n\right\|^2}e[n] = \mathbf{u}_{n-1} + \frac{\mathbf{u}_n}{\left\|\mathbf{u}_n\right\|^2}\varepsilon[n]. \tag{5.56}$$

By defining the step size as

$$\overline{\mu}_n = \begin{cases} 1/\left\|\mathbf{u}_n\right\|^2 & \mathbf{u}_n \neq 0 \\ 0 & \mathbf{u}_n = 0, \end{cases} \tag{5.57}$$

and taking the quadratic norm of the members of both sides of (5.56) is the following *energy conservation theorem*.

5.1.3.5 Energy Conservation Theorem

For each AF of the form (5.42), for each input type $d[n]$ and $\mathbf{x}_n$, applies

$$\left\|\mathbf{u}_n\right\|^2 + \overline{\mu}[n]\left|e[n]\right|^2 = \left\|\mathbf{u}_{n-1}\right\|^2 + \overline{\mu}[n]\left|\varepsilon[n]\right|^2, \tag{5.58}$$

where $e[n] = \mathbf{x}_n^T\mathbf{u}_{n-1}$, $\varepsilon[n] = \mathbf{x}_n^T\mathbf{u}_n$, and $\mathbf{u}_n = \mathbf{w}_n - \mathbf{w}_{\text{opt}}$, or equivalently the form

$$\left\|\mathbf{x}_n\right\|^2 \cdot \left\|\mathbf{u}_n\right\|^2 + \left|e[n]\right|^2 = \left\|\mathbf{x}_n\right\|^2 \cdot \left\|\mathbf{u}_{n-1}\right\|^2 + \left|\varepsilon[n]\right|^2. \tag{5.59}$$

5.2 Method of Descent Along the Gradient: The Steepest-Descent Algorithm

The SDA method can be defined considering a recursive solution of the normal equations in the Wiener form $\mathbf{R}\mathbf{w}_{\text{opt}} = \mathbf{g}$. The algorithm has no memory and is of the first order (the Hessian is not estimated) and, in its general form (5.2), can be written as[1]

$$\mathbf{w}_n = \mathbf{w}_{n-1} + \tfrac{1}{2}\mu\left(-\nabla J(\mathbf{w}_{n-1})\right), \tag{5.60}$$

where the multiplication by 1/2 is only for further simplifications. Denoting the error expectation as $J_n = E\left\{\left|e[n]\right|^2\right\}$ (for the sake of simplicity consider $J_n \equiv J(\mathbf{w}_n)$), the

[1] Note that the subscript n represents an iteration index not necessarily temporal.

explicit gradient expression ∇J_n of the CF can be easily obtained from the quadratic form (3.44) as

$$J_n = \sigma_d^2 - 2\mathbf{g}\mathbf{w}_n^T + \mathbf{w}_n^T\mathbf{R}\mathbf{w}_n, \tag{5.61}$$

for which, deriving respects to weights $\mathbf{w}_n$, applies

$$\nabla J_n = \frac{\partial J_n}{\partial \mathbf{w}_n} = 2(\mathbf{R}\mathbf{w}_n - \mathbf{g}). \tag{5.62}$$

Substituting the latter, evaluated at the step n–1, in (5.60), the explicit SDA form of the algorithm becomes

$$\begin{aligned}\mathbf{w}_n &= \mathbf{w}_{n-1} - \mu(\mathbf{R}\mathbf{w}_{n-1} - \mathbf{g}) \\ &= (\mathbf{I} - \mu\mathbf{R})\mathbf{w}_{n-1} + \mu\mathbf{g},\end{aligned} \tag{5.63}$$

that is precisely a recursive multidimensional finite difference equation (FDE) with IC $\mathbf{w}_{-1}$.

Remark As already mentioned in Chap. 3 (Sect. 3.3.4), the quadratic form (5.61) can be represented in canonical form as

$$J_n = \sigma_d^2 - \mathbf{g}^T\mathbf{R}^{-1}\mathbf{g} + \left(\mathbf{w}_n - \mathbf{R}^{-1}\mathbf{g}\right)^T\mathbf{R}\left(\mathbf{w}_n - \mathbf{R}^{-1}\mathbf{g}\right). \tag{5.64}$$

Note that, by definition place $\mathbf{w}_{\text{opt}} = \mathbf{R}^{-1}\mathbf{g}$, the *error surface* can be written as

$$\begin{aligned}J_n &= J_{\min} + \left(\mathbf{w}_n - \mathbf{w}_{\text{opt}}\right)^T\mathbf{R}\left(\mathbf{w}_n - \mathbf{w}_{\text{opt}}\right) \\ &= J_{\min} + \mathbf{u}_n^T\mathbf{R}\mathbf{u}_n,\end{aligned} \tag{5.65}$$

where $J_{\min} \triangleq J(\mathbf{w}_{\text{opt}})$ and $\mathbf{u}_n = \mathbf{w}_n - \mathbf{w}_{\text{opt}}$.

5.2.1 Multichannel Extension of the SDA

Considering the *composite form* 1 (Sect. 3.2.2.1), and the MIMO Wiener normal equations (Sect. 3.3.8), the multichannel SDA extension is written as

$$\begin{aligned}\mathbf{W}_n &= \mathbf{W}_{n-1} - \mu(\mathbf{R}\mathbf{W}_{n-1} - \mathbf{P}) \\ &= (\mathbf{I} - \mu\mathbf{R})\mathbf{W}_{n-1} + \mu\mathbf{P},\end{aligned} \tag{5.66}$$

with the composite weights matrix defined as $\mathbf{W} \in \mathbb{R}^{P(M)\times Q}$ and where the correlations are $\mathbf{R} \in \mathbb{R}^{P(M)\times P(M)}$ and $\mathbf{P} \in \mathbb{R}^{P(M)\times Q}$.

The reader can easily verify that using the *composite form* 2, the expression of adaptation is completely equivalent.

5.2.2 Convergence and Stability of the SDA

An algorithm is called stable if it converges to a minimum regardless of the choice of IC. To study the properties of convergence and stability of the SDA, consider the weights error vector (5.31) for which, recalling that $\mathbf{g} = \mathbf{R}\mathbf{w}_{\text{opt}}$, from (5.63) is[2]

$$\mathbf{u}_n = (\mathbf{I} - \mu\mathbf{R})\mathbf{u}_{n-1}. \tag{5.67}$$

Decoupling the equations with the *similarity unitary transformation* (Sect. 3.3.6) of the correlation matrix $\mathbf{R}$

$$\mathbf{R} = \mathbf{Q}\mathbf{\Lambda}\mathbf{Q}^T = \sum_{i=0}^{M-1} \lambda_i \mathbf{q}_i \mathbf{q}_i^T, \tag{5.68}$$

and rewriting (5.67) considering the decomposition (5.68) we have that $\mathbf{u}_n = \left(\mathbf{I} - \mu\mathbf{Q}\mathbf{\Lambda}\mathbf{Q}^T\right)\mathbf{u}_{n-1}$. Placing $\hat{\mathbf{u}}_n = \mathbf{Q}^T\mathbf{u}_n$ ($\hat{\mathbf{u}}_n$ represents the vector $\mathbf{u}_n$ rotated), it follows that

$$\hat{\mathbf{u}}_n = (\mathbf{I} - \mu\mathbf{\Lambda})\hat{\mathbf{u}}_{n-1}. \tag{5.69}$$

Because $\mathbf{\Lambda} = \text{diag}[\lambda_0 \quad \lambda_1 \quad \cdots \quad \lambda_{M-1}]$, (5.69) is a set of M decoupled first-order FDE, in the k index, of the type

$$\hat{u}_n(i) = (1 - \mu\lambda_i)\hat{u}_{n-1}(i), \qquad n \geq 0, \qquad i = 0, 1, \ldots, M-1. \tag{5.70}$$

This expression describes all of the M SDA's *natural modes*. The solution of the (5.70) can be determined starting from IC $\hat{u}_{-1}(i)$ for $i = 0, 1, \ldots, M-1$, so, with a simple back substitution, we can write

$$\hat{u}_n(i) = (1 - \mu\lambda_i)^n \hat{u}_{-1}(i), \qquad n \geq 0, \qquad i = 0, 1, \ldots, M-1. \tag{5.71}$$

Necessary condition because the algorithm does not diverge, and therefore for the stability, is that the argument of the exponent is $|1 - \mu\lambda_i| < 1$, or, equivalently,

$$0 < \mu < \frac{2}{\lambda_i}, \qquad \text{for} \quad i = 0, 1, \ldots, M-1. \tag{5.72}$$

This proves that, with a suitable choice of the step of adaptation μ such as to satisfy the (5.72), $\hat{u}_n(i)$ tends to zero for $n \to \infty$. This implies that

[2] Subtracting $\mathbf{w}_{\text{opt}}$ from both members of $\mathbf{w}_n = \mathbf{w}_{n-1} + \mu(\mathbf{g} - \mathbf{R}\mathbf{w}_{n-1}) = \mathbf{w}_{n-1} + \mu\mathbf{R}(\mathbf{w}_{\text{opt}} - \mathbf{w}_{n-1})$, we get $\mathbf{w}_n - \mathbf{w}_{\text{opt}} = \mathbf{w}_{n-1} - \mathbf{w}_{\text{opt}} + \mu\mathbf{R}(\mathbf{w}_{\text{opt}} - \mathbf{w}_{-1}) \Rightarrow \mathbf{u}_n = (\mathbf{I} - \mu\mathbf{R})\mathbf{u}_{n-1}$.

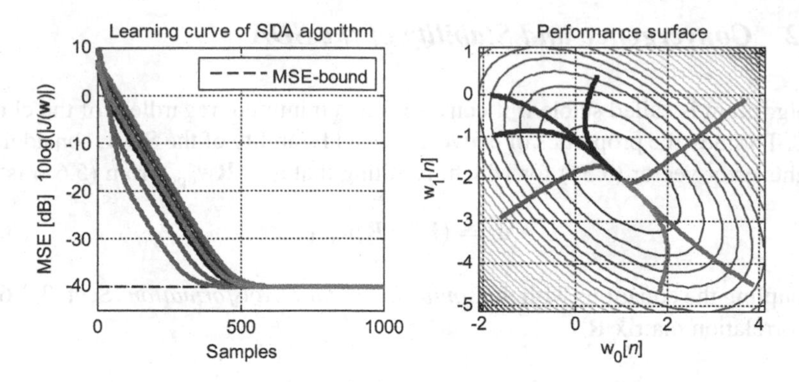

Fig. 5.6 Typical trends of the learning curve and the weights trajectories for the SDA for different IC in the case of an AF with only two coefficients: w_0 and w_1

$$\lim_{n\to\infty} \mathbf{w}_n = \mathbf{w}_{\text{opt}}, \ \forall \mathbf{w}_{-1} \ (\text{IC}). \tag{5.73}$$

It follows that the vector $\mathbf{w}_n$ converges exponentially and exactly the optimum.

Q.E.D.

To illustrate experimentally the convergence properties, Fig. 5.6 shows the typical behavior of the SDA learning curve. To obtain a coefficients graphical representation, it is considered a simple AF with only two adaptive parameters. Observe that the convergence is obtained for different IC values.

Remark Note that, or the SDA convergence proof, no particular assumptions were made about the nature of the input signal which, as for the Wiener filter, is simply described in terms of its second-order statistics.

5.2.2.1 SDA's Stability Condition

By the expression (5.72), the upper limit for the parameter μ is determined by the eigenvalues of the correlation $\mathbf{R} \in \mathbb{R}^{M\times M}$ which, by definition, is semi-positive defined matrix, for which it appears that $\lambda_i \in \mathbb{R}^+$ for $i = 0, 1, \ldots, M - 1$. So, for convergence, it is necessary that the learning rate is upper bounded by the maximum eigenvalue of $\mathbf{R}$, whereby, saying $\lambda_{\max} = \max(\lambda_0,\lambda_1, \ldots,\lambda_{M-1})$, we have that

$$0 < \mu < 2/\lambda_{\max}. \tag{5.74}$$

Many authors say that the maximum value for the step size $(\mu = 2/\lambda_{\max})$ is too large for ensuring the algorithm stability. Moreover, it is convenient to estimate this value before starting the adaptive procedure. One way to overcome this problem consists in considering the trace of the matrix $\mathbf{R}$, defined as $\text{tr}(\mathbf{R}) = \sum_{i=0}^{M-1}\lambda_i$, and, as a new upper limit for the convergence, the value for the step size as $0 < \mu < 2/\text{tr}(\mathbf{R})$. Indeed, given the Toeplitz nature of the autocorrelation matrix, it has

$\sum_{i=0}^{M-1}\lambda_i = \text{tr}(\mathbf{R}) = M \cdot r[0] = M \cdot E\{x^2[n]\}$. Then, the step size can assume the value

$$0 < \mu < \frac{2}{M}\frac{1}{E\{x^2[n]\}}. \tag{5.75}$$

The latter represents a more realistic condition that ensures the stability of the SDA. With this choice for the step size, the stability condition is much stronger than the condition (5.74).

5.2.3 Convergence Speed: Eigenvalues Disparities and Nonuniform Convergence

In AF applications, an important aspect is the convergence speed and, in this regard, the learning rate μ is the parameter that can be defined as an "accelerator" of the adaptation process. In general, the convergence of the SDA is not uniform, which is not identical for all the filter coefficients wi). To understand this phenomenon, let us consider (5.70) of the rotated expected error, relative to the ith element

$$\hat{u}_n(i) = (1 - \mu\lambda_i)\hat{u}_{n-1}(i), \qquad n \geq 0 \quad i = 0,\ 1, \ldots, M-1. \tag{5.76}$$

The decay of the expected error vector $\mathbf{u}_n$ has a rate that is determined by the constant $(0 < \mu < 2/\lambda_{\max})$. The decay of the ith element $\hat{u}_n(i)$ depends on its eigenvalue $|1 - \mu\lambda_i|$. In general, the eigenvalues can have very different values among them. This entails a *nonuniform convergence* of the vector $\mathbf{u}_n$; some elements of the vector converge before others. This problem is known as eigenvalues disparity or *eigenspread.*

In other words, the convergence speed as described by (5.61) depends on the performance surface nature $J(\mathbf{w})$. The most influential effect for the convergence speed is determined by the condition number of the correlation matrix that appears in the CF (5.61) that describes precisely the shape of the contour of $J(\mathbf{w})$. It is demonstrated, in fact, that for a $J(\mathbf{w})$ of the quadratic form that is

$$J(\mathbf{w}_n) \leq \left[\frac{\chi(\mathbf{R}) - 1}{\chi(\mathbf{R}) + 1}\right]^2 J(\mathbf{w}_{n-1}).$$

$\chi(\mathbf{R}) = \lambda_{\max}/\lambda_{\min}$ is the condition number that defines the eigenvalue spread of the matrix $\mathbf{R}$. From the geometry remember that the eigenvectors corresponding to the eigenvalues $\lambda_{\min}$ and $\lambda_{\max}$ are pointing, respectively, to the directions of maximum and minimum curvature of $J(\mathbf{w})$. Observe that the convergence slows down if the surface is more eccentric, or if the eigenspread is very high, i.e., if $\chi(\mathbf{R}) \equiv \lambda_{\max}/\lambda_{\min} \gg 1$. For a circular contour of $J(\mathbf{w})$ is $\chi(\mathbf{R}) = 1$ and the

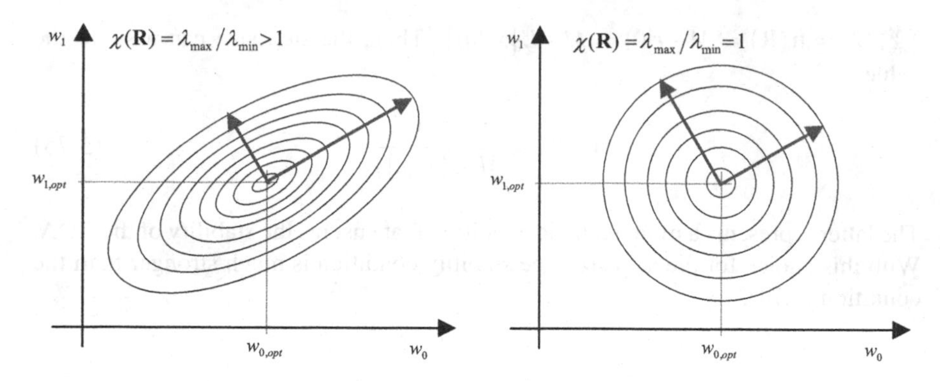

Fig. 5.7 Typical trends of the performance surface $J(\mathbf{w})$ for $M = 1$ (order 2)

convergence to the optimum point can be obtained (theoretically) in one adaptation step.

Figure 5.7 shows the typical behavior of the performance surface with indication of the main directions described by the maximum and minimum eigenvectors.

5.2.3.1 Signal Spectrum and Eigenvalues Spread

It is shown that the condition number [9] is upper bounded by the ratio between the maximum and minimum spectral components of input $x[n]$. Namely, considering the DTFT of the signal $X(e^{j\omega})$, it is possible to demonstrate that

$$1 \le \frac{\lambda_{\max}}{\lambda_{\min}} \le \frac{\left|X_{\max}(e^{j\omega})\right|^2}{\left|X_{\min}(e^{j\omega})\right|^2}.$$

With reference to Fig. 5.8, it also shows that when the filter length increases this inequality tends to equal

$$\lim_{M\to\infty} \Rightarrow \frac{\lambda_{\max}}{\lambda_{\min}} = \frac{\left|X_{\max}(e^{j\omega})\right|}{\left|X_{\min}(e^{j\omega})\right|}. \tag{5.77}$$

In the extreme case in which the input signal $x[n]$ is a white noise, the eigenvalues disparity is minimal $(\lambda_{\max}/\lambda_{\min}) = 1$, and there is uniform convergence. In the case of narrowband input signal, e.g., $x[n]$ is a sine wave (or a noise-free harmonic process), we have that $(\lambda_{\max}/\lambda_{\min}) = \infty$, for which convergence is no longer uniform and slowed down from the maximum eigenvalue.

Remark The ratio $\lambda_{\max}/\lambda_{\min}$ is a monotone nondecreasing function of the size M of the $\mathbf{R}$ matrix. This means that the problem of the disparities of the eigenvalues increases with the filter length M. This implies that increasing the length of the filter does not improve the convergence speed of the adaptation.

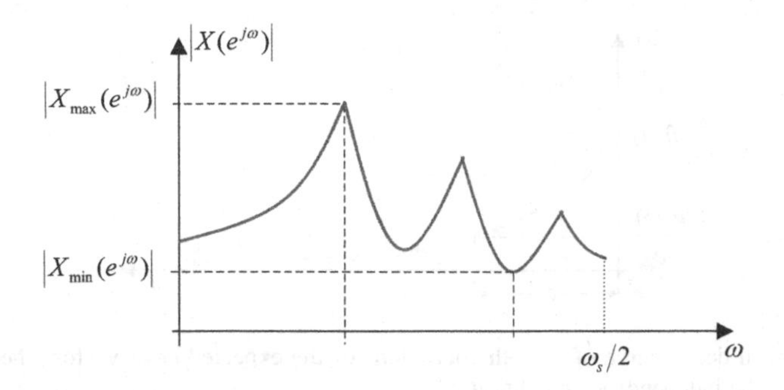

Fig. 5.8 A high eigenspread is characteristic of a process with "lines" spectrum

5.2.3.2 Convergence Time Constant and Learning Curve

The SDA convergence may be evaluated with a time constant. In particular, the adaptation time constant τ, for the ith coefficient of the expected error vector $\hat{u}_n(i)$, can be defined as the decay time for which the initial value decays by a factor equal to $1/e$, or $\tau.\because.\hat{u}_\tau(i) = \frac{1}{e}\hat{u}_{-1}(i)$.

Note that the decay equation (5.76) can be interpreted as that of a dynamic system with initial condition $\hat{u}_{-1}(i)$

$$\hat{u}_n(i) = (1 - \mu\lambda_i)^n \hat{u}_{-1}(i).$$

For $n = \tau$, it is then

$$\hat{u}_\tau(i) \equiv \hat{u}_{-1}(i)/e = (1 - \mu\lambda_i)^\tau \hat{u}_{-1}(i),$$

considering the logarithm is

$$\ln(1 - \mu\lambda_i)^\tau = \ln e^{-1} \quad \Rightarrow \quad \tau \ln(1 - \mu\lambda_i) = -1,$$

and assuming that $\mu\lambda_i \ll 1$ is $\ln(1 - \mu\lambda_i) \cong -\mu\lambda_i$, for which

$$\tau \cong 1/\mu\lambda_i.$$

The typical decay curve for a single element of the vector $\hat{u}_k(i)$ is shown in Fig. 5.9.

Each element of the error vector has a decay constant due to the relative eigenvalue of the correlation matrix. The overall convergence speed is thus determined by the slowest natural mode, i.e., by the smallest eigenvalue

$$\tau \cong 1/\mu\lambda_{\min}. \tag{5.78}$$

Figure 5.10 shows some typical learning curves of the SDA for some values of the learning rate μ. It should be noted, consistent with (5.78), that the convergence time,

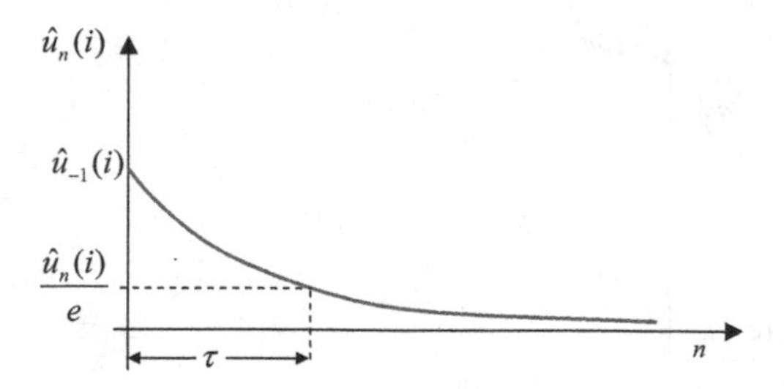

Fig. 5.9 Typical decay curve of the i-th coefficient of the expected error vector when the input signal satisfies the independence condition

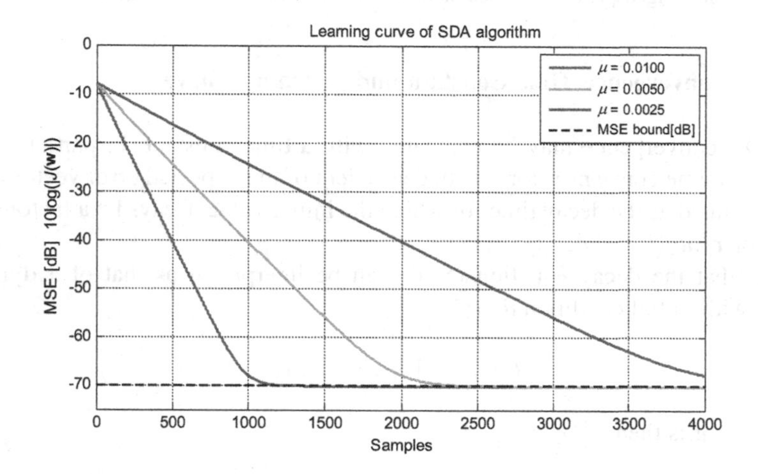

Fig. 5.10 Typical behavior of the learning curve in (dB) of the SDA, for some values of the learning rate μ, for identical initial conditions values

expressed by the number of iterations to the optimum value (indicated in the figure as MSE bound), is higher for lower values of μ.

Remark For the canonical error surface form (5.65), the steady-state excess mean square error takes the form

$$J_{\text{EMSE}} \triangleq J_{\infty} - J_{\min} = \mathbf{u}_n^T \mathbf{R} \mathbf{u}_n. \tag{5.79}$$

From (5.73), the SDA guarantees the optimal solution $\mathbf{w}_{\infty} = \mathbf{w}_{\text{opt}}$, the error function is, then, in an unique and absolute minimum point. It follows that $J(\mathbf{w}_{\text{opt}}) = J_{\infty} = J_{\min}$, for which, for the SDA the excess mean square error is zero, $J_{\text{EMSE}} = 0$.

5.3 First-Order Stochastic-Gradient Algorithm: The Least Mean Squares

Introduced by Widrow–Hoff in 1960 [10], the most popular memoryless SGA consists in considering, similar to the techniques LS, simply instantaneous squared error $e^2[n]$ instead of its expectation, so the CF is defined as $\hat{J}_n = |e[n]|^2$.

Note that this algorithm has already been introduced in Sect. 4.3.2, where it was obtained from the Lyapunov's attractor theorem and used for the determination of the recursive solution of the LS systems. Historically, in the adaptive filtering context, the LMS has been formulated as a stochastic version of the SDA method. Given the popularity of the method, in this section it is presented following this approach.

5.3.1 Formulation of the LMS Algorithm

The LMS algorithm differs from the SDA for the definition of CF that, in the LMS, is deterministic while in the SDA it is stochastic. In practice, the LMS can be interpreted as stochastic approximation of the SDA. Another important aspect concerns the iteration index algorithm that, in this case, always coincides with the time index n.

Calling $\nabla\hat{J}_{n-1} \approx \nabla J_{n-1}$ the estimate of the gradient vector, like for the SDA, see (5.60), the general expression of adaptation is

$$\mathbf{w}_n = \mathbf{w}_{n-1} + \frac{1}{2}\mu\left(-\nabla\hat{J}\left(\mathbf{w}_{n-1}\right)\right), \tag{5.80}$$

where the vector $\mathbf{w}_n$ is an RV.

Denoting the CF as the instantaneous error $\hat{J}_\mathrm{n} = |e[n]|^2$, where the *a priori* error or simply *error* (5.9) is defined as

$$\begin{aligned} e[n] &= d[n] - y[n] \\ &= d[n] - \mathbf{w}_{n-1}^T\mathbf{x}. \end{aligned} \tag{5.81}$$

The gradient vector $\nabla\hat{J}_{n-1}$ evaluated at step n is equal to

$$\nabla\hat{J}_{n-1} = \frac{\partial e^2[n]}{\partial \mathbf{w}_{n-1}} = 2e[n]\frac{\partial e[n]}{\partial \mathbf{w}_{n-1}} = 2e[n]\frac{\partial\left(d[n] - \mathbf{x}^T\mathbf{w}_{n-1}\right)}{\partial \mathbf{w}_{n-1}} = -2e[n]\mathbf{x}, \tag{5.82}$$

so, the adaptation formula (5.80) simply becomes

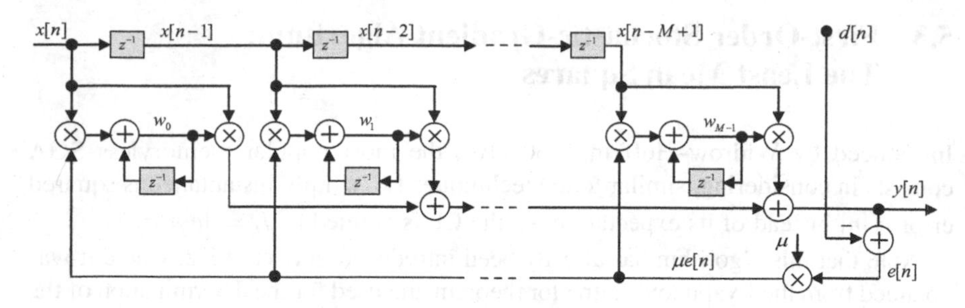

Fig. 5.11 DT circuit representation of the LMS algorithm

$$\mathbf{w}_n = \mathbf{w}_{n-1} + \mu e[n]\mathbf{x}_n. \tag{5.83}$$

The algorithm, whose discrete-time (TD) circuit is shown in Fig. 5.11, is regulated by the step size (or learning rate) μ, fixed or variable μ_n, which in the basic formulation is kept constant.

It is shown (the demonstration similar to that SDA is shown in the following) that the convergence of the algorithm is when

$$0 < \mu < 2/\lambda_{\max}, \tag{5.84}$$

where $\lambda_{\max}$ represents the maximum eigenvalue of the estimated autocorrelation matrix $\mathbf{R}_{xx} \in \mathbb{R}^{M\times M} = \mathbf{X}^T\mathbf{X}$.

In the LMS algorithms filter weights, calculated with (5.83), are estimations; thus $\mathbf{w}_n$ is a RV vector, whose expectation in the stationary case for $n \to \infty$, as will be shown later, tends to the Wiener filter optimal value.

The error minimized in the LMS method is often incorrectly referred to as MSE. What is actually minimized (Sect. 3.2) is, however, the *sum of squared error* (SSE), defined by a time average rather than the ensemble average. However, for the theoretical analysis of the algorithm performance, it always refers to the ensemble average.

5.3.1.1 LMS Formulation with Instantaneous SDA Approximation

The LMS algorithm can be formulated as an instantaneous approximation of the SDA [11]. In this case the following approximations are valid:

$$\mathbf{R} \approx \mathbf{x}_m\mathbf{x}_n^T \qquad \text{and} \qquad \mathbf{g} \approx \mathbf{x}_n d[n]. \tag{5.85}$$

The CF (5.61) $\hat{J}_n = e^2[n]$ takes the form

$$\hat{J}_n = d^2[n] - 2\mathbf{w}_{n-1}^T\mathbf{x}_n d[n] + \mathbf{w}_{n-1}^T\mathbf{x}_n\mathbf{x}_n^T\mathbf{w}_{n-1}, \tag{5.86}$$

and the expression of the gradient $\nabla\hat{J}_n$ can be easily obtained differentiating (5.86) with respect to the weights $\mathbf{w}_n$, for which we have

$$\nabla\hat{J}_n = \frac{\partial\hat{J}_n}{\partial\mathbf{w}_n} = -2\mathbf{x}_n\left(d[n] - \mathbf{x}_n^T\mathbf{w}_{n-1}\right). \tag{5.87}$$

Substituting the latter in (5.80), the explicit form of the algorithm is

$$\mathbf{w}_n = \mathbf{w}_{n-1} + \mu e[n]\mathbf{x}_n \qquad \text{with} \qquad \text{IC } \mathbf{w}_{-1} \tag{5.88}$$

which is exactly identical to (5.83).

5.3.1.2 Summary of the LMS Algorithm

The LMS algorithm consists in the iterative solution of the normal equations in the Yule–Walker form $\left(\mathbf{X}^T\mathbf{X}\mathbf{w} = \mathbf{X}^T\mathbf{d}\right)$. Calling $\mathbf{e} = \mathbf{d} - \mathbf{X}\mathbf{w}$, for the optimum is $\mathbf{w}_* \therefore \arg\min \left\|\mathbf{e}\right\|_2^2$ approximated iteratively by the following recursive form:

(i) Initialization $\mathbf{w}_{-1}$ (small random, all null, *a priori* known, ...)

(ii) For $n = 0, 1, \ldots\{$

$$\begin{aligned} y[n] &= \mathbf{w}_{n-1}^T\mathbf{x}_n, && \textit{input filtering} \\ e[n] &= d[n] - y[n], && \textit{a priori error} \\ \mathbf{w}_n &= \mathbf{w}_{n-1} + \mu e[n]\mathbf{x}_n, && \textit{adaptation} \end{aligned}$$

$\}$.

LMS and SDA Comparison

The LMS algorithm specified by the relations (5.81) and (5.83) has important similarities and differences with the SDA, some of which are shown in Table 5.2. The SDA contains deterministic quantity while the LMS operates with stochastic quantity. The SDA is not a true adaptive algorithm, since it depends only on the second-order moments $\mathbf{g}$ and $\mathbf{R}$ $\left(\text{not directly from the signals } x[n] \text{ and } d[n]\right)$ and n is not necessarily the time index. In practice, the SDA provides an iterative solution of the system $\mathbf{R}\mathbf{w} = \mathbf{g}$.

Table 5.2 Similarities and differences between the SDA and LMS algorithm (the comparison is only possible in average)

SDA	LMS
A priori note statistical functions	Approximated statistical functions
$\mathbf{R}$, $\mathbf{g}$	$\mathbf{R} \to \mathbf{x}\mathbf{x}^T$ $\mathbf{g} \to \mathbf{x}d[n]$
Learning rule	Learning rule
$\mathbf{w}_k = \mathbf{w}_{k-1} + \mu(\mathbf{g} - \mathbf{R}\mathbf{w}_{k-1})$	$\mathbf{w}_n = \mathbf{w}_{n-1} + \mu\mathbf{x}_n e[n]$
Deterministic convergence	Stochastic convergence
$\lim_{k\to\infty} \mathbf{w}_k = \mathbf{w}_{\text{opt}}$	$\lim_{n\to\infty} E\{\mathbf{w}_n\} = \mathbf{w}_{\text{opt}}$
If converges, converges to $\mathbf{w}_{\text{opt}}$	Converge on average. Fluctuation around $\mathbf{w}_{\text{opt}}$ with amplitude proportional to μ

5.3.1.3 LMS Algorithm Computational Cost

It is noted that the LMS computational cost is quite low. For each iteration it is necessary to assess the inner product $\mathbf{w}_{n-1}^T\mathbf{x}_n$ which consists of M multiplications and $M - 1$ additions. For the calculation of the error $e[n] = d[n] - y[n]$, there must be one addition. The product $\mu e[n]\mathbf{x}_n$ requires $M + 1$ multiplications. Finally, for the adaptation, other M additions are required. In total we have

$$(2M + 1) \text{ multiplications and } 2M \text{ additions for each iteration.} \tag{5.89}$$

5.3.2 *Minimum Perturbation Properties and Alternative LMS Algorithm Derivation*

Let us see now that the general minimum perturbation property of the adaptive algorithms, described in Sect. 5.1.3.1, can be considered in the specific case of the LMS. In particular, it is necessary to determine the specific form of the energy constraint $\varepsilon[n] = (1 - \alpha)e[n]$ for the LMS adaptation form. From the definitions of the *a priori* $e[n] = d[n] - \mathbf{w}_{n-1}^T\mathbf{x}$ and *a posteriori* $\varepsilon[n] = d[n] - \mathbf{w}_n^T\mathbf{x}$ errors and from the adaptation formula $\mathbf{w}_n = \mathbf{w}_{n-1} + \mu\mathbf{x}_n e[n]$, we can rewrite the constraint for the LMS adaptation as

$$\begin{aligned} \varepsilon[n] &= d[n] - \mathbf{x}_n^T\mathbf{w}_n \\ &= d[n] - \mathbf{x}_n^T\left(\mathbf{w}_{n-1} + \mu\mathbf{x}_n e[n]\right) \\ &= e[n] - \mu e[n]\mathbf{x}_n^T\mathbf{x}_n \\ &= \left(1 - \mu\|\mathbf{x}_n\|_2^2\right)e[n], \end{aligned} \tag{5.90}$$

which shows, that in the case LMS, we have that $\alpha = \mu\|\mathbf{x}_n\|_2^2$. The expression (5.46), specific to the LMS algorithm, is then

$$\mathbf{w}_{*} \therefore \arg\min_{\mathbf{w}} \|\delta\mathbf{w}\|_2^2 \qquad \text{s.t.} \qquad \varepsilon[n] = \left(1 - \mu\|\mathbf{x}_n\|_2^2\right)e[n], \tag{5.91}$$

(5.91) shows that the optimal LMS solution is the one that the vectors $\mathbf{w}_n$ and $\mathbf{w}_{n-1}$ are a minimum Euclidean distance and subject to the constraint (5.90) between $\varepsilon[n]$ and $e[n]$. Necessarily, it follows that $\left|1 - \mu\|\mathbf{x}_n\|_2^2\right| \leq 1$, and the constraint is more relevant when μ is small so as to satisfy $\left|1 - \mu\|\mathbf{x}_n\|_2^2\right| \ll 1$, namely for

$$0 < \mu \ll \frac{2}{\|\mathbf{x}_n\|_2^2}, \ \forall n. \tag{5.92}$$

Remark An alternative mode to derive the LMS algorithm is to minimize the CF $J(\mathbf{w}) = \|\delta\mathbf{w}\|_2^2$ imposing, *a priori* and *axiomatically*, the energy constraint between the errors $\varepsilon[n] = \left(1 - \mu\|\mathbf{x}_n\|_2^2\right)e[n]$. Proceeding as in (5.47), so as to express this constraint only in function of the *a priori* error, we get

$$\mathbf{x}_n^T\delta\mathbf{w} = \mu\|\mathbf{x}_n\|_2^2 e[n]. \tag{5.93}$$

Proceeding as in Sect. 5.1.3.1, it follows that the CF expression to optimize assumes the form

$$\delta\mathbf{w}_{*} \therefore \arg\min_{\delta\mathbf{w}} \|\delta\mathbf{w}\|_2^2 \qquad \text{s.t.} \qquad \mathbf{x}_n^T\delta\mathbf{w} = \mu\|\mathbf{x}_n\|_2^2 e[n]. \tag{5.94}$$

Note that from (5.93) and for $\|\mathbf{x}_n\|_2^2 \neq 0$, we have that

$$\begin{aligned}\delta\mathbf{w}_{*} &= \mathbf{x}_n\left(\mathbf{x}_n^T\mathbf{x}_n\right)^{-1}\mu\|\mathbf{x}_n\|_2^2 e[n] \\ &= \mu\mathbf{x}_n e[n],\end{aligned} \tag{5.95}$$

so, the adaptation formula

$$\mathbf{w}_n = \mathbf{w}_{n-1} + \mu\mathbf{x}_n e[n], \tag{5.96}$$

which coincides with the LMS (5.83), shows that the LMS algorithm turns out to be equivalent to the *exact solution of a local optimization problem.*

5.3.3 Extending LMS in the Complex Domain

In the case of complex domain signals [40], very common in telecommunications engineering, we consider the following notations: $\mathbf{x} = \mathbf{x}_{\text{Re}} + j\mathbf{x}_{\text{Im}}$ and $d[n] = d_{\text{Re}}[n] + jd_{\text{Im}}[n]$. For the filter coefficients applies the notation (3.8) $\mathbf{w} \triangleq \begin{bmatrix} w^*[0] & \cdots & w^*[M-1] \end{bmatrix}^T$, such that (3.9) the filter output can be computed as

$$\begin{aligned} y[n] &= \mathbf{w}^H \mathbf{x} \\ &= (\mathbf{w}_{\text{Re}} - j\mathbf{w}_{\text{Im}})^T (\mathbf{x}_{\text{Re}} + j\mathbf{x}_{\text{Im}}) \\ &= \mathbf{w}_{\text{Re}}^T \mathbf{x}_{\text{Re}} + j\mathbf{w}_{\text{Re}}^T \mathbf{x}_{\text{Im}} - j\mathbf{w}_{\text{Im}}{}^T \mathbf{x}_{\text{Re}} + \mathbf{w}_{\text{Im}}^T \mathbf{x}_{\text{Im}} \\ &= (\mathbf{w}_{\text{Re}}^T \mathbf{x}_{\text{Re}} + \mathbf{w}_{\text{Im}}^T \mathbf{x}_{\text{Im}}) + j(\mathbf{w}_{\text{Re}}^T \mathbf{x}_{\text{Im}} - \mathbf{w}_{\text{Im}}^T \mathbf{x}_{\text{Re}}). \end{aligned} \tag{5.97}$$

Separating the real and imaginary part of the error, defined as $e[n] = d[n] - y[n]$, we have that

$$\begin{aligned} e[n] &= e_{\text{Re}}[n] + je_{\text{Im}}[n] \\ &= d[n] - \mathbf{w}^H \mathbf{x} \\ &= (d_{\text{Re}}[n] - \mathbf{w}_{\text{Re}}^T \mathbf{x}_{\text{Re}}, -\mathbf{w}_{\text{Im}}^T \mathbf{x}_{\text{Im}}) + j(d_{\text{Im}}[n] - \mathbf{w}_{\text{Re}}^T \mathbf{x}_{\text{Im}}, +\mathbf{w}_{\text{Im}}^T \mathbf{x}_{\text{Re}}). \end{aligned} \tag{5.98}$$

The CF in the real case can written as $\hat{J}(\mathbf{w}) = e^2[n]$, while for the complex case is defined as $\hat{J}(\mathbf{w}) = |e[n]|^2 = e[n]e^*[n] = e_{\text{Re}}^2[n] + e_{\text{Im}}^2[n]$. For the calculation of the complex domain stochastic gradient we can separate the real and imaginary part as

$$\nabla \hat{J}(\mathbf{w}) = \frac{\partial \hat{J}(\mathbf{w})}{\partial \mathbf{w}_{\text{Re}}} + j\frac{\partial \hat{J}(\mathbf{w})}{\partial \mathbf{w}_{\text{Im}}}. \tag{5.99}$$

for which (5.80) can be rewritten as

$$\mathbf{w}_n = \mathbf{w}_{n-1} - \frac{1}{2}\mu\left(\frac{\partial \hat{J}(\mathbf{w})}{\partial \mathbf{w}_{\text{Re}}} + j\frac{\partial \hat{J}(\mathbf{w})}{\partial \mathbf{w}_{\text{Im}}}\right). \tag{5.100}$$

Calculating the partial derivative for the real part, we get

$$\begin{aligned} \frac{\partial |e[n]|^2}{\partial \mathbf{w}_{\text{Re}}} &= \frac{\partial \left[\left(d_{\text{Re}}[n] - \mathbf{w}_{\text{Re}}^T \mathbf{x}_{\text{Re}} - \mathbf{w}_{\text{Im}}^T \mathbf{x}_{\text{Im}}\right)^2 + \left(d_{\text{Im}}[n] - \mathbf{w}_{\text{Re}}^T \mathbf{x}_{\text{Im}} + \mathbf{w}_{\text{Im}}^T \mathbf{x}_{\text{Re}}\right)^2 \right]}{\partial \mathbf{w}_{\text{Re}}} \\ &= -2e_{\text{Re}}[n]\mathbf{x}_{\text{Re}} - 2e_{\text{Im}}[n]\mathbf{x}_{\text{Im}}, \end{aligned} \tag{5.101}$$

while for the imaginary, it is

$$\begin{aligned} \frac{\partial |e[n]|^2}{\partial \mathbf{w}_{\text{Im}}} &= \frac{\partial \left[\left(d_{\text{Re}}[n] - \mathbf{w}_{\text{Re}}^T \mathbf{x}_{\text{Re}} - \mathbf{w}_{\text{Im}}^T \mathbf{x}_{\text{Im}}\right)^2 + \left(d_{\text{Im}}[n] - \mathbf{w}_{\text{Re}}^T \mathbf{x}_{\text{Im}} + \mathbf{w}_{\text{Im}}^T \mathbf{x}_{\text{Re}}\right)^2 \right]}{\partial \mathbf{w}_{\text{Im}}} \\ &= -2e_{\text{Re}}[n]\mathbf{x}_{\text{Im}} + 2e_{\text{Im}}[n]\mathbf{x}_{\text{Re}}. \end{aligned} \tag{5.102}$$

Substituting (5.101) and (5.102) into (5.100) we obtain

$$\begin{aligned}\mathbf{w}_n &= \mathbf{w}_{n-1} + \mu\big(e_{\mathrm{Re}}[n]\mathbf{x}_{\mathrm{Re}} + e_{\mathrm{Im}}[n]\mathbf{x}_{\mathrm{Im}} + je_{\mathrm{Re}}[n]\mathbf{x}_{\mathrm{Im}} - je_{\mathrm{Im}}[n]\mathbf{x}_{\mathrm{Re}}\big)\\ &= \mathbf{w}_{n-1} + \mu\big(e_{\mathrm{Re}}[n] - je_{\mathrm{Im}}[n]\big)\big(\mathbf{x}_{\mathrm{Re}} + j\mathbf{x}_{\mathrm{Im}}\big)\\ &= \mathbf{w}_{n-1} + \mu e^*[n]\mathbf{x}_n.\end{aligned} \tag{5.103}$$

Note that in the complex case the notation is identical to the real one (5.83), except for the presence of the conjugate value [6].

The LMS complex convergence properties are very similar to those of the real case. For the convergence speed the expression (5.171) is still valid.

5.3.3.1 Computational Cost

A product in the complex domain is equivalent to four real multiplications and two sums, while the complex sum is equivalent to two real additions. Hence, for *t*(5.89), the computational cost per iteration of the complex LMS is

$$(8M + 2) \text{ real multiplications and } 8M \text{ real additions.} \tag{5.104}$$

5.3.4 LMS with Linear Constraints

As noted earlier in Chap. 4 (Sect. 4.2.5.5), some adaptive filtering applications may require the presence of external constraints, due to the specific nature of the problem. For example, consider the case that $e[n] = d[n] - \mathbf{w}_{n-1}^H\mathbf{x}_n$, and it is necessary to determine the CF minimum that, for some reason, is subject to the following constraint:

$$\mathbf{C}^H\mathbf{w}_n = \mathbf{b} \tag{5.105}$$

where $\mathbf{C} \in (\mathbb{R}, \mathbb{C})^{M\times N_c}$ and $\mathbf{b} \in (\mathbb{R}, \mathbb{C})^{N_c\times 1}$ with $M > N_c$ are the constraint matrix and vector, *a priori* fixed. For $\hat{J}_n = |e[n]|^2$ then the problem can be formulated with the following CF:

$$\mathbf{w}_c \therefore \min_{\mathbf{w}} |e[n]|^2 \qquad \text{s.t.} \qquad \mathbf{C}^H\mathbf{w}_n = \mathbf{b}, \tag{5.106}$$

for which the local Lagrangian is

$$L(\mathbf{w}, \boldsymbol{\lambda}) = \tfrac{1}{2}|e[n]|^2 + \boldsymbol{\lambda}^H\big(\mathbf{C}^H\mathbf{w}_n - \mathbf{b}\big). \tag{5.107}$$

5.3.4.1 The Linearly Constrained LMS Algorithm

In the presence of the linear constraint, the recursive solution, called linearly constrained LMS (LCLMS) algorithm, can be obtained from the standard LMS solution considering the local minimization of the Lagrangian (5.107).

For the determination the LCLMS recursion, we can consider the steepest descent directly of the Lagrangian gradient surface

$$\mathbf{w}_n = \mathbf{w}_{n-1} - \mu \nabla_{\mathbf{w}} L(\mathbf{w}, \boldsymbol{\lambda}), \tag{5.108}$$

where (5.82) $\nabla_{\mathbf{w}} L(\mathbf{w},\boldsymbol{\lambda}) = -e^*[n]\mathbf{x} + \mathbf{C}\boldsymbol{\lambda}$. For $N = 1$ (only one equation) (5.108) can be written as

$$\mathbf{w}_n = \mathbf{w}_{n-1} + \mu \mathbf{x}_n e^*[n] - \mu \mathbf{C}\boldsymbol{\lambda}. \tag{5.109}$$

Multiplying the last equation with $\mathbf{C}^H$, and for $\mathbf{C}^H\mathbf{w}_n = \mathbf{b}$, we get

$$\mathbf{C}^H\mathbf{w}_n \equiv \mathbf{b} = \mathbf{C}^H\mathbf{w}_{n-1} + \mu \mathbf{C}^H \mathbf{x}_n e^*[n] - \mathbf{C}^H\mathbf{C}\mu\boldsymbol{\lambda}.$$

Solving for $\mu\boldsymbol{\lambda}$, we obtain

$$\mu\boldsymbol{\lambda} = \mathbf{C}^H\left(\mathbf{C}^H\mathbf{C}\right)^{-1}\mathbf{w}_{n-1} + \mu\mathbf{C}^H\left(\mathbf{C}^H\mathbf{C}\right)^{-1}\mathbf{x}_n e^*[n] + \left(\mathbf{C}^H\mathbf{C}\right)^{-1}\mathbf{b}.$$

Substituting in (5.109) and rearranging we obtain

$$\begin{aligned}\mathbf{w}_n = &\left(\mathbf{I} - \mathbf{C}^H\left(\mathbf{C}^H\mathbf{C}\right)^{-1}\mathbf{C}\right)\mathbf{w}_{n-1} \\ &+\mu\left(\mathbf{I} + \mathbf{C}^H\left(\mathbf{C}^H\mathbf{C}\right)^{-1}\mathbf{C}\right)\mathbf{x}_n e^*[n] + \mathbf{C}\left(\mathbf{C}^H\mathbf{C}\right)^{-1}\mathbf{b}.\end{aligned} \tag{5.110}$$

5.3.4.2 Recursive Gradient Projection LCLMS

Proceeding as in Sect. 4.2.5.5, considering the following projection operators:

$$\begin{aligned}\tilde{\mathbf{P}} &\in (\mathbb{R},\mathbb{C})^{M\times M} \triangleq \mathbf{C}\left(\mathbf{C}^H\mathbf{C}\right)^{-1}\mathbf{C} \\ \mathbf{P} &\in (\mathbb{R},\mathbb{C})^{M\times M} \triangleq \left[\mathbf{I} - \tilde{\mathbf{P}}\right] \\ \mathbf{F} &\in (\mathbb{R},\mathbb{C})^{M\times 1} \triangleq \mathbf{C}\left(\mathbf{C}^H\mathbf{C}\right)^{-1}\mathbf{b},\end{aligned} \tag{5.111}$$

we have the recurrence equation (5.110) written as

$$\mathbf{w}_n = \mathbf{P}\left(\mathbf{w}_{n-1} + \mu\mathbf{x}_n e^*[n]\right) + \mathbf{F}, \tag{5.112}$$

where the projection matrix $\mathbf{P}$ and $\mathbf{F}$ can be *a priori* computed.

Remark The problem of adaptation in the presence of linear constraint is of fundamental importance in the area of space-time filtering (*array processing*). That argument will be reintroduced later in Chap. 9, where, in the problem of beamforming, a physical and geometrical interpretation of the constrained

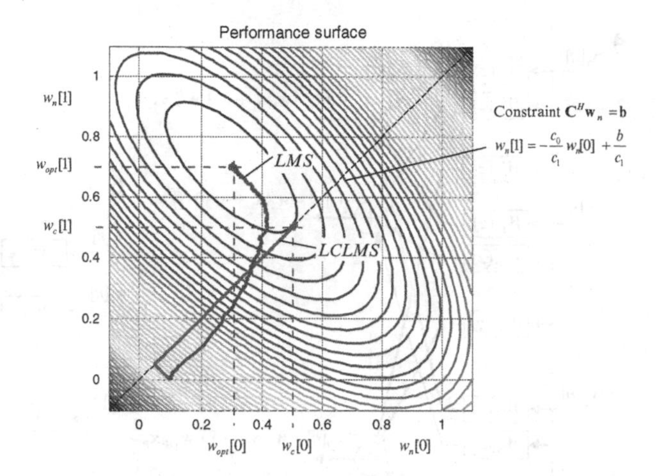

Fig. 5.12 Example of LCLMS. Weights trajectories during the of LMS and CLMS adaptation, in the performance surface $J(\mathbf{w}_n)$. The founded optimal constrained solution is $w_c[1] = w_c[0] = 0.5$

methodology will be given. For more details on the performance of the constrained LMS, see, for example, [12].

Example Consider an example of identifying a system with impulse response $\mathbf{h} = [0.3\ \ 0.7]$ (Sect. 3.4.1) in which is imposed for the optimal solution the constraint that the weights are identical, i.e., $w_c[1] = w_c[0]$. As in the example in Sect. 4.2.5.5, considering the expression (5.106), with $M = 2$, you can insert only one constraint $N_c = 1$, which can be formalized as

$$\mathbf{C}^H\mathbf{w}_n = \mathbf{b} \Rightarrow [1 \quad -1]\begin{bmatrix} w_n[0] \\ w_n[1] \end{bmatrix} = 0.$$

The unconstrained optimal solution is obviously $\mathbf{w}_{\text{opt}} = \mathbf{h}$. As illustrated in Fig. 5.12, the constrained solution is the closest (according to the metric choice) to the optimal solution $\mathbf{w}_{\text{opt}}$, which satisfies the constraint imposed, i.e., that lies in the plane of the constraint (in our case the line $w[1] = w[0]$). In other words, the optimal constrained solution corresponds to the point of tangency between the constraint line and the isolevel curve of the standard LMS CF $J(\mathbf{w})$.

5.3.4.3 Summary of the LCLMS Algorithm

(i) Initialization $\mathbf{w}_{-1} = \mathbf{0}$, $y[-1] = 0$, $\mathbf{P} = \mathbf{I} - \mathbf{C}(\mathbf{C}^H\mathbf{C})^{-1}\mathbf{C}^H$, $\mathbf{F} = \mathbf{C}(\mathbf{C}^H\mathbf{C})^{-1}\mathbf{b}$

(ii) For $n = 0, 1, \ldots \{$

$$y[n] = \mathbf{w}_n^H \mathbf{x}_n$$
$$e[n] = d[n] - y[n]$$
$$\mathbf{w}_n = \mathbf{P}(\mathbf{w}_{n-1} + \mu e^*[n]\mathbf{x}_n) + \mathbf{F}$$

$\}$.

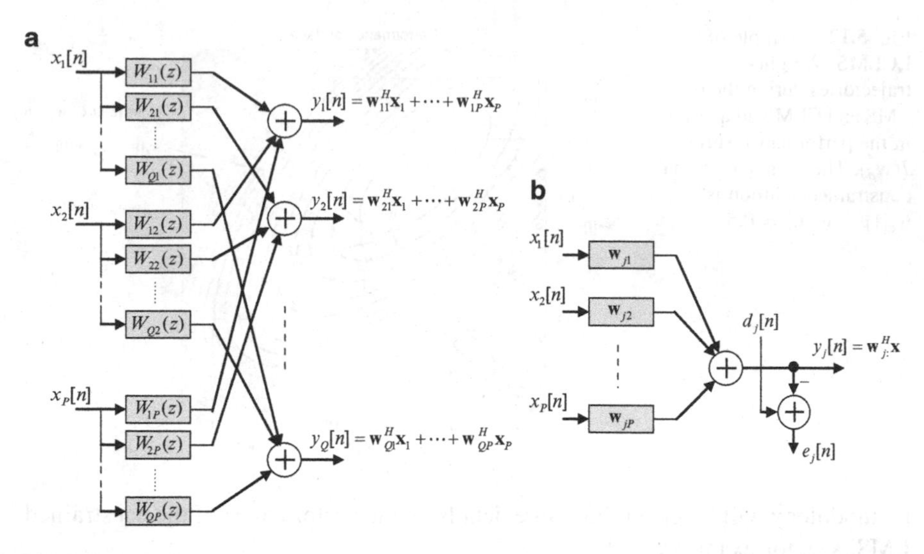

Fig. 5.13 MIMO adaptive filter: (**a**) formalism; (**b**) representation of the j-th MISO sub-system of the MIMO system

5.3.5 Multichannel LMS Algorithms

The generalization of the adaptation algorithms to the MIMO case with P inputs and Q outputs has already been introduced in Chap. 3 in particular with the definition of the MIMO Wiener–Hopf equations.

From the formalism already introduced in Chap. 3 and shown in Fig. 5.13 (Sect. 3.2.2), we recall that $\mathbf{w}_{ji} \in (\mathbb{R},\mathbb{C})^{M\times 1} \triangleq \begin{bmatrix} w_{ji}[0] & \cdots & w_{ji}[M-1] \end{bmatrix}^H$ is the impulse response between the ith input and the jth output, whereas the matrix $\mathbf{W} \in (\mathbb{R},\mathbb{C})^{Q\times P(M)}$ is defined as

$$\mathbf{W} \in (\mathbb{R},\mathbb{C})^{Q\times P(M)} = \begin{bmatrix} \mathbf{w}_{11}^H & \mathbf{w}_{12}^H & \cdots & \mathbf{w}_{1P}^H \\ \mathbf{w}_{21}^H & \mathbf{w}_{22}^H & \cdots & \mathbf{w}_{2P}^H \\ \vdots & \vdots & \ddots & \vdots \\ \mathbf{w}_{Q1}^H & \mathbf{w}_{Q2}^H & \cdots & \mathbf{w}_{QP}^H \end{bmatrix}_{Q\times P}. \tag{5.113}$$

Indicating with $\mathbf{w}_{j:}^H \in (\mathbb{R},\mathbb{C})^{1\times P(M)} \triangleq \begin{bmatrix} \mathbf{w}_{j1}^H & \cdots & \mathbf{w}_{jP}^H \end{bmatrix}$, the jth row of the $\mathbf{W}$ matrix, as shown in Fig. 5.13b, identifies the bank of P filters relating to the jth output of the system, also defined as

$$\mathbf{W} \in (\mathbb{R},\mathbb{C})^{Q\times 1(PM)} = \begin{bmatrix} \mathbf{w}_{1:} & \mathbf{w}_{2:} & \cdots & \mathbf{w}_{Q:} \end{bmatrix}^H. \tag{5.114}$$

Calling $\mathbf{x}_i \in (\mathbb{R},\mathbb{C})^{M\times 1} \triangleq \begin{bmatrix} x_i[n] & \cdots & x_i[n-M+1] \end{bmatrix}^H$ the ith input signal, for the output we get

$$\begin{aligned} y_j[n] &= \sum_{i=1}^{P} \mathbf{w}_{ji}^H \mathbf{x}_i \\ &= \mathbf{w}_{j:}^H \mathbf{x}. \end{aligned} \tag{5.115}$$

(5.115) indicates that the MIMO filter consists of a parallel of Q filters bank of P channels MISO; each of them is characterized by the weights vector $\mathbf{w}_{j:}$ and which can be adapted in a independent way to each other.

For the output snapshot $\mathbf{y}[n] \in (\mathbb{R}, \mathbb{C})^{Q\times 1} = \begin{bmatrix} y_1[n] & y_2[n] & \cdots & y_Q[n] \end{bmatrix}^H$

$$\mathbf{y}[n] = \mathbf{W}\mathbf{x}, \tag{5.116}$$

where, omitting the writing of the subscript n, the vector

$$\mathbf{x} \in (\mathbb{R}, \mathbb{C})^{P(M)\times 1} = \begin{bmatrix} \mathbf{x}_1^H & \mathbf{x}_2^H & \cdots & \mathbf{x}_P^H \end{bmatrix}^H, \tag{5.117}$$

contains the vectors of the input channels, all stacked at the instant n (we remind the reader the convention $\mathbf{x} \equiv \mathbf{x}_n$ and $\mathbf{x}_i \equiv \mathbf{x}_{i,n}$).

Indicating, respectively, with $(\mathbf{y}[n], \mathbf{d}[n]) \in (\mathbb{R},\mathbb{C})^{Q\times 1}$ the output and the desired output snapshots, for the *a priori* error vector $\mathbf{e}[n] \in (\mathbb{R},\mathbb{C})^{Q\times 1}$, we can write

$$\begin{aligned} \mathbf{e}[n] &= \mathbf{d}[n] - \mathbf{y}[n] \\ &= \mathbf{d}[n] - \mathbf{W}_{n-1}\mathbf{x}. \end{aligned} \tag{5.118}$$

Considering the jth output of the system, from the definition (5.114) holds

$$e_j[n] = d_j[n] - \mathbf{w}_{j:}^H \mathbf{x}, \qquad \text{for} \qquad j = 1, 2, \ldots, Q, \tag{5.119}$$

or, explaining all filters $\mathbf{w}_{ij}$, the above is equivalent to

$$e_j[n] = d_j[n] - \sum_{i=1}^{P} \mathbf{w}_{ji}^H \mathbf{x}_i, \qquad \text{for} \qquad j = 1, 2, \ldots, Q. \tag{5.120}$$

For the definition of the *multichannel least mean squares* or MIMO-LMS, we can refer to one of the error expressions (5.118)–(5.120).

5.3.5.1 MIMO-LMS by Global Adaptation

As a first case, we consider the development with the vector expression vector (5.118). In Sect. 3.3.8 we defined the stochastic CF as $J(\mathbf{W}) = E\left\{\mathbf{e}^H[n]\mathbf{e}[n]\right\}$, for which, extending the development in Sect. 5.3.1 to the multichannel case, by replacing the expectation operator with the instantaneous squared error, the MIMO-LMS cost function can be defined as

$$\hat{J}_{n-1} = \mathbf{e}^H[n]\mathbf{e}[n]. \tag{5.121}$$

The adaptation law is

$$\mathbf{W}_n = \mathbf{W}_{n-1} + \tfrac{1}{2}[-\nabla\hat{J}_{n-1}], \tag{5.122}$$

where, by generalizing (5.88), the stochastic gradient is a matrix $\nabla\hat{J}_{n-1} \in (\mathbb{R},\mathbb{C})^{Q\times PM}$ calculated by differentiating (5.121) with respect to $\mathbf{W}_{n-1}$

$$\nabla\hat{J}_{n-1} = \frac{\partial\left(^{\mathbf{e}H}[n]\mathbf{e}[n]\right)}{\partial\mathbf{W}_{n-1}} = 2\mathbf{e}^*[n]\frac{\partial\left(\mathbf{d}[n] - \mathbf{W}_{n-1}\mathbf{x}\right)}{\partial\mathbf{W}_{n-1}} = -2\mathbf{e}^*[n]\mathbf{x}^H.$$

A first vector form of the MIMO-LMS algorithm is then

$$\begin{aligned} \mathbf{e}[n] &= \mathbf{d}[n] - \mathbf{W}_{n-1}\mathbf{x} \\ \mathbf{W}_n &= \mathbf{W}_{n-1} + \mu\mathbf{e}^*[n]\mathbf{x}^H. \end{aligned} \tag{5.123}$$

5.3.5.2 MIMO-LMS by Filters Banks Adaptation

Considering the expression (5.119), the adaptation algorithm development can be made by considering Q independent filters banks (Fig. 5.13b); in other words the CF (5.121) is expressed as

$$\hat{J}_{n-1}(\mathbf{W}) = \left[\hat{J}_{1,n-1}(\mathbf{w}_{1:}) \quad \hat{J}_{2,n-1}(\mathbf{w}_{2:}) \quad \cdots \quad \hat{J}_{Q,n-1}(\mathbf{w}_{Q:})\right]^T. \tag{5.124}$$

By differentiating the jth component of the previous CF is then

$$\nabla\hat{J}_{j,n-1} = \frac{\partial e_j^2[n]}{\partial\mathbf{w}_{j:,n-1}} = 2e_j^*[n]\frac{\partial\left(d_j[n] - \mathbf{x}^H\mathbf{w}_{j:,n-1}\right)}{\partial\mathbf{w}_{j:,n-1}} = -2e_j^*[n]\mathbf{x}, \tag{5.125}$$

where $\nabla\hat{J}_{j,n-1} \in \mathbb{R}^{PM\times 1}$. For the adaptation we can write

$$\begin{aligned} e_j[n] &= d_j[n] - \mathbf{w}_{j:,n-1}^H\mathbf{x} \\ \mathbf{w}_{j:,n} &= \mathbf{w}_{j:,n-1} + \mu e_j^*[n]\mathbf{x}, \end{aligned} \qquad j = 1, 2, \ldots, Q. \tag{5.126}$$

Each of Q filters bank is interpreted as a unique single filter, of length equal to $(P\cdot M)$, with an input signal $\mathbf{x}$ containing, stacked, all inputs $\mathbf{x}_i$ for $i = 1, \ldots, P$.

5.3.5.3 Filter-by-Filter Adaptation

From the expression of the error (5.120), the CF is defined as

$$\hat{J}_{n-1}(\mathbf{W}) = \begin{bmatrix} \hat{J}_{11,n-1}(\mathbf{w}_{12}) & \hat{J}_{12,n-1}(\mathbf{w}_{12}) & \cdots & \hat{J}_{1P,n-1}(\mathbf{w}_{1P}) \\ \hat{J}_{21,n-1}(\mathbf{w}_{21}) & \hat{J}_{22,n-1}(\mathbf{w}_{22}) & \cdots & \hat{J}_{2P,n-1}(\mathbf{w}_{1P}) \\ \vdots & \vdots & \ddots & \vdots \\ \hat{J}_{Q1,n-1}(\mathbf{w}_{Q1}) & \hat{J}_{Q2,n-1}(\mathbf{w}_{Q1}) & \cdots & \hat{J}_{QP,n-1}(\mathbf{w}_{1P}) \end{bmatrix}. \tag{5.127}$$

Considering the element $\hat{J}_{ji,n-1}(\mathbf{w}_{ji})$,

$$\nabla\hat{J}_{ji,n-1} = \frac{\partial e_j^2[n]}{\partial \mathbf{w}_{ji,n-1}} = -2e_j^*[n]\mathbf{x}_i, \tag{5.128}$$

where $\nabla\hat{J}_{ij,n-1} \in \mathbb{R}^{M\times1}$. For the adaptation we can write

$$e_j[n] = d_j[n] - \sum_{i=1}^{P} \mathbf{x}_i^T \mathbf{w}_{ij}, \quad j = 1,2,\ldots,Q, \tag{5.129}$$

$$\mathbf{w}_{ij,n} = \mathbf{w}_{ij,n-1} + \mu e_j^*[n]\mathbf{x}_i. \quad i = 1,2,\ldots, \ P, \ j = 1,2,\ldots,Q. \tag{5.130}$$

Remark Being the output error uniquely defined, the formulations (5.123), (5.126), and (5.130) are algebraically completely equivalent. The reader can also easily verify that using the *composite notation* 2, the adaptation expressions are completely equivalent.

5.3.5.4 The MIMO-LMS as a MIMO-SDA Approximation

The multichannel LMS algorithm can be formulated as an instantaneous approximation of the multichannel SDA method, with P inputs and Q outputs (Sect. 5.2.1). So considering the *composite notation* 1 (Sect. 3.2.2.1), the following approximations are valid:

$$\mathbf{R} \approx \mathbf{x}_n\mathbf{x}_n^H \quad \text{and} \quad \mathbf{P} \approx \mathbf{x}_n\mathbf{d}[n], \tag{5.131}$$

where, in the MIMO case, the vector $\mathbf{x}_n \in (\mathbb{R},\mathbb{C})^{PM\times1} = \begin{bmatrix} \mathbf{x}_{1,n}^T & \mathbf{x}_{2,n}^T & \cdots & \mathbf{x}_{P,n}^T \end{bmatrix}^T$ represents the composite signal input in the composite notation 1. With similar reasoning presented in the previous section, we show that the MIMO-LMS adaptation rule can be written as

$$\mathbf{W}_n = \mathbf{W}_{n-1} - \mu\mathbf{e}^*[n]\mathbf{x}_n^H, \qquad \textit{LMS-MIMO algorithm.} \tag{5.132}$$

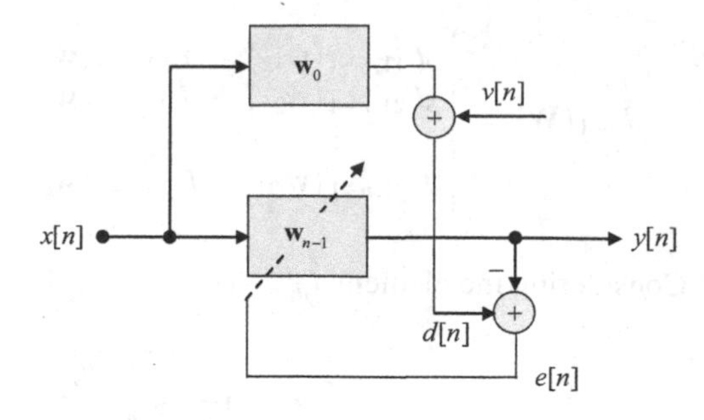

Fig. 5.14 Model for the statistical analysis of the performance of LMS (and other) algorithms

5.4 Statistical Analysis and Performance of the LMS Algorithm

The LMS performance is evaluated according to the fundamental aspects of stability, convergence speed, the accuracy of the result at steady state, the transient behavior, and the tracking capability, expressed in terms of opportune statistical functions of the error signal. The analysis reported in the following is carried out considering the stationary environment and the learning algorithm similar to a dynamic system represented by a stochastic difference equation (SDE).

5.4.1 Model for Statistical Analysis of the Adaptive Algorithms Performance

The dynamic system model considered for the statistical analysis of the algorithm performance is shown in Fig. 5.14. This consists in an identification problem of a dynamic system $\mathbf{w}_0$ when to the reference signal $d[n]$ is superimposed (added) a Gaussian noise, for which we have that

$$d[n] = \mathbf{w}_0^H \mathbf{x}_n + v[n].$$

In other words, the desired output $d[n]$ consists of a stationary *moving average* (MA) time series (Appendix C) with superimposed noise, where $\mathbf{v}_n \sim N(0,\sigma_v^2)$ is zero mean, for each n independent and identically distributed (iid) RV, with constant variance. In this situation, the Wiener optimal solution is, by definition, indicated as $\mathbf{w}_0 = \mathbf{R}^{-1}\mathbf{g}$.

It should be noted that the model of Fig. 5.14 is generic and allows, with specific variations, the analysis of all the adaptive algorithms characterized by a more general learning law of the type

$$\mathbf{w}_n = \mathbf{w}_{n-1} + \mu \mathbf{x}_n g\big(e[n]\big). \tag{5.133}$$

The convergence can be more easily demonstrated when the following assumptions are assumed:

1. The input sequence $\mathbf{x}_n$ is zero-mean WGN $\mathbf{x}_n \sim N(\mathbf{0},\mathbf{R})$
2. For each n, $\mathbf{w}_n$, $\mathbf{x}_n$ and $v[n]$ are iid sequences

Given that the quantity $\mathbf{w}_n$, as well as $\mathbf{x}_n$, also depends on its past values $\mathbf{x}_{n-1}, \mathbf{x}_{n-2}, \ldots$; the statistical independence assumption is equivalent to the condition that also applies to the vector $\mathbf{x}_n$ compared to previous instants, or, also applies that

$$E\left\{\mathbf{x}_n \mathbf{x}_m^H\right\} = 0 \qquad \forall n \neq m. \tag{5.134}$$

Note that this assumption is very strong and unrealistic. The vectors $\mathbf{x}_n$ and $\mathbf{x}_{n-1}$ have, in fact, $M-1$ common elements and belong to the same stochastic process. This assumption is, however, one of the few cases in which the average convergence of the LMS is explicitly proved and, furthermore, with a procedure similar to that of the SDA.

The transient and steady-state filter performances are evaluated by the solution of (5.133) with regard to the optimal Wiener solution that, in the case of Fig. 5.14, is precisely $\mathbf{w}_0$. In particular, for the convergence demonstration, is evaluated the first-order error statistic behavior $E\{\mathbf{u}_n\}$, while, for transient characteristic and tracking filter analysis, is considered the mean squares behavior, i.e., we consider the solution of the second-order statistics, or error vector mean square deviation $E\{\|\mathbf{u}_n\|_2^2\}$ (Sect. 5.1.2.3).

5.4.1.1 Minimum Energy Error, in the Performance Analysis's Model

In the dynamic system model identification, with the measurement noise superimposed on the desired output, such as that of Fig. 5.14, at the optimum solution, we know that the minimum error energy (Sect. 3.3.4.2) is equal to

$$J_{\min} = E\{d^2[n]\} - \mathbf{w}_0^H \mathbf{R} \mathbf{w}_0.$$

In the case of independent noise, we have that

$$\begin{aligned} E\{d^2[n]\} &= E\left\{\left(\mathbf{w}_0^H \mathbf{x} + v[n]\right)^2\right\} \\ &= \sigma_v^2 + \mathbf{w}_0^H \mathbf{R} \mathbf{w}_0. \end{aligned}$$

For the LMS algorithm, in the case of convergence in which for $n \to \infty \Rightarrow \mathbf{w}_n \to \mathbf{w}_0$, the determination of the minimum error energy is due to the Wiener's statistically optimal solution. Given variance σ_v^2, from the previous expressions, the minimum error energy is equal to

$$J_{\min} = \sigma_v^2. \tag{5.135}$$

Notice that this result was already implicitly discussed in the application example, of the dynamic system modeling (Sect. 3.4.1).

5.4.2 LMS Characterization and Convergence with Stochastic Difference Equation

From the adaptation formula (5.103), subtracting the optimal solution $\mathbf{w}_0$ from both members, and considering the weights error vector (5.31) $\mathbf{u}_n = \mathbf{w}_n - \mathbf{w}_0$, we get

$$\mathbf{u}_n = \mathbf{u}_{n-1} + \mu e^*[n]\mathbf{x}_n. \tag{5.136}$$

By defining the quantity, *error at optimal solution*, as

$$v[n] = d[n] - \mathbf{w}_0^H\mathbf{x}_n \tag{5.137}$$

we can express the relation of the *a priori* error as

$$\begin{aligned} e[n] &= d[n] - \mathbf{w}_{n-1}^H\mathbf{x}_n \\ &= d[n] - \mathbf{w}_{n-1}^H\mathbf{x}_n - \mathbf{w}_0^H\mathbf{x}_n + \mathbf{w}_0^H\mathbf{x}_n \\ &= v[n] - \mathbf{u}_{n-1}^H\mathbf{x}_n. \end{aligned} \tag{5.138}$$

By replacing this into (5.136) we obtain the following SDE:

$$\begin{aligned} \mathbf{u}_n &= \mathbf{u}_{n-1} + \mu\left(v[n] - \mathbf{u}_{n-1}^H\mathbf{x}_n\right)\mathbf{x}_n \\ &= \left(\mathbf{I} - \mu\mathbf{x}_n\mathbf{x}_n^H\right)\mathbf{u}_{n-1} + \mu v^*[n]\mathbf{x}_n, \end{aligned} \tag{5.139}$$

where, by definition, the variables $\mathbf{u}_n$, $\mathbf{w}_n$, and $\mathbf{x}_n$ are RV, for which the previous expression represents a nonuniform and time-variant SDE. The forcing term $\mu v^*[n]\mathbf{x}_n$ is due to the *irreducible noise* $v[n]$ whose causes are due to measurement error, quantization effects, and other disturbances.

Remark The determination of the statistical solution of the SDE is very complex because it requires the calculation of first- and second-order moments of its both members. For example, taking the expectation of (5.139), we can note the presence of the third-order moment $E\{\mathbf{x}_n\mathbf{x}_n^H\mathbf{u}_{n-1}\}$. This poses some mathematical–statistical difficulties, and it is for this reason that for the proof it is preferable to refer to the simple independence assumption.[3]

5.4.2.1 Study of Weak Convergence

A substantial simplification for the LMS statistical study and, as we shall see, other algorithms with more general learning law (5.133) is possible if we consider the *weak convergence*.

It is, in practice, to determine the boundary conditions for simplifying (5.139) such that it can be studied as a normal ordinary difference equation (ODE), and if these assumptions are met, it is possible to directly determine the solution (on *average*). In particular, the weak convergence study of (5.139) can be performed with the so-called direct-averaging method (DAM) [13], reported below.

Direct-Averaging Method

The condition that allows the simplification of the problem is to consider a very small adaptation step ($\mu \ll 1$), With such a condition, called DAM, we can consider the approximation $\left(\mathbf{I} - \mu\mathbf{x}_n\mathbf{x}_n^H\right) \sim (\mathbf{I} - \mu\mathbf{R})$. As a result, (5.139) can be rewritten as[4]

$$\mathbf{u}_n = (\mathbf{I} - \mu\mathbf{R})\mathbf{u}_{n-1} + \mu v^*[n]\mathbf{x}_n. \tag{5.140}$$

Considering the solution of the previous with first-order statistics, i.e., by making the expectation of both sides, for the independence between the quantities $\mathbf{x}_n$ and $v[n]$ is $E\{\mu v^*[n]\mathbf{x}_n\} = 0$. Then we have

$$E\{\mathbf{u}_n\} = (\mathbf{I} - \mu\mathbf{R})E\{\mathbf{u}_{n-1}\}. \tag{5.141}$$

Likewise to the SDA development (Sect. 5.2.2), decomposing the correlation with the similarity unitary transformation $\mathbf{\Lambda} = \mathbf{Q}^H\mathbf{R}\mathbf{Q}$, by placing $\hat{\mathbf{u}}_n = \mathbf{Q}^H\mathbf{u}_n$, we can write

$$E\{\hat{\mathbf{u}}_n\} = (\mathbf{I} - \mu\mathbf{\Lambda})E\{\hat{\mathbf{u}}_{n-1}\}, \tag{5.142}$$

which has the same form as the SDA (5.69), where the rotated vector error $\hat{\mathbf{u}}_n$ is replaced by its expectation $E\{\hat{\mathbf{u}}_n\}$; in other words, (5.142) is precisely an ODE.

[3] For independence, it holds that $E\{\mathbf{u}[n]\mathbf{v}[n]\} = E\{\mathbf{u}[n]\} \cdot E\{\mathbf{v}[n]\}$.

[4] The reader will note that, with the direct-averaging approximation, the assumption of independence is not strictly necessary, since it takes into account implicitly.

The previous development confirms that the LMS has the same average behavior of the SDA. From this point, in fact, the proof proceeds as in the SDA for which you get to a set of M first-order finite difference equations in the index n (analogous to (5.70)) of the type

$$E\{\hat{u}_n(i)\} = (1 - \mu\lambda_i)E\{\hat{u}_{n-1}(i)\}, \qquad n \geq 0,\ i = 0, 1, \ldots, M - 1. \tag{5.143}$$

The latter represents a set of M disjoint finite difference equations. The solution is determined by simple backwards substitution, from index n to 0. For which, expressing the result at the time index n, as a function of the IC, we get

$$E\{\hat{u}_n(i)\} = (1 - \mu\lambda_i)^n \hat{u}_{-1}(i), \qquad i = 0, 1, \ldots, M - 1, \tag{5.144}$$

which is stable for $|1 - \mu\lambda_i| < 1$ $\left(\text{or } 0 < \mu < \frac{2}{\lambda_i}\right)$. Clearly, the boundedness of the expected value of all modes is guaranteed by the following step size condition:

$$0 < \mu < \frac{2}{\lambda_{\max}}. \tag{5.145}$$

It follows that $\lim_{n\to\infty} E\{\hat{u}_n(i)\} = 0$, $\forall\, i \in [0, M - 1]$; then we have that

$$\lim_{n\to\infty} E\{\hat{\mathbf{u}}_n\} = \mathbf{0}, \tag{5.146}$$

or

$$\lim_{n\to\infty} E\{\mathbf{w}_n\} = \mathbf{w}_0. \tag{5.147}$$

It can be stated, then, that for $n \to \infty$ the vector $\mathbf{w}_n$ converges, on average, to the optimal solution.

Q.E.D.

5.4.2.2 Mean Square Convergence: Study of the Error Vector's Mean Square Deviation

For the study of LMS's transient and tracking characteristic, we proceed to the solution of the second-order SDE. By squaring and taking the expectation of (5.139) we can write

$$E\{\mathbf{u}_n\mathbf{u}_n^H\} = E\left\{\left(\mathbf{I} - \mu\mathbf{x}_n\mathbf{x}_n^H\right)\mathbf{u}_{n-1}\mathbf{u}_{n-1}^H\left(\mathbf{I} - \mu\mathbf{x}_n\mathbf{x}_n^H\right)\right\} + E\left\{\mu^2 v^2[n]\mathbf{x}_n\mathbf{x}_n^H\right\}.$$

For the independence assumption and the definition of the error vector correlation matrix (5.33), $\mathbf{K}_n = E\{\mathbf{u}_n\mathbf{u}_n^H\}$, it follows that

$$\mathbf{K}_n = (\mathbf{I} - \mu\mathbf{R})\mathbf{K}_{n-1}(\mathbf{I} - \mu\mathbf{R}) + \mu^2 J_{\min}\mathbf{R}. \tag{5.148}$$

This result can be expressed in a more convenient form. Decomposing the correlation with the transformation $\mathbf{\Lambda} = \mathbf{Q}^H\mathbf{R}\mathbf{Q}$, if we set $\hat{\mathbf{u}}_n = \mathbf{Q}^H\mathbf{u}_n$ and $\hat{\mathbf{K}}_{n-1} = E\{\hat{\mathbf{u}}_{n-1}\hat{\mathbf{u}}_{n-1}^H\}$, we have that $\hat{\mathbf{K}}_{n-1} = \mathbf{Q}^H\mathbf{K}_{n-1}\mathbf{Q}$. Accordingly, we can write

$$\hat{\mathbf{K}}_n = (\mathbf{I} - \mu\mathbf{\Lambda})\hat{\mathbf{K}}_{n-1}(\mathbf{I} - \mu\mathbf{\Lambda}) + \mu^2 J_{\min}\mathbf{\Lambda} \tag{5.149}$$

where $\hat{\mathbf{K}}_n = \operatorname{diag}[\hat{k}_n(i)]$ and $\mathbf{\Lambda} = \operatorname{diag}[\lambda_i]$. Decoupling, we get M difference equations of the type

$$\hat{k}_n(i) = (1 - \mu\lambda_i)^2\hat{k}_{n-1}(i) + \mu^2 J_{\min}\lambda_i, \qquad i = 0, 1, \ldots, M-1. \tag{5.150}$$

By back substituting for n, repeatedly, we write

$$\begin{aligned}
\hat{k}_0(i) &= (1 - \mu\lambda_i)^2\hat{k}_{-1}(i) + \mu^2 J_{\min}\lambda_i \\
\hat{k}_1(i) &= (1 - \mu\lambda_i)^4\hat{k}_{-1}(i) + (1 - \mu\lambda_i)^4\mu^2 J_{\min}\lambda_i + \mu^2 J_{\min}\lambda_i \\
&\vdots
\end{aligned}$$

By generalizing, we get

$$\hat{k}_n(i) = (1 - \mu\lambda_i)^{2n}\hat{k}_{-1}(i) + \mu^2\sum_{i=0}^{n/2}(1 - \mu\lambda_i)^{2i}J_{\min}\lambda_i,$$

so, for large n, choosing $0 < \mu < 2/\lambda_i$, the term due to $\hat{k}_{-1}(i)$ tends to zero.

Moreover, by definition $\hat{k}_n(i) = E\left\{|\hat{u}_n(i)|^2\right\}$. It follows that, for the generic component of the rotated error vector variance,

$$\begin{aligned}
\lim_{n\to\infty}\hat{k}_n(i) &= \mu^2\sum_{i=0}^{n/2}(1 - \mu\lambda_i)^{2i}J_{\min}\lambda_i \\
&= \mu^2\lambda_i J_{\min}\frac{1}{1 - (1 - \mu\lambda_i)^2} \\
&= \frac{\mu J_{\min}}{2 - \mu\lambda_i}.
\end{aligned} \tag{5.151}$$

Equation (5.151) provides an expression of the steady-state least mean squares error for the ith filter coefficient.

Remark The mean square convergence is proved because $\lim_{n\to\infty}E\left\{|\hat{u}_n(i)|^2\right\}$ $= constant, \forall i \in [0, M-1]$ for $0 < \mu < \frac{2}{\lambda_{\max}}$ (Sect. 5.1.2.2). Recalling that $\operatorname{tr}(\hat{\mathbf{K}}_n) \triangleq \sum_{i=0}^{M-1}\hat{k}_n(i)$, (5.151) can be generalized in vector form considering the mean square deviation (MSD) of the error vector $E\left\{\|\hat{\mathbf{u}}_n\|_2^2\right\}$ (Sect. 5.1.2.3); we can then write

$$\begin{aligned}\lim_{n\to\infty} E\left\{\|\hat{\mathbf{u}}_n\|_2^2\right\} &= \mathrm{tr}\left(\hat{\mathbf{K}}_n\right)\\ &= J_{\min}\sum_{i=0}^{M-1}\frac{\mu}{2-\mu\lambda_i} < \infty.\end{aligned} \tag{5.152}$$

One can therefore conclude that for $n \to \infty$, the vector $\mathbf{w}_n$ converges quadratically to the optimum solution.

5.4.2.3 LMS Steady-State Behavior with the Noisy Gradient Model

The *noisy gradient model* is an alternative way for the LMS convergence study that leads to results, at times, more physically interpretable. For a simplified analysis, in fact, it is convenient to consider the stochastic gradient modeled as the sum of the exact gradient and a noise contribution, formally

$$\nabla\hat{J}_{n-1}(\mathbf{w}) = \nabla J_{n-1}(\mathbf{w}) + 2\mathbf{N}_n, \tag{5.153}$$

in which the term $2\mathbf{N}_n$ represents the zero-mean noise of the estimated gradient.

From the stochastic-gradient equation (5.82) and by (5.153) we can write

$$-2\mathbf{x}_n e^*[n] = \nabla J(\mathbf{w}_{n-1}) + 2\mathbf{N}_n. \tag{5.154}$$

Replace, in the above, the gradient exact value $\nabla J(\mathbf{w}_{n-1}) = 2(\mathbf{R}\mathbf{w}_{n-1} - \mathbf{g})$, for $\mathbf{R}\mathbf{w}_{\mathrm{opt}} = \mathbf{g}$ and $\mathbf{u}_{n-1} = \mathbf{w}_{n-1} - \mathbf{w}_{\mathrm{opt}}$, we get

$$\begin{aligned}\mathbf{x}_n e^*[n] &= -\mathbf{R}\mathbf{w}_{n-1} + \mathbf{g} - \mathbf{N}_n\\ &= -\mathbf{R}\mathbf{u}_{n-1} - \mathbf{R}\mathbf{w}_{\mathrm{opt}} + \mathbf{g} - \mathbf{N}_n\\ &= -\mathbf{R}\mathbf{u}_{n-1} - \mathbf{N}_n.\end{aligned}$$

Substituting this into the adaptation equation in terms of error vector (5.136) $\left(\mathbf{u}_n = \mathbf{u}_{n-1} + \mu\mathbf{x}_n e^*[n]\right)$, we obtain the following difference equation[5]:

$$\begin{aligned}\mathbf{u}_n &= \mathbf{u}_{n-1} - \mu\mathbf{R}\mathbf{u}_{n-1} - \mu\mathbf{N}_n\\ &= (\mathbf{I} - \mu\mathbf{R})\mathbf{u}_{n-1} - \mu\mathbf{N}_n.\end{aligned} \tag{5.155}$$

Proceeding as usual to the decoupling of equation (5.155), setting $\left(\hat{\mathbf{u}}_n = \mathbf{Q}^H\mathbf{u}_n\right)$ and $\hat{\mathbf{N}}_n = \mathbf{Q}^H\mathbf{N}_n$, we can write

[5] Note that, by making the expectation of both sides of (5.155), we obtain a difference equation identical to (5.142).

$$\hat{\mathbf{u}}_n = (\mathbf{I} - \mu\boldsymbol{\Lambda})\hat{\mathbf{u}}_{n-1} - \mu\hat{\mathbf{N}}_n,$$

which, as seen above, is equivalent to a set of M independent equations of the type

$$\hat{u}_n(i) = (1 - \mu\lambda_i)\hat{u}_{n-1}(i) - \mu\hat{N}_n(i), \qquad i = 0, 1, \ldots, M-1.$$

By taking the expectation of the square and placing $\hat{k}_n(i) = E\left\{|\hat{u}_n(i)|^2\right\}$, we obtain

$$\hat{k}_n(i) = \left(1 - \mu\lambda_i\right)^2 \hat{k}_{n-1}(i) + \mu^2 E\left\{\hat{N}_n(i)\right\} - 2\mu(1 - \mu\lambda_i)E\left\{\hat{u}_{n-1}(i)\hat{N}_n(i)\right\}.$$

For the independence assumption $E\{\hat{u}_{n-1}(i)\hat{N}_n(i)\} = 0$, and the above becomes

$$\hat{k}_n(i) = \left(1 - \mu\lambda_i\right)^2 \hat{k}_{n-1}(i) + \mu^2 E\left\{\hat{N}_n(i)\right\}. \tag{5.156}$$

This expression describes how to propagate the MSE into the rotated error vector and is not of particular interest if you do not know the noise $\mathbf{N}_n$. Since we are interested in the steady-state solution, we can consider $\lim_{n\to\infty} E\{\mathbf{w}_n\} = \mathbf{w}_{\text{opt}}$, so the CF is at the minimum and absolute point and is $\nabla J(\mathbf{w}) = \mathbf{0}$. In such a situation we can write

$$e^*[n]\mathbf{x} = -\mathbf{N}_n.$$

Squaring and performing the expectation, for the independence assumption, it follows that

$$\begin{aligned} E\{\mathbf{N}_n\mathbf{N}_n^H\} &= E\left\{e^2[n,]\mathbf{x}_n\mathbf{x}_n^H\right\} \\ &= E\left\{e^2[n]\right\}E\left\{\mathbf{x}_n\mathbf{x}_n^H\right\} \\ &= J_{\min}\mathbf{R}. \end{aligned}$$

Using the rotated form, $E\left\{\hat{\mathbf{N}}_n\hat{\mathbf{N}}_n^H\right\} = J_{\min}\boldsymbol{\Lambda}$. Note also that, since $\mathbf{N}_n$ is a noise, the matrix $E\left\{\mathbf{N}_n\mathbf{N}_n^H\right\}$ has elements equal to $E\{\hat{N}_n^2(i)\} = J_{\min}\lambda_i$. It follows that we can write (5.156) as

$$\hat{k}_n(i) = (1 - \mu\lambda_i)^2\hat{k}_{n-1}(i) + \mu^2 J_{\min}\lambda_i, \qquad i = 0, 1, \ldots, M-1, \tag{5.157}$$

which is identical to the expression (5.150), obtained from the solution of the stochastic differential equation.

5.4.3 Excess of Error and Learning Curve

The CF effectively minimized by the LMS is the squared error $\hat{J}_n = |e[n]|^2$, which represents an estimate value of the MSE $J_n = E\{|e[n]|^2\}$. When the adaptation has stochastic nature, for the study and performance analysis, it is usual and convenient to refer to the ensemble averages. For this reason, as for the SDA, the LMS performance analysis is made considering the expectation $J_n = E\{\hat{J}_n\}$ and, consequently, the MSE. The LMS is characterized by a convergence, on average, of the weights vector towards the optimal filter, namely for $n \to \infty$, $E\{\mathbf{w}_n\} = \mathbf{w}_{\text{opt}}$. Therefore, at convergence the instantaneous value of the weights can deviate by the optimum and the instantaneous value of the gradient is nonzero. This situation entails a certain residual oscillation around the minimum value of the weights, which by definition coincides with the optimal Wiener CF $J_{\min} = J(\mathbf{w}_{\text{opt}})$, even after the convergence of the algorithm. This perturbation causes after convergence, a residual excess of steady-state error, called *excess of MSE* (EMSE), defined as $J_{EMSE\_\infty} \triangleq J_\infty - J_{\min}$ (Sect. 5.1.2.4).

5.4.3.1 Excess of Steady-State Error

By placing $\mathbf{u}_{n-1} = \mathbf{w}_{n-1} - \mathbf{w}_{\text{opt}}$, from (5.65) we can write

$$\hat{J}_n = \hat{J}_{\min} + \mathbf{u}_{n-1}^H \mathbf{x}\mathbf{x}_n^H \mathbf{u}_{n-1}. \tag{5.158}$$

The *excess of error* is, by definition, reportedly with respect to the MSE (EMSE) for which, performing the expectation of the previous, we get[6]

$$\begin{aligned} J_n &= J_{\min} + E\left\{\mathbf{u}_{n-1}^H \mathbf{x}\mathbf{x}^H \mathbf{u}_{n-1}\right\} \\ &= J_{\min} + E\left\{\mathbf{u}_{n-1}^H \mathbf{R}\mathbf{u}_{n-1}\right\}. \end{aligned} \tag{5.159}$$

The excess of error indicates the average amount of deviation, with respect to the Wiener's optimal solution. So, excess of error is equal to

$$J_{\text{EMSE}} = E\left\{\mathbf{u}_{n-1}^H \mathbf{R}\mathbf{u}_{n-1}\right\}. \tag{5.160}$$

Note that the EMSE represents the square of the term $E\{(\mathbf{u}_{n-1}^H \mathbf{x}_n)^2\}$, and that $\mathbf{u}_{n-1}^H \mathbf{x}_n$ is a scalar quantity for which the above can be written as

[6] In LMS the vector $\mathbf{u}_n$ is an RV, for which (5.65) $J_n = J_{\min} + \mathbf{u}_n^H \mathbf{R}\mathbf{u}_n$, related to the SDA, in this case must be written as $J_n = J_{\min} + E\{\mathbf{u}_{n-1}^H \mathbf{R}\mathbf{u}_{n-1}\}$.

$$E\{\mathbf{u}_{n-1}^H\mathbf{R}\mathbf{u}_{n-1}\} = E\Big\{\mathrm{tr}\big[\mathbf{u}_{n-1}^H\mathbf{R}\mathbf{u}_{n-1}\big]\Big\}. \tag{5.161}$$

From algebra, tr[**AB**] = tr[**BA**]; then we can write

$$\begin{aligned} E\Big\{\mathrm{tr}\big[\mathbf{u}_{n-1}^H\mathbf{R}\mathbf{u}_{n-1}\big]\Big\} &= E\Big\{\mathbf{R}, \mathrm{tr}\big[\mathbf{u}_{n-1}\mathbf{u}_{n-1}^H,\big]\Big\} \\ &= \mathrm{tr}\Big[\mathbf{R}, E\{\mathbf{u}_{n-1}\mathbf{u}_{n-1}^H\}\Big]. \end{aligned} \tag{5.162}$$

Recalling the definition of the error vector correlation $\mathbf{K}_n \triangleq E\{\mathbf{u}_n\mathbf{u}_n^H\}$ ((5.33) Sect. 5.1.2.3)

$$J_n = J_{\min} + \mathrm{tr}[\mathbf{R}\mathbf{K}_{n-1}]. \tag{5.163}$$

Remark Similarly to the case of optimal Wiener filter (Sect. 3.3.4.4 (3.81) and (3.82)), equation (5.163) shows that in the LMS algorithm, the mean square value of the estimated error is given by the contribution of two terms. The first term is the MMSE $J_{\min}$; the second term depends on the correlation **R** and the transient behavior of the matrix $\mathbf{K}_n$ (correlation of the error vector). By definition, both the matrix **R** and $\mathbf{K}_n$ are positive definite for all n. It follows that in the case of stochastic gradient it is possible to define the excess of the minimum square error (EMSE) as

$$J_{\mathrm{EMSE}} = \mathrm{tr}[\mathbf{R}\mathbf{K}_{n-1}]. \tag{5.164}$$

This result may be expressed in a more convenient form decomposing the correlations with the unitary transformation of similarity $\boldsymbol{\Lambda} = \mathbf{Q}^H\mathbf{R}\mathbf{Q}$; moreover, placing $\hat{\mathbf{u}}_n = \mathbf{Q}^H\mathbf{u}_n$ and $\hat{\mathbf{K}}_{n-1} = E\{\hat{\mathbf{u}}_{n-1}\hat{\mathbf{u}}_{n-1}^H\}$, it appears that $\hat{\mathbf{K}}_{n-1} = \mathbf{Q}^H\mathbf{K}_{n-1}\mathbf{Q}$. Therefore, the excess of error can be expressed as

$$J_{\mathrm{EMSE}} = \mathrm{tr}\big[\boldsymbol{\Lambda}\hat{\mathbf{K}}_{n-1}\big], \tag{5.165}$$

where $\hat{\mathbf{K}}_n = \mathrm{diag}\big[\hat{k}_n(i)\big]$ and $\boldsymbol{\Lambda} = \mathrm{diag}[\lambda_i]$. From (5.165), generalizing, the excess of error can be expressed as

$$J_{\mathrm{EMSE}} = \mathrm{tr}\big[\boldsymbol{\Lambda}\hat{\mathbf{K}}_n\big] = \sum_{i=0}^{M-1}\lambda_i\hat{k}_n(i), \qquad \text{for} \qquad n = 0, 1, \ldots, \infty. \tag{5.166}$$

For $n \to \infty$, we can consider the *excess of steady-state error* with the expression (5.151), for which the previous at steady state becomes

$$J_{\mathrm{EMSE\_}\infty} = J_\infty - J_{\min} = J_{\min}\sum_{i=0}^{M-1}\frac{\mu\lambda_i}{2-\mu\lambda_i}. \tag{5.167}$$

From the above it appears that the excess of steady-state error is less than MMSE if it is $J_{\text{EMSE}} < J_{\min}$, or

$$\sum_{i=0}^{M-1} \frac{\mu\lambda_i}{2 - \mu\lambda_i} < 1. \tag{5.168}$$

Regarding the misadjustment defined in (5.41) from (5.167), then

$$\mathcal{M} = \frac{J_{\text{EMSE}}}{J_{\min}} = \sum_{i=0}^{M-1} \frac{\mu\lambda_i}{2 - \mu\lambda_i}. \tag{5.169}$$

Remark You can give a more physically interpretable formulation of excess steady-state error, if we consider (5.167), solving for J_∞, so you have

$$J_\infty = J_{\min} + J_{\min} \sum_{i=0}^{M-1} \frac{\mu\lambda_i}{2 - \mu\lambda_i}.$$

Expanding on a Taylor series, the above can be written as

$$J_\infty = J_{\min} + J_{\min} \frac{\mu}{2} \sum_{i=0}^{M-1} \lambda_i \left(1 + \frac{\mu}{2}\lambda_i + \frac{\mu^2}{4}\lambda_i^2 + \cdots\right).$$

For $\mu \ll 2/\lambda_{\max}$, we can write

$$J_\infty \approx J_{\min} + J_{\min} \frac{\mu}{2} \sum_{i=0}^{M-1} \lambda_i = J_{\min}\left(1 + \frac{\mu}{2}\text{tr}\{\mathbf{R}\}\right),$$

or, in other words, you come to the relation

$$J_\infty \approx J_{\min}\left(1 + \frac{\mu}{2} \cdot M \cdot \text{input\_power}\right). \tag{5.170}$$

The above expression indicates that, at steady state, to have a result tending to optimum value, you must have a *sufficiently small* learning rate.

Remark For the stability condition of the LMS, the same considerations in Sect. 5.2.2.1 for the SDA are valid, and for the convergence we have to choose the learning rate μ properly. Since as for the SDA, $0 < \mu < \frac{2}{M}\left(1/E\{x^2[n]\}\right)$ (5.75), the learning rate can take a maximum value inversely proportional to the energy of the input signal which, as shown, is a necessary condition for the LMS stability.

The condition (5.75) is the basis of some variable learning rate techniques made in some LMS variants, such as, for example, in the NLMS algorithm, that will be discussed below (5.175).

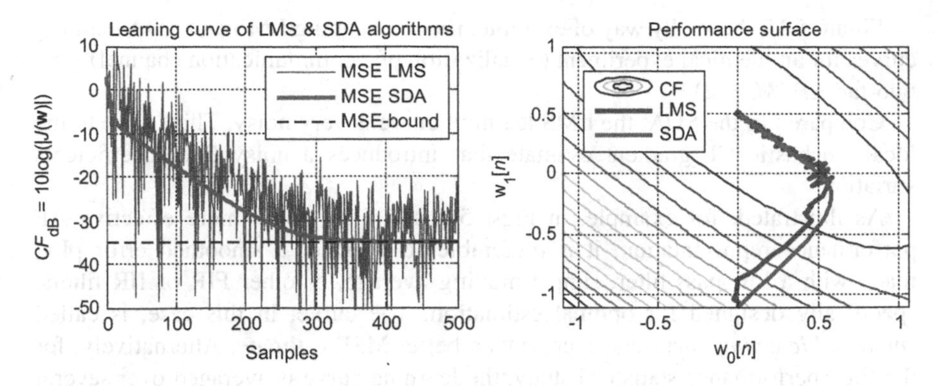

Fig. 5.15 Learning curves comparison between the LMS and the SDA

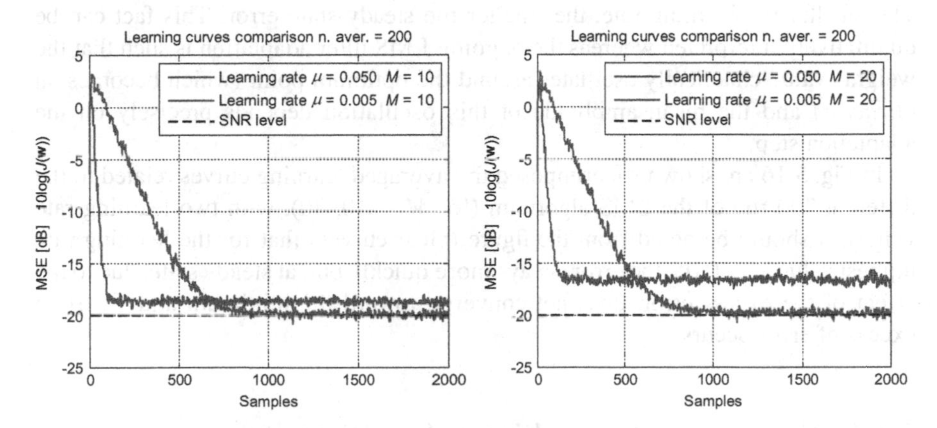

Fig. 5.16 Steady-state error for two different learning rate values and filter length reported in figure. For each experiment were considered the average of 200 runs with different IC

5.4.3.2 Learning Curve

To monitor the adaptation process, it is often useful to consider the plot of the squared error values $\hat{J}_n$ or, more properly, its ensemble average respect to the algorithm iterations (also called *learning epochs*). However, what is interesting to study is the CF without superimposed noise, i.e., $J_n(\mathbf{w})$. The error graph vs the learning iterations, whose typical behavior has already been shown in Fig. 5.4 (Sect. 5.1.2.5), is defined as *learning curve*.

Its trend is similar to the steepest-descent MSE, precisely because, as demonstrated by the theoretical development set out above, it represents its average trend. The noisy weights trajectories variation introduces additional error and pushes up the curve average trend. Note that the amplitude of the noise is small, if the parameter μ is small. Therefore, for small μ, as is usual in practice, the difference between the LMS square error (average) trend coincides with the MSE SDA trend.

Figure 5.15 shows, by way of example, the overlapping LMS and SDA learning curves for an identical experiment (equalization of a communication channel) for a two-tap AF ($M = 2$).

Compared to the SDA, the LMS learning curve is very noisy. This is due to the local stochastic CF gradient estimate that introduces a noisy filter coefficients variation.

As illustrated, for example, in Figs. 5.4 and 5.15, for a more effective AF performance representation, it is preferable to consider a smoothed error plot, made with a low-pass filter, simple moving average, or other FIR or IIR filters, specifically designed for optimal estimation. The curve, in this case, is called *smoothed learning curve* and represents a better MSE estimate. Alternatively, for the filter performance statistical study, the learning curve is averaged over several trials, starting from there different ICs.

Figure 5.16 indicates that the steady-state error depends on the learning rate μ. The smaller the learning rate, the smaller the steady-state error. This fact can be qualitatively interpreted whereas the ongoing LMS filter adaptation is such that the weights value chaotically oscillates around the optimum point (which becomes an *attractor*) and the mean amplitude of this oscillation depends precisely on the adaptation step.

In Fig. 5.16 are shown superimposed the averaged learning curves related to the different 200 run of the LMS algorithm (for $M = 10, 20$), with two learning rate values. It should be noted from the figure (clear curves) that for the learning rate highest value $\mu = 0.05$, the error decays more quickly but, at steady state, due to the effect of the excess error, does not converge toward the minimum and a certain excess of error occurs.

5.4.4 Convergence Speed: Eigenvalues Disparity and Nonuniform Convergence

For the LMS performance analysis, in terms of convergence speed, the same considerations of the case SDA analyzed in Sect. 5.2.3 are valid, with the difference of considering the variables as RV averages. In fact, even in the LMS, as in SDA, the convergence speed is determined by the slower *mode* of $\mathbf{R}$ matrix according to the expression (5.78) rewritten as

$$\tau \cong 1/\mu\lambda_{\min}. \tag{5.171}$$

However, in this case it should be noted that for (5.146), the convergence to optimal point is a convergence on average, or what tends to zero is the *expected error* (and not directly the error as in the SDA case). Moreover, it is precisely for this reason that the LMS learning curve has a rather irregular shape with sharp changes for each iteration.

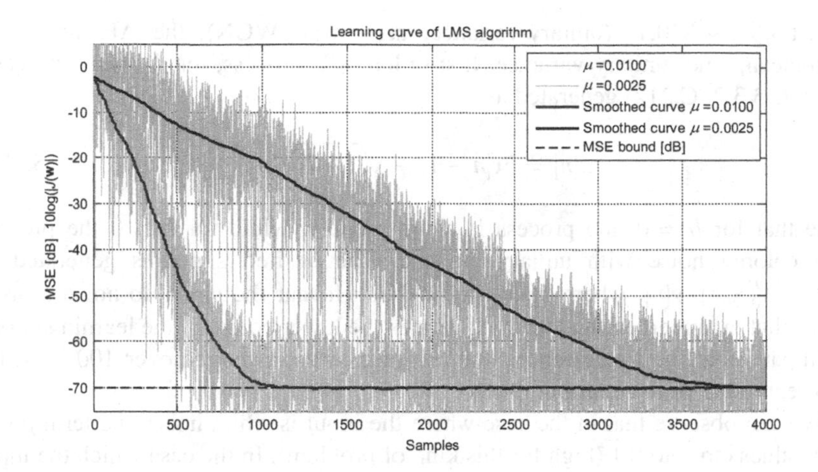

Fig. 5.17 Learning curve trend for different values of the learning rate μ, concerning the same experiment of Fig. 5.10. The *dark lines* shown the smoothed curves obtained by a zero-phase four-th order IIR Butterworth low-pass filter

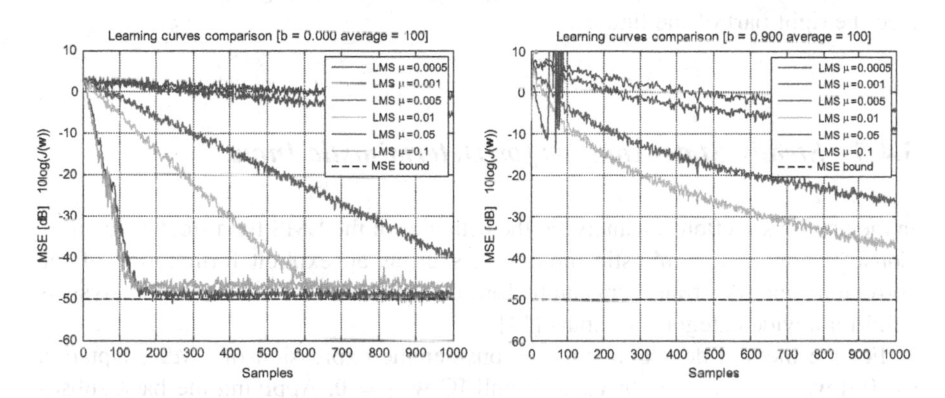

Fig. 5.18 Comparison of LMS learning curve averaged over 100 runs (left) white noise ($b = 0.0$) input; (right) narrowband MA colored process ($b = 0.9$)

For example, Fig. 5.17 shows learning curves related to similar SDA experiment of Fig. 5.10, from which it can be observed that the average performance (described by smooth curves) of the two algorithms is quite similar (only shows two curves for reasons of graphical clarity).

It should be noted, consistent with (5.171), that the convergence time, expressed by the number of iterations, to the optimal value (shown in the figure as MSE bound) is higher for smaller values of the adaptation step μ.

Figure 5.18 reports the results, in terms of learning curves, of an experiment of a dynamic system identification, of the type used for performance analysis just illustrated in Fig 5.13 (Sect. 5.4.1).

Let $\eta[n] \sim N(0,1)$ (unitary-variance, zero-mean WGN), the AF input is a zero-mean, and unitary-variance first-order AR process, or Markov process, (Sect. C.3.3.2, C.212) generated as

$$x[n] = bx[n-1] + \sqrt{1-b^2}\eta[n]. \tag{5.172}$$

Note that for $b = 0$, the process is simply WGN, while for $b > 0$ the process is a colored noise with unitary-variance. The desired signal is generated as $d[n] = \mathbf{w}_0^H \mathbf{x}_n + v[n]$, where $v[n]$ is a WGN such that the signal-to-noise ratio is 50 dB that in agreement with (5.135) defines the lower bound of the learning curve.

In particular the figure reports the learning curve, averaged over 100 runs, for different value of the learning rate μ.

We can observe that in the case where the input is white noise, the term μ can take values close to 0.1 (high for this kind of problem). In the case which the input process is narrowband, i.e., in (5.172) $0 \ll b < 1$, due to the high eigenspread $\chi(\mathbf{R}_{xx}) = \frac{1+b}{1-b}$ (C.214), in order to avoid the adaptation process divergence, it is necessary to maintain very low learning rate, so the convergence is much slower (see the right part of the figure).

5.4.5 Steady-State Analysis for Deterministic Input

In the previous section, we analyzed the behavior of the LMS for a stochastic input. Consider now a deterministic input $x[n]$, such as an explicit formulation of the correlation and its z-transform can be formulated. Note that this type of analysis is valid for a wide category of inputs [14].

For the method development, we consider the expression of LMS adaptation (5.103) $\mathbf{w}_n = \mathbf{w}_{n-1} + \mu e^*[n]\mathbf{x}_n$, with null IC $\mathbf{w}_{-1} = \mathbf{0}$. Applying the back substitutions we have that

$$\mathbf{w}_n = \mu \sum_{i=0}^{n-1} e^*[i]\mathbf{x}_i.$$

Multiplying the left-hand side by $\mathbf{x}^H$, and considering the filter output $y[n] = \mathbf{w}_n^H \mathbf{x}_n$, we get

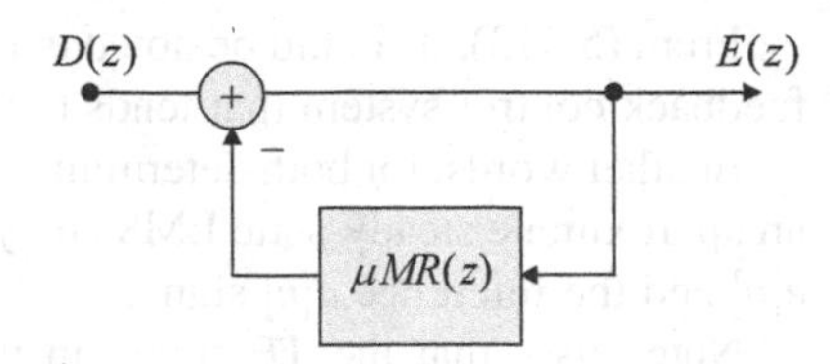

Fig. 5.19 Representation of an AF as a feedback control system

$$
\begin{aligned}
y[n] &= \mu\sum_{i=0}^{n-1} e[i]\mathbf{x}_i^H\mathbf{x}_n \\
&= \mu M\sum_{i=0}^{n-1} e[i]r_{i,n},
\end{aligned}
$$

where $r_{i,n} = \frac{1}{M}\mathbf{x}_i^H\mathbf{x}_n$. From the error definition $e[n] = d[n] - y[n]$, it follows

$$
e[n] + \mu M\sum_{i=0}^{n-1} e[i]r_{i,n} = d[n].
$$

This last is a finite difference equation between the time-varying error, considered as an output, and the desired output signal, considered as input. For $M \to \infty$, and for a finite energy signals, the correlation is such that $r_{i,n} \approx r[n-i]$, for which, for sufficiently long filters, the following approximation is valid:

$$
e[n] + \mu M\sum_{i=0}^{n-1} r[n-i]e[i] = d[n].
$$

By performing the z-transform we get

$$
E(z)\left[1 + \mu MR(z)\right] = D(z),
$$

or

$$
\frac{E(z)}{D(z)} = \frac{1}{1 + \mu MR(z)}, \tag{5.173}
$$

where $R(z) = r_1 z^{-1} + r_2 z^{-2} + \cdots$. In other words, as shown in Fig. 5.19, the adaptive filter is assimilated to a linear TF between $d[n]$ and $e[n]$. Note that, from the error definition, $y[n] = d[n] - e[n]$ and we can write

$$
\frac{Y(z)}{D(z)} = \frac{\mu MR(z)}{1 + \mu MR(z)}. \tag{5.174}
$$

The TF expresses a relationship between $e[n]$, $y[n]$, and $d[n]$ and represents only a simple approximation of the behavior of the steady-state optimal filter.

From (5.173), it should be noted that a steady-state AF can be treated as a simple feedback control system that tends to minimize the error $E(z)$.

In other words, for both deterministic or random signals, it is possible to perform an approximate steady-state LMS analysis, by considering the TF between the error $e[n]$ and the reference $d[n]$ signals.

Note, also, that the TF study, in the case of deterministic signals, allows an explicit analysis, both of convergence and stability, in terms of the pole–zero plot, while, in the case of random signals, it is used to describe the average system properties. In particular, the TF study is particularly useful in the case of colored inputs, as it highlights significant polarizations, compared to the optimal LS solution. For more details and examples, [14].

5.5 LMS Algorithm Variants

In the literature, there are many LMS algorithm variations. Some of these implementations are oriented to the real time and/or to simplify the necessary hardware or to have a low computational cost. While other variants, considering a certain increase in the computational cost, are oriented in order to have better convergence speed and/or better steady-state performance, still others tend to stabilize the weights trajectories, etc.

5.5.1 Normalized LMS Algorithm

The NLMS algorithm represents a very used variant to accelerate the convergence speed at the expense of a modest increase in the computational cost. The NLMS is characterized by a variable learning rate according to the following law:

$$\mu_n = \frac{\mu}{\delta + \|\mathbf{x}_n\|_2^2}. \tag{5.175}$$

Consequently, the update formula is

$$\mathbf{w}_n = \mathbf{w}_{n-1} + \mu \frac{e^*[n]\mathbf{x}_n}{\delta + \mathbf{x}_n^H \mathbf{x}_n}, \tag{5.176}$$

with $\mu \in (0, 2]$ and $\delta > 0$. Note that δ is the *regularization parameter* which also ensures the computability of (5.176) in the case of zero input. In the complex case the algorithm becomes

$$\mathbf{w}_n = \mathbf{w}_{n-1} + \mu \frac{e^*[n]\mathbf{x}_n}{\delta + \mathbf{x}_n^H \mathbf{x}_n}, \tag{5.177}$$

(5.176) and (5.177) indicate that the step size is inversely proportional to the energy of the input signal. This formula, although quite intuitive, has substantial theoretical reasons arising from (5.75) and (5.84). Given the implementative simplicity, the NLMS is one of the most used algorithms in the equalization applications, echo cancelation, active noise control, etc.

5.5.1.1 NLMS Algorithm's Computational Cost

Compared to the LMS, the NLMS requires the calculation of the hidden product to evaluate $\|\mathbf{x}_n\|_2^2$, a real addition and a real division to evaluate $\mu/\left(\delta + \|\mathbf{x}_n\|_2^2\right)$. To evaluate $\mathbf{x}_n^H \mathbf{x}_n$, since the two vectors are complex conjugate, one can easily verify that requests are only $2M$ real multiplications. Therefore, for the NLMS complex case the algorithm complexity for each iteration is equal to

$$(10M + 2) \text{ real mult.s, } 10M \text{ real adds, and one real division.} \tag{5.178}$$

For the NLMS real case, the computational cost is

$$(3M + 1) \text{ real mult.s, } 3M \text{ real adds, and one real division.} \tag{5.179}$$

Remark A simple way to reduce the number of multiplications for the $\|\mathbf{x}_n\|_2^2$ calculation is obtained by observing that the vector $\mathbf{x}_n$ contains $M - 1$ common values with the $\mathbf{x}_{n-1}$ vector. For this reason, the following relationship holds:

$$\|\mathbf{x}_n\|_2^2 = \|\mathbf{x}_{n-1}\|_2^2 - |x[n-M]|^2 + |x[n]|^2 \tag{5.180}$$

With this expedient the expressions (5.178) and (5.179) become, in the complex case,

$$(8M + 6) \text{ real mult.s, } 8M + 5 \text{ adds, and one real division,} \tag{5.181}$$

while in the real case

$$(2M + 3) \text{ real mult.s, } 2M + 3 \text{ adds, and one real division.} \tag{5.182}$$

The recursive computation (5.180) should be made with accuracy because the round-off error accumulation can lead to situations in which the nonnegativity of $\|\mathbf{x}_n\|_2^2$ is no longer true.

5.5.1.2 Minimal Perturbation Properties of NLMS Algorithm

In the case of NLMS adaptation the constraint (5.47), for the (5.177), is

$$\begin{aligned}\varepsilon[n] &= d[n] - \mathbf{w}_n^H \mathbf{x}_n \\ &= d[n] - \left(\mathbf{w}_{n-1} + \mu \frac{e^*[n]\mathbf{x}_n}{\delta + \mathbf{x}_n^T \mathbf{x}_n}\right)^H \mathbf{x}_n \\ &= \left(1 - \mu \frac{\|\mathbf{x}_n\|_2^2}{\delta + \|\mathbf{x}_n\|_2^2}\right) e[n],\end{aligned} \tag{5.183}$$

for which in (5.47) $\alpha = \mu\|\mathbf{x}_n\|_2^2/(\delta + \|\mathbf{x}_n\|_2^2)$, and the expression (5.46) was explicitly as

$$\mathbf{w}_* \therefore \arg\min_{\mathbf{w}} \|\delta\mathbf{w}\|_2^2 \quad \text{s.t.} \quad \varepsilon[n] = \left(1 - \mu \frac{\|\mathbf{x}_n\|_2^2}{\delta + \|\mathbf{x}_n\|_2^2}\right) e[n]. \tag{5.184}$$

As already discussed in the LMS case, also in this case the constraint (5.44) is more relevant when μ is small in such a way that $\left|1 - \mu\|\mathbf{x}_n\|_2^2/(\delta + \|\mathbf{x}_n\|_2^2)\right| < 1$, or

$$0 < \mu < \frac{2\left(\delta + \|\mathbf{x}_n\|_2^2\right)}{\|\mathbf{x}_n\|_2^2}, \quad \forall n. \tag{5.185}$$

In the case of NLMS, the expression (5.48) becomes

$$\delta\mathbf{w}_* \therefore \arg\min_{\delta\mathbf{w}} \|\delta\mathbf{w}\|_2^2 \quad \text{s.t.} \quad \mathbf{x}_n^H \delta\mathbf{w} = \mu \frac{\|\mathbf{x}_n\|_2^2}{\delta + \|\mathbf{x}_n\|_2^2} e[n]. \tag{5.186}$$

Note that in this case (5.49), for $\|\mathbf{x}_n\|_2^2 \neq 0$, is

$$\begin{aligned}\delta\mathbf{w}_* &= \mathbf{x}_n\left(\mathbf{x}_n^H \mathbf{x}_n\right)^{-1} \mu \frac{\|\mathbf{x}_n\|_2^2}{\delta + \|\mathbf{x}_n\|_2^2} e[n] \\ &= \frac{\mu}{\delta + \|\mathbf{x}_n\|_2^2} \mathbf{x}_n e^*[n].\end{aligned} \tag{5.187}$$

So, the update formula can be written as

$$\mathbf{w}_n = \mathbf{w}_{n-1} + \frac{\mu}{\delta + \|\mathbf{x}_n\|_2^2} \mathbf{x}_n e^*[n], \tag{5.188}$$

which coincides with the NLMS update rule (5.177). Therefore, it was shown that even the NLMS algorithm is equivalent to the exact solution of a local optimization problem.

5.5.2 Proportionate LMS Algorithms

The *proportionate NLMS* (PNLMS) algorithm, proposed in [15], is characterized by an adaptation rule of the type

$$\mathbf{w}_n = \mathbf{w}_{n-1} + \mu \frac{\mathbf{G}_{n-1}\mathbf{x}_n e^*[n]}{\delta_p + \mathbf{x}_n^H \mathbf{G}_{n-1}\mathbf{x}_n}, \tag{5.189}$$

where $0 < \mu < 1$ and $\mathbf{G}_n \in \mathbb{R}^{M\times M} = \text{diag}\left[g_n(0) \quad g_n(1) \quad \cdots \quad g_n(M-1) \right]$ is a diagonal matrix identified in order to adjust the step size of the filter weights in an individual mode. The $\mathbf{G}_n$ matrix is determined to have the step size proportional to the amplitude of the considered filter coefficient. In other words, the larger coefficients have a greater increase. Following this philosophy, a possible $\mathbf{G}_n$ matrix choice is the following:

$$\gamma_n[m] = \max\left\{ \rho \cdot \left(\max\left[\delta_p, |w_n[0]|, \ldots, |w_n[M-1]| \right] \right), |w_n[m]| \right\}, \tag{5.190}$$

$$g_n(m) = \frac{\gamma_n[m]}{\|\boldsymbol{\gamma}_n\|_1}, \qquad m = 0, 1, \ldots, M-1, \tag{5.191}$$

where $\boldsymbol{\gamma}_n \in \mathbb{R}^{M\times 1} = \left[\gamma_n[0] \quad \cdots \quad \gamma_n[M-1] \right]^T$ and $\delta_p, \rho \in \mathbb{R}^+$, called *precautionary constants*, have typical values $\rho = 0.01$ and $\delta_p = 0.01$. In practice δ_p is a regularization parameter that ensures the consistency of (5.189), also for null taps, while ρ serves to prevent stalling of the mth coefficient $w_n[m]$ when its amplitude is lower than the amplitude of the maximum coefficient.

In the algorithm called improved PNLMS (IPNLMS) [16], a more elegant $\mathbf{G}_n$ matrix choice is proposed

$$\gamma_n[m] = (1-\beta)\frac{\|\mathbf{w}_n\|_1}{M} + (1+\beta)|w_n[m]|, \tag{5.192}$$

$$\begin{aligned} g_n(m) &= \frac{\gamma_n[m]}{\|\boldsymbol{\gamma}_n\|_1} \\ &= \frac{(1-\beta)}{2M} + (1+\beta)\frac{|w_n[m]|}{2\|\mathbf{w}_n\|_1}, \qquad m = 0, 1, \ldots, M-1, \end{aligned} \tag{5.193}$$

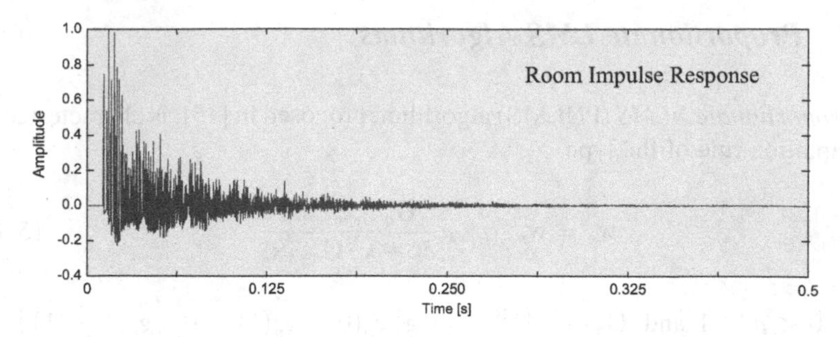

Fig. 5.20 Example of sparse impulse response. Trend of the impulse response of an acoustic path between two points for a room $(3.40 \times 5.10 \times 4.25)$ (m) (calculated using a simulator)

where $(-1 < \beta < 1)$ represents the proportionality control parameter. Note that for $\beta = -1$, the IPNLMS coincides with the NLMS. As reported in [16], a good choice of the parameter of proportionality β is –0.5 or 0. Furthermore, in the IPNMLS is usual to choose the regularization parameter with in the form

$$\delta_p = \frac{(1-\beta)}{2M}\delta, \tag{5.194}$$

where δ is the NLMS regularization parameter.

Remark The proportional algorithms are suitable in the case of systems identification with *sparse* impulse response. A simple definition of *sparsity* is the following: *an impulse response is called sparse if a large fraction of its energy is concentrated in a small fraction of its duration.* In more formal terms, a simple measure of an impulse response $\mathbf{w}$ sparseness is the following [17]:

$$\xi(\mathbf{w}) \triangleq \frac{M}{M-\sqrt{M}}\left(1 - \frac{\|\mathbf{w}\|_1}{\sqrt{M}\|\mathbf{w}\|_2}\right), \tag{5.195}$$

where, we remind the reader that the L_1 and L_2 norms are defined, respectively, as

$$\|\mathbf{w}\|_1 = \sum_{m=0}^{M-1}|w[m]| \qquad \text{and} \qquad \|\mathbf{w}\|_2 = \sqrt{\sum_{m=0}^{M-1}|w[m]|^2}, \tag{5.196}$$

for which $0 \le \xi(\mathbf{w}) < 1$ and for sparse $\mathbf{w}$, we have that $\xi(\mathbf{w}) \to 1$.

A typical example of sparse impulse response is the one that refers to TF of an acoustic path in a reverberating environment as, for example, the impulse response shown in Fig. 5.20.

A more consistent theoretical justification of the PNLMS and IPNLMS methods is shown in the following after the definition of the *general adaptation laws* (Sect. 6.8).

5.5.3 Leaky LMS

The leaky LMS algorithm has a cost function characterized by the sum of two contributions. At the LMS CF (5.87) $\hat{J}_n = |e[n]|^2$, is added a penalty contribution proportional to the inner product $\mathbf{w}^T\mathbf{w}$, for which the CF[7] is redefined as

$$J_n = \frac{1}{2}\Big(|e[n]|^2 + \delta \mathbf{w}_{n-1}^H \mathbf{w}_{n-1}\Big), \tag{5.197}$$

where the regularizing parameter δ is referred to as the "leak."

The square error minimization, together with the penalty function, limits the weights "*energy*" during the adaptation process. Therefore, equation (5.197) represents a regularized CF. It is easy to show that, called $\lambda_{\min}$ and $\lambda_{\max}$, respectively, the minimum and maximum correlation matrix $\mathbf{R}$ eigenvalues, for $\delta > 0$ we have that the eigenspread for the regularized form (5.197) is

$$\frac{\lambda_{\max} + \delta}{\lambda_{\min} + \delta} \le \frac{\lambda_{\max}}{\lambda_{\min}}. \tag{5.198}$$

In this case, the leaky LMS algorithm's worst-case transient performance will be better than that of the standard LMS algorithm [18].

Differentiating (5.197), we get

$$\begin{aligned} \nabla \hat{J}_n &= \frac{\partial\Big(|e[n]|^2 + \delta \mathbf{w}_{n-1}^H \mathbf{w}_{n-1}\Big)}{\partial \mathbf{w}_n} \\ &= -2e^*[n]\mathbf{x} + 2\delta \mathbf{w}_{n-1}, \end{aligned} \tag{5.199}$$

for which the adaptation law $\big(\mathbf{w}_n = \mathbf{w}_{n-1} - \mu\frac{1}{2}\nabla\hat{J}_n\big)$ is

$$\mathbf{w}_n = (1 - \mu\delta)\mathbf{w}_{n-1} + \mu e^*[n]\mathbf{x}. \tag{5.200}$$

In this case, proceeding as in Sect. 5.4.2.1, the step size upper bound is

$$0 < \mu < \frac{2}{\gamma + \lambda_{\max}}, \tag{5.201}$$

and, recalling that $\lambda_{\max} \le \mathrm{tr}(\mathbf{R}) = \sum_{i=0}^{M-1}\lambda_i = M \cdot E\{|x[n]|^2\}$, we have that the upper bound is

---

[7] Note that identical result can be reached considering a CF defined as

$\mathbf{w} \therefore \min\ \delta(\mathbf{w}_{n-1}^H \mathbf{w}_{n-1})$ s.t. $|e[n]|^2 = |d[n] - \mathbf{w}_{n-1}^H \mathbf{x}|^2$,

with Lagrangian $L(\mathbf{w},\boldsymbol{\lambda}) = \delta \mathbf{w}_{n-1}^H \mathbf{w}_{n-1} + \lambda|e[n]|^2$ and considering the descent in the stochastic gradient of the Lagrangian surface $\mathbf{w}_n = \mathbf{w}_{n-1} - \mu\frac{1}{2}\nabla_{\mathbf{w}} L(\mathbf{w},\boldsymbol{\lambda})$.

$$0 < \mu < \frac{2}{M \cdot \left(E\left\{ |x[n]|^2 \right\} + \delta \right)}. \tag{5.202}$$

Moreover, the ith component of the steady-state solution of the rotate vector is nonzero

$$\lim_{n \to \infty} E\{\hat{u}_n(i)\} = \frac{-\delta}{\lambda_i + \delta} w_{\text{opt}}(i), \qquad i = 0, 1, \ldots, M-1, \tag{5.203}$$

so, a nonzero leakage factor results in nonzero steady-state coefficient bias. However, the weak convergence (or convergence in the mean) does not guarantee mean square error convergence [19].

5.5.4 *Other Variants of the LMS Algorithm*

While variants such as that of *Leaky LMS* and the *momentum LMS* tend to regularize and/or stabilize the direction and speed of the stochastic-gradient descent, in the following, are reported some LMS's variants with reduced computational complexity at the expense of a modest performance degradation.

5.5.4.1 Signed-Error LMS Algorithm

The signed-error *LMS* (SE-LMS) algorithm is based on replacing the error $e[n]$ with its three-level (− 1, 0, 1) quantized version, defined by the function $\text{sign}(e[n])$. It is shown (see [5] for details) that in this case the block LS CF turns out to be of L_1 type. That is,

$$\arg\min_{\mathbf{w}} \hat{J}(\mathbf{w}) = \arg\min_{\mathbf{w}} \left\| \mathbf{d} - \mathbf{X}\mathbf{w} \right\|_1.$$

The recursive adaptation formula is equal to

$$\mathbf{w}_n = \mathbf{w}_{n-1} + \mu \mathbf{x}_n \,\text{sign}\big(e[n]\big). \tag{5.204}$$

In hardware realizations, to increase the calculation speed and/or simplify the circuit structure, the step size μ can be constrained to be a (negative) power of two. In this way you can replace the multiplication operation, which appears in the conventional version of LMS, with simple shift operation of the input signal.

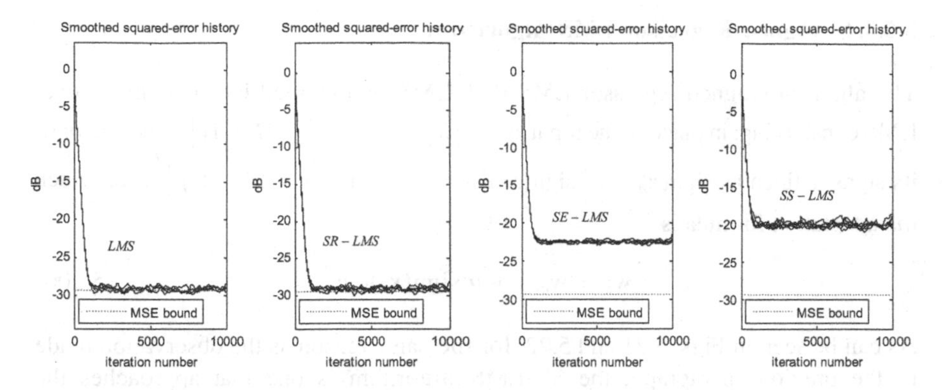

Fig. 5.21 Quantized variants of the LMS algorithm

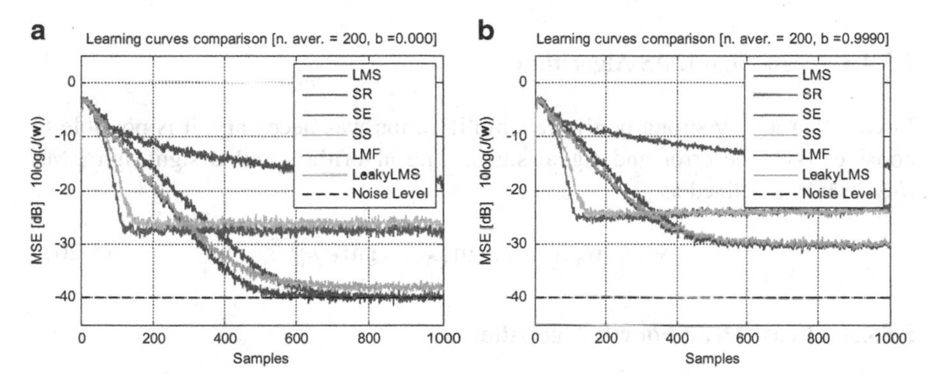

Fig. 5.22 Comparison among some LMS algorithm variants: (**a**) white noise ($b = 0.0$) input; (**b**) MA colored process ($b = 0.999$)

Remark The sign function can be defined as $\text{sign}(e[n]) = e[n]/|e[n]|$ for which the (5.204) can be rewritten as

$$\mathbf{w}_n = \mathbf{w}_{n-1} + \frac{\mu}{|e[n]|}\mathbf{x}_n e[n]. \tag{5.205}$$

The above expression can be interpreted as the conventional LMS in which the learning rate increases when the error decreases. This is in clear contrast with what was stated above in which it is demonstrated that the steady-state error should be inversely proportional to the learning rate. This observation is confirmed by the experimental results shown in Figs. 5.21 and 5.22.

5.5.4.2 Signed-Regressor LMS Algorithm

The algorithm signed-regressor LMS (SR-LMS) is obtained by the conventional LMS considering, in place of the input $\mathbf{x}_n = \begin{bmatrix} x[n] & \cdots & [n-M+1] \end{bmatrix}^T$, the vector of its signs defined as $\text{sign}(\mathbf{x}_n) = \begin{bmatrix} \text{sign}(x[n]) & \cdots & \text{sign}(x[n-M+1]) \end{bmatrix}^T$, for which the adaptation formula is

$$\mathbf{w}_n = \mathbf{w}_{n-1} + \mu \,\text{sign}(\mathbf{x}_n) e[n]. \tag{5.206}$$

As can be seen in Figs. 5.21 and 5.22, for the same reason as the observation made in the previous paragraph, the SR-LMS algorithm is one that approaches the performance of conventional LMS.

5.5.4.3 Sign–Sign LMS Algorithm

In case that a very strong hardware simplification was necessary, it is possible to consider both the error and signal signs. The algorithm, called sign–sign LMS (SS-LMS), is defined as

$$\mathbf{w}_n = \mathbf{w}_{n-1} + \mu \,\text{sign}(\mathbf{x}_n)\,\text{sign}(e[n]). \tag{5.207}$$

5.5.4.4 Least Mean Fourth Algorithm

Introduced by Walach–Widrow in [20], the Least Mean Fourth (LMF) algorithm represents an LMS variant in which is minimized the CF's $2N_\text{k}$-norm with $N_\text{k} = 1, 2, \ldots$. The CF assumes, therefore, the form

$$\underset{\mathbf{w}}{\arg\min}\, \hat{J}(\mathbf{w}) = \underset{\mathbf{w}}{\arg\min} \left\| \mathbf{d} - \mathbf{X}\mathbf{w} \right\|_2^{2N_\text{k}}, \tag{5.208}$$

where the choice of the $2N_k$ norm value influences the filter adaptation performance. The iterative approximation of (5.208) can be formulated as

$$\mathbf{w}_n = \mathbf{w}_{n-1} + \mu N_k e^{2N_k - 1}[n]\mathbf{x}_n,$$

that, for $N_\text{k} = 1$, becomes the standard LMS.

The most common form of the algorithm is for $N_\text{k} = 2$ and in this case is called least mean fourth (LMF). The adaptation formula then becomes

$$\mathbf{w}_n = \mathbf{w}_{n-1} + \mu \mathbf{x}_n e[n] \left| e[n] \right|^2. \tag{5.209}$$

The use of algorithms with norm greater than 2, such *higher order error algorithms*, must be done with care. In fact, as noted in [20], only in some specific operating conditions, such as the presence of non-Gaussian additive noise on the reference signal (desired output), it may have better performance than the standard LMS.

In Fig. 5.22, is reported a comparison among some LMS algorithm variants. The AF input is a stochastic MA process generated with the expression (5.172).

The experiment, similar to that described in Sect. 5.4.4, is carried out with a random IC coefficients filter of length $M = 6$, for $b = 0.0$ (white noise) and for $b = 0.999$ (colored noise). The learning curves are the average of 200 runs.

5.5.4.5 Least Mean Mixed Norm Algorithm

In case the minimization of a mixed norm is required, the CF can be defined as a linear combination of the type

$$\arg\min_{\mathbf{w}} \hat{J}(\mathbf{w}) = \arg\min_{\mathbf{w}} \left[\delta\|\mathbf{e}\|_2^2 + \tfrac{1}{2}(1-\delta)\|\mathbf{e}\|_2^4\right], \qquad \text{with} \qquad \mathbf{e} = \mathbf{d} - \mathbf{X}\mathbf{w}.$$

The latter, for $0 \le \delta \le 1$, can be approximated by the following recursion:

$$\mathbf{w}_n = \mathbf{w}_{n-1} + \mu\mathbf{x}_n e^*[n]\left(\delta + (1-\delta)\left|e[n]\right|^2\right). \tag{5.210}$$

5.5.4.6 LMS with Gradient Estimation Filter

In the LMS algorithm, the weights adaptation depends on the, rather noisy, local error surface gradient estimation. A simple trick to strengthen and improve the estimate is to use a smoothing filter. For example, in the technique called *average* LMS, the update formula is defined as

$$\mathbf{w}_n = \mathbf{w}_{n-1} + \mu\mathbf{v}_n, \tag{5.211}$$

in which $\mathbf{v}_n = \begin{bmatrix} v_n(0) & v_n(1) & \cdots & v_n(M-1) \end{bmatrix}^T$ represents a simple moving average of the last L instantaneous gradient estimates, i.e.,

$$\begin{aligned} \mathbf{v}_n &= \frac{1}{L}\sum_{k=n-L}^{n-1} \Delta\mathbf{w}_k \\ &= \frac{1}{L}\sum_{k=n-L}^{n-1} e^*[k]\mathbf{x}_k. \end{aligned} \tag{5.212}$$

More generally, the ith element $v_n(i)$ is a filtered version of the ith component of the gradient $\mathbf{v}_n$, in formal terms

$$v_n(i) = \text{LPF}\left\{e[n]x[n-i-1], e[n-1]x[n-i-2], \ldots\right\}, \quad i = 0, 1, \ldots, M-1, \tag{5.213}$$

where the operator LPF (low-pass filter) is the gradient optimal estimator made with a low-pass FIR filter.

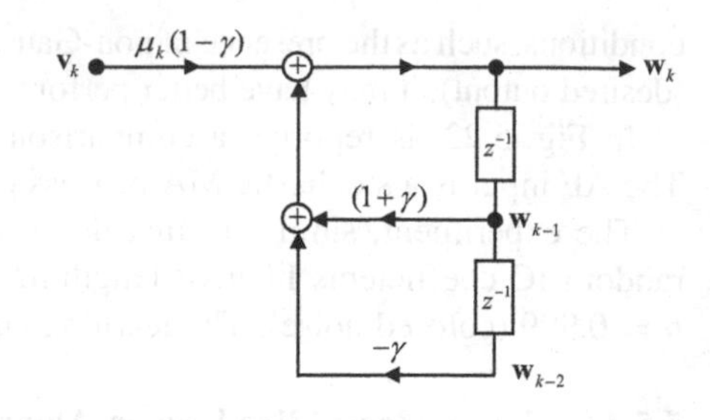

Fig. 5.23 The momentum LMS algorithm, modelled as second-order multi-channel IIR digital filter. Note that for $\gamma = 0$, the algorithm exactly coincides with the LMS

5.5.4.7 Momentum LMS

In the *momentum LMS* algorithm, to strengthen the estimate, the weights updating must depend, as well as by the vector $\mathbf{v}_k = -\nabla \hat{J}(\mathbf{w}_{k-1})$, on the difference from the weights of the previous iteration according to the following relation:

$$\mathbf{w}_k = \mathbf{w}_{k-1} + \mu_k(1-\gamma)\mathbf{v}_k + \gamma(\mathbf{w}_{k-1} - \mathbf{w}_{k-2}). \tag{5.214}$$

It is interesting to observe that the above relationship can be interpreted as a MIMO-IIR numerical filter with input $\mathbf{v}_k$ and output $\mathbf{w}_k$, as shown in Fig. 5.23, governed by the following finite differences equation:

$$\mathbf{w}_k = (1+\gamma)\mathbf{w}_{k-1} - \gamma\mathbf{w}_{k-2} + \mu_k(1-\gamma)\mathbf{v}_k. \tag{5.215}$$

In fact, (5.215) corresponds to M numerical filters (one for each filter tap) which exerts a low-pass smoothing of the AF's weights trajectory with the effect, in certain conditions, of stabilizing the solution.

5.5.5 *Delayed Learning LMS Algorithms*

In many LMS practical applications, the signals of adaptation (reference signal) are available only after a certain delay due, essentially, to a further path of the output signal, before arriving at the comparison node relative to the specific application. This delay provides a mismatch between the filter output and the desired signal which results in an AF performance degradation. A typical example is illustrated in Fig. 5.24 in which the delay is defined by an "*in the air*" acoustic path.

For the delayed learning LMS algorithm definition, consider the dynamical system identification process with the general scheme shown in Fig. 5.25. The dynamic system TF $H(z)$ to be identified is modeled as an FIR filter characterized by the impulse response $\mathbf{h} \in (\mathbb{R},\mathbb{C})^{M_h \times 1} = \begin{bmatrix} h[0] & \cdots & h[M_h-1] \end{bmatrix}^T$. The additional path TF $C(z)$, at the adaptive filter output, is modeled with an impulse response $\mathbf{c} \in (\mathbb{R},\mathbb{C})^{M_c \times 1} = \begin{bmatrix} c[0] & \cdots & c[M_c-1] \end{bmatrix}^T$.

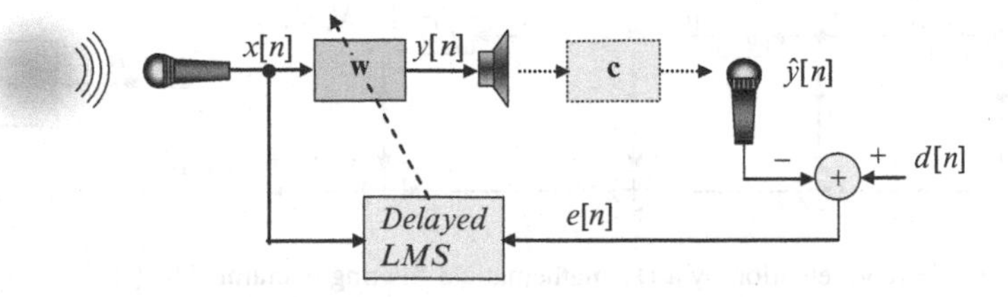

Fig. 5.24 Example of a typical scheme with delayed learning

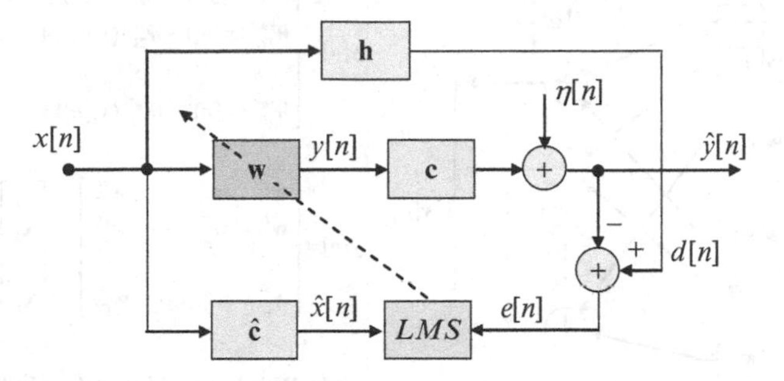

Fig. 5.25 System $H(z)$ identification block diagram, for the delayed LMS algorithms definition

In some applications such as, for example, in the predistortion (Sect. 2.3.3.2), the $C(z)$ indicates the distorting physical system TF, to be linearized or controlled, while the desired output path is a simple delay $H(z) = z^{-D}$. In active noise cancelation (Sect. 2.3.4.3), or more generally in the room acoustics active control, the $C(z)$ indicates the room transfer function (RTF), or the acoustic environment TF to equalize, while the $H(z)$ represents the optimal acoustic TF you want to obtain (target response) [36].

5.5.5.1 Definition of Discrete-Time Domain Filtering Operator

For a more compact and also more effective for the theoretical development representation is, in some situations, it is convenient to represent a numerical FIR filter as a discrete-time mathematical operator defined below.

Definition Denoting by q^{-1} the unit delay operator defines the discrete-time filtering operator, indicated as $W^{q^{-1}}(\cdot)$ and represented in Fig. 5.26, the time domain path representing the TF $W(z)$, such that following relations hold:

$$y[n] = W^{q^{-1}}\big(x[n]\big), \tag{5.216}$$

for an input sequence $\mathbf{x}_n = \begin{bmatrix} x[n] & x[n-1] & \cdots \end{bmatrix}^T$ is, by definition,

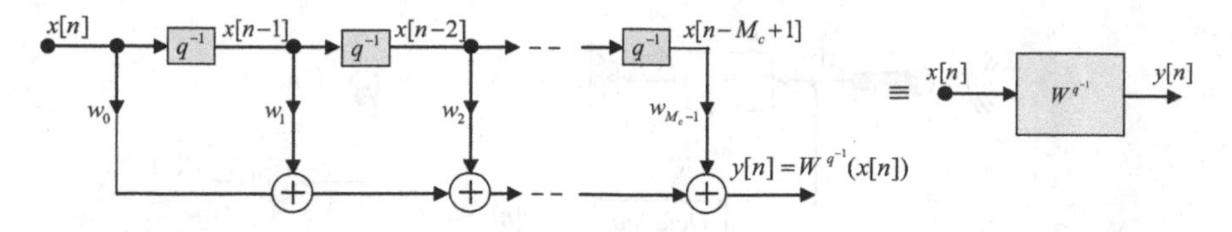

Fig. 5.26 TF representation by a DT mathematical filtering operator $W^{q^{-1}}(\cdot)$

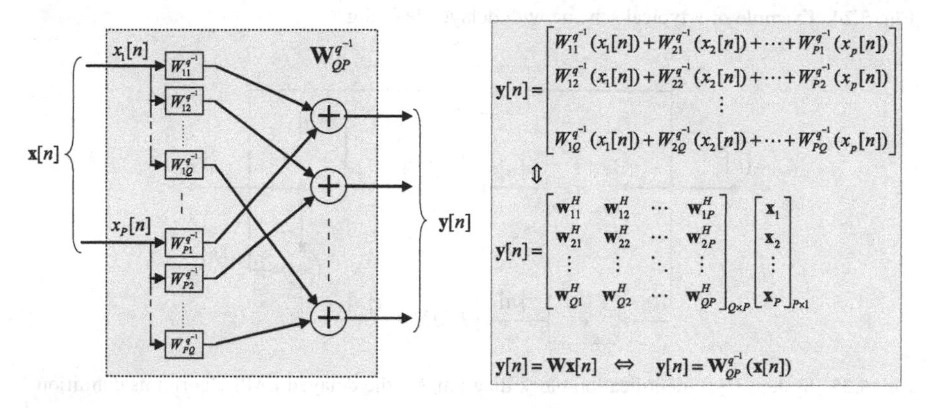

Fig. 5.27 MIMO filtering operator, in the case of P inputs and Q outputs

$$\mathbf{y}_n = W^{q^{-1}}(\mathbf{x}_n), \tag{5.217}$$

with $\mathbf{y}_n = \begin{bmatrix} y[n] & y[n-1] & \cdots \end{bmatrix}^T$.

The multichannel extension case, as depicted in Fig. 5.27, is such that for an input snapshot defined as $\mathbf{x}[n] \in (\mathbb{R}, \mathbb{C})^{P\times 1} = \begin{bmatrix} x_1[n] & \cdots & x_P[n] \end{bmatrix}^T$, and for the MIMO filter output snapshot defined as $\mathbf{y}[n] \in (\mathbb{R}, \mathbb{C})^{Q\times 1} = \begin{bmatrix} y_1[n] & \cdots & y_Q[n] \end{bmatrix}^T$, we have

$$\mathbf{y}[n] = \mathbf{W}_{\mathrm{QP}}^{q^{-1}}(\mathbf{x}[n]). \tag{5.218}$$

The formalism can be extended for a matrix of signals. In fact, in the case that at the filter input is present a signal matrix containing N-length time window snapshots, let $\mathbf{x}_{jn} = \begin{bmatrix} x_j[n] & \cdots & x_j[n-N+1] \end{bmatrix}^T$; we have that

$$\mathbf{X}_n \in \mathbb{R}^{1(N)\times P} = \begin{bmatrix} \mathbf{x}_{1n} & \mathbf{x}_{2n} & \cdots & \mathbf{x}_{Pn} \end{bmatrix}_{1\times P}, \tag{5.219}$$

and the output signal matrix is defined as

$$\mathbf{Y}_n = \mathbf{W}_{\mathrm{QP}}^{q^{-1}}(\mathbf{X}_n). \tag{5.220}$$

Therefore, the discrete-time filtering operator appears to be a very versatile formal instrument, which can be used for scalars, vectors, and matrices that, without loss of generality, can be very useful for adaptation algorithms representation.

5.5.5.2 Delayed LMS Algorithm

The delayed LMS (DLMS) algorithm [21–24] is defined by an output path characterized with a pure delay $C(z) = z^{-D}$. For error signal calculation, the output is available after a D delay, i.e., $y[n] = \mathbf{x}_{n-D}^H \mathbf{w}_{n-D}$. So we have that

$$e[n] = d[n] - \mathbf{x}_{n-D}^H \mathbf{w}_{n-D} + \eta[n]. \tag{5.221}$$

Denoting by $\hat{D}$ the estimated D delay, the adaptation rule is

$$\mathbf{w}_{n+1} = \mathbf{w}_n + \mu e^*[n] \mathbf{x}_{n-\hat{D}}. \tag{5.222}$$

For the algorithm analysis, proceeding as in [24], we substitute (5.221) into (5.222)

$$\mathbf{w}_{n+1} = \mathbf{w}_n + \mu\left(d[n]\mathbf{x}_{n-\hat{D}} + \eta[n]\mathbf{x}_{n-\hat{D}} - \mathbf{x}_{n-\hat{D}}\mathbf{x}_{n-D}^H \mathbf{w}_{n-D}\right). \tag{5.223}$$

By performing the expectation of the previous, a simplification for the algorithm performance analysis can be made considering the independence hypothesis true. For which we have that $E\left(\eta[n]\mathbf{x}_{n-\hat{D}}\right) = 0$ and $E\left(\mathbf{x}_{n-\hat{D}}\mathbf{x}_{n-D}^H \mathbf{w}_{n-D}\right) \cong E\left(\mathbf{x}_{n-\hat{D}}\mathbf{x}_{n-D}^H\right) E(\mathbf{w}_{n-D})$. Therefore, the performance analysis's stochastic difference equation is defined as

$$E(\mathbf{w}_{n+1}) = E(\mathbf{w}_n) + \mu \mathbf{g}_{n-\hat{D}} - \mathbf{R}_{D-\hat{D}} E(\mathbf{w}_{n-D}), \tag{5.224}$$

where $\mathbf{R}_{D-\hat{D}} = E\left(\mathbf{x}_{n-\hat{D}}\mathbf{x}_{n-D}^H\right)$ and $\mathbf{g}_{n-\hat{D}} = E\left(d[n]\mathbf{x}_{n-\hat{D}}\right)$.

For the convergence analysis, we can proceed as for the standard LMS (Sect. 5.4.2.1) and for the study of the mean quadratic behavior as in Sect. 5.4.2.2. It is shown (see [24] for details) that in the case of perfect estimation of the delay, i.e., $D \cong \hat{D}$, there is convergence to the optimal point for

$$0 < \mu < \frac{2}{2(D+1)\lambda_{\max} + \sum_{i=0}^{M-1} \lambda_i}. \tag{5.225}$$

This means that the step size upper bound is much the smaller, the greater the delay D.

5.5.5.3 Filtered-X LMS Algorithm

In the case of transfer function error path existence, one of the most widespread adaptation algorithms is the so-called filtered-x LMS (FX-LMS) [25–28].

Considering the general scheme of Fig. 5.25, for the adaptation it is necessary to estimate the TF $\hat{C}(z)$. The name "filtered-x" comes from the fact that to achieve adaptation, the input is filtered by this estimated TF.

Whereas the $C(z)$ path's model is of FIR type, characterized by the impulse response $\mathbf{c}$, the output error is defined as

$$e[n] = d[n] - \mathbf{c}^H \mathbf{y}_n. \tag{5.226}$$

By placing $\hat{x}[n] = \hat{\mathbf{c}}^H \mathbf{x}_n$, the update rule is

$$\mathbf{w}_{n+1} = \mathbf{w}_n + \mu e^*[n] \hat{\mathbf{x}}_n. \tag{5.227}$$

For the performance analysis, we can proceed as in the DLMS based on the independence processes assumption [29]. It is found that the algorithm performances are highly sensitive to the $C(z)$ path estimated goodness. The theoretical development is quite complex and for details on weak and quadratic convergence, please refer to the literature [29–33].

5.5.5.4 Adjoint LMS Algorithm

The adjoint LMS (AD-LMS) algorithm, developed by Eric Wan in [34], is an alternative way for the FX-LMS implementation. The AD-LMS algorithm exploits the linearity and the *adjoint network* definitions (Fig. 5.28).

For the algorithm presentation, as proposed by the author in [34], we proceed by representing the $C(z)$ path by the discrete-time operator $C^{q^{-1}}(\cdot)$. With this formalism the output error (5.226) can be rewritten in the time domain as

$$e[n] = d[n] - C^{q^{-1}}(y[n]), \tag{5.228}$$

and FX-LMS updated rule (5.227) is rewritten as

$$\mathbf{w}_{n+1} = \mathbf{w}_n + \mu e^*[n] \hat{C}^{q^{-1}}(\mathbf{x}_n). \tag{5.229}$$

Definition Given a DT circuit defined by a graph G, we define the *adjoint network* a circuit whose graph is determined by G with the following modifications: (1) the paths verses are reversed; (2) junction nodes are switched with sum nodes; and (3) delay elements are replaced with anticipation elements. For example, Fig. 5.29 shows a FIR filter graph and its adjoint network.

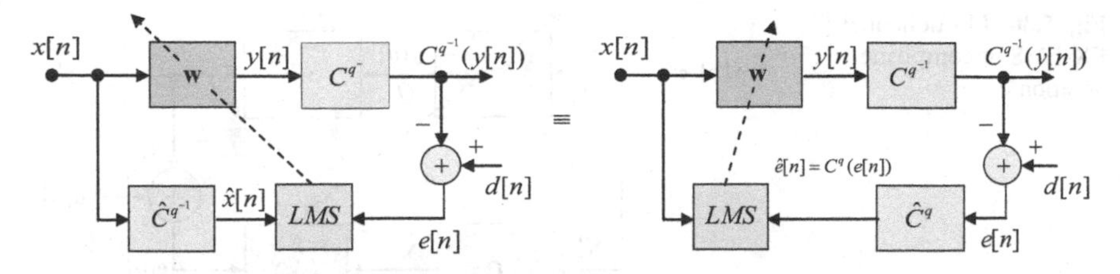

Fig. 5.28 Equivalence between FX-LMS (left) and AD-LMS (right) algorithms

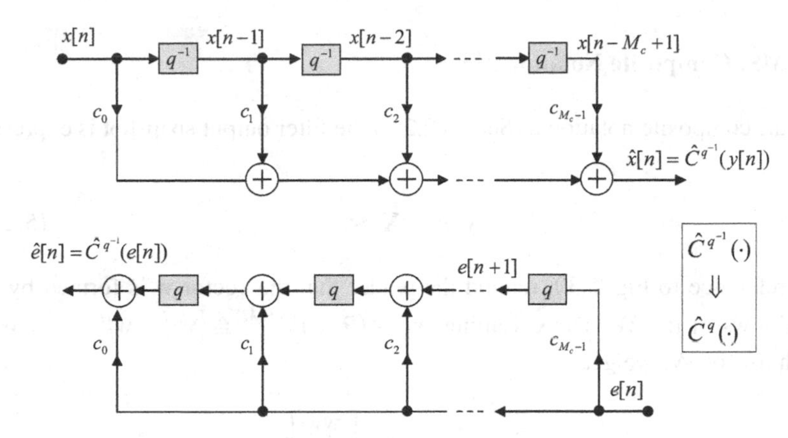

Fig. 5.29 DT filtering operator path of a M_c length FIR and corresponding *adjoint network*

By using the adjoint network paradigm, illustrated in Fig. 5.29, the update rule (5.229) can be rewritten as

$$\begin{aligned}\mathbf{w}_{n+1} &= \mathbf{w}_n + \mu\hat{e}^*[n - M_c]\mathbf{x}_{n-M_c}\\ \hat{e}[n] &= \hat{C}^q(e[n]).\end{aligned} \tag{5.230}$$

Note that, in (5.230) is filtered the error $e[n]$, rather than the input signal as in (5.229). The method is general and can be extended to paths modeled with IIR-FIR lattice structures.

The error filter defined by the adjoint network is characterized by the not causal operator $C^q(\cdot)$. Consequently, for the online algorithm feasibility, the sequences should be aligned by introducing a delay equal to the M_c filter length.

Note that, in the one-dimensional case, the algorithms described by (5.229) and (5.230) are characterized by the same computational complexity and have almost similar performance [34].

5.5.5.5 Multichannel FX-LMS Algorithm

For the FX-LMS MIMO development, we consider the composite notations 2 and 1.

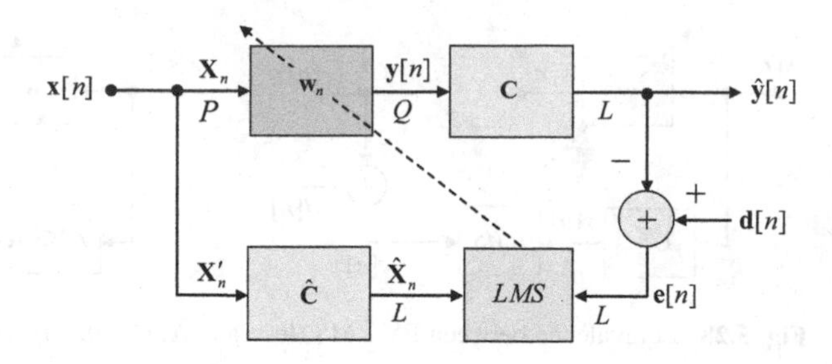

Fig. 5.30 Multichannel FX-LMS in composite notation 2

FX-LMS: Composite Notation 2

With the composite notation 2 (Sect. 3.2.2.2) the filter output snapshot is expressed as

$$\mathbf{y}[n] = \mathbf{X}^T\mathbf{w}. \tag{5.231}$$

With reference to Fig. 5.30, remind the reader that the vector $\mathbf{w}$ is formed by the staked rows of the $\mathbf{W}$ matrix. Calling $\mathbf{w}_{j:}^T \in (\mathbb{R},\mathbb{C})^{1\times P(M)} \triangleq \begin{bmatrix} \mathbf{w}_{j1}^H & \mathbf{w}_{j2}^H & \cdots & \mathbf{w}_{jP}^H \end{bmatrix}$ the jth row of $\mathbf{W}$, we get

$$\mathbf{w} \in (\mathbb{R},\mathbb{C})^{(\mathrm{PM})Q\times 1} \triangleq \begin{bmatrix} \mathbf{w}_{1:} \\ \vdots \\ \mathbf{w}_{Q:} \end{bmatrix}_{Q\times 1}. \tag{5.232}$$

In order to be true expression (5.231), the data matrix $\mathbf{X}^H \in (\mathbb{R},\mathbb{C})^{Q\times Q(\mathrm{PM})}$ is such that

$$\mathbf{y}[n] = \begin{bmatrix} \mathbf{x}^H & \mathbf{0} & \cdots & \mathbf{0} \\ \mathbf{0} & \mathbf{x}^H & \cdots & \mathbf{0} \\ \vdots & \vdots & \ddots & \vdots \\ \mathbf{0} & \mathbf{0} & \cdots & \mathbf{x}^H \end{bmatrix}_{Q\times Q} \begin{bmatrix} \mathbf{w}_{1:} \\ \vdots \\ \mathbf{w}_{Q:} \end{bmatrix}_{Q\times 1}, \tag{5.233}$$

for which $\mathbf{X}^H$ is formed by identical diagonal elements $\mathbf{x}^H \in (\mathbb{R},\mathbb{C})^{1\times \mathrm{PM}}$ that contain the PM inputs delay line samples.

Calling $\mathbf{C}$ matrix the AF's MIMO downstream path

$$\mathbf{C} \in (\mathbb{R},\mathbb{C})^{L\times Q(M_c)} = \begin{bmatrix} \mathbf{c}_{11}^H & \mathbf{c}_{12}^H & \cdots & \mathbf{c}_{1Q}^H \\ \mathbf{c}_{21}^H & \mathbf{c}_{22}^H & \cdots & \mathbf{c}_{2Q}^H \\ \vdots & \vdots & \ddots & \vdots \\ \mathbf{c}_{L1}^H & \mathbf{c}_{L2}^H & \cdots & \mathbf{c}_{LQ}^H \end{bmatrix}_{L\times Q}, \tag{5.234}$$

such that each element is a row vector containing the individual impulse responses $\mathbf{c}_{ij}^H \in (\mathbb{R},\mathbb{C})^{1\times M_c}$, the $\mathbf{C}$ path output snapshot, in composite notation 2, is

$$\hat{\mathbf{y}}[n] \in (\mathbb{R},\mathbb{C})^{L\times 1} = \mathbf{Y}^T\mathbf{c},$$

where the vector $\mathbf{c}$ is formed by $\mathbf{C}$ matrix rows, all in columns

$$\mathbf{c} \in (\mathbb{R},\mathbb{C})^{(QM_c)L\times 1} \triangleq \begin{bmatrix} \mathbf{c}_{1:} \\ \vdots \\ \mathbf{c}_{L:} \end{bmatrix}_{L\times 1}, \tag{5.235}$$

where $\mathbf{c}_{j:} \in (\mathbb{R},\mathbb{C})^{QM_cL\times 1} \triangleq \begin{bmatrix} \mathbf{c}_{j1}^H & \mathbf{c}_{j2}^H & \cdots & \mathbf{c}_{jL}^H \end{bmatrix}^H$, and similarly to the (5.233), the composite data matrix $\mathbf{Y} \in (\mathbb{R},\mathbb{C})^{(QM_c)L\times L}$ is defined as

$$\mathbf{Y}^H \in (\mathbb{R},\mathbb{C})^{L\times L(QM_c)} = \begin{bmatrix} \mathbf{y}^H & \mathbf{0} & \cdots & \mathbf{0} \\ \mathbf{0} & \mathbf{y}^H & \cdots & \mathbf{0} \\ \vdots & \vdots & \ddots & \vdots \\ \mathbf{0} & \mathbf{0} & \cdots & \mathbf{y}^H \end{bmatrix}_{L\times L}, \tag{5.236}$$

i.e., $\mathbf{Y}$ is a $L\times L$ matrix, where each diagonal element $\mathbf{y} \in (\mathbb{R},\mathbb{C})^{QM_c\times 1}$ contains, all stacked, the delay line filters samples $\mathbf{c}_{1:}, \mathbf{c}_{2:}, \ldots, \mathbf{c}_{L:}$.

We define the estimated path matrix $\hat{\mathbf{C}}$, as

$$\hat{\mathbf{C}} \in (\mathbb{R},\mathbb{C})^{L\times P(M_c)} = \begin{bmatrix} \hat{\mathbf{c}}^H{}_{11} & \hat{\mathbf{c}}^H{}_{12} & \cdots & \hat{\mathbf{c}}^H{}_{1P} \\ \hat{\mathbf{c}}^H{}_{21} & \hat{\mathbf{c}}^H{}_{22} & \cdots & \hat{\mathbf{c}}^H{}_{2P} \\ \vdots & \vdots & \ddots & \vdots \\ \hat{\mathbf{c}}^H{}_{L1} & \hat{\mathbf{c}}^H{}_{L2} & \cdots & \hat{\mathbf{c}}^H{}_{LP} \end{bmatrix}_{L\times P}, \tag{5.237}$$

while estimated path's output data matrix has the form

$$\hat{\mathbf{X}} = \hat{\mathbf{C}} \odot \mathbf{X}' = \begin{bmatrix} \hat{\mathbf{c}}^H{}_{11} & \hat{\mathbf{c}}^H{}_{12} & \cdots & \hat{\mathbf{c}}^H{}_{1P} \\ \hat{\mathbf{c}}^H{}_{21} & \hat{\mathbf{c}}^H{}_{22} & \cdots & \hat{\mathbf{c}}^H{}_{2P} \\ \vdots & \vdots & \ddots & \vdots \\ \hat{\mathbf{c}}^H{}_{L1} & \hat{\mathbf{c}}^H{}_{L2} & \cdots & \hat{\mathbf{c}}^H{}_{LP} \end{bmatrix}_{L\times P} \odot \begin{bmatrix} \mathbf{x}_1' & \mathbf{x}_2' & \cdots & \mathbf{x}_P' \\ \mathbf{x}_1' & \mathbf{x}_2' & \cdots & \mathbf{x}_P' \\ \vdots & \vdots & \ddots & \vdots \\ \mathbf{x}_1' & \mathbf{x}_2' & \cdots & \mathbf{x}_P' \end{bmatrix}_{L\times P}, \tag{5.238}$$

where $\odot$ is defined as the *Kronecker convolution*. The symbol $\odot$ indicates that each ij element of $\hat{\mathbf{X}}$ matrix is the convolution between the ij elements of the $\hat{\mathbf{C}}$ and $\mathbf{X}'$ matrices. With reference to Fig. 5.31, calling $\hat{\mathbf{c}}_{ij} \in (\mathbb{R},\mathbb{C})^{M_c\times 1}$ the estimated path impulse response between the input i and output j, $\mathbf{X}' \in (\mathbb{R},\mathbb{C})^{L(N_c)\times P}$ indicates the matrix in which each element of the ith column contains a signal block, of suitable N_c length, relative to $\hat{\mathbf{c}}_{ij}$ for each j. By defining the convolution between $\hat{\mathbf{c}}_{ji}$ and $\hat{\mathbf{x}}_i'$ as $\hat{\mathbf{x}}_{ji}' = \hat{\mathbf{c}}_{ji} * \mathbf{x}_i'$ so $\hat{\mathbf{x}}_i' \in \mathbb{R}^{(M_c+N_c-1)\times 1}$, (5.238) can be written as

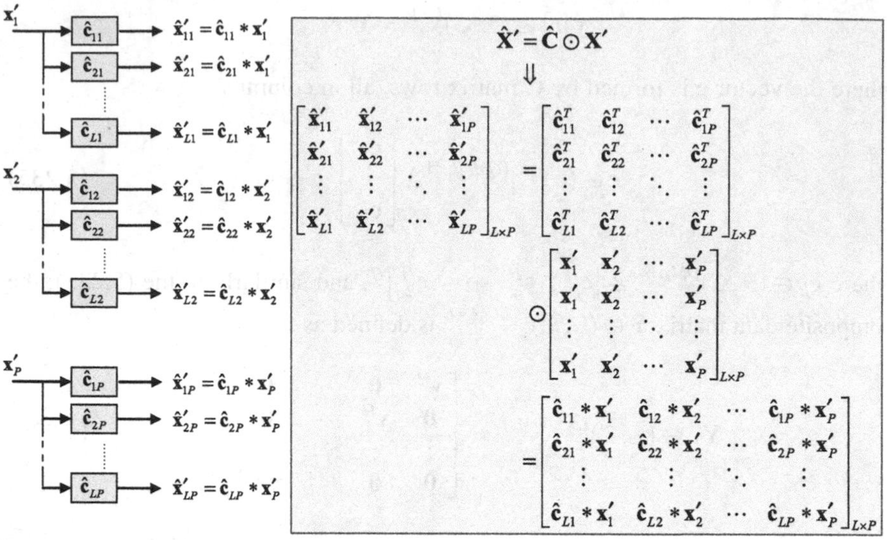

Fig. 5.31 Data matrix definition

$$\hat{\mathbf{X}} \in (\mathbb{R}, \mathbb{C})^{(N_c+M_c-1)L\times P} = \begin{bmatrix} \hat{\mathbf{c}}_{11} * \mathbf{x}_1' & \hat{\mathbf{c}}_{12} * \mathbf{x}_2' & \cdots & \hat{\mathbf{c}}_{1P} * \mathbf{x}_P' \\ \hat{\mathbf{c}}_{21} * \mathbf{x}_1' & \hat{\mathbf{c}}_{22} * \mathbf{x}_2' & \cdots & \hat{\mathbf{c}}_{2P} * \mathbf{x}_P' \\ \vdots & \vdots & \ddots & \vdots \\ \hat{\mathbf{c}}_{L1} * \mathbf{x}_1' & \hat{\mathbf{c}}_{L2} * \mathbf{x}_2' & \cdots & \hat{\mathbf{c}}_{LP} * \mathbf{x}_P' \end{bmatrix}_{L\times P}. \quad (5.239)$$

The adaptation rule may be defined by extending the update FX-LMS SISO (5.227), to the MIMO case. The gradient in the composite notation 2 is $\Delta \hat{J}_n^{C2} = -\hat{\mathbf{X}}_n \mathbf{e}^*[n]$, so we have that

$$\mathbf{e}[n] = \mathbf{d}[n] - \hat{\mathbf{y}}[n], \quad (5.240)$$

$$\mathbf{w}_n = \mathbf{w}_{n-1} + \mu \hat{\mathbf{X}}_n \mathbf{e}^*[n]. \quad (5.241)$$

Remark Rewriting (5.241) indicating the vectors and matrices size, we get

$$\underset{(\mathrm{PM})Q\times 1}{\mathbf{w}_n} = \underset{(\mathrm{PM})Q\times 1}{\mathbf{w}_{n-1}} + \mu \cdot \underset{[(N_c+M_c-1)L\times P]}{\hat{\mathbf{X}}_n} \cdot \underset{(L\times 1)}{\mathbf{e}^*[n]}, \quad (5.242)$$

we observe that, for the product $\hat{\mathbf{X}}_n \mathbf{e}^*[n]$ computation, it is necessary that the $\hat{\mathbf{X}}_n$ columns number is equal to the $\mathbf{e}[n]$ rows, i.e., $P \equiv L$. In this case, for the computability of the sum, it is necessary that $(\mathrm{PM})Q \equiv (N_c + M_c - 1)P$ for which the data block length at the $\hat{\mathbf{C}}$ MIMO filter input must be equal to $N_c = MQ - M_c + 1$.

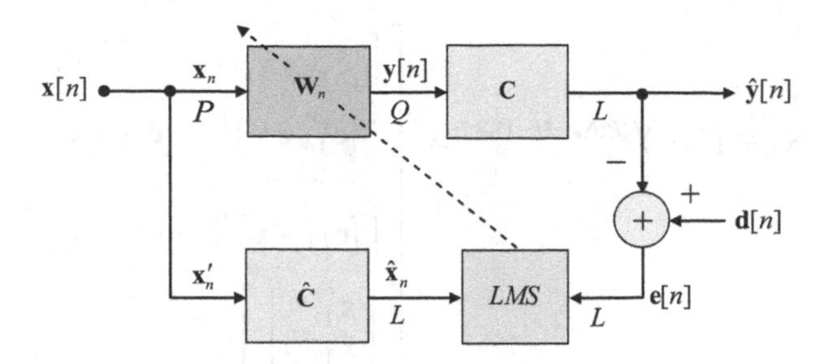

Fig. 5.32 Multichannel FX-LMS in composite notation 1

FX-LMS: Composite Notation 1

With the *composite notation* 1 (Sect. 3.2.2.1), with reference to Fig. 5.32, the filter's output snapshot $\mathbf{y}[n] \in (\mathbb{R},\mathbb{C})^{Q\times 1}$ is expressed as

$$\mathbf{y}[n] = \mathbf{W}\mathbf{x}, \tag{5.243}$$

where the $\mathbf{x}$ vector is defined as

$$\mathbf{x} \in (\mathbb{R},\mathbb{C})^{P(M)\times 1} = \begin{bmatrix} \mathbf{x}_1^H & \mathbf{x}_2^H & \cdots & \mathbf{x}_P^H \end{bmatrix}^H_{P\times 1}, \tag{5.244}$$

and the matrix $\mathbf{W} \in (\mathbb{R},\mathbb{C})^{Q\times \mathrm{PM}}$ as

$$\mathbf{W} \in (\mathbb{R},\mathbb{C})^{Q\times P(M)} = \begin{bmatrix} \mathbf{w}_{11}^H & \mathbf{w}_{12}^H & \cdots & \mathbf{w}_{1P}^H \\ \mathbf{w}_{21}^H & \mathbf{w}_{22}^H & \cdots & \mathbf{w}_{2P}^H \\ \vdots & \vdots & \ddots & \vdots \\ \mathbf{w}_{Q1}^H & \mathbf{w}_{Q2}^H & \cdots & \mathbf{w}_{QP}^H \end{bmatrix}_{Q\times P}. \tag{5.245}$$

For the adaptation rule, the composite notation 1 gradient is equal to $\Delta\hat{J}_n^{C1} = -\mathbf{e}[n]\hat{\mathbf{x}}_n^H$. It is therefore

$$\mathbf{W}_n = \mathbf{W}_{n-1} + \mu\mathbf{e}[n]\hat{\mathbf{x}}_n^H. \tag{5.246}$$

Where the data vector $\hat{\mathbf{x}}_n \in (\mathbb{R},\mathbb{C})^{N_x\times 1}$ is built as the vector containing all convolutions $\mathbf{x}'_i * \hat{\mathbf{c}}_{ij}$ (between the inputs $\hat{\mathbf{x}}_n \in (\mathbb{R},\mathbb{C})^{N_x\times 1}$ and the impulse responses $\hat{\mathbf{c}}_{ij} \in \mathbb{R}^{M_c\times 1}$ for $j = 1, \ldots, L$), all staked, and where each convolution has $(N_c + M_c - 1)$ length. Formally

$$\hat{\mathbf{x}}_n \in (\mathbb{R},\mathbb{C})^{L[P(N_c+M-1)]\times 1} = \begin{bmatrix} \left[[\hat{\mathbf{c}}_{11} * \mathbf{x}_1']^T & [\hat{\mathbf{c}}_{12} * \mathbf{x}_2']^T & \cdots & [\hat{\mathbf{c}}_{1P} * \mathbf{x}_P']^T \right]^T \\ \left[[\hat{\mathbf{c}}_{12} * \mathbf{x}']^T & [\hat{\mathbf{c}}_{22} * \mathbf{x}_2']^T & \cdots & [\hat{\mathbf{c}}_{2P} * \mathbf{x}_P']^T \right]^T \\ & \vdots & & \\ \left[[\hat{\mathbf{c}}_{L1} * \mathbf{x}']^T & [\hat{\mathbf{c}}_{L2} * \mathbf{x}_2']^T & \cdots & [\hat{\mathbf{c}}_{LP} * \mathbf{x}_P']^T \right]^T \end{bmatrix}_{L\times 1}$$

$$= \begin{bmatrix} \begin{bmatrix} \hat{\mathbf{x}}_{11}'^H \\ \hat{\mathbf{x}}_{12}'^H \\ \vdots \\ \hat{\mathbf{x}}_{1P}'^H \end{bmatrix} \\ \vdots \\ \begin{bmatrix} \hat{\mathbf{x}}_{L1}'^H \\ \hat{\mathbf{x}}_{L2}'^H \\ \vdots \\ \hat{\mathbf{x}}_{LP}'^H \end{bmatrix} \end{bmatrix}_{L\times 1}. \tag{5.247}$$

For which the $\hat{\mathbf{x}}_n$ vector length is equal to $LP(N_c + M_c - 1)$.

Remark Rewriting (5.246), indicating the size of the vectors and matrices,

$$\underset{[Q\times P(M)]}{\mathbf{W}_n} = \underset{[Q\times P(M)]}{\mathbf{W}_{n-1}} + \mu \cdot \underset{(L\times 1)}{\mathbf{e}^*[n]} \cdot \underset{[1\times LP(N_c+M_c-1)]}{\hat{\mathbf{x}}_n^H}, \tag{5.248}$$

we observe that, for consistency, it is necessary that $Q \equiv L$. In this case, for the update rule it is necessary that $PM \equiv Q[P(N_c + M_c - 1)]$, so the input data block length must be equal to $N_c = (M/Q) - M_c + 1$.

Remark The composite formulations 1 and 2, while being algebraically equivalent, have a different computational cost.

In composite notation 1 from (5.248), for the gradient estimate $\Delta\hat{J}_n^{C1} = -\mathbf{e}^*[n]\hat{\mathbf{x}}_n^T$ calculation, for $Q = L$, $Q \times QP(N_c + M_c - 1)$ multiplications are needed and the required input buffer length is equal to $N_c = (M/Q) - M_c + 1$. The total computational cost is equal to MQP.

With similar reasoning, in composite notation 2, from (5.242), for the gradient estimate $\Delta\hat{J}_n^{C2} = -\hat{\mathbf{X}}_n\mathbf{e}^*[n]$ calculation, for $P = L$, $(N_c + M_c - 1)P^3$ multiplications are needed and the required input buffer length is equal to $N_c = MQ - M_c + 1$. In this case the total computational cost is equal to MQP^3.

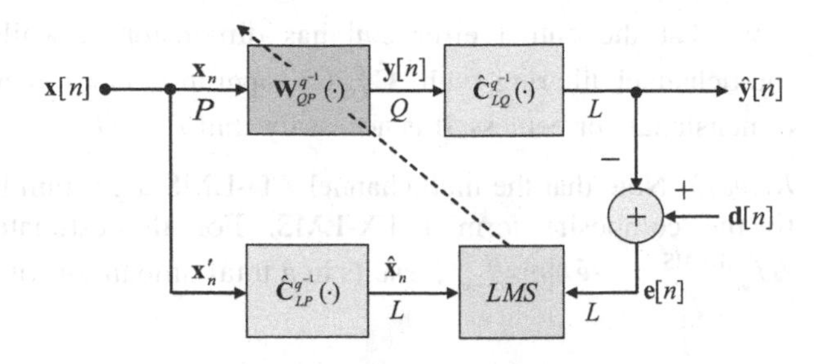

Fig. 5.33 FX-LMS MIMO in multichannel delay operators

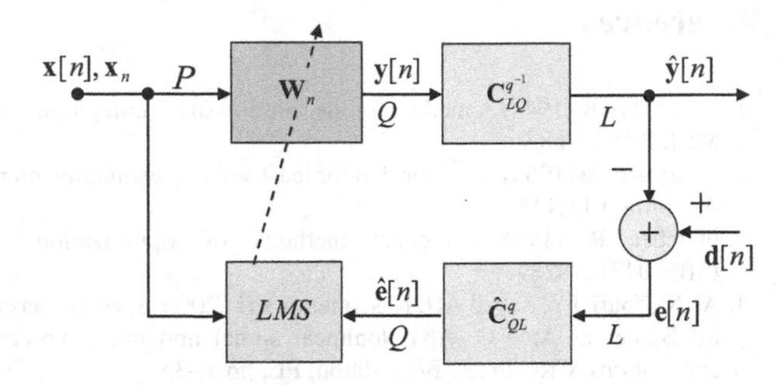

Fig. 5.34 Multichannel AD-LMS algorithm

MIMO FX-LMS in Multichannel Delay Operators Notation

The FX-LMS MIMO notation can be simplified by considering the DT filtering operators formalism. Note that by operators nesting, the output of the entire system can be written as $\hat{\mathbf{y}}[n] = \mathbf{C}_{LQ}^{q^{-1}}\left[\mathbf{W}_{QP}^{q^{-1}}(\mathbf{x}[n])\right]$ (Fig. 5.33).

The adaptation rule (5.246) for $Q = L$ can be written as

$$\mathbf{e}[n] = \mathbf{d}[n] - \hat{\mathbf{y}}[n], \tag{5.249}$$

$$\mathbf{W}_n = \mathbf{W}_{n-1} + \mu \mathbf{e}^*[n] \cdot \left[\hat{\mathbf{C}}_{LP}^{q^{-1}}\left(\mathbf{x}_n'\right)\right]^H. \tag{5.250}$$

5.5.5.6 Multichannel AD-LMS

Considering the multichannel AD-LMS, illustrated in Fig. 5.34, the adaptation algorithm can be simply implemented in the following mode:

$$\mathbf{e}[n] = \mathbf{d}[n] - \mathbf{C}_{\mathrm{LQ}}^{q^{-1}}\left(\mathbf{y}[n]\right), \tag{5.251}$$

$$\hat{\mathbf{e}}[n] = \hat{\mathbf{C}}_{\mathrm{QL}}^{q}(\mathbf{e}[n]), \tag{5.252}$$

$$\mathbf{W}_n = \mathbf{W}_{n-1} + \mu \hat{\mathbf{e}}^*[n - M_c]\mathbf{x}_{n-M_c}^H. \tag{5.253}$$

Note that the output error $\mathbf{e}[n]$ has dimension L while the error $\hat{\mathbf{e}}[n]$, after multichannel filtering with $\hat{\mathbf{C}}^{q}_{\text{QL}}(\cdot)$ operator, has dimension Q for which, for dimensional correctness, it is necessary that $L = Q$.

Remark Note that the multichannel AD-LMS algorithm has a similar complexity to the composite form 1 FX-LMS. For the estimated gradient calculation $\Delta\hat{J}^{\,\text{ADLMS}}_{n} = -\hat{\mathbf{e}}^{*}[n]\mathbf{x}^{H}_{n-M_c}$, results in a total amount of MQP computational cost.

References

1. Levenberg K (1944) A method for the solution of certain problems in least squares. Quart Appl Math 2:164–168
2. Marquardt D (1963) An algorithm for least squares estimation on nonlinear parameters. Siam J Appl Math 11:431–441
3. Fletcher R (1986) Practical methods of optimization. Wiley, New York, NY. ISBN 0471278289
4. Al-Naffouri TY, Sayed AH, Nascimento VH (2003) Energy conservation in adaptive filtering. In: Barner E, Arce G (eds) Nonlinear signal and image processing: theory, methods, and applications. CRC Press, Boca Raton, FL, pp 1–35
5. Sayed AH (2003) Fundamentals of adaptive filtering. IEEE Wiley - Interscience, New York, NY
6. Yousef NR, Sayed AH (2001) A unified approach to the steady-state and tracking analyses of adaptive filters. IEEE Trans Signal Proc 49(2)
7. Al-Naffouri TY, Sayed AH (2003) Transient analysis of adaptive filters with error nonlinearities. IEEE Trans Signal Proc 51(3):653–663
8. Al-Naffouri TY, Sayed AH (2003) Transient analysis of data-normalized adaptive filters. IEEE Trans Signal Proc 51(3):639–652
9. Haykin S (1996) Adaptive filter theory, Third Editionth edn. Prentice Hall, Upper Saddle River, NJ
10. Widrow B, Hoff ME (1960) Adaptive switching circuits. IRE WESCON, Conv. Rec., pt. 4:96–104
11. Widrow B (1966) Adaptive filters I: fundamentals. Stanford Electron. Labs, Stanford, CA, SEL-66-126
12. Godara LC, Cantoni A (1986) Analysis of constrained LMS algorithm with application to adaptive beamforming using perturbation sequences. IEEE Trans Antennas Propagat AP-34 (3):368–379
13. Kushner HJ (1984) Approximation and weak convergence methods for random processes, with applications to stochastic systems theory. MIT Press, Cambridge, MA. ISBN 0262110903, 9780262110907
14. Clarkson PM, White PR (1987) Simplified analysis of the LMS adaptive filter using a transfer function approximation. IEEE Trans Acoustics Speech Signal Proc ASSP-35(7):987–933
15. Duttweiler DL (2000) Proportionate normalized least-mean-squares adaptation in echo cancelers. IEEE Trans Speech Audio Proc 8:508–518
16. Benesty J, Gay SL (2002) An improved PNLMS algorithm. IEEE International Conference on Acoustics, Speech, and Signal Processing, ICASSP '02:1881–1884
17. Huang Y, Benesty J, Chen J (2006) Acoustic MIMO signal processing. Springer Series on Signal and Communication Technology, ISBN 10 3-540-37630-5
18. Kamenetsky M, Widrow B (2004) A variable leaky lms adaptive algorithm. IEEE Thirty-Eighth Asilomar Conference on Signals, Systems and Computers 1:125–128

19. Mayyas K, Tyseer A (1997) Leaky LMS algorithm: MSE analysis for Gaussian data. IEEE Trans Signal Proc 45:927–934
20. Walach E, Widrow B (1984) The least mean fourth (LMF) adaptive algorithm and its family. IEEE Trans Inform Theor IT-30(2):215
21. Long G, Ling F, Proakis J (1989) The LMS with delayed coefficient adaptation. IEEE Trans Acoustics Speech Signal Proc 37:1397–1405
22. Long G, Ling F, Proakis J (1992) Corrections to the LMS with delayed coefficient adaptation. IEEE Trans Signal Proc 40:230–232
23. Rupp M, Frenzel R (1994) Analysis of LMS and NLMS algorithms with delayed coefficient update under the presence of spherically invariant processes. IEEE Trans Signal Proc 42: 668–672
24. Tobias OJ, Bermudez JCM, Bershad NJ (2000) Stochastic analysis of the delayed LMS algorithm for a new model. In: Proceedings of IEEE International Conference on Acoustics, Speech, and Signal Processing, ICASSP '00. Vol. 1, pp 404–407, 15-9 June 2000
25. Widrow B, Stearns SD (1985) Adaptive signal processing. Prentice Hall, Upper Saddle River, NJ
26. Morgan DR (1980) An analysis of multiple correlation cancellation loops with a filter in the auxiliary path. IEEE Trans Acoust Speech Signal Proc ASSP-28(4):454–467
27. Widrow B, Shur D, Shaffer S (1981) On adaptive inverse control. In 15th Asilomar Conference Circuits, Systems, and Components, 185–189
28. Elliott SJ, Stothers IM, Nelson PA (1987) A multiple error LMS algorithm and its application to the active control of sound and vibration. IEEE Trans Acoust Speech Signal Proc ASSP-35 (10):1423–1434
29. Tobias OJ, Bermudez JCM, Bershad NJ (2000) Mean weight behavior of the filtered-X LMS algorithm. IEEE Trans Signal Proc 48:1061–1075
30. Snyder SD, Hansen CH (1994) Effect of transfer function estimation errors on the filtered-X LMS algorithm. IEEE Trans Signal Proc 42(4):950–953
31. Bjarnason E (1995) Analysis of the filtered-X LMS algorithm. IEEE Trans Speech Audio Proc 3:504–514
32. Douglas S, Pan W (1995) Exact expectation analysis of the LMS adaptive filter. IEEE Trans Signal Proc 43:2863–2871
33. Boucher CC, Elliott SJ, Nelson PA (1991) Effect of errors in the plant model on the performance of algorithms for adaptive feed forward control. Proc Inst Elect Eng F 138: 313–319
34. Wan EA (1996) Adjoint LMS: an efficient alternative to the filtered-X LMS and multiple error LMS algorithms. Proc IEEE ICASSP-1996:1842–1845
35. Widrow B (1971) Adaptive filters, from aspect of networks and system theory. Kalman – De Claris (ed.) Holt, Rinehart and Winston
36. Widrow B et al (1975) Adaptive noise cancellation: principles and applications. Proc IEEE 63:1691–1717
37. Wiener N (1949) Extrapolation, interpolation and smoothing of stationary time series, with engineering applications. Wiley, New York, NY
38. Bode HW, Shannon CE (1950) A simplified derivation of linear least squares smoothing and prediction theory. Proc IRE 38:417–425
39. Kivinen J, Warmuth MK (1997) Exponential gradient versus gradient descent for linear prediction. Inform Comput 132:1–64
40. Widrow B, McCool J, Ball M (1975) The complex LMS algorithm. Proc IEEE 63(4):719–720
41. Feuer A, Weinstein E (1985) Convergence analysis of LMS filters with uncorrelated Gaussian data. IEEE Trans Acoustics Speech Signal Proc 33(1):222–230
42. Farhang-Boroujeny B (1998) Adaptive filters: theory and applications. Wiley, New York, NY
43. Manolakis DG, Ingle VK, Kogon SM (2000) Statistical and adaptive signal processing. McGraw-Hill, New York, NY

Chapter 6
Second-Order Adaptive Algorithms

6.1 Introduction

This chapter introduces the second-order algorithms for the solution of the Yule–Walker normal equations with online recursive methods, such as error sequential regression (ESR) algorithm [1–3].

In the LS standard method, presented in Chap. 4, the solution is calculated considering that the entire signal block is known, without taking into account any estimates previously calculated of the same process. In the ESR class, the estimate of the LS optimal at time n, of a (at limit) infinite length sequence, is calculated starting from the estimates made in the previous instants: $n - 1, n - 2, \ldots, 0$. In other words, what is calculated at time n is just an optimal solution update due to new information present at the input.

Although not strictly necessary, it is preferred to derive these algorithms in the *classical* mode, or as approximate version of the Newton's algorithm (or second-order SDA).

In the first part of this chapter, the Newton method and its version with estimated time-average correlations which define the class of adaptive methods such as the sequential regression algorithms are briefly exposed. It is subsequently presented as a variant of the NLMS algorithm, the said affine projection algorithm (APA), in the context of the second-order algorithms [4, 5, 24].

Then we introduce the family of algorithms called recursive least squares (RLS) and their convergence properties are studied [2, 3, 6, 7, 21]. In Sect. 6.4.6 are presented some variants and RLS generalizations as, for example, the Kalman filter, with optimal performance in the case of nonstationary environment. Moreover, some criteria for the study of the adaptive algorithms performance, operating in nonstationary environments, are exposed [8, 9, 11, 12, 25].

Finally, the fundamental criteria for the definition of more general adaptation laws are presented. In particular methods based on non-Euclidean CF, that is, based on the *natural gradient* approach, and other methods in the presence of sparsity

A. Uncini, *Fundamentals of Adaptive Signal Processing*, Signals and Communication Technology, DOI 10.1007/978-3-319-02807-1_6,

constraints are presented and discussed. Moreover, in the final part of the chapter is presented the class of exponentiated gradient algorithms (EGA) [15, 17–20].

6.2 Newton's Method and Error Sequential Regression Algorithms

The ESR algorithms derivation can be made considering the approximate solution of second-order SDA methods (or Newton algorithms). A different presentation form can be derived, as seen previously, with the iterative LS system solution with Lyapunov's attractor (see Sect. 4.3.2), i.e., from the recursive solution of the Yule–Walker equations.

6.2.1 Newton's Algorithm

The Newton's methods for the adaptive filtering represent a class of recursive steepest-descent second-order algorithms, based on *a priori* knowledge of both the gradient and the Hessian matrix (see Sect. 5.1.1.3). The general algorithm formulation is described by the following expression of adaptation:

$$\mathbf{w}_k = \mathbf{w}_{k-1} + \mu\left[\nabla^2 J(\mathbf{w}_{k-1})\right]^{-1}\left(-\nabla J(\mathbf{w}_{k-1})\right) \tag{6.1}$$

if $\det\nabla^2 J(\mathbf{w}_k) \neq \mathbf{0}$. For a quadratic form, with the usual cost function (CF) $J(\mathbf{w}) = \sigma_d^2 - \mathbf{w}^T\mathbf{g} - \mathbf{g}^T\mathbf{w} + \mathbf{w}^T\mathbf{R}\mathbf{w}$, the gradient, and the Hessian matrix, the kth iteration takes, respectively, the form

$$\nabla J(\mathbf{w}_{k-1}) = \frac{\partial J(\mathbf{w})}{\partial \mathbf{w}_{k-1}} = 2(\mathbf{R}\mathbf{w}_{k-1} - \mathbf{g}) \tag{6.2}$$

$$\nabla^2 J(\mathbf{w}_{k-1}) = \frac{\partial^2 J(\mathbf{w})}{\partial \mathbf{w}_{k-1}^2} = 2\mathbf{R}. \tag{6.3}$$

Substituting the latter in (6.1), the Newton's algorithm, or II order SDA, is equivalent to the following expression:

$$\mathbf{w}_k = \mathbf{w}_{k-1} - \mu\mathbf{R}^{-1}(\mathbf{R}\mathbf{w}_{k-1} - \mathbf{g}). \tag{6.4}$$

It is noted that for $\mu = 1$, and simplifying the expression, it appears that $\mathbf{w}_k = \mathbf{R}^{-1}\mathbf{g} \equiv \mathbf{w}_{\mathrm{opt}}$ coincides with the Wiener optimal solution obtained in a single iteration. In this case, the convergence proof is immediate.

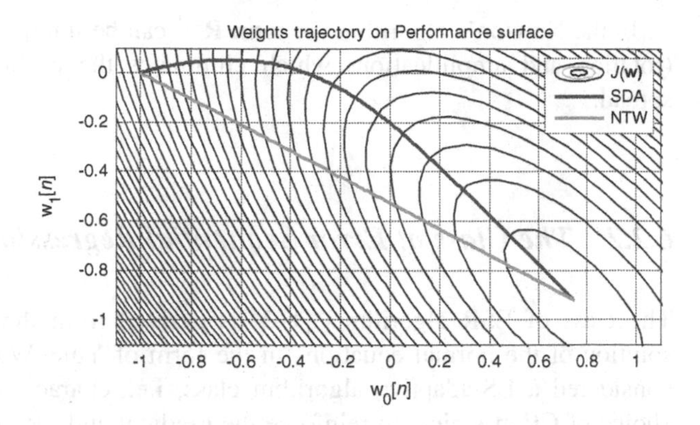

Fig. 6.1 Typical weights trajectories behavior for the SDA and Newton algorithms

Remark Newton's method is not a *true* adaptive algorithm, as it is based on *a priori* knowledge of the second-order statistics of the adaptive filter (AF) input processes. In fact, as to the exact Wiener formulation and as for the SDA, Newton's method has mainly theoretical implications that are used as a reference for the study of real adaptive algorithms (Fig. 6.1).

6.2.1.1 Study of the Convergence

In the case that $\mu \neq 1$, it is possible to express (6.4) as a simple FDE of the type

$$\begin{aligned}\mathbf{w}_k &= \mathbf{w}_{k-1} - \mu\mathbf{R}^{-1}\mathbf{R}\mathbf{w}_{k-1} + \mu\mathbf{R}^{-1}\mathbf{g} \\ &= (1-\mu)\mathbf{w}_{k-1} + \mu\mathbf{R}^{-1}\mathbf{g}.\end{aligned} \tag{6.5}$$

The study of the convergence is immediate and can be done, as usual, considering the *weights error vector* (WEV) $\mathbf{u}_n = \mathbf{w}_n - \mathbf{w}_{\text{opt}}$ (see Sect. 5.1.2.3) for which, for $\mathbf{w}_{\text{opt}} = \mathbf{R}^{-1}\mathbf{g}$, we can write

$$\mathbf{u}_n = (1-\mu)\mathbf{u}_{n-1},$$

and substituting back up to the initial conditions (ICs), indicated as $\mathbf{u}_{-1}$, we get

$$\mathbf{u}_n = (1-\mu)^n\mathbf{u}_{-1}$$

from which we observe that

1. the algorithm for $|1-\mu| < 1$ *exactly* converges for $n \to \infty \Rightarrow \mathbf{w}_n \to \mathbf{w}_{\text{opt}}$,
2. the convergence rate depends on $1-\mu$,
3. the convergence is identical for all the filter coefficients and is independent from the $\mathbf{R}$ eigenspread,
4. is possible to demonstrate that $J_{\text{EMSE}} \approx \frac{\mu}{2}\text{tr}\{\mathbf{R}\}$.

In the Newton's algorithm the term $\mathbf{R}^{-1}$ can be interpreted as a transformation (rotation and amplification) which eliminates the problem of the eigenvalues spread.

6.2.2 The Class of Error Sequential Regression Algorithms

The class of ESR algorithms [1–3] is derived from the iterative (or adaptive) solution of the normal equations in the form of Yule–Walker. The ESR may be considered a LS-adaptive algorithm class, i.e., characterized by a deterministic choice of CF in which, to reinforce the gradient and the Hessian matrix estimate, all the available information up to the instant n are used. Unlike the LS method, in which the solution $\mathbf{w}_{\text{LS}}$ is provided by processing the entire data block, the ESR solution depends on all the available data and the filter weights vector are updated, at every instant time, with the rule (6.1), in which the (true) statistics quantities are replaced with their estimates.

6.2.2.1 Definitions and Notation

The ESR algorithms notation is similar to that introduced in the LS methodology (see Sect. 4.2.2.1) and here it is briefly recalled.

Consider a measure interval $k \in [n_1, n_2]$, that is, the convention that n_2 is equal to the last available sample $n_2 = n$, while n_1 is the first or $n_1 = 0$. The analysis window has a length equal to $N = n + 1$ samples.

For the various algorithms derivation consider a LS system, with $N > M$, for which the following definitions, of *a priori* and *a posteriori* errors, shall apply:

$$\mathbf{e}_n = \mathbf{d}_n - \mathbf{X}_n \mathbf{w}_{n-1}, \qquad \textit{a priori error} \tag{6.6}$$

$$\boldsymbol{\varepsilon}_n = \mathbf{d}_n - \mathbf{X}_n \mathbf{w}_n, \qquad \textit{a posteriori error} \tag{6.7}$$

with $\mathbf{w}_n \in (\mathbb{R},\mathbb{C})^{M\times 1}$ and data matrix $\mathbf{X}_n \in (\mathbb{R},\mathbb{C})^{N\times M}$, defined as (see Sect. 4.2.2.1)

$$\mathbf{X}_n \triangleq \begin{bmatrix} \mathbf{x}_n^T \\ \mathbf{x}_{n-1}^T \\ \vdots \\ \mathbf{x}_0^T \end{bmatrix} = \begin{bmatrix} \mathbf{x}^T[n] \\ \mathbf{x}^T[n-1] \\ \vdots \\ \mathbf{x}^T[n-M+1] \end{bmatrix}^T$$

$$= \begin{bmatrix} x[n] & x[n-1] & \cdots & x[n-M+1] \\ x[n-1] & x[n-2] & \cdots & \vdots \\ \vdots & \vdots & \ddots & x[-M-2] \\ x[0] & x[-1] & \cdots & x[-M-1] \end{bmatrix}. \quad (6.8)$$

where the elements, $x[-1]$, $x[-2]$, ..., $x[-M-1]$, represent the recurrence ICs and, unless otherwise specified, shall be considered null. The vectors in (6.6), (6.7), and (6.8) are defined as

$$\mathbf{x}_n = \begin{bmatrix} x[n] & x[n-1] & \cdots & x[n-M+1] \end{bmatrix}^T, \qquad \textit{AF filter input} \quad (6.9)$$

$$\mathbf{x}[n] = \begin{bmatrix} x[n] & x[n-1] & \cdots & x[0] \end{bmatrix}^T, \qquad \textit{analysis window} \quad (6.10)$$

$$\mathbf{e}_n = \begin{bmatrix} e[n] & e[n-1] & \cdots & e[0] \end{bmatrix}^T, \qquad \textit{a priori error} \quad (6.11)$$

$$\boldsymbol{\varepsilon}_n = \begin{bmatrix} \varepsilon[n] & \varepsilon[n-1] & \cdots & \varepsilon[0] \end{bmatrix}^T, \qquad \textit{a posteriori error} \quad (6.12)$$

$$\mathbf{d}_n = \begin{bmatrix} d[n] & d[n-1] & \cdots & d[0] \end{bmatrix}^T, \qquad \textit{desired output.} \quad (6.13)$$

With this formalism the CF $\hat{J}(\mathbf{w}_n)$ assumes, for this algorithms class, an expression of the type

$$\begin{aligned} \hat{J}(\mathbf{w}_n) &= \sum_{i=0}^{n} |e[i]|^2 = \sum_{i=0}^{n} |d[i] - \mathbf{w}_{n-1}^H \mathbf{x}_{n-i}|^2 = \mathbf{e}_n^H \mathbf{e}_n \\ &= \|\mathbf{d}_n - \mathbf{X}_n \mathbf{w}_{n-1}\|_2^2. \end{aligned} \quad (6.14)$$

Note that some algorithm classes are derived from a CF defined considering the *a posteriori* error or a combination of *a priori* and *a posteriori* errors.

6.2.2.2 Derivation of ESR Algorithms

The ESR algorithms can be derived from the Newton's method, where instead of the *a priori* known correlations, their time-average estimates are used, and from the iterative solution of weighted LS (see Sect. 4.2.5.1). In both derivations, the matrix $\mathbf{R}$ and the vector $\mathbf{g}$ are replaced with time-average estimates at time n, indicated respectively as $\mathbf{R}_{xx,n}$ and $\mathbf{R}_{xd,n}$, calculated, for example, with the expressions (4.23) and (4.24), rewritten as

$$\mathbf{R}_{xx,n} \in (\mathbb{R},\mathbb{C})^{M\times M} = \mathbf{X}_n^H\mathbf{X}_n; \qquad \mathbf{R}_{xd,n} \in (\mathbb{R},\mathbb{C})^{M\times 1} = \mathbf{X}_n^H\mathbf{d}_n. \tag{6.15}$$

In fact, in the case of ergodic processes, applies

$$\mathbf{R} \approx \frac{1}{n}\mathbf{R}_{xx}, \quad \mathbf{g} \approx \frac{1}{n}\mathbf{R}_{xd}. \tag{6.16}$$

Typically in these cases, the adaptation formula (6.4), to avoid possible matrix inversion problems, is rewritten in the Levenberg–Marquardt form (5.18), as

$$\mathbf{w}_n = \mathbf{w}_{n-1} - \mu\left[\delta\mathbf{I} + \mathbf{R}_{xx,n}\right]^{-1}(\mathbf{R}_{xx,n}\mathbf{w}_{n-1} - \mathbf{R}_{xd,n}). \tag{6.17}$$

By placing, for simplicity $\delta = 0$, the adaptation formula is

$$\begin{aligned}\mathbf{w}_n &= \mathbf{w}_{n-1} - \mu\left[\mathbf{X}_n^H\mathbf{X}_n\right]^{-1}\left(\mathbf{X}_n^H\mathbf{X}_n\mathbf{w}_{n-1} - \mathbf{X}_n^H\mathbf{d}_n\right) \\ &= (1-\mu)\mathbf{w}_{n-1} - \mu\mathbf{R}_{xx,n}^{-1}\mathbf{R}_{xd,n}.\end{aligned} \tag{6.18}$$

The above expression is formally identical to (6.5). However, $\mathbf{w}_n$ being a RV, (6.18) is a stochastic difference equation (SDE) whose solution, expressed in terms of mean and mean square, provides the basic analysis tool for the study of the algorithm characteristics and its performance.

The expression (6.18) coincides with the iterative solution weighted LS presented above and obtained through the Lyapunov's attractor [see Sect. 4.3.2.1, (4.123)].

Remark The expression (6.18) can be written as

$$\mathbf{w}_n = \mathbf{w}_{n-1} + \mu\mathbf{X}_n^{\#}\mathbf{e}_n \tag{6.19}$$

where the term $\mathbf{X}_n^{\#} = [\mathbf{X}_n^H\mathbf{X}_n]^{-1}\mathbf{X}_n^H$ is, by definition, the Moore–Penrose pseudoinverse of the data matrix $\mathbf{X}_n$.

6.2.2.3 Average Convergence Study of ESR

The average solution of (6.18) can be derived by performing the expectation of both members, for which we can write

$$E\{\mathbf{w}_n\} = (1-\mu)E\{\mathbf{w}_{n-1}\} + \mu\mathbf{w}_{\text{opt}}. \tag{6.20}$$

For $\mu \neq 1$, considering the expected WEV $E\{\mathbf{u}_n\} = E\{\mathbf{w}_n\} - \mathbf{w}_{\text{opt}}$ with the hypothesis that applies $\mathbf{w}_{\text{opt}} = \mathbf{R}^{-1}\mathbf{g}$ (optimal Wiener solution), we can write

$$E\{\mathbf{u}_n\} = (1 - \mu)E\{\mathbf{u}_{n-1}\},$$

and back substituting up to the ICs, we get

$$E\{\mathbf{u}_n\} = (1 - \mu)^n \mathbf{u}_{-1}.$$

Similarly to the exact Newton case, we see that

1. the algorithm *converges in average* for $|1 - \mu| < 1$; for $n \to \infty \Rightarrow E\{\mathbf{w}_n\} \to \mathbf{w}_{\text{opt}}$,
2. the rate of convergence depends on $1 - \mu$,
3. convergence is identical for all the filter coefficients and is independent from the eigenvalues spread of the of $\mathbf{R}_{xx}$,
4. is possible to demonstrate that $J_{\text{EMSE}} \approx \frac{\mu}{2}\text{tr}\{\mathbf{R}_{xx}\}$.

6.2.3 LMS–Newton Algorithm

Equating the expression (6.19) with the general definition (6.1), it can be observed that the product $\mathbf{X}_n^H \mathbf{e}_n$ is an estimate of the CF gradient. From (6.1), considering a more simpler gradient approximation, for example, the same used for the LMS algorithm (see Sect. 5.3.1), namely, $\nabla \hat{J}_{n-1} = -2e^*[n]\mathbf{x}_n$, the adaptation equation can be expressed as

$$\mathbf{w}_n = \mathbf{w}_{n-1} - 2\mu \mathbf{R}^{-1} e^*[n]\mathbf{x}_n, \quad \textit{LMS} - \textit{Newton algorithm} \tag{6.21}$$

known as LMS/Newton algorithm [1]. The expression (6.21) has only a theoretical value because, in general, the knowledge of the input process (true) correlation is not available. For the inverse Hessian matrix it is possible to use the estimate $\mathbf{R}_{xx,n}^{-1}$. The algorithm that results is written as

$$\mathbf{w}_n = \mathbf{w}_{n-1} - 2\mu \mathbf{R}_{xx,n}^{-1} e^*[n]\mathbf{x}_n, \quad \textit{approximate LMS} - \textit{Newton algorithm}. \tag{6.22}$$

However, note that even this solution is in practice never used as $\mathbf{R}_{xx,n}^{-1}$ should be calculated for each iteration with great computational resources expenditure. In fact, in the ESR algorithm the estimate of the inverse matrix correlation is recursively performed with the method described in the following paragraph.

Remark As for the Newton's algorithm, even in the LMS–Newton the matrix $\mathbf{R}_{xx,n}^{-1}$ performs a rotation and gain, which allows the vector $\mathbf{w}_n$ to follow a more direct way toward the CF minimum.

6.2.4 *Recursive Estimation of the Time-Average Autocorrelation*

In the methods derived from the approximate sequential regression Newton form, one of the most important aspects concerns the calculation of the time-average autocorrelation matrix $\mathbf{R}_{xx,n}$ that at n instant is calculated as

$$\mathbf{R}_{xx,n} = \sum_{k=0}^{n} \mathbf{x}_k \mathbf{x}_k^H, \qquad \text{for} \qquad n = 0, 1, \ldots. \tag{6.23}$$

For the $\mathbf{R}_{xx,n}$ determination we can proceed in a recursive way by observing that the above expression is equivalent to

$$\mathbf{R}_{xx,n} = \mathbf{R}_{xx,n-1} + \mathbf{x}_n \mathbf{x}_n^H \tag{6.24}$$

for which the correlation can be simply recursively updated with the new input vectors outer product $\mathbf{x}_n \mathbf{x}_n^H$. We shall now see how, with the *matrix inversion lemma*, it is possible to determine a recursive relationship for the direct estimation of the inverse correlation matrix.

6.2.4.1 Recursive Estimation of $\mathbf{R}_{xx,n}^{-1}$ with Matrix Inversion Lemma

The matrix inversion lemma (MIL) or Sherman–Morrison–Woodbury formula (see Sect. A.3.4) [22, 23] asserts that, given the matrices $\mathbf{A} \in \mathbb{C}^{M \times M}$, $\mathbf{B} \in \mathbb{C}^{M \times N}$, $\mathbf{C} \in \mathbb{C}^{N \times N}$, and $\mathbf{D} \in \mathbb{C}^{N \times M}$, if $\mathbf{A}^{-1}$ and $\mathbf{C}^{-1}$ exist, the following equality is algebraically verified:

$$[\mathbf{A} + \mathbf{BCD}]^{-1} = \mathbf{A}^{-1} - \mathbf{A}^{-1}\mathbf{B}\left[\mathbf{C}^{-1} + \mathbf{D}\mathbf{A}^{-1}\mathbf{B}\right]^{-1}\mathbf{D}\mathbf{A}^{-1}. \tag{6.25}$$

A variant useful in AF is when $\mathbf{B}$ and $\mathbf{D}$ are vectors defined as $\mathbf{B} \rightarrow \mathbf{x} \in \mathbb{C}^{M \times 1}$, $\mathbf{D} \rightarrow \mathbf{x}^H \in \mathbb{C}^{1 \times M}$, and $\mathbf{C} = \mathbf{I}$, for which (6.25) can be written as

$$\left(\mathbf{A} + \mathbf{x}\mathbf{x}^H\right)^{-1} = \mathbf{A}^{-1} - \frac{\mathbf{A}^{-1}\mathbf{x}\mathbf{x}^H\mathbf{A}^{-1}}{1 + \mathbf{x}^H\mathbf{A}^{-1}\mathbf{x}}. \tag{6.26}$$

Denote the inverse of the correlation matrix with $\mathbf{P}_n$, for which ($\mathbf{P}_n \triangleq \mathbf{R}_{xx,n}^{-1}$), and applying the MIL to the $\mathbf{P}_n$ matrix, by (6.24) and (6.26), we get

$$\mathbf{P}_n = \left(\mathbf{R}_{xx,n-1} + \mathbf{x}_n \mathbf{x}_n^H\right)^{-1} = \mathbf{P}_{n-1} - \frac{1}{\alpha_n}\mathbf{P}_{n-1}\mathbf{x}_n\mathbf{x}_n^H\mathbf{P}_{n-1} \tag{6.27}$$

where $\alpha_n = 1 + \mathbf{x}_n^H \mathbf{P}_{n-1} \mathbf{x}_n$. Note that the $\mathbf{P}_n$ estimate does not require matrix inversions since α is a scalar. The complexity of the MIL formula is M^2 rather than M^3 of the direct matrix inversion.

6.2.4.2 Sequential Regression Algorithm with MIL

The ESR algorithm that derives from MIL, originally developed in [2], can be summarized in the following way:

(i) Initialization $\mathbf{w}_{-1}$, $\mathbf{P}_{-1} = \delta^{-1}\mathbf{I}$
(ii) For $n = 0,1, \ldots\{$

$$\begin{aligned} \mathbf{P}_n &= \mathbf{P}_{n-1} - \frac{\mathbf{P}_{n-1}\mathbf{x}_n\mathbf{x}_n^H\mathbf{P}_{n-1}}{1 + \mathbf{x}_n^H\mathbf{P}_{n-1}\mathbf{x}_n} \\ e[n] &= d[n] - \mathbf{w}_{n-1}^H\mathbf{x}_n \\ \mathbf{w}_n &= \mathbf{w}_{n-1} + \mu\mathbf{P}_n\mathbf{x}_n e^*[n]. \end{aligned} \tag{6.28}$$

}

In practice, the algorithm is formally identical to the LMS, described by (5.103) (see Sect. 5.3.3), in which the $\mathbf{P}_n$ weighing matrix is inserted to recursively estimate the inverse Hessian. In other words, the weighing $\mathbf{P}_n$ performs a transformation that tends to eliminate the problem of the spread of the eigenvalues of the correlation $\mathbf{R}$.

6.2.4.3 Algorithm Initialization

The choice of the initial value of the correlation is $\mathbf{P}_{-1} = \delta^{-1}\mathbf{I}$, with a small positive constant ($\delta \sim 10^{-1} - 10^{-4}$), or explicitly pre-calculating the $\mathbf{P}_{-1}$ from the first signal window and starting the iteration (ii). Note that the IC value affects the bias of the correlation matrix estimate.

6.3 Affine Projection Algorithms

The NLMS algorithm (see Sect. 5.5.1), due to its implementative simplicity and low computational cost, is widely used in the filters adaptation. It is known, however, that colored input signals can appreciably deteriorate its convergence speed [1].

Introduced in 1984 in [4, 5, 24], the algorithms class called affine projection algorithms (APA) is a NLMS generalization which improves its performance in the case of colored and correlated inputs. In the literature there are numerous APA versions and, in the following, we will refer to this type of algorithms as class of APA.

The NLMS can be seen as a one-dimensional affine projection. The APA adapts the filter, assuming length M, considering multiple projections in a subspace of

$K < M$ dimension. Increasing the projections order K there is an increase of the convergence speed but, unfortunately, also results in an increase of computational complexity. In practice, in the NLMS, the weights are adapted taking into account the only current input, i.e., $K = 1$, while APA updates the weights considering the earlier K input–output pairs.

Remark The APA is not an exact second-order algorithm, as in the adaptation it is used as an estimate of the correlation matrix $\mathbf{R}_{xx}$ projected onto a subspace of appropriate dimension. In the one-dimensional case the algorithm takes the form of the NLMS.

For the derivation of the APA consider a LS system, whereas the window index $k \in [n_1, n_2]$ is defined over the extremes $n_2 = n$ and $n_1 = n - K + 1$, i.e., consider only the last K sequence samples for which the data matrix $\mathbf{X}_n \in (\mathbb{R},\mathbb{C})^{K \times M}$ is defined as

$$\mathbf{X}_n \triangleq \begin{bmatrix} \mathbf{x}_n^T \\ \mathbf{x}_{n-1}^T \\ \vdots \\ \mathbf{x}_{n-K+1}^T \end{bmatrix} = \begin{bmatrix} \mathbf{x}^T[n] \\ \mathbf{x}^T[n-1] \\ \vdots \\ \mathbf{x}^T[n-M+1] \end{bmatrix}^T$$
$$= \begin{bmatrix} x[n] & x[n-1] & \cdots & x[n-M+1] \\ x[n-1] & x[n-2] & \cdots & x[n-M] \\ \vdots & \vdots & \ddots & \vdots \\ x[n-K+1] & x[n-K] & \cdots & x[n-K-M+2] \end{bmatrix}. \qquad (6.29)$$

Therefore, apply the definitions of the vectors (6.10), (6.11), (6.12), and (6.13) in which the lower bound index is not zero and assumes the value equal to $n - K + 1$.

Remark For $K < M$ the LS system is underdetermined, and the index K defines the number of projections, or the number of signal-reference pairs, for the K-order APA calculation.

6.3.1 APA Derivation Through Minimum Perturbation Property

The APA class methods can be derived from the properties of minimal perturbation already discussed above with the consideration, which for any adaptive algorithm, at convergence, apply the properties (i)–(iii) described in Sect. 5.1.3.2.

Considering the *a priori* error vector $\mathbf{e}_n$ and the *a posteriori* error vector $\boldsymbol{\varepsilon}_n$, defined in (6.6) and (6.7), the property (i) is, in this context, rewritten as $|\boldsymbol{\varepsilon}_n| < |\mathbf{e}_n|$ for which the property (ii) may be generalized as

$$\boldsymbol{\varepsilon}_n = (\mathbf{I} - \boldsymbol{\alpha})\mathbf{e}_n \tag{6.30}$$

where $(\mathbf{I} - \boldsymbol{\alpha}) < \mathbf{I}$. By defining the quantity

$$\delta\mathbf{w}_n = \mathbf{w}_n - \mathbf{w}_{n-1} \tag{6.31}$$

such that the CF $J(\mathbf{w}) = \|\delta\mathbf{w}\|_2^2$ represents the minimal perturbation property (ii) (i.e., near the optimum point, the weights do not change during the adaptation), the APA can be defined as a *constrained exact local minimization problem.* In practice, it is formulated as a constrained optimization problem of the type

$$\mathbf{w}_{\text{opt}} \therefore \underset{\mathbf{w}}{\operatorname{argmin}} \|\delta\mathbf{w}\|_2^2 \quad \text{s.t.} \quad \boldsymbol{\varepsilon}_n = (\mathbf{I} - \boldsymbol{\alpha})\mathbf{e}_n. \tag{6.32}$$

Multiplying the left-hand side of (6.31) for data matrix $\mathbf{X}_n$, we can express the constraint (6.30) in the following form:

$$\begin{aligned} \mathbf{X}_n\delta\mathbf{w} &= \mathbf{X}_n\mathbf{w}_n - \mathbf{X}_n\mathbf{w}_{n-1} = (\mathbf{X}_n\mathbf{w}_n - \mathbf{d}_n) - (\mathbf{d} - \mathbf{X}_n\mathbf{w}_{n-1}) \\ &= -\boldsymbol{\varepsilon}_n + \mathbf{e}_n \\ &= \boldsymbol{\alpha}\mathbf{e}_n. \end{aligned} \tag{6.33}$$

From the above, the constraint in (6.32) can be expressed as a function of only the *a priori* error. Therefore, the optimization problem (6.32) becomes

$$\mathbf{w}_{\text{opt}} \therefore \underset{\mathbf{w}}{\operatorname{argmin}} \|\delta\mathbf{w}\|_2^2 \quad \text{s.t.} \quad \mathbf{X}_n\delta\mathbf{w}_n = \boldsymbol{\alpha}\mathbf{e}_n. \tag{6.34}$$

Given the simplicity of the formulation, the adaptation equation can be directly obtained by solving the system relative to the constraint, $\mathbf{X}_n\delta\mathbf{w}_n = \boldsymbol{\alpha}\mathbf{e}_n$; it is then

$$\delta\mathbf{w}_n = \mathbf{X}_n^{\#}\boldsymbol{\alpha}\mathbf{e}_n. \tag{6.35}$$

Note that in (6.35) it is $K < M$ and, therefore, the expression represents a linear underdetermined system. It follows that for the pseudoinverse matrix definition, explaining the term $\delta\mathbf{w}_n$, we can write that

$$\mathbf{w}_n - \mathbf{w}_{n-1} = \mathbf{X}_n^H\left(\mathbf{X}_n\mathbf{X}_n^H\right)^{-1}\boldsymbol{\alpha}\mathbf{e}_n \tag{6.36}$$

so by inserting the adaptation constant μ such that $\boldsymbol{\alpha} = \operatorname{diag}(\mu)$, and by setting the regularization parameter δ, the standard APA updating formula appears to be

$$\mathbf{w}_n = \mathbf{w}_{n-1} + \mu\mathbf{X}_n^H\left[\delta\mathbf{I} + \mathbf{X}_n\mathbf{X}_n^H\right]^{-1}\mathbf{e}_n \tag{6.37}$$

Note that for $K = 1$ the former becomes

Table 6.1 Estimation of the computational cost of the real-domain APA

Term	Multiplications	Sums
$\mathbf{d}_n - \mathbf{X}_n\mathbf{w}_{n-1}$	KM	$K(M-1)+K$
$\delta\mathbf{I}+\mathbf{X}_n\mathbf{X}_n^T$	K^2M	$K^2(M-1)+K$
$(\delta\mathbf{I}+\mathbf{X}_n\mathbf{X}_n^T)^{-1}$	K^3	K^3
$(\delta\mathbf{I}+\mathbf{X}_n\mathbf{X}_n^T)^{-1}(\mathbf{d}_n - \mathbf{X}_n\mathbf{w}_{n-1})$	K^2	$K(K-1)$
$\mathbf{X}_n^T(\delta\mathbf{I}+\mathbf{X}_n\mathbf{X}_n^T)^{-1}(\mathbf{d}_n - \mathbf{X}_n\mathbf{w}_{n-1})$	KM	$(K-1)M$
$\mathbf{w}_n = \mathbf{w}_{n-1}+\mu(....)$	M	M
Total per iteration	$(K^2+2K+1)M+K^3+K$	$(K^2+2K)M+K^3+K^2$

$$\mathbf{w}_n = \mathbf{w}_{n-1} + \mu\frac{\mathbf{x}_n}{\delta + \|\mathbf{x}_n\|_2^2}e^*[n] \tag{6.38}$$

which coincides with the NLMS algorithm described in the previous chapter.

6.3.1.1 APA Derivation as Approximate Newton's Method

It is possible to derive the APA class, directly considering the iterative LS solution (6.18). By entering the regularization parameter (6.18) is

$$\mathbf{w}_n = \mathbf{w}_{n-1} + \mu\left[\delta\mathbf{I} + \mathbf{X}_n^H\mathbf{X}_n\right]^{-1}\mathbf{X}_n^H(\mathbf{d}_n - \mathbf{X}_n\mathbf{w}_{n-1}) \tag{6.39}$$

For $\delta > 0$, considering the matrix equality (algebraically provable) (see 4.116),

$$\left[\delta\mathbf{I} + \mathbf{X}_n^H\mathbf{X}_n\right]^{-1}\mathbf{X}_n^H = \mathbf{X}_n^H\left[\delta\mathbf{I} + \mathbf{X}_n\mathbf{X}_n^H\right]^{-1}, \tag{6.40}$$

the standard APA is directly formulated as

$$\mathbf{w}_n = \mathbf{w}_{n-1} + \mu\mathbf{X}_n^H\left[\delta\mathbf{I} + \mathbf{X}_n\mathbf{X}_n^H\right]^{-1}(\mathbf{d}_n - \mathbf{X}_n\mathbf{w}_{n-1}). \tag{6.41}$$

6.3.2 *Computational Complexity of APA*

The matrix $\mathbf{X}_n\mathbf{X}_n^H$ has dimension $(K \times K)$ for which the complexity of its inversion depends on the depth of the projection. In fact, the parameter K defines just the number of projections.

In Table 6.1 is shown an estimate of the complexity, for the real signal case, in which the inversion of the $(K \times K)$ symmetric matrix has a cost of $O(K^2)$ operations. As a result, the overall computational cost of the APA is equal to $O(K^2M)$ operations per iteration. In the complex case, considering four real multiplications for each complex multiplication and two real sums for a complex sum, the number of operations per iteration is $4(K^2+2K+1)M+4K^3+4K$ multiplications and $4(K^2+2K)M+4K^3+2K^2$ sums.

Table 6.2 The APA family with positive $\{K,\delta,\alpha,D\}$

Algorithm	K	δ	α	D
APA	$K \leq M$	$\delta = 0$	$\alpha = 0$	$D = 1$
BNDR-LMS	$K = 2$	$\delta = 0$	$\alpha = 0$	$D = 1$
R-APA	$K \leq M$	$\delta \neq 0$	$\alpha = 0$	$D = 1$
PRA	$K \leq M$	$\delta \neq 0$	$\alpha = 1$	$D = 1$
NMLS-OCF	$K \leq M$	$\delta = 0$	$\alpha = 0$	$D \geq 1$

6.3.3 The APA Class

In the literature numerous APA variants have been developed. To take account of some of them, as reported in [4], (6.41) can be rewritten in more general form as

$$\mathbf{w}_n = \mathbf{w}_{n-1-\alpha(K-1)} + \mu \mathbf{X}_n^H \left(\delta \mathbf{I} + \mathbf{X}_n \mathbf{X}_n^H\right)^{-1} \mathbf{e}_n \tag{6.42}$$

where the vectors and matrices that appear are redefined as

$$\begin{aligned}
\mathbf{X}_n &\triangleq \begin{bmatrix} x[n] & x[n-D] & \cdots & x[n-(M+1)D] \\ x[n-D] & x[n-2D+1] & \cdots & x[n-MD] \\ \vdots & \vdots & \iddots & \vdots \\ x[n-(K+1)D] & x[n-KD] & \cdots & x[n-(K-M+2)D] \end{bmatrix}, \\
\mathbf{e}_n &\triangleq \begin{bmatrix} e[n] & e[n-D] & \cdots & e[n-(K-1)D] \end{bmatrix}^T, \\
\mathbf{d}_n &\triangleq \begin{bmatrix} d[n] & d[n-D] & \cdots & d[n-(K-1)D] \end{bmatrix}^H.
\end{aligned} \tag{6.43}$$

The step size is such that $0 < \mu < 2$; the index D, in (6.43), is defined as the *delay input vector*, which takes into account the temporal depth with which past inputs samples should be considered. In practice, different choices of the parameters $\{K,\delta,\alpha,D\}$ in (6.42) define a specific APA. For example, for $\delta = 0$, $\alpha = 0$, and $D = 1$, we obtain the standard APA described by (6.42).

The APA family is particularly suitable in the acoustic echo cancelation problems, where the filter can reach the size of thousands of coefficients, as it has better performance than the NLMS: (1) in case $K \ll M$, namely with temporal depth much less than the length of the impulse response; (2) as already indicated above, in the case of colored inputs. Note, also, that in (6.41) [or in (6.42)] the size of the matrix to be inverted is equal to K and that this index can be chosen compatibly with the available computing resources.

Among the most common APA variants, with reference to Table 6.2, we can cite: the regularized APA (R-APA); the partial rank algorithm (PRA); the decorrelating algorithm (DA); NLMS with the orthogonal correction factor or orthogonal correction factors (NLMS-OCF), the fast APA, etc. [5]. For example in the PRA, to reduce the average computational cost, the filter coefficients updating is performed every K samples. In the case of particularly colored inputs PRA has lower performance than APA, while in the case where the input is a speech signal, the performance is quite similar. The main disadvantage of the PRA consists in the fact that, although the computational average has lower cost, peak is unchanged and the processor speed is calculated on the peak. In the algorithm NLMS-OCF the update formula is

$$\mathbf{w}_n = \mathbf{w}_{n-1} + \mu_0 \mathbf{x}_n + \mu_1 \mathbf{x}_n^1 + \cdots + \mu_K \mathbf{x}_n^K \tag{6.44}$$

where $\mathbf{x}_n$ is the input at time n, and $\mathbf{x}_n^i$, for $i = 1, 2, \ldots, K$, is the orthogonal component relative to the delayed inputs, where D is the delay between the input vectors used in the adaptation. The term μ_i, for $i = 0, 1, \ldots, K$, is chosen as

$$\mu_i = \begin{cases} \mu e_n^* / \mathbf{x}_n^H \mathbf{x}_n & \text{for } n = 0, \qquad \text{if } |\mathbf{x}_n| \neq 0 \\ \mu e_n^{i*} / \mathbf{x}_n^{iH} \mathbf{x}_n^i & \text{for } \; i = 1, 2, \ldots, K, \quad \text{if } |\mathbf{x}_n^i| \neq 0 \\ 0 & \text{otherwise.} \end{cases} \tag{6.45}$$

In order to further reduce the computational cost, and to avoid matrix inversion, the matrix inversion lemma (6.27) can be used. Although in the case APA the size of the matrix is equal to the size of the projections $K \ll M$ the computational advantage is evident.

6.4 The Recursive Least Squares

Known in the literature as a recursive least squares (RLS), the algorithm differs from the previously described ESR, for the correlation matrix which is estimated considering a certain *forgetting factor*. In this way, in the case of time-varying processes, the estimation of the correlation is improved by considering more the recent data samples available.

6.4.1 Derivation of the RLS Method

The CF $\hat{J}_n(\mathbf{w})$ for this algorithm class has an expression of the type

$$\begin{aligned} \hat{J}_n(\mathbf{w}) &= \sum_{i=0}^{n} \lambda^{n-i} \left| e[i] \right|^2 \\ &= \sum_{i=0}^{n} \lambda^{n-i} \left| d[i] - \mathbf{w}_{n-1}^H \mathbf{x}_n \right|^2 \end{aligned} \tag{6.46}$$

in which the constant $0 \ll \lambda \leq 1$, defined as *forgetting factor* which, with typical trend illustrated in Fig. 6.2, takes into account of the algorithm memory. In other words, the CF depends on both the instantaneous error and the past errors value with

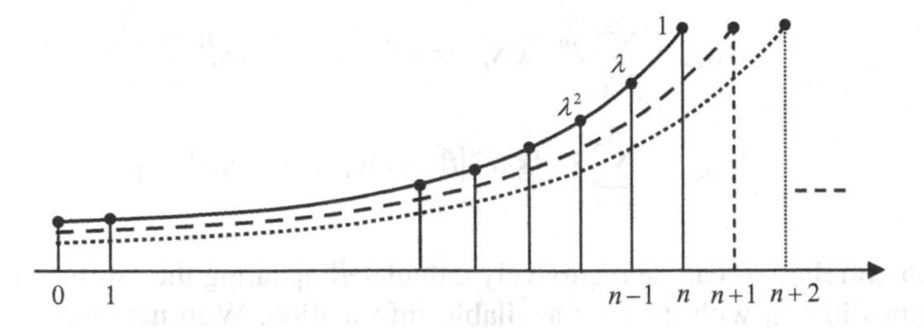

Fig. 6.2 Typical trend of the forgetting factor

weight, over time, more and more small. It is noted that for $\lambda = 1$, the past errors are taken into account with the same weight. In this case the algorithm is said *growing memory* RLS.

Considering the ESR notation Sect. 6.2.2.1, (6.46) can be written as

$$\hat{J}_n(\mathbf{w}) = \mathbf{e}_n^H \mathbf{\Lambda}_n \mathbf{e}_n = \mathbf{\Lambda}_n \|\mathbf{d}_n - \mathbf{X}_n \mathbf{w}_n\|_2^2. \tag{6.47}$$

Note that the above expression corresponds to the weighed LS, indicated as $\hat{J}_n(\mathbf{w}) = \|\mathbf{d}_n - \mathbf{X}_n \mathbf{w}_n\|_{\mathbf{\Lambda}}^2$ (see Sect. 4.2.5.1), with the weighing matrix $\mathbf{\Lambda}_n$ defined as

$$\mathbf{\Lambda}_n = \begin{bmatrix} 1 & 0 & \cdots & 0 \\ 0 & \ddots & \cdots & 0 \\ \vdots & \vdots & \lambda^{n-1} & \vdots \\ 0 & 0 & \cdots & \lambda^n \end{bmatrix}. \tag{6.48}$$

For the method development, we can refer to the weighed LS with weighing matrix $\mathbf{\Lambda}_n$, for which the normal equations at instant n, called in this case *regression equations*, take the form

$$\mathbf{X}_n^H \mathbf{\Lambda}_n \mathbf{X}_n \mathbf{w}_n = \mathbf{X}_n^H \mathbf{\Lambda}_n \mathbf{d}_n. \tag{6.49}$$

6.4.2 Recursive Class of the Correlation Matrix with Forgetting Factor and Kalman Gain

Indicating the correlations estimates, performed with weighted temporal averages, as

$$\mathbf{R}_{xx,n} = \mathbf{X}_n^H \mathbf{\Lambda}_n \mathbf{X}_n \quad \text{and} \quad \mathbf{R}_{xd,n} = \mathbf{X}_n^H \mathbf{\Lambda}_n \mathbf{d}_n \tag{6.50}$$

from the data matrix $\mathbf{X}_n$ definition in (6.8), for each instant n we observe that the time averages for the correlations estimates (6.50) can be written as

$$\mathbf{R}_{xx,n} = \sum_{i=0}^{n} \lambda^{n-i} \mathbf{x}_i \mathbf{x}_i^H = \lambda \mathbf{R}_{xx,n-1} + \mathbf{x}_n \mathbf{x}_n^H \tag{6.51}$$

$$\mathbf{R}_{xd,n} = \sum_{i=0}^{n} \lambda^{n-i} \mathbf{x}_i d^*[i] = \lambda \mathbf{R}_{xd,n-1} + \mathbf{x}_n d^*[n] \tag{6.52}$$

for which correlations can be recursively calculated updating the estimate made at the previous instant with the new available information. With notation similar to LS, we can write the solution of the sequential regression (6.49), at nth instant, as

$$\mathbf{R}_{xx,n} \mathbf{w}_n = \mathbf{R}_{xd,n}. \tag{6.53}$$

Applying the MIL (6.27) to the matrix (6.51) (with $\mathbf{P}_n \triangleq \mathbf{R}_{xx,n}^{-1}$), we get

$$\mathbf{P}_n = \lambda^{-1} \mathbf{P}_{n-1} - \frac{\lambda^{-1} \mathbf{P}_{n-1} \mathbf{x}_n \lambda^{-1} \mathbf{x}_n^H \mathbf{P}_{n-1}}{1 + \lambda^{-1} \mathbf{x}_n^H \mathbf{P}_{n-1} \mathbf{x}_n} \tag{6.54}$$

in which for computational convenience, it is usual to define the vector

$$\mathbf{k}_n = \frac{\lambda^{-1} \mathbf{P}_{n-1} \mathbf{x}_n}{1 + \lambda^{-1} \mathbf{x}_n^H \mathbf{P}_{n-1} \mathbf{x}_n} \tag{6.55}$$

for which the recurrence (6.54) can be rewritten as

$$\mathbf{P}_n = \lambda^{-1} \mathbf{P}_{n-1} - \lambda^{-1} \mathbf{k}_n \mathbf{x}_n^H \mathbf{P}_{n-1} \tag{6.56}$$

known as the *Riccati equation*.

Remark Note that the expression (6.55) can be written as

$$\begin{aligned} \mathbf{k}_n &= \lambda^{-1} \mathbf{P}_{n-1} \mathbf{x}_n - \lambda^{-1} \mathbf{k}_n \mathbf{x}_n^H \mathbf{P}_{n-1} \mathbf{x}_n \\ &= \left[\lambda^{-1} \mathbf{P}_{n-1} - \lambda^{-1} \mathbf{k}_n \mathbf{x}_n^H \mathbf{P}_{n-1} \right] \mathbf{x}_n \end{aligned} \tag{6.57}$$

where the part in brackets for (6.56) is equal to $\mathbf{P}_n$. Then the vector $\mathbf{k}_n$ can be defined in an equivalent way as

$$\mathbf{k}_n = \mathbf{P}_n \mathbf{x}_n \tag{6.58}$$

and, given the Toeplitz nature of $\mathbf{P}_n$ matrix, is also true that $\mathbf{k}_n^H = \mathbf{x}_n^H \mathbf{P}_n$. In other words, $\mathbf{k}_n$ is defined as the input vector transformed by the inverse correlation matrix $\mathbf{R}_{xx,n}^{-1}$. The vector $\mathbf{k}_n$ is said *Kalman gain vector*.

6.4.3 *RLS Update with* A Priori *and* A Posteriori *Error*

The solution of the regression equations (6.53) can be carried out through the so-called *a priori* or *a posteriori* formulation, considering the errors definition (see Sect. 5.1.1.2):

$$e[n] = d[n] - \mathbf{w}_{n-1}^H \mathbf{x}_n, \qquad a\ priori\ error \tag{6.59}$$

$$\varepsilon[n] = d[n] - \mathbf{w}_n^H \mathbf{x}_n, \qquad a\ posteriori\ error \tag{6.60}$$

depending on whether the error calculation is made with the old filter coefficients or with the current ones.

6.4.3.1 Weights Update with *A Priori* Error

In the *a priori* update, we consider the normal equations at instant $n-1$. In this case the adaptation takes the form

$$\mathbf{R}_{xx,n-1}\mathbf{w}_{n-1} = \mathbf{R}_{xd,n-1}. \tag{6.61}$$

Substituting (6.51) and (6.52) into (6.61), we write

$$\left[\mathbf{R}_{xx,n} - \mathbf{x}_n\mathbf{x}_n^H\right]\mathbf{w}_{n-1} = \mathbf{R}_{xd,n} - \mathbf{x}_n d^*[n] \tag{6.62}$$

from which it follows that

$$\mathbf{R}_{xx,n}\mathbf{w}_{n-1} + \mathbf{x}_n e^*[n] = \mathbf{R}_{xd,n} \tag{6.63}$$

where the error is *a priori* calculated by (6.59).

Multiplying both members of (6.63) by $\mathbf{P}_n$, where by definition $\mathbf{w}_n = \mathbf{P}_n\mathbf{R}_{xd,n}$, we can write

$$\mathbf{w}_n = \mathbf{w}_{n-1} + \mathbf{P}_n\mathbf{x}_n e^*[n] \tag{6.64}$$

which basically coincides with the LMS error sequential regression algorithm with MIL (6.28) (see Sect. 6.2.4.2). Considering the Kalman gain vector in (6.58), the update formula (6.64) is rewritten as

$$\mathbf{w}_n = \mathbf{w}_{n-1} + \mathbf{k}_n e^*[n]. \tag{6.65}$$

6.4.3.2 Weights Update with *A Posteriori* Error

For the *a posteriori* update, the normal equations are solved at the time n, as

$$\mathbf{R}_{xx,n}\mathbf{w}_n = \mathbf{R}_{xd,n}. \tag{6.66}$$

Substituting (6.51) and (6.52) into (6.66), we get

$$\left[\lambda\mathbf{R}_{xx,n-1} + \mathbf{x}_n\mathbf{x}_n^H\right]\mathbf{w}_n = \left[\lambda\mathbf{R}_{xd,n-1} + \mathbf{x}_n d^*[n]\right] \tag{6.67}$$

so by the definition of *a posteriori* error (6.60), (6.67) can be written as

$$\lambda\mathbf{R}_{xx,n-1}\mathbf{w}_n - \mathbf{x}_n\varepsilon^*[n] = \lambda\mathbf{R}_{xd,n-1}. \tag{6.68}$$

Multiplying both sides of (6.67) for $\mathbf{P}_{n-1}$, with $\mathbf{w}_{n-1} = \mathbf{P}_{n-1}\mathbf{R}_{xd,n-1}$, we obtain

$$\mathbf{w}_n = \mathbf{w}_{n-1} + \lambda^{-1}\mathbf{P}_{n-1}\mathbf{x}_n\varepsilon^*[n]. \tag{6.69}$$

Note that the above expression is noncausal since the vector $\mathbf{w}_n$ depends on the $\varepsilon[n]$ which also depends on $\mathbf{w}_n$, namely $\varepsilon[n]$ represents the error related to the future sample. Moreover, similarly to what was done previously, if we define the *alternative vector gain* or *alternative Kalman vector gain*, the vector $\tilde{\mathbf{k}}_n$ such that

$$\tilde{\mathbf{k}}_n = \lambda^{-1}\mathbf{P}_{n-1}\mathbf{x}_n \quad \text{or} \quad \tilde{\mathbf{k}}_n^H = \lambda^{-1}\mathbf{x}_n^H\mathbf{P}_{n-1} \tag{6.70}$$

we can write

$$\mathbf{w}_n = \mathbf{w}_{n-1} + \tilde{\mathbf{k}}_n\varepsilon^*[n]. \tag{6.71}$$

Remark Substituting the latter in (6.60), we get

$$\begin{aligned}\varepsilon[n] &= d[n] - \left(\mathbf{w}_{n-1}^H + \varepsilon[n]\tilde{\mathbf{k}}_n^H\right)\mathbf{x}_n \\ &= e[n] - \varepsilon[n]\tilde{\mathbf{k}}_n^H\mathbf{x}_n.\end{aligned}$$

By defining the conversion factor $\tilde{\alpha}_n$, as

$$\tilde{\alpha}_n \triangleq 1 + \lambda^{-1}\mathbf{x}_n^H\mathbf{P}_{n-1}\mathbf{x}_n = 1 + \tilde{\mathbf{k}}_n^H\mathbf{x}_n \tag{6.72}$$

which coincides with the denominator of (6.55), now we can relate the *a priori* and the *a posteriori* error energy with a simple relationship of the type

$$\varepsilon[n] = \frac{e[n]}{\tilde{\alpha}_n}. \tag{6.73}$$

The error $\varepsilon[n]$ calculation can be estimated by (6.72) and (6.73) before updating the filter weights with (6.69). This mode is causal and allows calculating the adaptive LS with *a posteriori* error. Furthermore, since $\mathbf{P}_{n-1}$ is, by definition, positive definite, it follows that the conversion factor is $\tilde{\alpha}_n < 1$, for which it appears that $|\varepsilon[n]| < |e[n]|$ for every n, i.e.,

$$\sum_n |\varepsilon[n]|^2 < \sum_n |e[n]|^2 \tag{6.74}$$

Note that the latter result is consistent with the general minimal perturbation properties previously described (see Sect. 5.1.3.2).

Moreover, combining (6.65), (6.71), and (6.73) we have that

$$\mathbf{k}_n = \frac{\tilde{\mathbf{k}}_n}{\tilde{\alpha}_n} \tag{6.75}$$

for which the gain vectors $\mathbf{k}_n$ and $\tilde{\mathbf{k}}_n$ have the same direction but different lengths. In addition, it is easy to show that the following relation holds: $\mathbf{k}_n\varepsilon^*[n] = \tilde{\mathbf{k}}_n e^*[n]$.

Remark From previous expressions we can see that the *adaptive gain vector* is a function of the input signal, while the desired output changes only the amplitude and the sign of the filter coefficients correction.

You can define a different conversion factor, called *likelihood variable*, as

$$\alpha_n \triangleq 1 - \mathbf{x}_n^H\mathbf{P}_n\mathbf{x}_n = 1 - \mathbf{k}_n^H\mathbf{x}_n \tag{6.76}$$

so from (6.75) is $\alpha_n = 1 - \mathbf{x}_n^H\tilde{\mathbf{k}}_n/\tilde{\alpha}_n$ and, with simple steps, is $\alpha_n = 1/\tilde{\alpha}_n$; moreover, given that by definition $\mathbf{x}_n^H\mathbf{P}_n\mathbf{x}_n \geq 0$, (6.76) implies that

$$0 < \alpha_n \leq 1. \tag{6.77}$$

It is also demonstrated, see [6] for details, that $\alpha_n = \lambda^M[\det(\mathbf{R}_{n-1})/\det(\mathbf{R}_n)]$.

Table 6.3 shows a RLS algorithm summary with *a priori* and *a posteriori* formulation.

6.4.4 Conventional RLS Algorithm

From the previous development we have seen that the most expensive part for calculating the RLS consists in the Kalman vectors gain determination $\mathbf{k}_n = \mathbf{P}_n\mathbf{x}_n$ or its alternative form $\tilde{\mathbf{k}}_n = \lambda^{-1}\mathbf{P}_{n-1}\mathbf{x}_n$. In fact, by the previous definition of the Kalman gain, the Riccati equation (6.56) can be expressed in several equivalent forms. Taking also into account the Toeplitz nature of the $\mathbf{P}_n$ matrix, (6.54) calculated at the index n is then[1]

[1] Recall that, for the symmetrical nature of the matrix $\mathbf{P}_n$, it holds that $[\mathbf{P}_{n-1}\mathbf{x}_n]^H = \mathbf{x}_n^H\mathbf{P}_{n-1}$.

Table 6.3 Summary of the RLS algorithms

RLS	*A priori* update	*A posteriori* update
Correl. matrix estimate	$\mathbf{R}_{xx,n} = \lambda\mathbf{R}_{xx,n-1} + \mathbf{x}_n\mathbf{x}_n^H$	$\mathbf{R}_{xx,n} = \lambda\mathbf{R}_{xx,n-1} + \mathbf{x}_n\mathbf{x}_n^H$
Kalman gain	$\mathbf{k}_n = \mathbf{P}_n\mathbf{x}_n$	$\tilde{\mathbf{k}}_n = \lambda^{-1}\mathbf{P}_{n-1}\mathbf{x}_n$
A priori error	$e[n] = d[n] - \mathbf{w}_{n-1}^H\mathbf{x}_n$	$\varepsilon[n] = d[n] - \mathbf{w}_n^H\mathbf{x}_n$
Conversion factor	$\alpha_n = 1 - \mathbf{k}_n^H\mathbf{x}_n$	$\tilde{\alpha}_n = 1 + \tilde{\mathbf{k}}_n^H\mathbf{x}_n$
A posteriori error	$\varepsilon[n] = \alpha_n e[n]$	$\varepsilon[n] = \tilde{\alpha}_n^{-1}e[n]$
Coefficients update	$\mathbf{w}_n = \mathbf{w}_{n-1} + \mathbf{k}_n e^*[n]$	$\mathbf{w}_n = \mathbf{w}_{n-1} + \tilde{\mathbf{k}}_n\varepsilon^*[n]$

$$
\begin{aligned}
\mathbf{P}_n &= \lambda^{-1}\mathbf{P}_{n-1} - \frac{1}{\tilde{\alpha}_n}\tilde{\mathbf{k}}_n\tilde{\mathbf{k}}_n^H \\
&= \lambda^{-1}\mathbf{P}_{n-1} - \mathbf{k}_n\tilde{\mathbf{k}}_n^H \\
&= \lambda^{-1}\left[\mathbf{I} - \mathbf{k}_n\mathbf{x}_n^H\right]\mathbf{P}_{n-1}
\end{aligned}
\tag{6.78}
$$

where $\tilde{\alpha}_n = 1 + \lambda^{-1}\mathbf{x}_n^H\mathbf{P}_{n-1}\mathbf{x}_n$, $\tilde{\mathbf{k}}_n = \lambda^{-1}\mathbf{P}_{n-1}\mathbf{x}_n$, $\tilde{\mathbf{k}}_n^H = \lambda^{-1}\mathbf{x}_n^H\mathbf{P}_{n-1}$, and $\mathbf{k}_n = \mathbf{P}_n\mathbf{x}_n$.

The algorithm that derives from (6.78) is said recursive LS or *Conventional RLS* (CRLS) or, simply RLS, and is characterized by the following equations:

$$
\begin{aligned}
&\tilde{\mathbf{k}}_n = \lambda^{-1}\mathbf{P}_{n-1}\mathbf{x}_n, && \textit{a priori Kalman gain o whitening,} \\
&\tilde{\alpha}_n = 1 + \tilde{\mathbf{k}}_n^H\mathbf{x}_n, && \textit{convention factor,} \\
&\mathbf{k}_n = \tilde{\alpha}_n^{-1}\tilde{\mathbf{k}}_n, && \textit{a posteriori Kalman gain,} \\
&\mathbf{P}_n = \lambda^{-1}\mathbf{P}_{n-1} - \mathbf{k}_n\tilde{\mathbf{k}}_n^H, && \textit{Riccati equation.}
\end{aligned}
$$

For the output and error calculation, and the weights update, we have that

$$
\begin{aligned}
e[n] &= d[n] - \mathbf{w}_{n-1}^H\mathbf{x}, && \textit{filtering and a priori error,} \\
\mathbf{w}_n &= \mathbf{w}_{n-1} + \mathbf{k}_n e^*[n], && \textit{filter weights update.}
\end{aligned}
$$

In practice, the CRLS algorithm can be written, as shown below, by introducing small changes to save some multiplications (for the parameter λ).

6.4.4.1 Summary of CRLS Algorithm

(i) Initialization $\mathbf{w}_{-1} = \mathbf{0}$, $\mathbf{P}_{-1} = \delta^{-1}\mathbf{I}$, $y[0] = 0$ // *Conventional RLS (CRLS)*
(ii) For $n = 0,1, \ldots$ {

$$
\begin{aligned}
\hat{\mathbf{k}}_n &= \mathbf{P}_{n-1}\mathbf{x}_n \\
\hat{\alpha}_n &= \lambda + \hat{\mathbf{k}}_n^H\mathbf{x}_n
\end{aligned}
$$

$$\mathbf{k}_n = \hat{\alpha}_n^{-1}\hat{\mathbf{k}}_n$$
$$\mathbf{P}_n = \lambda^{-1}\left[\mathbf{P}_{n-1} - \mathbf{k}_n\hat{\mathbf{k}}_n^H\right]$$
$$e[n] = d[n] - y[n]$$
$$\mathbf{w}_n = \mathbf{w}_{n-1} + \mathbf{k}_n e^*[n]$$
$$y[n] = \mathbf{w}_n^H\mathbf{x}.$$

}

Remark In some texts, the RLS algorithm is called *growing memory*, for $\lambda = 1$, while for $0 \ll \lambda \leq 1$, the algorithm is called exponentially weighted RLS (EWRLS).

6.4.4.2 Alternative CRLS Formulation

To complete the above, for compatibility with other texts on the subject and for further study by the reader, the following is an alternative formulation, but exactly equivalent RLS algorithm. In this formulation it does not take into account the symmetry of the $\mathbf{P}_n$. matrix. In practice, the CRLS is reformulated as

$$\begin{aligned}
y[n] &= \mathbf{w}_{n-1}^H\mathbf{x}, && \textit{output},\\
e[n] &= d[n] - y[n], && \textit{a priori error},\\
\mathbf{k}_n &= \frac{\lambda^{-1}\mathbf{P}_{n-1}\mathbf{x}_n}{1 + \lambda^{-1}\mathbf{x}_n^H\mathbf{P}_{n-1}\mathbf{x}_n}, && \textit{gain vector},\\
\mathbf{w}_n &= \mathbf{w}_{n-1} + \mathbf{k}_n e^*[n], && \textit{filter weights update},\\
\mathbf{P}_n &= \lambda^{-1}\left(\mathbf{P}_{n-1} - \mathbf{k}_n\mathbf{x}_n^H\mathbf{P}_{n-1}\right), && \textit{Riccati equation}.
\end{aligned}$$

Regarding the CF value, it is easily demonstrated that it applies the update

$$\begin{aligned}
\hat{J}_n &= \lambda\hat{J}_{n-1} + e^*[n]\varepsilon[n]\\
&= \lambda\hat{J}_{n-1} + \alpha_n\left|e[n]\right|^2.
\end{aligned} \tag{6.79}$$

6.4.4.3 Computational Complexity of RLS

The conventional RLS computational load is much higher than the LMS that is $O(M)$. For the RLS, in fact, the computational cost is equal to $O(M^2)$ despite the use of the MIL. For the EWRLS, we have a total of $(4M^2 + 4M)$ multiplications and $(3M^2 + M - 1)$ additions. The complexity can be reduced by using special symmetries of the matrix $\mathbf{R}_{xx,n}^{-1}$, but the total is always $O(M^2)$.

In literature, as will be introduced later in Chap. 8, there are faster versions of CRLS as, for example, the *Fast RLS* in which by developing the symmetry and redundancy and adopting an *recursive order algorithm* approach, we can get to a complexity equal to $O(7M)$.

6.4.5 Performance Analysis and Convergence of RLS

To the RLS behavior study, we proceed to the definition of a dynamic learning model based on a stochastic difference equation (SDE). For the analysis, as in the LMS case described in Sect. 5.4, we consider the desired output $d[n]$ defined by a moving average stationary model with superimposed noise, of the type illustrated in Fig. 6.3, and defined as

$$d[n] = \mathbf{w}_0^H \mathbf{x}_n + v[n]. \tag{6.80}$$

The term $\mathbf{w}_0$, constant and *a priori* fixed, represents the regression model vector; $v[n]$ indicates the zero-mean Gaussian measurement noise. Note that, considering the entire regression, the above can be expressed in matrix form as

$$\mathbf{d}_n = \mathbf{X}_n \mathbf{w}_0 + \mathbf{v}_n. \tag{6.81}$$

The input $\mathbf{x}_n$ is applied to both the model and the AF. The difference between the filter output $y[n] = \mathbf{w}^H \mathbf{x}_n$ and that of the model is minimal when, writing explicitly the regression equations (6.49), we have

$$\begin{aligned} \mathbf{w}_n &= \left(\mathbf{X}_n^H \mathbf{\Lambda}_n \mathbf{X}_n\right)^{-1} \mathbf{X}_n^H \mathbf{\Lambda}_n \mathbf{d}_n \\ &= \mathbf{P}_n \mathbf{R}_{xd,n}. \end{aligned} \tag{6.82}$$

Substituting in the first of the previous equation (6.81), we can write

$$\begin{aligned} \mathbf{w}_n &= \left(\mathbf{X}_n^H \mathbf{\Lambda}_n \mathbf{X}_n\right)^{-1} \mathbf{X}_n^H \left(\mathbf{\Lambda}_n \mathbf{X}_n \mathbf{w}_0 + \mathbf{v}_n\right) \\ &= \mathbf{w}_0 + \mathbf{P}_n \mathbf{X}_n^H \mathbf{\Lambda}_n \mathbf{v}_n. \end{aligned} \tag{6.83}$$

Taking the expectation of the above, for independence and because the noise has zero mean, we can write

$$\begin{aligned} E\{\mathbf{w}_n\} &= \mathbf{w}_0 + E\{\mathbf{P}_n \mathbf{X}_n^H\} \mathbf{\Lambda}_n E\{\mathbf{v}_n\} \\ &= \mathbf{w}_0. \end{aligned} \tag{6.84}$$

which proves the algorithm convergence (on average) for null ICs.

For a convergence, as for the LMS (see Sect. 5.4.1.1), the minimum error energy is

$$J_{\min} = \sigma_v^2.$$

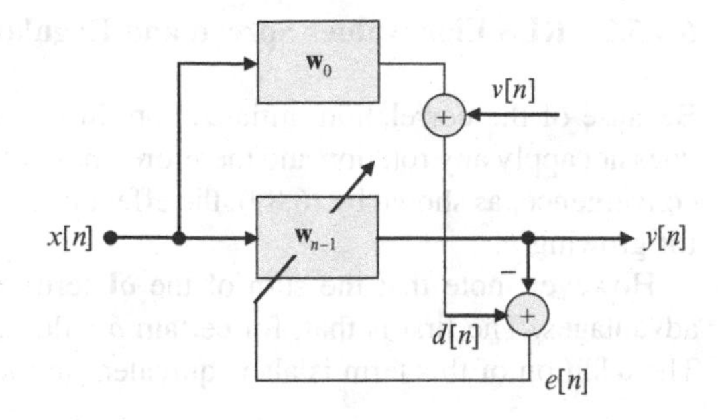

Fig. 6.3 Model for the study of the adaptive filter performance

6.4.5.1 Convergence of the Growing Memory RLS

For a complete study we must consider not null ICs in the form $\mathbf{R}_{xx,-1} = \delta\mathbf{I}$, for which it is necessary to (1) include them in the recursive expression of correlations computation and (2) determine the *bias*, from the optimal solution, in function of the parameters λ and δ.

To take into account the nonzero ICs for simplicity, we set $\lambda = 1$ (growing memory algorithm) and write correlations in expanded form as in

$$\mathbf{R}_{xx,n} = \sum_{i=0}^{n} \mathbf{x}_i \mathbf{x}_i^H + \mathbf{R}_{xx,-1} \tag{6.85}$$

$$\mathbf{R}_{xd,n} = \sum_{i=0}^{n} \mathbf{x}_i d^*[i]. \tag{6.86}$$

With this position, substituting (6.80) in (6.86) and using (6.85), we get

$$\begin{aligned} \mathbf{R}_{xd,n} &= \mathbf{w}_0 \sum_{i=0}^{n} \mathbf{x}_i \mathbf{x}_i^H + \sum_{i=0}^{n} \mathbf{x}_i v^*[i] \\ &= \mathbf{R}_{xx,n}\mathbf{w}_0 - \mathbf{R}_{xx,-1}\mathbf{w}_0 + \mathbf{X}_n^H \mathbf{v}_n. \end{aligned} \tag{6.87}$$

Substituting the latter in the current solution (6.82), we have that

$$\mathbf{w}_n = \mathbf{w}_0 - \mathbf{P}_n \mathbf{P}_{-1} \mathbf{w}_0 + \mathbf{P}_n \mathbf{X}_n^H \mathbf{v}_n. \tag{6.88}$$

By placing the expectation and considering the ergodicity $\mathbf{R} \approx \mathbf{R}_{xx,n}/n$, it is

$$\begin{aligned} E\{\mathbf{w}\} &= \mathbf{w}_0 - E\{\mathbf{P}_n \mathbf{P}_{-1}\}\mathbf{w}_0 + E\{\mathbf{P}_n \mathbf{X}_n^H \mathbf{v}_n\} \\ &= \mathbf{w}_0 - \frac{\delta}{n}\mathbf{P}\mathbf{w}_0. \end{aligned} \tag{6.89}$$

Equation (6.89) shows that the solution is biased and that the bias effect is proportional to δ and tends to zero for $n \to \infty$.

6.4.5.2 RLS Eigenvalues Spread and Regularization

Because of the correlation initialization, in the first iterations the inverse of $\mathbf{R}_{xx,n}$ does not apply any rotation and therefore does not reduce the eigenvalues spread. At convergence, as shown by (6.89), the effect is a solution bias that tends to disappear for growing n.

However, note that the sum of the $\delta\mathbf{I}$ term, with $\delta > 0$, also presents certain advantages. The first is that, for certain δ values, the matrix is not always unique. The addition of this term is also equivalent to the following CF definition:

$$\hat{J}_n(\mathbf{w}) = \sum_{i=0}^{n} \lambda^{n-i} \left|e[i]\right|^2 + \delta\lambda^n \|\mathbf{w}_n\|^2 \tag{6.90}$$

in which $\delta\lambda^n\|\mathbf{w}_n\|^2$ can be seen as a *Tikhonov regularization parameter*, of the type already studied in Sect. 4.2.5.2, which makes CF smooth so as to stabilize the solution and make it easier to search for the minimum.

Remark The regularization term transforms an *ill-posed problem* in a *well-posed problem* by adding *a priori* knowledge about the problem structure (for example, a smooth mapping in the least-squares sense, between $x[n]$ and $d[n]$). However, by (6.89), the regularization effect decays with time.

The regularization parameter δ is usually selected in a way inversely proportional to the SNR. In the case of low SNR (very noisy environment) it can assume higher values. In practice, the smoothing consists in a kind of CF low-pass filtering (CF's *smooth operator*).

6.4.5.3 Study of the Mean Square Convergence

For the mean square convergence study, it is necessary to analyze the behavior of the error vector correlation matrix $\mathbf{K}_n = E\{\mathbf{u}_n\mathbf{u}_n^H\}$ (see Sect. 5.1.2.3), where the WEV is $\mathbf{u}_n = \mathbf{w}_n - \mathbf{w}_0$. From (6.83) we can write

$$\mathbf{u}_n = \mathbf{P}_n\mathbf{X}_n^H\mathbf{\Lambda}_n\mathbf{v}_n \quad \text{and} \quad \mathbf{u}_n\mathbf{u}_n^H = \left(\mathbf{P}_n\mathbf{X}_n^H\mathbf{\Lambda}_n\mathbf{v}_n\right)\left(\mathbf{P}_n\mathbf{X}_n^H\mathbf{\Lambda}_n\mathbf{v}_n\right)^H. \tag{6.91}$$

Recalling that $\mathbf{\Lambda}_n = \mathbf{\Lambda}_n^H$ and that $\mathbf{R}_n$ is Toeplitz (for which $\mathbf{P}_n = \mathbf{P}_n^H$), we can write

$$E\{\mathbf{u}_n\mathbf{u}_n^H\} = E\{\mathbf{P}_n\mathbf{X}_n^H\mathbf{\Lambda}_n\mathbf{v}_n\mathbf{v}_n^H\mathbf{\Lambda}_n\mathbf{X}_n\mathbf{P}_n\} \tag{6.92}$$

since, by definition $E\{\mathbf{v}_n\mathbf{e}_n^H\} = \sigma_v^2\mathbf{I}$, and $\mathbf{P}_n \equiv \mathbf{R}_{xd,n}^{-1} = (\mathbf{X}_n^H\mathbf{\Lambda}_n\mathbf{X}_n)^{-1}$, considering the statistical independence, is

$$\mathbf{K}_n = \sigma_v^2 E\{\mathbf{P}_n \mathbf{X}_n^H \mathbf{\Lambda}_n^2 \mathbf{X}_n \mathbf{P}_n\}. \tag{6.93}$$

Before proceeding, let us consider the expectation of the term $\mathbf{R}_{xx,n}$, recalling that the following relation $\mathbf{R}_{xx,n} = \sum_{i=0}^{n} \lambda^{n-i} \mathbf{x}_i \mathbf{x}_i^H$ holds, we can write that

$$\begin{aligned} E\{\mathbf{R}_{xx,n}\} &= \sum_{i=0}^{n} \lambda^{n-i} E\{\mathbf{x}_i \mathbf{x}_i^H\} \\ &= \mathbf{R}\left(1 + \lambda + \lambda^2 + \cdots + \lambda^{n-1}\right) \\ &= \frac{1-\lambda^n}{1-\lambda}\mathbf{R} \end{aligned} \tag{6.94}$$

where $\mathbf{R} \triangleq E\{\mathbf{x}_i \mathbf{x}_i^H\}$. In other words, with the approximation $\mathbf{R}_{xx,n} \approx E\{\mathbf{R}_{xx,n}\}$, the relationship between true and estimated correlation can be expressed as

$$\mathbf{R}_{xx,n} = \frac{1-\lambda^n}{1-\lambda}\mathbf{R} \tag{6.95}$$

or

$$\mathbf{P}_n = \frac{1-\lambda}{1-\lambda^n}\mathbf{R}^{-1}. \tag{6.96}$$

In the case that the input vectors $\mathbf{x}_1$, $\mathbf{x}_2$, …, $\mathbf{x}_n$ are iid and the forgetting factor $0 \ll \lambda < 1$, for $n \gg M$, substituting (6.96) in (6.93) we have that

$$\mathbf{K}_n = \sigma_v^2 \left(\frac{1-\lambda}{1-\lambda^n}\mathbf{R}^{-1} \frac{1-\lambda^{2n}}{1-\lambda^2}\mathbf{R} \frac{1-\lambda}{1-\lambda^n}\mathbf{R}^{-1} \right). \tag{6.97}$$

Therefore, at steady state, for $n \to \infty$ $\lambda^n \to 0$, we have that

$$\begin{aligned} \mathbf{K}_\infty &= \sigma_v^2 \frac{(1-\lambda)^2}{1-\lambda^2}\mathbf{R}^{-1} \\ &= \sigma_v^2 \frac{1-\lambda}{1+\lambda}\mathbf{R}^{-1}. \end{aligned} \tag{6.98}$$

6.4.5.4 Convergence Speed and Learning Curve of RLS

Recall that (see Sect. 5.1.2.3)

$$J_n = J_{\min} + \mathrm{tr}[\mathbf{R}\mathbf{K}_{n-1}]. \tag{6.99}$$

Substituting (6.97) in (6.99) we get

$$J_n \approx J_{\min}\left(1 + \frac{1-\lambda}{1+\lambda} \cdot M\right). \tag{6.100}$$

From the previous expression we observe that in the RLS algorithm the convergence speed depends on the exponential term λ^{n-1}. In fact, according to (6.100) for RLS time constant τ_{RLS} we have that $\lambda^n = e^{-n/\tau_{\text{RLS}}}$, i.e., solving for we obtain

$$\tau_{\text{RLS}} = -\frac{1}{\ln\lambda} \tag{6.101}$$

and for $0 \ll \lambda < 1$

$$\tau_{\text{RLS}} \approx \frac{1}{1-\lambda}. \tag{6.102}$$

In the LMS algorithm, the convergence speed is determined by the slower mode of **R** matrix. Otherwise, for the RLS the convergence speed is independent from the eigenvalues of the correlation matrix and convergence is controlled only by the forgetting factor λ.

6.4.5.5 Excess of Steady-State Error of RLS

Form expression (6.100), the excess of MSE for $n \to \infty$ is

$$J_{\text{EMSE}\_\infty} = J_\infty - J_{\min} = MJ_{\min}\frac{1-\lambda}{1+\lambda} \tag{6.103}$$

and regarding the misadjustment we have that

$$\mathcal{M}_{\text{RLS}} = \frac{J_{\text{EMSE}}}{J_{\min}} = J_{\min}\frac{1-\lambda}{1+\lambda}. \tag{6.104}$$

Note that, as for the convergence speed, the forgetting factor affects also the excess of MSE and the misadjustment.

In Fig. 6.4 is reported an experiment of the identification of two random systems $\mathbf{w}_k$ generated with a uniform distribution as $w_k[n] = U(-0.5, 0.5)$ for $k = 0, 1$ and $n = 0, \ldots, M-1$, with $M = 6$, according to the scheme of study of Fig. 6.3. The learning curve, averaged over 200 trials, was evaluated for different values of λ (shown in the figure). The system input is a unitary-variance zero-mean colored noise generated by the expression (5.172) with $b = 0.9$.

In the first part of the experiment is identified the system $\mathbf{w}_0$ and for $n \geq \frac{N}{2}$ the system became $\mathbf{w}_1$. Note that, in agreement with (6.102), a high value of the forgetting factor corresponds to a slower transient behavior.

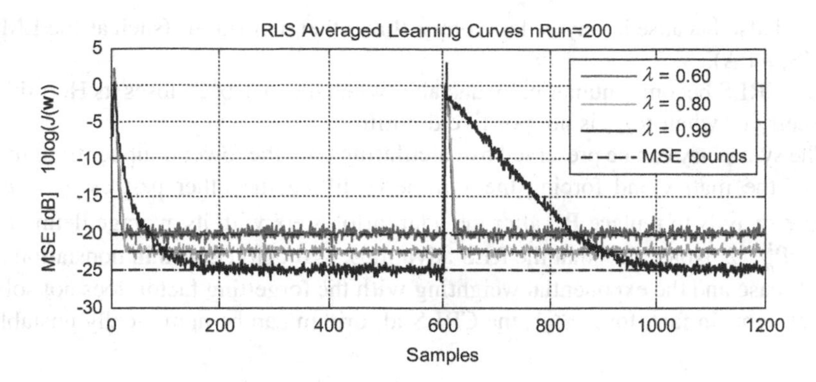

Fig. 6.4 Steady-state and convergence performance of the RLS algorithm for different values of forgetting factor λ in the presence of an abrupt change of the system to be identified. The SNR is 25 dB and IC $\mathbf{P}_{-1} = 100 \cdot \mathbf{I}$

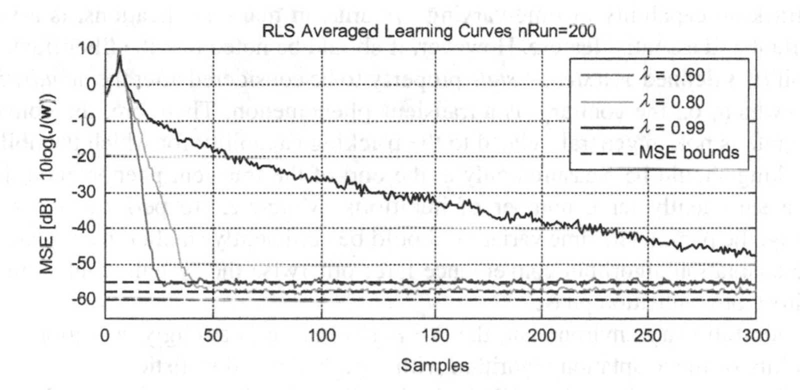

Fig. 6.5 Transient performance of the RLS algorithm for different values of forgetting factor λ and SNR 60 dB

In agreement with (6.100), it can be observed that the lower limit of the learning curves depends on the level of noise and on the parameter λ and does depend on the statistical characteristic of the input. Moreover, as also shown in Fig. 6.5 for similar experiment of Fig. 6.4, you do not have optimal transient performance for $0 \ll \lambda < 1$ $(\lambda \approx 1)$.

6.4.5.6 On the CRLS Robustness

The CRLS algorithm is extensively used in parameter estimation and identification problems. In the *online* DSP is less used, beyond that due to the high computational cost and also because it may be less robust than other algorithms (such as the LMS, NLMS, APA).

The CRLS becomes numerically unstable when the matrix $\mathbf{P}_n$ loses its Hermitian symmetry or when $\mathbf{R}_{xx,n}$ is not positive definite.

The symmetry can be preserved by calculating only the lower or upper triangular part of the matrix and forcing the symmetry filling the other part as $p_{ij} = p_{ij}^*$. Another way is to replace $\mathbf{P}_n$, after the adaptation step, with its average defined as $[\mathbf{P}_n + \mathbf{P}_n^H]/2$. Note, also, that the RLS advantage is much reduced in nonstationary signals case and the exponential weighting with the forgetting factor does not solve the problem. In fact, for $\lambda \ll 1$, the CRLS algorithm can be numerically unstable.

6.4.6 Nonstationary RLS Algorithm

The tracking capability of time-varying systems, in many applications, is a very important and essential feature. However, it should be noted that the filter tracking capability is defined as a *steady-state* property to be considered after the *acquisition phase* which, on the contrary, is a transient phenomenon. Therefore, the convergence rate is not, in general, related to the tracking capability for which the ability of tracking should be measured only at the end of the transient phenomenon, i.e., after a sufficiently large number of iterations. Moreover, to perform a correct tracking, the parameters time variation should be sufficiently smaller in comparison to the adaptation algorithm convergence rate; otherwise the system would still be transitory or acquisition phase.

In nonstationary environment, the AF performance is strongly conditioned by the ability of the adaptation algorithm with locally defined statistics.

In the exponential weighting RLS, the locally defined statistics are emphasized by the weight function that reduces the influence of the past data. In fact, the CF to minimize is of the type

$$J(\mathbf{w}_n) = \sum_{i=0}^{n} \lambda^{n-i} \left| d[i] - \mathbf{w}^H \mathbf{x}_i \right|^2 = \lambda J(\mathbf{w}_{n-1}) + \left| d[n] - \mathbf{w}^H \mathbf{x}_n \right|^2 \tag{6.105}$$

where $0 < \lambda < 1$. For which the analysis window *effective length* is expressed by the relation

$$L_{\text{eff}} \triangleq \sum_{n=0}^{\infty} \lambda^n / \lambda^0 = \frac{1}{1-\lambda}. \tag{6.106}$$

For good tracking capability the forgetting factor λ must be in the range $0.6 < \lambda < 0.8$. Note that for $\lambda = 1$ la the window has increasing length and is of rectangular type; in this case, it is considered the entire signal statistic for which the tracking capability is compromised.

A second way to emphasize the current system statistics is to use finite-length analysis windows. In this case, the CF is

$$J(\mathbf{w}_n) = \sum_{i=n-L+1}^{n} \left| d[i] - \mathbf{w}^H \mathbf{x}_i \right|^2 \tag{6.107}$$

where the window length is $L > M$.

6.5 Kalman Filter

The Kalman filter (KF) represents an alternative approach to the adaptive filtering formulation with MMSE criterion which, in some way, generalizes and provides a unified version of the RLS methods [1, 7–9]. The KF algorithms, even though they represent a special case of optimal linear filtering, are used in numerous applications such as maritime and aerospace navigation, where the correct prediction and smooth of the vehicle trajectory have a value of great importance.

One of the main KF prerogatives consists in the formulation and solution of the adaptive filtering problem in the context of the *theory of dynamical systems*. In other words, the AF's coefficients $\mathbf{w}_n$ are seen as the *state* of a linear dynamic system with random inputs and able to recursively update itself according to new data presented at its input.

The KF is suitable for stationary and nonstationary contexts and presents a recursive solution in which, at every step, it produces an estimate of the new state which depends only on the previous state and on new input data. The no need to memorize all the past states may lead to high computational efficiency.

For the KF development, we consider a linear system defined in state-space form as shown in Fig. 6.6. The *state vector* or simply *state*, at instant n, indicated with $\mathbf{w}_n$, is defined as the minimum data set for the system dynamic description, in the absence of external excitation. In other words, the state represents the minimum amount of data to describe the past and for the future prediction of the system behavior. Typically, the state $\mathbf{w}_n$ is unknown and its estimate is used for a set of observed data, called *observation vector* or simply *observation*, indicated with the vector $\mathbf{y}_n$.

Mathematically, the DT-linear dynamic system is described by two equations in which the first, which represents the *process*, has the form

$$\mathbf{w}_{n+1} = \mathbf{F}_{n+1,n}\mathbf{w}_n + \mathbf{B}_n \boldsymbol{\eta}_n, \quad \textit{process equation} \tag{6.108}$$

where $\mathbf{F}_{n+1,n} \in \mathbb{R}^{M \times M}$, defined as a *state-transition matrix*, links the states $\mathbf{w}_n$ and $\mathbf{w}_{n+1}$, and $\mathbf{B}_n \in \mathbb{R}^{M \times M}$ is the *input matrix* in the absence of external forcing. The input process $\boldsymbol{\eta}_n \in \mathbb{R}^{M \times 1}$, also called *driving noise*, is zero-mean *white Gaussian noise* (WGN), i.e., $\eta[n] \sim N(0, \sigma_\eta^2)$, with covariance matrix $\mathbf{Q}_n$.

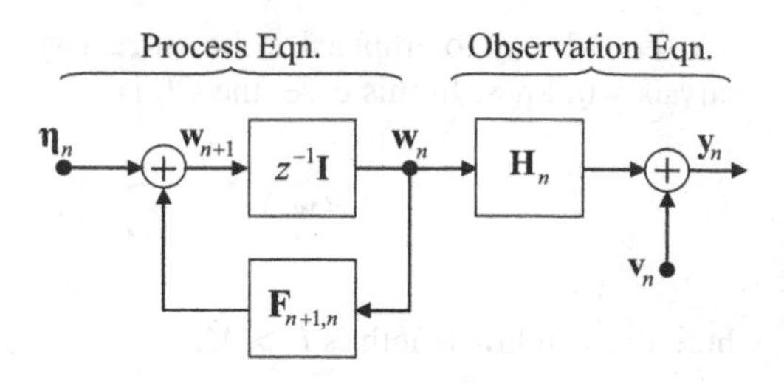

Fig. 6.6 State-space representation of a discrete-time linear dynamic system

The second equation, which represents the *observation* or the *measure*, has the form

$$\mathbf{y}_n = \mathbf{H}_n\mathbf{w}_n + \mathbf{v}_n, \quad \textit{observation equation} \tag{6.109}$$

where $\mathbf{H}_n \in \mathbb{R}^{N\times M}$, that is, the *observation or measurement matrix*, links the state $\mathbf{w}_n$ to the vector $\mathbf{y}_n$ observation. The process $\mathbf{v}_n$, which represents the observation noise, is zero-mean WGN $v[n] \sim N(0,\sigma_v^2)$, with covariance matrix $\mathbf{R}_n$.

6.5.1 Discrete-Time Kalman Filter Formulation

The Kalman filtering role is the optimal state variables estimation, which in general terms represents the trajectories to be tracked through the process and measurement equations joint solution. Considering, for simplicity $\mathbf{B}_n = \mathbf{I}$, the dynamic system is described as

$$\mathbf{w}_{n+1} = \mathbf{F}_{n+1,n}\mathbf{w}_n + \boldsymbol{\eta}_n \tag{6.110}$$

$$\mathbf{y}_n = \mathbf{H}_n\mathbf{w}_n + \mathbf{v}_n. \tag{6.111}$$

Formally, the problem consists in estimating the vector state[2] $\hat{\mathbf{w}}_i$ in light of all the observations $[\mathbf{y}_j]_1^n \triangleq [\mathbf{y}_1, \mathbf{y}_2, \ldots, \mathbf{y}_n]$ and, in general terms, we have $\hat{\mathbf{w}}_i = k\left([\mathbf{y}_j]_1^n\right)$ where with $k(\cdot)$ is indicated the prediction function, *a priori* known or to be determined in some way, called estimator. In the case where the time index i of the state to estimate is internal to the time window of available measures, namely, $1 \le i \le n$, the problem is that of the *classical filtering*. For $i < n$, it is also referred to as *smoothing*, while for $i > n$ the problem is that of the *linear prediction*.

In the KF, the basic assumptions for estimating the state are as follows:

1. the matrices **F**, **H** are known;
2. the input and the observation noise are independent zero-mean WGN, $\boldsymbol{\eta}_n \sim N(\mathbf{0},\mathbf{Q}_n)$ and $\mathbf{v}_n \sim N(\mathbf{0},\mathbf{R}_n)$ with known statistics $\mathbf{Q}_n$ and $\mathbf{R}_n$;
3. the estimator is of type linear MMSE and consists of a simple linear combination of the measures (see Sect. C.3.2.8).

[2] In this context $\hat{\mathbf{v}}$ indicates an RV that represents an estimate of a deterministic vector $\mathbf{v}$.

In particular in the KF the state estimator is modeled as

$$\hat{\mathbf{w}}_n = \mathbf{K}_n^{(1)}\hat{\mathbf{w}}_n^- + \mathbf{K}_n\mathbf{y}_n \tag{6.112}$$

where $\hat{\mathbf{w}}_n^-$ indicates the *a priori state estimate* and the matrices $\mathbf{K}_n^{(1)}$ and $\mathbf{K}_n$ represent the unknown parameters of the linear estimator. The determination of these matrices is accomplished through the *principle of orthogonality*, for which by defining the *state error vector* as

$$\tilde{\mathbf{w}}_n = \mathbf{w}_n - \hat{\mathbf{w}}_n \tag{6.113}$$

and by imposing the orthogonality, we get

$$E\left[\tilde{\mathbf{w}}_n\mathbf{y}_i^T\right] = \mathbf{0}, \quad \text{for } i = 1, 2, \ldots, n-1. \tag{6.114}$$

Using (6.111), (6.112), and (6.113) in (6.114), we get

$$E\left[\left(\mathbf{w}_n - \mathbf{K}_n^{(1)}\hat{\mathbf{w}}_n^- - \mathbf{K}_n\mathbf{H}_n\mathbf{w}_n - \mathbf{K}_n\boldsymbol{\eta}_n\right)\mathbf{y}_i^T\right] = \mathbf{0}. \tag{6.115}$$

The noise processes are independent also from the observation, for which is worth that $E[\boldsymbol{\eta}_n\mathbf{y}_i^T] = \mathbf{0}$, and therefore, rearranging the previous expression, we have that

$$E\left[\left(\mathbf{I} - \mathbf{K}_n\mathbf{H}_n - \mathbf{K}_n^{(1)}\right)\mathbf{w}_n\mathbf{y}_i^T - \mathbf{K}_{nn}^{(1)}\left(\mathbf{w}_n - \hat{\mathbf{w}}_n^-\right)\mathbf{y}_i^T\right] = \mathbf{0}.$$

Always for the principle of orthogonality, observe that $\mathbf{K}_n^{(1)}\left(\mathbf{w}_n - \hat{\mathbf{w}}_n^-\right)\mathbf{y}_i^T = \mathbf{0}$, for which the above can be simplified as

$$\left(\mathbf{I} - \mathbf{K}_n\mathbf{H}_n - \mathbf{K}_n^{(1)}\right)E\left[\mathbf{w}_n\mathbf{y}_i^T\right] = \mathbf{0}, \quad \text{for } i = 1, 2, \ldots, n-1. \tag{6.116}$$

For arbitrary values of the state $\mathbf{w}_n$ and observations $\mathbf{y}_n$, (6.116) can be satisfied only if $\mathbf{I} - \mathbf{K}_n\mathbf{H}_n - \mathbf{K}_n^{(1)} = \mathbf{0}$ or, equivalently, if it is possible to relate the matrices $\mathbf{K}_n^{(1)}$ and $\mathbf{K}_n$, as

$$\mathbf{K}_n^{(1)} = \mathbf{I} - \mathbf{K}_n\mathbf{H}_n. \tag{6.117}$$

Substituting (6.117) into (6.112) we can express the *a posteriori* state estimate at the time n as

$$\hat{\mathbf{w}}_n = \hat{\mathbf{w}}_n^- + \mathbf{K}_n\left(\mathbf{y}_n - \mathbf{H}_n\hat{\mathbf{w}}_n^-\right) \tag{6.118}$$

where the matrix $\mathbf{K}_n$ is defined as *Kalman gain* matrix.

It is possible to derive the matrix **K**, still applying the principle of orthogonality. Therefore, we have that

$$E\Big[(\mathbf{w}_n - \hat{\mathbf{w}}_n)\mathbf{y}_n^T\Big] = \mathbf{0} \quad \text{and} \quad E\Big[(\mathbf{w}_n - \hat{\mathbf{w}}_n)\hat{\mathbf{y}}_n^T\Big] = \mathbf{0} \tag{6.119}$$

where $\hat{\mathbf{y}}_n$ indicates the $\mathbf{y}_n$ estimate, obtained from the previous measurements $[\mathbf{y}_i]_1^{n-1}$. We define *innovation process*

$$\tilde{\mathbf{y}}_n = \mathbf{y}_n - \hat{\mathbf{y}}_n \tag{6.120}$$

which represents a measure of the new information contained in $\mathbf{y}_n$; this can be expressed as

$$\begin{aligned} \tilde{\mathbf{y}}_n &= \mathbf{y}_n - \mathbf{H}_n\hat{\mathbf{w}}_n^- \\ &= \mathbf{H}_n\mathbf{w}_n + \mathbf{v}_n - \mathbf{H}_n\hat{\mathbf{w}}_n^- \\ &= \mathbf{H}_n\tilde{\mathbf{w}}_n^- + \mathbf{v}_n \end{aligned} \tag{6.121}$$

where $\tilde{\mathbf{w}}_n^- = \mathbf{w}_n - \hat{\mathbf{w}}_n^-$ represents the *state error estimate vector*. From (6.119) and the definition (6.120) it is shown that the orthogonality principle also applies to the innovation process and therefore we can write

$$E\Big[(\mathbf{w}_n - \hat{\mathbf{w}}_n)\tilde{\mathbf{y}}_n^T\Big] = \mathbf{0}. \tag{6.122}$$

Using (6.111) and (6.118) it is possible to express the state error vector as

$$\begin{aligned} \mathbf{w}_n - \hat{\mathbf{w}}_n &= \tilde{\mathbf{w}}_n^- - \mathbf{K}_n\big(\mathbf{H}_n\tilde{\mathbf{w}}_n^- + \mathbf{v}_n\big) \\ &= (\mathbf{I} - \mathbf{K}_n\mathbf{H}_n)\tilde{\mathbf{w}}_n^- - \mathbf{K}_n\mathbf{v}_n \end{aligned} \tag{6.123}$$

and substituting (6.121) and (6.123) in (6.122), we obtain

$$E\Big\{\big[(\mathbf{I} - \mathbf{K}_n\mathbf{H}_n)\tilde{\mathbf{w}}_n^- - \mathbf{K}_n\mathbf{v}_n\big]\big(\mathbf{H}_n\tilde{\mathbf{w}}_n^- + \mathbf{v}_n\big)\Big\} = \mathbf{0} \tag{6.124}$$

because the noise $\mathbf{v}_n$ è is independent of the state $\mathbf{w}_n$ and therefore also of the error $\tilde{\mathbf{w}}_n^-$; it appears that the expectation (6.124) reduces to

$$(\mathbf{I} - \mathbf{K}_n\mathbf{H}_n)\mathbf{P}_n^-\mathbf{H}_n^T - \mathbf{K}_n\mathbf{R}_n = \mathbf{0} \tag{6.125}$$

where $\mathbf{R}_n = E[\mathbf{v}_n\mathbf{v}_n^T]$ is the covariance matrix of the observation noise and

$$\mathbf{P}_n^- = E\Big[\big(\mathbf{w}_n - \hat{\mathbf{w}}_n^-\big)\big(\mathbf{w}_n - \hat{\mathbf{w}}_n^-\big)^T\Big] = E\Big[\tilde{\mathbf{w}}_n^-\tilde{\mathbf{w}}_n^{-T}\Big] \tag{6.126}$$

is defined as the *a priori* covariance matrix.

Solving (6.125) with respect to $\mathbf{K}_n$, it is possible to define Kalman gain matrix as

$$\mathbf{K}_n = \mathbf{P}_n^- \mathbf{H}_n^T \left[\mathbf{H}_n \mathbf{P}_n^- \mathbf{H}_n^T + \mathbf{R}_n\right]^{-1}. \tag{6.127}$$

To complete the recursive estimation procedure, consider the *covariance error propagation* that describes the covariance matrix error estimation starting from its *a priori* estimate.

We define the *a posteriori covariance matrix* as the estimated quantity equal to

$$\mathbf{P}_n = E\left[\tilde{\mathbf{w}}_n \tilde{\mathbf{w}}_n^T\right] = E\left[(\mathbf{w}_n - \hat{\mathbf{w}}_n)(\mathbf{w}_n - \hat{\mathbf{w}}_n)^T\right], \tag{6.128}$$

so from the old value of *a posteriori* covariance $\mathbf{P}_{n-1}$, it is possible to estimate the *a priori* covariance $\mathbf{P}_n$. In fact, substituting (6.123) in (6.128) and for $\mathbf{v}_k$ independent of $\tilde{\mathbf{w}}_n^-$, we get

$$\mathbf{P}_n = (\mathbf{I} - \mathbf{K}_n \mathbf{H}_n)\mathbf{P}_n^-(\mathbf{I} - \mathbf{K}_n \mathbf{H}_n)^T + \mathbf{K}_n \mathbf{R}_n \mathbf{K}_n^T. \tag{6.129}$$

Further expanding the latter and using (6.127) it is possible with simple steps to reformulate the *a posteriori* and *a priori* covariance dependence, in the following ways:

$$\begin{aligned} \mathbf{P}_n &= (\mathbf{I} - \mathbf{K}_n \mathbf{H}_n)\mathbf{P}_n^- - (\mathbf{I} - \mathbf{K}_n \mathbf{H}_n)\mathbf{P}_n^- \mathbf{K}_n^T \mathbf{H}_n^T + \mathbf{K}_n \mathbf{R}_n \mathbf{K}_n^T \\ &= (\mathbf{I} - \mathbf{K}_n \mathbf{H}_n)\mathbf{P}_n^-. \end{aligned} \tag{6.130}$$

For the second stage of the error covariance propagation, it is noted that the state *a priori* estimate can be defined in terms of the old *a posteriori* estimate using the expression (6.110), and defining the matrix $\mathbf{F}_{n,n-1}$ for null $\mathbf{v}_n$, as

$$\hat{\mathbf{w}}_n^- = \mathbf{F}_{n,n-1} \hat{\mathbf{w}}_{n-1}. \tag{6.131}$$

From the above and from (6.110), the *a priori* estimate can be written as

$$\begin{aligned} \tilde{\mathbf{w}}_n^- = \mathbf{w}_n - \hat{\mathbf{w}}_n^- &= (\mathbf{F}_{n,n-1}\mathbf{w}_{n-1} + \boldsymbol{\eta}_{n-1}) - (\mathbf{F}_{n,n-1}\hat{\mathbf{w}}_{n-1}) \\ &= \mathbf{F}_{n,n-1}(\mathbf{w}_{n-1} - \hat{\mathbf{w}}_{n-1}) + \boldsymbol{\eta}_{n-1} \\ &= \mathbf{F}_{n,n-1}\tilde{\mathbf{w}}_{n-1} + \boldsymbol{\eta}_{n-1}. \end{aligned} \tag{6.132}$$

Using the above expression in the definition of the *a priori* covariance (6.126) and for the independence between $\boldsymbol{\eta}_{n-1}$ and $\tilde{\mathbf{w}}_{n-1}$, we can write

$$\begin{aligned} \mathbf{P}_n^- &= \mathbf{F}_{n,n-1} E(\tilde{\mathbf{w}}_{n-1}\tilde{\mathbf{w}}_{n-1}^T)\mathbf{F}_{n,n-1}^T + E(\boldsymbol{\eta}_{n-1}\boldsymbol{\eta}_{n-1}^T) \\ &= \mathbf{F}_{n,n-1}\mathbf{P}_{n-1}\mathbf{F}_{n,n-1}^T + \mathbf{Q}_{n-1} \end{aligned} \tag{6.133}$$

which defines the dependence of the *a priori* covariance $\mathbf{P}_n^-$ from the previous value of the *a posteriori* covariance $\mathbf{P}_{n-1}$.

6.5.2 The Kalman Filter Algorithm

The previous development, described by (6.131), (6.133), (6.127), (6.118), and (6.130), represents a set of equations for the recursive estimation of the state and is defined as *Kalman filter*. The results of the state estimation algorithm may be summarized in the following way:

1. *Knowledge of the process model*—matrix $\mathbf{F}_{n+1,n}$, covariance $\mathbf{Q}_n$, such that $\boldsymbol{\eta}_n \sim N(0,\mathbf{Q}_n)$

$$\mathbf{w}_{n+1} = \mathbf{F}_{n+1,n}\mathbf{w}_n + \boldsymbol{\eta}_n, \qquad n = 0, 1, \ldots$$

2. *Knowledge of the observation mode*—matrix $\mathbf{H}_n$, covariance $\mathbf{R}_n$, so $\mathbf{v}_n \sim N(0,\mathbf{R}_n)$

$$\mathbf{y}_n = \mathbf{H}_n\mathbf{w}_n + \mathbf{v}_n, \qquad n = 0, 1, \ldots$$

(i) Initialization $\hat{\mathbf{w}}_{-1} = E(\mathbf{w}_{-1}), \mathbf{P}_{-1} = E\Big[\big(\mathbf{w}_{-1} - E[\mathbf{w}_{-1}]\big)\big(\mathbf{w}_{-1} - E[\mathbf{w}_{-1}]\big)^T\Big]$

(ii) For $n = 0,1, \ldots$ {

$$\begin{aligned}
\hat{\mathbf{w}}_n^- &= \mathbf{F}_{n,n-1}\hat{\mathbf{w}}_{n-1}^-, && \textit{state estimation prediction}\\
\mathbf{P}_n^- &= \mathbf{F}_{n,n-1}\mathbf{P}_{n-1}\mathbf{F}_{n,n-1}^T + \mathbf{Q}_{n-1}, && \textit{covariance error prediction}\\
\mathbf{K}_n &= \mathbf{P}_n^-\mathbf{H}_n^T\left[\mathbf{H}_n\mathbf{P}_n^-\mathbf{H}_n^T + \mathbf{R}_n\right]^{-1}, && \textit{optimal Kalman gain}\\
\hat{\mathbf{w}}_n &= \hat{\mathbf{w}}_n^- + \mathbf{K}_n\left(\mathbf{y}_n - \mathbf{H}_n\hat{\mathbf{w}}_n^-\right), && \textit{state estimation update}\\
\mathbf{P}_n &= (\mathbf{I} - \mathbf{K}_n\mathbf{H}_n)\mathbf{P}_n^-, && \textit{covar.estimate update (Riccati Eqn.)}
\end{aligned}$$

}

The ICs choice indicated, in addition to being "*reasonable*," produces an unbiased state estimate $\mathbf{w}_n$.

Remark The KF can be considered as a kind of feedback control. The filter estimates the process state at a certain instant and gets a feedback in the form of measurement error (see Fig. 6.7). Therefore, the Kalman equations may be considered as belonging to two groups: the first group consists of the temporal update equations, and the second in the update measure equations. The time update equations are responsible for projecting forward in time the current state and error covariance estimates, to obtain the *a priori* estimate for the next time instant. The measurement update is responsible for the feedback because it incorporates the new measurement into the *a priori* estimate, to obtain an *a posteriori* estimate improvement.

In other words, the *a priori* estimate can be seen as a predictor, while the measurement update can be seen as a set of correction term equations. Therefore,

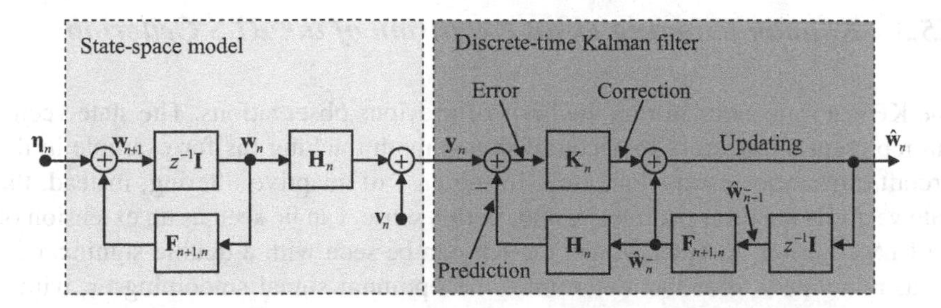

Fig. 6.7 Discrete-time Kalman filter scheme (modified from [6])

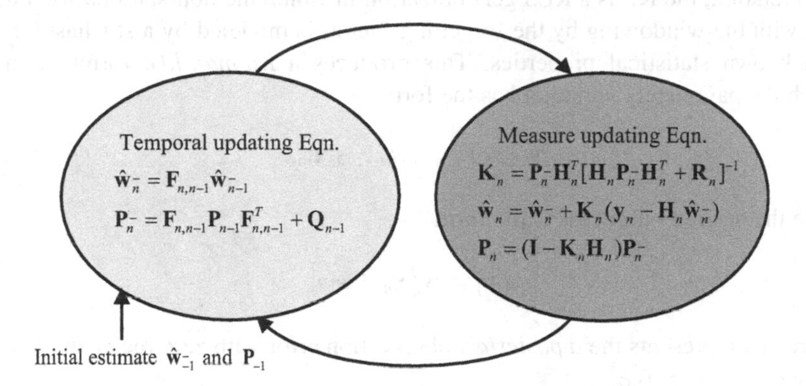

Fig. 6.8 The cyclic representation of KF. The time-update projects forward the current state estimate at time n. The measurement update projects the current measure estimate at time n

the KF can be thought of as a cyclic two-state algorithm, of type prediction correction as Fig. 6.8 describes.

The recursive nature of the KF is just one of its strengths that allow a practical and efficient implementability and consequent applicability to a wide class of problems.

Remark The recursive structure of the Kalman filter is similar to the *chains Markov model* with hidden or internal state or *Hidden Markov Model* (HMM) built on a linear operator and perturbed by Gaussian noise. The system state is represented by a real vector to which, for each time instant, a linear operator is applied to generate a new state to which is added, if known, the input and measurement additive noise contribution. Similarly, for the visible output, we consider a linear operator applied to the internal states with additive noise. In other words, the KF can be considered as the analogue of the HMM with the difference that the internal state variables belong to a continuous space differently by Markov models in which the space is discrete. In addition, the value of the future state for the HMM can be represented by an arbitrary distribution while for the KF we consider only the Gaussian distribution [10].

6.5.3 Kalman Filtering as an Extension of the RLS Criterion

The KF is a state estimator on the basis of previous observations. The state vector can represent a trajectory to accomplish a smooth tracking (as for example in the aircraft trajectories estimation, etc.). In the case of adaptive filtering, instead, the state vector is the filter parameters and, in this sense, can be seen as an extension of the RLS criterion. In other words, the KF can be seen with a double significance: (1) as a low-pass filter that determines the optimum signal smoothing or, with a different interpretation, (2) as optimal estimator of AF parameters, considered as the state trajectory of a linear dynamic system. Moreover, according to the latter interpretation, the KF is a RLS generalization in which the non-stationarity, rather than with the windowing by the forgetting factor, is modeled by a stochastic FDE with known statistical properties. This produces a *Kalman-like* formulation in which the parameters variation has the form

$$\mathbf{w}_n = \mathbf{F}_{n,n-1}\mathbf{w}_{n-1} + \boldsymbol{\eta}_n,$$

while the desired output takes the form

$$d[n] = \mathbf{w}_n^T\mathbf{x}_n + \varepsilon[n]$$

where $\varepsilon[n]$ represents the *a posteriori* observation error with zero mean and known variance $\varepsilon[n] \sim N(0,\sigma_\varepsilon^2)$.

In the KF scenario, the best unbiased linear estimate $\hat{\mathbf{w}}_n$ of the state $\mathbf{w}_n$, based on past observations $\left[d[i]\right]_{i=0}^{n}$, can be obtained from the following recursive equations:

$$\hat{\mathbf{w}}_n = \mathbf{F}_{n,n-1}\hat{\mathbf{w}}_{n-1} + \mathbf{k}_n\left[d[n] - \left[\hat{\mathbf{w}}_{n-1}\mathbf{F}_{n,n-1}\right]^T\mathbf{x}_n\right], \quad \textit{state estimate} \tag{6.134}$$

$$\mathbf{k}_n = \frac{\mathbf{F}_{n,n-1}\mathbf{P}_{n-1}\mathbf{x}_n}{\sigma_\varepsilon^2 + \mathbf{x}_n^T\mathbf{P}_{n-1}\mathbf{x}_n}, \quad \textit{Kalman gain} \tag{6.135}$$

$$\begin{aligned}\mathbf{P}_n &= \mathbf{F}_{n,n-1}\mathbf{P}_{n-1}\mathbf{F}_{n,n-1}^T + \mathbf{Q}_n \\ &\quad - \mathbf{F}_{n,n-1}\mathbf{P}_{n-1}\frac{\mathbf{x}_n\mathbf{x}_n^T}{\sigma_\varepsilon^2 + \mathbf{x}_n^T\mathbf{P}_{n-1}\mathbf{x}_n}\mathbf{P}_{n-1}\mathbf{F}_{n,n-1}^T, \quad \textit{cov.est.}\end{aligned} \tag{6.136}$$

where, as in RLS, $\mathbf{k}_n$ is the vector of the Kalman gain and $\mathbf{P}_n$ represents the error covariance matrix. The KF, in fact, is identical to the exponentially weighted RLS (EWRLS) for the following substitutions:

$$\mathbf{F}_{n,n-1} = \mathbf{I}; \quad \lambda = \sigma_\varepsilon^2; \quad \mathbf{Q}_n = \frac{1-\lambda}{\lambda}\left[\mathbf{I} - \mathbf{k}_n\mathbf{x}_n^T\right]\mathbf{P}_{n-1}$$

and to growing memory RLS algorithm for

$$\mathbf{F}_{n,n-1} = \mathbf{I}; \quad \sigma_{\varepsilon}^2 = 1; \quad \mathbf{Q}_n = \mathbf{0}$$

previously reported in Sect. 6.4.

6.5.4 Kalman Filter Robustness

The KF implementation poses a series of numerical problems well documented in the literature, which are mainly related to the computer arithmetic with finite word length. For example, the *a posteriori* estimate of the covariance matrix $\mathbf{P}_n$ defined in (6.130) as the difference $\mathbf{P}_n = \mathbf{P}_n^- - \mathbf{K}_n\mathbf{H}_n\mathbf{P}_n^-$ is such that it could have not semidefinite positive matrix, and, $\mathbf{P}_n$ being a covariance, this result would be unacceptable. As previously indicated in the RLS case, you can work around these problems by using unitary transformations in order to emphasize the algorithm robustness. One of these expedients is to propagate the matrix $\mathbf{P}_n$ with a root-square form using the Cholesky factorization, for which the covariance can be defined by the product $\mathbf{P}_n = \mathbf{P}_n^{1/2}\mathbf{P}_n^{T/2}$, where $\mathbf{P}_n^{1/2}$ is the lower triangular matrix and $\mathbf{P}_n^{T/2}$ is transposed. From algebra, in fact, every product of a square matrix for its transpose is always positive definite for which, even in the presence of numerical or rounding error, the condition of nonnegativity of the matrix $\mathbf{P}_n$ is respected.

6.5.5 KF Algorithm in the Presence of an External Signal

In the KF implementation, the noise observation covariance $\mathbf{R}_n$, such that $\mathbf{v}_n \sim N(\mathbf{0},\mathbf{R}_n)$, is supposed to be known and prior determined to the filtering procedure itself. This estimation is possible because it is generally able to measure the $\mathbf{v}_n$ process and (externally) calculate the measurement error variance.

The $\mathbf{Q}_n$ determination, which represents the input noise covariance matrix such that $\boldsymbol{\eta}_n \sim N(\mathbf{0},\mathbf{Q}_n)$, is generally more difficult since, typically, it is not able to have direct process observations.

As said, for a correct filter parameters initial tuning, it is convenient to determine $\mathbf{Q}$ and $\mathbf{R}$ with external identification procedure. Also, note that for stationary process, the parameters $\mathbf{R}_n$, $\mathbf{Q}_n$, and $\mathbf{K}_n$ quickly stabilize and remain almost constant.

For greater generality, consider the case in which in addition to the noise of $\boldsymbol{\eta}_n$ is also present as an external input, indicated as $\mathbf{u}_n$, for which the process equation (6.110), to take account of this external signal, takes the form

$$\mathbf{w}_{n+1} = \mathbf{F}_{n+1,n}\mathbf{w}_n + \mathbf{B}_n\mathbf{u}_n + \boldsymbol{\eta}_n$$

where $\mathbf{B}_n$ represents the input model applied to the control of the signal $\mathbf{u}_n$.

In this case the equation of state propagation estimate (6.131) is modified as

$$\hat{\mathbf{w}}_n^- = \mathbf{F}_{n,n-1}\hat{\mathbf{w}}_{n-1}^- + \mathbf{B}_n\mathbf{u}_{n-1}.$$

In the presence of external input $\mathbf{u}_n$, by introducing the intermediate variable $\tilde{\mathbf{z}}_n$, called *innovation* or *residual measure* and its covariance matrix $\mathbf{S}$, the set of equations that describe the KF algorithm is reformulated as

$$\hat{\mathbf{w}}_{n^-} = \mathbf{F}_{n,n-1}\hat{\mathbf{w}}_{n-1^-} + \mathbf{B}_n\mathbf{u}_{n-1}, \qquad \textit{state estimate prediction}$$
$$\mathbf{P}_n^- = \mathbf{F}_{n,n-1}\mathbf{P}_{n-1}\mathbf{F}_{n,n-1}^T + \mathbf{Q}_{n-1}, \qquad \textit{covariance error prediction}$$
$$\tilde{\mathbf{z}}_n = \mathbf{y}_n - \mathbf{H}_n\hat{\mathbf{w}}_n^-, \qquad \textit{innovation or residual measure}$$
$$\mathbf{S}_n = \mathbf{H}_n\mathbf{P}_n^-\mathbf{H}_n^T + \mathbf{R}_n, \qquad \textit{innovation covariance } cov(\tilde{\mathbf{z}}_n)$$
$$\mathbf{K}_n = \mathbf{P}_n^-\mathbf{H}_n^T\mathbf{S}_n^{-1}, \qquad \textit{optimal Kalman gain}$$
$$\hat{\mathbf{w}}_n = \hat{\mathbf{w}}_n^- + \mathbf{K}_n\tilde{\mathbf{z}}_n, \qquad \textit{state estimate update}$$
$$\mathbf{P}_n = (\mathbf{I} - \mathbf{K}_n\mathbf{H}_n)\mathbf{P}_n^-, \qquad \textit{covariance estimate (Riccati Eqn.)}$$

For further information, please refer to the vast literature on the subject (for example, [8, 9, 11]).

6.6 Tracking Performance of Adaptive Algorithms

In the previous chapters we have analyzed the AF properties considering stationary environment, i.e., the statistics $\mathbf{R}$ and $\mathbf{g}$ (or their estimates $\mathbf{R}_{xx}$ and $\mathbf{R}_{xd}$) of the processes involved in the algorithms are constant. In this case, the performance surface is fixed and algorithm optimization tends towards the optimal Wiener point $\mathbf{w}_{opt}$. In particular, the transient property of the algorithm in terms of the *average performance* of the *learning curve*, and the steady-state properties in terms of *excess of error* have been highlighted.

An environment is not stationary when the signals involved in the process are nonstationary. The non-stationarity can affect the input $\mathbf{x}_n$, the desired output $d[n]$, or both. In the case of time-variant input process, both correlation and cross-correlation are time varying. In the case of non-stationarity of the only reference $d[n]$ then only cross-correlation is time varying. Since the adaptation algorithm requires the invertibility of the correlation matrix $\mathbf{R}_n$, this means that the more critical is the input non-stationarity.

In this section we want to examine the behavior of adaptive algorithms in the case of nonstationary environment. The performance surface and the minimum point, denoted by $\mathbf{w}_{0,n}$, are variable in time and the adaptation algorithm must exhibit characteristics that, rather than the achievement, are aimed for tracking the minimum point. As mentioned in the AF general properties (see Sect. 5.1.2.6), in contrast to the convergence phase, which is a transitory phenomenon, tracking is a steady-state phenomenon. The convergence speed and tracking features are distinct properties. In fact, it is not always guaranteed that algorithms with high convergence speed also exhibit good tracking capability and vice versa. The two properties are different and are characterized with different performance indices. The tracking is possible only if the degree of non-stationarity is slower than the AF acquisition speed. The general characterization of the tracking properties is dependent on the algorithm type. In the following we will analyze the LMS and RLS algorithm performance.

6.6.1 *Tracking Analysis Model*

The non-stationarity is a specific problem and the systematic study of the tracking properties is generally quite complicated. Nevertheless, in this section we will extend the concepts discussed in Sect. 5.1.2.6, and already discussed above for the LMS (see Sect. 5.4), in the simplest case in which only the reference $d[n]$ is not stationary. In this situation, the correlation is static while the cross-correlation is time variant $\mathbf{g} \rightarrow \mathbf{g}_n$.

This section introduces a general methodology for the AF tracking performance analysis with a generic type adaptation law $\mathbf{w}_n = \mathbf{w}_{n-1} + \mu g\big(e[n]\big)\mathbf{x}_n$, when the only reference is a signal generated by a nonstationary stochastic system. In particular, $d[n]$ is a moving average time series, characterized by an FDE with time-varying coefficients. To define a general mode that allows a meaningful analysis and available in closed form, the parameters variation law of the reference generation equation consists of a first-order Markov process.

6.6.1.1 Generation of Nonstationary Process with Random Walk Model

The model for the nonstationary stochastic process generation $d[n]$, illustrated in Fig. 6.9, is defined by the law

$$d[n] = \mathbf{w}_{0,n-1}^H \mathbf{x}_n + v[n] \tag{6.137}$$

in which the *moving average* (MA) vector $\mathbf{w}_{0,n}$ is time variant and where $v[n]$ represents zero-mean WGN, independent of $\mathbf{x}_n$, with constant variance. Note that the (6.137) represents the time-variant generalization of the expression (6.80) used for the RLS analysis (see Sect. 6.4.5, Fig. 6.3).

In addition to the signal generation model (6.137), we must also define the time-varying MA process for generating the $\mathbf{w}_{0,n}$ coefficients. A wide use paradigm in the AF literature for this purpose is to said *random walk* in which the parameters generation $\mathbf{w}_{0,n}$ is considered to be the output of a MIMO linear dynamic system described by the following FDE:

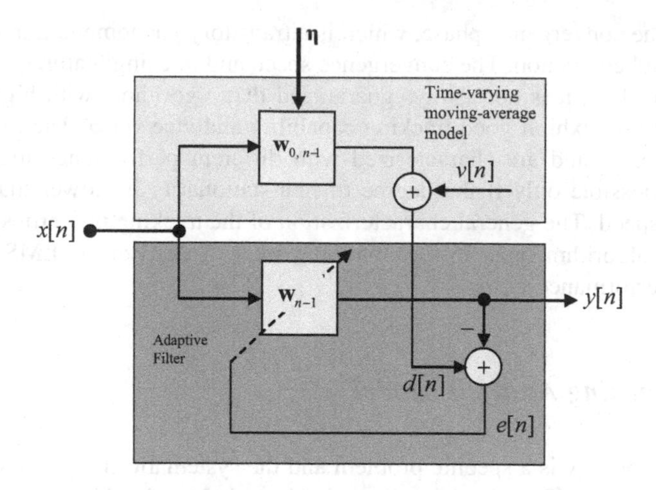

Fig. 6.9 Nonstationary model for AF tracking properties analysis

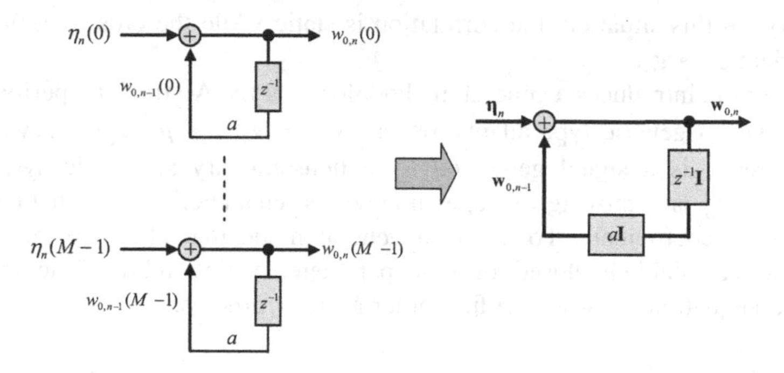

Fig. 6.10 Random walk model, with the first-order Markov process, for the generation of time-varying filter coefficients $\mathbf{w}_{0,n}$

$$\mathbf{w}_{0,n} = a\mathbf{w}_{0,n-1} + \boldsymbol{\eta}_n, \qquad \textit{First-order Markov process} \tag{6.138}$$

which represents a *first-order Markov process* illustrated in Fig. 6.10. In (6.138), the term a represents a fixed model parameter and $\boldsymbol{\eta}_n$ zero-mean WGN, independent of $\mathbf{x}_n$ and $v[n]$, with correlation matrix $\mathbf{R}_\eta = E\{\boldsymbol{\eta}\boldsymbol{\eta}^H\}$.

In practice, the vector $\mathbf{w}_{0,n}$ is generated by noise source $\boldsymbol{\eta}_n$ low-pass filtered with a single pole, with TF $1/(1 - az^{-1})$, filter bank. To obtain a very slow rate change of the model parameters, or to produce significant changes in the vector $\mathbf{w}_{0,n}$, only after several adaptation iterations, the filter TF is chosen with a very low cutoff frequency. For this reason, the parameter a has a value $0 \ll a < 1$, i.e., the pole is close to the unit circle so as to ensure a bandwidth much less than the bandwidth of the process $\boldsymbol{\eta}_n$.

In summary, basic assumptions of an analysis model of an AF tracking properties are

A. the input sequence $\mathbf{x}_n$ is a zero-mean WGN $\mathbf{x}_n \sim N(\mathbf{0},\mathbf{R})$;
B. for the desired output is $d[n] = \mathbf{w}_{0,n-1}^H \mathbf{x}_n + v[n]$, where $v[n] \sim N(0,\sigma_v^2)$ with constant variance;
C. the non-stationarity is modeled as $\mathbf{w}_{0,n} = a\mathbf{w}_{0,n-1} + \boldsymbol{\eta}_n$ and $\boldsymbol{\eta}_n \sim N(\mathbf{0},\mathbf{R}_\eta)$ with a close to 1;
D. the sequences $\mathbf{x}_n$, $v[n]$, and $\boldsymbol{\eta}_n$ are mutually independent (iid).

With these assumptions, the system non-stationarity is due to the sole presence of the time-variant vector $\mathbf{w}_{0,n}$.

6.6.1.2 Minimum Error Energy

In the case of statistically optimum stationary filter (see Sect. 5.4.1.1), we know that the minimum error energy is identical to the measurement noise variance σ_0^2. In the case of time varying, where $\mathbf{w}_0 \to \mathbf{w}_{0,n}$, the determination of minimum error is also attributable to the Wiener theory. Accordingly, if $\mathbf{w}_n \approx \mathbf{w}_{0,n}\ \forall n$, given the constancy of variance σ_v^2, even if not stationary, the minimum energy of error is

$$J_{\min} \approx \sigma_v^2. \tag{6.139}$$

6.6.2 *Performance Analysis Indices and Fundamental Relationships*

The nonstationary AF performance analysis is carried out by generalizing the standard methodologies previously defined (see Sect. 5.1.2). It is therefore necessary to redefine some variables already used for the stationary case. In this case the WEV $\mathbf{u}_n$ is redefined as a

$$\mathbf{u}_n = \mathbf{w}_n - \mathbf{w}_{0,n}. \tag{6.140}$$

We define also the *optimal solution a priori error* as

$$e_a[n] = \mathbf{x}_n^H \mathbf{w}_{n-1} - \mathbf{x}_n^H \mathbf{w}_{0,n-1} = \mathbf{x}_n^H \mathbf{u}_{n-1} \tag{6.141}$$

while the *optimal solution a posteriori error* is defined as

$$e_p[n] = \mathbf{x}_n^H \mathbf{w}_n - \mathbf{x}_n^H \mathbf{w}_{0,n} = \mathbf{x}_n^H \mathbf{u}_n. \tag{6.142}$$

6.6.2.1 Excess Error

The *a priori* error $e[n] = d[n] - \mathbf{x}_n^H \mathbf{w}_{n-1}$, considering the generation model (6.137) and the (6.140), can be expressed as

$$\begin{aligned} e[n] &= v[n] + \mathbf{x}_n^H \mathbf{w}_{0,n-1} - \mathbf{x}_n^H \mathbf{w}_{n-1} \\ &= v[n] - e_a[n]. \end{aligned} \tag{6.143}$$

For the independence hypothesis we have that $E\left\{|e[n]|^2\right\} = E\left\{|v[n]|^2\right\} - E\left\{|e_a[n]|^2\right\}$ and, since $J_{\min} = \sigma_v^2$, we get

$$E\left\{|e[n]|^2\right\} - J_{\min} = E\left\{|e_a[n]|^2\right\}.$$

It follows that for the excess of MSE (EMSE) in the nonstationary case we can write that

$$J_{EMSE} = \lim_{n \to \infty} E\left\{|e_a[n]|^2\right\}. \tag{6.144}$$

For which, in the nonstationary case, the EMSE can be calculated by evaluating the steady-state variance of the *a priori* error estimation.

6.6.2.2 Misalignment and Non-stationarity Degree

The EMSE lower limit can be determined as follows. From the definition of the WEV, considering $\mathbf{w}_{0,n} = \mathbf{w}_{0,n-1} + \boldsymbol{\eta}_n$, we can write

$$\begin{aligned} e_a[n] &= \mathbf{x}_n^H \mathbf{w}_{n-1} - \mathbf{x}_n^H \mathbf{w}_{0,n-1} \\ &= \mathbf{x}_n^H \mathbf{w}_{n-1} - \mathbf{x}_n^H (\mathbf{w}_{0,n-2} + \boldsymbol{\eta}_{n-1}) \\ &= \mathbf{x}_n^H (\mathbf{w}_{n-1} - \mathbf{w}_{0,n-2}) + \mathbf{x}_n^H \boldsymbol{\eta}_{n-1}. \end{aligned} \tag{6.145}$$

Taking the second-order moment and considering the independence, we obtain

$$\begin{aligned} E\left\{|e_a[n]|^2\right\} &= E\left\{\left|\mathbf{x}_n^H (\mathbf{w}_{n-1} - \mathbf{w}_{0,n-2}) + \mathbf{x}_n^H \boldsymbol{\eta}_{n-1}\right|^2\right\} \\ &= E\left\{\left|\mathbf{x}_n^H (\mathbf{w}_{n-1} - \mathbf{w}_{0,n-2})\right|^2\right\} + E\left\{\left|\mathbf{x}_n^H \boldsymbol{\eta}_{n-1}\right|^2\right\} \\ &\approx E\left\{\left|\mathbf{x}_n^H \boldsymbol{\eta}_{n-1}\right|^2\right\} \\ &= \mathrm{tr}(\mathbf{R}\mathbf{R}_\eta). \end{aligned} \tag{6.146}$$

The misalignment (see Sect. 5.1.2.4) is therefore

$$\mathcal{M} \triangleq \frac{J_{\text{EMSE}}}{J_{\min}} \geq \frac{\mathrm{tr}(\mathbf{R}\mathbf{R}_\eta)}{\sigma_v^2}. \tag{6.147}$$

It defines the *non-stationarity degree* as the square root of the previous expression:

$$\mathcal{DN} \triangleq \sqrt{\frac{\mathrm{tr}(\mathbf{R}\mathbf{R}_\eta)}{\sigma_v^2}}. \tag{6.148}$$

For small values, $(\mathcal{DN} \ll 1)$, it has a high degree of traceability of nonstationary environments. On the contrary, for $\mathcal{DN} > 1$, the statistical variation of the environment is too fast to be properly tracked.

6.6.2.3 Weights Error Vector Mean Square Deviation and Correlation Matrix

The scalar quantity $\mathcal{D}_n$ called WEV mean square deviation (MSD) is defined as

$$\mathcal{D}_n \triangleq E\left\{\|\mathbf{u}_n\|_2^2\right\} = E\left\{\|\mathbf{w}_n - \mathbf{w}_{0,n}\|_2^2\right\}. \tag{6.149}$$

The MSD, although is not a measurable quantity, represents a very important paradigm for the theoretical analysis of the statistical adaptive algorithms. It is also noted that (see Sect. 5.1.2.3) the WEV correlation matrix is defined as

$$\mathbf{K}_n \triangleq E\{\mathbf{u}_n\mathbf{u}_n^H\} \tag{6.150}$$

for which, in order to have good tracking properties, $\mathbf{K}_n$ must also be small.

To perform a more detailed analysis, it is necessary to separate the effects of non-stationarity from those due to measurement noise [12]. In this regard, it is useful to express the WEV as the sum of two independent terms

$$\begin{aligned}\mathbf{u}_n &= \mathbf{w}_n - \mathbf{w}_{0,n}\\ &= [\mathbf{w}_n - E\{\mathbf{w}_n\}] + [E\{\mathbf{w}_n\} - \mathbf{w}_{0,n}]\\ &\triangleq \mathbf{u}_n^{\mathrm{wen}} + \mathbf{u}_n^{\mathrm{lag}}\end{aligned} \tag{6.151}$$

where

$$\mathbf{u}_n^{\mathrm{wen}} = \mathbf{w}_n - E\{\mathbf{w}_n\} \tag{6.152}$$

defined as *weight error noise* (WEN) is the term due to measurement noise, while the term

$$\mathbf{u}_n^{\mathrm{lag}} = E\{\mathbf{w}_n\} - \mathbf{w}_{0,n} \tag{6.153}$$

defined as *weight error lag* (LAG) represents the degree of non-stationarity due to the change of the coefficients $\mathbf{w}_{0,n}$. For the independence of the two terms we have that

$$E\{\mathbf{u}_n^{\text{wen}H}\mathbf{u}_n^{\text{lag}}\} = E\{\mathbf{u}_n^{\text{lag}H}\mathbf{u}_{1n}^{\text{wen}}\} = 0 \tag{6.154}$$

and defining $\mathcal{D}_n^{\text{wen}} = E\left\{\left\|\mathbf{u}_n^{\text{wen}}\right\|_2^2\right\}$ and $\mathcal{D}_n^{\text{lag}} = E\left\{\left\|\mathbf{u}_n^{\text{lag}}\right\|_2^2\right\}$, we get

$$\mathcal{D}_n = \mathcal{D}_n^{\text{wen}} + \mathcal{D}_n^{\text{lag}}.$$

From the previous decomposition also the EMSE can be expressed as the sum of two contributions $J_{\text{EMSE}} = J_{\text{ESME}}^{\text{wen}} + J_{\text{ESME}}^{\text{lag}}$. The first term is due to WEN $\mathbf{u}_{1n}$ and is called *estimation noise*. The second term is related to the term $\mathbf{u}_{2n}$ and is said *delay noise*. The presence of the contribution $J_{2\text{EMSE}}$ is due to the nonstationary nature of the problem.

Correspondingly also the misalignment can be decomposed as the sum of two terms

$$\begin{aligned} \mathcal{M} &\triangleq \frac{J_{\text{ESME}}^{\text{wen}}}{\sigma_v^2} + \frac{J_{\text{ESME}}^{\text{lag}}}{\sigma_v^2} \\ &= \mathcal{M}^{\text{wen}} + \mathcal{M}^{\text{lag}}. \end{aligned} \tag{6.155}$$

6.6.3 Tracking Performance of LMS Algorithm

For the behavior characterization of the LMS in nonstationary environment, it is necessary to redefine the SDE (see Sect. 5.4.2) in the specific model described in Fig. 6.9.

Consider the LMS adaptation equation

$$\mathbf{w}_n = \mathbf{w}_{n-1} + \mu e^*[n]\mathbf{x}_n \tag{6.156}$$

From the error expression $e[n] = d[n] - y[n]$ and from the WEV definition, we can write

$$\begin{aligned} e[n] &= d[n] - \mathbf{x}_n^H\mathbf{w}_{n-1} \\ &= d[n] - \mathbf{x}_n^H\mathbf{w}_{n-1} + \mathbf{x}_n^H\mathbf{w}_{0,n-1} - \mathbf{x}_n^H\mathbf{w}_{0,n-1} \\ &= v[n] - \mathbf{x}_n^H\mathbf{u}_{n-1} \end{aligned} \tag{6.157}$$

where $v[n] = d[n] - \mathbf{x}_n^H\mathbf{w}_{0,n-1}$. Substituting in (6.156), (6.157), and (6.138), for $a = 1$, taking into account the fundamental assumptions of the analysis model (A.-D.), we get the SDE (5.144) $\left(\mathbf{u}_n = \left(\mathbf{I} - \mu\mathbf{x}_n\mathbf{x}_n^H\right)\mathbf{u}_{n-1} + \mu v^*[n]\mathbf{x}_n\right)$. In the case of nonstationary environment, we have that

$$\mathbf{u}_n = \left(\mathbf{I} - \mu\mathbf{x}_n\mathbf{x}_n^H\right)\mathbf{u}_{n-1} + \mu v^*[n]\mathbf{x}_n + \boldsymbol{\eta}_n. \tag{6.158}$$

The *weak convergence analysis* can be made by proceeding to the SDE solution with the DAM (see Sect. 5.4.2.1). The solution is studied in average in the condition of very small learning rate. In fact, for $\mu \ll 1$ the term $(\mathbf{I} - \mu\mathbf{x}_n\mathbf{x}_n^H)$, in (6.158), can be approximated as $(\mathbf{I} - \mu\mathbf{R})$ and, with this hypothesis, (6.158) is rewritten as

$$\mathbf{u}_n = (\mathbf{I} - \mu\mathbf{R})\mathbf{u}_{n-1} + \mu v^*[n]\mathbf{x}_n + \boldsymbol{\eta}_n. \tag{6.159}$$

For the tracking properties definition is necessary to consider the average second-order solution or evaluate the trend of the term $\mathbf{K}_n = E\{\mathbf{u}_n\mathbf{u}_n^H\}$.

6.6.3.1 Mean Square Convergence of Nonstationary LMS: MSD Analysis

Multiplying both sides of the above for the respective Hermitian (remembering that $\mathbf{K}_n = E\{\mathbf{u}_n\mathbf{u}_n^H\}$), taking the expectation, and considering the independence (for which the cross-products expectations are zero), we obtain

$$\begin{aligned}\mathbf{K}_n &= E\left\{\left[(\mathbf{I} - \mu\mathbf{R})\mathbf{u}_{n-1} + \mu v^*[n]\mathbf{x}_n + \boldsymbol{\eta}_{0,n}\right]\left[(\mathbf{I} - \mu\mathbf{R})\mathbf{u}_{n-1} + \mu v^*[n]\mathbf{x}_n + \boldsymbol{\eta}_{0,n}\right]^H\right\}\\ &= E\left\{(\mathbf{I} - \mu\mathbf{R})\mathbf{u}_{n-1}\mathbf{u}_{n-1}^H(\mathbf{I} - \mu\mathbf{R})\right\} + E\left\{\mu^2 v^*[\mathrm{n}]\mathbf{x}_n\mathbf{x}_n^H\right\} + E\left\{\boldsymbol{\eta}_{0,n}\boldsymbol{\eta}_{0,n}^H\right\}\\ &= (\mathbf{I} - \mu\mathbf{R})\mathbf{K}_{n-1}(\mathbf{I} - \mu\mathbf{R}) + \mu^2\sigma_v^2\mathbf{R} + \mathbf{R}_\eta.\end{aligned} \tag{6.160}$$

At steady state, for large n, we can assume $\mathbf{K}_n \approx \mathbf{K}_{n-1}$ and the previous results

$$\begin{aligned}\mathbf{K}_n &= (\mathbf{I} - \mu\mathbf{R})\mathbf{K}_n(\mathbf{I} - \mu\mathbf{R}) + \mu^2\sigma_v^2\mathbf{R} + \mathbf{R}_\eta\\ &= \mathbf{K}_n - \mu\mathbf{R}\mathbf{K}_n - \mu\mathbf{K}_n\mathbf{R} + \mu^2\mathbf{R}\mathbf{K}_n\mathbf{R} + \mu^2\sigma_v^2\mathbf{R} + \mathbf{R}_\eta.\end{aligned}$$

For $\mu \ll 1$, the term $\mu^2\mathbf{R}\mathbf{K}_n\mathbf{R}$ can be neglected. With this simplification, the above is rewritten as

$$\mathbf{R}\mathbf{K}_n + \mathbf{K}_n\mathbf{R} \approx \mu\sigma_v^2\mathbf{R} + \frac{1}{\mu}\mathbf{R}_\eta.$$

Multiplying both sides by $\mathbf{R}^{-1}$ and recalling that $\mathrm{tr}(\mathbf{K}_n) = \mathrm{tr}(\mathbf{R}^{-1}\mathbf{K}_n\mathbf{R})$ and $\mathrm{tr}(\mathbf{I}) = M$ we can write

$$\mathrm{tr}(\mathbf{K}_n) \approx \mu\sigma_v^2\frac{M}{2} + \frac{\mathrm{tr}(\mathbf{R}^{-1}\mathbf{R}_\eta)}{2\mu}.$$

For $n \to \infty$ we have that $\mathcal{D}_n = \mathrm{tr}(\mathbf{K}_n)$ for which the MSD can be written as

$$\mathcal{D}_n = \mu\sigma_v^2 \frac{M}{2} + \frac{\mathrm{tr}\left(\mathbf{R}^{-1}\mathbf{R}_\eta\right)}{2\mu}.$$

Note that the MSD is given by the sum of two contributions. The first, called *estimation deviation*, is due to the measurement noise variance and directly proportional to μ. The other, referred to as *lag deviation*, is dependent and inversely proportional to μ.

Equating the two contributions we can define an optimal step size μ_{opt} as

$$\mu_{\mathrm{opt}} = \sqrt{\frac{\mathrm{tr}\left(\mathbf{R}^{-1}\mathbf{R}_\eta\right)}{\sigma_v^2 M}}$$

or

$$\mathcal{D}_\infty = \sqrt{\mathrm{tr}\left(\mathbf{R}^{-1}\mathbf{R}_\eta\right)\sigma_v^2 M}.$$

6.6.4 RLS Performance in Nonstationary Environment

To determine the nonstationary RLS performance, note that the update equation with *a priori* error RLS [(6.64), see Sect. 6.4.3], we have that $\mathbf{w}_n = \mathbf{w}_{n-1} + \mathbf{P}_n\mathbf{x}_n e^*[n]$ where the error is defined as $e^*[n] = d^*[n] - \mathbf{x}_n^H\mathbf{w}_{n-1}$. It follows that RLS update expression is

$$\mathbf{w}_n = \mathbf{w}_{n-1} + \mathbf{P}_n\mathbf{x}_n\left(d^*[n] - \mathbf{x}_n^H\mathbf{w}_{n-1}\right). \tag{6.161}$$

With reference to Fig. 6.9, the desired output is $d^*[n] = \mathbf{x}_n^H\mathbf{w}_{0,n-1} + v^*[n]$ for which, substituting in (6.161), we can write

$$\begin{aligned}\mathbf{w}_n &= \mathbf{w}_{n-1} + \mathbf{P}_n\mathbf{x}_n\left(\mathbf{x}_n^H\mathbf{w}_{0,n-1} + v^*[n] - \mathbf{x}_n^H\mathbf{w}_{n-1}\right)\\ &= \mathbf{w}_{n-1} + \mathbf{P}_n\mathbf{x}_n\mathbf{x}_n^H\mathbf{w}_{0,n-1} - \mathbf{P}_n\mathbf{x}_n\mathbf{x}_n^H\mathbf{w}_{n-1} + \mathbf{P}_n\mathbf{x}_n v^*[n].\end{aligned}$$

Subtracting in both members of the term $\mathbf{w}_{0,n}$, and from the WEV definition, we have that

$$\begin{aligned}\mathbf{u}_n &= \mathbf{w}_{n-1} - \mathbf{w}_{0,n} - \mathbf{P}_n\mathbf{x}_n\mathbf{x}_n^H\mathbf{w}_{n-1} + \mathbf{P}_n\mathbf{x}_n\mathbf{x}_n^H\mathbf{w}_{0,n-1} + \mathbf{P}_n\mathbf{x}_n v^*[n]\\ &= -\mathbf{P}_n\mathbf{x}_n\mathbf{x}_n^H\mathbf{u}_{n-1} + \mathbf{w}_{n-1} - \mathbf{w}_{0,n} + \mathbf{P}_n\mathbf{x}_n v^*[n].\end{aligned}$$

From (6.138), place for simplicity $a = 1$, that is, $\mathbf{w}_{0,n} = \mathbf{w}_{0,n-1} + \boldsymbol{\eta}_n$, and replacing in the above expression, the SDE in terms of RLS error vector is

$$\begin{aligned}\mathbf{u}_n &= -\mathbf{P}_n\mathbf{x}_n\mathbf{x}_n^H\mathbf{u}_{n-1} + \mathbf{w}_{n-1} - \mathbf{w}_{0,n-1} - \boldsymbol{\eta}_n + \mathbf{P}_n\mathbf{x}_n v^*[n] \\ &= \left(\mathbf{I} - \mathbf{P}_n\mathbf{x}_n\mathbf{x}_n^H\right)\mathbf{u}_{n-1} - \boldsymbol{\eta}_n + \mathbf{P}_n\mathbf{x}_n v^*[n].\end{aligned} \quad (6.162)$$

6.6.4.1 Mean Square Convergence of Nonstationary RLS: MSD Analysis

Let us consider (6.162), and with the approximation $E\{\mathbf{R}_{xx,n}\} = \frac{1-\lambda^n}{1-\lambda}\mathbf{R} \approx \frac{1}{1-\lambda}\mathbf{R}$ [see (6.94)] we have that

$$\mathbf{u}_n = \left(\mathbf{I} - (1-\lambda)\mathbf{R}^{-1}\mathbf{x}_n\mathbf{x}_n^H\right)\mathbf{u}_{n-1} + (1-\lambda)\mathbf{R}^{-1}\mathbf{x}_n v^*[n] - \boldsymbol{\eta}_n. \quad (6.163)$$

For $(1-\lambda) \ll 1$ we can use the DAM discussed above (see Sect. 5.4.2.1) for which considering the approximation $\mathbf{x}_n\mathbf{x}_n^H \sim \mathbf{R}$, it follows that the SDE (6.163) takes the form

$$\mathbf{u}_n = \lambda\mathbf{u}_{n-1} + (1-\lambda)\mathbf{R}^{-1}\mathbf{x}_n v^*[n] - \boldsymbol{\eta}_n. \quad (6.164)$$

Multiplying both sides of the above for the respective Hermitian, taking the expectation, and considering the independence (for which the expectations of cross-products are zero), we obtain

$$E\{\mathbf{u}_n\mathbf{u}_n^H\} = \lambda^2 E\{\mathbf{u}_{n-1}\mathbf{u}_{n-1}^H\} + (1-\lambda)^2 E\{\mathbf{R}^{-1}\mathbf{x}_n v^*[n]v[n]\mathbf{R}^{-1}\mathbf{x}_n^H\} - \mathbf{R}_\eta$$

In terms of MSD

$$\mathbf{K}_n = \lambda^2\mathbf{K}_{n-1} + (1-\lambda)^2\sigma_v^2\mathbf{R}^{-1} - \mathbf{R}_\eta. \quad (6.165)$$

For large n $\mathbf{K}_n \approx \mathbf{K}_{n-1}$, for which

$$\left(1-\lambda^2\right)\mathbf{K}_n = (1-\lambda)^2\sigma_v^2\mathbf{R}^{-1} - \mathbf{R}_\eta.$$

Furthermore, $(1-\lambda) \ll 1$, the following approximation applies $(1-\lambda)^2 \approx 2(1-\lambda)$

$$\mathbf{K}_n \approx \frac{(1-\lambda)}{2}\sigma_v^2\mathbf{R}^{-1} - \frac{1}{2(1-\lambda)}\mathbf{R}_\eta, \quad n \to \infty.$$

For $n \to \infty$ we have that $\mathcal{D}_n = \mathrm{tr}(\mathbf{K}_n)$, for which the MSD

$$\mathcal{D}_n \approx \frac{(1-\lambda)}{2}\sigma_v^2\mathrm{tr}\left[\mathbf{R}^{-1}\right] - \frac{1}{2(1-\lambda)}\mathrm{tr}\left[\mathbf{R}_\eta\right],$$

is given by the sum of two contributions. The first, called *estimation deviation*, is due to the variance of the measurement noise $v[n]$ and directly proportional to

$(1 - \lambda)$. The other, referred to as *lag deviation*, depends on the noise process $\mathbf{R}_\eta$ and inversely proportional to $(1 - \lambda)$.

Equating the two contributions we can define an optimal forgetting factor λ_{opt} as

$$\lambda_{\text{opt}} \approx 1 - \frac{1}{\sigma_v}\sqrt{\frac{\text{tr}(\mathbf{R}_\eta)}{\text{tr}(\mathbf{R})}}$$

or

$$\mathcal{D}_\infty \approx \sigma_v\sqrt{\text{tr}\left(\mathbf{R}^{-1}\mathbf{R}_\eta\right)}.$$

6.7 MIMO Error Sequential Regression Algorithms

From the formalism already defined in Chap. 3 and the LMS MIMO introduced in Chap. 5, we extend to the case of multi-channel some ESR algorithms.

Considering the formalism already introduced in Chap. 3 and briefly illustrated in Fig. 6.11, indicating respectively with $\mathbf{y}[n], \mathbf{d}[n] \in (\mathbb{R},\mathbb{C})^{Q\times 1}$ the output and desired output snapshot and with $\mathbf{e}[n] \in (\mathbb{R},\mathbb{C})^{Q\times 1}$ the *a priori* error vector, we can write

$$\begin{aligned}\mathbf{e}[n] &= \mathbf{d}[n] - \mathbf{y}[n] \\ &= \mathbf{d}[n] - \mathbf{W}_{n-1}\mathbf{x}.\end{aligned} \tag{6.166}$$

Considering the jth system output (see Sect. 3.2.2.3), that is,

$$e_j[n] = d_j[n] - \mathbf{x}^H\mathbf{w}_{j,} \qquad \text{for} \qquad j = 1, 2, ..., Q \tag{6.167}$$

where we remind the reader that $\mathbf{w}_{j:}^T \in (\mathbb{R},\mathbb{C})^{1\times P(M)}$ indicates the jth row of the matrix $\mathbf{W}$.

6.7.1 *MIMO RLS*

The MIMO RLS algorithm, with *a priori* update, can be easily formulated by considering the Q filters bank (each of P-channels), or Q independent of each other MISO systems as described by (6.167) (see Fig. 6.11). Considering the composite input $\mathbf{x}$, the correlation matrix for multi-channel RLS is defined as $\mathbf{R}_{xx,n} \in (\mathbb{R},\mathbb{C})^{P(M)\times P(M)} = \sum_{k=0}^{n}\lambda^{n-k}\mathbf{x}_k\mathbf{x}_k^H$ (also see Sect. 3.3.8). So we have that

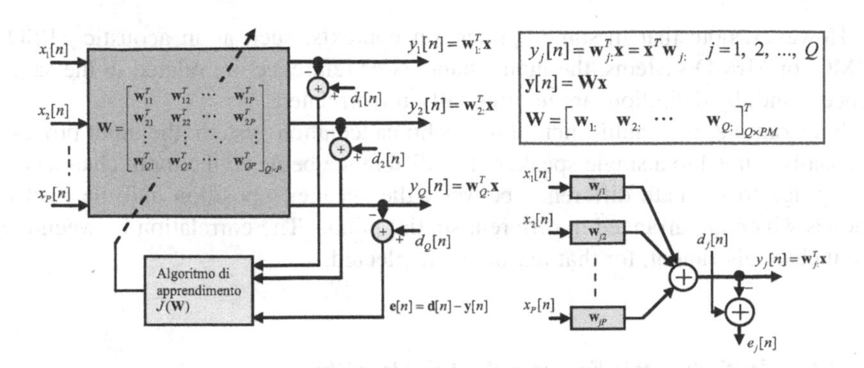

Fig. 6.11 The notation MIMO adaptive filter

$$
\mathbf{R}_{xx,n} = \sum_{k=0}^{n} \lambda^{n-k} \begin{bmatrix} \mathbf{x}_{1,k} \\ \vdots \\ \mathbf{x}_{P,k} \end{bmatrix} \begin{bmatrix} \mathbf{x}_{1,k}^H & \cdots & \mathbf{x}_{P,k}^H \end{bmatrix}
= \begin{bmatrix} \mathbf{R}_{x_1x_1,n} & \mathbf{R}_{x_1x_2,n} & \cdots & \mathbf{R}_{x_1x_P,n} \\ \mathbf{R}_{x_2x_1,n}^* & \mathbf{R}_{x_2x_2,n} & \cdots & \mathbf{R}_{x_2x_P,n} \\ \vdots & \vdots & \ddots & \vdots \\ \mathbf{R}_{x_Px_1,n}^* & \mathbf{R}_{x_Px_2,n}^* & \cdots & \mathbf{R}_{x_Px_P,n} \end{bmatrix}_{P\times P} \tag{6.168}
$$

with $\mathbf{R}_{x_ix_j,n} \in (\mathbb{R},\mathbb{C})^{M\times M} = \sum_{k=0}^{n} \lambda^{n-k}\mathbf{x}_{i,k}\mathbf{x}_{j,k}^H$.

By extending the RLS adaptation rule (see Table 6.3), we get

$$
\begin{aligned}
e_j[n] &= d_j[n] - \mathbf{w}_{j:,n-1}^H\mathbf{x}_n \\
\mathbf{k}_n &= \mathbf{R}_{xx,n}^{-1}\mathbf{x}_n, \qquad\qquad j = 1, 2, ..., Q \\
\mathbf{w}_{j:,n} &= \mathbf{w}_{j:,n-1} + \mathbf{k}_n e_j^*[n]
\end{aligned} \tag{6.169}
$$

where the Kalman gain is identical for all sub-MISO systems.

6.7.2 Low-Diversity Inputs MIMO Adaptive Filtering

In adaptive MISO systems an important aspect concerns the correlation between the input processes. In fact, as an extreme case, if we feed MISO AF, with an identical process on all inputs it is obvious that the MISO system is equivalent to a single filter SISO, with no advantage in using multiple channels. Similar reasoning can be done in the SIMO case when all desired outputs are identical.

The *diversity* between the input processes is, therefore, an essential feature to have an actual benefit in the use of multi-channel adaptive systems.

However, note that in some application contexts, such as in acoustic MISO, SIMO, or MIMO systems, the input channels or reference are related to the same process and, by definition, are not mutually uncorrelated.

For example, in a multi-microphone echo cancelation system, the input process is usually related to a single speaker. The difference between the input channels is solely due to the path difference between the speaker's position and the microphones which are arranged in different spatial points. The correlation between the input channels cannot, for that reason, be neglected.

6.7.2.1 Multi-channels Factorized RLS Algorithm

In many practical situations, the assumption of independent inputs is very strong and, for the adaptation, it is not possible to ignore the correlation between the input channels. In [13–16], to take account of these cross-correlations, it is proposed an improved version of RLS-MISO based on a particular factorization of the inverse cross-correlation matrix. For this purpose we define the vectors $\mathbf{z}_{i,n} \in (\mathbb{R},\mathbb{C})^{M\times 1}$ and matrices $\mathbf{C}_{ij} \in (\mathbb{R},\mathbb{C})^{M\times M}$ such that

$$\begin{aligned}\mathbf{z}_{i,n} &= \sum_{j=1}^{P} \mathbf{C}_{ij,n}\mathbf{x}_{j,n} \\ &= \mathbf{C}_{ii}\mathbf{x}_{i,n} + \sum_{j=1, j\neq i}^{P} \mathbf{C}_{ij,n}\mathbf{x}_{j,n} \\ &= \mathbf{x}_{i,n} - \hat{\mathbf{x}}_{i,n}, \quad i = 1, \ldots, P\end{aligned} \tag{6.170}$$

where $\mathbf{C}_{ii} = \mathbf{I}_{M\times M}$, $\hat{\mathbf{x}}_{i,n} = -\sum_{j=1, i\neq j}^{P} \mathbf{C}_{ij,n}\mathbf{x}_{j,n}$, and the $\mathbf{C}_{ij,n}$ matrices, called *cross-interpolation matrices*, are obtained by minimizing the CF

$$J_n(\mathbf{z}_i) = \sum_{k=0}^{n} \lambda^{n-k}\mathbf{z}_{i,k}^H\mathbf{z}_{i,k}, \quad i = 1, \ldots, P \tag{6.171}$$

where $\mathbf{z}_{i,n}$ are the *interpolation error vectors* and λ is the forgetting factor.

From the above definitions, it is possible to demonstrate that the $\mathbf{R}_{xx,n}^{-1} \in (\mathbb{R},\mathbb{C})^{P(M)\times P(M)}$ matrix can be factorized as

$$\mathbf{R}_{xx,n}^{-1} = \begin{bmatrix} \mathbf{R}_{1,n}^{-1} & \mathbf{0} & \cdots & \mathbf{0} \\ \mathbf{0} & \mathbf{R}_{2,n}^{-1} & \mathbf{0} & \vdots \\ \vdots & \vdots & \ddots & \mathbf{0} \\ \mathbf{0} & \cdots & \mathbf{0} & \mathbf{R}_{P,n}^{-1} \end{bmatrix} \begin{bmatrix} \mathbf{I} & \mathbf{C}_{12,n} & \cdots & \mathbf{C}_{1P,n} \\ \mathbf{C}_{21,n} & \mathbf{I} & \cdots & \vdots \\ \vdots & \vdots & \ddots & \mathbf{C}_{(P-1)P,n} \\ \mathbf{C}_{P1,n} & \cdots & \mathbf{C}_{P(P-1),n} & \mathbf{I} \end{bmatrix} \tag{6.172}$$

where the matrices on the diagonal are defined as

$$\mathbf{R}_{i,n} \in (\mathbb{R}, \mathbb{C})^{M \times M} = \sum_{j=1}^{P} \mathbf{C}_{ij,n} \mathbf{R}_{x_j x_i, n}, \quad i = 1, ..., P. \tag{6.173}$$

Note that the demonstration of the factorization of $\mathbf{R}_{xx,n}^{-1}$ can be made by multiplying both sides of (6.172) by $\mathbf{R}_{xx,n}$ and checking, with the help of (6.171), that the right side is equivalent to an identity matrix (see [16] for details). As an example, consider the two channels case ($P = 2$). From (6.170), we have that

$$\mathbf{z}_1 = \mathbf{x}_1 + \mathbf{C}_{12}\mathbf{x}_2$$
$$\mathbf{z}_2 = \mathbf{x}_2 + \mathbf{C}_{21}\mathbf{x}_1$$

where

$$\mathbf{C}_{12} = -\mathbf{R}_{x_1x_2}\mathbf{R}_{x_2x_2}^{-1}$$
$$\mathbf{C}_{21} = -\mathbf{R}_{x_2x_1}\mathbf{R}_{x_1x_1}^{-1}$$

are the cross-interpolators obtained by minimization, respectively, of the CFs $\sum_{k=0}^{n} \lambda^{n-k}\mathbf{z}_{1,k}^{H}\mathbf{z}_{1,k}$, $\sum_{k=0}^{n} \lambda^{n-k}\mathbf{z}_{2,k}^{H}\mathbf{z}_{2,k}$. It is then that

$$\mathbf{R}_{xx,n}^{-1} = \begin{bmatrix} \mathbf{R}_{1,n}^{-1} & \mathbf{0} \\ \mathbf{0} & \mathbf{R}_{2,n}^{-1} \end{bmatrix} \begin{bmatrix} \mathbf{I} & -\mathbf{R}_{x_1x_2}\mathbf{R}_{x_2x_2}^{-1} \\ -\mathbf{R}_{x_2x_1}\mathbf{R}_{x_1x_1}^{-1} & \mathbf{I} \end{bmatrix}$$

where

$$\mathbf{R}_{1,n} = \mathbf{R}_{x_1x_1} - \mathbf{R}_{x_1x_2}\mathbf{R}_{x_2x_2}^{-1}\mathbf{R}_{x_2x_1}$$
$$\mathbf{R}_{2,n} = \mathbf{R}_{x_2x_2} - \mathbf{R}_{x_2x_1}\mathbf{R}_{x_1x_1}^{-1}\mathbf{R}_{x_1x_2}$$

are the *Schur complement matrices* of $\mathbf{R}_{xx,n}$ with respect to $\mathbf{R}_{x_2x_2}$ and $\mathbf{R}_{x_1x_1}$.

Finally, it appears that from (6.172) the adaptation rule, of the so-called *factorized multi-channel* RLS (6.169), can be written as

$$\mathbf{w}_{ij,n} = \mathbf{w}_{ij,n-1} + \mathbf{R}_{i,n}^{-1}\mathbf{z}_{i,n}e_j[n], \qquad i = 1, 2, ..., P, \ j = 1, 2, ..., Q. \tag{6.174}$$

In practice, the filters of $\mathbf{w}_{ij}$ of the $\mathbf{W}$ matrix are individually adapted, one at a time.

6.7.2.2 Channels Dependent MIMO LMS Algorithm

In the MIMO LMS adaptation, the dependence between the channels, as proposed in [16], can be taken into account in the error gradient. For the vector $\nabla \hat{J}_{j,n-1}(\mathbf{w}_{ji})$ calculation, in addition to the dependence from $\mathbf{w}_{ij}$, is considered the dependence to all its neighboring channels, i.e., to the $\mathbf{w}_{j:}$ filter of the jth row of the matrix

$\mathbf{W}$ adjacent to $\mathbf{w}_{ji}$. In formal terms, considering that the expectation operator $E\{\cdot\}$, for each element of the $\mathbf{W}$ matrix, is imposed, the solution is

$$\begin{aligned}\nabla \hat{J}_{j,n-1} &= \frac{\partial \hat{J}_{j,n-1}}{\partial \mathbf{w}_{ji}}(\mathbf{w}_{j:,n-1}) \\ &= -E\{\mathbf{z}_{j,n}[d_i[n] - \mathbf{x}^H \mathbf{w}_{j:,n-1}\}, \quad j = 1, ..., P\end{aligned} \tag{6.175}$$

where

$$\begin{aligned}\mathbf{z}_{j,n} &= \sum_{k=1}^{P} [\partial \mathbf{w}_{ki} / \partial \mathbf{w}_{ji}]^H \mathbf{x}_{k,n} \\ &= \sum_{k=1}^{P} \mathbf{C}_{jk} \mathbf{x}_{k,n}, \qquad j = 1, ..., P\end{aligned} \tag{6.176}$$

Note, also, the following orthogonality properties:

$$\begin{aligned}E\{\mathbf{x}_j^H \mathbf{z}_{k,n}\} &= 0 \\ E\{\mathbf{z}_{j,n} \mathbf{x}_k^H\} &= \mathbf{0}\end{aligned} \qquad \forall j \quad k = 1, ..., P, \; j \neq k, \; j = 1, ..., P. \tag{6.177}$$

From the previous development, the LMS adaptation takes the form

$$\mathbf{w}_{j:,n} = \mathbf{w}_{j:,n-1} + \mu e_j^*[n] \mathbf{z}, \qquad j = 1, 2, ..., Q \tag{6.178}$$

with

$$\mathbf{z} \in (\mathbb{R}, \mathbb{C})^{PM \times 1} = \begin{bmatrix} \mathbf{z}_{1,n}^T & \mathbf{z}_{2,n}^T & \cdots & \mathbf{z}_{P,n}^T \end{bmatrix}^T. \tag{6.179}$$

Finally, note that the adaptation rule (6.178) can be obtained from the (6.174) by substituting in place of $\mathbf{R}_{i,n}^{-1}$ the $\mathbf{I}$ matrix.

6.7.3 *Multi-channel APA Algorithm*

The multi-channel APA algorithm derivation can be accomplished with minimal perturbation property, by generalizing the SISO method in Sect. 6.3.1.

By defining the vectors $\mathbf{e}_{j,n}$ and $\boldsymbol{\varepsilon}_{j,n}$

$$\mathbf{e}_{j,n} \in (\mathbb{R},\mathbb{C})^{K\times 1} \triangleq \begin{bmatrix} e_j[n] & e_j[n-1] & \cdots & e_j[n-K+1] \end{bmatrix}^T \tag{6.180}$$
$$\mathbf{d}_{j,n} \in (\mathbb{R},\mathbb{C})^{K\times 1} \triangleq \begin{bmatrix} d_j[n] & d_j[n-1] & \cdots & d_j[n-K+1] \end{bmatrix}^T,$$

respectively, as the *a priori* and *a posteriori* error vectors, for the jth channel of the MISO bank, we have that

$$\mathbf{e}_{j,n} = \mathbf{d}_{j,n} - \mathbf{X}_n \mathbf{w}_{j:,n-1} \tag{6.181}$$
$$\boldsymbol{\varepsilon}_{j,n} = \mathbf{d}_{j,n} - \mathbf{X}_n \mathbf{w}_{j:n}. \tag{6.182}$$

The input data matrix is, in this case, defined as

$$\mathbf{X}_n \in \mathbb{R}^{K\times P(M)} \triangleq \begin{bmatrix} \mathbf{X}_{1,n}^H & \mathbf{X}_{2,n}^H & \cdots & \mathbf{X}_{P,n}^H \end{bmatrix} \tag{6.183}$$

where

$$\mathbf{X}_{j,n} \in (\mathbb{R},\mathbb{C})^{M\times K} \triangleq \begin{bmatrix} \mathbf{x}_{j,n} & \mathbf{x}_{j,n-1} & \cdots & \mathbf{x}_{j,n-K+1} \end{bmatrix}^H \tag{6.184}$$

From the minimal perturbation property $\delta\mathbf{w}_{j\ :,n} = \mathbf{X}_n^{\#}\boldsymbol{\alpha}\mathbf{e}_{j,n}$ (see Sect. 6.3.1), it is

$$\mathbf{w}_{j:,n} = \mathbf{w}_{j:,n-1} + \mu \mathbf{X}_{j,n}^H \left[\delta\mathbf{I} + \mathbf{X}_{j,n}\mathbf{X}_{j,n}^H\right]^{-1} \mathbf{e}_{j,n}. \tag{6.185}$$

6.8 General Adaptation Law

In Chap. 4 we have seen how some available *a priori* knowledge can be exploited for the determination of new classes of adaptive algorithms, which allow a more accurate solution. For example, in Sect. 4.2.5.2, the confidence on the solution hypothesis $\overline{\mathbf{w}}$ led to the *regularized LS* algorithm definition, formulated by the inclusion in the CF of a constraint derived from prior knowledge. Even in adaptive algorithms case, the insertion of any *a priori* knowledge can be translated to learning rule redrafting, more appropriate to the problem under consideration.

A first example, already discussed in Sect. 4.3.2.2, is the *iterative weighted LS* algorithm, in which, starting by the standard weighted LS, can be defined its recursive version.

Here, in light of the previous three chapters, we present a new more general adaptive paradigm that makes it more feasible for the inclusion, into adaptation rule, of any prior knowledge.

As is known, the adaptation algorithm is treated as a dynamic system in which the weights represent a state variable. Starting from this point of view, by generalizing the form of such a system, it is possible to identify new algorithms classes. As introduced in Chap. 5 (see Sect. 5.1.1.3), recursive approach to optimal filtering, the dynamic system model related to the adaptation procedure, can have a form of the following type:

$$\mathbf{w}_k = \mathbf{w}_{k-1} + \mu_k \mathbf{H}_k \mathbf{v}_k \tag{6.186}$$

where, in the case of stochastic gradient, $\mathbf{v}_k \equiv -\nabla \hat{J}(\mathbf{w}_{k-1})$ and $\mathbf{H}_k \equiv \left[\nabla^2 \hat{J}(\mathbf{w}_{k-1})\right]^{-1}$ are the estimates gradient and the inverse Hessian of the CF. So, by extending the model (6.186), we can identify new paradigms of adaptation.

A first adaptation law, more general than (6.186), is a rule in which the weights $\mathbf{w}_k$ linearly depend on the weights of the instantly $(k-1)$. In formal terms, we can write

$$\mathbf{w}_k = \mathbf{M}_k \mathbf{w}_{k-1} + \hat{\mathbf{v}}_k \tag{6.187}$$

where $\mathbf{M}_k$ and $\hat{\mathbf{v}}_k$ are independent of $\mathbf{w}_k$. For example, in (6.186) $\hat{\mathbf{v}}_k = \mu_k \mathbf{H}_k \mathbf{v}_k$ and $\mathbf{M}_k = \mathbf{I}$. A second, even more general, model consists in the definition of a nonlinear relationship of the type

$$\mathbf{w}_k = \mathbf{M}(\mathbf{w}_{k-1}) + \hat{\mathbf{v}}_k \tag{6.188}$$

where $\mathbf{M}(\cdot)$ is a nonlinear operator of the weights $\mathbf{w}_{k-1}$, determined by any *a priori* knowledge on the processes or on the type of desired solution.

Remark In the previous sections, were presented primarily algorithms of the class described by (6.187) with $\mathbf{M}_k = \mathbf{I}$ and $\hat{\mathbf{v}}_k$ that consists in the gradient (and inverse Hessian) estimate. Classical algorithms such as the LMS, NLMS, APA, RLS etc., can be deduced from a general approach described by (6.187). Note, also, that in the algorithms PNLMS and IPNLMS with the law of adaptation, characterized by an adaptation rule of the type

$$\mathbf{w}_n = \mathbf{w}_{n-1} + \mu \frac{\mathbf{G}_{n-1}\mathbf{x}_n e[n]}{\delta + \mathbf{x}_n^T \mathbf{G}_{n-1}\mathbf{x}_n} \tag{6.189}$$

(see Sect. 5.5.2), the matrix $\mathbf{G}_n$ is a sparsity constraint. In other words, $\mathbf{G}_n$ takes account of *a priori* knowledge and is a function of the weights $\mathbf{w}_{n-1}$ and, in this sense, may be considered as the general algorithm class described by the expression (6.188).

6.8.1 Adaptive Regularized Form, with Sparsity Constraints

Proceeding as in the regularized LS (see Sect. 4.2.5.2), we can consider a CF, to which is added a *stabilizer* or *regularization term*, referred to as $J_s(\mathbf{w}_n)$, which takes into account of available *a priori* knowledge. The regularized CF takes the form

$$J(\mathbf{w}_n) = J_s(\mathbf{w}_n) + \hat{J}(\mathbf{w}_n). \tag{6.190}$$

The above expression, together with the model (6.188), translated into more explicit mode, can be used to derive different classes of adaptation algorithms. The *stabilizing function* is generally a distance $\delta(\mathbf{w}_n, \mathbf{w}_{n-1})$ with a metric that defines the adaptation rule and which can be linear or nonlinear.

A possible choice for the regularization term is represented by a weighted norm of the type

$$\begin{aligned} J_s(\mathbf{w}_n) &\triangleq \delta(\mathbf{w}_n, \mathbf{w}_{n-1}) \\ &= [\mathbf{w}_n - \mathbf{w}_{n-1}]^T \mathbf{Q}_n [\mathbf{w}_n - \mathbf{w}_{n-1}]^T \\ &= \|\mathbf{w}_n - \mathbf{w}_{n-1}\|_{\mathbf{Q}_n}^2. \end{aligned} \tag{6.191}$$

where $\mathbf{Q}_n$ is a positive definite matrix. A further constraint able to mitigate the possible presence of disturbances due to noise, can be expressed as a *minimum energy perturbation constraint*, applied to the weights trajectory and defined as

$$\|\mathbf{w}_n - \mathbf{w}_{n-1}\|_2^2 \leq \delta_{n-1} \tag{6.192}$$

where δ_{n-1} is a positive sequence whose choice influences the algorithm dynamics. In other words, the (6.192) ensures that the noise can perturb the quadratic norm at most by a factor equal to δ_{n-1}.

For the definition of a new class of adaptive algorithms, as suggested in [17], also considering the constraint (6.192), a possible CF $J(\mathbf{w})$ choice is as follows:

$$\mathbf{w}_* = \underset{\mathbf{w}}{\operatorname{argmin}} \left(\|\mathbf{w}_n - \mathbf{w}_{n-1}\|_{\mathbf{Q}_n}^2 + \left(\mathbf{X}_n \mathbf{G}_n \mathbf{X}_n^T\right)^{-1} \boldsymbol{\varepsilon}_n^T \boldsymbol{\varepsilon}_n \right) \tag{6.193}$$

subject to the constraint (6.192), where $\boldsymbol{\varepsilon}_n = \mathbf{d}_n - \mathbf{X}_n \mathbf{w}_n$ is the *a posteriori* error defined in (6.7). The matrices $\mathbf{Q}_n$ and $\mathbf{G}_n$ are positive definite and their choice defines the algorithms class.

In the case in which these matrices depended on the weights $\mathbf{w}_{n-1}$, the parameter space could have Riemannian nature; in other words we would be in the presence of a *differentiable manifold* or *curved manifold* and where the distance properties are not uniform but functions of the point. As we shall see, the use of the *Riemannian manifolds* can allow the insertion of some *a priori* knowledge.

In the simplest case, without the imposition of the constraint (6.192), the CF (6.190) can be written as

$$J(\mathbf{w}_n) = [\mathbf{w}_n - \mathbf{w}_{n-1}]^T \mathbf{Q}_n [\mathbf{w}_n - \mathbf{w}_{n-1}] + \left(\mathbf{X}_n \mathbf{G}_n \mathbf{X}_n^T\right)^{-1} \boldsymbol{\varepsilon}_n^T \boldsymbol{\varepsilon}_n. \tag{6.194}$$

Considering $\nabla J(\mathbf{w}_n) \to \mathbf{0}$ and placing

$$\mathbf{P}_n = \mathbf{X}_n^T \left[\mathbf{X}_n \mathbf{G}_n \mathbf{X}_n^T\right]^{-1} \mathbf{X}_n \tag{6.195}$$

it follows that

$$\begin{aligned}
\frac{\partial J(\mathbf{w}_n)}{\partial \mathbf{w}_n} &= 2\mathbf{Q}_n(\mathbf{w}_n - \mathbf{w}_{n-1}) - 2\mathbf{X}_n^T(\mathbf{X}_n \mathbf{G}_n \mathbf{X}_n^T)^{-1} \boldsymbol{\varepsilon}_n \\
&= \mathbf{Q}_n(\mathbf{w}_n - \mathbf{w}_{n-1}) - \mathbf{X}_n^T(\mathbf{X}_n \mathbf{G}_n \mathbf{X}_n^T)^{-1}(\mathbf{d}_n - \mathbf{X}_n \mathbf{w}_n - \mathbf{X}_n \mathbf{w}_{n-1} + \mathbf{X}_n \mathbf{w}_{n-1}) \\
&= (\mathbf{Q}_n - \mathbf{P}_n)\mathbf{w}_n - (\mathbf{Q}_n + \mathbf{P}_n)\mathbf{w}_{n-1} - \mathbf{X}_n^T(\mathbf{X}_n \mathbf{G}_n \mathbf{X}_n^T)^{-1} \mathbf{e}_n \equiv \mathbf{0}.
\end{aligned} \tag{6.196}$$

Equation (6.196) is characterized by a single minimum for which it is possible to define the adaptation rule, which can be expressed as

$$\mathbf{w}_n = \mathbf{w}_{n-1} + (\mathbf{Q}_n + \mathbf{P}_n)^{-1} \mathbf{X}_n^T \left(\mathbf{X}_n \mathbf{G}_n \mathbf{X}_n^T\right)^{-1} \mathbf{e}_n. \tag{6.197}$$

The reader can observe that for $\mathbf{G}_n = \mathbf{I}$, $\mathbf{P}_n$ is a projection operator (see Sect. A.6.5). The matrices $\mathbf{Q}_n$ and $\mathbf{G}_n$ in (6.197) can be chosen in function of any *a priori* knowledge on the AF application domain.

Below we see how (6.197) can be used for the simple derivation of already known algorithms.

6.8.1.1 Linear Adaptation: The APA and RLS Classes

A class of adaptation algorithms is that in which $\mathbf{G}_n = \mathbf{I}$ and the distance $\delta(\mathbf{w}_n, \mathbf{w}_{n-1})$ is characterized by a symmetric positive definite matrix $\mathbf{Q}_n$ dependent on the signal $x[n]$. In this case, the update equation (6.197) takes the form

$$\mathbf{w}_n = \mathbf{w}_{n-1} + [\mathbf{P}_n + \mathbf{Q}_n]^{-1} \mathbf{X}_n^T \left[\mathbf{X}_n \mathbf{X}_n^T\right]^{-1} \mathbf{e}_n \tag{6.198}$$

for which, considering

$$\mathbf{Q}_n = \mu^{-1}\mathbf{I} - \mathbf{P}_n, \tag{6.199}$$

the adaptation law can be rewritten as

$$\mathbf{w}_n = \mathbf{w}_{n-1} + \mu \mathbf{X}_n^T \left[\mathbf{X}_n \mathbf{X}_n^T\right]^{-1} \mathbf{e}_n \tag{6.200}$$

that appears to be precisely the APA (see Sect. 6.3). While, for $K = 1$, and choosing the matrix $\mathbf{Q}_n$ as,

$$\mathbf{Q}_n = \frac{\mathbf{R}_{xx,n}}{\mathbf{x}_n^T\mathbf{x}_n} - \mathbf{P}_n, \tag{6.201}$$

(6.198) turns out to be a second-order algorithm (see Sect. 6.4)

$$\mathbf{w}_n = \mathbf{w}_{n-1} + \mathbf{R}_{xx,n}^{-1}\mathbf{x}_n e[n]. \tag{6.202}$$

Note that the above adaptation law is the so-called LMS–Newton algorithm (see Sect. 6.2.3).

6.8.1.2 Nonlinear Adaptation with Gradient Descent Along the Natural Gradient: The PNLMS Class

To derive new nonlinear adaptation algorithms classes, place $\mathbf{T}_n = \mathbf{Q}_n\mathbf{G}_n$; we express the distance (6.191) as

$$\delta(\mathbf{w}_n, \mathbf{w}_{n-1}) = [\mathbf{w}_n - \mathbf{w}_{n-1}]^T\mathbf{T}_n[\mathbf{w}_n - \mathbf{w}_{n-1}]^T \tag{6.203}$$

where $\mathbf{T}_n$, symmetric positive definite matrix, is a function of the input $x[n]$ and, being $\mathbf{G}_n \triangleq \mathbf{G}_n(\mathbf{w}_{n-1})$, and of the impulse response $\mathbf{w}_{n-1}$. Equation (6.190) minimization, with the definition (6.203), allows to write an adaptation formula of the type

$$\mathbf{w}_n = \mathbf{w}_{n-1} + [\mathbf{P}_n + \mathbf{T}_n]^{-1}\mathbf{X}_n^T\left[\mathbf{X}_n\mathbf{X}_n^T\right]^{-1}\mathbf{e}_n. \tag{6.204}$$

Equation (6.204) is nonlinear as it appears in the product $\mathbf{Q}_n\mathbf{G}_n$ and the matrix $\mathbf{G}_n$ depends on the impulse response $\mathbf{w}_{n-1}$.

Remark For the presence of the product $\mathbf{Q}_n\mathbf{G}_n$, the distance measure (6.203), is not defined on a Euclidean, but on a curved space, said also *Riemannian space*. The matrix $\mathbf{T}_n = \mathbf{Q}_n\mathbf{G}_n$, is defined as *Riemann metric tensor*, which is a function of the point where the measurement is performed.[3]

From the $\mathbf{Q}_n$ and $\mathbf{G}_n$ matrices definition, it is possible to define certain adaptive algorithm classes. For example, considering the error vector defined as

---

[3] We remind the reader that in Riemannian geometry, for two vectors $\mathbf{w}$ and $\mathbf{w}+\delta\mathbf{w}$ the *metric distance* $d\mathbf{w}(\cdot,\cdot)$, which by definition depends on the space point in which it is located $\mathbf{w}$, is defined as $d(\mathbf{w}, \mathbf{w}+\delta\mathbf{w}) = \sqrt{\sum_{i=0}^{M-1}\sum_{j=0}^{M-1}\delta w_i \delta w_j g_{ij}(\mathbf{w})} = \sqrt{\delta\mathbf{w}^T\mathbf{G}(\mathbf{w})\delta\mathbf{w}}$ where $\mathbf{G}(\mathbf{w}) \in \mathbb{R}^{M\times M}$ is a positive definite matrix representing the Riemann metric tensor. The $\mathbf{G}(\mathbf{w})$ characterizes the curvature of the particular *manifold* of the M-dimensional space. Namely, $\mathbf{G}(\mathbf{w})$ represents a "correction" of Euclidean distance defined for $\mathbf{G}(\mathbf{w}) = \mathbf{I}$.

$\mathbf{e}_n = \begin{bmatrix} e[n] & \cdots & e[n-K+1] \end{bmatrix}^T$, for $K = 1$, and $\mathbf{Q}_n = \mu^{-1}\mathbf{G}_n^{-1} - \mathbf{P}_n$ the adaptation formula (6.204) takes the form

$$\mathbf{w}_n = \mathbf{w}_{n-1} + \mu \frac{\mathbf{G}_n \mathbf{x}_n}{\mathbf{x}_n^T \mathbf{G}_n \mathbf{x}_n} e[k] \tag{6.205}$$

defined as natural gradient algorithm (NGA) proposed in 1998 by Amari (see [18]). In addition, from the specific definition of the matrix $\mathbf{G}_n$, is possible to derive *proportional* algorithms such as PNLMS and IPNLMS (see Sect. 5.5.2).

For $K > 1$, the algorithm (6.205) appears to be

$$\mathbf{w}_n = \mathbf{w}_{n-1} + \mu \frac{\mathbf{G}_n \mathbf{X}_n^T}{\mathbf{X}_n \mathbf{G}_n \mathbf{X}_n^T} \mathbf{e}_n \tag{6.206}$$

defined as natural APA (NAPA) and, depending on $\mathbf{G}_n$ matrix choice, can be derived other proportional algorithms such as the proportional APA (PAPA).

Following the same philosophy, you can derive the *Natural RLS* (NRLS) [13] algorithm, defined as

$$\mathbf{w}_n = \mathbf{w}_{n-1} + \mathbf{G}_n^{1/2} \mathbf{R}_{w,n}^{-1} \mathbf{G}_n^{1/2} \mathbf{x}_n e[n] \tag{6.207}$$

where the matrix $\mathbf{R}_w$ is estimated with the expression

$$\mathbf{R}_{w,n} = \lambda \mathbf{R}_{w,n-1} + \left[\mathbf{G}_n^{1/2} \mathbf{x}_n\right] \left[\mathbf{G}_n^{1/2} \mathbf{x}_n\right]^T. \tag{6.208}$$

6.8.2 Exponentiated Gradient Algorithms Family

The class of exponentiated gradient algorithms (EGA) derives from the particular metric choice in the distance $\delta(\mathbf{w}_n, \mathbf{w}_{n-1})$ measurement. As suggested in [19] and [17] as a distance measure is proposed the *relative entropy* or Kullback–Leibler divergence (KLD) indicated as $\delta_{\text{re}}(\mathbf{w}_n, \mathbf{w}_{n-1})$. Note that the KLD is not a true distance and should be used with care. In practice, for the algorithms development we have to consider: (1) the filter weights always positive, and (2) a minimal perturbation constraint, in terms of L_1 norm.

6.8.2.1 Positive Weights Exponentiated Gradient Algorithm

The KLD is always positive by definition and in the case of all positive weights is a consistent measure. For $K = 1$ and $\mathbf{G}_n = \mathbf{I}$, the general criterion (6.194) is simplified as

$$J(\mathbf{w}_n) = \delta_{\text{re}}(\mathbf{w}_n, \mathbf{w}_{n-1}) + \left[\mathbf{x}_n^T \mathbf{x}_n\right]^{-1} \varepsilon^2[n] \tag{6.209}$$

where for $\mu > 0$

$$\delta_{\text{re}}(\mathbf{w}_n, \mathbf{w}_{n-1}) = \mu^{-1} \sum_{j=0}^{M-1} w_n[j] \ln \frac{w_n[j]}{w_{n-1}[j]}. \tag{6.210}$$

With this formalism vectors $\mathbf{w}_n$ and $\mathbf{w}_{n-1}$ are probabilities vectors, with no negative components, and such that $\|\mathbf{w}_n\|_1 = \|\mathbf{w}_{n-1}\|_1 = u > 0$ where u represents a scale factor. Therefore, for $u = 1$, we consider a CF $J(\mathbf{w}_n)$ with the constraint $\sum_{j=0}^{M-1} w_n[j] = 1$, i.e., substituting (6.210) in (6.209) and considering the constraint, we get

$$\begin{aligned} &\mathbf{w}_* \therefore \underset{\mathbf{w}}{\operatorname{argmin}} \left(\sum_{j=0}^{M-1} w_n[j] \ln \frac{w_n[j]}{w_{n-1}[j]} + \mu \left[\mathbf{x}_n^T \mathbf{x}_n\right]^{-1} \varepsilon^2[n] \right) \\ &\text{s.t.} \\ &\|\mathbf{w}_n\|_1 = \|\mathbf{w}_{n-1}\|_1 = 1. \end{aligned} \tag{6.211}$$

It is shown that the Lagrangian (see Sect. B.3.2) for the constrained problem (6.211) in a scalar form is equal to

$$\left(\ln \frac{w_n[j]}{w_{n-1}[j]} + 1 \right) - 2\mu \left[\mathbf{x}_n^T \mathbf{x}_n\right]^{-1} x[n-j] \varepsilon[n] + \lambda_j = 0, \qquad j = 0, \ldots, M-1 \tag{6.212}$$

where λ_j is the jth Lagrange multiplier. The above expression is rather complex and difficult to solve. Assuming small variations between weights ($\mathbf{w}_n \sim \mathbf{w}_{n-1}$), it is possible to consider the error *a priori* in place of the *a posteriori*. With this assumption, place $\mu_n = 2\mu[\mathbf{x}_n^T \mathbf{x}_n]^{-1}$; (6.212) is approximated as

$$\left(\ln \frac{w_n[j]}{w_{n-1}[j]} + 1 \right) - \mu_n \cdot x[n-j] e[n] + \lambda - 0, \qquad j = 0, \ldots, M-1 \tag{6.213}$$

In this case, solution for $w_n[j]$ is

$$w_n[j] = \frac{w_{n-1}[j] r_n[j]}{\sum_{k=0}^{M-1} w_{n-1}[k] r_n[k]}, \qquad j = 0, \ldots, M-1 \tag{6.214}$$

where

$$r_n[j] = \exp\left(\mu_n \cdot x[n-j]e[n]\right), \quad j = 0, ..., M-1 \tag{6.215}$$

with ICs $\mathbf{w}_0[j] = c > 0, \forall\, j$. In vector form the EGA adaptive algorithm is defined by the relation

$$\mathbf{w}_n = \frac{\mathbf{w}_{n-1} \odot \mathbf{r}_n}{\mathbf{w}_{n-1}^T \mathbf{r}_n} \tag{6.216}$$

in which the operator $\odot$ denotes the Hadamard (or the entrywise) product, i.e., the point-to-point vectors multiplication $\mathbf{r}_n$ and $\mathbf{w}_{n-1}$, and

$$\mathbf{r}_n = \exp\left(\mu_n \mathbf{x}_n e[n]\right). \tag{6.217}$$

Note that the name *exponentiated gradient* derives from expression (6.215) in which the estimate of the *j*th component of the gradient vector $\nabla \hat{J}(\mathbf{w}_n) \triangleq \mu_n \mathbf{x}_n e[n]$ appears as an argument of the exponential function.

6.8.2.2 Positive and Negative Weights Exponentiated Gradient Algorithm

Generalizing the EGA, also for negative weights, is sufficient to express the weight vector as the difference of two positive quantities

$$\mathbf{w}_n = \mathbf{w}_n^+ - \mathbf{w}_n^- \tag{6.218}$$

allowing to express the *a priori* and *a posteriori* errors, respectively, as

$$e[n] = y[n] - \left[\mathbf{w}_{n-1}^+ - \mathbf{w}_{n-1}^-\right]^T \mathbf{x}_n \tag{6.219}$$

$$\varepsilon[n] = y[n] - \left[\mathbf{w}_n^+ - \mathbf{w}_n^-\right]^T \mathbf{x}_n. \tag{6.220}$$

Thus, the CF (6.209) takes the form

$$J\left(\mathbf{w}_n^{\pm}\right) = \delta_{\text{re}}\left(\mathbf{w}_n^+, \mathbf{w}_{n-1}^+\right) + \delta_{\text{re}}\left(\mathbf{w}_n^-, \mathbf{w}_{n-1}^-\right) + \frac{1}{u}\left[\mathbf{x}_n^T \mathbf{x}_n\right]^{-1} \varepsilon^2[n] \tag{6.221}$$

where u represents a scaling constant. Using the KLD the constant u takes the form of constraint of the type $\|\mathbf{w}_n^+\|_1 + \|\mathbf{w}_n^-\|_1 = u > 0$ for which (6.213) is transformed into the pair of expressions

$$\left.\begin{array}{l}\left(\ln\dfrac{w_n^+[j]}{w_{n-1}^+[j]}+1\right)-\dfrac{2\mu_n}{u}\cdot x[n-j]e[n]+\lambda=0\\[2ex]\left(\ln\dfrac{w_n^-[j]}{w_{n-1}^-[j]}+1\right)-\dfrac{2\mu_n}{u}\cdot x[n-j]e[n]+\lambda=0\end{array}\right., \quad j=0,...,M-1. \tag{6.222}$$

Proceeding as in the case of positive weights

$$\begin{aligned}\mathbf{w}_n^+ &= u\frac{\mathbf{w}_{n-1}^+\odot\mathbf{r}_n^+}{\mathbf{w}_{n-1}^{+T}\mathbf{r}_n^+ + \mathbf{w}_{n-1}^{-T}\mathbf{r}_n^-}\\ \mathbf{w}_n^- &= u\frac{\mathbf{w}_{n-1}^-\odot\mathbf{r}_n^-}{\mathbf{w}_{n-1}^{+T}\mathbf{r}_n^+ + \mathbf{w}_{n-1}^{-T}\mathbf{r}_n^-}\end{aligned} \tag{6.223}$$

in which the vectors $\mathbf{r}_n^+$ and $\mathbf{r}_n^-$ take values

$$\mathbf{r}_n^+ = \exp\left(\frac{\mu_n}{u}\mathbf{x}_n e[n]\right) \tag{6.224}$$

$$\mathbf{r}_n^- = \exp\left(-\frac{\mu_n}{u}\mathbf{x}_n e[n]\right) = \frac{1}{\mathbf{r}_n^+}. \tag{6.225}$$

Note that it is worth the expression $u = \|\mathbf{w}_n^+\|_1 + \|\mathbf{w}_n^-\|_1 \geq \|\mathbf{w}_n^+\|_1 - \|\mathbf{w}_n^-\|_1 = \|\mathbf{w}_n\|_1$. It follows that, for convergence, it is necessary to choose the scaling factor such that $u \geq \|\mathbf{w}_n\|_1$.

6.8.2.3 Exponentiated RLS Algorithm

The *a priori* RLS algorithm update is characterized by the formula (see Sect. 6.4.3)

$$\mathbf{w}_n = \mathbf{w}_{n-1} + \mathbf{k}_n e[n] \tag{6.226}$$

where $\mathbf{k}_n$ is the Kalman gain defined as

$$\mathbf{k}_n = \mathbf{R}_{xx,n}^{-1}\mathbf{x}_n \tag{6.227}$$

and the *a priori* error $e[n]$ is defined by (6.219). With the above assumptions the RLS adaptation formulas are identical to (6.224) and (6.225) in which vectors $\mathbf{r}_n^+$ and $\mathbf{r}_n^-$ depend on the Kalman gain and take values

$$\mathbf{r}_n^+ = \exp\left(\frac{\mathbf{k}_n}{u}e[n]\right) \\ = \frac{1}{\mathbf{r}_n^-} \tag{6.228}$$

For further developments and investigations, in the case of sparse adaptive filters and on the natural gradient, refer to the literature [13–20].

References

1. Widrow B, Stearns SD (1985) Adaptive signal processing. Prentice Hall, Englewood Cliffs, NJ
2. Ahmed N, Soldan DL, Hummels DR, Parikh DD (1977) Sequential regression considerations of adaptive filter. IEE Electron Lett 13(15):446–447
3. Ahmed N, Hummels DR, Uhl M, Soldan DL (1979) A short term sequential regression algorithm. IEEE Trans Acoust Speech Signal Process ASSP-27:453
4. Shin HC, Sayed AH (2004) Mean-square performance of a family of affine projection algorithms. IEEE Trans Signal Process 52(1):90–102
5. Sankaran SG, (Louis) Beex AA (2000) Convergence behavior of affine projection algorithms. IEEE Trans Signal Process 48:1086–1096
6. Manolakis DG, Ingle VK, Kogon SM (2000) Statistical and adaptive signal processing. McGraw-Hill, New York, NY
7. Sayed AH (2003) Fundamentals of adaptive filtering. Wiley, New York, NY
8. Haykin S (2001) Kalman filter. In: Haykin S (ed) Kalman filtering and neural networks. Wiley. ISBN 0-471-36998-5
9. Kalman RE (1960) A new approach to linear filtering and prediction problems. J Basic Eng 82:34–45
10. Roweis S, Ghahramani Z (1999) A unifying review of linear Gaussian models. Neural Comput 11(2):305–345
11. Welch G, Bishop G (2006) An introduction to the Kalman filter. TR 95-041, Department of Computer Science, University of North Carolina at Chapel Hill (NC 27599-3175), July
12. Macchi O (1996) The theory of adaptive filtering in a random time-varying environment. In: Figueiras-Vidal AR (ed) Digital signal processing in telecommunications. Springer, London
13. Huang Y, Benesty J, Chen J (2006) Acoustic MIMO signal processing. Springer series on signal and communication technology. ISBN 10 3-540-37630-5
14. Benesty J, Gänsler T, Eneroth P (2000) Multi-channel sound, acoustic MIMO echo cancellation, and multi-channel time-domain adaptive filtering. In: Acoustic signal processing for telecommunication. Kluwer. ISBN 0-7923-7814-8
15. Martin RK, Sethares WA, Williamson RC, Johnson CR Jr (2002) Exploiting sparsity in adaptive filters. IEEE Trans Signal Process 50(8):1883–1894
16. Benesty J, Gänsler T, Huang Y, Rupp M (2004) Adaptive algorithms for MIMO acoustic echo cancellation. In: Audio signal processing for next-generation multimedia communication systems. Kluwer. ISBN 1-4020-7768-8
17. Vega LR, Rey H, J. Benesty J, Tressens S (2009) A family of robust algorithms exploiting sparsity in adaptive filters. IEEE Trans Audio Speech Lang Process 17(4):572–581
18. Amari S (1998) Natural gradient works efficently in learning. Neural Comput 10:251–276
19. Kivinen J, Warmuth MK (1997) Exponentiated gradient versus gradient descent for linear predictors. Inform Comput 132:1–64

20. Benesty J, Gänsler T, Gay L, Sondhi MM (2000) A robust proportionate affine projection algorithm for network echo cancellation. In: Proceedings of IEEE international conference on acoustics, speech, and signal processing, ICASSP '00, pp II-793–II-796
21. Haykin S (1996) Adaptive filter theory, 3rd edn. Prentice Hall, Upper Saddle River, NJ
22. Golub GH, Van Loan CF (1989) Matrix computation. John Hopkins University Press, Baltimore, MD. ISBN 0-80183772-3
23. Sherman J, Morrison WJ (1950) Adjustment of an inverse matrix corresponding to a change in one element of a given matrix. Ann Math Stat 21(1):124–127
24. Ozeki K, Umeda T (1984) An adaptive filtering algorithm using an orthogonal projection to an affine subspace and its properties. Electron Commun Jpn J67-A(5):126–132
25. Rupp M, Sayed AH (1996) A time-domain feedback analysis of filtered error adaptive gradient algorithms. IEEE Trans Signal Process 44(6):1428–1439

Chapter 7
Block and Transform Domain Algorithms

7.1 Introduction

In this chapter structures and algorithms for the implementation of adaptive filters (AF) with the purpose of improving the convergence speed and reducing the computational cost are presented. In particular, they are classified as *block* and *online* methods, operating in the time domain, in the transformed domain (typically the frequency domain), and in frequency subbands mode.

In adaptive filtering algorithms such as LMS, APA, and RLS, the parameters update is performed for each time instant n in the presence of a new sample at the filter input. The filter impulse response of $\mathbf{w}_n$ is time variant, and the convolution algorithm is implemented directly in the time domain, i.e., the AF output is calculated as a linear combination $y[n] = \mathbf{w}_{n-1}^H \mathbf{x}_n$. The computational complexity, proportional to the filter length, can become prohibitive for considerable length filters.

The *block* algorithms are defined by a periodic update law. The filter coefficients are constant and updated only every L samples. Calling k the block index, as for LS systems described above (see Chap. 4), the output is calculated in blocks of length L as the convolution sum $\mathbf{y}_k = \mathbf{X}_k \mathbf{w}_k$, where $\mathbf{w}_k$ represents a static filter for all rows of the signal matrix $\mathbf{X}_k$ (see Sect. 1.6.2.1). This formalism facilitates the implementation in the frequency domain.

The transform domain algorithms, usually the frequency domain, are defined starting from the same theoretical assumptions already widely discussed in earlier chapters of this volume. In general, however, these are almost never a simple redefinition "in frequency" of the same algorithms operating "in time." The frequency domain algorithms have peculiarities that determine structures and properties, sometimes, also very different from similar time-domain algorithms.

The block algorithms nature, especially those operating in frequency, requires an appropriate mechanism of *memory buffers* filling, hereinafter simply *buffers*, containing the input signal block to be processed and the filtered output. In addition, the *transformation operator* $\mathbf{F}$ requires the variables redefinition in the new domain.

A. Uncini, *Fundamentals of Adaptive Signal Processing*, Signals and Communication Technology, DOI 10.1007/978-3-319-02807-1_7,

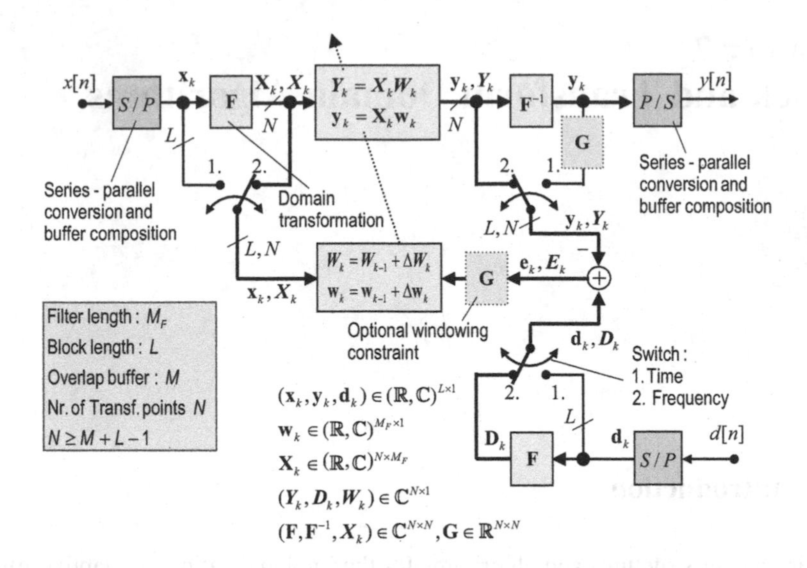

Fig. 7.1 General indicative framework for block algorithms in time and transformed domain by the operator **F**. For $\mathbf{F} = \mathbf{I}$, the algorithm is entirely in the time domain and the switches 1. or 2. position is indifferent. For $\mathbf{F} \neq \mathbf{I}$, the weights adaptation can be done in the time domain (switch position 1.) or in the transformed domain (switch position 2.)

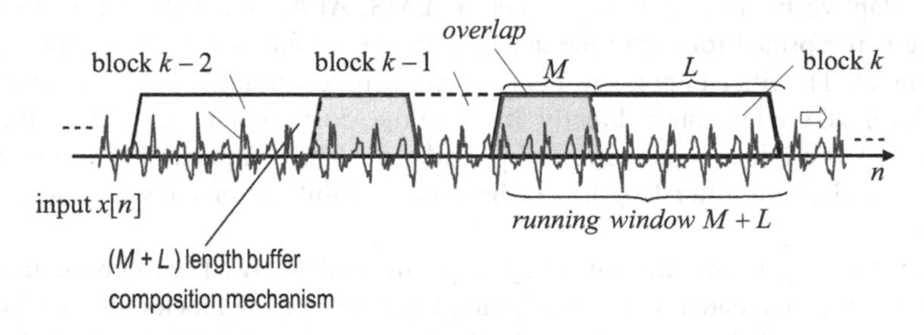

Fig. 7.2 Input signal buffer composition mechanism

These aspects, along with others discussed below, involve the proliferation of indices, symbols, and new variables that can, sometimes, burden the formalism.

A general representation framework for block algorithms described in this chapter is reported in Fig. 7.1, while Fig. 7.2 shows a possible mechanism example for buffer composition. Bearing in mind these figures, we define in this context the following quantities:

- $M_F \triangleq$ Adaptive filter length[1];
- $M + L \triangleq$ Analysis window length in the case of domain transformation;

[1] In some sections of this chapter, for reasons of notation clarified in the following, the length of the filter is referred to as M_F. The reader will note that generally the length of the filter is denoted by M, implying $M_F = M$.

- $N \geq M + L - 1 \triangleq$ Number transformed domain points. For example, number of DFT/FFT frequencies or other transformation;
- $L \triangleq$ New signal block length to be processed. Note that, L determines the algorithm latency;
- $M \triangleq$ Old data block length that overlap with the new;
- $\frac{M}{M+L} \times 100 \triangleq$ overlap percentage;
- $k \triangleq$ Block index and $i = 0, 1, \ldots, L - 1$ time index inside the block;
- $\mathbf{F} \triangleq$ Linear domain transform operator;
- $\mathbf{F}^{-1} \triangleq$ Inverse linear domain transform operator;
- $\mathbf{G} \triangleq$ Windowing constraint of the output signal, error, or weights;
- $\mathbf{x}_k$, $\mathbf{y}_k$, $\mathbf{w}_k \triangleq$ Vectors sequence block, respectively, of the input, output, and the filter weights;
- $\mathbf{X}_k \triangleq$ Time domain blocks input data matrix;
- $\boldsymbol{W}_k$, $\boldsymbol{Y}_k \triangleq$ Frequency domain output and the filter weights vectors;
- $\boldsymbol{X}_k \triangleq$ Frequency domain input data block diagonal matrix.

Again with reference to Fig. 7.1, the output and error signal windowing constraint $\mathbf{G}$ and that of the weights (the latter not shown in the figure) are necessary for the proper implementation of the inverse transformation operator. Note, also, the presence of the switches with positions 1. and 2. This presence indicates that the adaptive filtering algorithm can be implemented in mixed mode: the output calculation in the transformed domain and weights update in time domain. For $\mathbf{G} \equiv \mathbf{F} = \mathbf{I}$, the algorithm operates entirely in the time domain and, as the reader can observe from the figure, in this case the switches positions are indifferent.

7.1.1 Block, Transform Domain, and Online Algorithms Classification

The block algorithms [1, 2] operate, by definition, on a L-length signal block, but the (possible) domain transformation can be made by considering buffer of greater length. In general terms, the transformation can be performed on a signal segment (or *running window*) composed by L new samples (block) and possibly by M past samples. In this case, as shown in Fig. 7.2, the composition mechanism of the input buffer of length $M + L$ includes the presence of the new L samples block and M samples belonging to the previous block. Calling M_F the filter length, for the so-called frequency domain adaptive filters (FDAF) algorithms class, the block length is generally chosen as $L = M_F$; the FDAF buffer composition choice commonly used is such that $L = M \equiv M_F$. To operate a correct domain transformation, for example, with a DFT/FFT, and in particular for the filter output calculation, it is necessary to choose a number of FFT points $N \geq L + M - 1$. A usual choice for FDAF class is $N = L + M$.

In the case of very long filters (with thousands of coefficients), very common in AF's audio applications, the block length turns out to be necessarily $L \ll M_F$ and,

Table 7.1 Block and online algorithms operating in the time and/or in the transformed domain

Filter class	Block L	Overlap M (%)	F	G	N
LMS	1	$M_F - 1$	$\mathbf{I}$	$\mathbf{I}$	–
BLMS	$L = M_F$	≥ 0	$\mathbf{I}$	$\mathbf{I}$	–
FDAF	$L = M \equiv M_F$	M (50 %)	DFT/FFT	$\neq \mathbf{I}$	$\geq L + M - 1$
UFDAF	$L = M \equiv M_F$	0	DFT/FFT	$\mathbf{I}$	$N = M_F$
PFDAF	$L < M$	$M = pL$	DFT/FFT	$=, \neq \mathbf{I}$	$\geq L + M - 1$
TDAF	$L = 1$	$M_F - 1$	$\neq \mathbf{I}$	$\mathbf{I}$	$N = M_F$
SAF	$L = 1$	$M_F - 1$	B.F.	–	–

in this case, for the transform domain filter implementation, it is necessary to perform a impulse response partition. This partition enables the AF implementation with more contained *latencies*. As we shall see in the following, a very common choice is to consider P partitions of length M, such that the filter length is equal to the product $M_F = M \cdot P$, and a block length such that $M = pL$ with p integer; namely, the buffer length is equal to $(p + 1)L$. This class of algorithms is called partitioned frequency domain adaptive filters (PFDAF).

In the extreme case, where $L = 1$, a block of one sample length, the algorithm is defined as transform-domain adaptive filters (TDAF). The input window, in this context called *sliding window*, is simply defined by the filter delay-line length (see Fig. 3.1). The operator $\mathbf{F}$ performs a linear transformation just to orthogonalize the input signal so as to facilitate the *uniform convergence* of the adaptive algorithm.

The domain change can be of varied nature. Although, in theory, the operator $\mathbf{F}$ can be any orthonormal transformation, it is usual to choose transformations that allow, in addition to the input signal orthogonalization, a computational complexity reduction. Choices rather common are the DFT (implemented as FFT) the DCT, or other transformations tending to input signal orthogonalization (see Sect. 1.3).

Note that for $L = 1$, the transformation $\mathbf{F}$ can be replaced by a suitably designed parallel filters bank, uniformly or not non-uniformly spaced. In addition, to obtain a computational cost reduction, it is possible to perform a signal decimation/interpolation. The AF's class is in this case called subband adaptive filter (SAF).

A possible classification of the methods described in this chapter, refer to the formalism shown in Figs. 7.1 and 7.2, is reported in Table 7.1.

In the first part of this chapter, the block-LMS algorithm is introduced. Subsequently, two paragraphs concerning algorithms in the frequency domain, the constrained FDAF (CFDAF), the unconstrained FDAF (UFDAF), and the partitioned FDAF (PFDAF) are introduced. In the third paragraph, the TDAFs are presented. Note, that some authors introduce FDAF algorithms as a generalization of transform domain algorithms. Herein it is preferred the opposite, i.e., define the TDAF class as a particular case of the FDAF class.

In the last part of the chapter, after a brief reference to the multi-rate methods and filters, some architectures of SAF are presented.

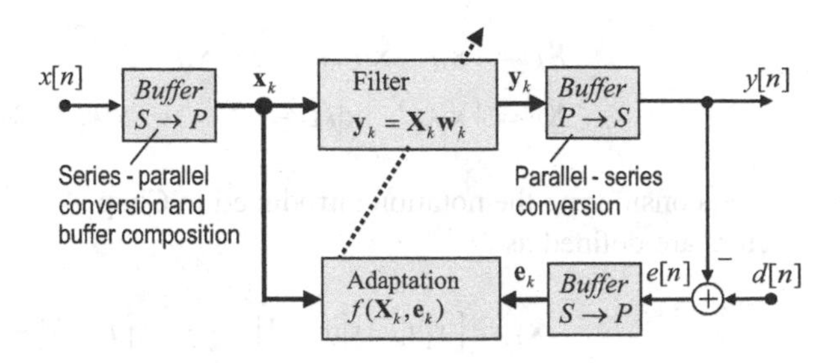

Fig. 7.3 General scheme of a *block adaptive filter*

7.2 Block Adaptive Filter

In the block algorithms class, represented schematically in Fig. 7.3, the input signal is stored in a L-length buffer (*block length*) to allow the output and weights update to be periodically calculated, with a period equal to L.

Calling k the block index, M the filter length, and $\mathbf{w}_k \in (\mathbb{R},\mathbb{C})^{M\times 1}$ the filter weights vector, the parameter update is characterized by a relation of the type

$$\mathbf{w}_{k+1} = \mathbf{w}_k + \frac{1}{L}\Delta\mathbf{w}_k \tag{7.1}$$

in which $\Delta\mathbf{w}_k$, defined as a *block update parameter*, is given by the sum of the instantaneous variations $\Delta\mathbf{w}_i$, i.e.,

$$\Delta\mathbf{w}_k = \sum_{i=0}^{L-1}\Delta\mathbf{w}_i. \tag{7.2}$$

With this definition, calling i the time index inside the block, the input sequence time index n is defined as

$$n = kL + i \quad \begin{matrix} i = 0, 1, \ldots, \; L-1 \\ k = 1, 2, \ldots \end{matrix} . \tag{7.3}$$

The term $\Delta\mathbf{w}_i$ is linked to the instantaneous estimate of the CF gradient $\nabla\hat{J}_{\mathrm{i}}$, and its calculation is performed for every block index, while keeping fixed filter coefficients.

The input signal, as in the LS methodology, is stored in the data matrix $\mathbf{X}_k$, indicated as *block matrix*, such that the kth block output is calculated as the convolution sum expressed in terms of the matrix-vector product as

$$\mathbf{y}_k = \mathbf{X}_k\mathbf{w}_k. \tag{7.4}$$

For the above equation, the block data matrix $\mathbf{X}_k \in (\mathbb{C},\mathbb{R})^{L\times M}$ can defined, by row or column, as

$$\mathbf{X}_k = \begin{bmatrix} \mathbf{x}_{kL} & \mathbf{x}_{kL-1} & \cdots & \mathbf{x}_{kL-L+1} \end{bmatrix}^T, \tag{7.5}$$

$$\mathbf{X}_k = \begin{bmatrix} \mathbf{x}[kL] & \mathbf{x}[kL-1] & \cdots & \mathbf{x}[kL-M+1] \end{bmatrix}, \tag{7.6}$$

where, considering the notation introduced in Chap. 4 (see Sect. 4.2.2.1), the signal vectors are defined as

$$\mathbf{x}_n = \begin{bmatrix} x[n] & x[n-1] & \cdots & x[n+M-1] \end{bmatrix}^T, \tag{7.7}$$

$$\mathbf{x}[n] = \begin{bmatrix} x[n] & x[n-1] & \cdots & x[n+L-1] \end{bmatrix}^T. \tag{7.8}$$

Note that the matrix $\mathbf{X}_k$ contains the input signal samples arranged in columns/rows shifted of one sample. For example, in the case of $L = 4$ and $M = 3$ for k and $k-1$, (7.4) is

$$k \rightarrow \begin{bmatrix} y[4k] \\ y[4k-1] \\ y[4k-2] \\ y[4k-3] \end{bmatrix} = \begin{bmatrix} x[4k] & x[4k-1] & x[4k-2] \\ x[4k-1] & x[4k-2] & x[4k-3] \\ x[4k-2] & x[4k-3] & x[4k-4] \\ x[4k-3] & x[4k-4] & x[4k-5] \end{bmatrix} \begin{bmatrix} w_k[0] \\ w_k[1] \\ w_k[2] \end{bmatrix},$$

$$k-1 \rightarrow \begin{bmatrix} y[4k-4] \\ y[4k-5] \\ y[4k-6] \\ y[4k-7] \end{bmatrix} = \begin{bmatrix} x[4k-4] & x[4k-5] & x[4k-6] \\ x[4k-5] & x[4k-6] & x[4k-7] \\ x[4k-6] & x[4k-7] & x[4k-8] \\ x[4k-7] & x[4k-8] & x[4k-9] \end{bmatrix} \begin{bmatrix} w_{k-1}[0] \\ w_{k-1}[1] \\ w_{k-1}[2] \end{bmatrix}.$$

Note that for $L = M$, the matrix $\mathbf{X}_k$ is Toeplitz.

For other vectors, similar to LS, we have the following definitions:

$$\begin{aligned} \mathbf{d}_k &\in (\mathbb{R}, \mathbb{C})^{L\times 1} \triangleq \begin{bmatrix} d[kL] & d[kL-1] & \cdots & d[kL-L+1] \end{bmatrix}^T \\ \mathbf{y}_k &\in (\mathbb{R}, \mathbb{C})^{L\times 1} \triangleq \begin{bmatrix} y[kL] & y[kL-1] & \cdots & y[kL-L+1] \end{bmatrix}^T \\ \mathbf{e}_k &\in (\mathbb{R}, \mathbb{C})^{L\times 1} \triangleq \begin{bmatrix} e[kL] & e[kL-1] & \cdots & e[kL-L+1] \end{bmatrix}^T \end{aligned} \tag{7.9}$$

for which the error vector can be defined as

$$\mathbf{e}_k = \mathbf{d}_k - \mathbf{y}_k. \tag{7.10}$$

From (7.4), the filter coefficients $\mathbf{w}_k$ remain constant for all L output samples $\mathbf{y}_k$, and the convolution can be performed with a block algorithm.

As regards the block length, we can identify three distinct situations: $L = M$, $L < M$, and $L > M$. The most common choice is that in which the block length is equal to (or less) the filter length and, in this case, the possibility to compute the convolution in the frequency domain suggests that filter lengths are equal to powers of two.

7.2.1 Block LMS Algorithm

In the LMS algorithm (see Sect. 5.3.1), the instantaneous parameters adaptation, at the ith instant time, equal to the local gradient estimate, is $\Delta\mathbf{w}_i \triangleq \nabla\hat{J}_i = e^*[i]\mathbf{x}_i$. So, considering the relation (7.1) and (7.2), the block LMS (BLMS) algorithm is characterized by a filter adaptation that occurs periodically every L iterations with a relation of the type

$$\begin{aligned}\mathbf{w}_{k+1} &= \mathbf{w}_k + \mu_B\frac{\sum_{i=0}^{L-1}\nabla\hat{J}_i}{L}\\ &= \mathbf{w}_k + \frac{\mu_B}{L}\nabla\hat{J}_k\end{aligned} \tag{7.11}$$

in which $\mu_B = L\cdot\mu$ is defined as *block learning rate* and represents $\nabla\hat{J}_k$ the *estimate gradient block* defined as

$$\nabla\hat{J}_k = \sum_{i=0}^{L-1} e^*[kL+i]\mathbf{x}_{kL+i} \tag{7.12}$$

interpretable as an approximation of the CF $J_k = E\{\mathbf{e}_k^H\mathbf{e}_k\}$ differentiation, as $J_k = L\cdot J_i$.

Remark Note that the expression (7.12) is formally identical to the cross-correlation estimate between the input vector and the error signal and, from the definition of the input data matrix (7.5), can be written in matrix form as

$$\nabla\hat{J}_k = \mathbf{X}_k^H\mathbf{e}_k^*. \tag{7.13}$$

7.2.1.1 Summary of BLMS Algorithm

The BLMS algorithm is then defined by the following iterative procedure

$$\mathbf{y}_k = \mathbf{X}_k\mathbf{w}_k, \qquad \textit{filtering}, \tag{7.14}$$

$$\mathbf{e}_k = \mathbf{d}_k - \mathbf{y}_k, \qquad \textit{error}, \tag{7.15}$$

$$\mathbf{w}_{k+1} = \mathbf{w}_k + \frac{\mu_B}{L}\mathbf{X}_k^H\mathbf{e}_k^*, \qquad \textit{adaptation}. \tag{7.16}$$

Remark The expression (7.14) represents a convolution, while the (7.16) a cross-correlation. In order to obtain greater computational efficiency and, moreover, better convergence characteristics, as we shall see in the following, both expressions can be implemented in the frequency domain.

7.2.2 Convergence Properties of BLMS

The BLMS algorithm minimizes the same CF of the LMS and, in addition, the block gradient estimation can be more accurate than the LMS because it is averaged on L values. It follows that BLMS steady-state solution, the misalignment, and the time constants for stationary signals are identical to those of the LMS. In fact, the adaptive algorithms convergence characteristics depend on the input correlation $\mathbf{R}$; thus, BLMS has the convergence behavior similar to the LMS. In particular it appears that the M modes decay time constant is defined as

$$\tau_{B,i} = \frac{1}{\mu_B \lambda_i}, \tag{7.17}$$

where λ_i is the ith eigenvalue of the matrix $\mathbf{R}$.

In the BLMS algorithm, the weight vector update is made by considering the average of the instantaneous perturbations (7.1). For which the weights have a mean trajectory which coincides with that of the SDA (see Sect. 5.1.1.1). Because of this averaging effect, the learning curve has smaller variance and is more smooth than the LMS [2, 3].

Remark The main difference between LMS and BLMS with regard to the maximum learning rate permissible value such as the algorithm is stable. In the case of BLMS, in fact, this is scaled by a factor L and, in the case of colored input sequence, i.e., input's correlation matrix with high eigenspread (or $\mathbf{R}$ with high condition number), the BLMS may converge more slowly.

7.3 Frequency Domain Block Adaptive Filtering

The subject of frequency domain adaptive filtering is a very broad topic, which presents many variations and specializations, evidenced by the numerous contributions, including recent ones, in the scientific literature (see for example [1, 4–17]). These algorithms have a high usability in applications in which the filter length is very high and is also required for high computational efficiency. In this section are presented, in particular, some known algorithms such as FDAF, which has a recursive formulation similar to BLMS. Also known in the literature as *fast LMS* (FLMS), it was presented for the first time by Ferrara [6] and, independently, by Clark, Mitra, and Parker [1].

In the BLMS algorithm the input filtering, by (7.4), is calculated by the convolution between the input and the filter coefficients. The block gradient estimate $\nabla \hat{J}_k$, for the definition (7.12), is similar to a cross-correlation between the input and the error signals. Both operations can, then, be effectively implemented in the frequency domain. In fact, both the output and the gradient can be evaluated on

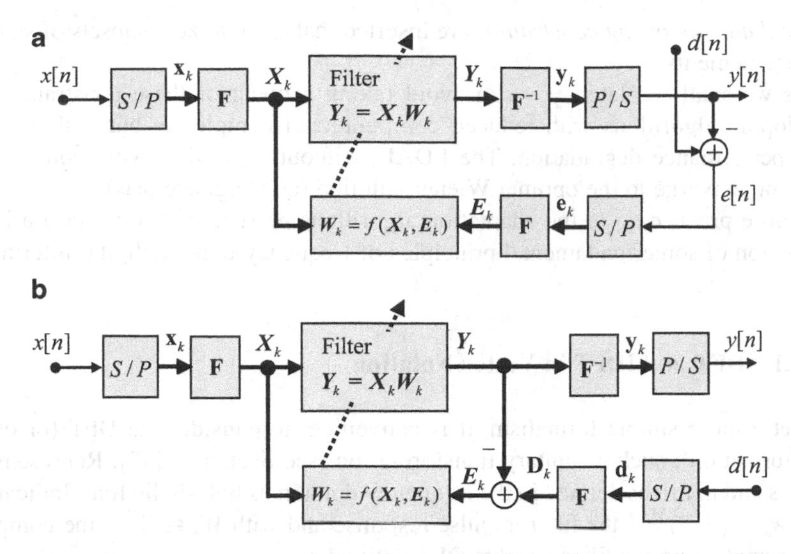

Fig. 7.4 Scheme of the frequency domain adaptive filters (FDAF), derived from the general structure of Fig. 7.1. Error calculation (**a**) in the time domain; (**b**) in the transformed domain

signal blocks; it is possible to obtain a considerable computational saving implementing the required operations in the frequency domain by the FFT. Indeed, the calculation of the N-length DFT, or of its inverse, requires N^2 multiplications while with the FFT algorithm are required only $N \log_2 N$ multiplications [18].

A block AF algorithms schematization, operating in the transformed domain, is shown in Fig. 7.4. In the figure, the operator **F** is a matrix that performs the transformation and in the case of the frequency domain, **F** represents the DFT matrix (see Sect. 1.3.1). The error calculation may be performed in the time domain, as shown in Fig. 7.4a, or, with proper precautions, in the frequency domain as shown in Fig. 7.4b.

7.3.1 Linear Convolution and Filtering in the Frequency Domain

The FDAF algorithm, as shown in Fig. 7.4, has a recurrent structure similar to the BLMS. The BLMS algorithm extension in the frequency domain is not, however, as immediate. Indeed, antitransforming the product of two DFT sequences, this corresponds to a *circular convolution* in the time domain, while the filtering operations are implemented with *linear convolution* [10]. The circular convolution is different from the linear one. Therefore, it is a necessary method for the determination of the linear convolution starting from the circular. In the FDAF, to obtain the linear convolution starting from the circular one, particular constraints, called *data windowing constraints*, are inserted that force to zero subsets of signals vectors elements.

As we shall see later, you can avoid taking account of these constraints, by developing algorithms with reduced computational complexity but with steady-state performance degradation. The FDAF, without data windowing constraints, may not converge to the optimal Wiener solution (convergence bias).

Before proceeding to the adaptation algorithms presentation, we report a brief discussion of some fundamental principles of frequency-domain digital filtering.

7.3.1.1 DFT and IDFT in Vector Notation

To get a more simple formalism, it is convenient to consider the DFT (or other transformation), such as unitary transformation (see Sect. 1.3) [12]. Representing vectors and matrices defined in the frequency domain as bold-italic font. Indicating with $\mathbf{w}_k \in (\mathbb{R},\mathbb{C})^{M\times 1}$ the filter impulse response and with $\boldsymbol{W}_k \in \mathbb{C}^{N\times 1}$ the complex vector containing the filter weights DFT defined as

$$\boldsymbol{W}_k = \mathbf{F}\mathbf{w}_k = \begin{bmatrix} w_k[0] & w_k[1] & \cdots & w_k[M-1] & \underbrace{0 \;\; \cdots \;\; 0}_{\text{zero-padding}} \end{bmatrix}^T, \tag{7.18}$$

where, being generally $N > M$, we must append to the weight vector $w_k[i]$ $(N - M)$ zeros. For the DFT definition, let $F_N = \frac{1}{\sqrt{N}}e^{-j2\pi/N}$, the matrix $\mathbf{F}$ (see Sect. 1.3.2, (1.17)) is defined as $\mathbf{F} \triangleq \left\{f_{kn} = F_N^{kn}\;\; k, n \in [0, N-1]\right\}$. In addition, the vector $\mathbf{w}_k$ $(N\times 1)$ appearing in (7.18) is an *augmented form* defined as

$$\mathbf{w}_k = \left[\hat{\mathbf{w}}_k\, \mathbf{0}_{N-M}\right]^T. \tag{7.19}$$

The actual filter weights are indicated in this context, as a *normal form* or *not augmented form:*

$$\hat{\mathbf{w}}_k = \left[w_k[0] \quad w_k[1] \quad \cdots \quad w_k[M-1]\right]^T. \tag{7.20}$$

Performing the IDFT of $\boldsymbol{W}_k$, i.e., left-multiplying by $\mathbf{F}^{-1}$ both members of the (7.18), we get the filter weights augmented form:

$$\mathbf{w}_k = \mathbf{F}^{-1}\boldsymbol{W}_k. \tag{7.21}$$

Therefore only the first M elements of the vector $\mathbf{w}_k$ are *significant*. In other words, the normal form can be indicated as

$$\hat{\mathbf{w}}_k = \left[\mathbf{F}^{-1} \boldsymbol{W}_k\right]^{\lceil M \rceil} \tag{7.22}$$

in which the symbol $[\mathbf{w}_k]^{\lceil M \rceil}$ indicates the selection of first M elements of the vector $\mathbf{w}_k$.

7.3.1.2 Convolution in the Frequency Domain with Overlap-Save Method

Consider the convolution between an infinite duration sequence (filter input) and another one of finite duration (the filter impulse response). For the determination of the linear convolution by the product of the respective FFT, proceed by sectioning the input sequence into finite length blocks and impose appropriate windowing constraints. Indeed, antitransforming the product of the two FFT, in time domain, a circular convolution/correlation is produced. In practice, there are two distinct sequence sectioning methods called, respectively, *overlap-save* and *overlap-add* [10].

To understand the overlap-save technique, we analyze a simple filtering problem of a sequence of infinite duration with an FIR filter. For the output determination, the frequency domain convolution calculation is performed on blocks of the input signal.

Consider, for simplicity, a M-length filter and a signal block of length L, with $L \leq M$. In order to generate the actual L output samples, you should have FFT of length N such that $N \geq M + L - 1$. As usual in adaptive filtering, and also for formal simplicity, we analyze the case where the input block length is smaller than that of the filter impulse response, i.e., $L < M$ and $N = M + L$. Denoting by k the index block, by (7.18), the DFT of the impulse response, is defined as

$$\begin{aligned} \boldsymbol{W}_k &= \mathbf{F}[\,\hat{\mathbf{w}}_k \; \mathbf{0}_L]^T \\ &= \mathbf{F}\mathbf{w}_k, \end{aligned} \tag{7.23}$$

where $\hat{\mathbf{w}}_k = \left[w[0], w[1], ..., w[M-1]\right]^T$ contains the filter impulse response, and $\mathbf{w}_k = [\,\hat{\mathbf{w}}_k \; \mathbf{0}_L]^T$ represents its augmented form.

Remark For presentation consistency reasons, it is appropriate to have a similar formalism in time and frequency domains. For example, in (7.14) the output vector is calculated as matrix-vector product; in frequency domain similar formalism can be maintained by expressing the output signal as $\boldsymbol{Y}_k = \boldsymbol{X}_k \boldsymbol{W}_k$, where $\boldsymbol{X}_k$ denotes a diagonal matrix containing the input signal DFT.

Let L be the input block length. The FFT of length equal to $N = M + L$ can be calculated by considering the L samples input block to which M past samples are appended. In formal terms, for $k = 0$, we can write[2]

---

[2] The symbols $\mathbf{x}^{\lceil M \rceil}$ and $\mathbf{x}^{\lfloor L \rfloor}$ denote, respectively, the first M and the last L samples of the vector $\mathbf{x}$.

$$X_0 = \text{diag}\mathbf{F}\Big[\underbrace{0 \quad \cdots \quad 0}_{\text{IC } \lceil M \rceil \text{ samples}} \ \underbrace{x[0] \quad \cdots \quad x[L-1]}_{\text{block } \lfloor L \rfloor \text{ samples}}\Big]^T$$

$$= \text{diag}\Big\{\mathbf{F}\big[\,\mathbf{0}_M \quad \mathbf{x}_0^L\big]^T\Big\} \tag{7.24}$$

for $k > 0$ is

$$X_k = \text{diag}\mathbf{F}\Big[\underbrace{x[kL-M+1] \cdots x[kL-1]}_{\text{overlap } \lceil M \rceil \text{ samples}} \ \underbrace{x[kL] \cdots x[kL+L-1]}_{\text{block } \lfloor L \rfloor \text{ samples}}\Big]^T$$

$$= \text{diag}\Big\{\mathbf{F}\big[\mathbf{x}_{\text{old}}^M \, \mathbf{x}_k^L\big]^T\Big\}. \tag{7.25}$$

This formalism allows the expression of the output signal as $\boldsymbol{Y}_k = \boldsymbol{X}_k \boldsymbol{W}_k$. It should be noted that the form of matrix-vector product, of the type $\boldsymbol{X}_k \boldsymbol{W}_k$, is possible only by inserting the DFT of the input sequence, in a diagonal matrix $\boldsymbol{X}_k$. In fact, considering the DFT vector, for example indicated as $\hat{\boldsymbol{X}}_k = \mathbf{F}\big[\mathbf{x}_{\text{old}}^M \, \mathbf{x}_k^L\big]^T$, the output takes the form $\boldsymbol{Y}_k = \hat{\boldsymbol{X}}_k \odot \boldsymbol{W}_k$ in which the operator $\odot$ denotes the Hadamard product, i.e., the point-to-point multiplication of the vectors $\hat{\boldsymbol{X}}_k$ and $\boldsymbol{W}_k$.

With the overlap-save method, the time-domain output samples are determined by selecting only the last L samples of the vector IDFT$(\boldsymbol{Y}_k)$. Formally, we get

$$\hat{\mathbf{y}}_k = \big[\mathbf{F}^{-1}\boldsymbol{X}_k \boldsymbol{W}_k\big]^{\lfloor L \rfloor}. \tag{7.26}$$

In fact, by performing the IDFT of the product $\boldsymbol{X}_k \boldsymbol{W}_k$, it is not guaranteed that the first M values are zero. The output augmented form is, then, obtained by constraining to zero the first M samples. In formal terms, we can write

$$\mathbf{y}_k = \mathbf{g}_{0,L}\mathbf{F}^{-1}\boldsymbol{X}_k \boldsymbol{W}_k, \tag{7.27}$$

where $\mathbf{g}_{0,L} \in \mathbb{R}^{(M+L)\times(M+L)}$ is a square matrix, called *weighing matrix* or *output projection matrix*, defined as

$$\mathbf{g}_{0,L} \in \mathbb{R}^{(M+L)\times(M+L)} = \begin{bmatrix} \mathbf{0}_{M,M} & \mathbf{0}_{M,L} \\ \mathbf{0}_{L,M} & \mathbf{I}_{L,L} \end{bmatrix}, \tag{7.28}$$

where $\mathbf{0}_{M,M}$ is a matrix of zeros and $\mathbf{I}_{L,L}$ is a diagonal unitary matrix. In practice, the multiplication by $\mathbf{g}_{0,L}$ forces to zero the first M samples of the vector $\hat{\mathbf{y}}_k$, leaving unchanged the last L.

In other words, the DFT of the output $\hat{\mathbf{y}}_k$ does not coincide with the product $\boldsymbol{X}_k \boldsymbol{W}_k$, i.e., $\mathbf{F}\big[\,\mathbf{0}_M \quad \hat{\mathbf{y}}_k\big] \neq \boldsymbol{X}_k \boldsymbol{W}_k$. Note that for the correct output DFT calculation, we must enforce the constraint (7.27) and we get

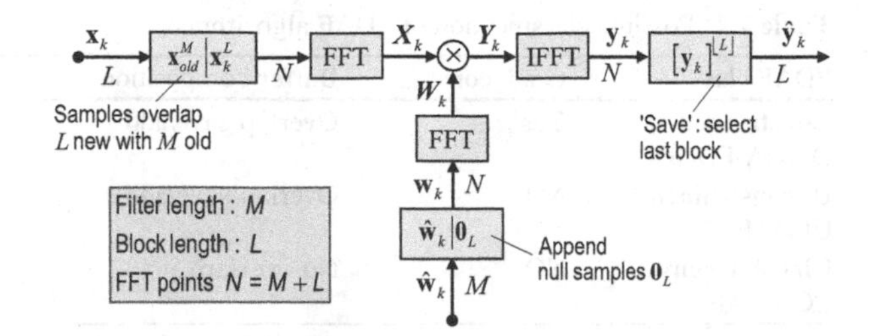

Fig. 7.5 The overlap-save sectioning method representation, in the case of $\hat{\mathbf{w}}_k \in (\mathbb{R}, \mathbb{C})^{M \times 1}$ filter, block length equal to L, and FFT of $N = M + L$ points

$$Y_k' = \mathbf{F}\mathbf{g}_{0,L}\mathbf{F}^{-1}X_k W_k. \tag{7.29}$$

In the previous term $\mathbf{F}\mathbf{g}_{0,L}\mathbf{F}^{-1}$, it is often referred to as

$$\mathbf{G}_{0,L} = \mathbf{F}\mathbf{g}_{0,L}\mathbf{F}^{-1} \tag{7.30}$$

and is defined as the *windowing constraint*.

Note that the input FFT is defined by considering a window of $M + L$ samples. With reference to Figs. 7.2 and 7.5, advancing the input sequence of one block forward (*running window*), the new FFT is calculated considering also the old M samples. In other words, the new FFT window contains L new and M old samples. Generally, this is referred to as an overlap of $(100L/(M + L))\%$.

7.3.2 Introduction of the FDAF

In the time-domain BLMS learning rule (7.12), the gradient estimate is determined by the cross-correlation between the data vector $\mathbf{x}_k$ and the error $\mathbf{e}_k$. Transforming the rule in the frequency domain, the weights update equation (7.16) can be rewritten, as suggested in [12], in a compact and general notation of the type

$$W_{k+1} = W_k + \mathbf{G}\boldsymbol{\mu}_k X_k^H E_k \tag{7.31}$$

in which the matrix $\boldsymbol{\mu}_k = \text{diag}\{\mu_k(0) \quad \mu_k(1) \quad \cdots \quad \mu_k(N-1)\}$ contains the learning rates or step size that can be of different value for each frequency bin. The matrix $\mathbf{G}$ represents the windowing or gradient constraint, necessary to impose the linearity of the correlation in the gradient calculation $X_k^H E_k$ and can be interpreted as a particular signal pre-windowing in the time domain. The matrix $\mathbf{G}$ is inserted in the learning rule only in order to generalize the FDAF formalism.

Remark In the class of the frequency domain adaptive algorithms, the error calculation can be performed directly in the time or frequency domain. In the case where the error is calculated in frequency domain, the gradient constraint can be chosen unitary $\mathbf{G} = \mathbf{I}$, and the FDAF is called UFDAF. In this case, the computational complexity is reduced but the convergence to the Wiener solution is biased.

Table 7.2 Possible classification of FDAF algorithms

FDAF class	Grad. const.	Buffer composition rule	Nr of FFT points
Constrained OS/OA-FDAF	Yes	Overlap save/add	$N \geq M + L - 1$ Typical $L = M$ $N = 2M$
Unconstrained UFDAF	NO	Overlap save/add	$N \geq M + L - 1$ Typical $L = M$ $N = 2M$
Circular conv. CC-FDAF	NO	No overlap	$L = M$ $N = M$

7.3.2.1 FDAF Algorithms Class

The FDAF algorithms class is very wide and, as already anticipated in the Chapter introduction, can be defined in relation to the input block length (*running window*), from the data buffer composition rule, the number of FFT points, the calculation mode of the error, and the presence or absence of the gradient constraint.

Indicating, respectively, with M, L, and N the filter length, the block length, and the FFT points, we can define the FDAF class according to Table 7.2.

7.3.2.2 Frequency Domain Step Size Normalization

One of the main advantages of the frequency approach is that the adaptation equations (7.31) are decoupled, i.e., in the frequency domain, the convergence of each filter coefficient is not dependent from the others. It follows that, to increase the convergence speed, the step size for each frequency, denoted as $\mu_k(m)$, can be determined independently from the others. For example, in an inversely proportional way to the relative power of the mth frequency component of the input signal (*frequency bin*). Indicating with $P_k(m)$ the estimation power of the mth frequency bin and let μ be a suitable predetermined scalar, the step size can be chosen as

$$\mu_k(m) = \mu / P_k(m), \quad m = 0, 1, ..., N - 1. \tag{7.32}$$

Another possible choice, recalling the normalized LMS, is as follows:

$$\mu_k(m) = \frac{\mu}{\alpha + P_k(m)}, \quad m = 0, 1, ..., N - 1 \tag{7.33}$$

with α and μ usually evaluated in an experimental way. This procedure, indicated in the literature also as *step-size normalization* procedure, allows to accelerate the AF's slower modes.

Note that, in the case of white and stationary input processes, the powers are identical for all frequency bins and we have $\boldsymbol{\mu}_k = \mu \mathbf{I}$.

Remark To avoid significant step-size discontinuity that could destabilize the adaptation, as suggested by some authors (see, for example, [8]), it is appropriate to estimate mth power frequency bin $P_k(m)$ with a one-pole low-pass smoothing filter, implemented by the following FDE:

$$P_k(m) = \lambda P_{k-1}(m) + (1-\lambda)\left|X_k(m)\right|^2, \quad m = 0, 1, ..., N-1, \tag{7.34}$$

where λ represents a *forgetting parameter* and $|X_k(m)|^2$ the mth measured energy bin.

7.3.3 Overlap-Save FDAF Algorithm

In adaptive filtering, in addition to the output calculation, it is necessary to calculate the update block parameter that, in practice, consists in the correlation calculation (7.13). Regarding the time domain, error for the kth block is $e[kL+i] = d[kL+i] - y[kL+i]$ for $i = 0, 1, \ldots, L-1$; indicating with

$$\hat{\mathbf{d}}_k = \begin{bmatrix} d[kL] & d[kL+1] & \cdots & d[kL+L-1] \end{bmatrix}^T,$$

the desired output vector, in the not augmented form, we have that

$$\hat{\mathbf{e}}_k = \hat{\mathbf{d}}_k - \hat{\mathbf{y}}_k \tag{7.35}$$

that, with appropriate zero-padding, is transformed in the frequency domain with the following DFT transformation:

$$\boldsymbol{E}_k = \mathbf{F}\begin{bmatrix} \mathbf{0}_M & \hat{\mathbf{e}}_k \end{bmatrix}^T. \tag{7.36}$$

The correlation can be seen as a reversed-sequence convolution. So, the linear correlation coefficients can be determined only selecting the first M samples of the vector $\mathbf{F}^{-1}\boldsymbol{X}_k^H\boldsymbol{E}_k$, formally

$$\nabla\hat{J}_k = \left[\mathbf{F}^{-1}\boldsymbol{X}_k^H\boldsymbol{E}_k\right]^{\lceil M \rceil}. \tag{7.37}$$

In fact, the last L samples are those to be discarded and relative to the circular correlation. Moreover, it should be noted that even in this case, it is not guaranteed that the last L elements of the vector $\boldsymbol{X}_k^H\boldsymbol{E}_k$ are zero.

7.3.3.1 Weight Update and Gradient's Constraint

To the weights update we can proceed in the time domain with the expression (7.11), considering a unique learning rate μ_B. To this solution, however, a frequency domain update of the type (7.31) is preferred, which allows the definition of a specific learning rate for each frequency bin. Therefore, we must transform again the frequency domain estimated gradient vector (7.37), considering the gradient augmented form by inserting L null terms, i.e., $\left[\nabla\hat{J}_k\,\mathbf{0}_L\right]^T$, namely,

$$\nabla\hat{J}_k^{\mathrm{F}} = \mathbf{F}\left[\nabla\hat{J}_k \quad \mathbf{0}_L\right]^T \tag{7.38}$$

and add it to the vector $\boldsymbol{W}_k$; as a result, the update with the overlap-save method can be written as

$$\boldsymbol{W}_{k+1} = \boldsymbol{W}_k + \boldsymbol{\mu}_k\nabla\hat{J}_k^{\mathrm{F}}. \tag{7.39}$$

For a better understanding of the algorithm and windowing constraint, it is convenient to express the OS-FDAF in matrix notation. Similarly to (7.28) the $(N{\times}N)$ windowing matrix $\mathbf{g}_{M,0}$ is defined as

$$\mathbf{g}_{M,0} \in \mathbb{R}^{(M+L)\times(M+L)} = \begin{bmatrix} \mathbf{I}_{M,M} & \mathbf{0}_{M,L} \\ \mathbf{0}_{L,M} & \mathbf{0}_{L,L} \end{bmatrix}. \tag{7.40}$$

With this formalism, the expression (7.37) can be rewritten in augmented form as

$$\left[\nabla\hat{J}_k\,\mathbf{0}_L\right]^T = \mathbf{g}_{M,0}\mathbf{F}^{-1}\boldsymbol{X}_k^H\boldsymbol{E}_k \tag{7.41}$$

and, consequently, the (7.39) can be rewritten as

$$\begin{aligned} \boldsymbol{W}_{k+1} &= \boldsymbol{W}_k + \mathbf{F}\mathbf{g}_{M,0}\mathbf{F}^{-1}\boldsymbol{\mu}_k\boldsymbol{X}_k^H\boldsymbol{E}_k \\ &= \boldsymbol{W}_k + \mathbf{G}_{M,0}\boldsymbol{\mu}_k\boldsymbol{X}_k^H\boldsymbol{E}_k. \end{aligned} \tag{7.42}$$

Comparing the latter with the general form (7.31), it appears that the *windowing constraint matrix* is defined as

$$\mathbf{G}_{M,0} = \mathbf{F}\mathbf{g}_{M,0}\mathbf{F}^{-1} \tag{7.43}$$

which is a full matrix with rank $< N$.

For the output computation, the expression (7.27) can be rewritten as

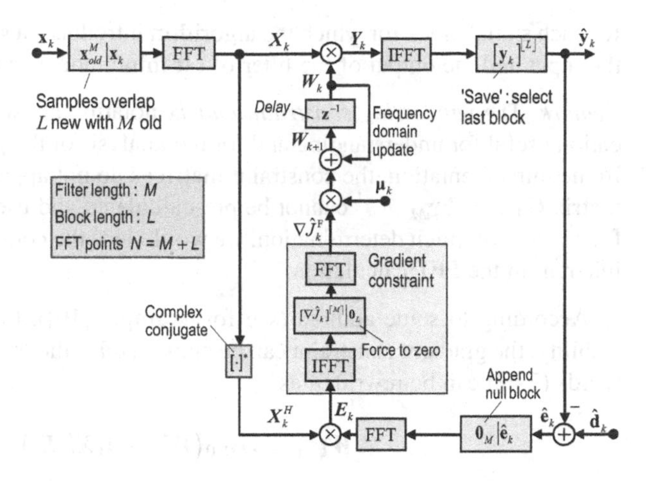

Fig. 7.6 Overlap-save FDAF (OS-FDAF) algorithm structure, also known as fast block LMS (FBLMS). The FFT is calculated for each signal block for which the algorithm introduces a systematic delay of (at least) L samples. In total, the OS-FDAF requires five N-points FFT calculation

$$\begin{aligned}\mathbf{y}_k &= \begin{bmatrix}\mathbf{0}_M & \hat{\mathbf{y}}_k\end{bmatrix}\\ &= \begin{bmatrix}\mathbf{0}_M & \left[\mathbf{F}^{-1}\boldsymbol{Y}_k\right]^{\lfloor L\rfloor}\end{bmatrix}\\ &= \mathbf{g}_{0,L}\mathbf{F}^{-1}\boldsymbol{X}_k\boldsymbol{W}_k.\end{aligned} \tag{7.44}$$

For the frequency domain error we have the expression (7.36).

The expression (7.42) is identical to (7.16) except that for convolution and correlation calculation the DFT was used. The complete algorithm structure is illustrated in Fig. 7.6 where we can observe the presence of five FFT/IFFT calculation blocks. This implementation has been independently derived by Ferrara [6] and by Clark et al. [1].

Remark The box illustrated in Fig. 7.6, which contains the IDFT (7.37) and the DFT (7.38), represents a *windowing constraint*, that in this case is a *gradient constraint*. From the previous development, it is clear that the constraint is necessary since the filter is of M-length and in performing the IDTF (7.37), only the first M values should be different from zero. Actually, the last L terms of the vector $\mathbf{F}^{-1}\boldsymbol{X}_k^H\boldsymbol{E}_k$ are not at all different from zero and, consequently, the gradient constraint forces such terms to zero ensuring proper weights update. Note, also, to avoid a biased solution, the initial weights value $\mathbf{w}_0$ must be chosen, necessarily, in such a way that the last L terms of its IDFT are zero [12].

Remark The overlap-save FDAF (OS-FDAF) algorithm, commonly also referred to as fast LMS (FLMS) or as fast block LMS (FBLMS), is the frequency domain equivalent of the BLMS; it has the same convergence characteristics in terms of speed, stability, misalignment, etc., and the algorithm converges, in average, to the optimum Wiener filter. The possibility of choosing learning rates different for each frequency bin, as with (7.39), allows a convergence speed improvement without, however, improving the reachable minimum MSE. The OS-FDAF presents, compared to BLMS, the dual advantage of having reduced complexity and higher convergence speed exploiting the step-size normalization. The FFT is calculated for each signal block for which the algorithm introduces a systematic delay between the input and the output of the filter of (at minimum) L samples.

Remark The *windowing constraint matrix* notation allows only a formal simplification useful for understanding and for the analysis of the properties of the method. In the implementation, the constraint matrices do not appear explicitly. In fact, the matrix $\mathbf{G}_{M,0} = \mathbf{F}\mathbf{g}_{M,0}\mathbf{F}^{-1}$ cannot be pre-calculated and used instead of the FFT. In fact, with its explicit determination, we would lose the computational cost reduction inherent in the FFT calculation.

According to some authors (see for example [19]), to have greater numerical stability, the gradient constraint can be applied *after* the weights $\boldsymbol{W}_k$ update. In other words (7.42) can be rewritten as

$$\boldsymbol{W}_{k+1} = \mathbf{G}_{M,0}\left(\boldsymbol{W}_k + \boldsymbol{\mu}_k \boldsymbol{X}_k^H \boldsymbol{E}_k\right). \tag{7.45}$$

From the implementation point of view, the algorithm can be realized as follows.

7.3.3.2 OS-FDAF Algorithm Summary

(i) Initialization $\boldsymbol{W}_0 = \mathbf{0}$, $P_0(m) = \delta_m$ for $m = 0, 1, \ldots, N-1$;
(ii) For $k = 0,1, \ldots$ { (for each L-samples block)

$$\boldsymbol{X}_k = \text{diag}\left\{\text{FFT}\left[\mathbf{x}_{old}^M \quad \mathbf{x}_k\right]^T\right\}$$

$$\hat{\mathbf{y}}_k = [\text{IFFT}(\boldsymbol{X}_k \boldsymbol{W}_k)]^{\lfloor L \rfloor}$$

$$\boldsymbol{E}_k = \text{FFT}\left(\left[\mathbf{0}_M \quad \hat{\mathbf{d}}_k - \hat{\mathbf{y}}_k\right]^T\right)$$

$$P_k(m) = \lambda P_{k-1}(m) + (1-\lambda)|X_k(m)|^2 \quad m = 0, 1, \ldots, N-1;$$

$$\boldsymbol{\mu}_k = \mu \text{diag}\left[P_k^{-1}(0), \ldots, P_k^{-1}(N-1)\right]$$

$$\boldsymbol{W}_{k+1} = \boldsymbol{W}_k + \boldsymbol{\mu}_k \boldsymbol{X}_k^H \boldsymbol{E}_k$$

$$\boldsymbol{W}_{k+1} = \text{FFT}\left(\begin{bmatrix} \text{IFFT}[\boldsymbol{W}_{k+1}]^{\lceil M \rceil} \\ \mathbf{0}_L \end{bmatrix}\right)$$

}

A more oriented scheme to the development of computer codes is presented in Fig. 7.7.

7.3.4 *UFDAF Algorithm*

In the so-called UFDAF algorithm [11], the gradient constraint is omitted, i.e., $\mathbf{G}_{M,0} = \mathbf{I}$. With this choice the configuration of the algorithm shown in Fig. 7.8 is

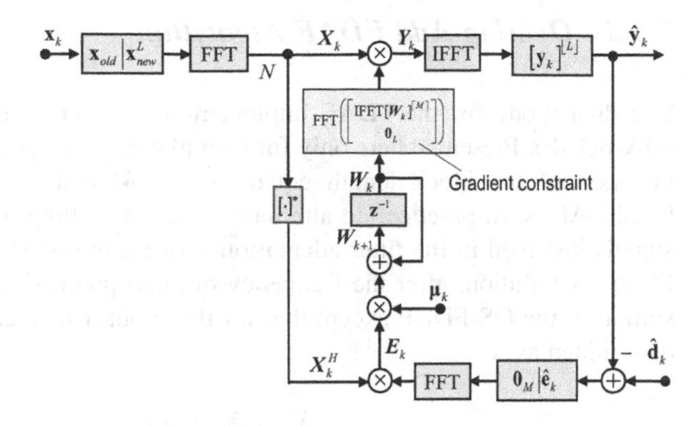

Fig. 7.7 Implementative scheme of the OS-FDAF algorithm

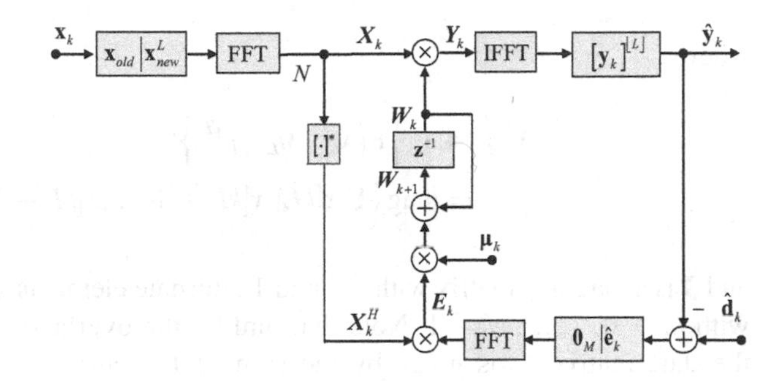

Fig. 7.8 Scheme of the algorithm (overlap-save) *unconstrained FDAF* (UFDAF). The UFDAF requires the calculation of three FFT of length $M + L$

identical to the OS-FDAF but with the gradient constraint block removed. The update rule is simplified as

$$\boldsymbol{W}_{k+1} = \boldsymbol{W}_k + \boldsymbol{\mu}_k \boldsymbol{X}_k^H \boldsymbol{E}_k. \tag{7.46}$$

The product $\boldsymbol{X}_k^H \boldsymbol{E}_k$ in (7.46) corresponds to a circular correlation over time (similar to the circular convolution $\boldsymbol{Y}_k = \boldsymbol{X}_k \boldsymbol{W}_k$). The output constraint (7.44) is instead maintained.

In general, the unconstrained algorithms have a biased convergence so they do not converge to the Wiener optimal solution and present a high steady-state error. In the case of systems identification, the algorithm tends to convergence to the optimum solution only in the case that the filter length M is greater than the order of the system to identify.

Although the convergence speed of the unconstrained algorithms can grow by optimizing the learning rate for each frequency bin (step size normalization), the misalignment due to the absence of constraints compensates for this improvement. Comparing experimentally constrained and unconstrained algorithms, it is seen that the latter requires approximately twice the iterations number to achieve the same misalignment level.

7.3.5 Overlap-Add FDAF Algorithm

The dual mode for the FDAF implementation is one called overlap-add FDAF (OA-FDAF). Presented here only for formal completeness, for simplicity, consider the case of the block length equal to $L = M$ and $N = 2M$ FFT points. The OA-FDAF is, in practice, an alternative way of cutting and reaggregation of the signals involved in the filter adaptation process in order to obtain a time-domain linear convolution, after the frequency domain processing [7]. The OA-FDAF is similar to the OS-FDAF except that for the input data vector which in this case is determined as

$$\boldsymbol{X}_k = \boldsymbol{X}_k^{'} + \mathbf{J}\boldsymbol{X}_{k-1}^{'}, \tag{7.47}$$

where

$$\begin{aligned} \boldsymbol{X}_k^{'} &= \text{diag}\left\{\mathbf{F}[\mathbf{x}_k \quad \mathbf{0}_{L=M}]^T\right\} \\ &= \text{diag}\left\{\mathbf{F}[x[kL], x[kL+1], ..., x[kL+1], 0, ..., 0]^T\right\} \end{aligned} \tag{7.48}$$

and $\mathbf{J}$ is a diagonal matrix with -1 and 1 alternate elements defined as $J_{mm} = (-1)^m$, with $m = 0, 1, \ldots, N - 1$. Note that, unlike the overlap-save method, in this case the data matrix $\boldsymbol{X}_k$ is given by the sum of the current block matrix, with zero-padding up to N, and the previous block matrix of with elements taken with alternate signs.

The filter output is calculated in accordance with the sectioning (7.47). For which we have

$$\hat{\mathbf{y}}_k = \left[\mathbf{F}^{-1}\boldsymbol{Y}_k\right]^{\lceil L \rceil}. \tag{7.49}$$

Even for the error the zero-padding is performed as

$$\boldsymbol{E}_k = \mathbf{F}[\,\hat{\mathbf{e}}_k\,\mathbf{0}_M]^T. \tag{7.50}$$

Regarding the learning rule this is entirely identical to that of OS-FDAF.

The algorithm structure is shown in Fig. 7.9 and comparing the overlap-save/add techniques one can observe that the only differences concern the vectors $\boldsymbol{X}_k$, $\boldsymbol{E}_k$, and $\mathbf{y}_k$ definition while, for the rest, the algorithms are identical.

Remark In the original formulation reported in [7], the sum of the current and previous blocks is performed in the time domain for which it is necessary to calculate two other DFT, i.e., $\mathbf{F}(\mathbf{F}^{-1}\boldsymbol{X}_{k-1}^{'} + \mathbf{F}^{-1}\boldsymbol{X}_k^{'})$. This is required because the time sequence, associated to the block $\boldsymbol{X}_{k-1}^{'}$, must be circularly shifted before being added to the IDFT $(\boldsymbol{X}_k^{'})$ sequence. Therefore, in total the original algorithm requires the calculation of seven DFT. One can easily see that in the expression (7.47), the

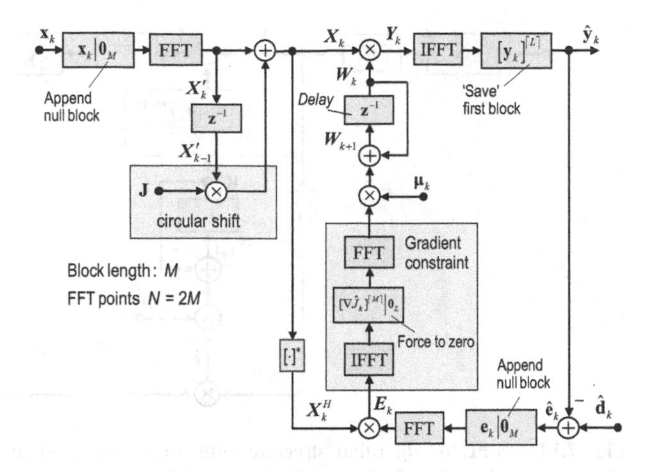

Fig. 7.9 Overlap-add FADF (OA-FDAF) algorithm structure

addition operation is carried out in the frequency domain. The multiplication by the matrix **J** is, in practice, the frequency domain operation equivalent to the time-domain circular shifting. Thus, the implementation (7.47), reported in [12], allows to save the calculation of two DFT (a direct and inverse).

7.3.6 Overlap-Save FDAF Algorithm with Frequency Domain Error

The overlap-save algorithm can be formulated in an alternative way, than presented previously in Sect. 7.3.3, performing the error calculation directly in the frequency domain.

From (7.27), (7.28), (7.29), and (7.30) the output DFT is defined as

$$\mathbf{Y}_k' = \mathbf{G}_{0,L}\mathbf{Y}_k. \tag{7.51}$$

For the $\mathbf{E}_k$ error calculation, define the *frequency domain desired response* of the amount

$$\mathbf{D}_k = \mathbf{F}\begin{bmatrix} \mathbf{0}_M & \hat{\mathbf{d}}_k \end{bmatrix}^T \tag{7.52}$$

for which the error in the frequency domain can be written as

$$\begin{aligned} \mathbf{E}_k &= \mathbf{D}_k' - \mathbf{Y}_k' \\ &= \mathbf{G}_{0,L}(\mathbf{D}_k - \mathbf{Y}_k). \end{aligned} \tag{7.53}$$

Note that the error is calculated by considering the constraint (7.51) and not as erroneously could be expected, from $\mathbf{E}_k = \mathbf{D}_k - \mathbf{X}_k\mathbf{W}_k$ directly. In Fig. 7.10, the

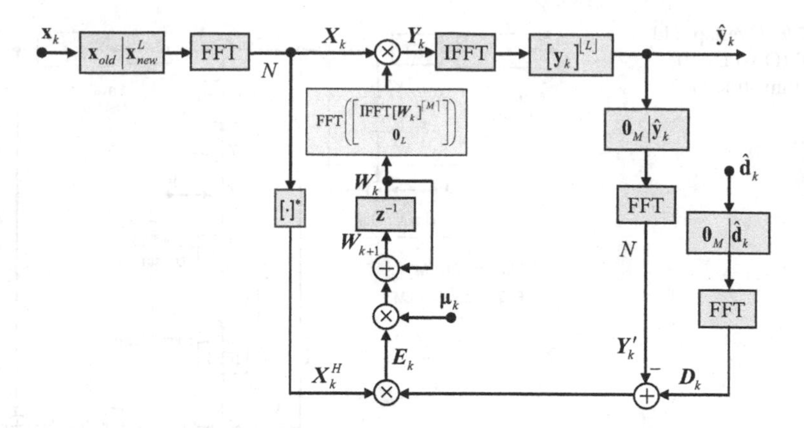

Fig. 7.10 OS-FDAF algorithm structure with error calculated in the frequency domain. The algorithm involves the calculation of six FFT and is, therefore, less efficient than other algorithms previously presented

algorithm diagram is shown, where we can observe the necessity of the calculation of six FFT, one more than the OS-FDAF with the time-domain error calculation.

Remark In theory, it is possible to define other FDAF implementation methods with different types of constraints that can improve performance in specific contexts (see for example [20] and [21]). However, in general terms other forms of implementation, while presenting interesting characteristics, do not always lead to accurate block adaptive algorithms, i.e., to the adaptation rule (7.16). For example, in [20], it is shown that using a full rank diagonal windowing matrix **g**, defined as $g_{mm} = (1/2)\cos(\pi m/N)$, $m = 0, 1, \ldots, N - 1$, may, in some situations, improve the convergence speed. In this case, moreover, the FDAF may be reformulated with reduced complexity with only three DFT.

7.3.7 UFDAF with $N = M$: Circular Convolution Method

The unconstrained algorithms for $N = M$ are characterized by the absence of constraints on both the input data windows and the gradient computation. The algorithm has a computational complexity approximately halved compared to the algorithms UFDAF at the expense, however, of a further convergence performance deterioration and misalignment. In fact, the absence of windowing constraints allows a 0 % overlap, whereas the absence of gradient constraint allows the direct frequency domain error calculation.

Before the algorithm description we present a brief review on *circulant matrices*.

7.3.7.1 Circulant Toeplitz Matrix

A circulant matrix $\mathbf{X}_\text{C}$ is a Toeplitz matrix with the form

$$\mathbf{X}_\text{C} = \begin{bmatrix} x_0 & x_{N-1} & x_{N-2} & \cdots & & x_1 \\ x_1 & x_0 & x_{N-1} & x_{N-2} & \cdots & x_2 \\ x_2 & x_1 & x_0 & x_{N-1} & \ddots & \\ \vdots & x_2 & x_1 & \ddots & \ddots & \vdots \\ & \vdots & \vdots & \vdots & & x_{N-1} \\ x_{N-1} & x_{N-2} & \cdots & & x_1 & x_0 \end{bmatrix}, \tag{7.54}$$

where, given the vector $\mathbf{x} = \begin{bmatrix} x_0 & x_1 & \cdots & x_{N-1} \end{bmatrix}^T$, each column (row) is constructed with a cyclic rotation of the previous column (row) element [22]. From the above definition, we have that $\mathbf{X}_\text{C}^H\mathbf{X}_\text{C} = \mathbf{X}_\text{C}\mathbf{X}_\text{C}^H$.

An important property, useful for the development below explained, is that each circulant matrix is diagonalizable with the DFT transformation, or with any other unitary transformation, such as

$$\mathbf{X}_d = \mathbf{F}\mathbf{X}_\text{C}\mathbf{F}^{-1}, \tag{7.55}$$

where the diagonal elements of $\mathbf{X}_d$ are constituted by the DFT of the first column of $\mathbf{X}_\text{C}$:

$$\mathbf{X}_d = \text{diag}\{\mathbf{F}\mathbf{x}\} = \text{diag}\begin{bmatrix} X(0) & X(1) & \cdots & X(N-1) \end{bmatrix}. \tag{7.56}$$

Applying the Hermitian transposition-conjugation operator in both sides of (7.55), since for the DFT matrix is $\mathbf{F}^{-1} = \mathbf{F}^H$, we can write

$$\begin{aligned} \mathbf{X}_d^H &= \mathbf{F}^{-H}\mathbf{X}_\text{C}^H\mathbf{F}^H \\ &= \mathbf{F}\mathbf{X}_\text{C}^H\mathbf{F}^{-1}. \end{aligned} \tag{7.57}$$

Left multiplying (7.55) by $\mathbf{F}^{-1}$ and right multiplying by $\mathbf{F}$, we have that

$$\mathbf{X}_\text{C} = \mathbf{F}^{-1}\mathbf{X}_d\mathbf{F} \tag{7.58}$$

in other words, the DFT transformation of a diagonal matrix produces always a circulant matrix.

For other properties on circulant matrices, see, for example, [23].

7.3.7.2 FDAF with Circulant Convolution

In the UFDAF algorithm the DFT length is equal to $N = M + L$ with M samples overlap of the input data window and computing needs of three FFT. An FDAF computational gain can be obtained, at the expense of performance deterioration, by

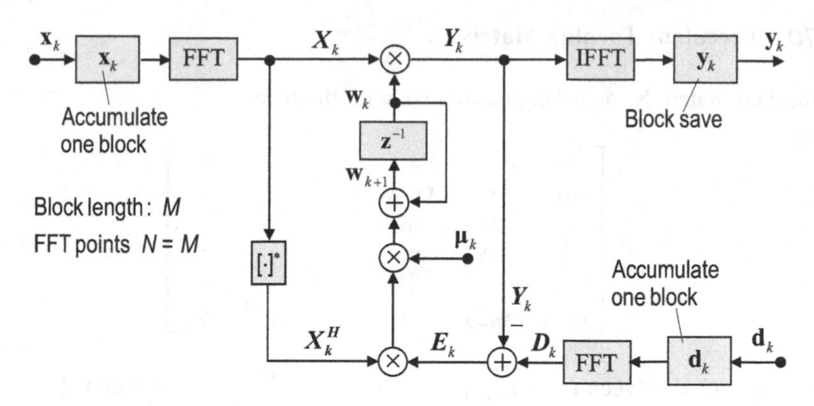

Fig. 7.11 The circular convolution FDAF (CC-FDAF) algorithm scheme

considering the DFT block length equal to the filter length, i.e., $N = L \equiv M$. In this case the augmented vectors are not needed and the DFT of the quantities $\mathbf{w}_k$ and $\mathbf{x}_k$ are defined, respectively, as

$$\boldsymbol{W}_k = \mathbf{F}\mathbf{w}_k, \tag{7.59}$$

$$\boldsymbol{X}_k = \text{diag}\{\mathbf{F}\mathbf{x}_k\}. \tag{7.60}$$

Also for the output no constraint is considered and it is simply

$$\mathbf{y}_k = \mathbf{F}^{-1}\boldsymbol{Y}_k, \tag{7.61}$$

where $\boldsymbol{Y}_k = \boldsymbol{X}_k\boldsymbol{W}_k$. The gradient constraints elimination is such that the output components are the result of a circular convolution. Note that input blocks since they are non-overlapping (0 % overlap), the error, unlike previous approaches, is a simple linear function of the output and the desired output. The error can therefore be directly calculated in the frequency domain without DFT and additional constraints. In other words, taking the desired output DFT, $\boldsymbol{D}_k = \mathbf{F}\mathbf{d}_k$, the frequency domain error in is simply

$$\boldsymbol{E}_k = \boldsymbol{D}_k - \boldsymbol{Y}_k \tag{7.62}$$

and the weights adaptation has the same UFDAF form (7.46). The circular convolution FDAF (CC-FDAF) algorithm, derived for the first time in [4], is shown in Fig. 7.11.

Although the algorithm does not require any data and gradient constraint, the CC-FDAF is, essentially, a block algorithm with adaptation similar to BLMS (7.16). Substituting the general form (7.31) in (7.61) and using the weights vector (7.59), the output can be expressed as

$$\begin{aligned}\mathbf{y}_k &= \mathbf{F}^{-1}X_k\mathbf{F}\mathbf{w}_k \\ &= \mathbf{X}_{\mathrm{C},k}\mathbf{w}_k,\end{aligned} \tag{7.63}$$

where $\mathbf{X}_{\mathrm{C},k} = \mathbf{F}^{-1}X_k\mathbf{F}$ and since by definition X_k is diagonal, it follows that, for the (7.58), $\mathbf{X}_{\mathrm{C},k}$ is a circulant matrix. For that reason, every column (row) of $\mathbf{X}_{\mathrm{C},k}$ entirely defines the matrix itself. In other words, the first column of $\mathbf{X}_{\mathrm{C},k}$ contains the M samples of the input block $x[kM], \ldots, x[kM + M - 1]$. So, considering the learning rate μ constant for all frequency, taking the IDFT of the UFDAF adaptation (7.46), we get

$$\mathbf{w}_{k+1} = \mathbf{w}_k + \mu\mathbf{X}_{\mathrm{C},k}^H\mathbf{e}_k. \tag{7.64}$$

Developing the matrix-vector product of the previous, the gradient estimate $\nabla\hat{J}_k = \mathbf{X}_{\mathrm{C},k}^H\mathbf{e}_k$ appears to be

$$\nabla\hat{J}_k = \sum_{i=0}^{L-1}\mathbf{x}_{\mathrm{C}i,k}e^*[kM + i], \tag{7.65}$$

where $\mathbf{x}_{\mathrm{C}i,k}$ indicates the ith matrix row of $\mathbf{X}_{C,k}^T$. Note that (7.65) has the same form of the adaptation block (7.12), except that, in this case, the error is correlated with the circulant version of the same input signal block. Similarly, the output vector (7.63) is the result of the circular convolution between TD filter weights and the input signal.

The obvious advantage of the method consists in the calculation of only three M points DFT that, together with the gradient constraint removal, allows a significant computational load reduction. The main disadvantage of the method is to have degraded performance because the method is only an approximate version of the BLMS. As a result of the distortions due to the circulant matrix, the convergence properties are quite different from the OS-FDAF methods.

The adaptation law (7.64) is quite different from (7.46), where each weight is updated by minimizing the MSE relative to its frequency rather than the MSE corresponding to the overall filter output performance. Only in the case in which the frequency bins are not correlated among them the two algorithms can converge in similar mode. Normally, however, there is a lot of spectral overlap and (7.64) has a steady-state performance lower than the linear convolution. A possible exception is in the adaptive line enhancer (ALE) applications (see Sect. 3.4.7.2) in which the signal to be cleaned generally has very narrow band or the process is constituted by well spatially separated sinusoids and therefore uncorrelated.

Table 7.3 FDAF vs. LMS computational efficiency ratio: $C_{\text{FDAF}}/C_{\text{LMS}}$ (from [12])

FDAF alg.	Filter length M					
	32	64	128	256	1024	2048
OS-FDAF	1.19	0.67	0.37	0.20	0.062	0.033
UFDAF	0.81	0.45	0.25	0.14	0.040	0.021
CC-FDAF	0.36	0.20	0.11	0.062	0.019	0.010

7.3.8 *Performance Analysis of FDAF Algorithms*

For performance analysis we consider the computational cost and the convergence analysis [5, 20, 24, 25].

7.3.8.1 Computational Cost Analysis

The real LMS algorithm (see Sect. 5.3.1.4) requires $(2M + 1)$ multiplications per sample. Thus, for N samples approximately $C_{\text{LMS}} = 2MN$ real multiplications are required. For an N-point FFT is required about $N\log_2 N$ multiplication. In the case of FDAF, N is, in general, chosen as $N = M$ or $N = 2M$, i.e., with a 50 % overlap. Therefore, the filter output and the gradient calculation require $4N$ real multiplications. Calling N_F the number of FFTs that correspond to the algorithm type, the computational cost for signal block processing, in terms of real multiplications, is approximately equal to

$$C_{\text{FDAF}} = N_F N\log_2 N + 10N. \tag{7.66}$$

An indicative summary of the FDAF algorithms computational costs is reported in Table 7.3 that indicates the relationship between the complexity of (7.66) and that for the LMS filters of equal length.

From the table it can be observed that the computational efficiency ratio increases with the filter length.

7.3.8.2 UFDAF Convergence Analysis

The OS-FDAF algorithm is exactly identical to BLMS with the only difference that the OS-FDAF is implemented in the frequency domain and, for this reason, it has identical convergence properties. The unconstrained algorithms have instead different characteristics. In this section we analyze the UFDAF performance with L-length block, M filter coefficients, and $N = M + L$ points FFT window.

For the convergence properties study, as already done in the time domain, consider a dynamic system identification problem of the type illustrated in Fig. 7.12. The frequency domain desired output is

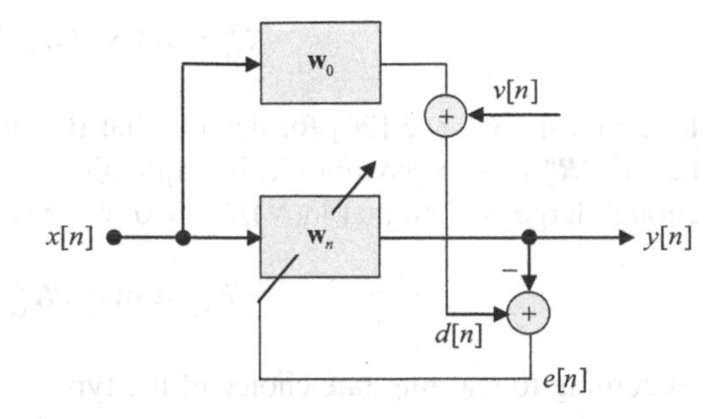

Fig. 7.12 Model for the statistical study of the UFDAF performance

$$\boldsymbol{D}_k = \mathbf{G}_{0,L}(\boldsymbol{X}_k \boldsymbol{W}_0 + \boldsymbol{V}_k), \tag{7.67}$$

where, indicating with $\mathbf{w}_0$ the Wiener solution, $\boldsymbol{W}_0 = \mathbf{F}[\mathbf{w}_0 \quad \mathbf{0}]^T$ is the optimal solution defined in the DFT domain and $\boldsymbol{V}_k$ indicates the frequency domain error at the optimal solution. In other words, if the filter weights are optimum, i.e., $\boldsymbol{W}_k = \boldsymbol{W}_0$, then the error is $\boldsymbol{E}_{0k} \equiv \boldsymbol{V}_k$.

For performance analysis proceeds as in the time domain (see Sect. 5.4.2). Using the above definitions, the error can be expressed as

$$\begin{aligned} \boldsymbol{E}_k &= \mathbf{G}_{0,L}(\boldsymbol{X}_k \boldsymbol{W}_0 + \boldsymbol{V}_k - \boldsymbol{X}_k \boldsymbol{W}_k) \\ &= \mathbf{G}_{0,L}\boldsymbol{X}_k \boldsymbol{W}_0 + \mathbf{G}_{0,L}\boldsymbol{V}_k - \mathbf{G}_{0,L}\boldsymbol{X}_k \boldsymbol{W}_k. \end{aligned} \tag{7.68}$$

Combining the above equation with the unconstrained adaptation law, we can write

$$\begin{aligned} \boldsymbol{W}_{k+1} &= \boldsymbol{W}_k + \boldsymbol{\mu}_k \boldsymbol{X}_k^H \boldsymbol{E}_k \\ &= \boldsymbol{W}_k - \boldsymbol{\mu}_k \boldsymbol{X}_k^H \mathbf{G}_{0,L}\boldsymbol{X}_k \boldsymbol{W}_k + \boldsymbol{\mu}_k \boldsymbol{X}_k^H \mathbf{G}_{0,L}\boldsymbol{X}_k \boldsymbol{W}_0 + \boldsymbol{\mu}_k \boldsymbol{X}_k^H \mathbf{G}_{0,L}\boldsymbol{V}_k. \end{aligned} \tag{7.69}$$

In addition, by defining the *frequency domain error vector* as

$$\boldsymbol{U}_k = \boldsymbol{W}_k - \boldsymbol{W}_0$$

the (7.69) can be written as

$$\boldsymbol{U}_{k+1} = \left(\mathbf{I} + \boldsymbol{\mu}_k \boldsymbol{X}_k^H \mathbf{G}_{0,L}\boldsymbol{X}_k\right)\boldsymbol{U}_k + \boldsymbol{\mu}_k \boldsymbol{X}_k^H \mathbf{G}_{0,L}\boldsymbol{V}_k \tag{7.70}$$

which is a SDE in the RV $\boldsymbol{U}_k$, $\boldsymbol{W}_k$, and $\boldsymbol{X}_k$, similar to the time-domain SDE $\mathbf{u}_n = (\mathbf{I} - \mu\mathbf{x}_n^H\mathbf{x}_n)\mathbf{u}_{n-1} + \mu v^*[n]\mathbf{x}_n$, already defined in the (see Sect. 5.4.2). Taking the expectation of the previous, the *weak convergence analysis* is made according to the orthogonality principle for which, $E\{\boldsymbol{X}_k^H\mathbf{G}_{0,L}\boldsymbol{V}_k\} = \mathbf{0}$. So, we get

$$E\{\boldsymbol{U}_{k+1}\} = \left(\mathbf{I} - \boldsymbol{\mu}_k \boldsymbol{R}_{xx}^u\right)E\{\boldsymbol{U}_k\}, \tag{7.71}$$

where $\boldsymbol{R}_{xx}^u$, which determines the various convergence modes (without learning rate normalization), is defined as

$$\boldsymbol{R}_{xx}^{u} = E\{\boldsymbol{X}_k^H \mathbf{G}_{0,L} \boldsymbol{X}_k\}. \tag{7.72}$$

It can be shown, see [26] for details, that the time-domain equivalent expression, i.e., $\mathbf{F}^{-1}\boldsymbol{R}_{xx}^{u}\mathbf{F}$, it is asymptotically equivalent to a circulant matrix. Such that, for enough large N such that $\ln(N)/N \to 0$, we have that

$$\boldsymbol{R}_{xx}^{u} \approx \text{diag}\{\boldsymbol{R}_{xx}^{u}\}. \tag{7.73}$$

According to learning rate choice of the type:

$$\begin{aligned}\boldsymbol{\mu}_k &= \mu \text{diag}\left[P_k^{-1}(0), \ldots, P_k^{-1}(N-1)\right] \\ &= \mu \mathbf{P}_k^{-1},\end{aligned}$$

it appears that the $\boldsymbol{\mu}_k$ elements tend to equalize the convergence modes such that it holds that $\boldsymbol{\mu}_k \boldsymbol{R}_{xx}^{u} = \mathbf{I}$. In other words

$$\text{diag}\{\boldsymbol{R}_{xx}^{u}\} = \mathbf{P}_k \tag{7.74}$$

so we can write

$$\boldsymbol{R}_{xx}^{u} \approx \frac{L}{N}\mathbf{P}_k. \tag{7.75}$$

7.3.8.3 Normalized Correlation Matrix

Equation (7.75) shows that the UFDAF convergence is regulated by the diagonal elements of the matrix $\mathbf{P}_k$ containing the various frequency-bin energy. With the choice (7.75) the product $\boldsymbol{\mu}_k \boldsymbol{R}_{xx}^{u} \approx \mathbf{I}$, for which the adaptation (7.71) has a single convergence mode. From the physical point of view, this is equivalent to the uniform sampling of the filter input power spectral density at $\omega_i = 2\pi i/N$, for $i = 0, 1, \ldots, N-1$. In other words, by defining the *normalized correlation matrix* as $\boldsymbol{R}_{xx}^{uN} = \boldsymbol{\mu}_k \boldsymbol{R}_{xx}^{u}$, for large N is

$$\boldsymbol{R}_{xx}^{uN} = \mathbf{P}_k^{-1}\boldsymbol{R}_{xx}^{u} = \text{diag}\{\boldsymbol{R}_{xx}^{u}\}\boldsymbol{R}_{xx}^{u} \approx \frac{L}{N}\mathbf{I}. \tag{7.76}$$

Note, finally, that indicating with $\boldsymbol{R}_{xx}^{c} = E\{\boldsymbol{X}_k^H \boldsymbol{X}_k\}$, the expression (7.72) can be written as

$$\boldsymbol{R}_{xx}^{u} = \boldsymbol{R}_{xx}^{c} \odot \mathbf{G}_{0,L}, \tag{7.77}$$

where the symbol $\odot$ indicates the point-to-point multiplication.

7.4 Partitioned Impulse Response FDAF Algorithms

The advantage of the block frequency domain algorithms depends both on the high computational efficiency and convergence properties. The latter are due to the intrinsic adaptation equations decoupling, namely the noncorrelation of the various frequency bins that determines an approximately diagonal correlation matrix.

However, the main disadvantage is related to the delay introduced, required for the preliminary acquisition of the entire block of signal, before processing. Even in the case of implementation with a certain degree of parallelism the systematic delay, also referred to as *latency*, introduced between the input and the output is at least equal to the block length L. A simple solution consists in defining short block lengths $(L \ll N)$. However, this choice may not be compatible with the filter length M and, in addition, the computational advantage may not be significant.

An alternative solution to decrease of the block length, given in [27] and later reproposed and modified by several other authors, see for example [19, 28–32], is to partition the filter impulse response in P subfilters. Thus, the convolution is implemented in P smaller convolutions, each of these implemented in the frequency domain. With this type of implementation, the frequency domain approach advantages are associated with a significant latency reduction. It should be noted that this class of algorithms is indicated in the literature as partitioned FBLMS (PBLMS), or as partitioned block FDAF (PBFDAF) and also with other names. In [29], for example, is indicated as multi-delay adaptive filter (MAF).

7.4.1 The Partitioned Block FDAF

Let us consider the implementation of a filter of length equal to $M_F = PM$ taps[3] where M is the length of the partition and P the number of partitions. The output of the filter is equal to

$$y[n] = \sum_{i=0}^{PM-1} w_n[i]x[n-i] \tag{7.78}$$

for the linearity of the convolution, the sum (7.78) can be partitioned as

$$y[n] = \sum_{l=0}^{P-1} y_l[n], \tag{7.79}$$

where

[3] In this section the filter length is referred to as M_F.

$$y_l[n] = \sum_{i=0}^{M-1} w_n[i + lM]x[n - i - lM]. \tag{7.80}$$

As schematically illustrated in Fig. 7.13, by inserting appropriate delay lines between the partitions, the filter is implemented with P separate M-length convolutions, each of which can be simply implemented in the frequency domain. The overall output is the sum (7.79).

Consider the case in which the block length is $L \leq M$. Let k be the block index, denoting with $\mathbf{x}_k^l \in (\mathbb{R},\mathbb{C})^{(M+L)\times 1}$ the lth partition of the input sequence vectors and with $\mathbf{w}_k^l \in (\mathbb{R},\mathbb{C})^{(M\times 1)}$ the augmented form of the filter weights, respectively, defined as

$$\mathbf{x}_k^l = \underbrace{\begin{bmatrix} \mathbf{x}_{\text{old}}^{l,M} & \mathbf{x}_{k,\text{new}}^{l,L} \end{bmatrix}^T}_{(M+L)} = \begin{bmatrix} x[kL - lM - M] \\ \vdots \\ x[kL - lM - 1] \\ x[kL - lM] \\ \vdots \\ x[kL - lM + L - 1] \end{bmatrix} \begin{matrix} \left.\vphantom{\begin{matrix} a \\ b \\ c \end{matrix}}\right\} \text{old samples } \lceil M \rceil \\ \left.\vphantom{\begin{matrix} a \\ b \\ c \end{matrix}}\right\} \text{new block } \lfloor L \rfloor \end{matrix} \tag{7.81}$$

and

$$\mathbf{w}_k^l = \begin{bmatrix} w_k[lM] \\ \vdots \\ w_k[lM + M - 1] \\ 0 \\ \vdots \\ 0 \end{bmatrix} \begin{matrix} \left.\vphantom{\begin{matrix} a \\ b \\ c \end{matrix}}\right\} \lceil M \rceil \text{ subfilter weights} \\ \left.\vphantom{\begin{matrix} a \\ b \\ c \end{matrix}}\right\} \text{zero padding } \lfloor L \rfloor \text{ samples} \end{matrix} . \tag{7.82}$$

Note that there is an L taps overlap between two successive filter partitions and the insertion of L zeros in the weights vector definition. The overlap and zero-padding of the weights vector are necessary for the algorithm implementation with overlap-save technique. The input data frequency domain representation for the single partition is defined by a diagonal matrix $X_k^l \in \mathbb{C}^{(M+L)\times(M+L)}$, with DFT elements of $\mathbf{x}_k^l$, i.e.,

$$X_k^l = \text{diag}\{\mathbf{F}\mathbf{x}_k^l\} \tag{7.83}$$

while the frequency domain representation of the impulse response partition $\mathbf{w}_k^l$ is defined as

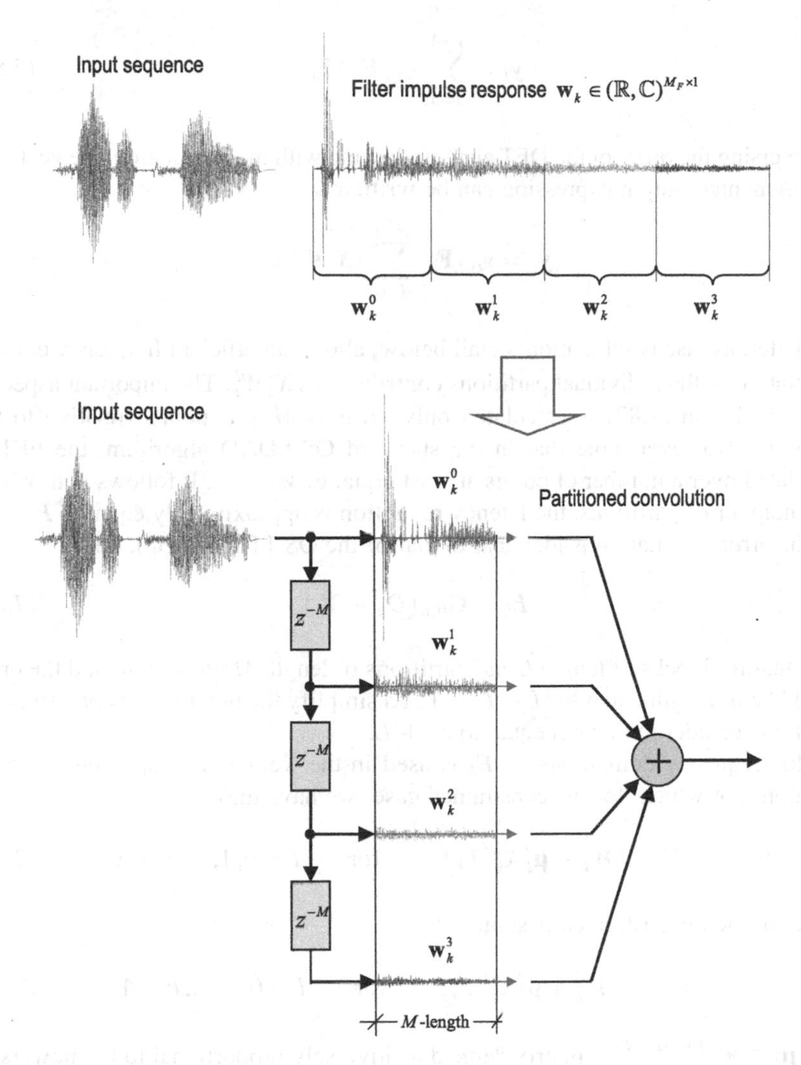

Fig. 7.13 Time-domain partitioned convolution schematization

$$\boldsymbol{W}_k^l = \mathbf{F}\mathbf{w}_k^l. \tag{7.84}$$

Calling $\boldsymbol{Y}_k^l = \boldsymbol{X}_k^l \boldsymbol{W}_k^l$ the output augmented form, for the lth partition, the time-domain output is defined as [see (7.27)]

$$\begin{aligned} \mathbf{y}_k^l &= \left[\mathbf{0} \quad \left[\mathbf{F}^{-1} \boldsymbol{Y}_k^l \right]^{\lfloor L \rfloor} \right] \\ &= \mathbf{g}_{0,L} \mathbf{F}^{-1} \boldsymbol{Y}_k^l \end{aligned} \tag{7.85}$$

whereby the filter overall output is defined by the sum of all partitions

$$\mathbf{y}_k = \sum_{l=0}^{P-1} \mathbf{g}_{0,L} \mathbf{F}^{-1} \boldsymbol{Y}_k^l. \tag{7.86}$$

By reversing the order of the DFT and windowing with the summation, the vector of the augmented output expression can be written as

$$\mathbf{y}_k = \mathbf{g}_{0,L} \mathbf{F}^{-1} \sum_{l=0}^{P-1} \left[\boldsymbol{X}_k^l \boldsymbol{W}_k^l \right]. \tag{7.87}$$

The latter, as discussed in more detail below, allows an efficient frequency domain calculation of the individual partitions contributions $\boldsymbol{X}_k^l \boldsymbol{W}_k^l$. The important aspect is that the FFT in (7.87) is calculated only on $N = M + L$ points (relative to the partition). However, note that in the standard OS-FDAD algorithm, the FFT is calculated over a number of points at least equal to $M_F + L$. It follows that, with a high number of partitions, the latency reduction is approximately equal to P.

The error calculation is identical to that of the OS-FDAF (7.53), i.e.,

$$\boldsymbol{E}_k = \mathbf{G}_{0,L}(\boldsymbol{D}_k - \boldsymbol{Y}_k). \tag{7.88}$$

Note that for blocks of length L and partitions of length M, the output and the error should be of length equal to $M + L - 1$. To simplify the notation it is preferred, as usual, to consider the length equal to $M + L$.

The frequency domain vector $\boldsymbol{E}_k$ is used in the weights law updating for each partition. For which, for the constrained case, we have that

$$\boldsymbol{W}_{k+1}^l = \mathbf{G}_{M,0}\left(\boldsymbol{W}_k^l + \boldsymbol{\mu}_k^l \boldsymbol{X}_k^{lH} \boldsymbol{E}_k\right), \qquad \text{for} \qquad l = 0, 1, \ldots, P-1 \tag{7.89}$$

while for the unconstrained is simply

$$\boldsymbol{W}_{k+1}^l = \boldsymbol{W}_k^l + \boldsymbol{\mu}_k^l \boldsymbol{X}_k^{lH} \boldsymbol{E}_k, \qquad \text{for} \qquad l = 0, 1, \ldots, P-1 \tag{7.90}$$

with $\boldsymbol{\mu}_k^l \in \mathbb{R}^{(M+L)\times(M+L)}$ matrix defined as inversely proportional to the powers of the relative frequency-bins $\{\mathbf{F}\mathbf{x}_k^l\}$ and updated with the mechanism of the type previously described in (7.34).

Remark As for the not PFDAF algorithms also in this case it is possible to implement constrained or unconstrained adaptation forms.

7.4.1.1 PBFDAF Algorithm Development

In PBFDAF algorithms the block length is always less than or equal to the length of the filter partition, i.e., $L \le M$. For the algorithm implementation is necessary, in practice, to choose the block length L submultiple of M or $L = M/p$ with p integer.

With this position, the partition is equal to $M = pL$, and from the definition (7.81) and (7.83), we can write

$$\mathbf{x}_k^l = [\underbrace{\mathbf{x}_{\text{old}}^{l,pL} \quad \cdots \quad \mathbf{x}_{\text{old}}^{l,2L} \quad \mathbf{x}_{\text{old}}^{l,L}}_{M=pL} \; \underbrace{\mathbf{x}_{k,\text{new}}^{l,L}}_{L}]^T$$

$$= \begin{bmatrix} x[kL - lpL - pL] \\ \vdots \\ x[kL - lpL - 1] \\ x[kL - lpL] \\ \vdots \\ x[kL - lpL + L - 1] \end{bmatrix} \begin{matrix} \left.\vphantom{\begin{matrix}a\\b\\c\end{matrix}}\right\} [pL] \\ \left.\vphantom{\begin{matrix}a\\b\\c\end{matrix}}\right\} [L] \end{matrix} . \tag{7.91}$$

For the adaptation algorithm development, we consider the following cases:

Case $M = L$ In case we have $p = 1, M = L$, and for $l = 0, 1, 2, \ldots$, we can write

$$\mathbf{x}_k^0 = \begin{bmatrix} \vdots \\ x[kL + L - 1] \end{bmatrix}, \mathbf{x}_k^1 = \begin{bmatrix} \vdots \\ x[kL - L + L - 1] \end{bmatrix}, \mathbf{x}_k^2 = \begin{bmatrix} \vdots \\ x[kL - 2L + L - 1] \end{bmatrix}, \ldots \tag{7.92}$$

It is easy to verify that for $M = L$ is

$$\boldsymbol{X}_k^l = \boldsymbol{X}_{k-l}^0. \tag{7.93}$$

For which (7.89) can be expressed as

$$\boldsymbol{W}_{k+1}^l = \mathbf{G}_{M,0}\left(\boldsymbol{W}_k^l + \boldsymbol{\mu}_k^l \boldsymbol{X}_{k-l}^{0H} \boldsymbol{E}_k\right), \quad \text{for } l = 0, 1, \ldots, P - 1. \tag{7.94}$$

The last property and the expression of the output calculation (7.87) allow an algorithm structure as shown in Fig. 7.14. An interesting interpretation can be made observing the figure in which there is a bank of P filters, of $N = 2M$ order, called frequency-bin filters [27, 32]. In addition, considering the (7.92) and (7.93), the delays z^{-1} are intended in unit of block size.

Case $M = pL$ For example, for $p = 2$ we have that $M = 2L$ and for $l = 0, 1, 2, \ldots$; so, we can write

$$\mathbf{x}_k^0 = \begin{bmatrix} \vdots \\ x[kL + L - 1] \end{bmatrix}, \mathbf{x}_k^l = \begin{bmatrix} \vdots \\ x[kL - 2L + L - 1] \end{bmatrix}, \mathbf{x}_k^2 = \begin{bmatrix} \vdots \\ x[kL - 4L + L - 1] \end{bmatrix}, \ldots$$

and from the above it is easy to generalize as

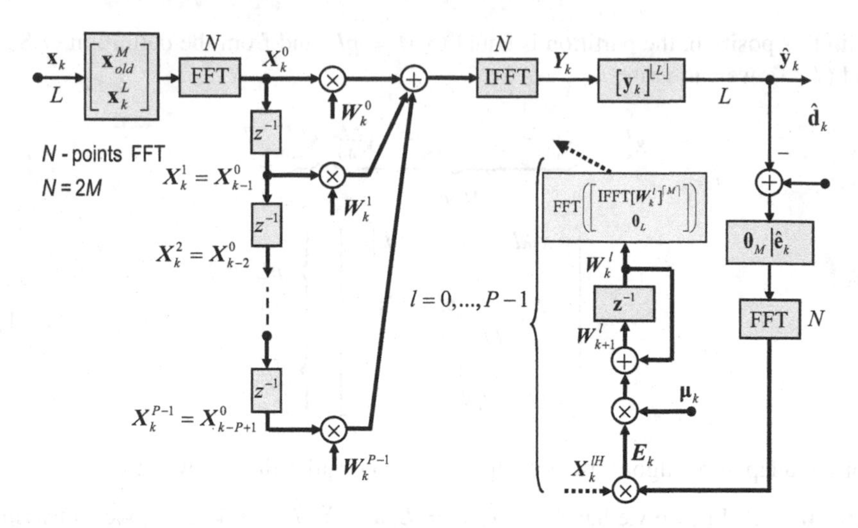

Fig. 7.14 Structure of the PFDAF algorithm, also known as PFBLMS, for $L = M$ developed in [27–32]

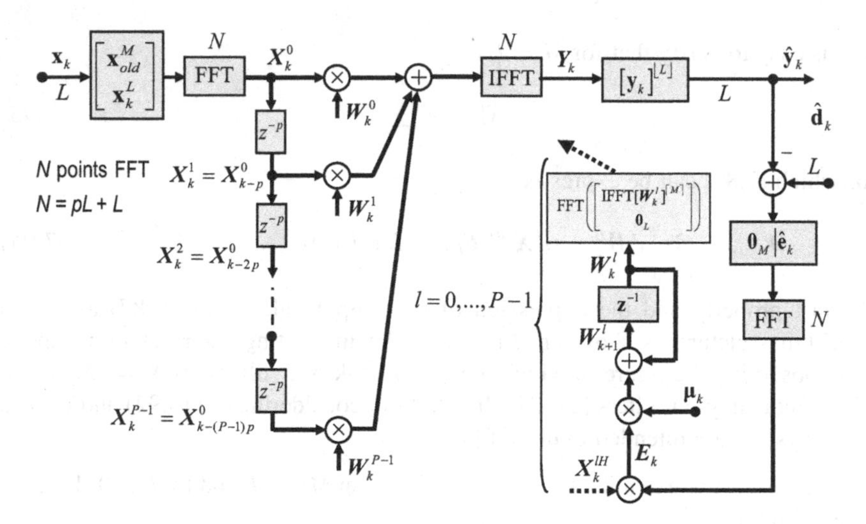

Fig. 7.15 Structure of the algorithm PFBLMS, for $M = pL$, developed in [19]. The delays z^{-p} are intended in unit of block size

$$X^l_k = X^0_{k-pl}. \tag{7.95}$$

For which (7.89) can be written as

$$W^l_{k+1} = \mathbf{G}_{M,0}\Big(W^l_k + \boldsymbol{\mu}^l_k X^{0H}_{k-pl} E_k\Big), \qquad \text{for} \qquad l = 0, 1, ..., P-1. \tag{7.96}$$

The algorithm structure is illustrated in Fig. 7.15, where it can be observed that in this case there are $N = pL + L$ frequency bins and the unit delay element z^{-1} of Fig. 7.14 is replaced with a delay element z^{-p} (also intended in unit of block size).

7.4.1.2 PFDAF Algorithm Summary

(i) Initialization $\mathbf{W}_0 = \mathbf{0}$, $P_0(m) = \delta_m$ for $m = 0, 1, \ldots, M + L - 1$;
(ii) For $k = 0,1, \ldots$ { *// for block of L samples*

$$\mathbf{X}_k^0 = \text{diag}\left\{\text{FFT}\left[\mathbf{x}_{old}^{0,M} \quad \mathbf{x}_{k,new}^{0,L}\right]^T\right\}$$

$$\hat{\mathbf{y}}_k = \left[\text{IFFT}\left(\sum\nolimits_{l=0}^{P-1} \mathbf{X}_{k-pl}^0 \mathbf{W}_k^l\right)\right]^{\lfloor L \rfloor}$$

$P_k(m) = \lambda P_{k-1}(m) + (1 - \lambda)|X_k^0(m)|^2$ for $m = 0, 1, \ldots, M + L-1$

$\boldsymbol{\mu}_k = \mu \, \text{diag}[P_k^{-1}(0), \ldots, P_k^{-1}(M + L - 1)]$

For $l = 0, \ldots, P - 1$ {

$$\mathbf{E}_k = \text{FFT}\left(\left[\mathbf{0}_M \quad \hat{\mathbf{d}}_k - \hat{\mathbf{y}}_k\right]^T\right)$$

$$\mathbf{W}_{k+1} = \mathbf{W}_k + \boldsymbol{\mu}_k \mathbf{X}_{k-pl}^{0H} \mathbf{E}_k$$

$$\mathbf{W}_{k+1} = \text{FFT}\left(\begin{bmatrix} \text{IFFT}[\mathbf{W}_{k+1}]^{\lceil M \rceil} \\ \mathbf{0}_L \end{bmatrix}\right)$$

}

}

7.4.2 Computational Cost of the PBFDAF

The complexity analysis of the PBFDAF depends on the used FFT type and many other factors. The exact calculation of the computational cost, in addition to being difficult, it is not strictly necessary and, in general, see for example [19], it is preferred to perform a macroscopic level analysis. In the unconstrained case each data block processing requires three FFT of $N = (p + 1)L$ points, and five FFT, in the case that a gradient constraint is added (see above figures).

Considering, for simplicity, the unconstrained case and real values input sequence, for each FFT (using a standard algorithm with power of two lengths) are required $N/4 \log_2 N/2$ butterflies calculations. Considering that the processing of each sample, in the frequency domain, requires three real multiplications, each gradient vector element requires a real multiplication with the learning rate, and other operations are required for the learning rate normalization; the computational cost for each signal sample will assume the expression (see [19] for details)

$$C_{\text{PBFDAF}} = \frac{(p+1)LP + \frac{3}{4}(p+1)\log_2 \frac{(p+1)L}{2}}{L}$$
$$= (p+1)P + \frac{3}{4}(p+1)\log_2 \frac{(p+1)L}{2}.$$

The required amount of memory for intermediate frequency domain data storage (delay lines z^{-P}, etc.), the filter coefficients, plus other intermediate buffers for the algorithm implementation is about $(p+1)^2 LP$ [19].

7.4.3 PFDAF Algorithm Performance

For the performance analysis we proceed with the same method used in the FBLMS case (see Sect. 7.3.8.2), and it is noted that even for the PFDAF we consider almost uncorrelated frequency bins and thus a correlation matrix approximately diagonal.

7.4.3.1 Performance of PFDAF for $L = M$

For the development proceed as in [19] considering, for simplicity, the unconstrained algorithm and $L = M$.

Recalling that $\mathbf{X}_k^l = \mathbf{X}_{k-l}^0$,

$$\begin{aligned} \mathbf{x}_{i,k} &= \begin{bmatrix} X_k^0(i) & X_k^1(i) & \cdots & X_k^{P-1}(i) \end{bmatrix}^T \\ &= \begin{bmatrix} X_k^0(i) & X_{k-1}^0(i) & \cdots & X_{k-P+1}^0(i) \end{bmatrix}^T \end{aligned}$$

the vector containing the ith frequency bin of the kth block, i.e., considering Fig. 7.14; $\mathbf{x}_{i,k}$ contains the P values of the input delay line of the ith *frequency-bin filter*. It follows that the behavior and the convergence properties of the ith frequency-bin filter depend on the $(P \times P)$ correlation matrix eigenvalues of its input that is defined as

$$\mathbf{R}_{xx,i} \triangleq E\{\mathbf{x}_{i,k}\mathbf{x}_{i,k}^H\} \tag{7.97}$$

or, equivalently, of its normalized version defined as

$$\mathbf{R}_{xx,i}^N \triangleq \left(\text{diag}[\mathbf{R}_{xx,i}]\right)^{-1}\mathbf{R}_{xx,i}. \tag{7.98}$$

For the correlation matrix determination, to simplify the analysis, we consider a white input sequence $x[n]$ and, recalling the (2M-points) DFT definition, for the ith frequency bin of the element $X_k^0(i)$, we have that

$$X_k^0(i) = \sum_{n=0}^{2M-1} x[kM - M + n]e^{-j\frac{2\pi}{2M}in}. \tag{7.99}$$

For which, from previous assumptions, it appears that

$$E\{X_k^0(i)X_{k-l}^{0*}(i)\} = \begin{cases} 2M\sigma_x^2 & \text{for } l = 0 \\ (-1)^i \times M\sigma_x^2 & \text{for } l = \pm 1 \\ 0 & \text{otherwise,} \end{cases} \tag{7.100}$$

where σ_x^2 is the variance of $x[n]$. Generalizing the result (7.100), for input white process, the normalized correlation matrix (7.98) is defined as

$$\mathbf{R}_{xx,i}^N = (\text{diag}[\mathbf{R}_{xx,i}])^{-1}\mathbf{R}_{xx,i} = \begin{bmatrix} 1 & \alpha_i & 0 & \cdots & 0 \\ \alpha_i & 1 & \alpha_i & 0 & \cdots & 0 \\ 0 & \alpha_i & 1 & \alpha_i & \ddots & \vdots \\ \vdots & 0 & \alpha_i & 1 & \ddots & 0 \\ & \vdots & \ddots & \ddots & \ddots & \alpha_i \\ 0 & 0 & \cdots & 0 & \alpha_i & 1 \end{bmatrix}, \tag{7.101}$$

where $\alpha_i = (-1)^i \times 0.5$. Note that the parameter α_i depends on the overlap level between two successive frames and in the case of 50 % overlap, its value is $\alpha_i = \pm 0.5$.

The $\mathbf{R}_{xx,i}^N$ matrix nature allows the calculation of its eigenvalues necessary for the convergence properties evaluation. From (7.100), we can observe that the condition number $\chi(\mathbf{R}_{xx,i}^N) = \lambda_{\max}/\lambda_{\min}$ (i) it does not depend on the frequency index i, (ii) it increases with the number of partitions P increase, (iii) it decreases with decreasing $|\alpha_i|$. As reported in [19], it can easily calculate that for $P = 2, \chi(\mathbf{R}_{xx,i}^N) = 3$ and for $P = 10, \chi(\mathbf{R}_{xx,i}^N) = 48.374$. Therefore, to increase the convergence speed it is convenient to implement the algorithm with overlap less than 50 %, for which $L < M$.

7.4.3.2 Performance of PFDAF for $L < M$

Let us consider the case where $L = M/p$, with p a positive integer. In this case, for the ith frequency bin, the $(M + L)$-points DFT expression is defined as

$$X_k^0(i) = \sum_{n=0}^{M+L-1} x[kL - M + n]e^{-j\frac{2\pi}{(M+L)}in}. \tag{7.102}$$

From this, it is immediate to show that, for white $x[n]$, it is

Table 7.4 Value of $\chi(\boldsymbol{R}_{xx,i}^{N})$ for $P = 10$ for different p values [19]

p	1	2	3	4	5	6	7	8	9	10
$\chi(\boldsymbol{R}_{xx,i}^{N})$	48.37	5.55	2.84	2.25	1.94	1.75	1.63	1.54	1.47	1.42

$$\alpha_i = \frac{1}{p+1} e^{j\left(2\pi p i / p+1\right)}$$

so, it is evident that, by increasing the overlap, α_i tends to decrease. In Table 7.4 a series of values of the condition number for $P = 10$ for various p values is shown.

Remark The convergence problems due to the overlap level disappear in the case where the filter weights update is performed with the constrained gradient. In the case of constraint gradient algorithm, in fact, the convergence is identical to that of not partitioned implementation.

7.5 Transform-Domain Adaptive Filters

The adaptive algorithms convergence properties depend on the input correlation matrix eigenvalues. In fact, the high condition number $\chi(\mathbf{R})$ in the colored processes determines the increase of the time-constant convergence. Online linear unitary transformations, as whitening pre-filtering and/or unitary orthogonalizing transformations, together with a step-size power normalization procedure, determine a new eigenvalues distribution lowering the condition number, with a consequent increase in the adaptation convergence speed.

With the TDAF, we refer to filters adapted with the LMS when the input is preprocessed by a unitary, usually data independent, transformation followed by step-size power normalization stage. The chosen transformation is, most of the time, the DFT, although other transformations operating on real data, as the DST, DCT, DHT, the Walsh–Hadamard transform (WHT), etc., have been proposed and used in the literature. The resulting algorithm takes the name of LMS-DFT, DCT-LMS, etc [33–35].

With reference to Fig. 7.16, the TDAF methods may be viewed as a special case of FDAF in which the block length is equal to 1. These algorithms are also called sliding window FDAF. Note, also, that the nickname TDAF, introduced in [33], is not entirely appropriate, as pointed out in [12], because also the FDAF operate in the transformed domain.

7.5.1 TDAF Algorithms

The TDAF, represented schematically in Fig. 7.16, can be viewed as FDAF in which the block length is $L = 1$. In this case the linear transformation $\mathbf{F}$ is performed in the presence of a new signal sample $x[n]$. In other terms, TDAF are

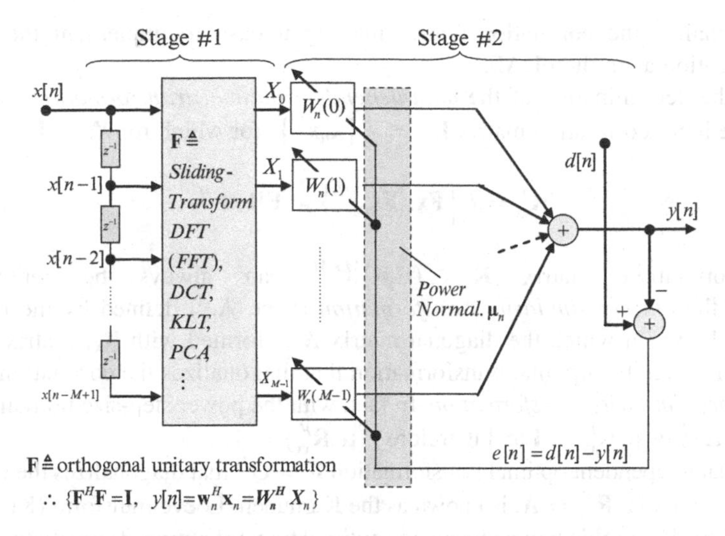

Fig. 7.16 Transform-domain adaptive filtering. The AF is realized in two stages: the first makes a sliding window domain transformation, while the second implements the LMS filtering algorithm with step-size power normalization

normal transversal AF characterized by an unitary orthogonal transformation $\mathbf{F}$ made on the input signal, i.e., such that $\mathbf{F}^H\mathbf{F} = \mathbf{I}$, tending to orthogonalize the input signal itself. The operator $\mathbf{F}$ is applied to the input $\mathbf{x}_n$ and to the weights $\mathbf{w}_n$, that, in the transformed domain, are denoted, respectively, as

$$X_n = \mathbf{F}\mathbf{x}_n,$$
$$W_n = \mathbf{F}\mathbf{w}_n.$$

As regards the time-domain output, it appears that $y[n] = \mathbf{w}_n^H\mathbf{x}_n$ or, given the nature of the transformation, we can also write

$$y[n] = \left[\mathbf{F}^{-1}W_n\right]^H\mathbf{F}^{-1}X_n = W_n^H\mathbf{F}\mathbf{F}^{-1}X_n = W_n^H X_n. \tag{7.103}$$

Note, that for (7.103), the time-domain output does not require the calculation of the inverse transformation.

7.5.1.1 TDAF with Data-Dependent Optimal and *A Priori* Fixed Sub-optimal Transformations

The LMS performance can be improved through a unitary transformation that tends to orthogonalize the input sequence $\mathbf{x}_n$ [12, 36]. In fact, the transformation $\mathbf{F}$ tends to diagonalize the correlation matrix making it easy to implement the power normalization as in the FDAF.

For the determination of the *data-dependent optimal transformation*, we consider the input correlation matrix $\mathbf{R}_{xx} = E\{\mathbf{x}_n\mathbf{x}_n^H\}$, for which for $\boldsymbol{X}_n = \mathbf{F}\mathbf{x}_n$

$$\mathbf{R}_{xx}^F = E\left\{\mathbf{F}\mathbf{x}_n[\mathbf{F}\mathbf{x}_n]^H\right\} = \mathbf{F}\mathbf{R}_{xx}\mathbf{F}^H. \tag{7.104}$$

The correlation matrix $\mathbf{R}_{xx} \in (\mathbb{C},\mathbb{R})^{M\times M}$ can always be represented through the *unitary similarity transformation* (Sect. A.9) defined by the relation $\boldsymbol{\Lambda} = \mathbf{Q}^H\mathbf{R}_{xx}\mathbf{Q}$, in which the diagonal matrix $\boldsymbol{\Lambda}$ is formed with $\mathbf{R}_{xx}$ matrix eigenvalues λ_k. Then, the optimal transformation that diagonalizes the correlation is just the *unitary similarity transformation.* In fact, with the power step-size normalization is $\boldsymbol{\mu}_n = \boldsymbol{\Lambda}^{-1}$ or $\boldsymbol{\mu}_n\mathbf{R}_{xx}^F = \mathbf{I}$ and therefore $\chi(\boldsymbol{\mu}_n\mathbf{R}_{xx}^F) = 1$.

The data-dependent optimal transformation $\mathbf{F} = \mathbf{Q}^H$ that diagonalizes the correlation, i.e., such that $\mathbf{R}_{xx}^F = \boldsymbol{\Lambda}$, is known as the Karhunen–Loeve transform (KLT) (see Sect. 1.3.6). The problem of choosing the optimal transformation is essentially related to the computational cost required for its determination. The optimal transformation, $\mathbf{Q}^H$, depends on the signal itself and its determination has complexity $O(M^2)$.

By choosing transformations not dependent on the input signal, i.e., signal representations related to a predetermined and *a priori* fixed orthogonal vectors base, such as DFT and DCT, the computational cost can be reduced to $O(M)$. Such transformations represent, moreover, in certain conditions, a KLT good approximation.

For example, in case of *lattice filters* we proceed in a rather different way. The input orthogonalization is performed with a lower triangular matrix $\mathbf{F}$ which is computed run-time for each new input sample (see Sect. 8.3.5).

In case of *a priori* fixed sub-optimal transformations, although there are infinite possibilities for the choice of the matrix $\mathbf{F}$, in signal processing, the DFT and DCT are among the most used (see Sect. 1.3).

Calling $f_{m,n}$ the elements of $\mathbf{F}$, for the DFT it is

$$f_{m,n}^{\mathrm{DFT}} = Ke^{-j\frac{2\pi}{M}mn}, \qquad \text{for} \qquad n,m = 0,1,...,M-1, \tag{7.105}$$

where to get $\mathbf{F}\mathbf{F}^{-1} = \mathbf{I}$ it results in $K = 1/\sqrt{M}$. The DFT has a wide range of uses as, distinguishing between positive and negative frequencies, it is applicable to both real signals as well as those complex.

For real domain signal it is possible, and often convenient, to use transformations defined only in the real domain. In this case, the complex arithmetic is not strictly necessary. In the following, some transformations definitions that can be used for TDAF algorithms implementation are given.

The DHT (see Sect. 1.3.3) is defined as

$$f_{m,n}^{\text{DHT}} = K\left(\cos\frac{2\pi}{M}mn + \sin\frac{2\pi}{M}mn\right), \quad \text{for } n,m = 0,1,...,M-1 \tag{7.106}$$

with $K = 1/\sqrt{M}$. In practice, the DHT coincides with the DFT for real signals.

Unlike the DFT, which is uniquely defined, real transformations, such as DCT and DST (see Sect. 1.3.4), may be defined in different ways. In literature (at least) four variants are given and *Type II*, which is based on a periodicity $2M$, appears to be one most used.

The *Type II discrete cosine transform* DCT-II is defined as

$$f_{m,n}^{\text{DCT}} = K_m \cos\frac{\pi(2n+1)m}{2M}, \quad \text{for } n,m = 0,1,...,M-1, \tag{7.107}$$

where, in order to have $\mathbf{F}\mathbf{F}^{-1} = \mathbf{I}$,

$$K_0 = 1/\sqrt{M} \text{ and } K_m = \sqrt{2/M} \quad \text{for } m > 0. \tag{7.108}$$

The *Type II discrete sine transform* (DST-II) is defined as

$$f_{m,n}^{\text{DST}} = K_m \sin\frac{\pi(2n+1)(m+1)}{2M}, \quad \text{for } n,m = 0,1,...,M-1 \tag{7.109}$$

with K_m defined as in (7.108). Note that the DCT, the DST, and other transformations can be computed with fast FFT-like algorithms. Other types of transformations can be found in literature [18, 26, 34, 37–39].

7.5.1.2 Transformed Domain LMS

The algorithm structure is independent of the transformation choice. The filter input and weights are transformed as in the circular-convolution FDAF in which it is placed $L = 1$ (see Sect. 7.3.7.2). The block index is identical to that of the input sequence ($k = n$) and the sliding transform computation does not require an augmented vector definition. Indicating the generic transforms of variables $\mathbf{w}_n$ and $\mathbf{x}_n$ with the notation $\text{FFT}(\cdot)$, we can write

$$\boldsymbol{W}_n = \text{FFT}(\mathbf{w}_n), \tag{7.110}$$

$$\boldsymbol{X}_n = \text{FFT}(\mathbf{x}_n). \tag{7.111}$$

For the time-domain output it is (7.103), and the error can be represented as

$$e[n] = d[n] - \boldsymbol{W}_n^H \boldsymbol{X}_n. \tag{7.112}$$

In practice each weight is updated with the same error.

Remark The transform domain LMS algorithm, also known as sliding DFT–DCT–DST–..., LMS, is formally identical to the LMS and requires, with respect to it, an

M-points FFT calculation for each new input sample. To the complexity of LMS, therefore, the FFT complexity must be added.

The availability of the transformed input allows the definition of a normalization step-size procedure, as that described by the relations (7.33) and (7.34), which represents a necessary part of this class of algorithms. In this case, the convergence appears to be rather uniform even for colored inputs.

7.5.1.3 Sliding Transformation LMS Algorithm Summary

(i) Initialization $\mathbf{W}_0 = \mathbf{0}$, $P_0(m) = \delta_m$ or $m = 0, 1, \ldots, M-1$;
(ii) For $n = 0,1, \ldots$ { *// for each new input sample*

$\mathbf{X}_n = \text{FFT}[\mathbf{x}_n]$
$y[n] = \mathbf{W}_n^H \mathbf{X}_n$ *// Eqn. (7.103)*
$e[n] = d[n] - y[n]$ *// Time-domain error*
// Step-size normalization for each frequency bin
$P_n(m) = \lambda P_{n-1}(m) + (1-\lambda)|X_n(m)|^2 \quad m = 0, 1, \ldots, M-1;$
$\boldsymbol{\mu}_n = \mu[P_n^{-1}(0)\ P_n^{-1}(1)\ \cdots\ P_n^{-1}(M-1)]^T$
// LMS up-date
$\mathbf{W}_{n+1} = \mathbf{W}_n + e^*[n]\boldsymbol{\mu}_n \odot \mathbf{X}_n.$

}

Note that the algorithm structure is identical for all transformation types that in this context, for formalism uniformity with the previous paragraphs, has been indicated with FFT(·).

7.5.2 Sliding Transformation LMS as Sampling Frequency Interpretation with Bandpass Filters Bank

The analysis sliding window, which determines the transformation, *sees* a time-variant process and consequently also the transformed signal is time variant. It thus appears that the frequency domain transformation of the input $x[n]$ is not stationary, and it is also a function of time. In this case the signal spectrum, indicated as $X(n, m)$, is a function of two variables: the time, understood as the time index n, and the frequency, represented by the index m. In the case of the frequency transformation, the spectrum is defined by a so-called short-time Fourier transform (STFT) which has the form (for details see [10]):

$$X(n,m) = \sum_{m=-\infty}^{\infty} w[n-m]x[n]e^{-j\frac{2\pi}{M}mn}, \tag{7.113}$$

where $w[n-m] = 1$ for $0 \leq n \leq M-1$ indicates the finite duration sliding window (short time) that *runs* on the signal $x[n]$.

For (7.113), it is possible to process the signal in the two-dimensional domain (n, m) in a dual manner: (1) fixing the time and considering the frequency variable or (2) by fixing the frequency and considering the time variable. The first mode is the STFT, defined by the expression (7.113), that fixes (or samples) the time variable n. In this case it is usual to indicate $X(n,m)$ as $X_n(m)$ which highlights the variability in frequency. The second mode may be interpreted as filters bank fixing (or sampling) the frequency m, and in this case it is usual to indicate the spectrum as $X_m(n)$ such as to highlight the time variability.

Remark The DFT and other transformations can be interpreted as a bank of M bandpass filters. At the bank input there is the sequence $x[n]$ while at the output we have the frequency bins of its mth frequency. In other words, the bank fixes m through a uniform M-points frequency-domain signal sampling.

Considering $x[n]$ as the input and $X_m(n)$ as the output of the mth filter of the bank, from the definition of DFT (7.105) and for (7.113), we can write

$$X_m(n) = K\sum_{p=0}^{M-1} x[n-p]e^{-j\frac{2\pi}{M}mp}, \quad m = 0, 1, ..., M-1. \tag{7.114}$$

In this case, the explicit definition of the window $w[n]$ is not necessary, since the summation has by definition finite duration.

From the above equation, $X_m(n)$ can be interpreted as the output of a FIR filter with impulse response defined as

$$h_m^{\mathrm{DFT}}[n] = Ke^{-j\frac{2\pi}{M}mn}. \tag{7.115}$$

By performing the z-transform of M bandpass filters, the corresponding TF are defined as

$$\begin{aligned} H_m^{\mathrm{DFT}}(z) &= \frac{X_m(z)}{X(z)} \\ &= K\sum_{p=0}^{M-1} e^{-j\frac{2\pi}{M}mp}z^{-p}, \quad \text{for } m = 0, 1, ..., M-1. \\ &= K\frac{1-z^{-M}}{1-e^{-j\frac{2\pi}{M}m}z^{-1}} \end{aligned} \tag{7.116}$$

It should be noted in particular that, for $p = 0$, not considering the gain factor K, the TF (7.116) is equal to

$$H_0^{\mathrm{DFT}}(z) = 1 + z^{-1} + \cdots + z^{-(M-1)} \tag{7.117}$$

which corresponds to a simple moving average filter with a rectangular window. The TF of the remaining filters can then be expressed as

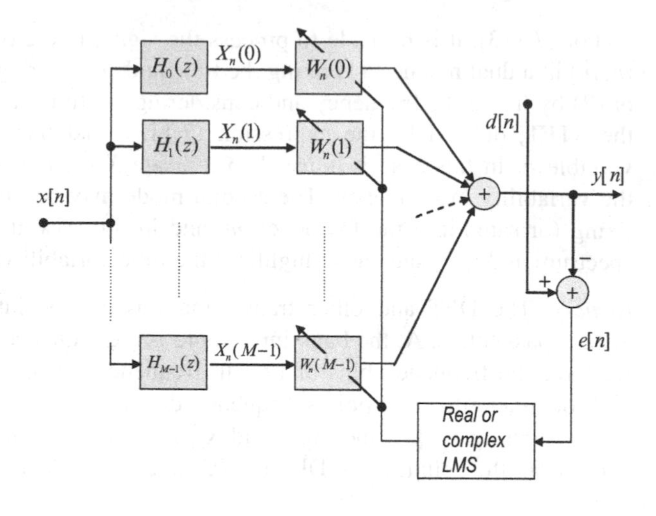

Fig. 7.17 Equivalence between DFT/DCT and a bank of M bandpass filters used for frequency sampling

$$H_m^{\mathrm{DFT}}(z) = e^{-j2\pi m/M} \cdot H_0^{\mathrm{DFT}}(z), \quad \text{for } m = 0, 1, \ldots, M-1 \tag{7.118}$$

or, in terms of frequency response, as

$$H_m^{\mathrm{DFT}}\left(e^{-j\omega}\right) = H_0^{\mathrm{DFT}}\left(e^{-j\left(\omega - \frac{2\pi}{M}m\right)}\right), \quad \text{for } m = 0, 1, \ldots, M-1. \tag{7.119}$$

For which the DFT filter bank can be interpreted as a moving-average low-pass filter, called *low-pass prototype*, that for each m for (7.119), it is shifted to the right as $(\omega - (2\pi/M)m)$ along the ω axis (i.e., around the unit circle) in order to generate the M channel bank.

The representation of the DFT bank is presented in Fig. 7.17.

Remark The TF (7.116) is characterized by a numerator with M zeros uniformly distributed around the unit circle (and a denominator containing only one pole that, by varying m, exactly cancels the respective zero).

As for the DFT also the DCT can be interpreted as a bank of M filters. From the definition (7.107), proceeding as in the DFT case, the impulse response is

$$h_m^{\mathrm{DCT}}(n) = K_m \cos\left[\left(\frac{\pi}{M}m\right)\left(n + \frac{1}{2}\right)\right].$$

It is demonstrated, using the relationship $\cos x = (e^{jx} + e^{-jx})/2$, (see [34, 35]) that the z-transform of the previous expression is

$$H_m^{\mathrm{DCT}}(z) = K_m \cos\left(\frac{\pi}{2M}m\right) \frac{(1 - z^{-1})[1 - (-1)^m z^{-M}]}{1 - 2\cos\left(\frac{\pi}{M}m\right)z^{-1} + z^{-2}}, \quad m = 0, 1, \ldots, M-1. \tag{7.120}$$

For other types of transformations see, for example, [18, 26, 34, 37–39].

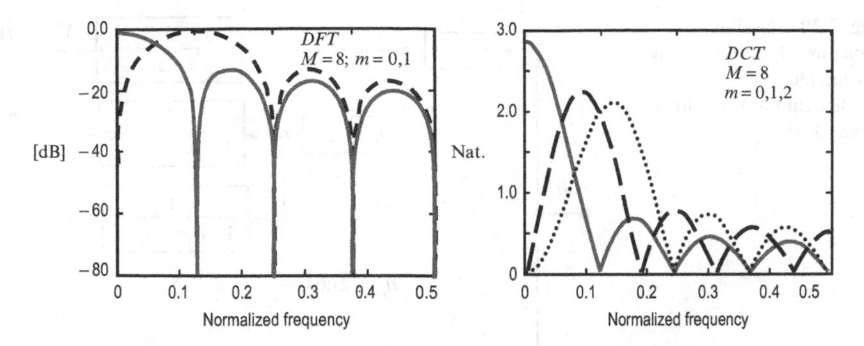

Fig. 7.18 Frequency responses of the DFT and DCT transformations, in dB and natural values, for $M = 8$, seen as bank of FIR filters with impulse response (7.115) and (7.120), respectively

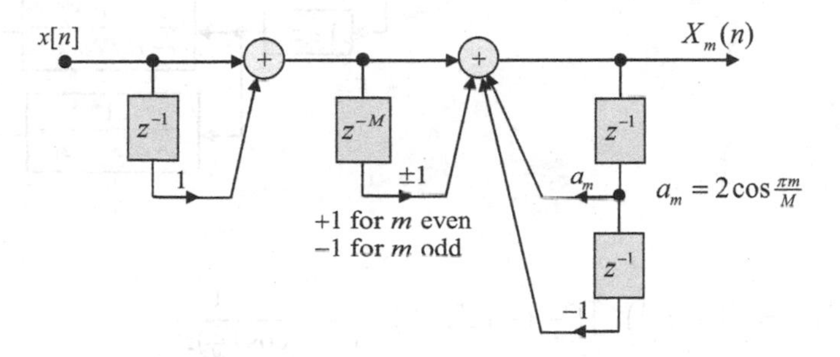

Fig. 7.19 Recursive IIR circuit that implements the mth filter of the DCT bank

Figure 7.18 shows the frequency responses of the first filters of the bank in the DFT/DCT case for $M = 8$. Note the high degree of frequency response overlap for filters with adjacent bands.

7.5.2.1 Implementation Notes

The interpretation of TDAF as filters bank, with the corresponding TF, suggests the use of appropriate circuit structures for the real-time transformations implementation.

DCT with Recursive Filters Bank

The TF $H_m^{\mathrm{DCT}}(z)$ defined in (7.120), neglecting for simplicity the gain factor $K_m \cos(\pi m/2M)$, can be expressed as the product of three terms

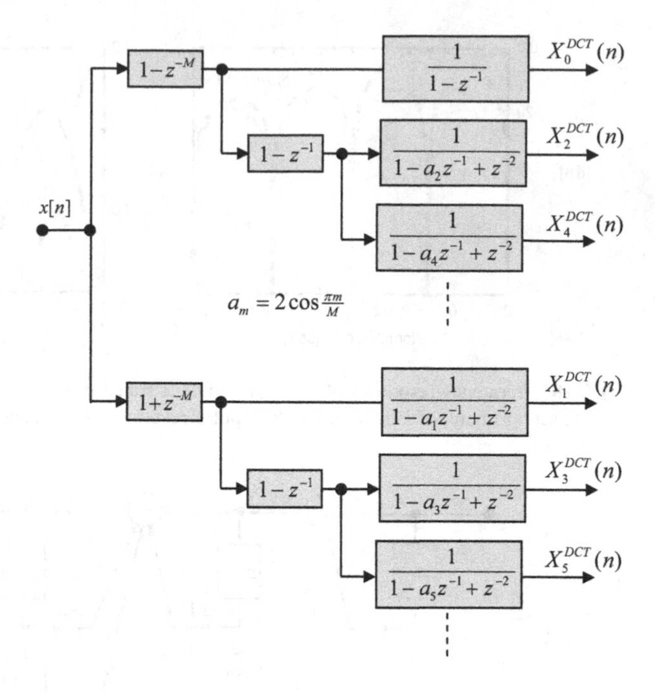

Fig. 7.20 Possible structure of the filter bank for the DCT implementation (modified from [35])

$$H_m^{\mathrm{DCT}}(z) = \left(1 - z^{-1}\right) \cdot \left[1 - (-1)^m z^{-M}\right] \cdot \frac{1}{1 - 2\cos\left(\frac{\pi m}{M}\right) z^{-1} + z^{-2}}. \tag{7.121}$$

The above factorization of $H_m^{\mathrm{DCT}}(z)$ corresponds to a recursive (IIR) circuit structure of the type shown in Fig. 7.19 (similar argument can be made for (7.116) or other transforms).

Following the development reported in [35], noting that for $m = 0$ the equality holds:

$$\frac{1}{1 - 2\cos\left(\frac{\pi m}{M}\right) z^{-1} + z^{-2}} = \frac{1}{\left(1 - z^{-1}\right)^2}.$$

It appears that

$$H_0^{\mathrm{DCT}}(z) = \frac{1 - z^{-M}}{1 - z^{-1}}. \tag{7.122}$$

In addition, we have that

$$1 - (-1)^m z^{-M} = \begin{cases} 1 - z^{-M} & \text{for } m \text{ even} \\ 1 + z^{-M} & \text{for } m \text{ odd} \end{cases}.$$

By grouping common terms, the entire bank resulting in circuit structure is of the type illustrated in Fig. 7.20.

Remark The structure of the bank of Fig. 7.20 presents a stability problem because the poles of the second-order recursive filters are located just around the unit circle. The errors accumulation due to the round-off error can bring the circuit to saturation. In addition, the coefficients quantization may cause some poles fall outside the unit circle. To overcome these drawbacks, it is possible to replace z^{-1} with βz^{-1} in which $\beta < 1$. This solution maps all the poles inside the circle ensuring the stability at the expense, however, of a non-negligible increase in the computational cost.

Non-recursive DFT Filter Bank: The Bruun's Algorithm

The Bruun's algorithm [40] derives from the following powers-of-two factorizations

$$1 - z^{-2N} = \left(1 - z^{-N}\right)\left(1 + z^{-N}\right) \tag{7.123}$$

and

$$1 - \alpha z^{-2N} + z^{-4N} = \left(1 + \sqrt{2 - \alpha} z^{-N} + z^{-2N}\right)\left(1 - \sqrt{2 - \alpha} z^{-N} + z^{-2N}\right) \tag{7.124}$$

whereby for $\alpha = 0$

$$1 + z^{-4N} = \left(1 + \sqrt{2} z^{-N} + z^{-2N}\right)\left(1 - \sqrt{2} z^{-N} + z^{-2N}\right). \tag{7.125}$$

To understand how the above factorization can be used for the DFT implementation as a non-recursive filter bank, we apply iteratively the Bruun factorization for $N = M, \ldots, 1$.

For better understanding we proceed to the development for $M = 8$. Unless of a gain factor, (7.116) can be written as

$$H_m^{\mathrm{DFT}}(z) = \frac{1 - z^{-8}}{1 - e^{-j\frac{\pi}{4}m} z^{-1}}. \tag{7.126}$$

Applying (7.123) to the DFT numerator, we have that[4]

$$1 - z^{-8} = \left(1 - z^{-4}\right)\left(1 + z^{-4}\right),$$

where the terms $(1 - z^{-4})$ and $(1 + z^{-4})$ for (7.123) and (7.125) are factorizable as

[4] The roots of the polynomial $(1 - z^{-M})$ are uniformly placed around the unit circles exactly like the frequency-bins of a M points DFT.

$$\left(1-z^{-4}\right)=\left(1-z^{-2}\right)\left(1+z^{-2}\right)$$

and

$$\left(1+z^{-4}\right)=\left(1+\sqrt{2}z^{-1}+z^{-2}\right)\left(1-\sqrt{2}z^{-1}+z^{-2}\right).$$

For the terms $\left(1+\sqrt{2}z^{-1}+z^{-2}\right)$ and $\left(1-\sqrt{2}z^{-1}+z^{-2}\right)$ for (7.124) is

$$\left(1+\sqrt{2}z^{-1}+z^{-2}\right)=\left(1+e^{j\frac{\pi}{4}}z^{-1}\right)\left(1+e^{-j\frac{\pi}{4}}z^{-1}\right)=\left(1+e^{j\frac{\pi}{4}}z^{-1}\right)\left(1+e^{j\frac{7\pi}{4}}z^{-1}\right)$$
$$\left(1-\sqrt{2}z^{-1}+z^{-2}\right)=\left(1-e^{j\frac{3\pi}{4}}z^{-1}\right)\left(1-e^{j\frac{5\pi}{4}}z^{-1}\right)$$

while for the terms $(1+z^{-2})$ and $(1-z^{-2})$ is, respectively,

$$\left(1+z^{-2}\right)=\left(1+jz^{-1}\right)\left(1-jz^{-1}\right)\ \text{and}\ \left(1-z^{-2}\right)=\left(1+z^{-1}\right)\left(1-z^{-1}\right).$$

This reasoning suggests the possibility of implementing (7.126) with M TF, one for each m, considering the cascade connection of filters made with the terms of the factorization (7.123), (7.124), and (7.125). From the development, in fact, we observe that (7.126) is factored in eight terms of first degree $(1-cz^{-1})$ with $c=e^{-j\pi m/4}$, for $m=0, 1, \ldots, M-1$, each of which is coming from the factorization of a second degree term $(1+bz^{-1}+z^{-2})$, with $b=0, \pm\sqrt{2}$; this term is coming, in turn, from the factorization of a term of the fourth degree and so on.

It follows that the DFT can be implemented with a binary tree structure with $N_s=\log_2 M$ stages in which each filter is formed by series connection of N_s elementary structures (sub-filters) that implement the various factors of development (7.123), (7.124), and (7.125). In practice, each bank channel has in common with one another N_s-1 sub-filters, with two other N_s-2 sub-filters, with four other N_s-3 sub-filters, and so on.

In Fig. 7.21, the filters tree structure with three stages that implements the DFT with $M=8$ is shown. Note that the filters of the structure are quite simple and consist, most of the times, of delay lines shared with multiple bank channels.

Remark With reference to Fig. 7.22, the reader can verify that the eight factors of first degree $(1-e^{-j\pi m/4}z^{-1})$, derived from the power-of-two factorization of $(1-z^{-8})$, have a coefficient that coincides with one (and only one) of the eight terms $e^{-j\pi m/4}$ that appears in the denominator of (7.126). The roots of the denominator of (7.126) are cancelled one at a time, and therefore the DFT filters bank TFs are $M-1$ common zeros.

In [35], to which the reader is referred for further details, it has been defined as a generalization of the previous structure for real domain transformations as the DCT and DST.

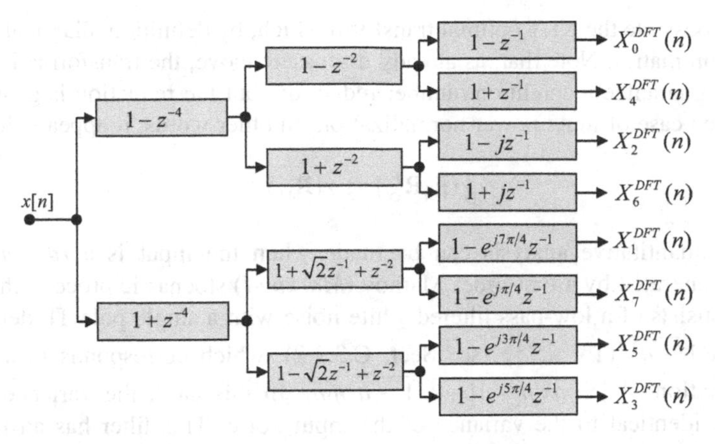

Fig. 7.21 Three stages ($M = 8$) Bruun's tree, consisting of non-recursive filter for the DFT implementation (modified from [35])

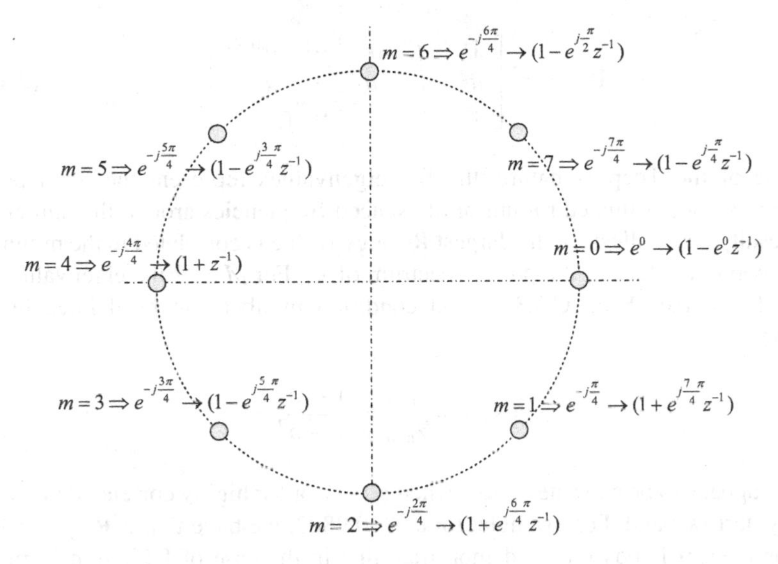

Fig. 7.22 Unitary circle zeros distribution for a DFT with $M = 8$

7.5.3 Performance of TDAF

The TDAF performance analysis can be performed by evaluating the correlation matrix condition number before and after the application of the input transformation. From a geometric point of view, the transformation $\mathbf{F}$ on the input signal produces a correlation $\mathbf{R}_{xx}^{F} = \mathbf{F}\mathbf{R}_{xx}\mathbf{F}^{H}$ that appears to be *more diagonal* with respect to $\mathbf{R}_{xx}$. This statement derives from the fact that $\mathbf{F}$ is chosen so as to approximate as much as possible the KLT optimal transform which, by definition, diagonalizes the correlation matrix. Note that, as already discussed above, the transformation itself does not guarantee a condition number reduction, but the reduction is guaranteed only in the case of input power normalization. In other words, it appears that

$$\chi\left(\boldsymbol{\mu}_n \mathbf{R}_{xx}^F\right) < \chi(\mathbf{R}_{xx}).$$

A more quantitative analysis can be made when the input is a *random walk model*. Generated by a first-order Markov (*Markov-I*) stochastic process, the input signal consists of a low-pass filtered white noise with a single pole TF defined as $H(z) = \sqrt{1-a^2}/(1-az^{-1})$ (see Sect. C.3.3.2), which corresponds to a difference equation $x[n] = ax[n-1] + \sqrt{1-a^2}\eta[n]$. In this case, the variance of the output is identical to the variance of the input noise. This filter has an impulse response that decreases geometrically with a rate a determined by the position of the pole on the z-plane. The autocorrelation is $r[k] = \sigma_\eta^2 a^k$ for $k = 0, 1, \ldots, M$, so the autocorrelation matrix is (C.213):

$$\mathbf{R}_{xx} = \sigma_\eta^2 \begin{bmatrix} 1 & a & a^2 & \cdots & a^{M-1} \\ a & 1 & a & \cdots & a^{M-2} \\ \vdots & \vdots & \ddots & \cdots & \vdots \end{bmatrix}. \tag{7.127}$$

Because of the Toeplitz nature, the $\mathbf{R}_{xx}$ eigenvalues represent the input power spectrum value, evaluated for uniformly spaced frequencies around the unit circle. As a result, the smallest and the largest $\mathbf{R}_{xx}$ eigenvalue is correlated to the minimum and maximum values of the power spectrum of $\mathbf{x}_n$. For $M = 2$ the eigenvalues are $\lambda_{1,2} = 1 \pm a$ (see Sect. C.3.3.2), and condition number can be defined by the relation:

$$\chi(\mathbf{R}_{xx}) = \frac{\lambda_{\max}}{\lambda_{\min}} = \frac{1+a}{1-a},$$

which happens to be extremely large when $a \to 1$ or for highly correlated processes or very narrow band. For example, for $a \to 0.9802$, we have that $\chi(\mathbf{R}_{xx}) = 100.0$.

Moreover, it is possible to demonstrate that in the case of DCT transform, we have that

$$\lim_{M\to\infty} \chi\left(\boldsymbol{\mu}_n \mathbf{R}_{xx}^{\mathrm{DCT}}\right) = (1+a),$$

whereby with the DCT transformation result $\lim_{M\to\infty} \chi\left(\boldsymbol{\mu}_n \mathbf{R}_{xx}^{\mathrm{DCT}}\right) \le 2$. For details and proofs see [34].

7.6 Subband Adaptive Filtering

The subband adaptive filtering (SAF) can be considered as a TDAF extension in which the DFT bank is replaced with a more selective filter bank (FB) that makes possible the signal decimation. Therefore, the SAF is a multirate system, i.e., that works with multiple sampling frequencies. The input signal is divided into usually uniform subbands from the FB and further decimated. Each subband is processed with a specific AF, only for that band, much shorter than the AF necessary in the case of full-band signal. Unlike the TDAF, for the output subband sequences, a complementary interpolation stage is necessary. A general scheme of SAF is illustrated in Fig. 7.23.

7.6.1 On the Subband-Coding Systems

The FBs are circuits constituted by low-pass, bandpass, and high-pass filters, combined in appropriate architectures, that act to decompose the input signal spectrum in a number of contiguous bands. For the required subband decomposition (SBD), you should consider if you want to use FIR or IIR filters. Indeed, the SBD characteristics vary depending on the context and for each specific application more methodologies are available.

The most used SBDs are the *uniform subdivisions* in which the contiguous signal bands have the same width. Another common subdivision is the *octave subdivision* in which the contiguous signal bands are doubled as, for example, in the so-called constant-Q filter banks. Other features often desired or imposed to the FB design can relate to the filters transition bands width, the stop-band attenuation, the ripple, aliasing level, the degree of regularity, etc.

The SBD, as shown in Fig. 7.23, is made by the so-called analysis-FB while the reconstruction is done through the synthesis-FB. A global subband-coding (SBC) is therefore made of an analysis FB followed by synthesis FB.

In the SBC design philosophy, rather than analyzing separately the individual filters of the analysis-synthesis FBs, you should consider the global SBC specific as, for example, the acceptable aliasing level, the group delay, the reconstruction error, etc. In other words, the specification is given in terms of global relationships between the input and output signals, regardless of the local characteristics of the analysis-synthesis filters. Sometimes you can trade computational efficiency or other characteristics (such as the group delay) by introducing acceptable signal reconstruction error. In general, imposing a certain quality of the reconstructed signal, it is possible to obtain more control and freedom degrees in the filters design.

In the FB design, for a given application, you should identify the cost function and the free parameters to be optimized with respect to them. Typically, the analysis-synthesis FBs lead to very complex input–output relations and complicated compromises between group delay, quality of filters, the quality of

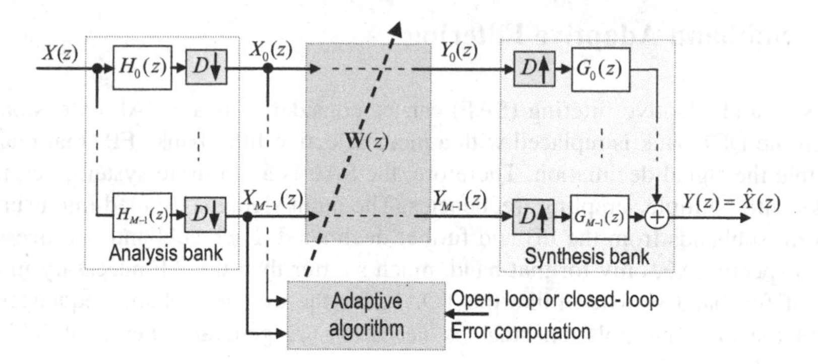

Fig. 7.23 General scheme of a subband adaptive filtering (SAF)

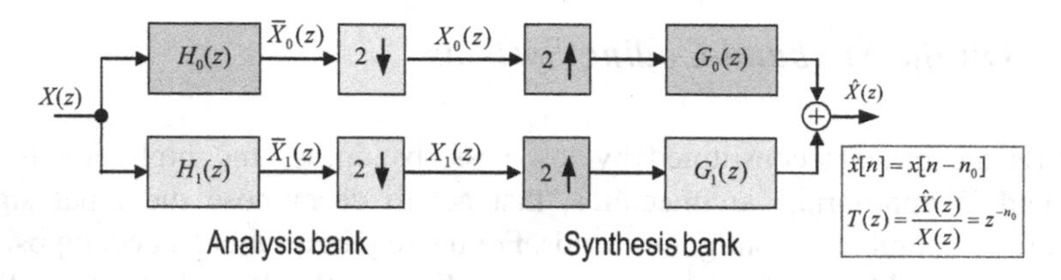

Fig. 7.24 Two-channel SBC with critical sample rate

reconstruction, processing speed, and, in the case of SAF, the convergence speed of the adaptation algorithm.

The design of the SBC is beyond the scope of this text, so for further information please refer to the extensive literature [41–48]. But before proceeding, some basic concepts referred to the two-channel SBC cases are presented.

7.6.2 Two-Channel Filter Banks

For simplicity, consider the two-channel SBC shown in Fig. 7.24. In fact, it is known that the two-channel SBC can be used to determine certain fundamental characteristics extensible to the case of M channels.

7.6.2.1 SBC in Modulation Domain z-Transform Representation

For $M = 2$, the analysis side TFs $H_0(z)$ and $H_1(z)$ are, respectively, a low-pass and high-pass symmetrical and power-complementary filters, with a cutoff frequency equal to $\pi/2$. As illustrated in Fig. 7.25, the high-pass filter $H_1(z)$ is constrained to be a π -rotated version of $H_0(z)$ on the unit circle, therefore is $H_1(z) = H_0(-z)$. The TFs of this type are usually called *half-band* filters.

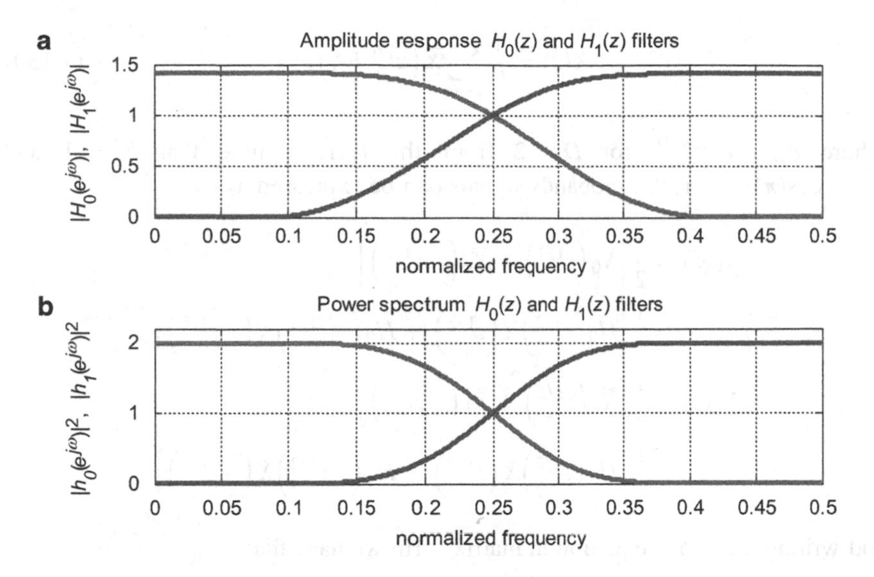

Fig. 7.25 Typical half-band filters response (**a**) symmetric amplitude response; (**b**) power complementary response (constant sum)

Similarly, synthesis FB is composed of two TFs $G_0(z)$ and $G_1(z)$, also low-pass and high-pass symmetrical and power complementary filters, with a cutoff frequency equal to $\pi/2$ and linked through some relation with $H_0(z)$ and $H_1(z)$.

To determine the so-called perfect reconstruction conditions (PRC) of the output signal (less of a delay), we consider the overall input–output TF. Considering the cascade connection of the analysis and synthesis FBs, as in Fig. 7.24, the circuit is defined as perfect reconstruction SBC, if the input–output relationship is a simple delay, so $\hat{x}[n] = x[n - n_0]$. The overall TF, relative to the input signal sampling frequency, is equal to

$$\frac{\hat{X}(z)}{X(z)} = z^{-n_0}. \tag{7.128}$$

In the analysis FB, the spectrum of the signal $X(e^{j\omega})$, for $0 \le \omega \le \pi$, is divided into two subbands. For which we have that

$$\begin{aligned} \overline{X}_0(z) &= H_0(z)X(z), \\ \overline{X}_1(z) &= H_1(z)X(z). \end{aligned} \tag{7.129}$$

Recalling that the z-transform of the D-decimated signal is (see for example [38])

$$X(z) = \frac{1}{D}\sum_{k=0}^{D-1}\overline{X}\left(z^{1/D}F_D^k\right), \tag{7.130}$$

where $F_D = e^{-j2\pi/D}$, for $D = 2$ from the (7.129), note that $F_2^0 = 1$ and $F_2^1 = \cos(\pi) = -1$; the subbands signals can be expressed as

$$\begin{aligned}
X_0(z) &= \frac{1}{2}\left[\overline{X}_0\left(z^{1/2}\right) + \overline{X}_0\left(-z^{1/2}\right)\right] \\
&= \frac{1}{2}\left[H_0\left(z^{1/2}\right)X\left(z^{1/2}\right) + H_0\left(-z^{1/2}\right)X\left(-z^{1/2}\right)\right] \\
X_1(z) &= \frac{1}{2}\left[\overline{X}_1\left(z^{1/2}\right) + \overline{X}_1\left(-z^{1/2}\right)\right] \\
&= \frac{1}{2}\left[H_1\left(z^{1/2}\right)X\left(z^{1/2}\right) + H_1\left(-z^{1/2}\right)X\left(-z^{1/2}\right)\right]
\end{aligned}$$

and writing the above equation in matrix form we have that

$$\begin{bmatrix} X_0(z) \\ X_1(z) \end{bmatrix} = \frac{1}{2}\begin{bmatrix} H_0(z^{1/2}) & H_0(-z^{1/2}) \\ H_1(z^{1/2}) & H_1(-z^{1/2}) \end{bmatrix}\begin{bmatrix} X(z^{1/2}) \\ X(-z^{1/2}) \end{bmatrix}. \tag{7.131}$$

Moreover, with similar reasoning, regarding the synthesis FB we have that

$$\hat{X}(z) = G_0(z)X_0(z^2) + G_1(z)X_1(z^2) \quad = \begin{bmatrix} G_0(z) & G_1(z) \end{bmatrix}\begin{bmatrix} X_0(z^2) \\ X_1(z^2) \end{bmatrix}. \tag{7.132}$$

For the analysis and definition of the FB specifications design, you must define a global transfer relationship that combines the input and output signals. To simplify the discussion, you can use the *modulation domain z-transform representation*, or simply *modulation representation*, that is defined by an array $\mathbf{x}^{(m)}(z)$ whose elements are the *modulated z-transform components* defined as $X_k^{(m)}(z) = X(z \cdot F_D^k)$, for $k = 0, 1, \ldots, D - 1$. So, for $D = 2$, the signal defined according to its modulated components can be written as

$$\mathbf{x}^{(m)}(z) = \begin{bmatrix} X_0^{(m)}(z) & X_1^{(m)}(z) \end{bmatrix}^T = \begin{bmatrix} X(z) & X(-z) \end{bmatrix}^T. \tag{7.133}$$

In fact, for $m = 0$, $X_0^{(m)}(z) = X(z)$ is the *baseband component*, while for $m = 1$ we have that $F_2^1 = -1$, and the *modulated component* $X_1^{(m)}(z) = X(-z)$ corresponds to that translated around $\omega = \pi$. From the expressions (7.131) and (7.132), you can also define the output modulation expansion representation as

$$\hat{\mathbf{x}}^{(m)}(z) = \left[\hat{X}^{(0)}(z)\hat{X}^{(1)}(z)\right]^T = \left[\hat{X}(z)\hat{X}(-z)\right]^T. \tag{7.134}$$

Indeed, for the output baseband component $\hat{X}(z)$, we have that

$$\hat{X}(z) = \frac{1}{2}\begin{bmatrix} G_0(z) & G_1(z) \end{bmatrix}\begin{bmatrix} H_0(z) & H_0(-z) \\ H_1(z) & H_1(-z) \end{bmatrix}\begin{bmatrix} X(z) \\ X(-z) \end{bmatrix} \tag{7.135}$$

while for the modulated component $\hat{X}(-z)$, we can write

$$\hat{X}(-z) = \frac{1}{2}\begin{bmatrix} G_0(-z) & G_1(-z) \end{bmatrix}\begin{bmatrix} H_0(z) & H_0(-z) \\ H_1(z) & H_1(-z) \end{bmatrix}\begin{bmatrix} X(z) \\ X(-z) \end{bmatrix}. \tag{7.136}$$

Combining the earlier you can define the matrix expression:

$$\begin{bmatrix} \hat{X}(z) \\ \hat{X}(-z) \end{bmatrix} = \frac{1}{2}\begin{bmatrix} G_0(z) & G_1(z) \\ G_0(-z) & G_1(-z) \end{bmatrix}\begin{bmatrix} H_0(z) & H_0(-z) \\ H_1(z) & H_1(-z) \end{bmatrix}\begin{bmatrix} X(z) \\ X(-z) \end{bmatrix}. \tag{7.137}$$

Finally, defining the matrices

$$\mathbf{H}^{(m)}(z) = \begin{bmatrix} H_0(z) & H_0(-z) \\ H_1(z) & H_1(-z) \end{bmatrix} \tag{7.138}$$

and

$$\mathbf{G}^{(m)}(z) = \begin{bmatrix} G_0(z) & G_1(z) \\ G_0(-z) & G_1(-z) \end{bmatrix} \tag{7.139}$$

as *modulated component matrices* of the analysis and synthesis FB, respectively, the compact form of modulation representation can be rewritten as

$$\hat{\mathbf{x}}^{(m)}(z) = \frac{1}{2}\mathbf{G}^{(m)}(z) \cdot \mathbf{H}^{(m)}(z) \cdot \mathbf{x}^{(m)}(z). \tag{7.140}$$

This last expression provides a global description of the two-channel SCB TF in terms of the input–output modulation representation.

7.6.2.2 PRC for Two-Channel SBC

The no aliasing PRC occurs when the FB output is exactly the same at the input less than a delay. Whereas both modulation components $X(z)$ and $X(-z)$, the PRC (7.128), appear to be $\hat{X}(z)/X(z) = z^{-n_0}$ and $\hat{X}(-z)/X(-z) = (-z)^{-n_0}$ that in matrix form can be written as

$$\begin{bmatrix} \hat{X}(z) \\ \hat{X}(-z) \end{bmatrix} = \begin{bmatrix} z^{-n_0} & 0 \\ 0 & (-z)^{-n_0} \end{bmatrix} \begin{bmatrix} X(z) \\ X(-z) \end{bmatrix} \tag{7.141}$$

and, by considering the (7.137) in an extended form, the PRC can be written as

$$\begin{aligned} G_0(z)H_0(z) + G_1(z)H_1(z) &= 2z^{-n_0} \\ G_0(z)H_0(-z) + G_1(z)H_1(-z) &= 0 \end{aligned} \tag{7.142}$$

and

$$\begin{aligned} G_0(-z)H_0(z) + G_1(-z)H_1(z) &= 0 \\ G_0(-z)H_0(-z) + G_1(-z)H_1(-z) &= 2(-z)^{-n_0}. \end{aligned} \tag{7.143}$$

Let

$$\left[\mathbf{H}^{(m)}(z)\right]^{-1} = \frac{1}{\Delta_H} \begin{bmatrix} H_1(-z) & -H_0(-z) \\ -H_1(z) & H_0(z) \end{bmatrix}, \tag{7.144}$$

where $\Delta_H = H_0(z)H_1(-z) - H_0(-z)H_1(z)$ is the determinant of $\mathbf{H}^{(m)}(z)$, considering (7.140) and (7.141); it is easy to derive the relationship between the matrices $\mathbf{G}^{(m)}(z)$ and $\mathbf{H}^{(m)}(z)$ which, for n_0 odd, is equal to

$$\begin{aligned} \mathbf{G}^{(m)}(z) &= 2z^{-n_0} \begin{bmatrix} 1 & 0 \\ 0 & (-1)^{-n_0} \end{bmatrix} \left[\mathbf{H}^{(m)}(z)\right]^{-1} \\ &= \frac{2z^{-n_0}}{\Delta_H} \begin{bmatrix} H_1(-z) & -H_0(-z) \\ H_1(z) & -H_0(z) \end{bmatrix} \end{aligned} \tag{7.145}$$

and hence, the connection between the analysis and synthesis FB and, because the PRC are verified, we have that

$$\begin{aligned} G_0(z) &= \frac{2z^{-n_0}}{\Delta_H} \cdot H_1(-z), \\ G_1(z) &= -\frac{2z^{-n_0}}{\Delta_H} \cdot H_0(-z). \end{aligned} \tag{7.146}$$

The TF of the synthesis bank can be implemented with IIR or FIR filters. However, in many applications the use of the FIR filters is more appropriate. Moreover, even if the $H_0(z)$ and $H_1(z)$ are of FIR type, from the presence of the denominator in (7.146), $G_0(z)$ and $G_1(z)$ are of the IIR type. The only possibility for which $G_0(z)$ and $G_1(z)$ are of FIR type is that the denominator is equal to a pure delay, i.e.,

$$\Delta_H = \alpha \cdot z^{-k} \quad \alpha \in \mathbb{R}, k \in \mathbb{Z}. \tag{7.147}$$

In this case we have that

$$\begin{aligned} G_0(z) &= +\tfrac{2}{\alpha} z^{-n_0+k} H_1(-z), \\ G_1(z) &= -\tfrac{2}{\alpha} z^{-n_0+k} H_0(-z). \end{aligned} \tag{7.148}$$

These conditions are rather simple and generic and easily verifiable in different ways. Below are presented the two most intuitive and common solution.

7.6.2.3 Quadrature Mirror Filters

The first solution suggested in the literature (see [44–51]) is the so-called quadrature mirror filters (QMF). Given $H(z)$ the TF of a *half-band* low-pass FIR filter, with a cutoff frequency equal to $\pi/2$, called *low-pass prototype*, determined according to some optimality criterion, then you can easily prove that the PRC (7.142) and/or (7.143) are verified, if the following conditions are met

$$\begin{aligned} H_0(z) &= H(z) & &\Leftrightarrow & h_0[n] &= h[n] \\ H_1(z) &= H(-z) & &\Leftrightarrow & h_1[n] &= (-1)^n h[n] \end{aligned} \tag{7.149}$$

and

$$\begin{aligned} G_0(z) &= 2H(z) & &\Leftrightarrow & g_0[n] &= 2h[n], \\ G_1(z) &= -2H(-z) & &\Leftrightarrow & g_1[n] &= -2(-1)^n h[n]. \end{aligned} \tag{7.150}$$

where the factor 2 in the synthesis FB is inserted to compensate for the factor 1/2 introduced by the decimation. Moreover, to obtain the PRC referred to only the low-pass prototype $H(z)$, we replace (7.149) and (7.150) in (7.142) and we get

$$H(z)H(z) - H(-z)H(-z) = z^{-n_0}, \tag{7.151}$$

$$H(z)H(-z) - H(-z)H(z) = 0. \tag{7.152}$$

Note that (7.151) is equivalent to

$$H^2(z) - H^2(-z) = z^{-n_0} \tag{7.153}$$

which has odd symmetry, for which $H(z)$ must necessarily be an FIR filter of even length. Whereby calling L_f the length of the low-pass prototype filter, the total delay of the analysis-synthesis FB pair is $n_0 = L_f - 1$.

In addition, the expression (7.153) explains the name QMF. Indeed, $H(z)$ is low-pass while $H(-z)$ is high-pass. The frequency response is just the *mirror* image of the axis of symmetry. Furthermore, the filters are also complementary in power. In fact, for $z = e^{j\omega}$,

$$\left|H\left(e^{j\omega}\right)\right|^2 + \left|H\left(e^{j(\omega-\pi)}\right)\right|^2 = 1. \tag{7.154}$$

To obtain the perfect reconstruction, the low-pass FIR prototype must fully satisfy the condition (7.154).

In literature many filter design techniques, which are able to determine the coefficients $h[n]$ in order to fine approximate this condition, are available.

Furthermore, note that the expression (7.152) is also indicated as *aliasing cancellation condition.* The (7.152) provides, in fact, the absence of cross components [see the diagonal matrix in (7.141)].

Remark The (7.149) indicates that the response of the high pass $h_1[n]$ is obtained by changing the sign to the odd samples of $h[n]$ (equivalent to a rotation of π on the unit circle). In terms of the z-transform, this is equivalent to the $H_1(z)$ zeros position, specular and conjugates whereas the vertical axis, compared to the zeros of $H_0(z)$. Indeed, indicating the ith zero of $H_0(z)$ as $z_i^{H_0} = \alpha_i \pm j\beta_i$, for $z \to -z$, the zeros of $H_1(z)$ are $z_i^{H_1} = -\alpha_i \mp j\beta_i$; then simply sign changes and conjugates.

7.6.2.4 Conjugate Quadrature Filters

A second solution for the two-channel PRC-FB design, similar to that suggested above, is to choose the high-pass filters in the conjugated form. In this case the FB is realized with conjugate quadrature filters (CQF). In this case, indicating with $h[n]$ the L_f-length low-pass prototype, the conditions (7.149) and (7.150) are rewritten as

$$\begin{aligned} H_0(z) &= H(z) &&\Leftrightarrow\quad h_0[n] = h[n] \\ H_1(z) &= z^{-(L_f-1)}H\left(-z^{-1}\right) &&\Leftrightarrow\quad h_1[n] = (-1)^{(L_f-1-n)}h\left[L_f-n-1\right] \end{aligned} \tag{7.155}$$

and

$$\begin{aligned} G_0(z) &= 2z^{-(L_f-1)}H\left(z^{-1}\right) &&\Leftrightarrow\quad g_0[n] = 2h\left[L_f-1-n\right] \\ G_1(z) &= -2H(-z) &&\Leftrightarrow\quad g_1[n] = -2(-1)^n h[n] \end{aligned} \tag{7.156}$$

so even for filters CQF it is easy to show that the PRC are met.

Remark Starting from the same low-pass prototype $h[n]$, we can observe that in the QMF case the zeros of $H_1(z)$ are a mirrored version with respect to the vertical symmetry axis of those of $H(z)$. In the case of CQF, however, they are a mirrored and reciprocal version. Indeed, indicating the ith zero of $H_0(z)$ as $z_i^{H_0} = \alpha_i \pm j\beta_i$, for $z \to -z^{-1}$ the $H_1(z)$ zeros are $z_i^{H_1} = -\frac{1}{\alpha_i^2+\beta_i^2}(\alpha_i \mp j\beta_i)$; then sign is changed and reciprocal. So the amplitude and power response of CQF is identical to the QMF bank.

This is due to CQF condition on $h_1[n]$ that in addition to the alternating sign change, also requires the time reversal of the impulse response. In fact, in the time domain the synthesis filters are equivalent to time-reversed version of the analysis filters (plus a gain that compensates for the decimation). In real situations, sometimes, instead of inserting a gain factor equal to 2 and in the synthesis FB, often the gain is distributed among the analysis filter synthesis and equal to $\sqrt{2}$.

7.6.3 Open-Loop and Closed-Loop SAF

We can define two types of SAF structures, called open-loop and closed-loop, which differ in the error calculation mode and for the definition of the update rule.

In the closed-loop structure, shown in Fig. 7.26, the error is calculated, at the output, in the usual way as $e[n] = d[n] - y[n]$. Thus, the error calculated is then divided into subbands, with the analysis filters bank, in such a way for each channel of the bank, it is defined as a decimated error $e_m^{CL}[k]$ with $k = nD$, related to the mth frequency band. This error is multiplied by AF input delay-line vector of the mth channel $\mathbf{x}_{m,k}$. The update rule is then

$$\mathbf{w}_{m,k+1} = \mathbf{w}_{m,k} + \mu_m \mathbf{x}_{m,k} e_m^{CL}[k]. \tag{7.157}$$

In the open-loop structure, shown in Fig. 7.27, it is the desired output signal $d[n]$ which is divided into subbands; the error is then calculated by considering the output of the mth filter channel $y_m[k]$ as

$$e_m[k] = d_m[k] - y_m[k]. \tag{7.158}$$

The update rule is therefore identical to (7.157) in which, instead of error, we consider the open-loop error $e_m[k]$.

From the formal point of view it is noted that, in general terms, $e_m^{CL}[k] \neq e_m[k]$, and that the correct error calculation is in a closed-loop, i.e., defined by the comparison of the full bandwidth signals $d[n]$ and $y[n]$, and subsequently divided into subbands. The two errors coincide only in the case of ideal filters bank and uncorrelated processes between contiguous bands.

From the application point of view, however, the SAF is usually implemented as the open-loop structure. In fact, the advantage of having a correct error calculation is thwarted by the latency introduced in the synthesis filter bank, needed to obtain the full-bandwidth output $y[n]$. This delay compromises, in many practical situations, the convergence speed of SAF implemented in a closed-loop scheme.

As shown in Fig. 7.27, in the open-loop SAF the signal $y_m[k]$ is taken before the synthesis bank. In practice, in the open-loop structure, the non optimality in the error calculation is compensated, in terms of performance, from the *zero-latency* between the input $x_m[k]$ and the output $y_m[k]$.

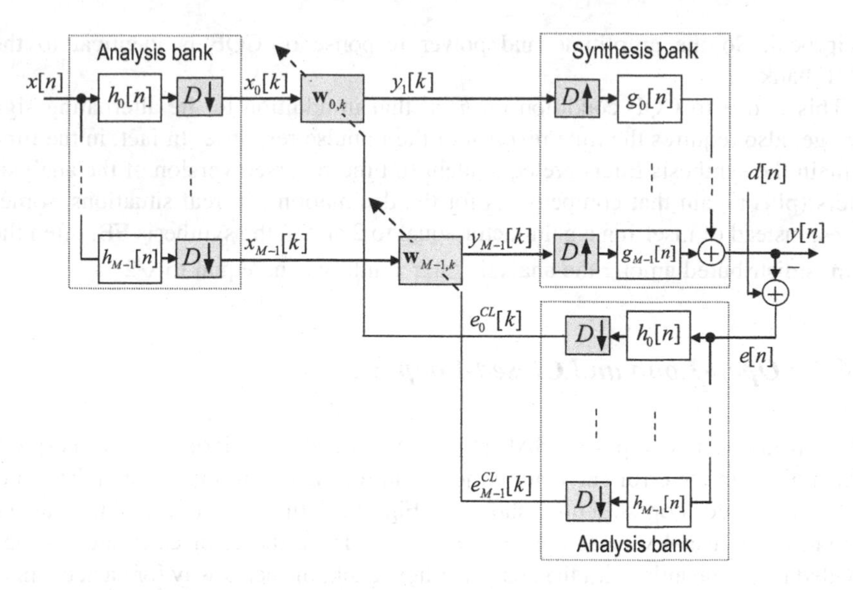

Fig. 7.26 Subband adaptive filtering with closed-loop error computation

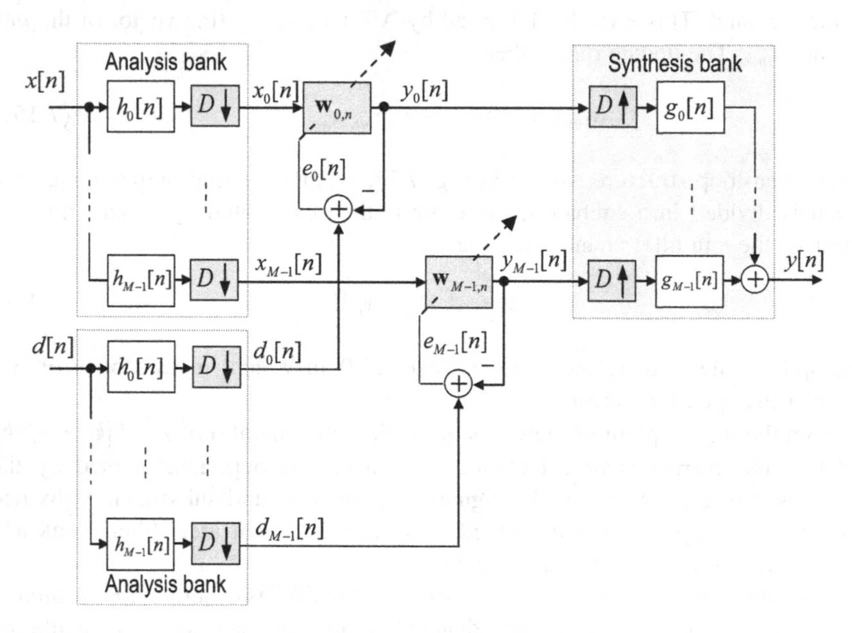

Fig. 7.27 Subband adaptive filtering with open-loop error computation

7.6.3.1 Condition for Existence of the Optimal Solution

For the determination of the existence conditions of the SAF optimal solution, we consider the problem of identifying a linear dynamic system with TF $S(z)$ described in Fig. 7.28. The reference subbands structure for the identification of $S(z)$, in the case of M channels, is instead shown in Fig. 7.29.

For the development consider, for simplicity, the case with only two channels, open-loop learning scheme, and consider the *modulation expansion* of signal and filters [41, 49]. So, the TF $S(z)$ represented in terms modulation expansion is defined as follows:

$$\mathbf{S}(z) = \begin{bmatrix} S(z) & 0 \\ 0 & S(-z) \end{bmatrix}. \tag{7.159}$$

For the output is

$$\begin{aligned} \begin{bmatrix} Y_0(z) \\ Y_1(z) \end{bmatrix} &= \frac{1}{2} \begin{bmatrix} H_0\left(z^{1/2}\right) & H_0\left(-z^{1/2}\right) \\ H_1\left(z^{1/2}\right) & H_1\left(-z^{1/2}\right) \end{bmatrix} \begin{bmatrix} S\left(z^{1/2}\right) & 0 \\ 0 & S\left(-z^{1/2}\right) \end{bmatrix} \begin{bmatrix} X\left(z^{1/2}\right) \\ X\left(-z^{1/2}\right) \end{bmatrix} \\ &= \frac{1}{2}\mathbf{H}\left(z^{1/2}\right)\mathbf{S}\left(z^{1/2}\right)\mathbf{x}\left(z^{1/2}\right). \end{aligned} \tag{7.160}$$

The identifier output is

$$\begin{aligned} \begin{bmatrix} \hat{Y}_0(z) \\ \hat{Y}_1(z) \end{bmatrix} &= \frac{1}{2} \begin{bmatrix} W_{0,0}(z) & W_{0,1}(z) \\ W_{1,0}(z) & W_{1,1}(z) \end{bmatrix} \begin{bmatrix} H_0\left(z^{1/2}\right) & H_0\left(-z^{1/2}\right) \\ H_1\left(z^{1/2}\right) & H_1\left(-z^{1/2}\right) \end{bmatrix} \begin{bmatrix} X\left(z^{1/2}\right) \\ X\left(-z^{1/2}\right) \end{bmatrix} \\ &= \frac{1}{2}\mathbf{W}(z)\mathbf{H}\left(z^{1/2}\right)\mathbf{x}\left(z^{1/2}\right). \end{aligned} \tag{7.161}$$

From the foregoing, in the case of open-loop learning, the *zero error condition* does not involve the synthesis filter bank and considering the M channels case, we can write:

$$\mathbf{W}\left(z^M\right)\mathbf{H}(z) = \mathbf{H}(z)\mathbf{S}(z). \tag{7.162}$$

The error can be cancelled and the adaptation algorithm can achieve the optimal solution, using only the subband signals. Each channel has an independent adaptation from the others and the algorithm converges to the optimum solution, with open-loop error determined according to the scheme of Fig. 7.27. From (7.162), the open-loop solution exists and is determined with the adaptive algorithm (LMS or other), if and only if, the analysis filters bank is *aliasing free*, i.e., $\mathbf{H}(z)\mathbf{H}^{-1}(z) = \mathbf{I}$.

From the scheme of Fig. 7.29, the most general identification condition is that of closed-loop adaptation, that also involves the synthesis filters bank. In this case, the output error is zero $E(z) = Y(z) - \hat{Y}(z) \equiv 0$, if applies

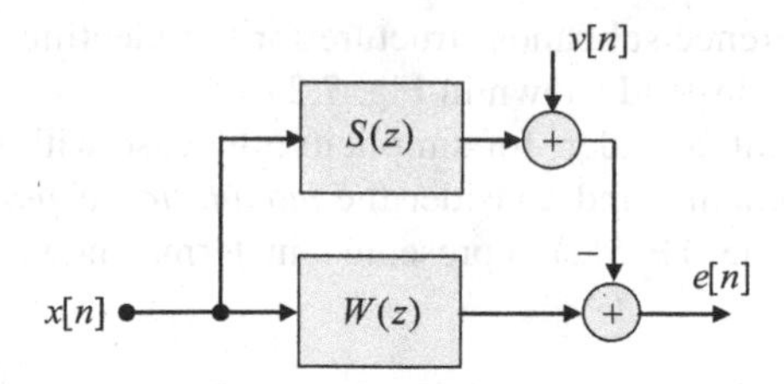

Fig. 7.28 Identification of a linear system $S(z)$

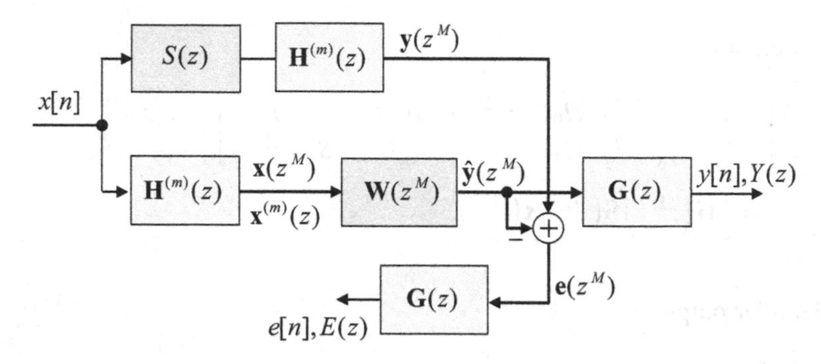

Fig. 7.29 Linear system identification $S(z)$

$$\mathbf{G}(z)\mathbf{W}(z^M)\mathbf{H}(z) = \mathbf{G}(z)\mathbf{H}(z)\mathbf{S}(z). \tag{7.163}$$

Note that, the subband errors are not necessarily zero and for the adaptation the knowledge of the global output error $E(z)$ is required. For the optimal solution determination, as suggested in [52], it is convenient to define the vector $\tilde{\mathbf{W}}(z^M)$ and the matrix $\tilde{\mathbf{G}}(z)$, respectively, as $\tilde{\mathbf{W}}(z^M) = \begin{bmatrix} W_0(z^M) & \cdots & W_{M-1}(z^M) \end{bmatrix}^T$ and $\tilde{\mathbf{G}}(z) = \text{diag}\begin{bmatrix} G_0(z) & \cdots & G_{M-1}(z) \end{bmatrix}$, such that $\tilde{\mathbf{W}}(z^M)\tilde{\mathbf{G}}(z) = \mathbf{G}(z)\mathbf{W}(z^M)$. Then, (7.163) can be rewritten as

$$\tilde{\mathbf{W}}(z^M)\tilde{\mathbf{G}}(z)\mathbf{H}(z) = \mathbf{G}(z)\mathbf{H}(z)\mathbf{S}(z) \tag{7.164}$$

which has a solution

$$\tilde{\mathbf{W}}(z^M) = \mathbf{G}(z)\mathbf{H}(z)\mathbf{S}(z)\mathbf{H}^{-1}(z)\tilde{\mathbf{G}}^{-1}(z).$$

From the previous development, it is possible to determine the solution $\tilde{\mathbf{W}}(z^M)$ if

$$\mathbf{G}(z)\mathbf{H}(z)\mathbf{H}^{-1}(z)\tilde{\mathbf{G}}^{-1}(z) = \mathbf{I}. \tag{7.165}$$

So, if for the analysis filters bank we have that $\mathbf{H}(z)\mathbf{H}^{-1}(z) = \mathbf{I}$, then, it is necessary that for the synthesis filters bank, it is $\mathbf{G}(z)\tilde{\mathbf{G}}^{-1}(z) = \mathbf{I}$.

7.6.4 Circuit Architectures for SAF

In general, the SAF approach is indicated in the case of very long AF impulse response. For example, in the identification of the acoustic paths, as in echo cancellation problems. In fact, with typical audio sampling frequencies, for reverberant environments, you may have impulse responses of length equal to tens of thousands of samples.

In such application contexts, for a correct implementability and AF effectiveness, one must necessarily use circuit architectures with:

1. Low computational complexity for real-time use;
2. Low latency, compatible with the type of application;
3. Remarkable convergence characteristics that allow a proper operability even in nonstationary environment.

In these cases, the SAF, when properly calibrated, are among the architectures that, in principle, allow to obtain a good compromise considering the above three specific requests.

7.6.4.1 The Gilloire–Vetterli's Tridiagonal SAF Structure

Consider the problem of identifying a linear dynamic system with TF $S(z)$, described in Fig. 7.29, with open-loop learning. Consider, the two-channel case ($M = 2$), for which the FB is composed by a low-pass and high-pass half-band complementary filters.

The condition for the optimal solution determination (7.162) is

$$\mathbf{W}(z^2)\mathbf{H}(z) = \mathbf{H}(z)\mathbf{S}(z). \tag{7.166}$$

Considering the QMF condition (7.149) and (7.150) and the modulation component matrix $\mathbf{H}^{(m)}$, (7.138) here rewritten as

$$\mathbf{H}^{(m)} \triangleq \begin{bmatrix} H_0(z) & H_0(-z) \\ H_1(z) & H_1(-z) \end{bmatrix} = \begin{bmatrix} H(z) & -H(-z) \\ -H(-z) & H(z) \end{bmatrix}. \tag{7.167}$$

Moreover, with the position (7.147), the determinant is a pure delay

$$\Delta_H \approx \alpha z^{-L_f+1}. \tag{7.168}$$

The PRC can be obtained considering the *paraunitary* condition for the composite analysis/synthesis TF. Let $\mathbf{G}^{(m)}(z)$ be the synthesis FB matrix; for the PRC [see (7.140)], we have that

$$\mathbf{T}^{(m)}(z) = \mathbf{G}^{(m)}(z)\mathbf{H}^{(m)}(z) \approx z^{-L_f+1} \tag{7.169}$$

whereby the $\mathbf{G}^{(m)}(z)$, for (7.145) considering the QMF conditions, takes the form:

$$\mathbf{G}^{(m)}(z) \approx z^{-L_f+1}\left[\mathbf{H}^{(m)}(z)\right]^{-1} = \frac{1}{\alpha}\begin{bmatrix} H(z) & -H(-z) \\ -H(-z) & H(z) \end{bmatrix}. \tag{7.170}$$

From (7.166) then

$$\begin{aligned} z^{-L_f+1}\mathbf{W}^{(m)}(z^2) &\approx \mathbf{H}^{(m)}(z)\mathbf{S}^{(m)}(z)\left[\mathbf{H}^{(m)}(z)\right]^{-1} \\ &\approx \frac{1}{\alpha}\begin{bmatrix} H^2(z)S(z) - H^2(-z)S(-z) & H(z)H(-z)\left[S(-z) - S(z)\right] \\ H(z)H(-z)\left[S(-z) - S(z)\right] & H^2(z)S(-z) - H^2(-z)S(z) \end{bmatrix} \end{aligned} \tag{7.171}$$

whereby $\mathbf{W}^{(m)}(z)$ is diagonal only if it is true, at least, one of the following conditions:

1. $H(z)H(-z) = 0$.
2. $S(-z) - S(z) = 0$.

The first condition is true only if $H(z)$ turns out to be an ideal filter with infinite attenuation in the stop band, namely $H(e^{j\omega}) = 0$ for $\pi/2 \le \omega \le \pi 3/4$, while the second condition does not correspond to a feasible physical system. In other words, $\mathbf{W}^{(m)}(z)$ is diagonal only in the case of ideal prototype low-pass filter, i.e., $H(z)$ is an ideal half-band filter.

As said, for the correct identifiability of a generic physical system $S(z)$, the matrix $\mathbf{W}^{(m)}(z)$ cannot have a pure diagonal structure, but must also contain the cross terms. In the case of a filter bank, with sufficient stop-band attenuation, in [49], a *tridiagonal* structure of $\mathbf{W}^{(m)}(z)$ is given, in which only for the adjacent bands the cross terms are present.

Formally,

$$\mathbf{W}^{(m)}\left(z^M\right) \approx z^{-K}\begin{bmatrix} W_{0,0}(z) & W_{0,1}(z) & 0 & \cdots & 0 & W_{0,M-1}(z) \\ W_{1,0}(z) & W_{1,1}(z) & W_{1,2}(z) & 0 & \cdots & 0 \\ 0 & W_{21}(z) & W_{2,2}(z) & W_{2,3}(z) & \vdots & \vdots \\ \vdots & 0 & 0 & \ddots & \ddots & 0 \\ 0 & \vdots & \vdots & W_{M-2,M-3}(z) & W_{M-2,M-2}(z) & W_{M-2,M-1}(z) \\ W_{M-1,0}(z) & 0 & 0 & \cdots & W_{M-1,M-2}(z) & W_{M-1,M-1}(z) \end{bmatrix}. \tag{7.172}$$

The inclusion of cross terms between the subband adaptive filter leads to slow convergence and of an increase in the computational cost.

The structure of the adaptive filter bank is shown in Fig. 7.30.

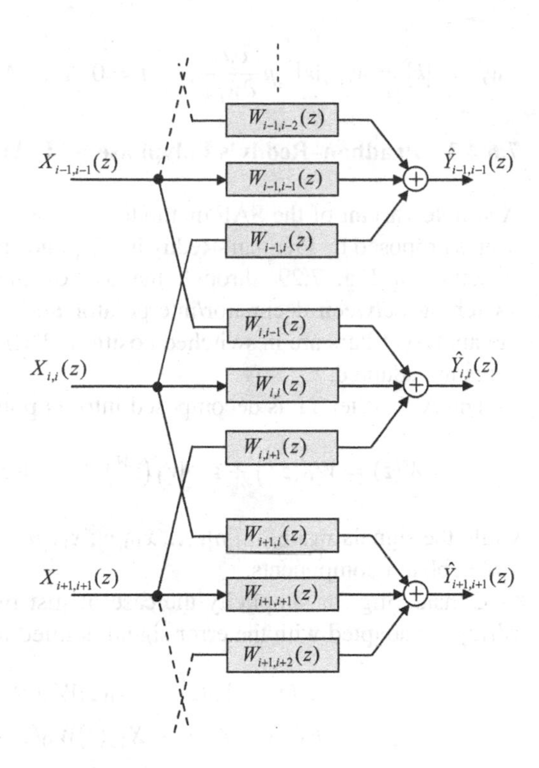

Fig. 7.30 Representation of the matrix $\mathbf{W}(z)$ in the tridiagonal SAF of Gilloire–Vetterli

7.6.4.2 LMS Adaptation

For the determination of LMS adaptation algorithm, proceed by minimizing the output error power and, for paraunitary $\mathbf{G}^{(m)}(z)$, the error power is equal to the sum of the subband errors powers, for which the cost function is given by the sums

$$J_n = \sum_{m=0}^{M-1} \alpha_m E\Big\{ |e_m[n]|^2 \Big\}, \tag{7.173}$$

where the coefficients α_m are inversely proportional to the mth band signal power (power normalization) and in the case of white input it has $\alpha_m = 1$ for $m = 0, \ldots, M - 1$.

Differentiating J_n with respect to the filter weights in scalar form (see Sect. 3.3.1), we obtain:

$$\frac{\partial J_n}{\partial w_m[k]} = 2\left[\alpha_0 e_0 \frac{\partial e_0}{\partial w_m[k]} + \alpha_1 e_1 \frac{\partial e_1}{\partial w_m[k]} + \cdots + \alpha_{M-1} e_{M-1} \frac{\partial e_{M-1}}{\partial w_m[k]}\right], \tag{7.174}$$

for $m = 0, \ldots, M - 1$, and let L_s be the AF length, for $k = 1, \ldots, L_s - 1$. Therefore, the adaptation takes the form:

$$w_{m,n+1}[k] = w_{m,n}[k] - \mu \frac{\partial J_n}{\partial w_{mk}}, \quad m = 0, 1, ..., M-1; k = 0, 1, ..., L_s - 1. \quad (7.175)$$

7.6.4.3 Pradhan–Reddy's Polyphase SAF Architecture

A simple variant of the SAF methodology for the dynamic system $S(z)$ identification is proposed by Pradhan–Reddy in [53] and shown in Fig. 7.31. Compared to the structure of Fig. 7.29, through the use of the noble identity (that allows the switching between decimator/interpolator and a TF [43, 44]), the decimator and the analysis filters are in switched position. Therefore, the AF's polyphase components are adapted.

The AF's filter TF is decomposed into its polyphase components as

$$W(z) = W_0(z^M) + z^{-1}W_1(z^M) + \cdots + z^{-(M-1)}W_{M-1}(z^M) \quad (7.176)$$

while the signals $x_{00}[n], x_{01}[n], \ldots, x_{10}[n], x_{11}[n], \ldots, x_{M-1,\, M-1}[n]$ represent the input $x[n]$ subband components.

Considering for simplicity the case of just two channels, the filters $W_0(z)$ and $W_1(z)$ are adapted with the error signal defined as

$$E_0(z) = Y_0(z) - X_{00}(z)W_0(z) - X_{01}(z)W_1(z), \quad (7.177)$$

$$E_1(z) = Y_1(z) - X_{10}(z)W_0(z) - X_{11}(z)W_1(z). \quad (7.178)$$

With the CF (7.173) that for $M = 2$ is

$$J_n = \alpha_0 E\left\{|e_0[n]|^2\right\} + \alpha_1 E\left\{|e_1[n]|^2\right\}. \quad (7.179)$$

From (7.179), differentiating with respect to the filter weights, we obtain

$$\frac{\partial J_n}{\partial w_{0k}} = 2\alpha_0 E\left\{e_0[n]\frac{\partial e_0[n]}{\partial w_{0k}}\right\} + 2\alpha_1 E\left\{e_1[n]\frac{\partial e_1[n]}{\partial w_{0k}}\right\}, \quad k = 0, 1, \ldots, \frac{L}{2} - 1, \quad (7.180)$$

$$\frac{\partial J_n}{\partial w_{1k}} = 2\alpha_0 E\left\{e_0[n]\frac{\partial e_0[n]}{\partial w_{1k}}\right\} + 2\alpha_1 E\left\{e_1[n]\frac{\partial e_1[n]}{\partial w_{1k}}\right\}, \quad k = 0, 1, \ldots, \frac{L}{2} - 1. \quad (7.181)$$

The partial derivatives of $E_0(z)$ and $E_1(z)$ respect to w_{0k} and w_{1k} are equal to

$$\frac{\partial E_0(z)}{\partial w_{0k}} = -X_{00}(z)z^{-k}, \quad (7.182)$$

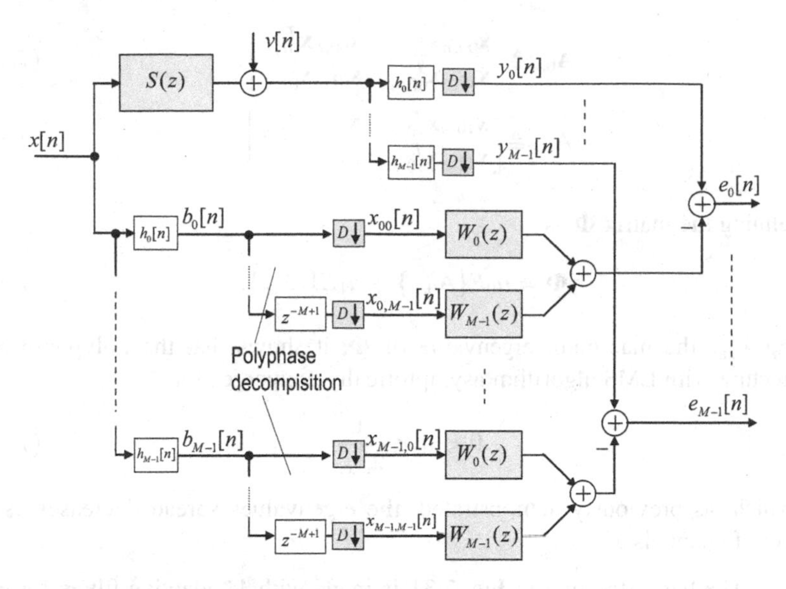

Fig. 7.31 *M*-channels Pradhan–Reddy's SAF structure

$$\frac{\partial E_1(z)}{\partial w_{0k}} = -X_{10}(z)z^{-k}, \tag{7.183}$$

$$\frac{\partial E_0(z)}{\partial w_{1k}} = -X_{01}(z)z^{-k}, \tag{7.184}$$

$$\frac{\partial E_1(z)}{\partial w_{1k}} = -X_{11}(z)z^{-k}. \tag{7.185}$$

Performing the inverse transform of the above equations, we have that

$$\begin{aligned} w_{0k}[n+1] &= w_{0k}[n] + 2\mu\left[\alpha_0 E\{e_0[n]x_{00}[n-k]\} + \alpha_1 E\{e_1[n]x_{10}[n-k]\}\right] \\ w_{1k}[n+1] &= w_{1k}[n] + 2\mu\left[\alpha_0 E\{e_0[n]x_{01}[n-k]\} + \alpha_1 E\{e_1[n]x_{11}[n-k]\}\right] \end{aligned} \tag{7.186}$$

for $k = 0, 1, \ldots, L/2 - 1$. By replacing the expectation with its instantaneous estimate, we get the LMS learning rule that is

$$\begin{aligned} w_{0k}[n+1] &= w_{0k}[n] + 2\mu\left[\alpha_0 e_0[n]x_{00}[n-k] + \alpha_1 e_1[n]x_{10}[n-k]\right], \\ w_{1k}[n+1] &= w_{1k}[n] + 2\mu\left[\alpha_0 e_0[n]x_{01}[n-k] + \alpha_1 e_1[n]x_{11}[n-k]\right]. \end{aligned} \tag{7.187}$$

Let $\mathbf{A}_{0,n}$ and $\mathbf{A}_{1,n}$ be the matrices related to the subband components of the input signal, defined as

$$\mathbf{A}_{0,n} \triangleq \begin{bmatrix} \mathbf{x}_{00,n}\mathbf{x}_{00,n}^T & \mathbf{x}_{00,n}\mathbf{x}_{01,n}^T \\ \mathbf{x}_{10,n}\mathbf{x}_{00,n}^T & \mathbf{x}_{01,n}\mathbf{x}_{01,n}^T \end{bmatrix}, \tag{7.188}$$

$$\mathbf{A}_{1,n} \triangleq \begin{bmatrix} \mathbf{x}_{10,n}\mathbf{x}_{10,n}^T & \mathbf{x}_{10,n}\mathbf{x}_{11,n}^T \\ \mathbf{x}_{11,n}\mathbf{x}_{10,n}^T & \mathbf{x}_{11,n}\mathbf{x}_{11,n}^T \end{bmatrix}. \tag{7.189}$$

By defining the matrix $\mathbf{\Phi}$ as

$$\mathbf{\Phi} = \alpha_0 E\{\mathbf{A}_{0,n}\} + \alpha_1 E\{\mathbf{A}_{1,n}\}. \tag{7.190}$$

Calling $\lambda_{\max}$ the maximum eigenvalue of $\mathbf{\Phi}$, it shows that the polyphase SAF architecture with LMS algorithm asymptotically converges for

$$0 < \mu < \frac{1}{\lambda_{\max}}. \tag{7.191}$$

For which, as previously demonstrated, the eigenvalues spread decreases as the number of channels M.

Remark The bank structure in Fig. 7.31 is made with M adaptive filters for each subband for which it has a computational complexity similar to or higher than the full-band adaptive filter.

7.6.5 Characteristics of Analysis-Synthesis Filter Banks in the SAF Structure

The proper SAF structure design is quite complex and highly dependent on the application context. As a first step, the analysis-synthesis filters bank structure is of crucial importance to have acceptable performance. Moreover, there is not a precise formal criterion for its optimization. The filters bank design is, therefore, difficult and should be carried out throughout several compromises to have a good balance between the required global specifications.

Because of the decimation, especially in the critical sampling case, the channel outputs of the bank are affected by aliasing that determines a quality degradation of the output. An obvious mode to reduce the aliasing effect is to use a decimation rate not critical ($D < M$) with an increase, however, in the computational load. Additionally, the use of analysis-synthesis symmetric FB determines a low value of the error signal around the crossover frequency between two contiguous bands that worsens significantly the speed of SAF convergence [49–51, 54].

A possible solution to these problems, that are most evident as the number of channels increase, is to widen the distance between the contiguous bands of the FB (method spectral-gap) or to work with decimation rate lower than the critical [52, 55]. However, in the case of large number of channels the spectral gap technique produces a not acceptable output quality.

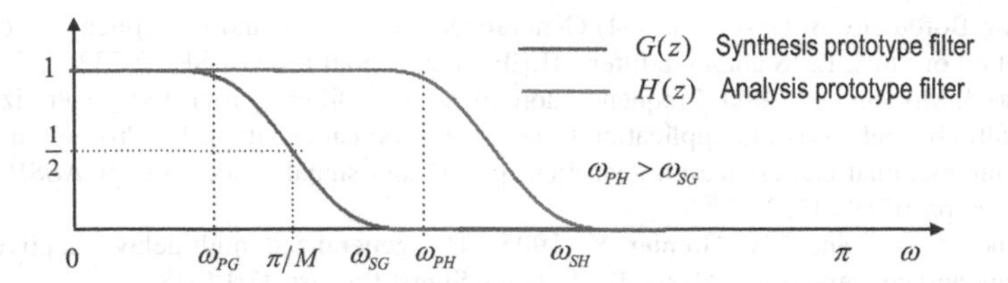

Fig. 7.32 Different low-pass prototypes for the analysis and synthesis FBs to increase the convergence speed around the crossover frequencies ($m\pi/M$)

A simple solution to increase the convergence speed, given in [50], is to choose the analysis FB prototype with wider bandwidth compared to the synthesis FB prototype, as shown in Fig. 7.32.

Other approaches to the reduction of aliasing make use of auxiliary adaptive channels [52].

Finally, in the literature there are a number of SAF architecture with uniform and nonuniform FBs. See, for example, [56–58].

References

1. Clark GA, Mitra SK, Parker SR (1981) Block implementation of adaptive digital filters. IEEE Trans Circuits Syst CAS-28(6):584–592
2. Widrow B, Stearns SD (1985) Adaptive signal processing. Prentice Hall, Englewood Cliffs, NJ
3. Feuer A (1985) Performance analysis of the block least mean square algorithm. IEEE Trans Circuits Syst CAS-32(9):960–963
4. Dentino M, McCool J, Widrow B (1978) Adaptive filtering in the frequency domain. Proc IEEE 66:1658–1660
5. Bershad NJ, Feintuch PD (1979) Analysis of the frequency domain adaptive fiter. Proc IEEE 67:1658–1659
6. Ferrara ER (1980) Fast implementation of LMS adaptive filters. IEEE Trans Acoust Speech Signal Process ASSP-28:474–475
7. Clark GA, Parker SR, Mitra SK (1983) A unified approach to time- and frequency-domain realization of fir adaptive digital filters. IEEE Trans Acoust Speech Signal Process ASSP-31:1073–1083
8. Narayan SS, Peterson AM (1981) Frequency domain least-mean square algorithm. Proc IEEE 69(1):124–126
9. Lee JC, Un CK (1989) Performance analysis of frequency-domain block LMS adaptive digital filters. IEEE Trans Circuits Syst 36:173–189
10. Oppenheim AV, Schafer RW, Buck JR (1999) Discrete-time signal processing, 2nd edn. Prentice Hall, Upper Saddle River, NJ
11. Mansour D, Gray AH (1982) Unconstrained frequency-domain adaptive filter. IEEE Trans Acoust Speech Signal Process ASSP-30(5):726–734
12. Shynk JJ (1992) Frequency domain and multirate adaptive filtering. IEEE Signal Process Mag 9:14–37
13. Bendel Y, Burshtein D, Shalvi O, Weinstein E (2001) Delayless frequency domain acoustic echo cancellation. IEEE Trans Speech Audio Process 9(5):589–597

14. Farhang-Boroujeny B, Gazor S (1994) Generalized sliding FFT and its application to implementation of block LMS adaptive filters. IEEE Trans Signal Process SP-42:532–538
15. Benesty J, Morgan DR (2000) Frequency-domain adaptive filtering revisited, generalization to the multi-channel case, and application to acoustic echo cancellation. In: Proceedings of the IEEE international conference on acoustics speech, and signal proceesing (ICASSP), vol 2, 5–9 June, pp II789–II792
16. Moulines E, Amrane OA, Grenier Y (1995) The generalized multidelay adaptive filter: structure and convergence analysis. IEEE Trans Signal Process 43:14–28
17. McLaughlin HJ (1996) System and method for an efficiently constrained frequency-domain adaptive filter. US Patent 5 526 426
18. Frigo M, Johnson SG (2005) The design and implementation of FFTW3. Proc IEEE 93 (2):216–231
19. Farhang-Boroujeny B (1996) Analysis and efficient implementation of partitioned block LMS adaptive filters. IEEE Trans Signal Process SP-44(11):2865–2868
20. Sommen PCW, Gerwen PJ, Kotmans HJ, Janssen AJEM (1987) Convergence analysis of a frequency-domain adaptive filter with exponential power averaging and generalized window function. IEEE Trans Circuits Syst CAS-34(7):788–798
21. Derkx RMM, Egelmeers GPM, Sommen PCW (2002) New constraining method for partitioned block frequency-domain adaptive filters. IEEE Trans Signal Process SP-50 (9):2177–2186
22. Golub GH, Van Loan CF (1989) Matrix computation. John Hopkins University Press, Baltimore. ISBN 0-80183772-3
23. Gray RM (2006) Toeplitz and circulant matrices: a review. Found Trends Commun Inf Theory 2(3):155–239
24. Farhang-Boroujeny B, Chan KS (2000) Analysis of the frequency-domain block LMS algorithm. IEEE Trans Signal Process SP-48(8):2332–2342
25. Chan KS, Farhang-Boroujeny B (2001) Analysis of the partitioned frequency-domain block LMS (PFBLMS) algorithm. IEEE Trans Signal Process SP-49(9):1860–1864
26. Lee JC, Un CK (1986) Performance of transform-domain LMS adaptive algorithms. IEEE Trans Acoust Speech Signal Process ASSP-34:499–510
27. Asharif MR, Takebayashi T, Chugo T, Murano K (1986) Frequency domain noise canceler: frequency-bin adaptive filtering (FBAF). In: Proceedings ICASSP, pp 41.22.1–41.22.4
28. Sommen PCW (1989) Partitioned frequency-domain adaptive filters. In: Proceedings of 23rd annual asilomar conference on signals, systems, and computers, Pacific Grove, CA, pp 677–681
29. Soo JS, Pang KK (1990) Multidelay block frequency domain adaptive filter. IEEE Trans Acoust Speech Signal Process 38:373–376
30. Sommen PCW (1992) Adaptive filtering methods. PhD dissertation, Eindhoven University of Technology, Eindhoven, The Netherlands
31. Yon CH, Un CK (1994) Fast multidelay block transform-domain adaptive flters based on a two-dimensional optimum block algorithm. IEEE Trans Circuits Syst II Analog Digit Signal Process 41:337–345
32. Asharif MR, Amano F (1994) Acoustic echo-canceler using the FBAF algorithm. Trans Commun 42:3090–3094
33. Narayan SS, Peterson AM, Marasimha MJ (1983) Transform domain lms algorithm. IEEE Trans Acoust Speech Signal Process ASSP-31(3):609–615
34. Beaufays F (1995) Transform domain adaptive filters: an analytical approach. IEEE Trans Signal Process SP-43(3):422–431
35. Farhan-Boroujeny B, Lee Y, Ko CC (1996) Sliding transforms for efficient implementation of transform domain adaptive filters. Elsevier, Signal Process 52: 83–96
36. Marshall DF, Jenkins WK, Murphy JJ (1989) The use of orthogonal transforms for improving performance of adaptive filters. IEEE Trans Circuits Syst 36(4):474–484
37. Ahmed N, Natarajan T, Rao KR (1974) Discrete cosine transform. IEEE Trans Comput C-23 (1):90–93

38. Feig E, Winograd S (1992) Fast algorithms for the discrete cosine transform. IEEE Trans Signal Process 40(9):2174–2193
39. Martucci SA (1994) Symmetric convolution and the discrete sine and cosine transforms. IEEE Trans Signal Process SP-42(5):1038–1051
40. Bruun G (1978) z-Transform DFT filters and FFTs. IEEE Trans Acoust Speech Signal Process 26(1):56–63
41. Vetterli M (1987) A theory of multirate filter banks. IEEE Trans Acoust Speech Signal Process ASSP-35:356–372
42. Johnston J (1980) A filter family designed for use in quadrature mirror filter banks. In: Proceedings of IEEE international conference on acoustics, speech, and signal processing, Denver, CO
43. Crochiere RE, Rabiner LR (1983) Multirate signal processing. Prentice Hall, Englewood Cliffs, NJ
44. Fliege NJ (1994) Multirate digital signal processing. Wiley, New York
45. Vaidyanathan PP (1993) Multirate systems and filterbanks. Prentice-Hall, Englewood Cliffs, NJ
46. Koilpillai RD, Vaidyanathan PP (1990) A new approach to the design of FIR perfect reconstruction QMF banks. IEEE international symposium on circuits and systems-1990, vol 1, 1–3 May 1990, pp 125–128
47. Nayebi K, Barnwell T, Smith M (1992) Time domain filter bank analysis: a new design theory. IEEE Trans Signal Process 40(6):1412–1429
48. Nguyen TQ (1994) Near perfect reconstruction pseudo QMF banks. IEEE Trans Signal Process 42(1):65–76
49. Gilloire A, Vetterli M (1992) Adaptive filtering in subbands with critical sampling: analysis, experiments, and application to acoustic echo cancellation. IEEE Trans Signal Process 40:1862–1875
50. De Léon PL II, Etter DM (1995) Experimental results with increased bandwidth analysis filters in oversampled, subband echo canceler. IEEE Trans Signal Process Lett 1:1–2
51. Croisier A, Esteban D, Galand C (1976) Perfect channel splitting by use of interpolation/decimation/tree decomposition techniques. Conference on information sciences and systems
52. Kellermann W (1988) Analysis and design of multirate systems for cancellation of acoustic echoes. In: Proceedings IEEE international conference on acoustics, speech, and signal processing, New York, NY, pp 2570–2573
53. Pradhan SS, Reddy VU (1999) A new approach to subband adaptive filtering. IEEE Trans Signal Process 47(3):65–76
54. Gilloire A (1987) Experiments with subband acoustic echo cancellation for teleconferencing. In: Proceedings IEEE ICASSP, Dallas, TX, pp 2141–2144
55. Yusukawa H, Shimada S, Furakawa I (1987) Acoustic echo with high speech quality. In: Proceedings IEEE ICASSP, Dallas, TX, pp 2125–2128
56. Petraglia MR, Alves RG, Diniz PSR (2000) New structures for adaptive filtering in subbands with critical sampling. IEEE Trans Signal Process 48(12):3316–3327
57. Petraglia MR, Batalheiro PB (2004) Filtre bank design for a subband adaptive filtering structure with critical sampling. IEEE Trans Signal Process 51(6):1194–1202
58. Kim SG, Yoo CD, Nguyen TQ (2008) Alias-free subband adaptive filtering with critical sampling. IEEE Trans Signal Process 56(5):1894–1904

Chapter 8
Linear Prediction and Recursive Order Algorithms

8.1 Introduction

The problem of optimal filtering consists in determining the filter coefficients $\mathbf{w}_{opt}$ through the normal equations solution in the Wiener stochastic or the Yule–Walker deterministic form. In practice this is achieved by inverting the correlation matrix $\mathbf{R}$ or its estimate $\mathbf{R}_{xx}$. Formally, the problem is simple. Basically, however, this inversion is most often of ill-posed nature. The classical matrix inversion approaches are not robust and in certain applications cannot be implemented. In fact, most of the adaptive signal processing problems concern the computational cost and robustness of the estimation algorithms.

Another important aspect relates to the parameters scaling of the calculation procedures. The adaptation algorithms produce a set of intermediate results whose values sometimes assume an important physical meaning. The online analysis of these parameters often allows the verification of important properties (stability, minimum phase, etc.) which are useful in some applications such as, for example, the speech coding and transmission. In connection with this point, an issue of central importance is the choice of the implementative circuit or algorithm structure.

Problems such as the noise control, scaling and efficient coefficients computation, other effects due to quantization, etc., are in fact difficult to solve and strongly influence the filter performance. Some implementative structures with equivalent transfer function (TF) may present, in addition to the typical static filtering advantages, also other interesting features that may determine a higher convergence speed and the possibility of more efficient adaptation methods.

This chapter introduces the linear prediction issue and the theme of the recursive order algorithms. Both of these topics are related to the implementative structures with particular robustness and efficiency properties.

A. Uncini, *Fundamentals of Adaptive Signal Processing*, Signals and Communication Technology, DOI 10.1007/978-3-319-02807-1_8,

8.2 Linear Estimation: Forward and Backward Prediction

Linear prediction plays an important role in signal processing in many theoretic-computational and applications areas. Although the linear prediction theory was initially formulated in the 1940s of the last century, its influence is still present [1, 2].

As already indicated in Chap. 2, the linear prediction problem can be formulated in very simple terms and can be defined in the more general context of linear estimation and linear filtering (understood as smoothing). In this section, the prediction and estimation arguments are formulated with reference to the formal aspects of the optimal filtering Wiener theory discussed in Chap. 3.

8.2.1 Wiener's Optimum Approach to the Linear Estimation and Linear Prediction

Suppose we know M samples of the sequence $x[n] \in (\mathbb{R}, \mathbb{C})$ between the extremes $[n, n-M]$ and that we want to estimate an unknown value of the sequence, indicated as $\hat{x}[n-i]$, not present in the known samples, using a linear combination of these known samples. In formal terms, indicating with $w[k]$, $k = 0, 1, \ldots, M$, the coefficients of the estimator, we can write

$$y[n] = \hat{x}[n-i] = \sum_{\substack{k=0 \\ k \neq i}}^{M} w^*[k]x[n-k]. \tag{8.1}$$

The estimation error can be defined considering the reference signal $d[n]$ defined as $d[n] = x[n-i]$, for which we have that

$$e^{\mathrm{i}}[n] = d[n] - y[n] = x[n-i] - \hat{x}[n-i], \tag{8.2}$$

where the superscript "i" indicates that the prediction error is relative to the sample $x[n-i]$ sample.

Depending on the sample to be estimated is internal or external to the analysis window, we can define three cases:

1. *Linear estimation*—for i inside the analysis window $0 < i < M$;
2. *Forward prediction*—for $i \leq 0$, prediction of the future signal known the past samples; in particular for $i = 0$, it has a *one-step forward prediction* or simply *forward prediction*;
3. *Backward prediction*—for $i \geq M$, prediction of the past signal known as the current samples, in particular for $i = M$ there is a *one-step backward prediction* also simply referred to as *backward prediction*.

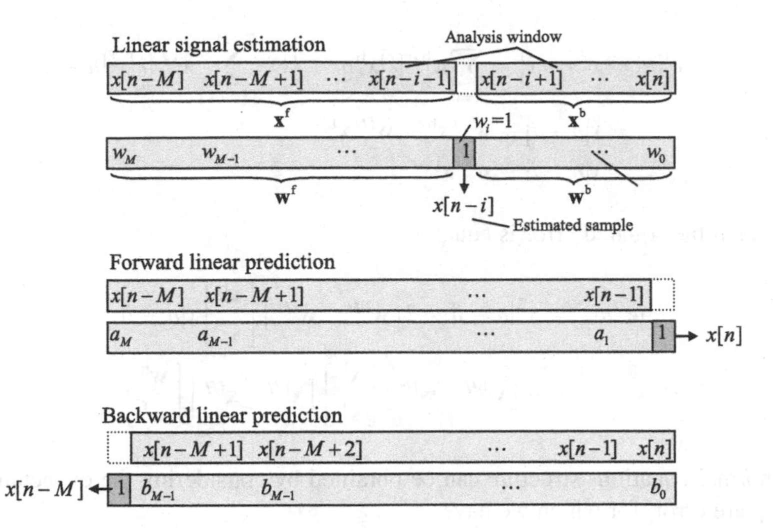

Fig. 8.1 Schematic representation of estimation and one-step forward and backward prediction process (modified from [3])

A general estimation and prediction process scheme is shown in Fig. 8.1. From the figure it is noted that the input signal can be partitioned as follows:

$$\begin{aligned}\mathbf{x}^{\mathrm{b}} &\in (\mathbb{R},\mathbb{C})^{i\times 1} \triangleq \left[x[n] \quad \cdots \quad x[n-i+1]\right]^T,\\ \mathbf{x}^{\mathrm{f}} &\in (\mathbb{R},\mathbb{C})^{(M-i)\times 1} \triangleq \left[x[n-i-1] \quad \cdots \quad x[n-M]\right]^T,\end{aligned} \tag{8.3}$$

where the superscript "f" and "b" stand for forward and backward.

Similarly, for the predictor vector we can write

$$\begin{aligned}\mathbf{w}^{\mathrm{b}} &\in (\mathbb{R},\mathbb{C})^{i\times 1} \triangleq \left[w[0] \quad w[1] \quad \cdots \quad w[i-1]\right]^H,\\ \mathbf{w}^{\mathrm{f}} &\in (\mathbb{R},\mathbb{C})^{(M-i)\times 1} \triangleq \left[w[i+1] \quad w[i+2] \quad \cdots \quad w[M]\right]^H.\end{aligned} \tag{8.4}$$

By defining the vectors

$$\mathbf{x}^{\mathrm{i}} \in (\mathbb{R},\mathbb{C})^{M\times 1} = \left[\mathbf{x}^{\mathrm{b}T} \quad \mathbf{x}^{\mathrm{f}T}\right]^T \tag{8.5}$$

$$\mathbf{w}^{\mathrm{i}} \in (\mathbb{R},\mathbb{C})^{M\times 1} = \left[\mathbf{w}^{\mathrm{b}H} \quad \mathbf{w}^{\mathrm{f}H}\right]^T \tag{8.6}$$

such that

$$y[n] = \mathbf{w}^{\mathrm{i}H}\mathbf{x}^{\mathrm{i}} = \left[\mathbf{w}^{\mathrm{b}H} \quad \mathbf{w}^{\mathrm{f}H}\right]\begin{bmatrix}\mathbf{x}^{\mathrm{b}}\\ \mathbf{x}^{\mathrm{f}}\end{bmatrix} \tag{8.7}$$

we have that the prediction error written as

$$
\begin{aligned}
e^{\mathrm{i}}[n] &= x[n-i] - \sum_{k=0}^{i-1} w^{\mathrm{b}*}[k]x^{\mathrm{b}}[n-k] - \sum_{k=i+1}^{M} w^{\mathrm{f}*}[k]x^{\mathrm{f}}[n-k] \\
&= x[n-i] - \mathbf{w}^{\mathrm{b}H}\mathbf{x}^{\mathrm{b}} - \mathbf{w}^{\mathrm{f}H}\mathbf{x}^{\mathrm{f}} \\
&= x[n-i] - \mathbf{w}^{\mathrm{i}H}\mathbf{x}^{\mathrm{i}}.
\end{aligned}
\tag{8.8}
$$

For which the squared error is equal to

$$
\begin{aligned}
\left|e^{\mathrm{i}}[n]\right|^2 &= x^2[n-i] - 2\begin{bmatrix}\mathbf{w}^{\mathrm{b}H} & \mathbf{w}^{\mathrm{f}H}\end{bmatrix}\begin{bmatrix}\mathbf{x}^{\mathrm{b}} \\ \mathbf{x}^{\mathrm{f}}\end{bmatrix}x[n-i] \\
&\quad + \begin{bmatrix}\mathbf{w}^{\mathrm{b}H} & \mathbf{w}^{\mathrm{f}H}\end{bmatrix}\begin{bmatrix}\mathbf{x}^{\mathrm{b}} \\ \mathbf{x}^{\mathrm{f}}\end{bmatrix}\begin{bmatrix}\mathbf{x}^{\mathrm{b}H} & \mathbf{x}^{\mathrm{f}H}\end{bmatrix}\begin{bmatrix}\mathbf{w}^{\mathrm{b}} \\ \mathbf{w}^{\mathrm{f}}\end{bmatrix}.
\end{aligned}
\tag{8.9}
$$

The normal equations structure can be obtained by considering the expectation of the square error, for which we have

$$
\begin{aligned}
J_n^{\mathrm{i}} &= E\left\{\left|e^{\mathrm{i}}[n]\right|^2\right\} \\
&= \sigma^2_{x[n-i]} - 2\begin{bmatrix}\mathbf{w}^{\mathrm{b}H} & \mathbf{w}^{\mathrm{f}H}\end{bmatrix}\begin{bmatrix}\mathbf{r}^{\mathrm{b}} \\ \mathbf{r}^{\mathrm{f}}\end{bmatrix} + \begin{bmatrix}\mathbf{w}^{\mathrm{b}H} & \mathbf{w}^{\mathrm{f}H}\end{bmatrix}\begin{bmatrix}\mathbf{R}^{\mathrm{bb}} & \mathbf{R}^{\mathrm{bf}} \\ \mathbf{R}^{\mathrm{bf}H} & \mathbf{R}^{\mathrm{ff}}\end{bmatrix}\begin{bmatrix}\mathbf{w}^{\mathrm{b}} \\ \mathbf{w}^{\mathrm{f}}\end{bmatrix} \\
&= \sigma^2_{x[n-i]} - 2\mathbf{w}^{\mathrm{i}H}\mathbf{r}^{\mathrm{i}} + \mathbf{w}^{\mathrm{i}H}\mathbf{R}^{\mathrm{i}}\mathbf{w}^{\mathrm{i}}.
\end{aligned}
\tag{8.10}
$$

From the previous, the correlation matrix $\mathbf{R}^{\mathrm{i}}$ is defined as

$$
\begin{aligned}
\mathbf{R}^{\mathrm{i}} &= E\left\{\begin{bmatrix}\mathbf{x}^{\mathrm{b}} \\ \mathbf{x}^{\mathrm{f}}\end{bmatrix}\begin{bmatrix}\mathbf{x}^{\mathrm{b}H} & \mathbf{x}^{\mathrm{f}H}\end{bmatrix}\right\} = \begin{bmatrix}E\left\{\mathbf{x}^{\mathrm{b}}\mathbf{x}^{\mathrm{b}H}\right\} & E\left\{\mathbf{x}^{\mathrm{b}}\mathbf{x}^{\mathrm{f}H}\right\} \\ E\left\{\mathbf{x}^{\mathrm{f}}\mathbf{x}^{\mathrm{b}H}\right\} & E\left\{\mathbf{x}^{\mathrm{f}}\mathbf{x}^{\mathrm{f}H}\right\}\end{bmatrix} \\
&= \begin{bmatrix}\mathbf{R}^{\mathrm{bb}} & \mathbf{R}^{\mathrm{bf}} \\ \mathbf{R}^{\mathrm{bf}H} & \mathbf{R}^{\mathrm{ff}}\end{bmatrix},
\end{aligned}
\tag{8.11}
$$

where

$$
\mathbf{R}^{\mathrm{bb}} = E\left\{\mathbf{x}^{\mathrm{b}}\mathbf{x}^{\mathrm{b}H}\right\} = \begin{bmatrix} r[0] & \cdots & r[i-1] \\ \vdots & \ddots & \vdots \\ r^*[i-1] & \cdots & r[0] \end{bmatrix},
$$

$$
\mathbf{R}^{\mathrm{ff}} = E\left\{\mathbf{x}^{\mathrm{f}}\mathbf{x}^{\mathrm{f}H}\right\} = \begin{bmatrix} r[0] & \cdots & r[M-i-1] \\ \vdots & \ddots & \vdots \\ r^*[M-i-1] & \cdots & r[0] \end{bmatrix},
$$

$$
\mathbf{R}^{\mathrm{bf}} = E\left\{\mathbf{x}^{\mathrm{b}}\mathbf{x}^{\mathrm{f}H}\right\} = \begin{bmatrix} r[i+1] & \cdots & r[M] \\ \vdots & \ddots & \vdots \\ r[2] & \cdots & r[M-i+1] \end{bmatrix}.
$$

For the cross-correlations vectors it is (see 3.55)

$$\begin{aligned}\mathbf{r}^{\mathrm{b}} &= E\{\mathbf{x}^{\mathrm{b}}x^*[n-i]\} = \begin{bmatrix} r[i] & r[i-1] & \cdots & r[1]\end{bmatrix}^T \\ \mathbf{r}^{\mathrm{f}} &= E\{\mathbf{x}^{\mathrm{f}}x^*[n-i]\} = \begin{bmatrix} r[1] & r[2] & \cdots & r[M-i]\end{bmatrix}^T\end{aligned} \tag{8.12}$$

and, furthermore,

$$\sigma^2_{x[n-i]} = E\left\{|x[n-i]|^2\right\}, \tag{8.13}$$

where note that in the stationary case it is $\sigma^2_{x[n-i]} = r[0]$.

Calculating the derivatives $\partial J_n^i(\mathbf{w})/\partial \mathbf{w}^{\mathrm{f}}$ and $\partial J_n^i(\mathbf{w})/\partial \mathbf{w}^{\mathrm{b}}$, and setting them to zero, we can write the normal equations in partitioned form, as

$$\begin{bmatrix} \mathbf{R}^{\mathrm{bb}} & \mathbf{R}^{\mathrm{bf}} \\ \mathbf{R}^{\mathrm{bf}H} & \mathbf{R}^{\mathrm{ff}} \end{bmatrix} \begin{bmatrix} \mathbf{w}^{\mathrm{b}} \\ \mathbf{w}^{\mathrm{f}} \end{bmatrix} = \begin{bmatrix} \mathbf{r}^{\mathrm{b}} \\ \mathbf{r}^{\mathrm{f}} \end{bmatrix} \tag{8.14}$$

or, in compact notation, as

$$\mathbf{R}^{\mathrm{i}}\mathbf{w}^{\mathrm{i}} = \mathbf{r}^{\mathrm{i}} \qquad \text{i.e.} \qquad \mathbf{w}^{i}_{\mathrm{opt}} = \mathbf{R}_i^{-1}\mathbf{r}_i. \tag{8.15}$$

The minimum energy error is equal to

$$J^{\mathrm{i}}\left(\mathbf{w}^{\mathrm{i}}_{\mathrm{opt}}\right) = \sigma^2_{x[n-i]} - \mathbf{r}^{\mathrm{b}H}\mathbf{w}^{\mathrm{b}} - \mathbf{r}^{\mathrm{f}H}\mathbf{w}^{\mathrm{f}}. \tag{8.16}$$

8.2.1.1 Augmented Normal Equations in the Wiener–Hopf Form

It is possible to formulate the normal equations in *extended notation*, by considering the *extended vectors coefficients* $\mathbf{w} \in (\mathbb{R},\mathbb{C})^{(M+1)\times 1}$ and the *extended sequence* $\mathbf{x} \in (\mathbb{R},\mathbb{C})^{(M+1)\times 1}$, defined as

$$\mathbf{w} = \begin{bmatrix} -\mathbf{w}^{\mathrm{f}H} & 1 & -\mathbf{w}^{\mathrm{b}H} \end{bmatrix}^T \tag{8.17}$$

and

$$\mathbf{x} = \begin{bmatrix} \mathbf{x}^{\mathrm{f}T} & x[n-i] & \mathbf{x}^{\mathrm{b}T} \end{bmatrix}^T \tag{8.18}$$

such that the prediction error (8.2) can be written as

$$e^{\mathrm{i}}[n] = \mathbf{w}^H\mathbf{x}. \tag{8.19}$$

For (8.17) and (8.18), considering the expressions (8.11), and (8.12), we can define the *extended correlation matrix*, expressed with the following partition, as

$$\mathbf{R} = E\{\mathbf{x}\mathbf{x}^H\}$$
$$= \begin{bmatrix} \mathbf{R}^{\mathrm{bb}} & \mathbf{r}^{\mathrm{b}} & \mathbf{R}^{\mathrm{bf}} \\ \mathbf{r}^{\mathrm{b}H} & \sigma^2_{x[n-i]} & \mathbf{r}^{\mathrm{f}H} \\ \mathbf{R}^{\mathrm{bf}H} & \mathbf{r}^{\mathrm{f}} & \mathbf{R}^{\mathrm{ff}} \end{bmatrix}$$
$$= \left[\begin{array}{ccc|c|ccc} r[0] & \cdots & r[i-1] & r[i] & r[i+1] & \cdots & r[M] \\ \vdots & \ddots & \vdots & \vdots & \vdots & \ddots & \vdots \\ r^*[i-1] & \cdots & r[0] & r[1] & r[2] & \cdots & r[M-i+1] \\ \hline r^*[i] & \cdots & r^*[1] & \sigma^2_{x[n-i]} & r^*[1] & \cdots & r^*[M-i] \\ \hline r^*[i+1] & \cdots & r^*[2] & r[1] & r[0] & \cdots & r[M-i-1] \\ \vdots & \ddots & \vdots & \vdots & \vdots & \ddots & \vdots \\ r^*[M] & \cdots & r^*[M-i+1] & r[M-i] & r^*[M-i-1] & \cdots & r[0] \end{array}\right] \tag{8.20}$$

for which the structure of the so-called *augmented normal equations* results in

$$\begin{bmatrix} \mathbf{R}^{\mathrm{bb}} & \mathbf{r}^{\mathrm{b}} & \mathbf{R}^{\mathrm{bf}} \\ \mathbf{r}^{\mathrm{b}H} & \sigma^2_{x[n-i]} & \mathbf{r}^{\mathrm{f}H} \\ \mathbf{R}^{\mathrm{bf}H} & \mathbf{r}^{\mathrm{f}} & \mathbf{R}^{\mathrm{ff}} \end{bmatrix} \begin{bmatrix} -\mathbf{w}^{\mathrm{f}} \\ 1 \\ -\mathbf{w}^{\mathrm{b}} \end{bmatrix} = \begin{bmatrix} \mathbf{0} \\ J^{\mathrm{i}}\left(\mathbf{w}^{\mathrm{i}}_{\mathrm{opt}}\right) \\ \mathbf{0} \end{bmatrix}. \tag{8.21}$$

With the above expression it is possible to determine both the prediction coefficients vector $\mathbf{w}^{\mathrm{i}}_{\mathrm{opt}}$ and the minimum error energy or MMSE $J^{\mathrm{i}}(\mathbf{w}^{\mathrm{i}}_{\mathrm{opt}})$.

Remark For $M = 2\,L$ and $i = L$, the filter is said to be *symmetric linear estimator*. In substance, the estimator is an odd length FIR filter and the signal is estimated by considering a window composed by the L past and L future samples of the sample to be predicted $\hat{x}[n-i]$. The augmented structure of the normal equations allows us to interpret the estimation of a sample inside of the analysis window, as a forward–backward prediction. The window to the left of $x[n-i]$ predicts forward, while that on its right predicts it backwards.

8.2.1.2 Forward Linear Prediction

The one-step forward prediction, commonly called forward linear prediction (FLP), can be developed in (8.2) for $i = 0$, i.e., when $d[n] = x[n]$. For (8.8) the filter output is

$$\hat{x}[n] = \sum_{k=1}^{M} w^{\mathrm{f}*}[k]x[n-k]. \tag{8.22}$$

The estimation error $e^{\mathrm{f}}[n] = x[n] - \hat{x}[n]$ appears to be

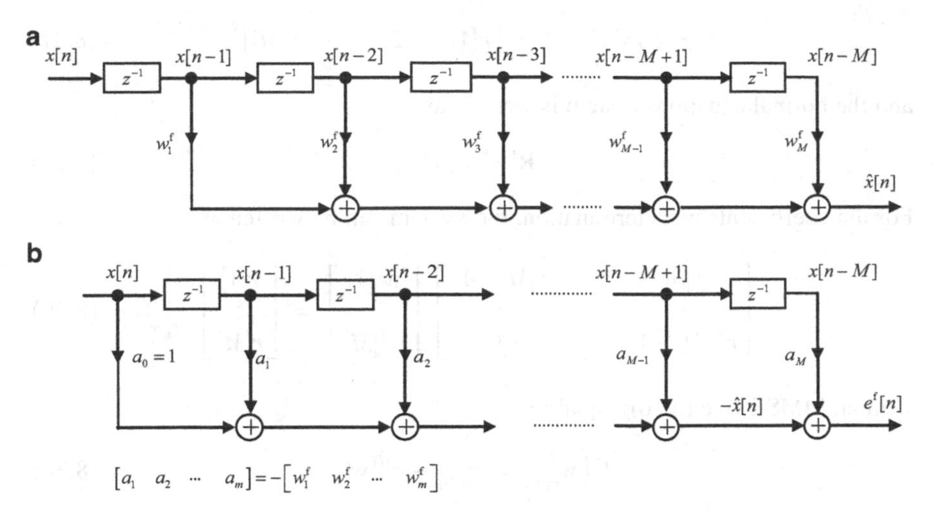

Fig. 8.2 Forward linear prediction: (**a**) one-step forward predictor or *forward predictor*; (**b**) forward error predictor filter

$$
\begin{aligned}
e^{\mathrm{f}}[n] &= x[n] - \sum_{k=1}^{M} w^{\mathrm{f}*}[k]x[n-k] \\
&= x[n] - \mathbf{w}^{\mathrm{f}H}\mathbf{x}^{\mathrm{f}} = \mathbf{a}^{H}\mathbf{x},
\end{aligned}
\tag{8.23}
$$

where the vectors $\mathbf{w}^{\mathrm{f}}$, $\mathbf{x}^{\mathrm{f}}$, $\mathbf{a}$, and $\mathbf{x}$ are defined, respectively, as

$$
\begin{aligned}
\mathbf{w}^{\mathrm{f}} \in (\mathbb{R},\mathbb{C})^{M\times 1} &= \begin{bmatrix} w^{\mathrm{f}}[1] & w^{\mathrm{f}}[2] & \cdots & w^{\mathrm{f}}[M] \end{bmatrix}^{H}, \\
\mathbf{x}^{\mathrm{f}} \in (\mathbb{R},\mathbb{C})^{M\times 1} &= \begin{bmatrix} x[n-1] & x[n-2] & \cdots & x[n-M] \end{bmatrix}^{T}, \\
\mathbf{a} \in (\mathbb{R},\mathbb{C})^{(M+1)\times 1} &= \begin{bmatrix} 1 & -w^{\mathrm{f}*}[1] & \cdots & -w^{\mathrm{f}*}[M] \end{bmatrix}^{T} = \begin{bmatrix} 1 & -\mathbf{w}^{\mathrm{f}H} \end{bmatrix}^{T}, \\
\mathbf{x} \in (\mathbb{R},\mathbb{C})^{(M+1)\times 1} &= \begin{bmatrix} x[n] & \mathbf{x}^{\mathrm{f}T} \end{bmatrix}^{T},
\end{aligned}
\tag{8.24}
$$

and note that $\mathbf{x}^{\mathrm{f}} = \mathbf{x}_{n-1}$. The prediction filter structure is that of Fig. 8.2a.

The CF has the form

$$
\begin{aligned}
J^{\mathrm{f}}(\mathbf{w}) &= E\left\{ \left| e^{\mathrm{f}}[n] \right|^{2} \right\} \\
&= E\left\{ \left| x[n] - \mathbf{w}^{\mathrm{f}H}\mathbf{x}^{\mathrm{f}} \right|^{2} \right\} \\
&= \sigma_{x[n]}^{2} - 2\mathbf{w}^{\mathrm{f}H}\mathbf{r}^{\mathrm{f}} + \mathbf{w}^{\mathrm{f}H}\mathbf{R}^{\mathrm{f}}\mathbf{w}^{\mathrm{f}}.
\end{aligned}
\tag{8.25}
$$

The correlation matrix $\mathbf{R}^{\mathrm{f}}$ is equal to

$$
\mathbf{R}^{\mathrm{f}} \equiv \mathbf{R}_{n-1} = E\{\mathbf{x}^{\mathrm{f}}\mathbf{x}^{\mathrm{f}H}\},
\tag{8.26}
$$

the correlation vector is

$$\mathbf{r}^{\mathrm{f}} = E\{\mathbf{x}^{\mathrm{f}} x^*[n]\} = \begin{bmatrix} r[1] & r[2] & \cdots & r[M] \end{bmatrix}^T \tag{8.27}$$

and the normal equations system is written as

$$\mathbf{R}^{\mathrm{f}}\mathbf{w}^{\mathrm{f}} = \mathbf{r}^{\mathrm{f}}. \tag{8.28}$$

For the coefficients $\mathbf{w}^{\mathrm{f}}$ determination, the system can be written as

$$\begin{bmatrix} r[0] & \cdots & r[M-1] \\ \vdots & \ddots & \vdots \\ r^*[M-1] & \cdots & r[0] \end{bmatrix} \begin{bmatrix} w^{\mathrm{f}}[1] \\ \vdots \\ w^{\mathrm{f}}[M] \end{bmatrix} = \begin{bmatrix} r[1] \\ \vdots \\ r[M] \end{bmatrix} \tag{8.29}$$

with an MMSE (see (8.16)) equal to

$$J^{\mathrm{f}}\left(\mathbf{w}_{\mathrm{opt}}^{\mathrm{f}}\right) = \sigma_{x[n]}^2 - \mathbf{r}^{\mathrm{f}H}\mathbf{w}^{\mathrm{f}}. \tag{8.30}$$

Extended Notation and Prediction Error Filter

From (8.23), the prediction error is equal to $e^{\mathrm{a}}[n] = \mathbf{a}^H\mathbf{x}$ where the coefficients $\mathbf{a}$, as shown in Fig. 8.2b, define the *forward prediction error filter*. The extended correlation, defined in (8.20) for $i = 0$, is rewritten as

$$\mathbf{R} = \left.\left[\begin{array}{ccc|c|ccc} r[0] & \cdots & r[i-1] & r[i] & r[i+1] & \cdots & r[M] \\ \vdots & \ddots & \vdots & \vdots & \vdots & \ddots & \vdots \\ r^*[i-1] & \cdots & r[0] & r[1] & r[2] & \cdots & r[M-i+1] \\ \hline r^*[i] & \cdots & r^*[1] & \sigma_{x[n-i]}^2 & r^*[1] & \cdots & r^*[M-i] \\ \hline r^*[i+1] & \cdots & r^*[2] & r[1] & r[0] & \cdots & r[M-i-1] \\ \vdots & \ddots & \vdots & \vdots & \vdots & \ddots & \vdots \\ r^*[M] & \cdots & r^*[M-i+1] & r[M-i] & r^*[M-i-1] & \cdots & r[0] \end{array}\right]\right|_{i=0} \Rightarrow$$

$$\mathbf{R} = E\{\mathbf{x}^{\mathrm{f}}\mathbf{x}^{\mathrm{f}H}\} = \begin{bmatrix} \sigma_{x[n]}^2 & \mathbf{r}^{\mathrm{f}H} \\ \mathbf{r}^{\mathrm{f}} & \mathbf{R}^{\mathrm{f}} \end{bmatrix} = \left[\begin{array}{c|ccc} \sigma_{x[n]}^2 & r^*[1] & \cdots & r^*[M] \\ \hline r[1] & r[0] & \cdots & r[M-1] \\ \vdots & \vdots & \ddots & \vdots \\ r[M] & r^*[M-1] & \cdots & r[0] \end{array}\right]. \tag{8.31}$$

The augmented normal equations (see (8.21)) assume the form

$$\begin{bmatrix} \sigma_{x[n]}^2 & \mathbf{r}^{\mathrm{f}H} \\ \mathbf{r}^{\mathrm{f}} & \mathbf{R}^{\mathrm{f}} \end{bmatrix} \begin{bmatrix} 1 \\ -\mathbf{w}^{\mathrm{f}} \end{bmatrix} = \begin{bmatrix} J^{\mathrm{f}}\left(\mathbf{w}_{\mathrm{opt}}^{\mathrm{f}}\right) \\ \mathbf{0} \end{bmatrix},$$

i.e.,

$$\mathbf{Ra} = \begin{bmatrix} J^{\mathrm{f}}\left(\mathbf{w}_{\mathrm{opt}}^{\mathrm{f}}\right) \\ \mathbf{0} \end{bmatrix} \tag{8.32}$$

with an MMSE equal to (8.30).

8.2.1.3 Backward Linear Prediction

In the problem of backward linear prediction (BLP), known as the samples of sequence $x[n-M+1]$, $x[n-M+2]$, …, $x[n-1]$, $x[n]$, we want to estimate the signal $x[n-M]$. The (8.1), for $i = M$, is written as

$$y[n] = \hat{x}[n-M] = \sum_{k=0}^{M-1} w^{\mathrm{b}*}[k]x[n-k]. \tag{8.33}$$

The estimation error $e^{\mathrm{b}}[n] = x[n-M] - y[n]$ appears to be

$$\begin{aligned} e^{\mathrm{b}}[n] &= x[n-M] - \sum_{k=0}^{M-1} w^{\mathrm{b}*}[k]x[n-k] \\ &= -\mathbf{w}^{\mathrm{b}H}\mathbf{x}^{\mathrm{b}} + x[n-M] = \mathbf{b}^H\mathbf{x}, \end{aligned} \tag{8.34}$$

where the vectors $\mathbf{w}^{\mathrm{b}}$, $\mathbf{x}^{\mathrm{b}}$, $\mathbf{a}$, and $\mathbf{x}$ are defined, respectively, as

$$\begin{aligned} \mathbf{w}^{\mathrm{b}} \in (\mathbb{R},\mathbb{C})^{M\times 1} &= \begin{bmatrix} w^{\mathrm{b}}[0] & w^{\mathrm{b}}[1] & \cdots & w^{\mathrm{b}}[M-1] \end{bmatrix}^H, \\ \mathbf{x}^{\mathrm{b}} \in (\mathbb{R},\mathbb{C})^{M\times 1} &= \begin{bmatrix} x[n] & x[n-1] & \cdots & x[n-M+1] \end{bmatrix}^T, \\ \mathbf{b} \in (\mathbb{R},\mathbb{C})^{(M+1)\times 1} &= \begin{bmatrix} w^{\mathrm{b}*}[0] & \cdots & w^{\mathrm{b}*}[M-1] & 1 \end{bmatrix}^T = \begin{bmatrix} -\mathbf{w}^{\mathrm{b}H} & 1 \end{bmatrix}^T, \\ \mathbf{x} \in (\mathbb{R},\mathbb{C})^{(M+1)\times 1} &= \begin{bmatrix} \mathbf{x}^{\mathrm{b}T} & x[n-M] \end{bmatrix}^T. \end{aligned} \tag{8.35}$$

The prediction filter structure is that of Fig. 8.3a.

The CF takes the form

$$\begin{aligned} J^{\mathrm{b}}(\mathbf{w}) &= E\left\{\left|e^{\mathrm{b}}[n]\right|^2\right\} \\ &= E\left\{\left|x[n-M] - \mathbf{w}^{\mathrm{b}H}\mathbf{x}^{\mathrm{b}}\right|^2\right\} \\ &= \sigma_{x[n-M]}^2 - 2\mathbf{w}^{\mathrm{b}H}\mathbf{r}^{\mathrm{b}} + \mathbf{w}^{\mathrm{b}H}\mathbf{R}^{\mathrm{b}}\mathbf{w}^{\mathrm{b}}. \end{aligned} \tag{8.36}$$

The correlation matrix $\mathbf{R}^{\mathrm{b}}$ is equal to

$$\mathbf{R}^{\mathrm{b}} \equiv \mathbf{R}_n = E\{\mathbf{x}^{\mathrm{b}}\mathbf{x}^{\mathrm{b}H}\} \tag{8.37}$$

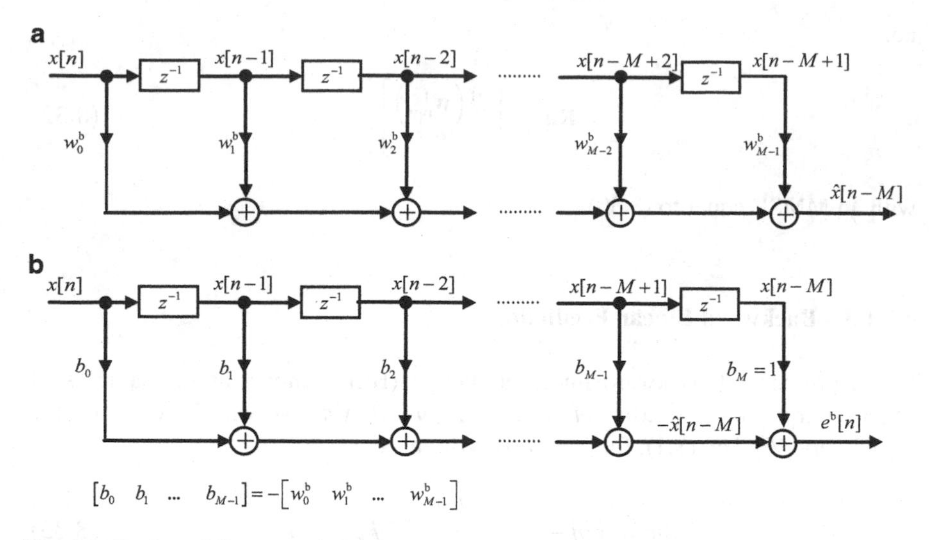

Fig. 8.3 Backward linear prediction: (**a**) one-step backward predictor or *backward predictor*; (**b**) backward error predictor filter

and

$$\mathbf{r}^{\mathrm{b}} = E\{\mathbf{x}^{\mathrm{b}}x^*[n-M]\} = \begin{bmatrix} r[M] & r[M-1] & \cdots & r[1] \end{bmatrix}^T. \tag{8.38}$$

The normal equations system

$$\mathbf{R}^{\mathrm{b}}\mathbf{w}^{\mathrm{b}} = \mathbf{r}^{\mathrm{b}} \tag{8.39}$$

assumes the form

$$\begin{bmatrix} r[0] & \cdots & r[M-1] \\ \vdots & \ddots & \vdots \\ r^*[M-1] & \cdots & r[0] \end{bmatrix} \begin{bmatrix} w^{\mathrm{b}}[0] \\ \vdots \\ w^{\mathrm{b}}[M-1] \end{bmatrix} = \begin{bmatrix} r[M] \\ \vdots \\ r[1] \end{bmatrix} \tag{8.40}$$

with an MMSE equal to (see (8.16))

$$J^{\mathrm{b}}\left(\mathbf{w}_{\mathrm{opt}}^{\mathrm{b}}\right) = \sigma_{x[n-m]}^2 - \mathbf{r}^{\mathrm{b}H}\mathbf{w}^{\mathrm{b}}. \tag{8.41}$$

Extended Notation and Prediction Error Filter

The predictor equations are $e^{\mathrm{b}}[n] = \mathbf{b}^T\mathbf{x}$ for which the *backward prediction error filter* is the one shown in Fig. 8.3b.

The extended correlation defined in (8.20) for $i = M$ is rewritten as

$$\mathbf{R} = \left[\begin{array}{ccc|c|ccc} r[0] & \cdots & r[i-1] & r[i] & r[i+1] & \cdots & r[M] \\ \vdots & \ddots & \vdots & \vdots & \vdots & \ddots & \vdots \\ r^*[i-1] & \cdots & r[0] & r[1] & r[2] & \cdots & r[M-i+1] \\ \hline r^*[i] & \cdots & r^*[1] & \sigma^2_{x[n-i]} & r^*[1] & \cdots & r^*[M-i] \\ \hline r^*[i+1] & \cdots & r^*[2] & r[1] & r[0] & \cdots & r[M-i-1] \\ \vdots & \ddots & \vdots & \vdots & \vdots & \ddots & \vdots \\ r^*[M] & \cdots & r^*[M-i+1] & r[M-i] & r^*[M-i-1] & \cdots & r[0] \end{array}\right]_{i=M} \Rightarrow$$

$$\mathbf{R} = E\{\mathbf{x}^{\mathrm{b}}\mathbf{x}^{\mathrm{b}H}\} = \begin{bmatrix} \mathbf{R}^{\mathrm{b}} & \mathbf{r}^{\mathrm{b}} \\ \mathbf{r}^{\mathrm{b}H} & \sigma^2_{x[n-M]} \end{bmatrix} = \left[\begin{array}{ccc|c} r[0] & \cdots & r[M-1] & r[M] \\ \vdots & \ddots & \vdots & \vdots \\ r^*[M-1] & \cdots & r[0] & r[1] \\ \hline r^*[M] & \cdots & r^*[1] & \sigma^2_{x[n-M]} \end{array}\right] \tag{8.42}$$

so, the augmented normal equations system assumes the form

$$\begin{bmatrix} \mathbf{R}^{\mathrm{b}} & \mathbf{r}^{\mathrm{b}} \\ \mathbf{r}^{\mathrm{b}H} & \sigma^2_{x[n-M]} \end{bmatrix} \begin{bmatrix} -\mathbf{w}^{\mathrm{b}} \\ 1 \end{bmatrix} = \begin{bmatrix} \mathbf{0} \\ J^{\mathrm{b}}\left(\mathbf{w}^{\mathrm{b}}_{\mathrm{opt}}\right) \end{bmatrix} \tag{8.43}$$

in compact notation

$$\mathbf{R}^{\mathrm{b}} = \begin{bmatrix} \mathbf{0} \\ J^{\mathrm{b}}\left(\mathbf{w}^{\mathrm{b}}_{\mathrm{opt}}\right) \end{bmatrix}. \tag{8.44}$$

8.2.1.4 Relationship Between Prediction Coefficients for Stationary Processes

In the case of stationary process the predictors autocorrelation matrices $\mathbf{R}^{\mathrm{f}}$ and $\mathbf{R}^{\mathrm{b}}$ are identical. For which we can write $\mathbf{R} = \mathbf{R}^{\mathrm{f}} = \mathbf{R}^{\mathrm{b}}$ (see (8.31) and (8.42)), i.e.,

$$\begin{aligned} \mathbf{R} &= \begin{bmatrix} \sigma^2_{x[n]} & \mathbf{r}^{\mathrm{f}H} \\ \mathbf{r}^{\mathrm{f}} & \mathbf{R}^{\mathrm{f}} \end{bmatrix} = \begin{bmatrix} \mathbf{R}^{\mathrm{b}} & \mathbf{r}^{\mathrm{b}} \\ \mathbf{r}^{\mathrm{b}H} & \sigma^2_{x[n-M]} \end{bmatrix} \\ &= \left[\begin{array}{c|ccc} r[0] & r^*[1] & \cdots & r^*[M] \\ \hline r[1] & r[0] & \cdots & r[M-1] \\ \vdots & \vdots & \ddots & \vdots \\ r[M] & r^*[M-1] & \cdots & r[0] \end{array}\right] = \left[\begin{array}{ccc|c} r[0] & \cdots & r[M-1] & r[M] \\ \vdots & \ddots & \vdots & \vdots \\ r^*[M-1] & \cdots & r[0] & r[1] \\ \hline r^*[M] & \cdots & r^*[1] & r[0] \end{array}\right]. \end{aligned} \tag{8.45}$$

Let $\mathbf{r}$ the vector defined as

$$\mathbf{r} = \begin{bmatrix} r[1] & r[2] & \cdots & r[M] \end{bmatrix}^H, \tag{8.46}$$

where $r[k] = E\{x[k]x^*[n + k]\}$ for $k = 1, \ldots, M$. Let us define the superscript "B" as the *reverse ordering* or *backward vector arrangement* operator; it is easy to see that the cross-correlation vectors $\mathbf{r}^{\mathrm{f}}$ and $\mathbf{r}^{\mathrm{b}}$ are related by

$$\mathbf{r}^{\mathrm{f}} = \mathbf{r}^*, \tag{8.47}$$

$$\mathbf{r}^{\mathrm{b}} = \mathbf{r}^{\mathrm{f}*B} = \mathbf{r}^B = \mathbf{Pr}, \tag{8.48}$$

whereby from the normal equations (8.28) and (8.39) rewritten (from (8.48)), respectively, as $\mathbf{Rw}^{\mathrm{f}} = \mathbf{r}^*$ and $\mathbf{Rw}^{\mathrm{b}} = \mathbf{r}^B$, we get

$$\mathbf{w}^{\mathrm{b}} = \mathbf{w}^{\mathrm{f}*B} = \mathbf{Pw}^{\mathrm{f}*}, \quad \text{i.e.} \quad \mathbf{b} = \mathbf{a}^{*B} = \mathbf{Pa}^*, \tag{8.49}$$

where $\mathbf{P}$, such that $\mathbf{P}^T\mathbf{P} = \mathbf{PP}^T = \mathbf{I}$, is the *permutation matrix operator* (that implements the reverse ordering), defined as

$$\mathbf{P} = \begin{bmatrix} 0 & 0 & \cdots & 1 \\ \vdots & \vdots & \ddots & \vdots \\ 0 & 1 & \cdots & 0 \\ 1 & 0 & \cdots & 0 \end{bmatrix}, \tag{8.50}$$

for which the forward and backward predictors coefficients are identical but in reverse order. It applies, of course, that the forward and backward error energies are identical

$$J\left(\mathbf{w}_{\mathrm{opt}}\right) \triangleq J^{\mathrm{b}}\left(\mathbf{w}_{\mathrm{opt}}^{\mathrm{b}}\right) = J^{\mathrm{f}}\left(\mathbf{w}_{\mathrm{opt}}^{\mathrm{f}}\right). \tag{8.51}$$

8.2.1.5 Combined and Symmetric Forward–Backward Linear Prediction

A case of particular interest, both theoretical and practical, is illustrated in Fig. 8.4 in which the same time-series window, in time-reversed mode, is used for the one-step forward and backward prediction. This prediction scheme is denoted as combined one-step forward–backward linear prediction (CFBLP).

Another case of interest, illustrated in Fig. 8.5, is denoted as symmetric (one-step) forward–backward linear prediction (SFBLP), in which two analysis windows, related to the same SP, predict the same sample.

In both cases, in order to have a more robust estimate, it is possible to impose a joint parameters measurement that, simultaneously, minimizes the forward and backward errors, i.e., defining a CF of the type

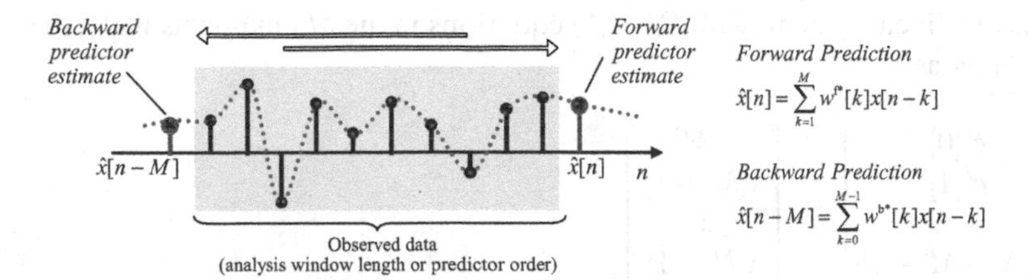

Fig. 8.4 Schematic of the combined one-step forward–backward linear prediction (CFBLP). By using the same time-series window, predict both the one-step forward and backward samples

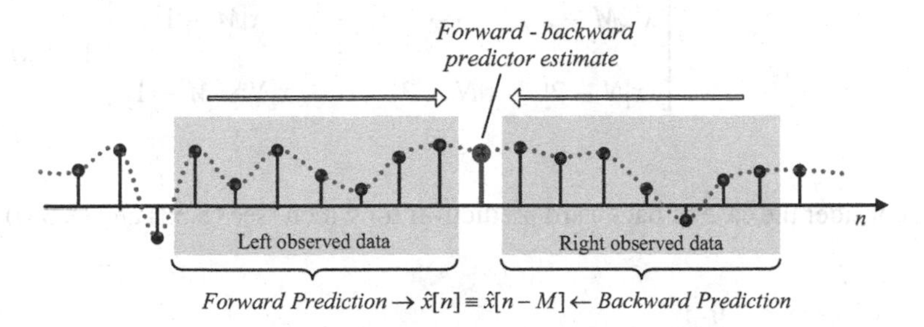

Fig. 8.5 Symmetric forward–backward linear prediction (SFBLP) method. Forward (*from left*) and backward (*from right*) prediction of the same signal sample

$$J^{\text{fb}}(\mathbf{w}) = E\left\{\left|e^{\text{f}}[n]\right|^2 + \left|e^{\text{b}}[n]\right|^2\right\}. \tag{8.52}$$

In the case of stationary processes, the combined/symmetric forward–backward predictor coefficients are conjugate and reversed, i.e., $\mathbf{w}_{\text{opt}}^{\text{b}} = \mathbf{w}_{\text{opt}}^{f*\mathsf{B}}$ $\left(\text{or } \mathbf{b}_{\text{opt}} = \mathbf{a}_{\text{opt}}^{*\mathsf{B}}\right)$.

8.2.2 Forward and Backward Prediction Using LS Approach

Consider the forward prediction previously discussed (see (8.22) and (8.23)), so that it is

$$\hat{x}[n] = \sum_{k=1}^{M} w^{\text{f}*}[k]x[n-k] = \mathbf{w}^{\text{f}H}\mathbf{x}, \qquad \text{for} \quad 0 \le n \le N-1$$

and

$$e^{\text{f}}[n] = x[n] - \hat{x}[n] = x[n] - \mathbf{w}^{\text{f}H}\mathbf{x}^{\text{f}}.$$

Writing the prediction error in explicit form for all the $(N-M)$ samples of the N-length sequence, for the covariance windowing method (see Sect. 4.2.3.1), we have a linear system, with $(N-M)$ equations in the M unknowns $w_f[k]$, that can be written as

$$\begin{bmatrix} e^{\mathrm{f}}[0] \\ e^{\mathrm{f}}[1] \\ \vdots \\ e^{\mathrm{f}}[N-M-1] \end{bmatrix} = \begin{bmatrix} x[M] \\ x[M+1] \\ \vdots \\ x[N-1] \end{bmatrix} - \begin{bmatrix} x[M-1] & x[M-2] & \cdots & x[0] \\ x[M] & x[M-1] & \cdots & x[1] \\ \vdots & \vdots & \ddots & \vdots \\ x[2M-2] & \cdots & \cdots & x[M-1] \\ \vdots & \vdots & \vdots & \vdots \\ x[N-2] & x[N-3] & \cdots & x[N-M-1] \end{bmatrix} \begin{bmatrix} w^{\mathrm{f}}[1] \\ w^{\mathrm{f}}[2] \\ \vdots \\ w^{\mathrm{f}}[M] \end{bmatrix}. \tag{8.53}$$

Now, consider the case of backward prediction for which (see (8.33) and (8.34)) we have

$$\hat{x}[n-M] = \sum_{k=0}^{M-1} w^{\mathrm{b}*}[k]x[n-k] = \mathbf{w}^{\mathrm{b}H}\mathbf{x}, \qquad \text{for} \qquad 0 \le n \le N-1$$

and

$$e^{\mathrm{b}}[n] = x[n-M] - \hat{x}[n-M] = x[n-M] - \mathbf{w}_b^H\mathbf{x}_b.$$

In this case the $(N-M)$ equations in the M unknowns $w_b[k]$ are

$$\begin{bmatrix} e^{\mathrm{b}}[0] \\ e^{\mathrm{b}}[1] \\ \vdots \\ e^{\mathrm{b}}[N-M-1] \end{bmatrix} = \begin{bmatrix} x[0] \\ x[1] \\ \vdots \\ x[N-M-1] \end{bmatrix} - \begin{bmatrix} x[1] & \cdots & x[M-1] & x[M] \\ x[2] & \cdots & x[M] & x[M-1] \\ \vdots & \iddots & \vdots & \vdots \\ x[M] & \cdots & \cdots & x[2M+1] \\ \vdots & \vdots & \vdots & \vdots \\ x[N-M] & \cdots & x[N-2] & x[N-1] \end{bmatrix} \begin{bmatrix} w^{\mathrm{b}}[0] \\ w^{\mathrm{b}}[1] \\ \vdots \\ w^{\mathrm{b}}[M-1] \end{bmatrix}. \tag{8.54}$$

The expressions (8.53) and (8.54) can be written, with obvious meaning of the used symbolism, as

$$\mathbf{e}^{\mathrm{f}} = \mathbf{d}^{\mathrm{f}} - \mathbf{X}^{\mathrm{f}}\mathbf{w}^{\mathrm{f}}, \tag{8.55}$$

$$\mathbf{e}^{\mathrm{b}} = \mathbf{d}^{\mathrm{b}} - \mathbf{X}^{\mathrm{b}}\mathbf{w}^{\mathrm{b}}. \tag{8.56}$$

By minimizing the energy of the prediction errors $E_e^{\mathrm{f}} = \{\mathbf{e}^{\mathrm{f}H}\mathbf{e}^{\mathrm{f}}\}$ and $E_e^{\mathrm{b}} = \{\mathbf{e}^{\mathrm{b}H}\mathbf{e}^{\mathrm{b}}\}$ the coefficient vectors of prediction $\mathbf{w}^{\mathrm{f}}$ and $\mathbf{w}^{\mathrm{b}}$ can be calculated by means of the LS normal equations (in the form of Yule–Walker; see Sect. 4.2.2.2). For which it is

$$\mathbf{w}^{\mathrm{f}} = \left(\mathbf{X}^{\mathrm{f}H}\mathbf{X}^{\mathrm{f}}\right)^{-1}\mathbf{X}^{\mathrm{f}H}\mathbf{d}^{\mathrm{f}}, \tag{8.57}$$

$$\mathbf{w}^{\mathrm{b}} = \left(\mathbf{X}^{\mathrm{b}H}\mathbf{X}^{\mathrm{b}}\right)^{-1}\mathbf{X}^{\mathrm{b}H}\mathbf{d}^{\mathrm{b}}. \tag{8.58}$$

8.2.2.1 Symmetric Forward–Backward Linear Prediction Using LS

In the stationary case, the coefficients forward and backward are identical but in reverse order, and so we can write

$$\mathbf{w} \equiv \mathbf{w}^{\mathrm{f}} = \mathbf{w}^{\mathrm{b}*B}. \tag{8.59}$$

For more robust prediction vector $\mathbf{w}$ estimate, we can think to jointly solve the expressions (8.57) and (8.58). In practice, it is not the single prediction error that is minimized, but their sum

$$E_e^{\mathrm{fb}} = \sum_{n=M}^{N}\left\{\left|e^{\mathrm{f}}[n]\right|^2 + \left|e^{\mathrm{b}}[n]\right|^2\right\} = \mathbf{e}^{\mathrm{f}H}\mathbf{e}^{\mathrm{f}} + \mathbf{e}^{\mathrm{b}H}\mathbf{e}^{\mathrm{b}}. \tag{8.60}$$

This can be interpreted, as illustrated in Fig. 8.5, such as writing the forward predictor (from right) and backward predictor (from the left) of the same sequence for a window of N samples.

In writing the equations, attention must be paid to the indices and formalism. Note that, although the sample to estimate is the same, this is indicated in the forward prediction with $\hat{x}[n]$ while in the backward prediction as $\hat{x}[n-M]$.

8.2.3 Augmented Yule–Walker Normal Equations

By combining the expressions (8.53) and (8.54), we can write a system of $2(N-M)$ equations in M unknowns $\mathbf{w}$, defined as

$$\begin{bmatrix} e^{\mathrm{f}}[0] \\ e^{\mathrm{f}}[1] \\ \vdots \\ e^{\mathrm{f}}[N-M-1] \\ e^{\mathrm{b}}[N-M-1] \\ \vdots \\ e^{\mathrm{b}}[1] \\ e^{\mathrm{b}}[0] \end{bmatrix} = \begin{bmatrix} x^{\mathrm{f}}[M] \\ \vdots \\ x^{\mathrm{f}}[N-1] \\ x^{\mathrm{b}}[N-M-1] \\ \vdots \\ x^{\mathrm{b}}[1] \\ x^{\mathrm{b}}[0] \end{bmatrix}$$
$$- \begin{bmatrix} x^{\mathrm{f}}[M-1] & x^{\mathrm{f}}[M-2] & \cdots & x^{\mathrm{f}}[0] \\ x^{\mathrm{f}}[M] & x^{\mathrm{f}}[M-1] & \cdots & x^{\mathrm{f}}[1] \\ \vdots & \vdots & \ddots & \vdots \\ x^{\mathrm{f}}[N-2] & x^{\mathrm{f}}[N-3] & \cdots & x^{\mathrm{f}}[N-M-1] \\ x^{\mathrm{b}}[N-M] & \cdots & x^{\mathrm{b}}[N-2] & x^{\mathrm{b}}[N-1] \\ \vdots & \cdots & \vdots & \vdots \\ x^{\mathrm{b}}[2] & \iddots & x^{\mathrm{b}}[M] & x^{\mathrm{b}}[M-1] \\ x^{\mathrm{b}}[1] & \cdots & x^{\mathrm{b}}[M-1] & x^{\mathrm{b}}[M] \end{bmatrix} \begin{bmatrix} w[1] \\ w[2] \\ \vdots \\ w[M] \end{bmatrix} \tag{8.61}$$

for which with the same number of unknowns, the number of equations is doubled. For stationary process, the estimate is more robust because the measurement error is averaged over a larger window. In compact form, with obvious symbolism, we can write the previous expression as

$$\mathbf{X}^H\mathbf{X}\mathbf{w} = \mathbf{X}^H\mathbf{d}. \tag{8.62}$$

Recalling that the optimal solution is the one with minimum error, so by (4.22) it is

$$E_{\min} \equiv \hat{J}(\mathbf{w}) = \mathbf{d}^H\mathbf{d} - \mathbf{d}^H\mathbf{X}\left(\mathbf{X}^H\mathbf{X}\right)^{-1}\mathbf{X}^H\mathbf{d}. \tag{8.63}$$

We can derive the augmented LS normal equations as

$$\begin{bmatrix} \mathbf{d}^H\mathbf{d} & \mathbf{d}^H\mathbf{X} \\ \mathbf{X}^H\mathbf{d} & \mathbf{X}^H\mathbf{X} \end{bmatrix} \begin{bmatrix} 1 \\ -\mathbf{w} \end{bmatrix} = \begin{bmatrix} E_{\min} \\ \mathbf{0} \end{bmatrix}. \tag{8.64}$$

Calling $\mathbf{\Phi} \in (\mathbb{R},\mathbb{C})^{(M+1)\times(M+1)}$ the Hermitian matrix in expression (8.64), the solution of system (8.64) determines a robust estimation of the M-order prediction error filter parameters of the type already illustrated in the previous sections.

Therefore, by defining the linear prediction coefficients vector as $\mathbf{a} \in (\mathbb{R}, \mathbb{C})^{(M+1)\times 1} = \begin{bmatrix} 1 & -\mathbf{w}^{fT} \end{bmatrix}^T$, the previous expression is generally written as

$$\mathbf{\Phi a} = \begin{bmatrix} E_{\min} \\ \mathbf{0} \end{bmatrix}, \tag{8.65}$$

where $\mathbf{\Phi}$ is denoted as *augmented correlation matrix* which is a persymmetric matrix (i.e., such that $\phi_{i,j} = \phi^*_{M+1-i,M+1-j}$) and for its inversion $O(M^2)$ order algorithms exist [4, 5]. In fact, note that the LS solution can be obtained using the LDL Cholesky decomposition (see Sect. 4.4.1).

8.2.4 Spectral Estimation of a Linear Random Sequence

The LS methods, as already indicated in Chap. 4, are based on a deterministic CF interpretation and on a precise stochastic model that characterizes the signal. From the theory of stochastic models (see Appendix C), a linear stochastic process is defined as the output of a LTI DT circuit, with a certain $H(z)$, when the input is a WGN $\eta[n]$, as illustrated in Fig. 8.6, where, without loss of generality, we assume $a_0 = 1$.

In the case where the model $H(z)$ is a FIR filter, which performs a weighted average of a certain time window of the input signal, the model is called moving average (MA). If the $H(z)$ is an all-pole IIR filter, for which the filter output depends only on the current input and the delayed outputs, the model is said autoregressive (AR). Finally, if there are poles and zeros, you would have the extended model called autoregressive moving average (ARMA). Calling q and p, respectively, the degree of the polynomial in the numerator and the denominator of $H(z)$, the order of the model is usually shown in brackets, for example, as ARMA(p, q).

Since the noise spectrum is white by definition, it follows that the spectral characteristics of the random sequence $x[n]$ at the filter output coincide with the spectral characteristics of the filter TF [24, 26, 27, 29]. Then, the estimate of TF $H(z)$ coincides with the $x[n]$ spectrum estimate. In practice, remembering that the *power spectral density* (PSD) of a sequence is equal to the DTFT of its correlation sequence, we have that, for a linear random process $x[n]$, described by the model of Fig. 8.6, with an autocorrelation $r_{xx}[n]$ such that $R_{xx}(e^{j\omega}) = \text{DTFT}\left(r_{xx}[n]\right)$, we get

ARMA(p,q) spectrum

$$\begin{aligned} R_{xx}\left(e^{j\omega}\right) &= \sigma_\eta^2 \left| H(z)\big|_{z=e^{j\omega}} \right|^2 \\ &= \sigma_\eta^2 \frac{\left| b_0 + b_1 e^{-j\omega} + b_2 e^{-j2\omega} + \cdots + b_q e^{-jq\omega} \right|^2}{\left| 1 + a_1 e^{-j\omega} + a_2 e^{-j2\omega} + \cdots + a_p e^{-jp\omega} \right|^2}. \end{aligned} \tag{8.66}$$

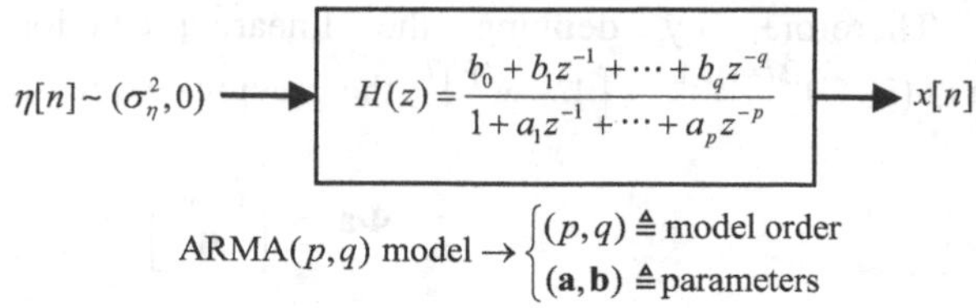

Fig. 8.6 Scheme for generating a linear random sequence $x[n]$

$$\eta[n] \sim (\sigma_\eta^2, 0) \rightarrow H(z) = \frac{b_0 + b_1 z^{-1} + \cdots + b_q z^{-q}}{1 + a_1 z^{-1} + \cdots + a_p z^{-p}} \rightarrow x[n]$$

$$\text{ARMA}(p,q)\text{ model} \rightarrow \begin{cases} (p,q) \triangleq \text{model order} \\ (\mathbf{a},\mathbf{b}) \triangleq \text{parameters} \end{cases}$$

MA(q) *spectrum*

$$R_{xx}\left(e^{j\omega}\right) = \sigma_\eta^2\left|b_0 + b_1 e^{-j\omega} + b_2 e^{-j2\omega} + \cdots + b_q e^{-jM\omega}\right|^2. \tag{8.67}$$

AR(p) *spectrum*

$$R_{xx}\left(e^{j\omega}\right) = \frac{\sigma_\eta^2}{\left|1 + a_1 e^{-j\omega} + a_2 e^{-j2\omega} + \cdots + a_p e^{-jp\omega}\right|^2}. \tag{8.68}$$

The model parameters estimation is therefore equivalent to the signal spectral estimate.

One of the central problems in the estimation of the linear random sequences parameters consists in choosing the correct model order. Typically, this is determined on the base of *a priori* known signal characteristics. However, in case these are not known, there are some (more or less empirical) criteria for determining that order. Note, also, that in the literature there are many estimators which work more or less accurately in dependence on the known sequence characteristics (length, statistic measurement noise, order, etc.).

In expressions (8.66), (8.67), and (8.68), for the correct spectrum scaling, it is also necessary to know the noise variance σ_η^2. In case of using an estimator based on the augmented normal equations as, for example, (8.65), the estimator would provide at the same time both the prediction filter coefficients and the error energy estimation which, of course, coincides with the noise variance.

8.2.5 Linear Prediction Coding of Speech Signals

One of the most powerful methods for the speech signal treatment, used in many real applications, is the linear prediction coding (LPC) [1, 6–8]. This methodology is predominant for the estimation of many speech signal fundamental parameters such as, for example, the fundamental frequency or pitch, the formants frequencies, the vocal tract modeling, etc. This method allows, in addition, also an efficient compressed speech signal encoding, with very low bit rate.

The general structure of the technique is illustrated in Fig. 8.7. The left part of the figure shows the *source coding*, while the right part reports the *source decoding*.

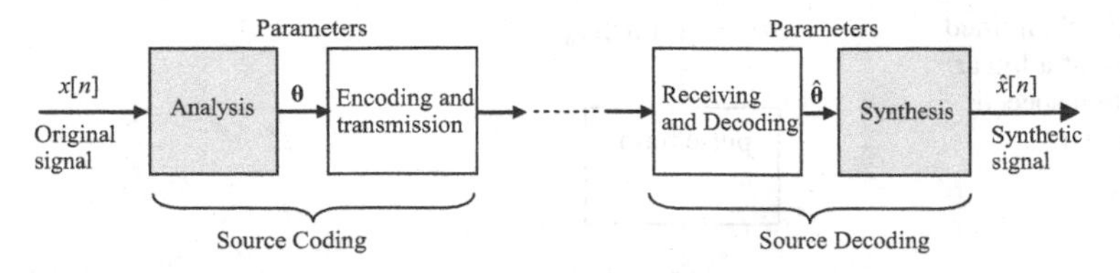

Fig. 8.7 General synthesis-by-analysis scheme. If the parameters estimate is correctly carried out and the parameters are sent to the synthesizer without further processing (compression etc.) the synthetic signal coincides with the original one

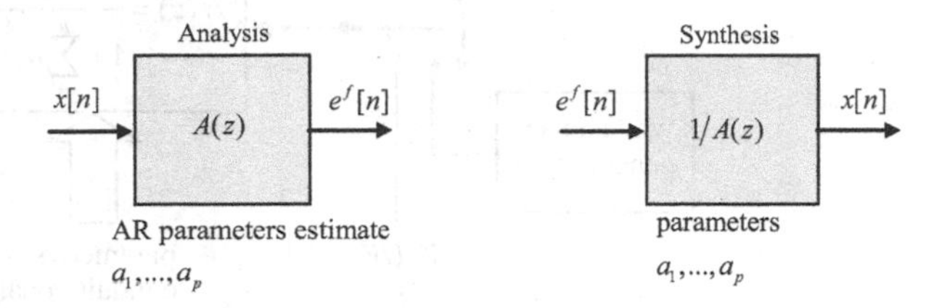

Fig. 8.8 Speech signals analysis–synthesis with AR model

The LPC speech encoding is based on a linear predictor, by means of which are estimated the filter parameters vector and the prediction error. As shown in Fig. 8.8, the speech synthesis is performed with the all-pole inverse filter, by feeding the inverse filter with the error signal.

In practice, the LPC technique is used for low-rate voice transmission (<2.4 kbit/s); for example, in the GSM it is used at 13.3 kbit/s. In the analysis phase is used an estimator that allows the estimation of both the model parameters and the signal variance which, in the context of LPC, is referred to as *gain* $G = \sigma^2_{x[n]}$. To decrease the number of transmitted bits, the error signal is (sometimes) not transmitted. The excitation signal, required for decoding, is generated directly at the synthesis side, based on some speech signal statistical characteristics as: discrimination of voiced/unvoiced sound and, in the case of voiced sounds, the fundamental frequency or pitch. The parameters of the synthesizer are updated approximately every 4–6 ms while the length of the analysis window is long, typically 15–30 ms. A simplified scheme of the LPC synthesizer is illustrated in Fig. 8.9.

In the analysis phase, it is necessary to determine, in addition to the model parameters, also other parameters such as the pitch and the voiced/unvoiced (V/UV) bit decision.

Note that the analysis and synthesis filters implementation almost never happen in the direct form.

In the case of vocal signal, is often used a *lattice structure* whose parameters, called reflection coefficients or PARCOR k_n, have the following property of: (1) directly representing the lossless model of the acoustic tubes representing the vocal tract, (2) determining a stable filter when $|k_m| < 1$, (3) being easily interpolated keeping the stable filter, (4) being easily calculated (for example with the

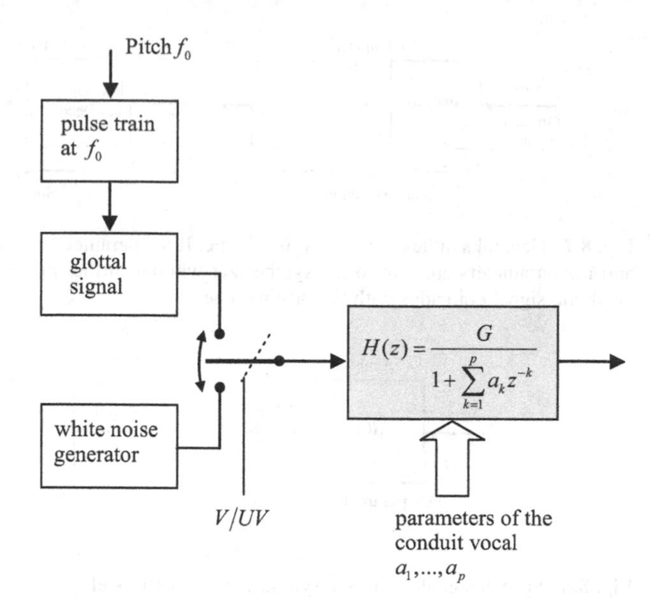

Fig. 8.9 Simplified diagram of a linear prediction speech synthesizer

Levinson algorithm described above), and (5) ensuring a minimum phase filter for which the inverse filter exists and is also stable and minimum phase.

Remark In MATLAB there is a function `A = LPC(X,p)` which allows the determination of the coefficients $\mathbf{a} = [1\ a_1\ \ldots\ a_p]^T$, of a p-order forward predictor such that

$$\hat{x}[n] = -a_1x[n-1] - a_2x[n-2] - \cdots - a_Mx[n-p].$$

The `X` variable can be either a vector or a matrix, in case it is a matrix containing the separated signals in each column and `[A, E] = LPC(X,p)` returns the estimated model of each column in each row of `A`, while `E` returns the variance of the prediction error (power of error). The LPC function uses the Levinson–Durbin algorithm to solve the normal equations that arise from the LS formulation with the autocorrelation method.

8.3 Recursive in Model Order Algorithms

In numerical methods to increase robustness and to reduce the computational complexity, one of the most used paradigms consists in the definition of a recursive mode for determining the solution of the given problem. In mathematics, the term indicated as *recursive solution* is an approach where, relative to a certain domain, the *current solution* is dependent on another solution in the neighborhood. In other words indicating the current solution $\mathbf{w}_k$, this is a function of solutions belonging to its neighborhood. Formally, $\mathbf{w}_k = h(\mathbf{w}_{k-1}, \ldots, \mathbf{w}_{k-p})$ where k is an index defined in a certain domain (such as time, space, order, etc.), and p is an index which defines the depth of the neighborhood and h is the estimator that, most of the times, is a linear MMSE (see Sect. C.3.2.8).

Note that, as seen in Chap. 6, the term *adaptive filtering* means precisely the optimal filtering implemented with recursive numerical methods defined in the time domain. In this case the current solution is a function of the past solutions.

In this section some recursive methods in which the domain of recurrence is the filter order are presented. In this case, the solution of order m is estimated starting from that of order $m - 1$, proceeding iteratively until it reaches the maximum filter order M. This type of recursive procedure, which defines the class of *recursive-in-model-order adaptive filter algorithms* or simply *recursive order filter* (ROF), is typically realized considering some algebraic–geometric properties of the correlation matrix as, for example, its Toeplitz structure. The recursive approach for the solution of the optimal filtering also presents a number of important properties which make it attractive in many application contexts. The ROFs are, in fact, important as they may allow (1) series–parallel algorithm decomposition, (2) a proper choice of the filter order, and (3) the production of intermediate results with particular physical mathematical evidence that allow a run-time evaluation of a certain filter properties (stability, minimum phase, order, etc.). Moreover, they sometimes permit, concerning the problem, to establish a suitable circuit architecture, for example, more adequate for hardware implementation.

8.3.1 Partitioned Matrix Inversion Lemma

The partitioned matrix inversion lemma allows the recursive inverse computation of a partitioned Hermitian matrix [9, 10]. Calling $\mathbf{R}_m \in (\mathbb{R},\mathbb{C})^{m\times m}$ a m-order matrix, the $(m + 1)$-order partitioned matrix $\mathbf{R}_{m+1} \in (\mathbb{R},\mathbb{C})^{(m+1)\times(m+1)}$, defined as

$$\mathbf{R}_{m+1} = \begin{bmatrix} \mathbf{R}_m & \mathbf{r}_m^{\mathrm{b}} \\ \mathbf{r}_m^{\mathrm{b}H} & \rho_m^{\mathrm{b}} \end{bmatrix} \tag{8.69}$$

admits the inverse $\mathbf{R}_{m+1}^{-1}$ which, by definition, is also Hermitian of the type

$$\mathbf{R}_{m+1}^{-1} = \begin{bmatrix} \mathbf{Q}_m & \mathbf{q}_m \\ \mathbf{q}_m^H & q_m \end{bmatrix} \tag{8.70}$$

such that

$$\begin{bmatrix} \mathbf{R}_m & \mathbf{r}_m^{\mathrm{b}} \\ \mathbf{r}_m^{\mathrm{b}H} & \rho_m^{\mathrm{b}} \end{bmatrix} \begin{bmatrix} \mathbf{Q}_m & \mathbf{q}_m \\ \mathbf{q}_m^H & q_m \end{bmatrix} = \begin{bmatrix} \mathbf{I}_m & \mathbf{0}_m \\ \mathbf{0}_m^T & 1 \end{bmatrix}. \tag{8.71}$$

To determine a recursive inversion formula, we express the $\mathbf{Q}_m$ matrix terms as a function of the known $\mathbf{R}_m$ matrix terms. From the product (8.71) we get

$$\mathbf{R}_m\mathbf{Q}_m + \mathbf{r}_m^{\mathrm{b}}\mathbf{q}_m^H = \mathbf{I}_m, \tag{8.72}$$

$$\mathbf{r}_m^{\mathrm{b}H}\mathbf{Q}_m + \rho_m^{\mathrm{b}}\mathbf{q}_m^H = \mathbf{0}_m^T, \tag{8.73}$$

$$\mathbf{R}_m\mathbf{q}_m + \mathbf{r}_m^{\mathrm{b}}q_m = \mathbf{0}_m, \tag{8.74}$$

$$\mathbf{r}_m^{\mathrm{b}H}\mathbf{q}_m + \rho_m^{\mathrm{b}}q_m = 1. \tag{8.75}$$

From (8.74) we get

$$\mathbf{q}_m = -\mathbf{R}_m^{-1}\mathbf{r}_m^{\mathrm{b}}q_m. \tag{8.76}$$

From (8.75) and from the previous, we can write

$$q_m = \frac{1}{\rho_m^{\mathrm{b}} - \mathbf{r}_m^{\mathrm{b}H}\mathbf{R}_m^{-1}\mathbf{r}_m^{\mathrm{b}}}. \tag{8.77}$$

For $(\rho_m^{\mathrm{b}} - \mathbf{r}_m^{bH}\mathbf{R}_m^{-1}\mathbf{r}_m^{\mathrm{b}}) \neq 0$ we get

$$\mathbf{q}_m = \frac{-\mathbf{R}_m^{-1}\mathbf{r}_m^{\mathrm{b}}}{\rho_m^{\mathrm{b}} - \mathbf{r}_m^{\mathrm{b}H}\mathbf{R}_m^{-1}\mathbf{r}_m^{\mathrm{b}}}, \tag{8.78}$$

which replaced in (8.72)

$$\begin{aligned}\mathbf{Q}_m &= \mathbf{R}_m^{-1} + \mathbf{R}_m^{-1}\mathbf{r}_m^{\mathrm{b}}\mathbf{q}_m^H\\ &= \mathbf{R}_m^{-1} + \frac{\mathbf{R}_m^{-1}\mathbf{r}_m^{\mathrm{b}}\left(\mathbf{R}_m^{-1}\mathbf{r}_m^{\mathrm{b}}\right)^H}{\rho_m^{\mathrm{b}} - \mathbf{r}_m^{\mathrm{b}H}\mathbf{R}_m^{-1}\mathbf{r}_m^{\mathrm{b}}}.\end{aligned} \tag{8.79}$$

From the previous development, we see that the inverse $\mathbf{R}_{m+1}^{-1}$ can be expressed in terms of known quantities. For a more suitable notation (see (8.35)), we define the quantities

$$\mathbf{w}_m^{\mathrm{b}} \in (\mathbb{R}, \mathbb{C})^{m\times 1} \triangleq \left[w_m^{\mathrm{b}}[0] \quad w_m^{\mathrm{b}}[1] \quad \cdots \quad w_m^{\mathrm{b}}[m-1]\right]^H = \mathbf{R}_m^{-1}\mathbf{r}_m^{\mathrm{b}}, \tag{8.80}$$

$$\alpha_m^{\mathrm{b}} \triangleq \rho_m^{\mathrm{b}} - \mathbf{r}_m^{\mathrm{b}H}\mathbf{R}_m^{-1}\mathbf{r}_m^{\mathrm{b}} = \rho_m^{\mathrm{b}} - \mathbf{r}_m^{\mathrm{b}H}\mathbf{w}_m^{\mathrm{b}}. \tag{8.81}$$

If the matrix $\mathbf{R}_m$ is invertible and $\alpha_m^{\mathrm{b}} \neq 0$, the (8.79) can be rewritten as

$$\mathbf{Q}_m = \mathbf{R}_m^{-1} + \frac{1}{\alpha_m^{\mathrm{b}}}\mathbf{w}_m^{\mathrm{b}}\mathbf{w}_m^{\mathrm{b}H}, \tag{8.82}$$

whereby

$$\mathbf{R}_{m+1}^{-1} = \begin{bmatrix} \mathbf{Q}_m & \mathbf{q}_m \\ \mathbf{q}_m^H & q_m \end{bmatrix}$$

$$= \begin{bmatrix} \mathbf{R}_m^{-1} + \dfrac{1}{\alpha_m^{\mathrm{b}}} \mathbf{w}_m^{\mathrm{b}} \mathbf{w}_m^{\mathrm{b}H} & -\dfrac{\mathbf{w}_m^{\mathrm{b}}}{\alpha_m^{\mathrm{b}}} \\ -\dfrac{\mathbf{w}_m^{\mathrm{b}H}}{\alpha_m^{\mathrm{b}}} & \dfrac{1}{\alpha_m^{\mathrm{b}}} \end{bmatrix}$$

$$= \begin{bmatrix} \mathbf{R}_m^{-1} & \mathbf{0}_m \\ \mathbf{0}_m^T & 0 \end{bmatrix} + \frac{1}{\alpha_m^{\mathrm{b}}} \begin{bmatrix} \mathbf{w}_m^{\mathrm{b}} \mathbf{w}_m^{\mathrm{b}H} & -\mathbf{w}_m^{\mathrm{b}} \\ -\mathbf{w}_m^{\mathrm{b}H} & 1 \end{bmatrix}$$

$$= \begin{bmatrix} \mathbf{R}_m^{-1} & \mathbf{0}_m \\ \mathbf{0}_m^T & 0 \end{bmatrix} + \frac{1}{\alpha_m^{\mathrm{b}}} \begin{bmatrix} -\mathbf{w}_m^{\mathrm{b}} \\ 1 \end{bmatrix} \begin{bmatrix} -\mathbf{w}_m^{\mathrm{b}H} & 1 \end{bmatrix} \tag{8.83}$$

and, from the definition (8.35), we have

$$\mathbf{R}_{m+1}^{-1} = \begin{bmatrix} \mathbf{R}_m^{-1} & \mathbf{0}_m \\ \mathbf{0}_m^T & 0 \end{bmatrix} + \frac{1}{\alpha_m^{\mathrm{b}}} \mathbf{b}_m \mathbf{b}_m^H \tag{8.84}$$

that, from $\mathbf{R}_m^{-1}$, allows the recursive computation of the $\mathbf{R}_{m+1}^{-1}$ matrix.(8.83) (or (8.84)) is also known as *partitioned matrix inversion lemma* [9]. It also demonstrates that the term α_m^{b} is equal to the determinants ratio

$$\alpha_m^{\mathrm{b}} = \frac{\det \mathbf{R}_{m+1}}{\det \mathbf{R}_m}. \tag{8.85}$$

It is shown, proceeding similarly to the previous mode, that we can write

$$\mathbf{R}_{m+1}^{-1} = \begin{bmatrix} \rho_m^{\mathrm{f}} & \mathbf{r}_m^{\mathrm{f}H} \\ \mathbf{r}_m^{\mathrm{f}} & \mathbf{R}_m \end{bmatrix}^{-1} = \begin{bmatrix} 0 & \mathbf{0}^T \\ \mathbf{0} & \mathbf{R}_m^{-1} \end{bmatrix} + \frac{1}{\alpha_m^{\mathrm{f}}} \begin{bmatrix} 1 \\ -\mathbf{w}_m^{\mathrm{f}} \end{bmatrix} \begin{bmatrix} 1 & -\mathbf{w}_m^{\mathrm{f}H} \end{bmatrix}, \tag{8.86}$$

where

$$\mathbf{w}_m^{\mathrm{f}} \in (\mathbb{R}, \mathbb{C})^{m \times 1} \triangleq \begin{bmatrix} w_m^{\mathrm{f}}[1] & w_m^{\mathrm{f}}[2] & \cdots & w_m^{\mathrm{f}}[m] \end{bmatrix}^H = \mathbf{R}_{m,n-1}^{-1} \mathbf{r}_m^{\mathrm{f}}, \tag{8.87}$$

$$\alpha_m^{\mathrm{f}} \triangleq \rho_m^{\mathrm{f}} - \mathbf{r}_m^{\mathrm{f}H} \mathbf{R}_m^{-1} \mathbf{r}_m^{\mathrm{f}} = \rho_m^{\mathrm{f}} - \mathbf{r}_m^{\mathrm{f}H} \mathbf{w}_m^{\mathrm{f}}. \tag{8.88}$$

8.3.2 Recursive Order Adaptive Filters

In the algorithms developed in the previous sections the order of the estimator is assumed known and *a priori* fixed. For this reason they are often referred to as fixed-order algorithms. In the case in which the order itself becomes a variable, as in the *recursive order algorithms*, the notation must also take into account the index order. Then, in the usual notation is added an index m defined in the recurrence order domain such that the input sequence can be indicated as

$$\begin{aligned}
\mathbf{x}_{m+1,n} &\in (\mathbb{R},\mathbb{C})^{(m+1)\times 1} \triangleq \begin{bmatrix} x[n] & x[n-1] & \cdots & x[n-m] \end{bmatrix}^T \\
\mathbf{x}_{m,n} &\in (\mathbb{R},\mathbb{C})^{m\times 1} \triangleq \begin{bmatrix} x[n] & x[n-1] & \cdots & x[n-m+1] \end{bmatrix}^T \\
\mathbf{x}_{m+1,n-1} &\in (\mathbb{R},\mathbb{C})^{(m+1)\times 1} \triangleq \begin{bmatrix} x[n-1] & x[n-2] & \cdots & x[n-m-1] \end{bmatrix}^T \\
\mathbf{x}_{m,n-1} &\in (\mathbb{R},\mathbb{C})^{m\times 1} \triangleq \begin{bmatrix} x[n-1] & x[n-2] & \cdots & x[n-m] \end{bmatrix}^T \\
&\vdots
\end{aligned} \tag{8.89}$$

If the time index is omitted, it is considered equal to n. In case that it is necessary to explicitly define the time dependence, the vector is indicated as $\mathbf{x}_{m,n}$.

In fact, considering the CFBLP scenario, in the ROF estimation the filter parameters of order $m+1$ is performed starting from the estimate of order m. Therefore, it appears that also the additional observation $x[n-m]$ must be added to the input vector. Therefore, for the mathematical development it is useful to consider the input vector $\mathbf{x}_{m+1,n}$ partitioned as

$$\mathbf{x}_{m+1,n} \triangleq \mathbf{x}_{m+1,n}^{\lceil m \rceil} \left\{ \begin{bmatrix} x[n] \\ x[n-1] \\ \vdots \\ x[n-m+1] \\ x[n-m] \end{bmatrix} \right\} \mathbf{x}_{m+1,n}^{\lfloor m \rfloor}, \tag{8.90}$$

where $\mathbf{x}_{m+1,n}^{\lceil m \rceil}$ and $\mathbf{x}_{m+1,n}^{\lfloor m \rfloor}$ are, respectively, the first and the last m samples of the $\mathbf{x}_{m+1,n}$ vector. In other words, $\mathbf{x}_{m+1,n}^{\lceil m \rceil} = \mathbf{x}_{m,n}$ and $\mathbf{x}_{m+1,n}^{\lfloor m \rfloor} = \mathbf{x}_{m,n-1}$ represent the one sample shifted versions of the sequence $\mathbf{x}_{m,n}$, for which we can write

$$\mathbf{x}_{m+1,n} = \begin{bmatrix} \mathbf{x}_{m,n} \\ x[n-m] \end{bmatrix} = \begin{bmatrix} x[n] \\ \mathbf{x}_{m,n-1} \end{bmatrix}. \tag{8.91}$$

Indicating with $\mathbf{w}_m$ the filter parameters vector of order m

$$\mathbf{w}_m \in (\mathbb{R},\mathbb{C})^{m\times 1} \triangleq \begin{bmatrix} w_{m,0} & w_{m,1} & \cdots & w_{m,m-1} \end{bmatrix}^H \tag{8.92}$$

we can write the relationships between vectors with the following equivalent notations:

$$e_m[n] \triangleq d[n] - y_m[n], \qquad \textit{mth order error}, \tag{8.93}$$

$$y_m[n] = \mathbf{w}_{m,n}^H \mathbf{x}_{m,n}, \qquad \textit{output at the time n for the order m}, \tag{8.94}$$

$$y_m[n-1] = \mathbf{w}_{m,n-1}^H \mathbf{x}_{m,n-1}, \qquad \textit{output at n} - \textit{1 for the order m}. \tag{8.95}$$

For the correlation matrix, recalling the definitions (8.31) and (8.42), we can write

$$\mathbf{R}_{m+1}^{\mathrm{f}} = E\left\{\begin{bmatrix} x[n] \\ \mathbf{x}_{m,n-1} \end{bmatrix}\begin{bmatrix} x^*[n] & \mathbf{x}_{m,n-1}^H \end{bmatrix}\right\} = \begin{bmatrix} \sigma_{x[n]}^2 & \mathbf{r}_m^{\mathrm{f}H} \\ \mathbf{r}_m^{\mathrm{f}} & \mathbf{R}_{m,n-1} \end{bmatrix}, \tag{8.96}$$

$$\mathbf{R}_{m+1}^{\mathrm{b}} = E\left\{\begin{bmatrix} \mathbf{x}_{m,n} \\ x[n-m] \end{bmatrix}\begin{bmatrix} \mathbf{x}_{m,n}^H & x^*[n-m] \end{bmatrix}\right\} = \begin{bmatrix} \mathbf{R}_m & \mathbf{r}_m^{\mathrm{b}} \\ \mathbf{r}_m^{\mathrm{b}H} & \sigma_{x[n-m]}^2 \end{bmatrix}, \tag{8.97}$$

in which the correlation vectors are defined as $\mathbf{r}_m^{\mathrm{f}} = E\{\mathbf{x}_{m,n-1}x^*[n]\}$ and $\mathbf{r}_m^{\mathrm{b}} = E\{\mathbf{x}_{m,n}x^*[n-m]\}$

Remark In case the vector $\mathbf{w}_{m,n}$ is already available, the calculation of the recursive estimation of order $m + 1$ ($\mathbf{w}_{m+1,n}$) starting from $\mathbf{w}_{m,n}$ would allow a high computational saving. In a similar way, as we shall see below, it is possible to develop a time domain recursive algorithms for which starting from $\mathbf{w}_{m,n-1}$, one calculates the estimate at the following instant $\mathbf{w}_{m,n}$.

Note, also, that the combination recurrences, in time n and in the order m, can coexist. This coexistence plays an important role in the development and implementation of fast and robust methodologies, and it is of central importance in adaptive filtering.

8.3.3 Levinson–Durbin Algorithm

A first example of fast and robust ROF algorithm, used in many real applications, is that we exploit thc Hermitian–Toeplitz symmetry of the correlation matrix for the normal equations solution. The solution proposed by Norman Levinson in 1947 and improved by Durbin in 1960 (see, for example, [6, 11, 12]) is of complexity $O(n^2)$, while the solution with Gauss elimination is of complexity $O(n^3)$.

The Levinson–Durbin algorithm (LDA) is a recursive procedure, which belongs to the ROF family, for the calculation of the solution of a linear equations system with Toeplitz coefficients matrix. Starting from the order $m - 1$, the estimator calculates the order m and so on up to order M. The calculation method is developed considering the combined forward and backward prediction filter coefficients of order m as a linear combination of the $m - 1$ order vectors. Therefore, we have that $\mathbf{a}_m \leftarrow (\mathbf{a}_{m-1},\mathbf{b}_{m-1})$ and $\mathbf{b}_m \leftarrow (\mathbf{b}_{m-1},\mathbf{a}_{m-1})$.

The algorithm can be developed in scalar or vector form. In vector form the recursion is defined as

$$\left.\begin{aligned} \mathbf{a}_m &= \begin{bmatrix} \mathbf{a}_{m-1} \\ 0 \end{bmatrix} + k_m^{\mathrm{f}}\begin{bmatrix} 0 \\ \mathbf{b}_{m-1} \end{bmatrix} \\ \mathbf{b}_m &= \begin{bmatrix} 0 \\ \mathbf{b}_{m-1} \end{bmatrix} + k_m^{\mathrm{b}}\begin{bmatrix} \mathbf{a}_{m-1} \\ 0 \end{bmatrix} \end{aligned}\right\}, \quad \text{for} \quad m = 1,2,\ldots,M. \tag{8.98}$$

The vectors $\mathbf{a}_m$ and $\mathbf{b}_m$, for (8.24) and (8.35), are defined as

$$\begin{aligned}\mathbf{a}_m &= \begin{bmatrix} a_{0,m} & a_{1,m} & \cdots & a_{m,m} \end{bmatrix}^H \\ \mathbf{b}_m &= \begin{bmatrix} b_{0,m} & b_{1,m} & \cdots & b_{m,m} \end{bmatrix}^H,\end{aligned} \tag{8.99}$$

where, by definition $a_{0,m} = b_{m,m} = 1$, the parameters k_m^{f} and k_m^{b}, as will be clarified in the following, are defined as *reflection coefficients*. Note that, in the scalar case (8.98), they are written as

$$\begin{aligned}a_{k,m} &= a_{k,m-1} + k_m^{\mathrm{f}} b_{k,m-1} \\ b_{k,m} &= b_{k,m-1} + k_m^{\mathrm{b}} a_{k,m-1}\end{aligned}, \quad \text{for} \quad k = 0, 1, \ldots, m. \tag{8.100}$$

For stationary process, for which it is (8.49), we have that $k_m^{\mathrm{f}} = k_m^*$, $k_m^{\mathrm{b}} = k_m$, and also that $\mathbf{b}_m = \mathbf{a}_m^{*B} = \begin{bmatrix} a_{m,m} & a_{m-1,m} & \cdots & 1 \end{bmatrix}^T$. Therefore, in the case of stationary process (8.100) can be rewritten in the following matrix form:

$$\begin{bmatrix} a_{k,m} \\ a_{m-k,m}^* \end{bmatrix} = \begin{bmatrix} 1 & k_m^* \\ k_m & 1 \end{bmatrix} \begin{bmatrix} a_{k,m-1} \\ a_{m-k,m-1}^* \end{bmatrix}, \quad \text{for} \quad k = 0, 1, \ldots, m. \tag{8.101}$$

8.3.3.1 Reflection Coefficients Determination

In the stationary case it is $\mathbf{b}_m = \mathbf{a}_m^{*\mathrm{B}}$ (or $\mathbf{b}_m = \mathbf{P}\mathbf{a}_m^*$) and $\mathbf{R}_{m+1} = \mathbf{R}_{m+1}^{\mathrm{f}} = \mathbf{R}_{m+1}^{\mathrm{b}}$, i.e., for (8.97) (see also (8.45)), we have that

$$\begin{aligned}\mathbf{R}_{m+1} &= \left[\begin{array}{ccc|c} r[0] & \cdots & r[m-1] & r[m] \\ \vdots & \ddots & \vdots & \vdots \\ r^*[m-1] & \cdots & r[0] & r[1] \\ \hline r^*[m] & \cdots & r^*[1] & r[0] \end{array}\right] \\ &= \left[\begin{array}{c|ccc} r[0] & r^*[1] & \cdots & r^*[m] \\ \hline r[1] & r[0] & \cdots & r[m-1] \\ \vdots & \vdots & \ddots & \vdots \\ r[m] & r^*[m-1] & \cdots & r[0] \end{array}\right], \\ &= \begin{bmatrix} \mathbf{R}_m & \mathbf{r}_m^B \\ \mathbf{r}_m^{BH} & r[0] \end{bmatrix} = \begin{bmatrix} r[0] & \mathbf{r}_m^H \\ \mathbf{r}_m & \mathbf{R}_{m,n-1} \end{bmatrix}\end{aligned} \tag{8.102}$$

where $\mathbf{R}_m$ and $\mathbf{R}_{m,n-1}$ are the $(m \times m)$ autocorrelation matrices and $\mathbf{r}_m$ is the $(m \times 1)$ correlation vector, as defined in (8.46), such that and $\mathbf{r}_m^{\mathrm{f}} = \mathbf{r}_m^*$ and $\mathbf{r}_m^{\mathrm{b}} = \mathbf{r}_m^{f*\mathrm{B}} = \mathbf{r}_m^B$.

For the determination of the parameters k_m consider the development of the forward predictor pre-multiplying both sides of the first of (8.98) for the correlation matrix of order $m + 1$. For which we have

$$\mathbf{R}_{m+1}\mathbf{a}_m = \mathbf{R}_{m+1}\begin{bmatrix}\mathbf{a}_{m-1}\\ 0\end{bmatrix} + k_m\mathbf{R}_{m+1}\begin{bmatrix}0\\ \mathbf{a}_{m-1}^{*B}\end{bmatrix}, \tag{8.103}$$

such that we can redefine the three terms of (8.103) as described below.

Considering the expression (8.32) the first term can be written as

$$\mathbf{R}_{m+1}\mathbf{a}_m = \begin{bmatrix}J_m\\ \mathbf{0}_m\end{bmatrix}, \tag{8.104}$$

while, for the (8.102), the other two terms can be rewritten as

$$\begin{aligned}
\mathbf{R}_{m+1}\begin{bmatrix}\mathbf{a}_{m-1}\\ 0\end{bmatrix} &= \begin{bmatrix}\mathbf{R}_m & \mathbf{r}_m^B\\ \mathbf{r}_m^{BH} & r[0]\end{bmatrix}\begin{bmatrix}\mathbf{a}_{m-1}\\ 0\end{bmatrix} = \begin{bmatrix}\mathbf{R}_m\mathbf{a}_{m-1}\\ \mathbf{r}_m^{BH}\mathbf{a}_{m-1}\end{bmatrix}\\
\mathbf{R}_{m+1}\begin{bmatrix}0\\ \mathbf{a}_{m-1}^{*B}\end{bmatrix} &= \begin{bmatrix}r[0] & \mathbf{r}_m^H\\ \mathbf{r}_m & \mathbf{R}_{m,n-1}\end{bmatrix}\begin{bmatrix}0\\ \mathbf{a}_{m-1}^{*B}\end{bmatrix} = \begin{bmatrix}\mathbf{r}_m^H\mathbf{a}_{m-1}^{*B}\\ \mathbf{R}_{m,n-1}\mathbf{a}_{m-1}^{*B}\end{bmatrix}.
\end{aligned} \tag{8.105}$$

Therefore, from (8.104) and (8.105) it follows that

$$\begin{bmatrix}J_m\\ \mathbf{0}_m\end{bmatrix} = \begin{bmatrix}\mathbf{R}_m\mathbf{a}_{m-1}\\ \mathbf{r}_m^{BH}\mathbf{a}_{m-1}\end{bmatrix} + k_m\begin{bmatrix}\mathbf{r}_m^H\mathbf{a}_{m-1}^{*B}\\ \mathbf{R}_{m,n-1}\mathbf{a}_{m-1}^{*B}\end{bmatrix}, \tag{8.106}$$

where the terms $\mathbf{R}_m\mathbf{a}_{m-1}$ and $\mathbf{R}_{m,n-1}\mathbf{a}_{m-1}^{*\mathrm{B}}$ (see (8.32)) can be rewritten as

$$\mathbf{R}_m\mathbf{a}_{m-1} = \begin{bmatrix}J_{m-1}\\ \mathbf{0}_{m-1}\end{bmatrix}, \quad \mathbf{R}_{m,n-1}\mathbf{a}_{m-1}^{*B} = \begin{bmatrix}\mathbf{0}_{m-1}\\ J_{m-1}\end{bmatrix}. \tag{8.107}$$

From the previous position the expression (8.106) can be rewritten as

$$\begin{bmatrix}J_m\\ \mathbf{0}_{m-1}\\ 0\end{bmatrix} = \begin{bmatrix}J_{m-1}\\ \mathbf{0}_{m-1}\\ \mathbf{r}_m^{BH}\mathbf{a}_{m-1}\end{bmatrix} + k_m\begin{bmatrix}\mathbf{r}_m^H\mathbf{a}_{m-1}^{*B}\\ \mathbf{0}_{m-1}\\ J_{m-1}\end{bmatrix}. \tag{8.108}$$

Let us define the scalar quantity β_{m-1}^* as

$$\beta_{m-1}^* \triangleq \mathbf{r}_m^H\mathbf{a}_{m-1}^{*B}, \tag{8.109}$$

(8.108), removing the $\mathbf{0}_{m-1}$ rows, can be rewritten in a compact form of the type

$$\begin{bmatrix}J_m\\ 0\end{bmatrix} = \begin{bmatrix}J_{m-1}\\ \beta_{m-1}\end{bmatrix} + k_m\begin{bmatrix}\beta_{m-1}^*\\ J_{m-1}\end{bmatrix}, \qquad \text{for} \qquad m = 1, 2, \ldots, M. \tag{8.110}$$

Finally, from the last of (8.110) ($0 = \beta_{m-1} + k_m J_{m-1}$), we get

$$k_m = -\frac{\beta_{m-1}}{J_{m-1}}, \qquad \text{for} \qquad m = 1, 2, \ldots, M-1. \tag{8.111}$$

Remark The computability of parameters β_m and k_m demonstrates that the recursive formulation (8.98) (or (8.100)) is consistent. Moreover, note that from the definitions (8.99) we have $\mathbf{a}_{m-1}^{*B} = \begin{bmatrix} a_{m-1,m-1} & a_{m-1,m-2} & \cdots & 1 \end{bmatrix}^T$. Therefore, the expression (8.109) can be rewritten as

$$\beta_{m-1} = \left(\beta_{m-1}^*\right)^* = \mathbf{r}^{TB[m-1]}\mathbf{a}_{m-1}^{[m-1]} + r[m] \tag{8.112}$$

and $\beta_0 = r[1]$.

8.3.3.2 Initialization of k and β Parameters

We observe that from the first of (8.110)

$$J_m = J_{m-1} + k_m\beta_{m-1}^*. \tag{8.113}$$

Therefore, replacing the expression of $\beta_{m-1} = -k_m J_{m-1}$ calculated with (8.111), we obtain the recursive expression:

$$J_m = J_{m-1}\left(1 - |k_m|^2\right). \tag{8.114}$$

It is recalled that the term J_m physically represents the prediction error energy and if the predictor order increases, the error decreases, for which

$$0 \le J_m \le J_{m-1}, \qquad \text{for} \qquad m \ge 1.$$

From (8.104) it follows that

$$J_0 = r[0]. \tag{8.115}$$

The zero-order prediction error energy is in fact the maximum possible, i.e., equal to the energy of the input signal. Initializing (8.114) with such value we have that the prediction error energy of a filter of order M is equal to

$$J_M = r[0]\prod_{m=1}^{M}\left(1 - |k_m|^2\right). \tag{8.116}$$

From the above and from (8.114) it is obvious that

$$|k_m| \le 1, \qquad \text{for} \qquad 1 \le m \le M. \tag{8.117}$$

The parameter k_m that appears in the LDA recurrence is defined as *reflection coefficient*, in analogy to the transmission lines theory where, at the interface between two media with different characteristic propagation impedance, part of the energy is transmitted and part is reflected.

From the first of (8.100), for a prediction filter of order m, the coefficient k_m is equal to the last coefficient $a_{m,m}$, i.e.,

$$k_m = a_{m,m}. \tag{8.118}$$

As regards the parameters β_{m-1} we can observe that, since the zero-order error is equal to the input, we get

$$e_0^{\mathrm{f}}[n] = e_0^{\mathrm{b}}[n] = x[n]. \tag{8.119}$$

It is worth also, in agreement with (8.112) for which $\beta_0 = r[1]$, and since by definition $J_0 = E\left\{|x[n]|^2\right\}$, the reflection coefficient k_0 for (8.111) is

$$k_1 = \frac{\beta_0}{J_0} = \frac{r^*[1]}{r[0]} \tag{8.120}$$

for which the Levinson–Durbin recurrence can be properly initialized.

The algorithm pseudo-code is reported below.

8.3.3.3 Summary of Levinson–Durbin Algorithm

Input $r[0], r[1], \ldots, r[M-1]$;

Initialization

$J_0 = r[0]$

$\beta_0 = r^*[1]$

$k_0 = -\beta_0 / J_0$

$a_0 = k_0$

$J_1 = J_0 + \beta_0 k_0^*$

For $m = 1, 2, \ldots, M-2$ {

$\beta_m = \mathbf{r}_{0:m}^{TB} \mathbf{a}_{0:m} + \mathbf{r}[m+1]$

$k_m = -\beta_m / J_m$

$$\mathbf{a}_m = \begin{bmatrix} \mathbf{a}_{m-1} \\ 0 \end{bmatrix} + k_m^* \begin{bmatrix} \mathbf{a}_{m-1}^{*B} \\ 1 \end{bmatrix}$$

$J_{m+1} = J_m + \beta_m k_m^*$

}

Output: $\mathbf{a}$; $k_0, k_1 \ldots, k_{M-1}$; J_{M-1}.

8.3.3.4 Reverse Levinson–Durbin Algorithm

In the reverse form of Levinson–Durbin algorithm we compute the reflection coefficients $\mathbf{k}$, based on the prediction error coefficients $\mathbf{a}$ and the final prediction error J, using an inverse recursion.

From the (8.118) we have that

$$a_{m,m} = k_m, \qquad \text{for} \qquad m = M, M-1, \ldots, 1. \tag{8.121}$$

The step-down formula can be derived considering the LDA forward–backward scalar recursion (8.101) solved for the filter coefficients $\mathbf{a}$. Therefore we have

$$a_{k,m-1} = \frac{a_{k,m} - k_m a^*_{m-k,m}}{1 - |k_m|^2}, \qquad \text{for} \qquad k = 0, 1, \ldots, m. \tag{8.122}$$

8.3.3.5 Summary of Reverse Levinson–Durbin Algorithm

Input $a_1^M,\ a_2^M,\ \ldots,\ a_{M-1}^M$;
Initialization
$a_{m,M} = a_m^M$
For $m = M,\ M-1,\ \ldots,\ 1$ {

$k_m = a^*_{m,m}$

For $k = 1,\ \ldots,\ m-1$ {

$$a_{k,m-1} = \frac{a_{k,m} - a_{m,m} a^*_{m-k,m}}{1 - |k_m|^2}$$

}

}

Output: $k_0, k_1 \ldots, k_{M-1}$.

8.3.3.6 Prediction Error Filter Structure

From the development carried out in the previous section we can express the forward–backward prediction error of order m in the following way:

$$e_m^{\mathrm{f}}[n] = x[n] + \sum_{k=1}^{m} a^*_{k,m} x[n-k] = \mathbf{a}_m^H \begin{bmatrix} x[n] & \mathbf{x}^T_{m,n-1} \end{bmatrix}^T, \tag{8.123}$$

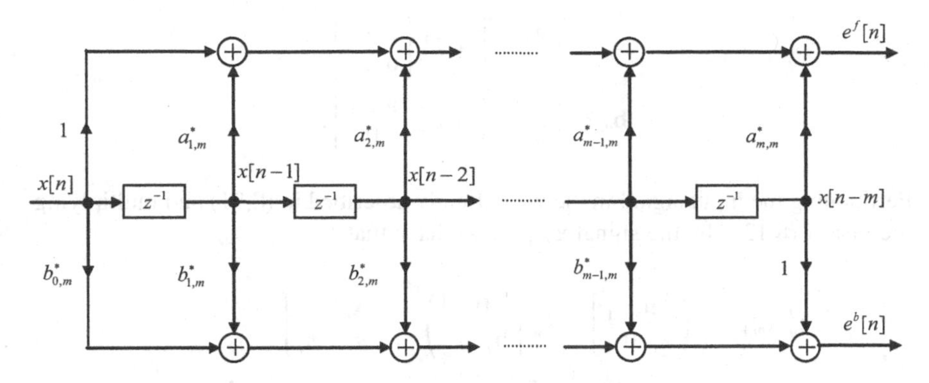

Fig. 8.10 Example of combined forward–backward prediction error filter structure in direct form

$$e_m^{\text{b}}[n] = x[n-m] + \sum_{k=0}^{m-1} b_{k,m}^* x[n-k] = \mathbf{b}_m^H \left[\mathbf{x}_m^T \quad x[n-m] \right]^T, \tag{8.124}$$

where $a_{0,m} = b_{m,m} = 1$ that corresponds to the filter structure in direct form, illustrated in Fig. 8.10.

8.3.4 Lattice Adaptive Filters and Forward–Backward Linear Prediction

The digital filters can be made with various structures: direct or inverse form-I and form-II, lattice, state space, etc. Among these, the lattice structure may not have the minimum number of multiplications/additions but has many advantages, including a block structure which also allows a modular hardware level, a immediate stability verifiability, low sensitivity to coefficient quantization, good performance in the case of finite-precision arithmetic, scalability, and, most important in the context of ROF, the possibility of nested structure (or pluggability), i.e., the possibility of increasing the filter order by simply adding a new lattice stage without having to recalculate the previous one.

These features have led to the use of such robust structures in many application areas such as, for example, the speech processing, the channel equalization, time-series prediction, etc. [6–8, 13, 28, 32]. Even in the case of adaptive filtering, the lattice structure has significant advantages including, a very important one, the reduced sensitivity to the eigenvalues spread of the input signal correlation matrix.

For the lattice structure determination, consider the partitions (8.98) used in the definition of the recursive filter and (see (8.123)) reformulate the forward prediction error $e_m^{\text{f}}[n] = \mathbf{a}_m^H \mathbf{x}_{m+1,\text{n}}^T$, in function of them. In practice, let us review the terms of the forward and backward order recursive filter (8.98) here rewritten

$$\mathbf{a}_m = \begin{bmatrix} \mathbf{a}_{m-1} \\ 0 \end{bmatrix} + k_m^{\mathrm{f}} \begin{bmatrix} 0 \\ \mathbf{b}_{m-1} \end{bmatrix}$$
$$\mathbf{b}_m = \begin{bmatrix} 0 \\ \mathbf{b}_{m-1} \end{bmatrix} + k_m^{\mathrm{b}} \begin{bmatrix} \mathbf{a}_{m-1} \\ 0 \end{bmatrix}. \tag{8.125}$$

Partitioning the input signal in the way already described in (8.91) and multiplying the first of (8.125) by the signal $\mathbf{x}_{m+1,n}$, we have that

$$\begin{aligned} \mathbf{a}_m^H \mathbf{x}_{m+1} &= \left(\begin{bmatrix} \mathbf{a}_{m-1} \\ 0 \end{bmatrix} + k_m^{\mathrm{f}} \begin{bmatrix} 0 \\ \mathbf{b}_{m-1} \end{bmatrix} \right)^H \begin{bmatrix} \mathbf{x}_{m,n} \\ x[n-m] \end{bmatrix} \\ &= [\,\mathbf{a}_{m-1} \quad 0\,] \begin{bmatrix} \mathbf{x}_{m,n} \\ x[n-m] \end{bmatrix} + k_m^{\mathrm{f}} [0 \quad \mathbf{b}_{m-1}\,] \begin{bmatrix} \mathbf{x}_{m,n} \\ x[n-m] \end{bmatrix}, \end{aligned}$$

where the terms are, by definition,

$$\begin{aligned} [\,\mathbf{a}_{m-1} \quad 0\,] \begin{bmatrix} \mathbf{x}_{m,n} \\ x[n-m] \end{bmatrix} &= \mathbf{a}_{m-1}^H \mathbf{x}_m = e_{m-1}^{\mathrm{f}}[n], \\ [0 \quad \mathbf{b}_{m-1}\,] \begin{bmatrix} x[n] \\ \mathbf{x}_{m,n-1} \end{bmatrix} &= \mathbf{b}_{m-1}^T \mathbf{x}_{m,n-1} = e_{m-1}^{\mathrm{b}}[n-1]. \end{aligned}$$

It follows that we can write

$$e_m^{\mathrm{f}}[n] = e_{m-1}^{\mathrm{f}}[n] + k_m^{\mathrm{f}} e_{m-1}^{\mathrm{b}}[n-1]. \tag{8.126}$$

With similar reasoning, multiplying the second of (8.125) by the signal $\mathbf{x}_{m+1,n}$, we get

$$\mathbf{b}_m^H \mathbf{x}_{m+1} = \left(\begin{bmatrix} 0 \\ \mathbf{b}_{m-1} \end{bmatrix} + k_m^{\mathrm{b}} \begin{bmatrix} \mathbf{a}_{m-1} \\ 0 \end{bmatrix} \right)^H \begin{bmatrix} x[n] \\ \mathbf{x}_{m,n} \end{bmatrix},$$

where the first and the second terms are by definition

$$\begin{aligned} [0 \quad \mathbf{b}_{m-1}\,] \begin{bmatrix} x[n] \\ \mathbf{x}_{m,n-1} \end{bmatrix} &= \mathbf{b}_{m-1}^H \mathbf{x}_{m,n-1} = e_{m-1}^{\mathrm{b}}[n-1], \\ [\,\mathbf{a}_{m-1}^H \quad 0\,] \begin{bmatrix} x[n] \\ \mathbf{x}_{m,n-m} \end{bmatrix} &= \mathbf{a}_{m-1}^H \mathbf{x}_{m,n} = e_{m-1}^{\mathrm{f}}[n]. \end{aligned}$$

It follows that, even in this case, we can write

$$e_m^{\mathrm{b}}[n] = e_{m-1}^{\mathrm{b}}[n-1] + k_m^{\mathrm{b}} e_{m-1}^{\mathrm{f}}[n]. \tag{8.127}$$

For stationary process we have that $k_m^* \equiv k_m^f = k_m^{b*}$, and the expressions (8.126) and (8.127) can be rewritten as

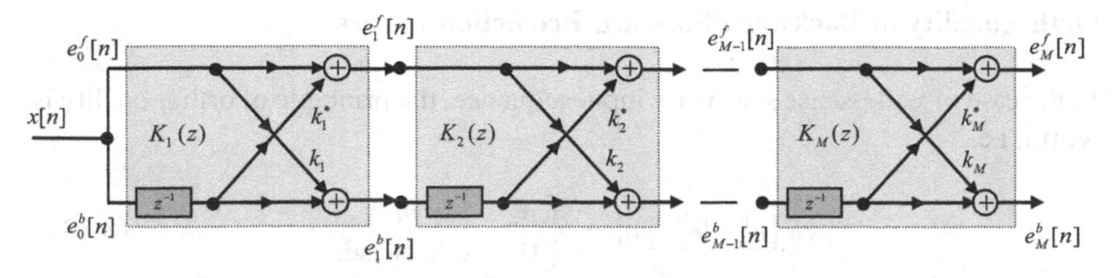

Fig. 8.11 Discrete-time two-port network structure of the combined forward–backward lattice prediction error filter derived from (8.129)

$$\begin{aligned} e_m^{\mathrm{f}}[n] &= e_{m-1}^{\mathrm{f}}[n] + k_m^* e_{m-1}^{\mathrm{b}}[n-1] \\ e_m^{\mathrm{b}}[n] &= k_m e_{m-1}^{\mathrm{f}}[n] + e_{m-1}^{\mathrm{b}}[n-1] \end{aligned} \tag{8.128}$$

or, in terms of two-port DT network (see [14, 15]), take the form

$$\begin{bmatrix} e_m^{\mathrm{f}}[n] \\ e_m^{\mathrm{b}}[n] \end{bmatrix} = \begin{bmatrix} 1 & k_m^* z^{-1} \\ k_m & z^{-1} \end{bmatrix} \begin{bmatrix} e_{m-1}^{\mathrm{f}}[n] \\ e_{m-1}^{\mathrm{b}}[n] \end{bmatrix}. \tag{8.129}$$

The latter, with the initial condition (8.119) $e_0^{\mathrm{f}}[n] = e_0^{\mathrm{b}}[n] = x[n]$, for $m = 1,2, \ldots, M$, is equivalent to the lattice structure shown in Fig. 8.11.

8.3.4.1 Properties of Lattice Filters

The main properties of the lattice structures are the (1) order selection; (2) easy verification of stability; and (3) orthogonality of backward/forward prediction errors.

Optimal Nesting

This property, which is the fundamental ORF's characteristic, allows us to vary the filter order by simply adding or removing a lattice stage, without having to fully solve the normal equations.

Stability

A lattice structure is stable for

$$0 \le |k_m| < 1, \qquad \text{for} \qquad m = 1, 2, \ldots, M. \tag{8.130}$$

This property is important in the case of inverse filtering and adaptive IIR filters where it allows an immediate verification of stability.

Orthogonality of Backward/Forward Prediction Errors

In the case of wide-sense stationary input sequence, the principle of orthogonality is worth, i.e.,

$$E\{e_m^{\mathrm{b}}[n]e_i^{\mathrm{b}*}[n]\} = \begin{cases} \sigma_m^2 & i = m \\ 0 & \text{otherwise.} \end{cases} \tag{8.131}$$

In fact, for $m \geq i$ substituting for $e_i^{\mathrm{b}*}[n]$ from (8.124), we have

$$E\Big\{e_m^{\mathrm{b}}[n]\big[x^*[n-i] + b_{1,i}x^*[n-i+1] + \cdots + b_{i,i}x^*[n]\big]\Big\}$$

and for orthogonality between input and error sequences, we have that

$$E\{e_m^{\mathrm{b}}[n]x^*[n-i]\} = 0, \qquad \text{for} \qquad i = 0, 1, \ldots, m-1$$

thus, for $m > i$, all terms in (8.131) are zero. Expanding $e_m^{\mathrm{b}}[n]$, with similar argument, we can prove that also for $m < i$ all terms in (8.131) are zero.

In the lattice structure the output of each stage is uncorrelated with that of the preceding stage. Unlike the standard delay lines (in which this is not done) the lattice equations represent a stage-by-stage orthogonalization section.

8.3.5 Lattice as Orthogonalized Transform: Batch Joint Process Estimation

In the previous sections the lattice structure has been introduced for CFBLP problems. In this section we want to extend the use of lattice structures for all typical adaptive filtering applications [16, 17, 25].

In the case of generic desired output $d[n]$, the relationships between the parameters of the adaptive filter $\mathbf{w}$ and AR coefficients $\mathbf{a}$ (or $\mathbf{b}$) are no longer those due to the previous sections that are defined in the case of one-step prediction.

Let us assume that the optimum lattice backward coefficients $\mathbf{b}_{\text{opt}}$ (or the related reflection coefficients k_m) are available, referring to Fig. 8.12; the output can be computed as

$$y[n] = \mathbf{h}^H \mathbf{e}_n^{\mathrm{b}}, \tag{8.132}$$

where

$$\mathbf{e}_n^{\mathrm{b}} \in (\mathbb{R}, \mathbb{C})^{(M+1)\times 1} = \begin{bmatrix} e_0^{\mathrm{b}}[n] & \cdots & e_M^{\mathrm{b}}[n] \end{bmatrix}^T$$

is the predetermined prediction error vector containing the output of each lattice stage for an input sequence $\mathbf{x}_n \in (\mathbb{R}, \mathbb{C})^{(M+1)\times 1} = \begin{bmatrix} x[n] & \cdots & x[n-M] \end{bmatrix}^T$.

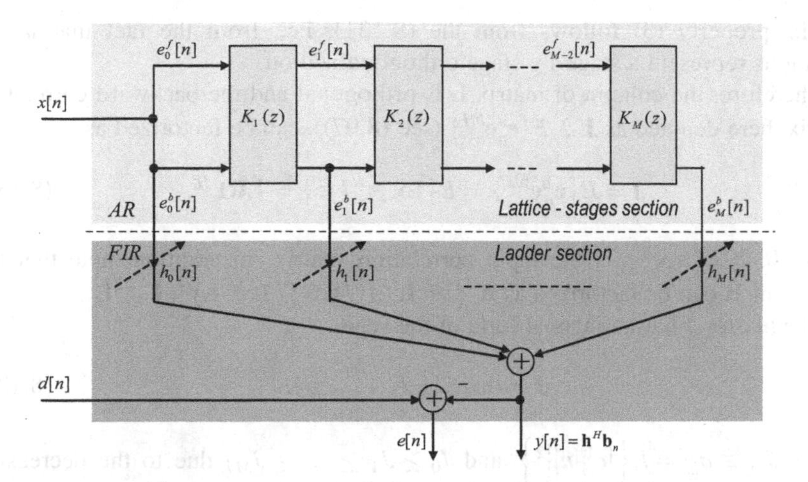

Fig. 8.12 Lattice-ladder filter structure for the joint process estimation. The lattice-stages performs an orthogonal transformation of the input sequence. The ladder-filter section **h** represents a simple transversal adaptive filter

8.3.5.1 Lattice Stages Section as Orthogonalized Transform

Considering (8.124), it is rewritten as

$$\begin{aligned}
e_0^{\mathrm{b}}[n] &= x[n] \\
e_1^{\mathrm{b}}[n] &= b_{0,1}^*[n]x[n] + x[n-1] \\
e_2^{\mathrm{b}}[n] &= b_{0,2}^*[n]x[n] + b_{1,2}^*[n]x[n-1] + x[n-2] \\
&\vdots \\
e_M^{\mathrm{b}}[n] &= b_{0,M}^*[n]x[n] + b_{1,M}^*[n]x[n-1] + \cdots + b_{M-1,M}^*[n]x[n-M+1] + x[n-M].
\end{aligned} \tag{8.133}$$

Let us define the lower triangular matrix **L** as

$$\mathbf{L} \in (\mathbb{R},\mathbb{C})^{(M+1)\times(M+1)} \triangleq \begin{bmatrix} 1 & 0 & \cdots & 0 & 0 & 0 \\ b_{0,1}^* & 1 & 0 & \cdots & \vdots & 0 \\ b_{0,2}^* & b_{1,2}^* & 1 & \cdots & 0 & 0 \\ \vdots & \vdots & \vdots & \ddots & 1 & \vdots \\ b_{0,M}^* & b_{1,M}^* & b_{2,M}^* & \cdots & b_{M-1,M}^* & 1 \end{bmatrix} \tag{8.134}$$

such that the expression (8.133) can be rewritten as

$$\mathbf{e}^{\mathrm{b}} = \mathbf{L}\mathbf{x}_n. \tag{8.135}$$

The matrix **L** has the following properties: (1) is lower triangular with unitary elements in the main diagonal, (2) has eigenvalue $\lambda_0 = \lambda_1 = \cdots = \lambda_M = 1$, and hence it is nonsingular, and (3) the column of **L** are orthogonal.

The property (3) follows from the (8.131), i.e., from the fact that lattice equations represent a stage-by-stage orthogonalization.

Therefore, the column of matrix **L** is orthogonal and the backward correlation matrix, here denoted as $\mathbf{J} \triangleq E\{\mathbf{e}_n^{\mathrm{b}}\mathbf{e}_n^{bH}\}$ (see (8.97)), can be factorized as

$$\mathbf{J} = E\{\mathbf{e}_n^{\mathrm{b}}\mathbf{e}_n^{\mathrm{b}H}\} = E\{\mathbf{L}\mathbf{x}_n\mathbf{x}_n^H\mathbf{L}^H\} = \mathbf{L}\mathbf{R}\mathbf{L}^H, \tag{8.136}$$

where $\mathbf{R} = E\{\mathbf{x}_n\mathbf{x}_n^H\}$ is the input correlation matrix. In addition, note that the inverse of **R** can be factorized as $\mathbf{R}^{-1} = \mathbf{L}^H\mathbf{J}^{-1}\mathbf{L} = (\mathbf{J}^{-1/2}\mathbf{L})^H(\mathbf{J}^{-1/2}\mathbf{L})$.

The matrix **J** has a diagonal form of the type

$$\mathbf{J} = \mathrm{diag}(J_0, J_1, \ldots, J_M), \tag{8.137}$$

where $J_m \equiv \sigma_m^2 = E\left\{\left|e_m^{\mathrm{b}}[n]\right|^2\right\}$ and $J_0 \geq J_1 \geq \ldots \geq J_M)$ due to the decreasing behavior of the prediction error energy with predictor order.

Remark The orthogonalization performed by the lattice stages, considering the (8.136), corresponds to Cholesky decomposition (see Sect. 4.4.1).

8.3.5.2 Adaptive Ladder Filter Parameters Determination

Figure 8.12 reminds the TDAF structures introduced in Sect. 7.5. In fact, the lattice ladder structure can be seen as an adaptive filter in which the delay line elements have been replaced with the lattice stages. Moreover, the orthogonal matrix **L** transforms the correlated input $\mathbf{x}_n$ into an uncorrelated sequence $\mathbf{e}^{\mathrm{b}} = \mathbf{L}\mathbf{x}_n$.

The optimal filter coefficients $\mathbf{h}_{\mathrm{opt}}$ can be determined in batch mode whereas the theory of Wiener, or in adaptively with online first or second order algorithms.

Proceeding with the Wiener's optimal approach, the cross-correlation vector between the filter input $\mathbf{e}^{\mathrm{b}}$ and the desired output $d[n]$ can be defined as

$$\mathbf{g}_{\mathrm{ed}} = E\{\mathbf{e}_n^{\mathrm{b}}d^*[n]\} = \mathbf{L}E\{\mathbf{x}_n d^*[n]\} = \mathbf{L}\mathbf{g}. \tag{8.138}$$

For (8.136) and (3.47), the normal equations take the form $\mathbf{L}^H\mathbf{J}\mathbf{h} = \mathbf{g}$, and the optimal ladder filter solution can be determined as

$$\mathbf{h}_{\mathrm{opt}} = \mathbf{J}^{-1}\mathbf{L}\mathbf{g}. \tag{8.139}$$

The output of the transversal ladder filter as shown in Fig. 8.12 can be obtained as a linear combination of the backward prediction error vector $\mathbf{e}^{\mathrm{b}}$. The lattice predictor is used to transform the input signals into the backward prediction errors. The linear combiner uses these backward prediction errors to produce an estimate of the desired signal $d[n]$.

Finally, equating the (8.139) with the Wiener optimal solution $\mathbf{w}_{\text{opt}} = \mathbf{R}^{-1}\mathbf{g}$, for (8.136) the one-to-one correspondence between the optimal FIR filter $\mathbf{w}_{\text{opt}}$ and the parameters of optimal ladder filter $\mathbf{h}_{\text{opt}}$ can be computed as

$$\mathbf{w}_{\text{opt}} = \mathbf{L}^H \mathbf{h}_{\text{opt}}. \tag{8.140}$$

8.3.5.3 Burg Estimation Formula

The batch or online reflection coefficients k_m estimation can be performed by Wiener/LS or SDA/LMS-like approach based on the minimization of a certain CF.

At the mth lattice stage the optimality criterion is represented to be the CF (see also (8.52))

$$J_{m,n}(k_m) = E\left\{\left|e_m^{\text{f}}[n]\right|^2 + \left|e_m^{\text{b}}[n]\right|^2\right\}. \tag{8.141}$$

Substituting (8.128) into (8.141) and taking the derivative respect to k_m we have that

$$\begin{aligned}\frac{\partial J_{m,n}(k_m)}{\partial k_m} &\triangleq \frac{\partial J}{\partial k_m} E\left\{\left|e_{m-1}^{\text{f}}[n] + k_m^* e_{m-1}^{\text{b}}[n-1]\right|^2 + \left|e_{m-1}^{\text{b}}[n-1] + k_m e_{m-1}^{\text{f}}[n]\right|^2\right\} \\ &= 2E\left\{\left|e_{m-1}^{\text{f}}[n]\right|^2 + \left|e_{m-1}^{\text{b}}[n-1]\right|^2\right\} k_m^* + 4E\left\{e_{m-1}^{\text{b}}[n-1] e_{m-1}^{\text{f}}[n]\right\} = 0.\end{aligned} \tag{8.142}$$

Therefore, considering the input $\mathbf{x}_n$ as an ergodic process and replacing the expectation operator $E(\cdot)$ with time average operator $\hat{E}(\cdot)$ we obtain the Burg formula:

$$k_m^* = -2 \frac{\sum_{n=0}^{N-1} \left\{e_{m-1}^{\text{b}}[n-1] e_{m-1}^{\text{f}*}[n]\right\}}{\sum_{n=0}^{N-1} \left\{\left|e_{m-1}^{\text{f}}[n]\right|^2 + \left|e_{m-1}^{\text{b}}[n-1]\right|^2\right\}}, \quad \text{for} \quad m = 1, \ldots, M, \tag{8.143}$$

which represent a LS-like blockwise formulation.

8.3.6 Gradient Adaptive Lattice Algorithm: Online Joint Process Estimation

The online estimation of the reflection coefficients k_m can be performed by LMS-like approach based on the CF minimization through the descent of its stochastic gradient. Therefore, as is usual, the CF can be chosen as the instantaneous version of (8.141), i.e.,

$$\hat{J}_{m,n}(k_m) = \left|e_m^{\mathrm{f}}[n]\right|^2 + \left|e_m^{\mathrm{b}}[n]\right|^2. \tag{8.144}$$

For the development of the algorithm, denoted as gradient adaptive lattice (GAL) [16, 17], we consider (8.128) with the initial condition $e_0^{\mathrm{f}}[n] = e_0^{\mathrm{b}}[n] = x[n]$ for $m = 1,2, \ldots, M$.

As for the LMS (see Sect. 5.3.1) the GAL algorithm can be implemented by the following finite difference equations:

$$k_{m,n} = k_{m,n-1} + \frac{1}{2}\mu_{m,n}\left(-\nabla\hat{J}_{m,n}(k_m)\right), \qquad \text{for} \qquad m = 1,\ldots,M. \tag{8.145}$$

Substituting (8.128) into (8.144), for the instantaneous gradient we have

$$\begin{aligned}\nabla\hat{J}_{m,n}(k_m) &\triangleq \frac{\partial\hat{J}}{\partial k_m}\left(\left|e_{m-1}^{\mathrm{f}}[n] + k_m^* e_{m-1}^{\mathrm{b}}[n-1]\right|^2 + \left|e_{m-1}^{\mathrm{b}}[n-1] + k_m e_{m-1}^{\mathrm{f}}[n]\right|^2\right)\\ &= 2e_m^{\mathrm{f}*}[n]e_{m-1}^{\mathrm{b}}[n-1] + 2e_m^{\mathrm{b}}[n]e_{m-1}^{\mathrm{f}*}[n].\end{aligned} \tag{8.146}$$

Substituting the latter in (8.145), we get

$$k_{m,n} = k_{m,n-1} - \mu_{m,n}\left(e_m^{\mathrm{f}*}[n]e_{m-1}^{\mathrm{b}}[n-1] + e_m^{\mathrm{b}}[n]e_{m-1}^{\mathrm{f}*}[n]\right). \tag{8.147}$$

Note that as in the NLMS algorithm (see Sect. 5.5.1), it is possible to determine the learning rate $\mu_{m,n}$ using an energy normalization. Therefore, we have that

$$\mu_{m,n} = \frac{\mu_0}{\delta + J_{m,n}}, \qquad \text{for} \qquad m = 1,\ldots,M.$$

To avoid significant step-size discontinuity that could destabilize the adaptation, as suggested in [17], it is appropriate to estimate the energy with a one-pole low-pass smoothing filter, implemented by the following FDE:

$$J_{m,n} = \gamma J_{m,n-1} + (1-\gamma)\left(\left|e_m^{\mathrm{f}}[n]\right|^2 + \left|e_m^{\mathrm{b}}[n-1]\right|^2\right), \qquad \text{for} \qquad m = 1,\ldots,M, \tag{8.148}$$

where $0 < \gamma < 1$ is a smoothing parameter, and where μ_0 and δ are small learning parameters empirically predetermined.

8.3.6.1 GAL Adaptive Filtering

Referring to Fig. 8.12, in the presence of a generic desired output $d[n]$, in addition to the estimation of the parametric k_m, we must also consider the estimation of the ladder filter coefficients **h**. Considering first-order stochastic gradient algorithm, the *joint process estimation* can be performed with the following adaptation rule:

$$\mathbf{h}_n = \mathbf{h}_{n-1} + \frac{1}{2}\mu_n\big(-\nabla\hat{J}(\mathbf{h})\big). \tag{8.149}$$

The CF $\hat{J}(\mathbf{h})$ in (8.149) is the instantaneous square error $\hat{J}(\mathbf{h}) = \left|d[n] - \mathbf{h}^H\mathbf{e}^{\mathrm{b}}\right|^2$ where its gradient is $\nabla\hat{J}(\mathbf{h}) = -2e^*[n]\mathbf{e}^{\mathrm{b}}$. Therefore, the LMS adaptation rule can be written as

$$\mathbf{h}_n = \mathbf{h}_{n-1} + \mu_n e^*[n]\mathbf{e}^{\mathrm{b}}. \tag{8.150}$$

Remark Due to orthogonality property of the lattice section transformation, compared with LMS algorithm, the GAL generally converges more quickly, and their convergence rate is independent of the eigenvalue spread of the input data covariance matrix.

In the case of uncorrelated input sequence, the reflection coefficients are zero, and the lattice stages become a simple delay line. No orthogonalization takes place, and the joint estimation process reduces to a simple transversal AF.

Numerical Example

Figure 8.13 reports the results, in terms of averaged learning curves, of an experiment of a dynamic system identification, of the type used for performance analysis just illustrated in the previous chapters (e.g., see Sects. 5.4.4 and 6.4.5.3). In particular, the experiment consists in the identification of two random system $\mathbf{w}_k$, for $k = 0,1$ and $M = 6$, according to the scheme of study of Fig. 5.14. The input of the system $\mathbf{w}_0$ is a unitary-variance zero-mean white noise, while for the system $\mathbf{w}_1$ the input is a colored noise generated by the expression (5.172) with $b = 0.995$. The learning curves are averaged over 200 runs. For all experiments the noise level was set at a level such that $SNR = 50$ dB.

In the first part of the experiment is identified the system $\mathbf{w}_0$ and for $n \geq \frac{N}{2}$ the system became $\mathbf{w}_1$. As you can see from the figure, for white noise input sequence, the performance of the three algorithms is rather similar. On the contrary, in the case of narrow band input sequence, the LMS algorithm does not converge and the GAL obtains the best performance.

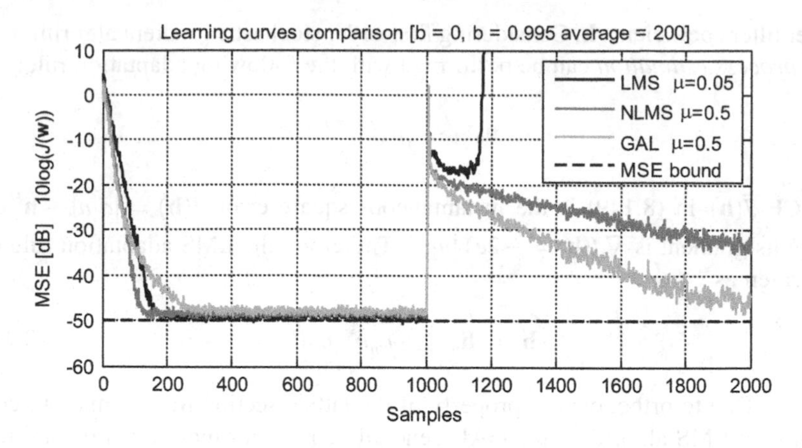

Fig. 8.13 Comparison of LMS, NLMS and GAL learning curve averaged over 200 runs. Left part identification of system $\mathbf{w}_0$ for white noise input; right part identification of system $\mathbf{w}_1$ for narrowband MA colored input

8.3.6.2 Summary of the GAL Algorithm

Input $M,\ \mu_0,\ \delta,\ \gamma,\ \mu_h$
Initialization $k(m)=0,\ m=1,2,\ldots,M;\ \mathbf{h}$
For $n=0,1,\ldots,$ {

$$e_0^{\mathrm{f}}[n]=e_0^{\mathrm{b}}[n]=x[n]$$

For $m=1,2,\ldots,M$ {

$$J_{m,n}=\gamma J_{m,n-1}+(1-\gamma)\left(\left|e_m^{\mathrm{f}}[n]\right|^2+\left|e_m^{\mathrm{b}}[n-1]\right|^2\right)$$

$$\begin{bmatrix} e_m^{\mathrm{f}}[n] \\ e_m^{\mathrm{b}}[n] \end{bmatrix}=\begin{bmatrix} 1 & k_m^* z^{-1} \\ k_m & z^{-1} \end{bmatrix}\begin{bmatrix} e_{m-1}^{\mathrm{f}}[n] \\ e_{m-1}^{\mathrm{b}}[n] \end{bmatrix}$$

$$k_{m,n}=k_{m,n-1}-\frac{\mu_0}{\delta+J_{m,n}}\left(e_m^{\mathrm{f}*}[n]e_{m-1}^{\mathrm{b}}[n-1]+e_m^{\mathrm{b}}[n]e_{m-1}^{\mathrm{f}*}[n]\right)$$

}

$$y[n]=\mathbf{h}_{n-1}^H\mathbf{e}^{\mathrm{b}}$$

$$e[n]=d[n]-y[n]$$

$$\mathbf{h}_n=\mathbf{h}_{n-1}+\mu_h e^*[n]\mathbf{e}^{\mathrm{b}}$$

}

Output: $k_1,k_2\ldots,k_M;\ \mathbf{h},\ y[n],\ e[n],\ J_M$.

8.3.7 Schür Algorithm

An alternative way for the development of adaptive lattice architectures, which allows a more appropriate understanding of the physical meaning, is that where the reflection coefficients $k_0, k_2, \ldots, k_{M-1}$ are directly estimated from the autocorrelation sequence $r[0], \ldots, r[m-1]$, without the explicit computation of the filter coefficients **a** and **b**.

For the method development, we multiply both members of (8.126) and (8.127) for $x^*[n-k]$ and taking the expectation we get

$$E\{e_m^{\mathrm{f}}[n]x^*[n-k]\} = E\{e_{m-1}^{\mathrm{f}}[n]x^*[n-k]\} + k_m^* E\{e_{m-1}^{\mathrm{b}}[n-1]x^*[n-k]\},$$
$$E\{e_m^{\mathrm{b}}[n]x^*[n-k]\} = E\{e_{m-1}^{\mathrm{b}}[n-1]x^*[n-k]\} + k_m E\{e_{m-1}^{\mathrm{f}}[n]x^*[n-k]\}.$$

Denoting the cross-correlations between signals and forward backward errors, respectively, as $q_m^{\mathrm{f}}[k] \triangleq E\{e^{\mathrm{f}}[n]x^*[n-k]\}$ and $q_m^{\mathrm{b}}[k] \triangleq E\{e^{\mathrm{b}}[n]x^*[n-k]\}$, the previous expression can be rewritten as

$$\begin{aligned} q_m^{\mathrm{f}}[k] &= q_{m-1}^{\mathrm{f}}[k] + k_m^* q_{m-1}^{\mathrm{b}}[k-1] \\ q_m^{\mathrm{b}}[k] &= q_{m-1}^{\mathrm{b}}[k-1] + k_m q_{m-1}^{\mathrm{f}}[k] \quad m=1,\ldots,M;\ k=m,\ldots,M. \end{aligned} \tag{8.151}$$

Considering the CFBLP (see Fig. 8.4), the algorithm is formulated by imposing the orthogonality between the prediction errors $e^{\mathrm{f}}[n]$, $e^{\mathrm{b}}[n]$ and the input signal. In fact, as seen in Sect. 8.2.1.2, the choice of optimal coefficients produces orthogonality between the error $e[n]$ and the input $x[n]$, for which we have that $q_m^{\mathrm{f}}[k] = 0$, for $k = 1, 2, \ldots, m$ and $q_m^{\mathrm{b}}[k]$, for $k = 0, 2, \ldots, m$. Therefore, considering (8.151), the reflection coefficient k_m can be computed as

$$k_m^* = -\frac{q_{m-1}^{\mathrm{f}}[m]}{q_{m-1}^{\mathrm{b}}[m-1]} \quad \text{or} \quad k_m = -\frac{q_{m-1}^{\mathrm{b}}[m-1]}{q_{m-1}^{\mathrm{f}}[m]}. \tag{8.152}$$

Finally the recurrence (8.151) is initialized as $J_0 = r[0]$, $q_0^{\mathrm{f}}[k] = q_0^{\mathrm{b}}[k] = r[k]$ for $k = 0, 1, \ldots, M-1$ and $k_0 = -q_0^{\mathrm{f}}[1]/q_0^{\mathrm{b}}[0]$.

Remark As for the LDA, the equations (8.151) describe a recursive procedure with autocorrelation sequence $r[k]$ as input. In other words, with the recurrence (8.151), you can determine the reflection coefficients, known as the autocorrelation samples.

8.3.8 All-Pole Inverse Lattice Filter

Inverse filtering or *deconvolution*[1] means the determination of the input signal $x[n]$ known the output $y[n]$ and the impulse response $h[n]$ of a system, such that $y[n] = x[n] * h[n]$.

The most intuitive way to determine the inverse of a given TF $H(z)$ consists in computing explicitly its reciprocal, i.e., $F(z) = 1/H(z)$. For example, given a TF $H(z) = 1 + az^{-1}$, which has a single zero at $z = -a$, the computation of $F(z)$, denoted as inverse or deconvolution filter, can be performed using a long division as

$$F(z) = \frac{1}{1 + az^{-1}} = 1 - az^{-1} + a^2z^{-2} - a^3z^{-3} + \cdots$$

Providing $|a| < 1$ (i.e., the $H(z)$ is minimum phase) the sequence converges to a stable TF.

Considering the recursion (8.129), since the lattice structure of Fig. 8.11 has minimum phase, denoting as $H(z)$ the TF of the lattice filter, as reported in the previous paragraph, it is possible to directly synthesize the inverse filter $1/H(z)$ such that if in input is placed the error sequence the filter output produces the signal $x[n]$. In fact, due to minimum phase characteristic, the solution of deconvolution problem can be solved by simply inverting the verses of the branches of the graph and exchanging the inputs and the outputs signals.

In practice, working backward, the *all-pole inverse filter* implementation for stationary process is just

$$\begin{aligned} e_M^{\mathrm{f}}[n] &= e[n] \\ \left.\begin{aligned} e_{m-1}^{\mathrm{f}}[n] &= e_m^{\mathrm{f}}[n] - k_m^* e_{m-1}^{\mathrm{b}}[n-1] \\ e_m^{\mathrm{b}}[n] &= k_m e_m^{\mathrm{f}}[n] + e_{m-1}^{\mathrm{b}}[n-1] \end{aligned}\right\}, &\quad \text{for} \quad m = M, M-1, \ldots, 1. \\ x[n] &= e_0^{\mathrm{f}}[n] = e_0^{\mathrm{b}}[n]. \end{aligned} \tag{8.153}$$

The structure, called the *inverse lattice*, is shown in Fig. 8.14.

Remark The role of the inverse filter is to estimate the input signal to a system, where its output is known. This process is also referred to as deconvolution and, as already noted above in the case of LPC speech synthesis (see Sect. 8.2.5), plays a central aspect of importance in many areas of great interest such as seismology, radio astronomy, optics and image-video processing, etc.

For example, in optics it is specifically used to refer to the process of reversing the optical distortion that takes place in an optical or electron microscope,

---

[1] The foundations for deconvolution and time-series analysis were largely laid by Norbert Wiener. The book [2] was based on work Wiener had done during World War II but that had been classified at the time. Some of the early attempts to apply these theories were in the fields of weather forecasting and economics.

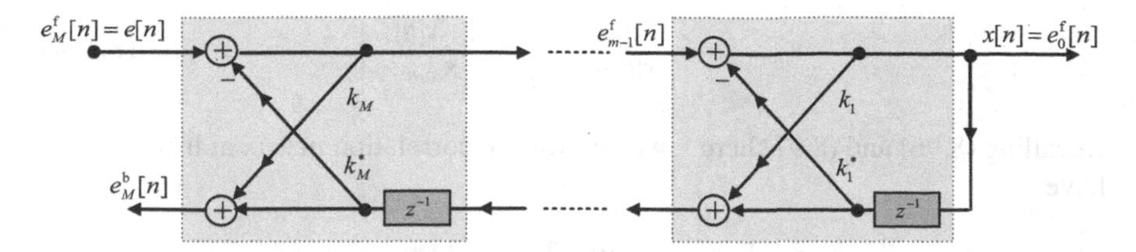

Fig. 8.14 All-pole lattice filter with TF $1/H(z)$, for the $x[n]$ signal reconstruction from the prediction error $e[n]$

telescope, or other imaging instrument, thus creating clearer images. Early *Hubble Space Telescope* images were distorted by a flawed mirror and could be sharpened by deconvolution.

As another example, in the geophysical signals analysis, the propagation model of a seismic trace is a convolution of the *reflectivity function* of the earth and an energy waveform referred to as the *seismic wavelet*. In this case, the objective of deconvolution is to extract the reflectivity function from the seismic trace.

8.4 Recursive Order RLS Algorithms

The RLS algorithm has complexity $O(M^2)$ and for high length filter the computational resources needed may be unacceptable. To overcome this drawback fast RLS (FRLS), with linear complexity $O(KM)$, have been studied. The basic idea for the FRLS algorithm development is to make use of the symmetries and redundancies, and developing recursive methods both in the order m and in time index n. In order to reduce the computational cost, the concepts of prediction and filtering are elegantly combined. In other words, you need to merge the concepts of filtering, forward–backward prediction, recursive order algorithms, and *a priori* and *a posteriori* updating.

In this paragraph are taken the basic concepts of the ROF already discussed in the previous paragraphs and the deterministic normal equations are reformulated in this context. The RLS implemented in lattice structure is discussed and the class of RLS lattice (RLSL) algorithms and fast transversal RLS (FTRLS or FTF) are introduced [14, 15, 18–23, 30, 31].

8.4.1 Fast Fixed-Order RLS in ROF Formulation

For the theoretical development, as previously introduced, defining the sequence $\mathbf{x}_{m,n} = \begin{bmatrix} x[n] & \cdots & x[n-m+1] \end{bmatrix}^T$, it is useful to consider the vector input data $\mathbf{x}_{m+1,n}$ with the partitioned notation (see Sect. 8.3.2). We can then write

$$\mathbf{x}_{m+1,n} = \begin{bmatrix} \mathbf{x}_{m,n} \\ x[n-m] \end{bmatrix} = \begin{bmatrix} x[n] \\ \mathbf{x}_{m,n-1} \end{bmatrix}. \tag{8.154}$$

Recalling (8.96) and (8.97) here rewritten for the correlation matrix at instant n, we have

$$\mathbf{R}_{m+1,n} = \begin{bmatrix} \sigma^2_{x[n]} & \mathbf{r}^{fH}_m \\ \mathbf{r}^f_m & \mathbf{R}_{m,n-1} \end{bmatrix} = \begin{bmatrix} \mathbf{R}_{m,n} & \mathbf{r}^b_m \\ \mathbf{r}^{bH}_m & \sigma^2_{x[n-M]} \end{bmatrix}. \tag{8.155}$$

The theoretical foundation for the definition of the FRLS algorithms class consists of an estimate of the correlation matrix as a temporal average considering the ROF notation (8.154) and the forgetting factor. Omitting to indicate, for the sake of simplicity, the subscript "xx", so $\mathbf{R}_{xx(m,n)} \rightarrow \mathbf{R}_{m,n}$, in this section $\mathbf{R}_{m+1,n}$ indicates time average correlation estimate calculated as

$$\mathbf{R}_{m+1,n} = \sum_{i=0}^{n} \lambda^{n-i} \mathbf{x}_{m+1,i} \mathbf{x}^H_{m+1,i} = \begin{bmatrix} E_{x[n]} & \mathbf{r}^{fH}_m \\ \mathbf{r}^f_m & \mathbf{R}_{m,n-1} \end{bmatrix} = \begin{bmatrix} \mathbf{R}_{m,n} & \mathbf{r}^b_m \\ \mathbf{r}^{bH}_m & E_{x[n-M]} \end{bmatrix} \tag{8.156}$$

with IC $\mathbf{x}_m(-1) = \mathbf{0}$, necessary to ensure the presence of the term $\mathbf{R}_{m,n-1}$ to estimate $\mathbf{R}_{m+1,n}$, and where the variance $\sigma^2_{x[n]}$ is simply replaced with energy $E_{x[n]}$.

The form of the estimator of the correlation (8.156), identical to the statistical form (8.155), enables the development of LS algorithms in recursive order mode. In particular, the notation (8.156), for the (6.51), the estimator of the recursive correlation, is expressed as

$$\mathbf{R}_{m,n} = \lambda \mathbf{R}_{m,n-1} + \mathbf{x}_{m,n} \mathbf{x}^H_{m,n} \tag{8.157}$$

that enables the development of RLS algorithms of complexity $O(KM)$.

8.4.1.1 Transversal RLS Filter

For the development of the method, consider the transversal filter of order m illustrated in Fig. 8.15. The filter input is the vector $\mathbf{x}_{m,n}$, while the desired response is equal to $d[n]$.

Calling $\mathbf{w}_m \triangleq \begin{bmatrix} w_m[0] & w_m[1] & \cdots & w_m[m-1] \end{bmatrix}^H$ the vector of unknown filter coefficients at time n, indicating $\mathbf{R}_{xd(m,n)} \rightarrow \mathbf{g}_{m,n}$, referring to Sect. 6.4.3 (see also Table 6.3), the recursive formulas of the m order RLS are

$$\mathbf{R}_{m,n} \mathbf{w}_{m,n} = \mathbf{g}_{m,n}, \qquad \textit{normal equation}, \tag{8.158}$$

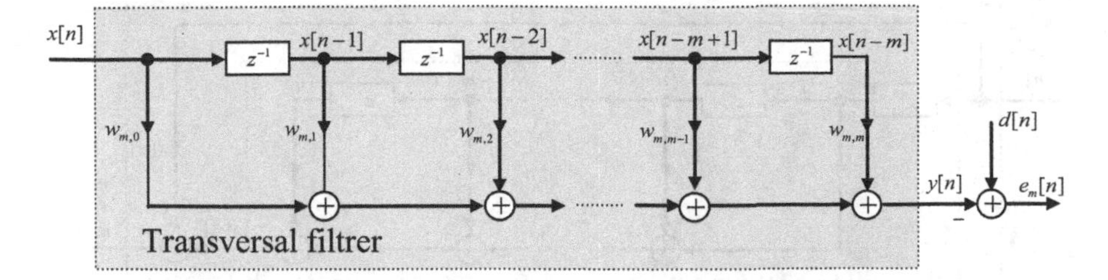

Fig. 8.15 Transversal filter of order m

$$\mathbf{R}_{m,n} = \sum_{k=0}^{n} \lambda^{n-k} \mathbf{x}_{m,k} \mathbf{x}_{m,k}^{H}, \qquad \textit{correlation}, \tag{8.159}$$

$$\mathbf{g}_{m,n} = \sum_{k=0}^{n} \lambda^{n-k} \mathbf{x}_{m,k} d^{*}[k], \qquad \textit{cross-correlation}, \tag{8.160}$$

$$J_{m,n} = E_{d[n]} - \mathbf{w}_{m,n}^{H} \mathbf{g}_{m,n}, \qquad \textit{LSE (error energy)}. \tag{8.161}$$

***A priori* error update**

$$\mathbf{k}_{m,n} = \mathbf{R}_{m,n}^{-1} \mathbf{x}_{m,n}, \qquad \textit{a priori Kalman gain}, \tag{8.162}$$

$$e_m[n] = d[n] - \mathbf{w}_{m,n-1}^{H} \mathbf{x}_m, \qquad \textit{a priori error}, \tag{8.163}$$

$$\mathbf{w}_{m,n} = \mathbf{w}_{m,n-1} + \mathbf{k}_{m,n} e_m^{*}[n], \qquad \textit{update}, \tag{8.164}$$

$$J_{m,n} = \lambda J_{m,n-1} + \alpha_{m,n} \left| e_m[n] \right|^2, \qquad \textit{error energy}. \tag{8.165}$$

***A posteriori* error update**

$$\tilde{\mathbf{k}}_{m,n} = \lambda^{-1} \mathbf{R}_{m,n-1}^{-1} \mathbf{x}_{m,n}, \qquad \textit{a posteriori Kalman gain}, \tag{8.166}$$

$$\varepsilon_m[n] = d[n] - \mathbf{w}_{m,n}^{H} \mathbf{x}_{m,n}, \qquad \textit{a posteriori error}, \tag{8.167}$$

$$\mathbf{w}_{m,n} = \mathbf{w}_{m,n-1} + \tilde{\mathbf{k}}_{m,n} \varepsilon_m^{*}[n], \qquad \textit{update}, \tag{8.168}$$

$$J_{m,n} = \lambda J_{m,n-1} + \tilde{\alpha}_{m,n}^{-1} \left| \varepsilon_m[n] \right|^2, \qquad \textit{error energy}. \tag{8.169}$$

8.4.1.2 Forward Prediction RLS Filter

Consider the forward predictor of order m illustrated in Fig. 8.16. The input of the filter consists in the vector $\mathbf{x}_{m,n-1} \triangleq \begin{bmatrix} x[n-1] & \cdots & x[n-m] \end{bmatrix}^T$ and the desired response is equal to $x[n]$.

Calling $\mathbf{w}_m^{\mathrm{f}} \triangleq \begin{bmatrix} w_m^{\mathrm{f}}[1] & w_m^{\mathrm{f}}[2] & \cdots & w_m^{\mathrm{f}}[m] \end{bmatrix}^H$ (see (8.87)) the coefficients vector of the forward predictor, defined as

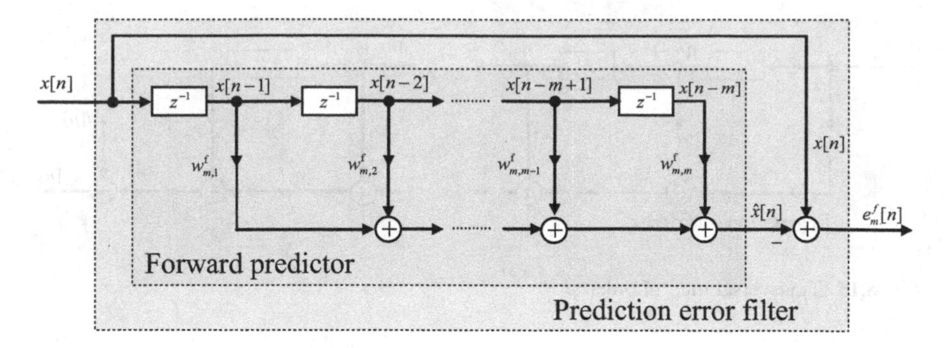

Fig. 8.16 Linear prediction and forward prediction error filter

$$\mathbf{R}_{m,n-1}\mathbf{w}^{\mathrm{f}}_{m,n} = \mathbf{r}^{\mathrm{f}}_{m,n}, \qquad \textit{normal equation} \tag{8.170}$$

the main relations for its estimation, in the LS sense, are

$$\mathbf{R}_{m,n-1} = \sum_{k=0}^{n} \lambda^{n-k}\mathbf{x}_{m,k-1}\mathbf{x}^{H}_{m,k-1}, \qquad \textit{correlation matrix}, \tag{8.171}$$

$$\mathbf{r}^{\mathrm{f}}_{m,n} = \sum_{k=0}^{n} \lambda^{n-k}\mathbf{x}_{m,k-1}x^{*}[k], \qquad \textit{correlation vector}. \tag{8.172}$$

A priori error update

By applying the standard RLS for the predictor coefficients $\mathbf{w}^{\mathrm{f}}_{m,n}$ calculation, derived from *a priori* forward error update, we have that

$$\mathbf{k}_{m,n} = \mathbf{R}^{-1}_{m,n}\mathbf{x}_{m,n}, \qquad \textit{a priori Kalman gain}, \tag{8.173}$$

$$e^{\mathrm{f}}_{m}[n] = x[n] - \mathbf{w}^{\mathrm{f}H}_{m,n-1}\mathbf{x}_{m,n-1}, \qquad \textit{a priori error}, \tag{8.174}$$

$$\mathbf{w}^{\mathrm{f}}_{m,n} = \mathbf{w}^{\mathrm{f}}_{m,n-1} + \mathbf{k}_{m,n-1}e^{\mathrm{f}*}_{m}[n], \qquad \textit{update}, \tag{8.175}$$

$$J^{\mathrm{f}}_{m,n} = \lambda J^{\mathrm{f}}_{m,n-1} + \alpha_{m,n-1}\left|e^{\mathrm{f}}_{m}[n]\right|^{2}, \qquad \textit{error energy}. \tag{8.176}$$

A posteriori error update

In the *a posteriori* error update, we have that

$$\tilde{\mathbf{k}}_{m,n} = \lambda^{-1}\mathbf{R}^{-1}_{m,n-1}\mathbf{x}_{m,n}, \qquad \textit{a posteriori Kalman gain}, \tag{8.177}$$

$$\varepsilon^{\mathrm{f}}_{m}[n] = x[n] - \mathbf{w}^{\mathrm{f}H}_{m,n}\mathbf{x}_{m,n}, \qquad \textit{a posteriori error}, \tag{8.178}$$

$$\mathbf{w}^{\mathrm{f}}_{m,n} = \mathbf{w}^{\mathrm{f}}_{m,n-1} + \tilde{\mathbf{k}}_{m,n-1}\varepsilon^{\mathrm{f}*}_{m}[n], \qquad \textit{update}, \tag{8.179}$$

$$J^{\mathrm{f}}_{m,n} = \lambda J^{\mathrm{f}}_{m,n-1} + \tilde{\alpha}^{-1}_{m,n}\left|\varepsilon^{\mathrm{f}}_{m}[n]\right|^{2}, \qquad \textit{error energy}. \tag{8.180}$$

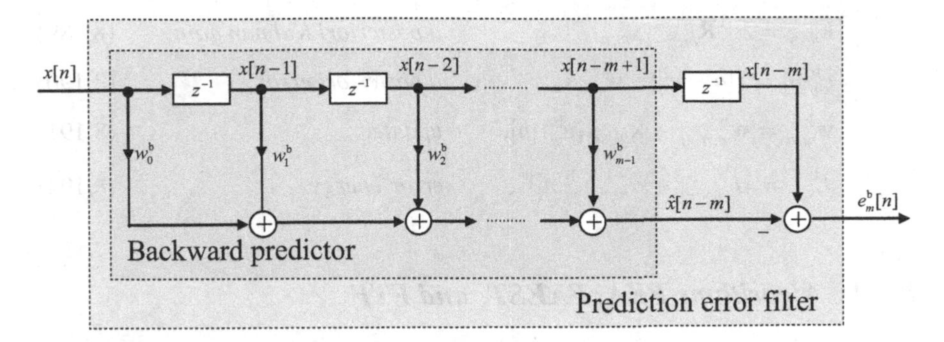

Fig. 8.17 Linear prediction and backward prediction error filter

8.4.1.3 Backward Prediction RLS Filter

Consider the order m backward predictor illustrated in Fig. 8.17. The filter input consists in the vector $\mathbf{x}_{m,n}$ and the desired response is equal $x[n-m]$.

Calling $\mathbf{w}_m^{\mathrm{b}} \triangleq \begin{bmatrix} w_m^{\mathrm{b}}[0] & w_m^{\mathrm{b}}[1] & \cdots & w_m^{\mathrm{b}}[m-1] \end{bmatrix}^H$ (see (8.80)) the coefficients vector of the backward predictor, defined as

$$\mathbf{R}_{m,n}\mathbf{w}_{m,n}^{\mathrm{b}} = \mathbf{r}_{m,n}^{\mathrm{b}}, \quad \textit{normal equations} \tag{8.181}$$

below are the main relations for its evaluation in the sense LS

$$\mathbf{R}_{m,n} = \sum_{k=0}^{n} \lambda^{n-k} \mathbf{x}_{m,k} \mathbf{x}_{m,k}^H, \quad \textit{correlation}, \tag{8.182}$$

$$\mathbf{r}_{m,n}^{\mathrm{b}} = \sum_{k=0}^{n} \lambda^{n-k} \mathbf{x}_{m,k} x^*[k-m], \quad \textit{cross-correlations}, \tag{8.183}$$

$$J_{m,n}^{\mathrm{b}} = E_{x[n-m]}^{\mathrm{b}} - \mathbf{w}_{m,n}^{\mathrm{b}H} \mathbf{r}_{m,n}^{\mathrm{b}}, \quad \textit{LSE}. \tag{8.184}$$

***A priori* error update**

For the *a priori* update we have that

$$\mathbf{k}_{m,n} = \mathbf{R}_{m,n}^{-1} \mathbf{x}_{m,n}, \quad \textit{a priori Kalman gain}, \tag{8.185}$$

$$e_m^{\mathrm{b}}[n] = x[n-m] - \mathbf{w}_{m,n-1}^{\mathrm{b}H} \mathbf{x}_{m,n}, \quad \textit{a priori error}, \tag{8.186}$$

$$\mathbf{w}_{m,n}^{\mathrm{b}} = \mathbf{w}_{m,n-1}^{\mathrm{b}} + \mathbf{k}_{m,n} e_m^{\mathrm{b}*}[n], \quad \textit{update}, \tag{8.187}$$

$$J_{m,n}^{\mathrm{b}} = \lambda J_{m,n-1}^{\mathrm{b}} + \alpha_{m,n} \left| e_m^{\mathrm{b}}[n] \right|^2, \quad \textit{error energy}. \tag{8.188}$$

***A posteriori* error update**

For the *a posteriori* update we have that

$$\tilde{\mathbf{k}}_{m,n} = \lambda^{-1}\mathbf{R}_{m,n-1}^{-1}\mathbf{x}_{m,n}, \qquad \textit{a posteriori Kalman gain}, \tag{8.189}$$

$$\varepsilon_m^{\mathrm{b}}[n] = x[n-m] - \mathbf{w}_{m,n}^{\mathrm{b}H}\mathbf{x}_{m,n}, \qquad \textit{a posteriori error}, \tag{8.190}$$

$$\mathbf{w}_{m,n}^{\mathrm{b}} = \mathbf{w}_{m,n-1}^{\mathrm{b}} + \tilde{\mathbf{k}}_{m,n-1}\varepsilon_m^{\mathrm{b}*}[n], \qquad \textit{update}, \tag{8.191}$$

$$J_{m,n}^{\mathrm{b}} = \lambda J_{m,n-1}^{\mathrm{b}} + \tilde{\alpha}_{m,n}^{-1}\left|\varepsilon_m^{\mathrm{b}}[n]\right|^2, \qquad \textit{error energy}. \tag{8.192}$$

8.4.2 Algorithms FKA, FAEST, and FTF

The class of FRLS algorithms is vast and in the scientific literature there are many variations and specializations. Below are just a few algorithms.

8.4.2.1 Fast Kalman Algorithm

In the RLS algorithm, the calculation of the vector of gain $\mathbf{k}_n$ or $\tilde{\mathbf{k}}_n$ assumes central importance since it provides for correlation matrix inversion (see Table 6.3). To reduce the complexity from $O(M^2)$ to $O(KM)$ we proceed to the calculation using the recursive order update. In the algorithm, developed in [18], it is supposed to know the Kalman gain at time $n-1$, for which it is

$$\mathbf{k}_{m,n-1} = \mathbf{R}_{m,n-1}^{-1}\mathbf{x}_{m,n-1} \tag{8.193}$$

and, using the new input data $\mathbf{x}_{m,n}$ e $d[n]$, suppose we want to calculate the gain at time n

$$\mathbf{k}_{m,n} = \mathbf{R}_{m,n}^{-1}\mathbf{x}_{m,n}. \tag{8.194}$$

From the partitioned matrix inverse formula (8.84), for the backward case we have

$$\mathbf{R}_{m+1,n}^{-1} = \begin{bmatrix} \mathbf{R}_{m,n}^{-1} & \mathbf{0}_m \\ \mathbf{0}_m^T & 0 \end{bmatrix} + \frac{1}{J_{m,n}^{\mathrm{b}}}\begin{bmatrix} -\mathbf{w}_{m,n}^{\mathrm{b}} \\ 1 \end{bmatrix}\begin{bmatrix} -\mathbf{w}_{m,n}^{\mathrm{b}H} & 1 \end{bmatrix} \tag{8.195}$$

while for the forward case (see (8.86)), we have that

$$\mathbf{R}_{m+1,n}^{-1} = \begin{bmatrix} 0 & \mathbf{0}_m^T \\ \mathbf{0}_m & \mathbf{R}_{m,n}^{-1} \end{bmatrix}^{-1} + \frac{1}{J_{m,n}^{\mathrm{f}}}\begin{bmatrix} 1 \\ -\mathbf{w}_{m,n}^{\mathrm{f}} \end{bmatrix}\begin{bmatrix} 1 & -\mathbf{w}_{m,n}^{\mathrm{f}H} \end{bmatrix}. \tag{8.196}$$

Using (8.195), the input sequence partition (8.154), and the definition of *a priori* error $e_m^{\mathrm{b}}[n]$, we get

$$\mathbf{k}_{m+1,n} = \begin{bmatrix} \mathbf{k}_{m,n} \\ 0 \end{bmatrix} + \frac{\varepsilon_m^{\mathrm{b}}[n]}{J_{m,n}^{\mathrm{b}}} \begin{bmatrix} -\mathbf{w}_{m,n}^{\mathrm{b}} \\ 1 \end{bmatrix}, \tag{8.197}$$

which allows the recursive update for the gain $\mathbf{k}_{m,n}$. Proceeding in a similar manner, from (8.196), from the partition (8.154), and by the definition of *a posteriori* error, we have that

$$\mathbf{k}_{m+1,n} = \begin{bmatrix} 0 \\ \mathbf{k}_{m,n-1} \end{bmatrix} + \frac{\varepsilon_m^{\mathrm{f}}[n]}{J_{m,n}^{\mathrm{f}}} \begin{bmatrix} 1 \\ -\mathbf{w}_{m,n}^{\mathrm{f}} \end{bmatrix} \tag{8.198}$$

that combines the recursive update in both the time n and the order m.

Given the gain $\mathbf{k}_{m,n-1}$, we first compute $\mathbf{k}_{m+1,n}$ with (8.198) then, from the first equation of (8.197), we calculate $\mathbf{k}_{m,n}$ as

$$\mathbf{k}_{m,n} = \mathbf{k}_{m+1,n}^{[m]} + k_{m+1}^{(m+1)}[n]\mathbf{w}_{m,n}^{\mathrm{b}}, \tag{8.199}$$

where

$$k_{m+1,n}^{(m+1)} = \frac{\varepsilon_{m,n}^{\mathrm{b}}}{J_{m,n}^{\mathrm{b}}}. \tag{8.200}$$

The (8.197) and (8.198) update requires the predictor $\mathbf{w}_{m,n}^{\mathrm{b}}$, $\mathbf{w}_{m,n}^{\mathrm{f}}$ and minimum energy errors $J_{m,n}^{\mathrm{b}}$, $J_{m,n}^{\mathrm{f}}$, adaptation.

For the calculation of the Kalman gain $\mathbf{k}_{m,n}$, we proceed substituting (8.187) in (8.199), for which it is

$$\mathbf{k}_{m,n} = \left(\frac{\mathbf{k}_{m+1,n}^{[m]} + k_{m+1}^{(m+1)}[n]\mathbf{w}_{m,n-1}^{\mathrm{b}}}{1 - k_{m+1}^{(m+1)}[n]e_m^{\mathrm{b}*}[n]} \right). \tag{8.201}$$

The algorithm organization for calculating fast fixed-order RLS or fast Kalman algorithm (FKA) is reported below.

FKA Algorithm Implementation

In the case of *fixed order*, we have $m = M$ for which the writing of the subscript m, where that is not expressly requested, may be omitted.

Suppose the estimates at the instant $(n-1)$ are known: $\mathbf{w}_{n-1}^{\mathrm{f}}$, $\mathbf{w}_{n-1}^{\mathrm{b}}$, $\mathbf{w}_{n-1}$, $\mathbf{k}_{n-1}$, J_{n-1}^{f}, the forward predictor algorithm structure $\big($i.e., $d[n] \equiv x[n]\big)$ is the following:

$$
\begin{aligned}
&e^{\mathrm{f}}[n] = x[n] - \mathbf{w}_{n-1}^{\mathrm{f}H}\mathbf{x}_{n-1},\\
&\mathbf{w}_n^{\mathrm{f}} = \mathbf{w}_{n-1}^{\mathrm{f}} + \mathbf{k}_{n-1}e^{\mathrm{f}*}[n],\\
&\varepsilon_m^{\mathrm{f}}[n] = x[n] - \mathbf{w}_n^{\mathrm{f}H}\mathbf{x}_{n-1},\\
&J_n^{\mathrm{f}} = \lambda J_{n-1}^{\mathrm{f}} + \varepsilon^{\mathrm{f}}[n]e^{\mathrm{f}*}[n],\\
&\mathbf{k}_{M+1,n} = \begin{bmatrix} 0 \\ \mathbf{k}_{n-1} \end{bmatrix} + \frac{\varepsilon_n^{\mathrm{f}}}{J_n^{\mathrm{f}}}\begin{bmatrix} 1 \\ -\mathbf{w}_n^{\mathrm{f}} \end{bmatrix},\\
&e^{\mathrm{b}}[n] = x[n-m] - \mathbf{w}_{n-1}^{\mathrm{b}H}\mathbf{x}_n,\\
&\mathbf{k}_n = \left(\frac{\mathbf{k}_{M+1,n}^{[M]} + k_{M+1}^{(M+1)}[n]\mathbf{w}_{n-1}^{\mathrm{b}}}{1 - k_{M+1}^{(M+1)}[n]e^{\mathrm{b}*}[n]} \right),\\
&\mathbf{w}_n^{\mathrm{b}} = \mathbf{w}_{n-1}^{\mathrm{b}} + \mathbf{k}_{n-1}e^{\mathrm{b}*}[n].
\end{aligned}
$$

For the transversal filter coefficients updating (the new input data are $\mathbf{x}_n$ and $d[n]$), we proceed as

$$
\begin{aligned}
&e[n] = d[n] - \mathbf{w}_{n-1}\mathbf{x}_n,\\
&\mathbf{w}_n = \mathbf{w}_{n-1} + \mathbf{k}_n e^*[n].
\end{aligned}
$$

The resulting algorithm has a complexity $O(9M)$ for each iteration.

8.4.2.2 Fast *a Posteriori* Error Sequential Technique

The fast *a posteriori* error sequential technique (FAEST) algorithm, developed in [14] and discussed below, is one of the fastest algorithms of the RLS class as it has a complexity $O(7M)$. The FAEST is based on the *a posteriori* error for which for the calculation of the Kalman gain is used the expression (8.189) $\left(\tilde{\mathbf{k}}_{m,n} = \lambda^{-1}\mathbf{R}_{m,n-1}^{-1}\mathbf{x}_{m,n}\right)$.

Using (8.195), the input partition (8.154) and the definition of the error we get the Levinson's recurrence

$$
\tilde{\mathbf{k}}_{m+1,n} = \begin{bmatrix} 0 \\ \tilde{\mathbf{k}}_{m,n-1} \end{bmatrix} + \frac{e_m^{\mathrm{f}}[n]}{\lambda J_{m,n-1}^{\mathrm{f}}}\begin{bmatrix} 1 \\ -\mathbf{w}_{m,n-1}^{\mathrm{f}} \end{bmatrix} \tag{8.202}
$$

and

$$
\tilde{\mathbf{k}}_{m+1,n} = \begin{bmatrix} \tilde{\mathbf{k}}_{m,n} \\ 0 \end{bmatrix} + \frac{e_m^{\mathrm{b}}[n]}{\lambda J_{m,n-1}^{\mathrm{b}}}\begin{bmatrix} -\mathbf{w}_{m,n-1}^{\mathrm{b}} \\ 1 \end{bmatrix} \tag{8.203}
$$

that determines a relationship between $\tilde{\mathbf{k}}_{m,n-1}$ and $\tilde{\mathbf{k}}_{m,n}$. Proceeding as for the FKA

$$\tilde{\mathbf{k}}_{m,n} = \tilde{\mathbf{k}}^{[m]}_{m+1,n} + \tilde{k}^{(m+1)}_{m+1}[n]\mathbf{w}^{\mathrm{b}}_{m,n-1}, \tag{8.204}$$

where

$$\tilde{k}^{(m+1)}_{m+1,n} = \frac{e^{\mathrm{b}}_{m}[n]}{\lambda J^{\mathrm{b}}_{m,n-1}}. \tag{8.205}$$

Note that, unlike the FKA, the filter weight vector appears with time index $(n-1)$, the latter also enables the simple calculation of the backward *a priori* error as

$$e^{\mathrm{b}}_{m,n} = \lambda J^{\mathrm{b}}_{m,n-1}\tilde{k}^{(m+1)}_{m+1,n}. \tag{8.206}$$

An important aspect of the FAEST algorithm regards the *conversion factor* $\tilde{\alpha}_{m,n}$, also known as *likelihood variable* (6.76, Sect. 6.4.3.2), which links the *a priori* and *a posteriori* errors. In fact, the FEAST algorithm proceeds with the recursive calculation of the conversion factor defined as

$$\tilde{\alpha}_{m,n} = 1 + \tilde{\mathbf{k}}^{H}_{m,n}\mathbf{x}_{m,n} \tag{8.207}$$

and which can be updated by combining the order m and the time index n, as

$$\tilde{\alpha}_{m+1,n} = \tilde{\alpha}_{m+1,n-1} + \frac{\left|e^{\mathrm{f}}_{m}[n]\right|^2}{\lambda J^{\mathrm{f}}_{m,n-1}}. \tag{8.208}$$

In addition, from (8.203) and the upper partition (8.154), it is possible to obtain

$$\tilde{\alpha}_{m,n} = \tilde{\alpha}_{m+1,n} + \tilde{k}^{(m+1)}_{m+1}e^{\mathrm{b}*}_{m}[n] \tag{8.209}$$

or

$$\tilde{\alpha}_{m,n} = \tilde{\alpha}_{m+1,n} + \frac{\left|e^{\mathrm{b}}_{m}[n]\right|^2}{\lambda J^{\mathrm{b}}_{m,n-1}}, \tag{8.210}$$

which together with the (8.208) provides the update of the sequence $\tilde{\alpha}_{m+1,n-1} \rightarrow \tilde{\alpha}_{m+1,n} \rightarrow \tilde{\alpha}_{m,n}$.

Fixed-Order FAEST Algorithm Implementation

Knowing the estimates at time $((n-1)$: $\mathbf{w}^{\mathrm{f}}_{n-1}$, $\mathbf{w}^{\mathrm{b}}_{n-1}$, $\mathbf{w}_{n-1}$, $\tilde{\mathbf{k}}_{n-1}$, J^{b}_{n-1}, J^{f}_{n-1}, $\tilde{\alpha}_{m,n-1}$, at the arrival of new input data $\mathbf{x}_{n-1}$ and $d[n] \equiv x[n]$, the predictor structure is

$$
\begin{aligned}
&e^{\mathrm{f}}[n] = x[n] - \mathbf{w}_{n-1}^{\mathrm{f}H}\mathbf{x}_{n-1},\\
&\mathbf{w}_n^{\mathrm{f}} = \mathbf{w}_{n-1}^{\mathrm{f}} + \tilde{\mathbf{k}}_{n-1}e^{\mathrm{f}*}[n],\\
&\varepsilon^{\mathrm{f}}[n] = \tilde{\alpha}_{n-1}^{-1}e^{\mathrm{f}}[n],\\
&J_n^{\mathrm{f}} = \lambda J_{n-1}^{\mathrm{f}} + \varepsilon^{\mathrm{f}}[n]e^{\mathrm{f}*}[n],\\
&\tilde{\mathbf{k}}_{M+1,n} = \begin{bmatrix} 0 \\ \tilde{\mathbf{k}}_{n-1} \end{bmatrix} + \frac{e_n^{\mathrm{f}}}{\lambda J_{n-1}^{\mathrm{f}}}\begin{bmatrix} 1 \\ -\mathbf{w}_{n-1}^{\mathrm{f}} \end{bmatrix},\\
&e^{\mathrm{b}}[n] = \lambda J_{n-1}^{\mathrm{b}}\tilde{g}_{M+1}^{(M+1)}[n],\\
&\tilde{\mathbf{k}}_n = \tilde{\mathbf{k}}_{M+1,n}^{[M]} + \tilde{k}_{M+1}^{(M+1)}[n]\mathbf{w}_{n-1}^{\mathrm{b}},\\
&\tilde{\alpha}_{M+1,n} = \tilde{\alpha}_{n-1} + \frac{\left|e^{\mathrm{f}}[n]\right|^2}{\lambda J_{n-1}^{\mathrm{f}}},\\
&\tilde{\alpha}_n = \tilde{\alpha}_{M+1,n} + \tilde{k}_{M+1}^{(M+1)*}e_m^{\mathrm{b}}[n],\\
&\mathbf{w}_n^{\mathrm{b}} = \mathbf{w}_{n-1}^{\mathrm{b}} + \tilde{\mathbf{k}}_n\varepsilon^{\mathrm{b}*}[n],\\
&\varepsilon^{\mathrm{b}}[n] = \tilde{\alpha}_n^{-1}e^{\mathrm{b}}[n],\\
&J_n^{\mathrm{b}} = \lambda J_{n-1}^{\mathrm{b}} + \varepsilon^{\mathrm{b}}[n]e^{\mathrm{b}*}[n].
\end{aligned}
$$

For the transversal filter updating (new input: $\mathbf{x}_n$ and $d[n]$), we proceed as

$$
\begin{aligned}
&e[n] = d[n] - \mathbf{w}_{n-1}^{H}\mathbf{x}_n,\\
&\varepsilon[n] = \tilde{\alpha}_n^{-1}e[n],\\
&\mathbf{w}_n = \mathbf{w}_{n-1} + \tilde{\mathbf{k}}_n\varepsilon^*[n].
\end{aligned}
$$

Note that the FAEST algorithm is very similar to the FKA but has a complexity $O(7M)$ for each iteration.

8.4.2.3 *A Priori* Error Fast Transversal Filter

The fast transversal filter (FTF) algorithm presented in [15], similar to the FKA and FAEST, is based on the *a priori* error. Similarly to the FEAST for the Kalman gain calculation, the relation (6.77) is used (see Sect. 6.4.1.2), which defines the likelihood variable, rewritten in recursive order notation as

$$
\alpha_{m,n} = 1 + \mathbf{k}_{m,n}^{H}\mathbf{x}_{m,n} \tag{8.211}
$$

such that $\tilde{\alpha}_{m,n} = 1/\alpha_{m,n}$; from (8.202), (8.203), and the upper–lower partition in (8.154), it is possible to write, respectively,

$$\alpha_{m+1,n} = \alpha_{m,n} - \frac{\left|e_m^{\rm b}[n]\right|^2}{J_{m,n}^{\rm b}}, \tag{8.212}$$

$$\alpha_{m+1,n} = \alpha_{m,n-1} - \frac{\left|e_m^{\rm f}[n]\right|^2}{J_{m,n}^{\rm f}}. \tag{8.213}$$

In FTF algorithm, the term $\tilde{\alpha}_{m,n}$ is replaced with $1/\alpha_{m,n}$ and the update equation of the likelihood variable $\tilde{\alpha}_{M+1,n} = \tilde{\alpha}_{n-1} + \left|e^{\rm f}[n]\right|^2/\lambda J_{n-1}^{\rm f}$ is replaced with (8.213). To get the term $\alpha_{m,n}$ from $\alpha_{m+1,n}$ the expression $\tilde{\alpha}_n = \tilde{\alpha}_{M+1} + \tilde{k}_{M+1}^{(M+1)*} e_m^{\rm b}[n]$ is not used, but the relationship

$$\alpha_{m,n} = \frac{\alpha_{m+1,n}}{1 + \alpha_{m+1,n}\tilde{k}_{m+1}^{(m+1)} e_m^{\rm b*}[n]} \tag{8.214}$$

obtained by combining the (8.210), (8.204), and $\tilde{\alpha}_{m,n} = 1/\alpha_{m,n}$. The algorithm has the same complexity of the FAEST.

Initialization

The FRLS algorithms class, in the case of implementation as a direct form transversal filter, is initialized considering the following IC:

$$J_{-1}^{\rm f} = J_{-1}^{\rm b} = \delta > 0 \tag{8.215}$$

$$\alpha_{-1} = 1 \quad \text{or} \quad \tilde{\alpha}_{-1} = 1 \tag{8.216}$$

with all other quantities void. The constant δ is positive with order of magnitude equal to $0.01\sigma_{x[n]}^2$. For a forgetting factor $\lambda < 1$, the effects of these ICs are quickly canceled.

References

1. Vaidyanathan PP (2008) The theory of linear prediction. In: Synthesis lectures on signal processing, vol 3. Morgan & Claypool, San Rafael, CA. ISBN 9781598295764, doi:0.2200/S00086ED1V01Y200712SPR03
2. Wiener N (1949) Extrapolation, interpolation and smoothing of stationary time series, with engineering applications. Wiley, New York
3. Manolakis DG, Ingle VK, Kogon SM (2005) Statistical and adaptive signal processing. Artech House, Boston, MA. ISBN 1-58053-610-7
4. Golub GH, Van Loan CF (1989) Matrix computation. John Hopkins University Press, Baltimore and London. ISBN 0-80183772-3

5. Strang G (1988) Linear algebra and its applications, 3rd edn. Thomas Learning, Lakewood, CO. ISBN 0-15-551005-3
6. Makhoul J (1975) Linear prediction: a tutorial review. Proc IEEE 63:561–580
7. Markel JD, Gray AH (1976) Linear prediction of speech. Springer, New York
8. Rabiner LB, Schafer RW (1978) Digital processing of speech signal. Prentice-Hall, Englewood Cliffs, NJ. ISBN 0-13-213603-1
9. Noble B, Daniel JW (1988) Applied linear algebra. Prentice-Hall, Englewood Cliffs, NJ
10. Petersen KB, Pedersen MS (2012) The matrix cookbook, Tech. Univ. Denmark, Kongens Lyngby, Denmark, Tech. Rep
11. Ammar GS, Gragg WB (1987) The generalized Schur algorithm for the superfast solution of Toeplitz systems. Rational Approx Appl Math Phys Lect Notes Math 1237:315–330
12. Levinson N (1947) The Wiener rms error criterion in filter design and prediction. J Math Phys 25:261–278
13. Atal BS, Schroeder MR (1979) Predictive coding of speech signals and subjective error criteria. IEEE Trans Acoust Speech Signal Process 27:247–254
14. Carayannis G, Manolakis DG, Kalouptsidis N (1983) A fast sequential algorithm for least-squares filtering and prediction. IEEE Trans Acoust Speech Signal Process ASSP-31:1394–1402
15. Cioffi JM, Kailath T (1984) Fast recursive least squares transversal filters for adaptive filtering. IEEE Trans ASSP 32:304–337
16. Griffiths LJ (1977) A continuously-adaptive filter implemented as a lattice structure. In: IEEE international acoustics, speech, and signal processing, conference (ICASSP'77), pp 683–68
17. Griffiths LJ (1978) An adaptive lattice structure for noise-cancelling applications. In: Proceedings of IEEE international acoustics, speech, and signal processing, conference (ICASSP'78), pp 87–90
18. Falconer DD, Ljung L (1978) Application of fast Kalman estimation to adaptive equalization. IEEE Trans Commun 26(10):1439–1446
19. Ling F (1991) Givens rotation based least-squares lattice and related algorithms. IEEE Trans Signal Process 39:1541–1551
20. Ling F, Manolakis D, Proakis JG (1896) Numerically robust least-squares lattice-ladder algorithm with direct updating of the reflection coefficients. IEEE Trans Acoust Speech Signal Process 34(4):837–845
21. Ling F, Proakis JG (1986) A recursive modified Gram–Schmidt algorithm with applications to least-squares and adaptive filtering. IEEE Trans Acoust Speech Signal Process 34(4):829–836
22. Ljung S, Ljung L (1985) Error propagation properties of recursive least-squares adaptation algorithms. Automatica 21:157–167
23. Slock DTM, Kailath T (1991) Numerically stable fast transversal filters for recursive least squares adaptive filtering. IEEE Trans Signal Process 39:92–114
24. Burg JP (1975) Maximum entropy spectral analysis. Ph.D. dissertation, Stanford University, Stanford
25. Chen S-J, Gibson JS (2001) Feedforward adaptive noise control with multivariable gradient lattice filters. IEEE Trans Signal Process 49(3):511–520
26. Kay SM (1988) Modern spectral estimation: theory and applications. Prentice-Hall, Englewood Cliffs, NJ
27. Kay SM, Marple SL (1981) Spectrum analysis—a modern perspective. Proc IEEE 69:1380–1419
28. Makhoul J (1978) A class of all-zero lattice digital filters: properties and applications. IEEE Trans Acoust Speech Signal Process ASSP-26:304–314
29. Marple SL (1987) Digital spectral analysis with applications. Prentice-Hall, Englewood Cliffs, NJ
30. Merched R (2003) Extended RLS lattice adaptive filters. IEEE Trans Signal Process 51 (9):2294–2309
31. Merched R, Sayed AH (2001) Extended fast fixed order RLS adaptive filtering. IEEE Trans Signal Process 49(12):3015–3031
32. Vaidyanathan PP (1986) Passive cascaded-lattice structures for low-sensitivity FIR filter design with applications to filter banks. IEEE Trans Circuits Syst CAS-33(11):1045–1064

Chapter 9
Discrete Space-Time Filtering

9.1 Introduction

In many scientific and technological areas, acquiring signals relating to the same stochastic process, with a multiplicity of homogeneous sensors and arranged in different spatial positions, is sometimes necessary or simply useful. For example, this is the case of the acquisition of biomedical signals, such as electroencephalogram (EEG), electrocardiogram (ECG), and tomography or of telecommunications signals such as those deriving from the antenna arrays and radars, the detection of seismic signals, the sonar, and the microphone arrays for the acquisition of acoustic signals. The phenomena measured in these applications may have different physical nature but, in any case, the array of sensors, or receivers, is made to acquire processes concerning the propagation of electromagnetic or mechanical waves coming from one or more radiation sources.

The arrangement of sensors illuminated from an energy field requires taking into account, in addition, to the temporal sampling, also the spatial sampling. The energy carried by a wave may be intercepted by a single receiver, of adequate size, or by a set of sensors (sensor array) which spatially sample the field. In the first case, the spatially acquired signal will have *continuous-space* nature, while *discrete-space* in the second one. In the electromagnetic case, the continuous-spatial sampling can be performed with an antenna, sized to be adequately illuminated by the impinging wave.

The discrete-space sampling occurs when as a field of acquisition a set of punctiform sensors is used. For example, in the case of acoustic field a set of microphones with omnidirectional characteristic can be used. Note that in both of the described situations, the geometry of the acquisition system plays a primary role. In the continuous case, the sensor size must be properly calculated as a wavelength function of the wave to be acquired. Similarly, as better described later in this chapter, in the case of array, the distance between sensors (or interdistance) must be such as to avoid the *spatial aliasing* phenomena.

A. Uncini, *Fundamentals of Adaptive Signal Processing*, Signals and Communication Technology, DOI 10.1007/978-3-319-02807-1_9,

The processing of signals from homogeneous and spatially distributed sensors array is referred to as *array signal processing* or simply array processing (AP) [1–6]. The purpose of the AP is, in principle, the same as the classical signal processing: the extraction of significant information from the acquired data. In the case of linear DSP, due to the time sampling nature, discrimination can take place in the frequency domain. The acquired signal can be divided and filtered according to its spectral characteristics.

In the case of spatial sampling, the distribution introduced by the array allows for some directional discrimination. Then, the possibility of discrimination of the signals exists, as well as in the usual time domain, even through the domain of the angle of arrival, named also *spatial frequency*.

9.1.1 Array Processing Applications

The main purpose of the AP can be summarized as:

- *Signal-to-noise ratio* (SNR) improvement of the received signal from one or more specific directions called look-directions (LD), with respect to the signal acquired with a single sensor.
- Determination of number, location, and waveform of the sources that propagate energy.
- Separation of independent sources.
- Motion tracking of sources emitting energy.

The SNR can be improved by considering the *radiation diagram* (or *spatial response* or *directivity pattern*) of the sensors array. The techniques referred to as *beamforming* (BF) allow, with appropriate procedures, to steer the array directivity pattern toward the source of interest (SOI). At the same time, such techniques also allow to mitigate the effects of any disturbing sources coming from other directions, through methods of sidelobe cancellation.

The estimation of the source position is performed by using methods based on the so-called direction of arrivals (DOA), through which it is possible to trace the angle of arrival of a wave. In practice, both BF and DOA methodologies are very often run at the same time allowing simultaneous *source tracking* and the spatial filtering.

Algorithms of *source separation* are defined in the case of multiple-independent sources operating at the same frequencies but in different spatial positions.

9.1.2 Types of Sensors

One of the basic AP assumptions is to have sensors with linear response, punctiform and omnidirectional, i.e., that respond in the same way to signals from all directions and all frequencies; such type of sensor is said to be *isotropic*. In acoustic and

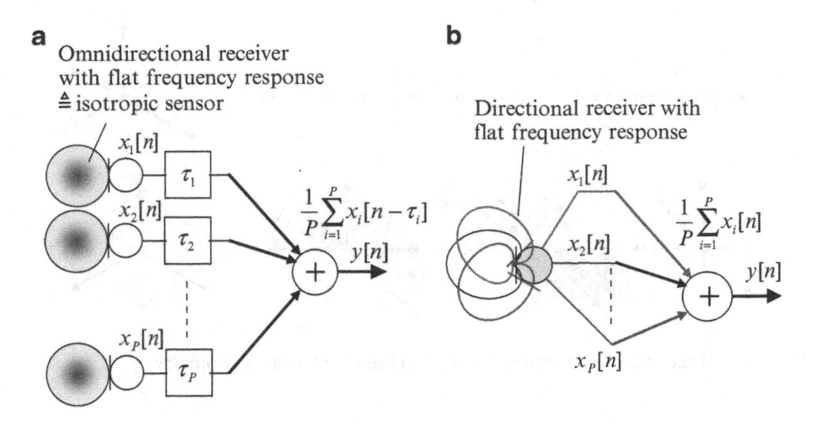

Fig. 9.1 Examples of sensor arrays: (**a**) non-coincident and (**b**) coincident

mechanical areas, these are microphones, hydrophones, geophones, accelerometers, vibrometers, etc. In the case of electromagnetic (EM) fields, we have electrodes, antennas, radar, etc. In the case of real sensors, the above ideal assumptions are almost never verified. Nevertheless, it is possible, within certain limits, to use correction filters, located downstream of the sensors themselves. These filters are able to perform appropriate space-frequency equalization.

In the case of isotropic sensors, as illustrated in Fig. 9.1a, the directivity pattern composition implies a certain spatial distribution of the receivers. Indeed, if they were arranged in the same point, the overall response would also be isotropic.

In some types of applications, very common in the audio sector, the array is *coincident*, i.e., all the sensors are positioned on the same point (*coincident microphone array*). In this case, as illustrated in Fig. 9.1b, the receivers are not isotropic but have a certain spatial directivity. The radiation pattern of the array is given by the additive combination of the individual microphones diagrams.

9.1.3 Spatial Sensors Distribution

The spatial array configuration is of fundamental importance, and it is highly dependent on the application context. The sensors geometry affects the processing methodology and the global array performance. The sensors position performs, in fact, sampling in the spatial domain.

In general, even if we can think the sensors distribution in three-dimensional space (3D), as illustrated in Fig. 9.2, the sensors most often are arranged on planes (2D) or lines (1D). For example, some typical configurations are below indicated:

- 1D: linear uniform distribution, harmonic nested sensors.
- 2D: *T*, *P*-Greek, *X*, circular, random.
- 3D: coincident spherical, cylindrical, with precise geometry, random, etc.

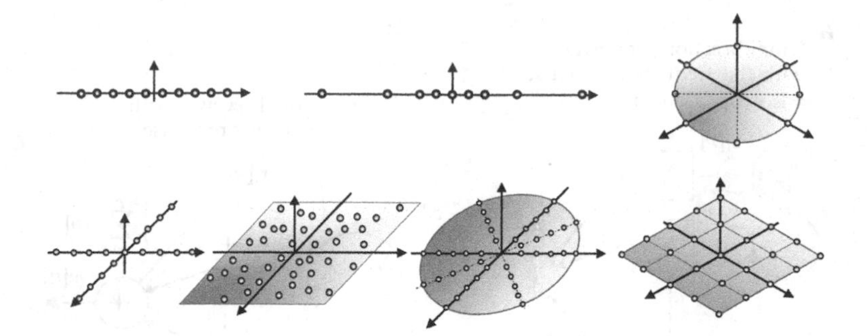

Fig. 9.2 Spatial receivers distribution. Typical 1D and 2D array's geometry

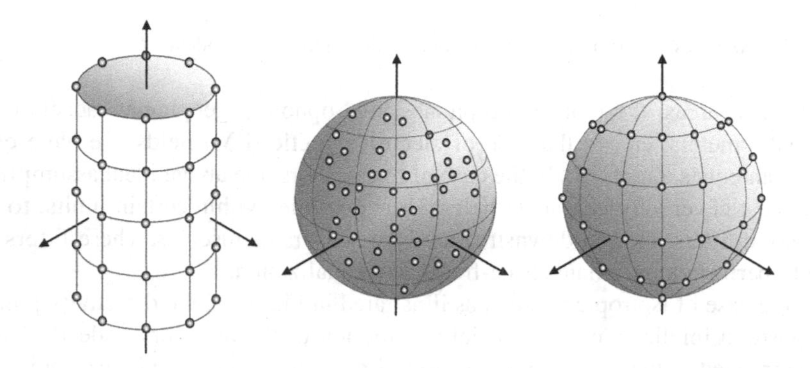

Fig. 9.3 Typical geometry of 3D array

In Fig. 9.3 are reported some typical architecture of 3D arrays.

9.1.4 AP Algorithms

A generic *array processor* is a MISO or MIMO system and its design consists in (1) determination of the geometry which is typically a priori chosen in accordance with considerations related to the application domain and (2) free parameters calculation, i.e., *discrete space-time filter synthesis*.

The filter synthesis can be performed with the same paradigms used in DT circuits and, in general, we proceed with an optimization criterion related to one or more quality's indices, subject to constraints of varied nature. The optimization procedure can be done by a static or an adaptive algorithm.

As illustrated in Fig. 9.1, the simplest processing form is the sum of the signals coming from the array. If the SOI appears to be in phase, with respect to the sensors, and the noise is different between the sensors (i.e., characterized by random phase), there is an improvement in output SNR directly proportional to the number of sensors. More generally, the signals from the array are processed with static or adaptive systems. In the BF case, in the presence of a single source, the filter is a MISO system. In the case of multiple sources, the system is MIMO.

Regarding processing techniques, we can think of paradigms based on desired frequency and spatial response of the array. If the BF is designed regardless of the signal statistic, the class of algorithms is *data independent*. The BF analysis and design are similar to that performed in digital filtering techniques.

In the case in which for the determination of the array response, the statistics of the input signals (known or estimated) are used, the algorithm class is *data dependent*. The analysis and design techniques derive, in this case, from the methodologies introduced in the adaptive filtering.

Note that the adaptive algorithms are always, by definition, data dependent, while for static algorithms they can be chosen according to both design philosophies. Regarding the processing methods, these can be batch or adaptive, of the first or second order and implemented in the time or in the frequency domain. In some specific applications, in addition to classical algorithms discussed in earlier chapters and extended to cases of MISO and MIMO systems, AP methodologies include specific optimization constraints due to the desired spatial response.

A further distinction is also related to the bandwidth of the signals to be acquired. In the case of EM signals, the antenna array capture modulated signals which, by definition, are narrowband. In other applications, as for example in speech signal capture, the process is considered broadband, even with respect to the array physical dimension.

9.2 Array Processing Model and Notation

The main AP's objective is to use a set of sensors, suitably distributed, to perform a space-time sampling of a travelling wave of an electromagnetic or acoustic-mechanical field, which propagates in a medium. The signal processing must be done in space-time domain with the purpose of extracting useful information in the presence of noise and interference. The array captures the energy, electromagnetic or mechanical, coming from one or more sources with a certain looking direction for simultaneous SNR increase and interference reduction.

9.2.1 Propagation Model

The signal model, due to the spatial propagation, is that resulting from the solution of the wave equation that can be written as

$$\nabla^2 s(t,\mathbf{r}) = \frac{1}{c^2}\frac{\partial^2 s(t,\mathbf{r})}{\partial t^2} \tag{9.1}$$

where the space-time function $s(t,\mathbf{r})$ represents the waveform quantity related to the specific area of interest and c indicates the propagation speed. In the acoustic case, $s(t,\mathbf{r})$ represents a pressure wave propagating in a fluid (air, water, etc.), with propagation speed defined by the relation $c = \partial P/\partial \rho$ (P is the pressure and ρ the fluid density). For example, in the air the propagation speed in standard conditions is equal to $c \approx 334\ \mathrm{ms}^{-1}$. In the EM case, the quantity can represent the electric field $s(t,\mathbf{r}) \equiv E(t,\mathbf{r})$, with propagation speed in vacuum equal to $c = 1/\sqrt{\mu_0\varepsilon_0} \approx 3\times 10^8\,\mathrm{ms}^{-1}$.

In (9.1), the terms $\mathbf{r}$ and ∇^2 represent, respectively, the *position vector* and the *Laplacian operator* that assume different definitions depending on the type of the used spatial coordinates. For example, in the Cartesian coordinate system they are $\mathbf{r} = [x \quad y \quad z]^T$ and $\nabla^2 \triangleq \frac{\partial^2}{\partial x^2} + \frac{\partial^2}{\partial y^2} + \frac{\partial^2}{\partial z^2}$, while for the spherical system, illustrated in Fig. 9.4, the position vector appears to be $\mathbf{r} = r[\sin\theta\cos\phi \quad \sin\theta\sin\phi \quad \cos\theta]^T$ and, for simplicity omitting the indices $(t,\mathbf{r})$, the wave equation can be written as (see [7])

$$\frac{\partial^2 s}{\partial r^2} + \frac{1}{r^2}\frac{\partial^2 s}{\partial \theta^2} + \frac{1}{r^2\sin^2\theta}\frac{\partial^2 s}{\partial \phi^2} + \frac{2}{r}\frac{\partial s}{\partial r} + \frac{1}{r^2\tan\theta}\frac{\partial s}{\partial \theta} = \frac{1}{c^2}\frac{\partial^2 s}{\partial t^2}. \tag{9.2}$$

The solution of (9.1) in Cartesian coordinates for monochromatic plane wave can be written as

$$s(t,\mathbf{r}) = S_0 e^{j(\omega t - \mathbf{r}^T\mathbf{k})}. \tag{9.3}$$

In (9.3), S_0 is the wave amplitude and, in case a modulation is present, without loss of generality, we consider a time-variant amplitude $S_0 \to S(t)$. The variable ω represents the angular frequency (or pulsatance) of the signal ($\omega = 2\pi f$, $f = 1/T$ represents the temporal frequency and T is the period). Note that, a spherical wave can be approximated as a plane wave on the receiver, only if its distance is much greater than the square of the maximum physical size of the source, divided by the wavelength (*far-field hypothesis*). For the electromagnetic waves the hypothesis of plane wave is almost always true, while for acoustic fields almost never. In the case of the *near field*, the assumptions of a plane wave is not verified for which the solution of the wave equation is of the type

$$s(t,\mathbf{r}) = \frac{S_0}{4\pi r} e^{j(\omega t - rk)}. \tag{9.4}$$

With reference to Fig. 9.4, the $\mathbf{k}$ vector, called *wavenumber vector*, indicates the *speed* and the wave *propagation direction*, is defined as

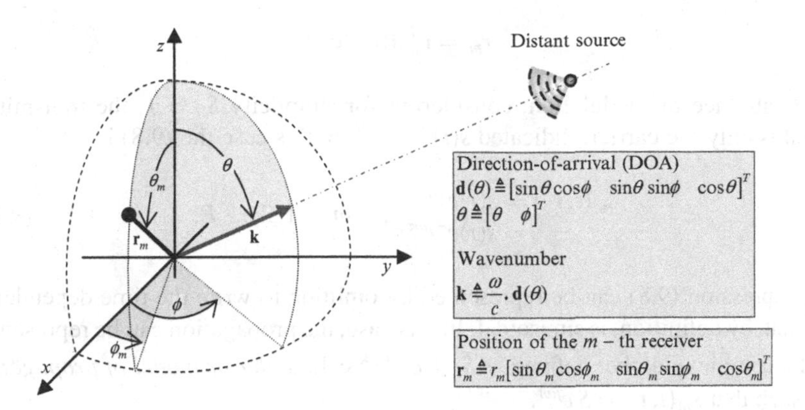

Fig. 9.4 Three-dimensional spatial description in spherical coordinates (*azimuth*, *elevation* and *range*) of the k-th source and the m-th receiver

$$\begin{aligned}\mathbf{k} &= k \cdot [\,\sin\theta\cos\phi \quad \sin\theta\sin\phi \quad \cos\theta]^T, \\ &= k \cdot \mathbf{d}(\theta)\end{aligned} \qquad \text{with} \quad k = |\mathbf{k}|. \tag{9.5}$$

where θ, for simplicity of notation, indicates the generic direction of arrival, i.e., in spherical coordinates, the pair $(\theta, \phi) \rightarrow \theta$.

The unit vector $\mathbf{d}(\theta)$, representing the direction of propagation of the wave, is referred to as direction of arrival (DOA). The amplitude of the vector wavenumber $\mathbf{k}$, or *scalar wavenumber* $k = |\mathbf{k}|$ along travel direction, is related to the propagation speed, as

$$k = 2\pi/\lambda = \omega/c \tag{9.6}$$

where $\lambda = c/f$ is the *wavelength.*

The receiver spatial position, by hypothesis punctiform, is indicated with the vector $\mathbf{r}_m$ defined, similarly to (9.5), in spherical or Cartesian coordinates, as

$$\begin{aligned}\mathbf{r}_m &= r_m[\,\sin\theta_m\cos\phi_m \quad \sin\theta_m\sin\phi_m \quad \cos\theta_m\,]^T \\ &= [\,x_m \quad y_m \quad z_m\,]^T.\end{aligned} \tag{9.7}$$

The signal at m-th sensor vicinity, for (9.3), (9.5), (9.6), and (9.7), can be written as

$$\begin{aligned}s_m(t, \mathbf{r}) &= S_0 e^{j\left(\omega t - \mathbf{r}_m^T \mathbf{k}\right)} \\ &= S_0 e^{j\omega t} e^{-j(\omega/c)\mathbf{r}_m^T \mathbf{d}(\theta)}.\end{aligned} \tag{9.8}$$

From the plane wave assumption, the wavenumber vector $\mathbf{k}$ does not depend on the sensor position for which the receivers are irradiated with the same but delayed signal. Indicating as origin the coordinates of the sensor selected as the reference $\mathbf{r}_1 = [0 \quad 0 \quad 0]^T$, the propagation delay between the sensors is obtained as

$$\tau_m = \mathbf{r}_m^T \mathbf{d}(\theta)/c. \tag{9.9}$$

In the absence of modulation, considering for simplicity $S_0 = 1$, the transmitted signal is only the carrier, indicated $s(t) = e^{j\omega t}$. In this case the (9.8) is

$$\begin{aligned} s_m(t, \mathbf{r})\big| &= s(t)e^{-j\mathbf{r}_m^T k} \\ &= s(t)e^{-j\omega\tau_m}, \end{aligned} \quad m = 1, 2, ..., P. \tag{9.10}$$

The expression (9.8) can be represented by omitting to write the time dependence $e^{j\omega t}$ that, by definition, is sinusoidal. In this case, the propagation can be represented by the complex number defined as $\bar{S} = e^{-j\mathbf{r}_m^T\mathbf{k}}$ called *phasor model of propagation* [1], such that $s_m(t, \mathbf{r}) = \bar{S}\, e^{j\omega t}$.

Note that for many theoretical developments, it is sometimes necessary to consider the spatial derivative of the wave equation solution (9.3) or (9.4). Whereas the model phasor, i.e., omitting writing of the temporal dependence in the case of far-field, takes the form

$$\frac{\partial^n}{\partial \mathbf{r}^n} s(t, \mathbf{r}) = S_0(-j\mathbf{k})^n e^{-j\mathbf{r}^T\mathbf{k}} \tag{9.11}$$

while from (9.4), in the case of near field we have that

$$\frac{\partial^n}{\partial r^n} s(t, r) = S_0 \frac{n!}{r^{n+1}} e^{-jr\mathbf{k}} (-1)^n \sum_{m=0}^{n} \frac{(jr\mathbf{k})^m}{m!}. \tag{9.12}$$

9.2.1.1 Steering

The *direction* or *steering* is defined as the phasor appearing in the solution of the wave equation $\bar{a}_m(\mathbf{r}, \mathbf{k}) = e^{-j\mathbf{r}_m^T\mathbf{k}}$, also referred to as $\bar{a}_m(\omega, \theta) = e^{-j\mathbf{r}_m^T\mathbf{k}}$. The variable $\bar{a}_m(\omega, \theta)$ contains all the geometric information about the wave that radiates the m-th sensor.

Whereas a single radiation source, in the case of sensors P, we define a *direction vector* or *steering vector* as

$$\bar{\mathbf{a}}(\omega, \theta) \in \mathbb{C}^{P\times 1} = \begin{bmatrix} e^{-j\mathbf{r}_1^T\mathbf{k}} & e^{-j\mathbf{r}_2^T\mathbf{k}} & \cdots & e^{-j\mathbf{r}_P^T\mathbf{k}} \end{bmatrix}^T. \tag{9.13}$$

The vector $\bar{\mathbf{a}}(\omega, \theta)$ incorporates all the spatial characteristic and propagation of the wave that illuminates the array. From the mathematical point of view, it represents a *differential manifold* or simply a *manifold*, i.e., in terms of differential geometry, a mathematical space that on a scale sufficiently small behaves like a Euclidean space. Formally, the manifold is introduced as a continuous set of *steering vectors* defined as

$$\mathcal{M} \triangleq \{\bar{\mathbf{a}}(\omega,\theta);\ \theta \in \Theta,\ \omega \in \Omega\} \tag{9.14}$$

where for a certain angular frequency ω, Θ represents the visual field or field-of-view (FOV) of the array. For example, for some 1D array the FOV is usually equal to $\Theta \in [-90°, 90°]$.

Note that in the case of a plane wave, by (9.10), the steering vector is

$$\bar{\mathbf{a}}(\omega,\theta) \in \mathbb{C}^{P\times 1} = \begin{bmatrix} 1 & e^{-j\omega\tau_2} & \cdots & e^{-j\omega\tau_P} \end{bmatrix}^T \tag{9.15}$$

where by definition $\tau_1 = 0$, i.e., the first sensor is the reference one.

9.2.1.2 Sensor Directivity Function and Steering Vector

A receiver is called *isotropic* if it has a flat frequency response in the band of interest and identical for all directions. In the case of receivers not isotropic, it is necessary to define a response function, in directivity and frequency $b_m(\omega,\theta)$, also called *sensor radiation diagram*, defined as

$$b_m(\omega,\theta) = B_m(\omega,\theta)e^{-j\gamma_m(\omega,\theta)}, \quad m = 1, 2, ..., P \tag{9.16}$$

where $B_m(\omega,\theta)$ is the gain and $\gamma_m(\omega,\theta)$ the phase for the m-th sensor. The (9.16) is a complex function which can be simply multiplied to the propagation phasor model, determining an attenuation or amplification, in function of frequency and angle. In the case of non-isotropic sensor, the steering $\bar{a}_m(\omega,\theta)$ must also include the radiation diagram. For which it is

$$a_m(\omega,\theta) = \bar{a}_m(\omega,\theta)\cdot b_m(\omega,\theta) = b_m(\omega,\theta)e^{-j\mathbf{r}_m^T\mathbf{k}}. \tag{9.17}$$

Indicating with

$$\mathbf{b}(\omega,\theta) \triangleq \begin{bmatrix} b_1(\omega,\theta) & b_2(\omega,\theta) & \cdots & b_P(\omega,\theta) \end{bmatrix}^T, \tag{9.18}$$

the vector with the radiation diagram of the receivers, the steering vector, referred to as $\mathbf{a}(\omega,\theta) \in \mathbb{C}^{P\times 1}$, is redefined as

$$\begin{aligned} \mathbf{a}(\omega,\theta) &= \bar{\mathbf{a}}(\omega,\theta) \odot \mathbf{b}(\omega,\theta) \\ &= \begin{bmatrix} b_1(\omega,\theta)e^{-j\mathbf{r}_1^T\mathbf{k}} & b_2(\omega,\theta)e^{-j\mathbf{r}_2^T\mathbf{k}} & \cdots & b_P(\omega,\theta)e^{-j\mathbf{r}_P^T\mathbf{k}} \end{bmatrix}^T \end{aligned} \tag{9.19}$$

where the symbol $\odot$ indicates the Hadamard product (point-to-point vectors multiplication). Note that for isotropic receivers $\mathbf{a}(\omega,\theta) \equiv \bar{\mathbf{a}}(\omega,\theta)$.

Figure 9.5 shows some typical examples of radiation diagram of electromagnetic or acoustic sensors.

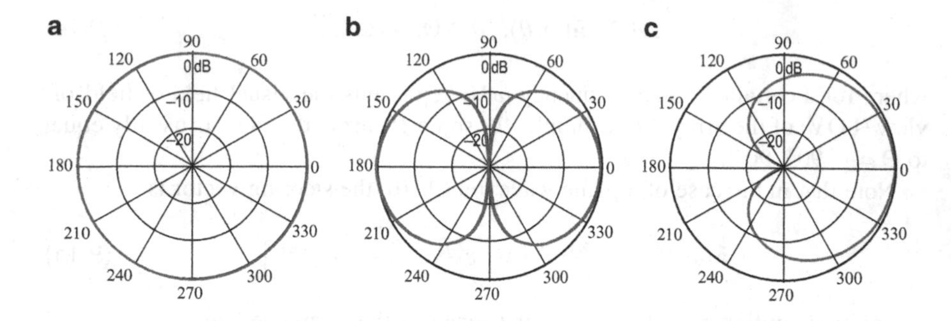

Fig. 9.5 Examples of *spatial radiation diagram* of sensors, evaluated for a specific frequency $b_m(\omega_0,\theta)$, (**a**) omnidirectional or isotropic; (**b**) "eight" diagram; (**c**) cardioid diagram

9.2.2 Signal Model

An array of sensors, as illustrated in Fig. 9.6, samples the propagated signal by the wave in space-time mode. The spatial sampling is due to the presence of multiple sensors in precise *geometric loci* while the temporal one is due to the analog to digital conversion of the acquired analog signal.

For the definition of a numerical model of the acquired signal from the array, for simplicity, we consider the case of sufficiently distant sources for which the propagated waves can be considered plane (*plane wave hypothesis*) and consider two separate cases:

(i) Free-field propagation model and no reflections (*anechoic or free-field model*).
(ii) Confined propagation model with reverb due to reflections of reflective surfaces (*echoic or confined model*).

9.2.2.1 Anechoic Signal Propagation Model

For the hypothesis (i), the received signal from the *m*-th sensor, considering the steering in (9.17), is defined as

$$\begin{aligned} x_m(t) &= s_m(t,\mathbf{r}) + n_m(t) \\ &= a_m(\omega,\theta)s(t) + n_m(t) \end{aligned} \tag{9.20}$$

where $n_m(t)$ is the *measurement noise* that is by hypothesis independent and different for each sensor. In addition, sometime, the noise can be subdivided in stationary and nonstationary components, i.e., $n_m(t) = n_m^s(t)+n_m^n(t)$.

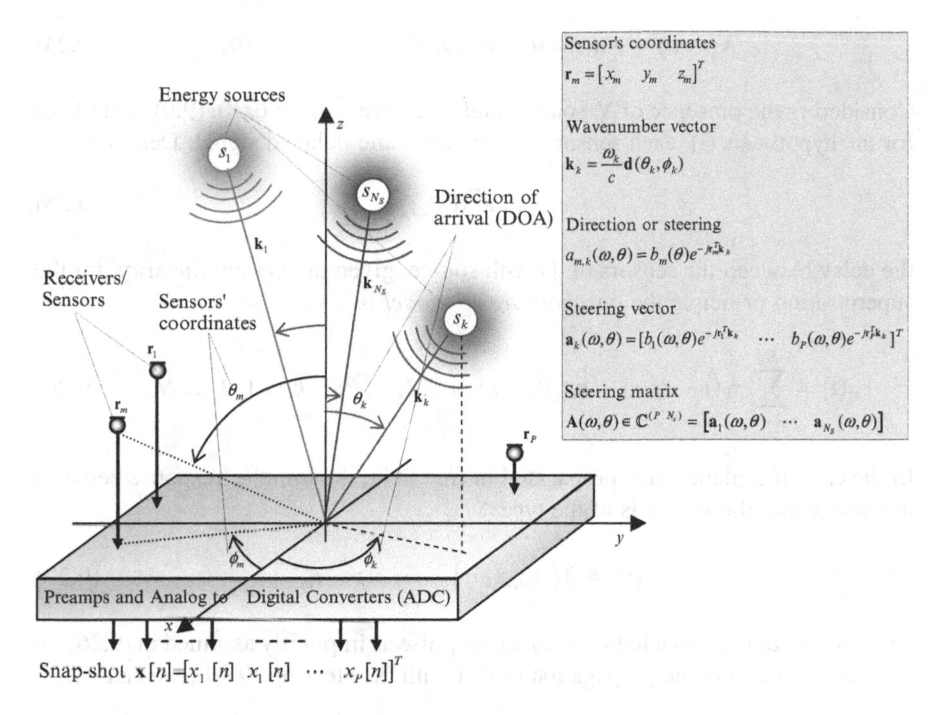

Fig. 9.6 Schematic of three-dimensional distribution of the sensors arrays. P represents the number of sensors, θ_k and ϕ_k the angles of arrival of the wave

Generalizing the previous vector notation we can write

$$\mathbf{x}(t) = \mathbf{a}(\omega, \theta)s(t) + \mathbf{n}(t) \tag{9.21}$$

where $\mathbf{x}(t) = \begin{bmatrix} x_1(t) & \cdots & x_P(t) \end{bmatrix}^T$ and $\mathbf{n}(t) = \begin{bmatrix} n_1(t) & \cdots & n_P(t) \end{bmatrix}^T$, $(P \times 1)$ vectors, which indicate, respectively, the output sensors *snap-shot* and the additive measurement noise. In case of linear propagation medium, the *superposition principle* is applied according to which, in the presence of N_S distant sources and incident on all the sensors, we can write

$$\mathbf{x}(t) = \sum_{k=1}^{N_S} \mathbf{a}_k(\omega, \theta)s_k(t) + \mathbf{n}(t). \tag{9.22}$$

In vector notation (9.22) becomes

$$\mathbf{x}(t) = \mathbf{A}(\omega, \theta)\mathbf{s}(t) + \mathbf{n}(t) \tag{9.23}$$

where $\mathbf{s}(t) = \begin{bmatrix} s_1(t) & \cdots & s_{N_S}(t) \end{bmatrix}^T$, and $\mathbf{A}(\omega, \theta) \in \mathbb{C}^{P \times N_S}$ is the *steering matrix* containing the P steering vectors related to the N_S sources. Therefore, we have

$$\mathbf{A}(\omega,\theta) = \begin{bmatrix} \mathbf{a}_1(\omega,\theta) & \mathbf{a}_2(\omega,\theta) & \cdots & \mathbf{a}_{N_S}(\omega,\theta) \end{bmatrix}. \tag{9.24}$$

Considering the presence of N_s sources and P receivers, from (9.8), (9.9), and (9.10) for the hypothesis (i), each sensor receives the same delayed signal. Defining

$$\tau_{m,k} = \mathbf{r}_m^T \mathbf{d}(\theta_k)/c \tag{9.25}$$

the delay between the sensors of the k-th source, given the system linearity, for the superposition principle the *anechoic signal model* is

$$x_m(t) = \sum_{k=1}^{N_S} s_k(t - \tau_{m,k}) + n_m(t), \quad m = 1, 2, \ldots, P; \quad k = 1, 2, \ldots, N_s. \tag{9.26}$$

In the case of a plane wave propagated in free field, the impulse response between the source and the sensor is of the type

$$a_{m,k}(t) \triangleq \Im\{A_{m,k}(\omega)\}^{-1} = \delta(t - \tau_{m,k}) \tag{9.27}$$

i.e., is a pure delay, modeled with a delayed pulse as implicitly assumed in (9.26). In this case, indicating the propagation model with the steering (9.19), we can write

$$A_{m,k}(\omega) \triangleq b_m(\omega,\theta) e^{-j\omega_k \tau_{m,k}}, \quad m = 1, 2, \ldots, P; \quad k = 1, 2, \ldots, N_s. \tag{9.28}$$

In the case of anechoic model for a plane wave coming from a certain direction, the steering vector models exactly the propagation delays on the sensors.

9.2.2.2 Echoic Signal Propagation Model

In the case of plane wave propagation in a confined environment, i.e., in the presence of reflections, with only one source of any form $s(t)$, the signal on the sensor can be expressed as

$$x_m(t) = a_m(t) * s(t) + n_m(t), \quad m = 1, 2, \ldots, P, \tag{9.29}$$

where $a_m(t)$ is the impulse response of the path between the source and the m-th sensor. The $a_m(t)$ impulse response and the relative TF, defined as $A_m(\omega) = \Im\{a_m(t)\}$, implicitly contains all deterministic-geometric information known about the array such as the direction, the propagation model, the directivity function, the spatial-frequency response of the sensor, the propagation delay between the source and the m-th sensor, and the possible presence of multiple paths due to reflections.

Such a propagation environment is said to be *reverberant* or *multipath* and is generally modeled as a discrete time with an FIR filter of length N_a, indicated as $\mathbf{a}_m = \begin{bmatrix} a_m[0] & \cdots & a_m[N_a-1] \end{bmatrix}^T$, different for each sensor.

In the discussion that follows, we consider the sensors' output directly in numerical form assuming an ideal analog-digital conversion.

As shown in Fig. 9.6, indicating with

$$\mathbf{a} \in \mathbb{R}^{P\times 1(N_a)} = \begin{bmatrix} \mathbf{a}_1 & \mathbf{a}_2 & \cdots & \mathbf{a}_P \end{bmatrix}^T \tag{9.30}$$

the matrix containing the P impulse responses of length N_a, and with $\mathbf{s}_n \in \mathbb{R}^{N_a\times 1} = \begin{bmatrix} s[n] & \cdots & s[n-N_a+1] \end{bmatrix}^T$ the signal vector, the sensors' snap-shot can be expressed in vector mode directly in the DT domain as

$$\mathbf{x}[n] = \mathbf{a}\mathbf{s}_n + \mathbf{n}[n]. \tag{9.31}$$

In the anechoic case, with $\mathbf{a}(\omega,\theta) \in \mathbb{C}^{P\times 1}$, (9.31) is reduced to the form (9.21) expressed in discrete time as

$$\mathbf{x}[n] = \mathbf{a}(\omega,\theta)s[n] + \mathbf{n}[n]. \tag{9.32}$$

In the presence of multiple sources, similarly to (9.23), we define $\mathbf{A}(\omega,\theta) \in \mathbb{R}^{P\times N_s(N_a)}$, the matrix with the impulse responses $a_{m,k}[n]$ between the sources and the sensors, as

$$\mathbf{A}(\omega,\theta) \in \mathbb{R}^{P\times N_s(N_a)} = \begin{bmatrix} \mathbf{a}_{1,1}^T & \mathbf{a}_{1,2}^T & \cdots & \mathbf{a}_{1,N_s}^T \\ \mathbf{a}_{2,1}^T & \mathbf{a}_{2,2} & \cdots & \mathbf{a}_{2,N_s} \\ \vdots & \vdots & \ddots & \vdots \\ \mathbf{a}_{P,1} & \mathbf{a}_{P,2} & \cdots & \mathbf{a}_{P,N_s} \end{bmatrix}_{P\times N_s}. \tag{9.33}$$

By defining the composite signal vector as $\mathbf{s} \in \mathbb{R}^{(N_a)N_s\times 1} = \begin{bmatrix} \mathbf{s}_1^T & \cdots & \mathbf{s}_{N_s}^T \end{bmatrix}^T$, for the output snap-shot, similar to (9.23), we can write

$$\mathbf{x}[n] = \mathbf{A}(\omega,\theta)\mathbf{s} + \mathbf{n}[n]. \tag{9.34}$$

The above equation represents the discrete-time MIMO model, for an array with P sensors of any type, illuminated by N_S sources.

9.2.3 Steering Vector for Typical AP Geometries

Most of AP literature is relative to the narrowband models in an anechoic environment, for which the steering is sufficient to describe the deterministic part of the

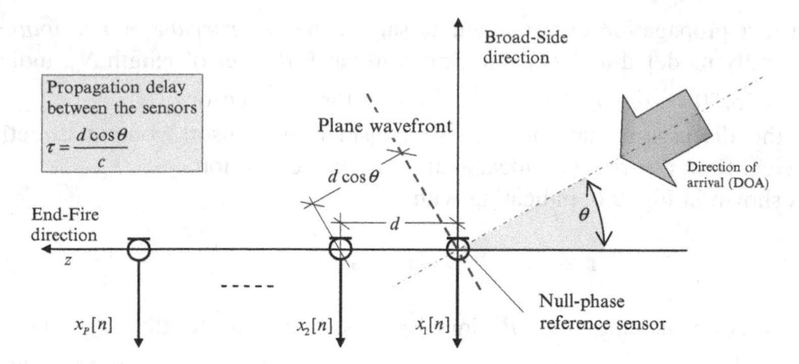

Fig. 9.7 Linear array geometry with uniform distribution of the sensors or *uniform linear array* (ULA). P represents the number of sensors, d the distance between them, and θ the angle of arrival

array. Very often, even in the case of reverberating environment, for the sake of simplicity, or for lack of knowledge of TF paths, we consider the anechoic steering vector. The steering vector plays, then, an important role in the array processing; therefore, we see in detail its definition for some of the most common geometries.

9.2.3.1 Uniform Linear Array

For a linear array it is usual to consider the sensors distribution along the z direction (see Fig. 9.6). With reference to Fig. 9.7, in the uniform distribution case or uniform linear array (ULA), the sensors position is defined by the coordinates $\mathbf{r}_m = \begin{bmatrix} 0 & 0 & (m-1)d \end{bmatrix}^T$, for $m = 1, 2, ..., P$. On the other hand, in the case of isotropic sensors in the presence of a single source, from (9.17) the steering is defined as

$$\begin{aligned} a_m(\omega, \theta) &= e^{-jk\begin{bmatrix} 0 & 0 & (m-1)d \end{bmatrix}\begin{bmatrix} \sin\theta\cos\phi & \sin\theta\sin\phi & \cos\theta \end{bmatrix}^T}, \quad m = 1, 2, ..., P. \\ &= e^{-j(\omega/c)(m-1)d\cos\theta} \end{aligned} \tag{9.35}$$

From (9.9), we define the propagation delay between the sensors as

$$\tau = \frac{d\cos\theta}{c}. \tag{9.36}$$

With simple geometric considerations on Fig. 9.7, τ represents the propagation delay between two adjacent sensors relative to an incident plane wave coming from the direction θ. By the ULA definition, the relative delays between sensors are identical. Therefore, the ULA steering vector (9.19) is defined as

$$\begin{aligned}\mathbf{a}_{\text{ULA}}(\omega,\theta) &= \begin{bmatrix}1 & e^{-jkd\cos\theta} & \cdots & e^{-j(P-1)kd\cos\theta}\end{bmatrix}^T \\ &= \begin{bmatrix}1 & e^{-j\omega\tau} & \cdots & e^{-j\omega(P-1)\tau}\end{bmatrix}^T.\end{aligned} \tag{9.37}$$

From (9.9), indicating with $\tau_m = (m-1)\tau$, the delay measured from the reference sensor for each sensor is $x_m(t) = s\big(t-(m-1)\tau\big)+n_m(t)$, for which in the discrete-time model we have that

$$x_m[n] = s[n-\tau_m] + n_m[n], \quad m = 1,2,...,P. \tag{9.38}$$

From (9.27) and (9.28), the impulse responses matrix, which in this case is a vector of delays, appears to be

$$\mathbf{a} \in \mathbb{R}^{P\times 1} \triangleq \mathfrak{I}\{\mathbf{a}(\omega,\theta)\} = \begin{bmatrix}1 & \delta[n-\tau] & \cdots & \delta[n-(P-1)\tau]\end{bmatrix}^T \tag{9.39}$$

where the filter length is $N_a = 1$ and in (9.31) $\mathbf{s}_n = s[n]$. It follows that for the ULA, the general form reduces to

$$\mathbf{x}[n] = \mathbf{a}^T s[n] + \mathbf{n}[n] \tag{9.40}$$

which coincides with (9.32).

Broadside and Endfire Directions

For an ULA, the orthogonal DOA with respect to the sensors alignment, i.e., $\theta = 90°$, is said to be *broadside direction.* While the DOA for an angle $\theta = 0°$ is referred to as *endfire direction.*

Note that in the endfire direction, the steering vector assumes the form $\mathbf{a}_{\text{ULA}}(\omega,\theta) = \begin{bmatrix}1 & e^{-j2\pi(d/\lambda)} & \cdots & e^{-j2\pi(d/\lambda)(P-1)}\end{bmatrix}^T$ for which for $d \ll \lambda$, and the vector is no longer dependent on the direction, i.e., $\mathbf{a}_{\text{ULA}}(\omega,\theta) \sim [1 \ \cdots \ 1]^T, \forall\, \theta$.

9.2.3.2 Uniform Circular Array

Another often used 2D architecture is the circular array with uniform distribution of sensors, called uniform circular array (UCA), as illustrated in Fig. 9.8.

The UCA spatial coordinate vector for the m-th sensor is defined as

$$\mathbf{r}_m = r\begin{bmatrix}\cos\frac{2\pi}{P}(m-1) & \sin\frac{2\pi}{P}(m-1) & 0\end{bmatrix}^T, \quad m = 1,2,...,P. \tag{9.41}$$

For the direction vector, relative to the direction of propagation (θ,ϕ), the steering vector is defined for isotropic sensors by considering the (9.5)

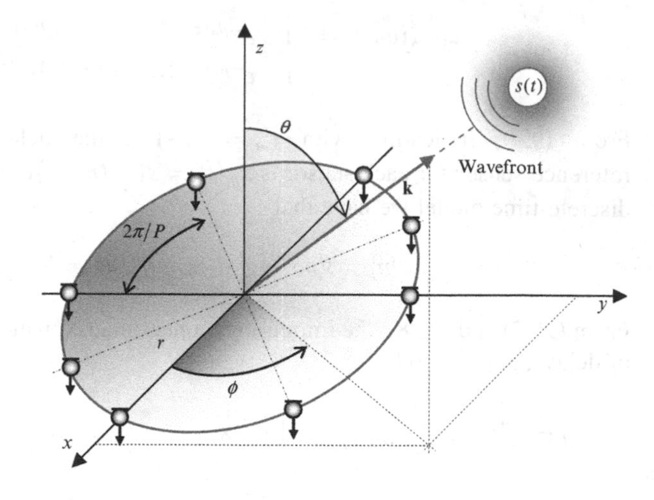

Fig. 9.8 Circular array with uniform distribution of sensors or uniform circular array (UCA)

$$\begin{aligned} a_m(\omega,\theta) &= e^{-jkr\left[\cos\frac{2\pi}{P}(m-1) \quad \sin\frac{2\pi}{P}(m-1) \quad 0\right]\left[\sin\theta\cos\phi \quad \sin\theta\sin\phi \quad \cos\theta\right]^T}, \quad m=1,\ldots,P. \\ &= e^{-j\omega\tau_m} \end{aligned} \tag{9.42}$$

where

$$\tau_m = \frac{\sin\theta}{c}\left[\cos\phi\cos\frac{2\pi}{P}(m-1) + \sin\phi\sin\frac{2\pi}{P}(m-1)\right], \quad m=1,\ldots,P \tag{9.43}$$

for which the discrete time UCA model is defined as

$$\mathbf{a} = \left[\delta[n-\tau_1] \quad \delta[n-\tau_2] \quad \cdots \quad \delta[n-\tau_P]\right]^T. \tag{9.44}$$

Note that, because of the array circular geometry, unlike the ULA case, propagation delays are different.

The steering vector definition, expressed in terms of delays, can be easily extended for any array geometry.

9.2.3.3 Harmonic Linear Array

A typical structure that allows acquisition in *spatial subbands* and a subsequent subband time processing is reported in Fig. 9.9.

This array shows a linear harmonic sensors distribution, or linear harmonic array (LHA), and it is much used in the microphone arrays for speech enhancement problems. In practice, the LHA can be seen as multiple ULA arrays, called sub-arrays, which partially share the same receivers. Each ULA sub-array is tuned to a specific wavelength, such that the distance between the sensors of each sub-array is selected to be $d \sim \lambda/2$.

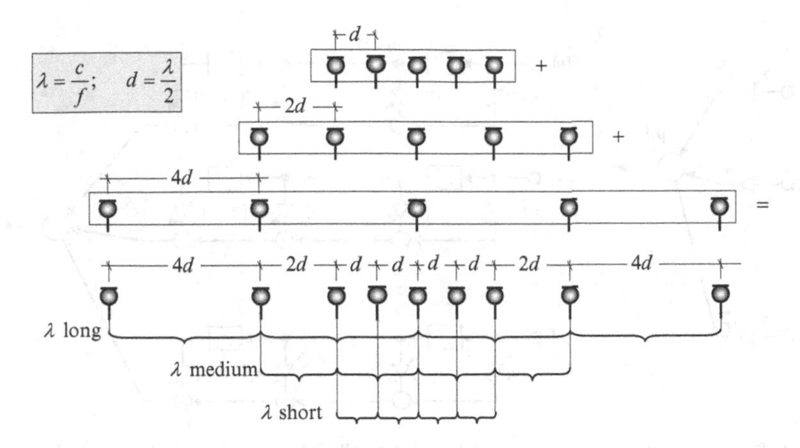

Fig. 9.9 Linear array of nine sensors with harmonic distribution

The determination of the steering vector can be performed by proceeding in a similar manner of the previously described cases.

9.2.4 *Circuit Model for AP and Space-Time Sampling*

An array processor can be viewed as a filter which operates simultaneously on both the spatial and temporal domains.

- The receivers arranged in certain geometric position, instantaneously acquire, with a certain range of spatial sampling, a portion of the incident wave. The ability of spatial discrimination is proportional to the maximum size of the array with respect to the source wavelength and is linked to the spatial sampling interval defined by the distance between the sensors.
- The temporal filtering, performed with appropriate methodologies, some of which are described later in this chapter, performs a frequency domain discrimination and is characterized by the choice of the sampling frequency.

9.2.4.1 AP Composite MIMO Notation

The BFs are multi-input circuits with one or more outputs, performing a filtering operation in the space-time domain. For narrowband sources, a situation very common in antenna arrays as shown in Fig. 9.10a, the beamforming consists of a simple linear combination, defined in the complex domain, of the signals present on

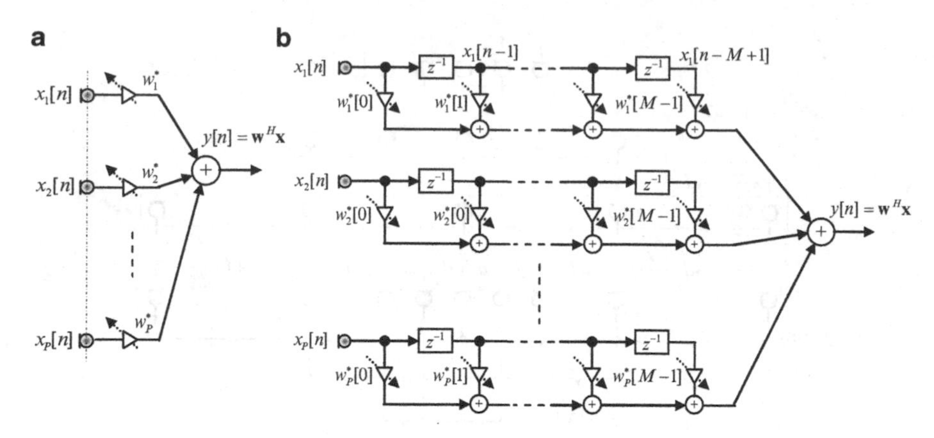

Fig. 9.10 Beamforming for narrowband and broadband sources: (**a**) the *phased array* or weighted-sum beamforming (WSBF) consists of a complex linear combination of signals on receivers; (**b**) the broadband BF or filter and sum BF (FSBF). The signal of each sensor is filtered with a M-length FIR filter. The filters outputs are summed: MISO system

the P receivers. In this case, the BF is said to be *phased array* or weighted-sum beamforming (WSBF), as the multiplication of the complex weights with the signals coming from the sensors determines a variation of their phase (or delay) before the sum. In this case the output is a simple linear combiner defined as

$$y[n] = \sum_{k=1}^{P} w_k^* x_k[n]. \tag{9.45}$$

For the ULA, the BF can be interpreted as an FIR filter operating in the discrete space domain, which can discriminate sinusoidal signals coming from different directions θ.

The weights w_k^* can be determined, as in the FIR filters design, imposing a desired response in the space domain and by minimizing a CF between the actual and the desired responses. This can be done through, for example, the windowing technique (usually Dolph–Chebyshev) or with the minimization of a certain norm $(L^1, L^2, \ldots)$ of the distance between the actual and the desired spatial response.

If the process (related to the field to be captured) is broadband, at the space domain sampling must be added that in time domain: the broadband BF performs a discrimination both in the space and in the time (or frequency) domains. As illustrated in Fig. 9.10b, the single weight is substituted with a delay line and for each delay element is defined a multiplication coefficient. In practice, at each sensor output is placed an FIR filter operating in the time domain. An array of P sensors with delay lines of length M consists of a MISO system, characterized by $P \cdot M$ free coefficients. These values are determinable on the basis of the philosophy design choice, some of which are discussed later in this chapter. The broadband beamforming is often referred to as filter and sum BF (FSBF).

The broadband BF output is calculated as for a MISO system, as

$$y[n] = \sum_{k=1}^{P}\sum_{j=0}^{M-1} w_k^*[j]x_k[n-j]. \tag{9.46}$$

Considering the composite notation for MIMO system (see Sect. 3.2.2.3), recalling the definition of the vectors $\mathbf{x}$ and $\mathbf{w}$ which contain the stacked inputs and weights of all filters

$$\mathbf{w} \in (\mathbb{R},\mathbb{C})^{P(M)\times 1} = \left[\mathbf{w}_1^T\ \mathbf{w}_2^T \cdots \mathbf{w}_P^T\right]^T \tag{9.47}$$

$$\mathbf{x} \in (\mathbb{R},\mathbb{C})^{P(M)\times 1} = \left[\mathbf{x}_1^T\quad \mathbf{x}_2^T\quad \cdots\quad \mathbf{x}_P^T\right]^T \tag{9.48}$$

with $\mathbf{w}_k = \left[w_k^*[0]\ \cdots\ w_k^*[M-1]^T\right.$ and $\mathbf{x}_k = \left[x_k[n]\ \ldots\ x_k[n-M+1]\right]^T$, $k = 1, 2, \ldots, P$, the output calculation is

$$y[n] = \mathbf{w}^H\mathbf{x}. \tag{9.49}$$

The expression (9.49) is such that the output array calculation for narrow or broad band appears formally identical. Defining $K = P \cdot M$ for broadband array and $K = P$ for those narrowband, we can determine the output as

$$y[n] = \sum_{j=0}^{K-1} w^*[j]x[j], \qquad \text{with} \qquad K = \begin{cases} P \cdot M & \text{broadband} \\ P & \text{narrowband}. \end{cases} \tag{9.50}$$

Note that defining $\boldsymbol{W} = \mathrm{DTFT}(\mathbf{w})$ and $\boldsymbol{X} = \mathrm{DTFT}(\mathbf{x})$, the frequency domain is

$$Y\left(e^{j\omega}\right) = \boldsymbol{W}^H\boldsymbol{X} \tag{9.51}$$

with similar formalism to that in the time domain (9.49).

9.2.4.2 Array Space-Time Aperture

Spatial *array aperture*, indicated with $r_{\max}$, for an ULA as shown in Fig. 9.11, is defined as the maximum size of the array measured in wavelengths. The term, $r_{\max}$, determines how many wavelengths of the front are simultaneously acquired from the array.

To avoid the *spatial aliasing* phenomenon, called $\lambda_{\min}$ the minimum wavelength of the source to acquire the spatial sampling interval must be $d < \lambda_{\min}/2$.

Consider a broadband FSBF. For an angle of arrival of the source $\theta \neq 0$, as illustrated in Fig. 9.12, the array in addition to seeing a spatial portion of the wave *sees* a certain time window $T(\theta)$ which is defined as *temporal array aperture*, which depends on the angle of arrival. From the figure we can observe that the temporal

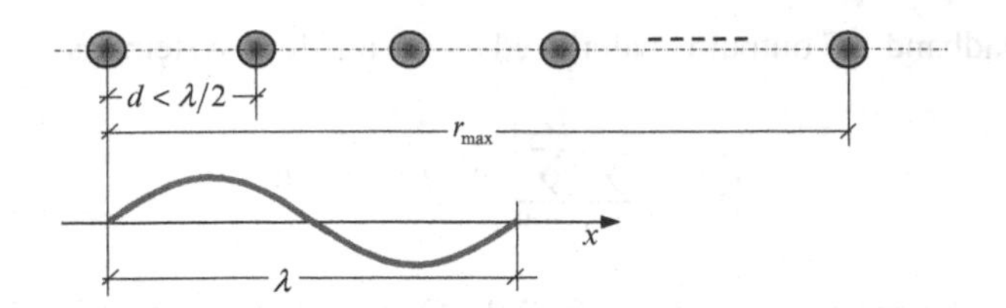

Fig. 9.11 Example of spatial sampling with an ULA, of an incident wavefront parallel to the array axis $d < \lambda/2$

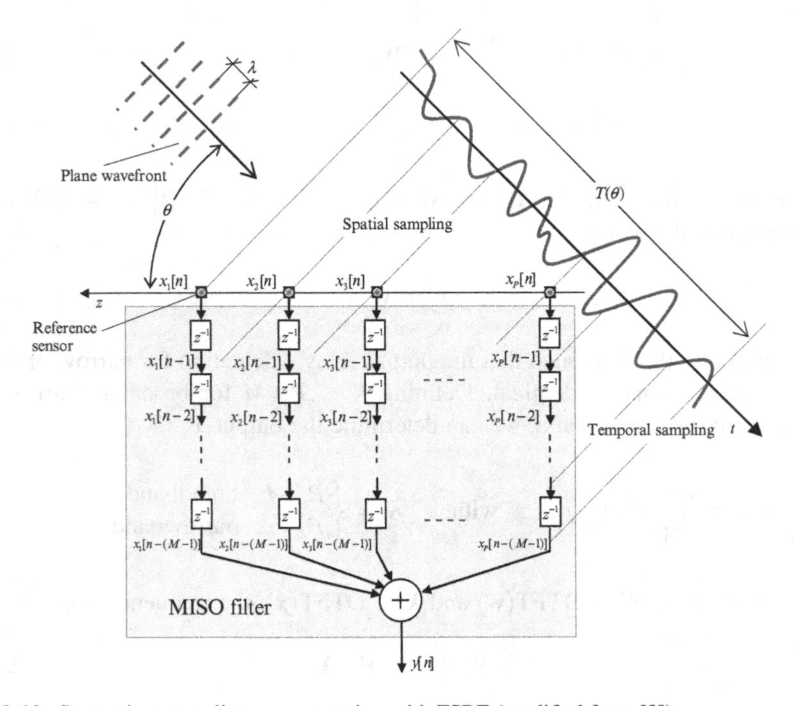

Fig. 9.12 Space-time sampling representation with FSBF (modified from [8])

component is stored in the filter delay line memories. Note that, both the space-aperture (together with the array geometry, and the relative distance between two neighboring sensors) and the temporal sampling frequency f_c must be such as to ensure the signal acquisition, free of spatial and temporal aliasing.

In the case of purely sinusoidal incident or narrowband wave, the discrimination in frequency is, of course, inconsistent and the array is a very simple FIR filter operating only in the spatial domain θ.

For non-sinusoidal source, the condition for which it can be considered narrowband depends on two factors (1) the source bandwidth (of the envelope if it is in the presence of modulation) and (2) the temporal array aperture $T(\theta)$.

Consider a narrowband process as a white noise processed with a bandpass filter with bandwidth $B = f_2 - f_1$ and center frequency f, for which the filter's *quality factor* or *Q factor* is $Q_f = f/B$. In the case in which the bandwidth is small compared to the center frequency, i.e., very high Q_f value, and the observation time such that $T(\theta) \ll 1/B$, then the observed wave has almost sinusoidal form. Note, also, that in the case where the observation time increases, so that the acquired wave appears as sinusoidal, the bandwidth B will proportionately decrease. It follows that the product between the observation time and the bandwidth, defined as *time band-width product* (TBWP) $T_{\mathrm{BWP}} \triangleq T(\theta) \cdot B$, is a parameter of fundamental importance to determine whether a source is narrowband or not. As said a source can be defined narrowband $T(\theta) \cdot B \ll 1$, for any direction of arrival.

9.2.4.3 Steering Vector for Filter and Sum Beamformer

For FSBF, with the composite MIMO notation (9.49), the composite input vector $\mathbf{x} \in \mathbb{C}^{P(M)\times 1}$ contains the signals of the array receivers, spatially and temporally sampled, all stacked in one column as illustrated in Fig. 9.12. Generalizing for the echoic model (9.32), called $\mathbf{a}(\omega,\theta) \in \mathbb{C}^{P(M)\times 1}$, the *composite steering vector*, considering the FSBF case, is then

$$\mathbf{x} = \mathbf{a}(\omega, \theta) \cdot \mathbf{s} + \mathbf{n} \tag{9.52}$$

where the composite model for the noise vector $\mathbf{n}$ is defined as

$$\mathbf{n} \in (\mathbb{R}, \mathbb{C})^{P(M)\times 1} = \begin{bmatrix} \mathbf{n}_1^T & \cdots & \mathbf{n}_P^T \end{bmatrix}^T \text{ with } \mathbf{n}_m = \begin{bmatrix} n_m[n] & \cdots & n_m[n-M+1] \end{bmatrix}^T$$

in the case of single source is $\mathbf{s} \in (\mathbb{R},\mathbb{C})^{M(P)\times 1}$, for which we have that

$$\mathbf{s} = \begin{bmatrix} \mathbf{s}_n^T & \cdots & \mathbf{s}_n^T \end{bmatrix}^T \quad \text{with } \mathbf{s}_n = \begin{bmatrix} s[n] & \cdots & s[n-M-1] \end{bmatrix}^T.$$

For the composite steering vector $\mathbf{a}^T(\omega,\theta)$ definition, it is necessary to extend the definition (9.19) taking into account the propagation in the filters' delay lines. For isotropic sensors, indicating with f_c the sampling frequency, $t_c = 1/f_c$ is the delay of each element of the delay line. The *composite isotropic steering vector* is formed by the ideal isotropic vector $\bar{\mathbf{a}}(\omega, \theta) \in \mathbb{C}^{P\times 1}$, as defined in (9.15), combined with the delays of the delay line

$$\bar{\mathbf{a}}_c(\omega, \theta) = \bar{\mathbf{a}}(\omega, \theta) \otimes \begin{bmatrix} 1 \\ e^{-j\omega t_s} \\ \vdots \\ e^{-j\omega(M-1)t_s} \end{bmatrix} = \begin{bmatrix} 1 \\ e^{-j\omega\tau_2} \\ \vdots \\ e^{-j\omega\tau_P} \end{bmatrix} \otimes \begin{bmatrix} 1 \\ e^{-j\omega t_s} \\ \vdots \\ e^{-j\omega(M-1)t_s} \end{bmatrix}. \tag{9.53}$$

where $\otimes$ indicates the Kronecker product.

Considering the radiation diagrams of the sensors, generalizing the expression (9.19), we get

$$\mathbf{a}(\omega,\theta) = \bar{\mathbf{a}}_c(\omega,\theta) \odot \mathbf{b}(\omega,\theta)$$
$$= \begin{bmatrix} \begin{bmatrix} 1 & e^{-j\omega t_s} & \cdots & e^{-j\omega(M-1)t_s} \end{bmatrix}^T \\ \begin{bmatrix} e^{-j\omega\tau_2} & e^{-j\omega(\tau_2+t_s)} & \cdots & e^{-j\omega\left(\tau_2+(M-1)t_s\right)} \end{bmatrix}^T \\ \vdots \\ \begin{bmatrix} e^{-j\omega\tau_P} & e^{-j\omega(\tau_P+t_s)} & \cdots & e^{-j\omega\left(\tau_P+(M-1)t_s\right)} \end{bmatrix}^T \end{bmatrix}_{P\times 1} \odot \begin{bmatrix} B_1(\omega,\theta)e^{-j\gamma_1(\omega,\theta)} \\ B_2(\omega,\theta)e^{-j\gamma_2(\omega,\theta)} \\ \vdots \\ B_P(\omega,\theta)e^{-j\gamma_P(\omega,\theta)} \end{bmatrix} \tag{9.54}$$

where, from (9.16) and (9.18), the vector $\mathbf{b}(\omega,\theta) \in \mathbb{C}^{P\times 1}$ contains the radiation diagrams of the receivers.

9.3 Noise Field Characteristics and Quality Indices

The prevalent use of sensor arrays is to identify the direction of arrival of the sources and, through the synthesis of a certain directivity patterns or radiation diagram (*beam*), the simultaneous extraction of information relating to them. In other words, the sensor array is used for the signal-to-noise ratio improvement by increasing the array gain toward the direction of arrival of the SOI and simultaneously decreasing it in other directions (at the limit all other) where the unwanted sources are both localized and diffused.

The BFs operate on a discrete space-time domain and, as for the traditional digital filters, it is possible to think of a static design independent from the input data or an adaptive approach which is, by definition, always data dependent. In the latter case, the BF coefficients, or part of them, are updated according to the SOI or noise statistics present at its input.

It is important, therefore, to determine the temporal and spatial statistical characteristics of the interfering signal, defined as a *noise field*, and performance indices useful in determining the topological and algorithmic specific of the array.

9.3.1 Spatial Covariance Matrix and Projection Operators

It is appropriate to define the space-time second-order statistical relations useful for theoretical analysis. The *spatial covariance matrix*[1] [10] is the matrix defined as $\mathbf{R}_{xx} \in (\mathbb{R},\mathbb{C})^{P\times P} \triangleq E\{\mathbf{x}\mathbf{x}^H\}$. Therefore, when considering a generic signal model, such as for example one of the models (9.31), (9.32), and (9.34) or model (9.52), for a generic beamformer (WSBF or FSBF), the spatial covariance matrix is

[1] In [9], the development is carried out in the case of anechoic model for which the matrix $\mathbf{A}$ is the steering matrix defined in (9.24). Here, the proposed study model is more general and also valid for reverberant propagation environments.

$$\begin{aligned}\mathbf{R}_{xx} &= E\{\mathbf{x}\mathbf{x}^H\} \\ &= \mathbf{A}(\omega,\theta)E\{\mathbf{s}\mathbf{s}^H\}\mathbf{A}^H(\omega,\theta) + E\{\mathbf{n}\mathbf{n}^H\} \\ &= \mathbf{A}(\omega,\theta)\mathbf{R}_{ss}\mathbf{A}^H(\omega,\theta) + \mathbf{R}_{nn}.\end{aligned} \tag{9.55}$$

Indicating $N = N_a N_s$, in (9.55) $\mathbf{A}(\omega,\theta) \in (\mathbb{R},\mathbb{C})^{P\times N}$ is the steering matrix of the echoic signal propagation model (9.34) and

$$\mathbf{R}_{ss} \in (\mathbb{R},\mathbb{C})^{N\times N} = E\{\mathbf{s}\mathbf{s}^H\} \tag{9.56}$$

$$\mathbf{R}_{nn} \in (\mathbb{R},\mathbb{C})^{N\times N} = E\{\mathbf{n}\mathbf{n}^H\} \tag{9.57}$$

represent, respectively, the source and noise covariance matrices.

Note that, for anechoic model, applies $N_a = 1$ and, in the case of a single source, is $N_s = 1$ and $\mathbf{A}(\omega,\theta) \rightarrow \mathbf{a}(\omega,\theta)$.

In addition, in the frequency domain, the covariance matrices $\mathbf{R}_{xx}$ and $\mathbf{R}_{nn}$, true or estimated sampling covariance matrix, are indicated, respectively, as

$$\mathbf{R}_{xx}\left(e^{j\omega}\right) = \text{DTFT}\left[E\{\mathbf{x}\mathbf{x}^H\}\right] = \begin{bmatrix} R_{x_1x_1}(e^{j\omega}) & \cdots & R_{x_1x_P}(e^{j\omega}) \\ \vdots & \ddots & \vdots \\ R_{x_Px_1}(e^{j\omega}) & \cdots & r_{x_Px_P}(e^{j\omega}) \end{bmatrix}, \tag{9.58}$$

$$\mathbf{R}_{nn}\left(e^{j\omega}\right) = \text{DTFT}\left[E\{\mathbf{n}\mathbf{n}^H\}\right] = \begin{bmatrix} R_{n_1n_1}(e^{j\omega}) & \cdots & R_{n_1n_P}(e^{j\omega}) \\ \vdots & \ddots & \vdots \\ R_{n_Pn_1}(e^{j\omega}) & \cdots & R_{n_Pn_P}(e^{j\omega}) \end{bmatrix}. \tag{9.59}$$

We remind the reader that the above two matrices are *power spectral density* (PSD) matrices, defined just as DTFT of the autocorrelation sequence.

Note, finally, that the signal covariance matrix is considered to be positive definite or non-singular (or almost non-singular for almost coherent signals).

9.3.1.1 Spatial White Noise

The noise is said to be *spatially white* if it is zero-mean, uncorrelated, and with the same power, or homogeneous, on all sensors. In this case, the covariance matrix is

$$\mathbf{R}_{nn} \in (\mathbb{R},\mathbb{C})^{P\times P} = E\{\mathbf{n}\mathbf{n}^H\} = \begin{bmatrix} r_{n_1n_1} & \cdots & r_{n_1n_P} \\ \vdots & \ddots & \vdots \\ r_{n_Pn_1} & \cdots & r_{n_Pn_P} \end{bmatrix} = \sigma_n^2\mathbf{I}. \tag{9.60}$$

In the case of homogeneous noise but not white you can proceed with a weighting called *whitening*. More specifically, the signal coming from the sensors are multiplied by $\mathbf{R}_{nn}^{-1/2}$ (square root of the Hermitian matrix $\mathbf{R}_{nn}^{-1}$) before processing.

9.3.1.2 Spatial Covariance Matrix Spectral Factorization

The *spectral factorization* of $\mathbf{R}_{xx} \in (\mathbb{R},\mathbb{C})^{P\times P}$ is of central importance to many theoretical developments and, for simplicity omitting the writing of the indices (ω,θ), can be expressed as

$$\mathbf{R}_{xx} = \mathbf{A}\mathbf{R}_{ss}\mathbf{A}^H + \mathbf{R}_{nn} = \mathbf{U}\boldsymbol{\Lambda}\mathbf{U}^H \tag{9.61}$$

with $\mathbf{U}$ unitary matrix and $\boldsymbol{\Lambda} = \text{diag}[\lambda_1, \lambda_2, \ \dots, \lambda_P]$ a diagonal matrix with real eigenvalues ordered as $\lambda_1 \geq \lambda_2 \geq \dots \geq \lambda_P > 0$.

Note that for $N < P$ and Gaussian noise ($\mathbf{R}_{nn} = \sigma_n^2\mathbf{I}$), you can partition the eigenvectors and eigenvalues belonging to the signal $\lambda_1, \lambda_2, \dots, \lambda_N \geq \sigma_n^2$ and belonging to the noise $\lambda_{N+1}, \lambda_{N+2}, \dots, \lambda_P = \sigma_n^2$. Therefore, it is possible to write

$$\mathbf{R}_{xx} = \mathbf{U}_s\boldsymbol{\Lambda}_s\mathbf{U}_s^H + \mathbf{U}_n\boldsymbol{\Lambda}_n\mathbf{U}_n^H \tag{9.62}$$

wherein $\boldsymbol{\Lambda}_n = \sigma_n^2\mathbf{I}$. Since the noise eigenvectors are orthogonal to $\mathbf{A}$, the columns of $\mathbf{U}_s$ represent a *span* for the *column space* of $\mathbf{A}$, referred to as $\mathcal{R}(\mathbf{A})$. While those of $\mathbf{U}_n$ are a span for its orthogonal complement, i.e., the *nullspace* of $\mathbf{A}^H$ indicated as $\mathcal{N}(\mathbf{A}^H)$ (see Sect. A.6).

9.3.1.3 Projection Operators

The *projection operators* (Sect. A.6.5) on the signal and the noise space are then defined as

$$\tilde{\mathbf{P}} = \mathbf{U}_s\mathbf{U}_s^H = \mathbf{A}\left(\mathbf{A}^H\mathbf{A}\right)^{-1}\mathbf{A}^H, \qquad \textit{projection on} \quad \Psi = \mathcal{R}(\mathbf{A}) \tag{9.63}$$

$$\mathbf{P} = \mathbf{U}_n\mathbf{U}_n^H = \mathbf{I} - \mathbf{A}\left(\mathbf{A}^H\mathbf{A}\right)^{-1}\mathbf{A}^H, \qquad \textit{projection on} \quad \Sigma = \mathcal{N}\left(\mathbf{A}^H\right) \tag{9.64}$$

for which $\mathbf{P} + \tilde{\mathbf{P}} = \mathbf{I}$.

Note that the projection operators are useful both in development and in the geometric interpretation of the adaptive AP's algorithms discussed below.

9.3.1.4 Isotropic Noise with Spherical and Cylindrical Symmetry

In general, *isotropic noise* is referred to the noise with spherical symmetry, i.e., radiating an isotropic sensor with uniform probability for all directions of the solid angle and for all frequencies. Therefore, we define with $N(e^{j\omega},\theta)$, the normalized noise power that must satisfy the following condition

$$N(e^{j\omega},\theta)\therefore\frac{1}{4\pi}\int_0^{2\pi}\int_0^{\pi}N(e^{j\omega},\theta)\sin\theta\cdot d\theta d\phi=1. \tag{9.65}$$

In some special situations, the noise is characterized by uniform probability only on the azimuth plane and is zero for the other directions. In this case, it is called *isotropic noise with cylindrical symmetry*. Similarly to (9.65), the normalized power indicated as $N_C(e^{j\omega},\phi)$ is defined as

$$N_C(e^{j\omega},\phi)\therefore\frac{1}{2\pi}\int_0^{2\pi}N_C(e^{j\omega},\phi)d\phi=1. \tag{9.66}$$

Remark The isotropic noise with cylindrical symmetry appears to be more appropriate in environments with a particular propagation. A typical example is the reverberating acoustics of a confined environment when the floor and the ceiling are treated with phono-absorbent materials. In this case, the noise component can be modeled only on the azimuth plane without taking into account the elevation.

9.3.2 Noise Field Characteristics

The design of the array geometry, the circuit topology, and the possible adaptation mechanisms depends heavily on the noise field characteristics in which they operate. Characteristics such as number and movement of sources, bandwidth, level, the presence of multiple paths or reverberation, and characteristics of the coherent or diffuse noise field are therefore of great interest for the correct definition of the beamformer type and the algorithm, static or adaptive, for determining its free parameters.

In particular, among the various APs' applications, more complex situations are those in the acoustic sectors. In fact, very often the microphones array operates in extremes noise and noise field conditions, at times even in the presence of precise design (and economic) constraints which limit the size, position, and number of sensors.

For the noise field characterization, we consider two spatially distinct stationary random processes, for example, as shown in Fig. 9.13, acquired by two sensors located in the coordinates $\mathbf{r}_i$ and $\mathbf{r}_j$ and indicated directly in discrete time as $n_i[n]$ and $n_j[n]$, with correlations $r_{n_k n_k}[n]=E\{n_k[n]n_k^*[n-l]\}$, for $k=i,j$. Consider the coherence function (see Sect. 3.3.3)

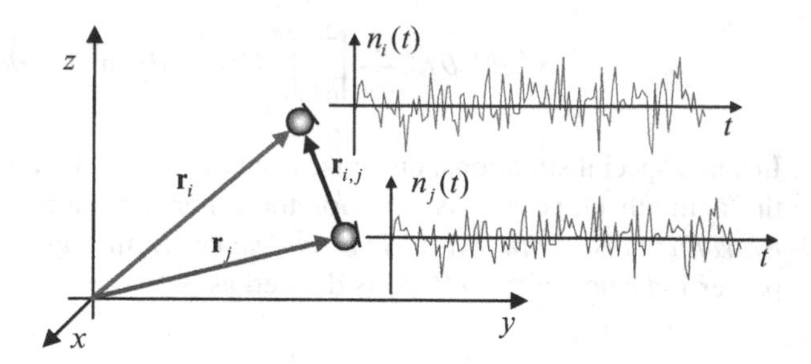

Fig. 9.13 Two nearby sensors immersed in a noise field may receive data more or less similar. In the case of strongly correlated signals, the field is said to be *coherent* and *incoherent* in the opposite case (modified from [9])

$$\gamma_{n_i n_j}(e^{j\omega}) \triangleq \frac{R_{n_i n_j}(e^{j\omega})}{\sqrt{R_{n_i n_i}(e^{j\omega}) R_{n_j n_j}(e^{j\omega})}} \tag{9.67}$$

where the terms $R_{n_k n_k}(e^{j\omega}) \triangleq \text{DTFT}\{r_{n_k n_k}[n]\}$, for $k = i, j$, are PSD while $R_{n_i n_j}(e^{j\omega}) \triangleq \text{DTFT}\{r_{n_i n_j}(e^{j\omega})\}$ is a cross-power spectral density (CPSD) and its amplitude squared, defined as $C_{n_i n_j}(e^{j\omega}) = |\gamma_{n_i n_j}(e^{j\omega})|^2$, is the magnitude square coherence (MSC) function.

Recall that the coherence function can be interpreted as a correlation in the space-frequency domain. In fact, if $n_i[n] \approx n_j[n]$, it follows that $\gamma_{n_i n_j}(e^{j\omega}) \approx 1$, that is, it has the highest correlation (similarity), and conversely if $n_j[n]$ is not correlated to $n_j[n]$ we have that $\gamma_{n_i n_j}(e^{j\omega}) \approx 0$. It has then $0 \leq \left|\gamma_{n_i n_j}(e^{j\omega})\right| \leq 1$ for each frequency.

9.3.2.1 Coherent Field

The noise field is said to be coherent when the sensors acquire signals strongly correlated consequently $\gamma_{n_i n_j}(e^{j\omega}) \approx 1$. This situation is typical when the radiated wave is not subject to reflections or strong dissipation due to the propagation. In the case of microphone arrays, the field is (almost) coherent in unconfined environments such as in open air or in anechoic chambers (confined environments in which the acoustic waves are absorbed by the walls for all wavelengths).

9.3.2.2 Incoherent Field

The noise field is said to be incoherent; in the case that the sensors acquire signals strongly uncorrelated it is $\gamma_{n_i n_j}(e^{j\omega}) \approx 0$. For example, the sensors' electrical noise appears to be almost always of incoherent nature. Note that the incoherent field is also spatially white.

9.3.2.3 Diffuse Field

The extreme conditions of completely coherent or incoherent fields are rare in real life situations. For example, for microphone arrays operating in confined spaces, where there is a certain reverberation due to the walls reflections, after a certain time the background noise, due to the numerous constructive and destructive interference, can be defined as a *diffuse noise field*. The diffused (or scattered) field can thus be generated by the plane wave superposition that propagates in a confined space with infinite reverberation time or from a number of infinite sources for each reverberation time. A diffuse field is characterized by:

- Weakly correlated signals on the sensors.
- Coming simultaneously with the same energy level, from all directions with spherical or cylindrical symmetry.

The dependence on the noise characteristics is particularly important in the case of microphone arrays used primarily for *speech enhancement*.

Typical acoustic environments such as offices and vehicles can be characterized by a diffuse field. In these cases the coherence between the noises acquired by two sensors i and j is a function of the distance between the sensors $d_{ij} = |\mathbf{r}_{i,j}|$ and the acquired frequency. In the case of isotropic sensor it is proved (see [11] for details) that the coherence function between the two sensors is a function equal to

$$\gamma_{n_i n_j}\left(e^{j\omega}\right) = \frac{\sin\left(kd_{ij}\right)}{kd_{ij}}. \tag{9.68}$$

Therefore, it follows that in the case of very close microphones, i.e., in terms of wavelength for $\frac{2\pi}{\lambda}d_{ij} \rightarrow 0$, the field is *approximately* coherent.

Figure 9.14, for example, shows typical MSC $C(\mathbf{r}_{i,j},\omega) = |\gamma(\mathbf{r}_{i,j},\omega)|^2$, for real acoustic environments superimposed on the ideal curves calculated with (9.68).

Similarly to (9.59), it is possible to define the *coherency matrix*, characterized by a diagonal unitary and Toeplitz symmetry, as

$$\boldsymbol{\Gamma}_{nn}\left(e^{j\omega}\right) = \begin{bmatrix} 1 & \gamma_{n_1 n_2}\left(e^{j\omega}\right) & \cdots & \gamma_{n_1 n_p}\left(e^{j\omega}\right) \\ \gamma_{n_2 n_1}\left(e^{j\omega}\right) & 1 & \cdots & \vdots \\ \vdots & \vdots & \ddots & \gamma_{n_{p-1} n_p}\left(e^{j\omega}\right) \\ \gamma_{n_P n_1}\left(e^{j\omega}\right) & \cdots & \gamma_{n_p n_{p-1}}\left(e^{j\omega}\right) & 1 \end{bmatrix}. \tag{9.69}$$

A noise field is called homogeneous if its characteristic PSD does not depend on the spatial position of the sensor. For example, the field is spatially incoherent or diffuse white is by definition homogeneous.

In the case of homogeneous noise all sensors are characterized by the same noise PSD, i.e., $R_{n_i n_i}(e^{j\omega}) = R_n(e^{j\omega})$, for $i = 1, 2, ..., P$. It follows that the coherence function is, in this case, defined as

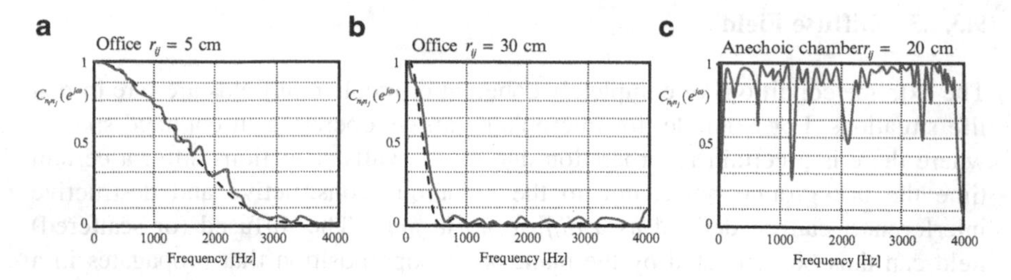

Fig. 9.14 Examples of *magnitude square coherence* for typical acoustic environments: (**a**) office with 5 cm distant microphones; (**b**) office with 30 cm distant microphones; (**c**) anechoic chamber (modified from [9])

$$\gamma_{n_i n_j}\left(e^{j\omega}\right) \triangleq \frac{R_{n_i n_j}\left(e^{j\omega}\right)}{R_n\left(e^{j\omega}\right)}. \tag{9.70}$$

9.3.2.4 Combined Noise Field

For microphone arrays operating in confined spaces, in general, the noise field can be defined as combined type. In fact, the distance between the noise source and the microphones and the walls reflection coefficients determine a direct noise path superimposed on an incoherent and diffuse noise. The array will, therefore, be designed so as to operate properly independently of the noise field characteristics.

In Fig. 9.15 an example of MSC for a typical confined environment (office) with microphones distance equal to 40 cm is reported.

9.3.3 Quality Indexes and Array Sensitivity

We define some characteristic parameters for defining the array quality for a generic architecture as that in Fig. 9.16. From the general models (9.34) and (9.51), the input and output signals in the frequency domain, by omitting writing the term $(e^{j\omega})$, are defined as

$$\begin{aligned} \boldsymbol{X} &= \mathbf{a}(\omega,\theta)\boldsymbol{S} + \boldsymbol{N} \\ Y &= \boldsymbol{W}^H \boldsymbol{X} \end{aligned} \tag{9.71}$$

where the signal model in the time domain is (9.52):

$$y[n] = \mathbf{w}^H \mathbf{a}(\omega,\theta)\mathbf{s} + \mathbf{w}^H \mathbf{n} \tag{9.72}$$

while in the frequency domain the output is

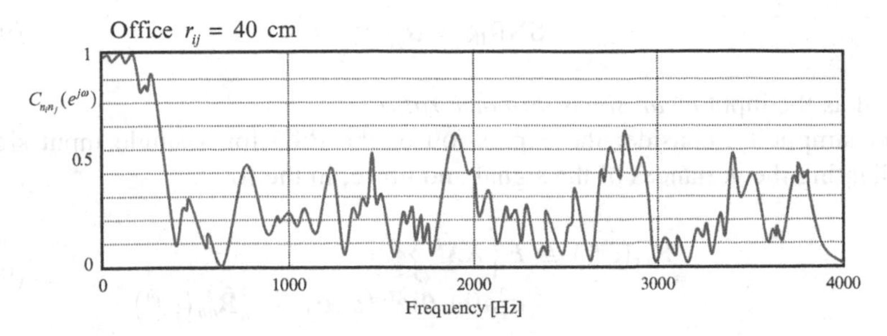

Fig. 9.15 Magnitude square coherence measured with 40 cm distant microphones, in a typical acoustic environment of an office, with the combined noise field (modified from [9])

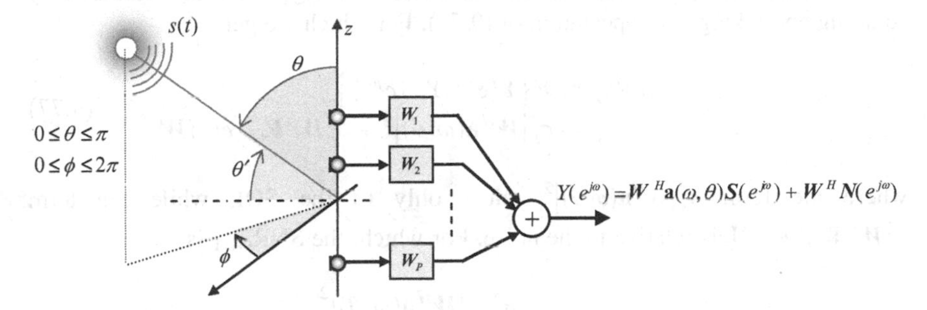

Fig. 9.16 Typical WSBF or FSBF used as a reference for the definition of quality specifications

$$Y = \boldsymbol{W}^H \mathbf{a}(\omega, \theta)\boldsymbol{S} + \boldsymbol{W}^H \boldsymbol{N} \tag{9.73}$$

with $\boldsymbol{W} \in \mathbb{C}^{P(M)\times 1} = \begin{bmatrix} W_1(e^{j\omega}) & \cdots & W_P(e^{j\omega}) \end{bmatrix}_{P\times 1}^T$. We remind the reader that for the WSBF it is $\mathbf{x}$, $\mathbf{w}$, $\mathbf{a}(\omega,\theta) \in (\mathbb{R},\mathbb{C})^{P\times 1}$, while in the case of FSBF we have the composite notation for which it is $\mathbf{x}$, $\mathbf{w}$, $\mathbf{a}(\omega,\theta) \in (\mathbb{R},\mathbb{C})^{P(M)\times 1}$.

9.3.3.1 Input and Output Beamformer Signal-to-Noise Ratio

From the definitions (9.56) and (9.59), the signal-to-noise ratio at the BF's input ($\mathrm{SNR_{IN}}$) is evaluated by considering the presence of a single isotropic sensor placed at the array center. Assuming stationary signals, the $\mathrm{SNR_{IN}}$ is defined as

$$\mathrm{SNR_{IN}} = \frac{E\{S(e^{j\omega})S^*(e^{j\omega})\}}{E\{N(e^{j\omega})N^*(e^{j\omega})\}} = \frac{R_s(e^{j\omega})}{R_n(e^{j\omega})} \tag{9.74}$$

where $R_s(e^{j\omega})$ is the PSD of the signal and $R_n(e^{j\omega})$ the noise PSD, measured in average at the BF's input. Note that indicating with σ_s^2 and σ_n^2, the variances, respectively, of signal and noise, (9.74) can be written as

$$\text{SNR}_{\text{IN}} = \sigma_s^2/\sigma_n^2 \tag{9.75}$$

defined as the input *mean signal-to-noise ratio*.

For simplicity, consider the expression of the PSD for a single input signal, highlighting the variances of the signal and noise, in the form

$$\begin{aligned} \mathbf{R}_{xx}(e^{j\omega}) &\triangleq E\left\{XX^H\right\} \\ &= \sigma_s^2\mathbf{a}(\omega,\theta)\mathbf{a}^H(\omega,\theta) + \sigma_n^2\hat{\mathbf{R}}_{nn}(e^{j\omega}) \end{aligned} \tag{9.76}$$

where $\hat{\mathbf{R}}_{nn}(e^{j\omega}) \triangleq \mathbf{R}_{nn}(e^{j\omega})/\sigma_n^2$ represents the normalized PSD covariance matrix of the noise so that its trace is equal to the number of sensors P.[2]

With a similar procedure, the output SNR (SNR_{OUT}) can be evaluated by squaring and taking the expectation of (9.73). For which we get

$$\begin{aligned} R_y(e^{j\omega}) &\triangleq E\left\{Y(e^{j\omega})Y^*(e^{j\omega})\right\} \\ &= \sigma_s^2\left|W^H\mathbf{a}(\omega,\theta)\right|^2 + \sigma_n^2 W^H\hat{\mathbf{R}}_{nn}(e^{j\omega})W \end{aligned} \tag{9.77}$$

where the term $\sigma_s^2\left|W^H\mathbf{a}(\omega,\theta)\right|^2$ refers only to the SOI, while the term $\sigma_n^2 W^H\hat{\mathbf{R}}_{nn}(e^{j\omega})W$ is relative to the noise. For which, the SNR_{OUT} is

$$\text{SNR}_{\text{OUT}} \triangleq \frac{\sigma_s^2}{\sigma_n^2}\cdot\frac{\left|W^H\mathbf{a}(\omega,\theta)\right|^2}{W^H\hat{\mathbf{R}}_{nn}(e^{j\omega})W}. \tag{9.78}$$

Denoting by A the matrix $A \triangleq \mathbf{a}(\omega,\theta)\mathbf{a}^H(\omega,\theta)$; the above is written as

$$\text{SNR}_{\text{OUT}} = \text{SNR}_{\text{IN}}\cdot\frac{W^H A W}{W^H\hat{\mathbf{R}}_{nn}(e^{j\omega})W}. \tag{9.79}$$

9.3.3.2 Radiation Functions

We define *radiation function*, assuming plane wave propagated in a homogeneous medium the TF defined as

$$\begin{aligned} R(e^{j\omega},\theta) &\triangleq \frac{Y(e^{j\omega},\theta)}{S(e^{j\omega})} \\ &= W^H\mathbf{a}(\omega,\theta) \end{aligned} \tag{9.80}$$

where θ indicates the pair $(\theta,\phi) \to \theta$, and $Y(e^{j\omega},\theta)$ is (θ,ϕ) variable. In (9.80), the term $S(e^{j\omega})$ represents the BF's input signal received by an isotropic virtual sensor

[2] Note that for $\mathbf{x} \in (\mathbb{R},\mathbb{C})^{P\times 1}$, we have $\text{tr}\left[E\{\mathbf{x}\mathbf{x}^T\}/(\frac{\mathbf{x}^T\mathbf{x}}{P})\right] \sim P$.

placed at the center of the array geometry, according to the anechoic propagation model.

Radiation Diagram

The *radiation diagram*, also called *beampattern*, is defined as the module of the radiation function, generally normalized with respect to the direction of maximum gain θ. Whereby, called $\theta_{\max} \triangleq \max\limits_{\theta} \left|R(e^{j\omega},\theta)\right|$, we have

$$R_d\left(e^{j\omega},\theta\right) = \left|R\left(e^{j\omega},\theta\right)\right| \Big/ \left|R\left(e^{j\omega},\theta_{\max}\right)\right|. \tag{9.81}$$

For example, the radiation diagrams of the sensors shown in Fig. 9.5 are evaluated with (9.81) expressed in dB.

Power Diagram: Spatial Directivity Spectrum

We define *power diagram* or spatial directivity spectrum (SDS) as the beampattern square amplitude that considering (9.80) is

$$\begin{aligned}\left|R(e^{j\omega},\theta)\right|^2 &= \left|\mathbf{W}^H\mathbf{a}(\omega,\theta)\right|^2 \\ &= \mathbf{W}^H\mathbf{A}\mathbf{W}.\end{aligned} \tag{9.82}$$

Note that, since $\mathbf{A}$ is complex, this can be decomposed into real and imaginary parts as $\mathbf{A} = \mathbf{A}_R + j\mathbf{A}_I$, where the imaginary part is anti-symmetric for which we have $\mathbf{w}^T\mathbf{A}_I(\omega,\theta)\mathbf{w} = 0$. Accordingly, we can write $\left|R(e^{j\omega},\theta)\right|^2 = \mathbf{w}^T\mathbf{A}_R(\omega,\theta)\mathbf{w}$.

Spatial Response at −3 dB

We define *main lobe width* as the region around the maximum of the response, i.e., $\theta_{\max}$, with amplitude >-3 dB.

9.3.3.3 Array Gain

The *array gain* or *directivity* is defined as the SNR improvement, for a certain direction θ_0, between the input and the output of the BF, i.e.,

$$G\left(e^{j\omega},\theta_0\right) \triangleq \frac{\mathrm{SNR_{OUT}}}{\mathrm{SNR_{IN}}} \tag{9.83}$$

therefore from (9.79), we have that

$$G(e^{j\omega},\theta_0) = \frac{\left|\boldsymbol{W}^H\mathbf{a}(\omega,\theta_0)\right|^2}{\boldsymbol{W}^H\hat{\mathbf{R}}_{nn}(e^{j\omega})\boldsymbol{W}} = \frac{\boldsymbol{W}^H\boldsymbol{A}\boldsymbol{W}}{\boldsymbol{W}^H\hat{\mathbf{R}}_{nn}(e^{j\omega})\boldsymbol{W}}. \tag{9.84}$$

The array gain depends therefore on the array characteristics, described by $\boldsymbol{A}$ and $\boldsymbol{W}$, and from those of the noise field defined by the matrix $\hat{\mathbf{R}}_{nn}(e^{j\omega})$.

Spherically Symmetric Isotropic-Noise Case

In the case of a symmetrical spherical isotropic noise, the array gain along the direction θ_0, explaining the noise expression (9.65), can be defined as

$$G(e^{j\omega},\theta_0) \triangleq \frac{\text{SNR}_{\text{OUT}}}{\text{SNR}_{\text{IN}}} = \frac{\left|R(e^{j\omega},\theta_0)\right|^2}{\frac{1}{4\pi}\int_0^{2\pi}\int_0^{\pi}\left|R(e^{j\omega},\theta)\right|^2 N(e^{j\omega},\theta)\sin\theta\cdot d\theta d\phi} \tag{9.85}$$

where θ_0 represents the steering direction indicated also as main response axis (MRA). Combining the latter with (9.84), we observe that the normalized noise correlation matrix can be defined as

$$\hat{\mathbf{R}}_{nn}(e^{j\omega}) = \frac{1}{4\pi}\int_0^{2\pi}\int_0^{\pi} A(e^{j\omega},\theta)N(e^{j\omega},\theta)\sin\theta\cdot d\theta d\phi. \tag{9.86}$$

In the case of transmitting antennas, $G(e^{j\omega},\theta)$ represents the maximum radiation intensity (power per solid angle) divided by the average radiation over the entire spherical angle. For a receiving antenna, the denominator of (9.85) represents the output power due to noise on the sensor with a certain spatial distribution $N(e^{j\omega},\theta)$ around the sphere.

Cylindrical Symmetry Isotropic-Noise Case

In the case of a symmetrical cylindrical isotropic noise, for (9.66), the array gain is defined as

$$G_C(e^{j\omega},\theta_0) \triangleq \frac{\text{SNR}_{\text{OUT}}}{\text{SNR}_{\text{IN}}} = \frac{\left|R(e^{j\omega},\theta_0)\right|^2}{\frac{1}{2\pi}\int_0^{2\pi}\left|R(e^{j\omega},\theta)\right|^2 N_C(e^{j\omega},\phi)d\phi} \tag{9.87}$$

for which, similarly to (9.86) we can write

$$\hat{\mathbf{R}}_{nn}(e^{j\omega}) = \frac{1}{2\pi}\int_0^{2\pi} A(e^{j\omega},\theta)N_C(e^{j\omega},\phi)d\phi. \tag{9.88}$$

Unless otherwise specified, the array gain is defined by the expression (9.85), i.e., noise with spherical symmetry.

Remark The expression (9.84) or (9.87) indicates that the array gain is much higher as the $\hat{\mathbf{R}}_{nn}(e^{j\omega})$ is small. This implies that the gain is large if the sensors receive uncorrelated as possible noise. In other words, the condition for which it is convenient to use an array, rather than a single receiver, is to have certain *spatial diversity* between the sensors. This is easily understandable considering that the BF makes a sum of the signals present on the sensors and, if the noise has zero mean, the sum of the uncorrelated noise tends to zero while the SOI, i.e., that is in phase on with the sensors, tends to be additive.

In beamforming practice, it is very important to consider the array gain as a function of the specific characteristics of the noise field. In various application contexts, in fact, the array gain is defined according to the noise field type: coherent, incoherent, or diffuse, as below described.

Homogeneous Noise Field

Note that for homogeneous noise field, for the (9.70), the normalized correlation matrix coincides with the coherence noise matrix [12]. Thus, indicating with $\mathbf{\Gamma}_{nn}(e^{j\omega})$ the coherence matrix, in the case of homogeneous noise, i.e., with identical powers for all the sensors, the expression (9.84) takes the form

$$G(e^{j\omega},\theta_0) = \frac{\left|\boldsymbol{W}^H\mathbf{a}(\omega,\theta_0)\right|^2}{\boldsymbol{W}^H\mathbf{\Gamma}_{nn}(e^{j\omega})\boldsymbol{W}}. \tag{9.89}$$

Directivity Index for Diffuse Noise Field

One of the most important parameters to define the quality and characteristics of an array of sensors is the *directivity*, defined in the presence of a diffuse noise field coming from all directions. The directivity index (DI) is defined as the quantity

$$DI\left(e^{j\omega}\right) \triangleq \frac{\left|\boldsymbol{W}^H\mathbf{a}(\omega,\theta_{\max})\right|^2}{\boldsymbol{W}^H\mathbf{\Gamma}_{nn}^{\text{diffuse}}(e^{j\omega})\boldsymbol{W}} \tag{9.90}$$

where the elements of the matrix $\mathbf{\Gamma}_{nn}^{\text{diffuse}}(e^{j\omega})$ are $\gamma_{n_in_j}(e^{j\omega}) \approx \text{sinc}\left(\frac{\omega d_{ij}}{c}\right)$, evaluated with (9.68). In general, we consider the evaluation in dB, $DI_{\text{dB}} = 10\log_{10}\left[DI(e^{j\omega})\right]$.

Uncorrelated Noise Field: White Noise Gain

In the case in which the noise is spatially white or uncorrelated, therefore it is $\boldsymbol{\Gamma}_{nn}(e^{j\omega}) \equiv \hat{\mathbf{R}}_{nn}(e^{j\omega}) = \mathbf{I}$; (9.84) takes the form

$$GW(e^{j\omega}) = \frac{\left|W^H\mathbf{a}(\omega,\theta)\right|^2}{W^H W} \tag{9.91}$$

where $GW(e^{j\omega})$ is defined as *white noise gain*. Note, as we shall see in the following, that in some types of beamformer the constraint $\mathbf{w}^H\mathbf{a}(\omega,\theta) = 1$ is assumed. In this case, the white noise gain is equal to $GW(e^{j\omega}) = \|\mathbf{w}\|^{-2}$ (i.e., the inverse of the weights' L_2 norm). For example, in the case of the WSB with all the same weights, it results as $GW(e^{j\omega}) = P$.

Geometric Gain

For spherically isotropic noise, the noise matrix is indicated as $\mathbf{Q}_g(e^{j\omega})$ to emphasize the dependence on the array geometry [13]. In this case, the corresponding gain, said *geometric gain*, is defined as

$$G_G\left(e^{j\omega},\theta\right) = \frac{\left|W^H\mathbf{a}(\omega,\theta)\right|^2}{W^H\mathbf{Q}_g(e^{j\omega})W}. \tag{9.92}$$

Supergain Ratio

The Q_a factor, or *supergain ratio*, which represents an alternative measure to the array sensibility, is defined as the ratio between the geometric gain and the white noise gain, i.e.,

$$Q_a\left(e^{j\omega},\theta\right) \triangleq \frac{G_G(e^{j\omega},\theta)}{GW(e^{j\omega},\theta)} = \frac{W^H W}{W^H\mathbf{Q}_g W}. \tag{9.93}$$

The scalar quantity $G(e^{j\omega},\theta)/GW(e^{j\omega},\theta)$ is defined as *generalized supergain ratio.*

9.3.3.4 Array Sensitivity

Consider an array perturbation, for example, a random movement of a sensor, such as an error signal, indicated as ξ, with zero mean and normalized variance $\mathbf{Q}_\xi(e^{j\omega})$, such that the covariance matrix of the SOI becomes $\sigma_s^2\left(\left|W^H\mathbf{a}(\omega,\theta)\right|^2 + \xi\mathbf{Q}_\xi(e^{j\omega})\right)$.

It is defined as the array gain sensitivity with respect to disturbance (*array sensitivity*) ξ as

$$S=\left(\frac{dG/d\xi}{G}\right)=\frac{W^H\mathbf{Q}_\xi(e^{j\omega})W}{\left|W^H\mathbf{a}(\omega,\theta)\right|^2}=\frac{1}{G_\xi}. \tag{9.94}$$

For uncorrelated disturbances for which $\mathbf{Q}_\xi(e^{j\omega})=\mathbf{I}$, by (9.91), the sensitivity is the reciprocal of white noise gain ($S_w=G_w^{-1}$) which is, for this reason, assumed as classical array sensitivity measure. The white noise gain is, therefore, the measure that is usually related to the array robustness.

9.4 Conventional Beamforming

A first BF category is the nonadaptive one, called *fixed beamforming*, in which both the topology and the circuit parameters are defined by minimizing a CF that does not depend on the statistics of the input data (SOI or noise field) to be acquired, but from a certain desired spatial-frequency response. In general terms, as previously noted, we can identify the following types of fixed beamforming:

- Delay and sum beamforming (DSBF)
- Delay and subtract (*differential beamforming*)
- Delay and weighted sum beamforming (DWSB)
- Filter and sum beamforming (FSBF)

The DSBF is the analog of the DT moving average filter. In practice, it does not perform any processing on the individual channels that are simply added together. In other cases, as for the digital filters, also for the narrowband or broadband array, it is possible to determine the parameters $\mathbf{w}$ in order to synthesize a desired frequency-spatial response, according to an appropriate optimization criterion.

In this section we present some very common types of fixed beamforming, often referred to as *conventional beamforming*, where the determination of the weights is performed in a similar way to the digital filters design with the windowing techniques or with approximation of a certain desired response. In practice, as for DT filters, the methods of polynomial approximation are used with various types of metrics like min–max, LS, weighed LS, etc.

9.4.1 Conventional Beamforming: DSBF-ULA

The uniform distribution linear array, called ULA and shown in Fig. 9.7, is among the most widely used applications in both electromagnetic and acoustic. Typically with ULA-DSBF it refers to a BF with identical weights.

9.4.1.1 Radiation Pattern

The array radiation function $R(e^{j\omega},\theta)$, defined in (9.80), represents the BF's spatial domain response for a given frequency sinusoidal signal, as a function of the angle of arrival. For an ULA the array radiation diagram, combining (9.37) and (9.80), is defined as

$$R\left(e^{j\omega},\theta\right)=\mathbf{w}^{H}\mathbf{a}_{ULA}(\omega,\theta)=\sum_{m=1}^{P}w_{m}^{*}e^{-jk(m-1)d\cos\theta} \tag{9.95}$$

the latter is, in fact, just the DTFT of the spatial response of the spatial filter. Note that, in the case of unitary weights, we can evaluate (9.95), in closed form as

$$R\left(e^{j\omega},\theta\right)=\sum_{m=1}^{P}\frac{1}{e^{jk(m-1)d\cos\theta}}=\frac{1-e^{-jkPd\cos\theta}}{1-e^{-jkd\cos\theta}}. \tag{9.96}$$

The radiation diagram for $\tau=(d\cos\theta)/c$ is

$$\left|R\left(e^{j\omega},\theta\right)\right|=\left|\frac{\sin\left(\frac{kPd}{2}\cos\theta\right)}{\sin\left(\frac{kd}{2}\cos\theta\right)}\right|=\left|\frac{\sin\left(\frac{1}{2}P\omega\tau\right)}{\sin\left(\frac{1}{2}\omega\tau\right)}\right|. \tag{9.97}$$

In Fig. 9.17, for example, the modules of the radiation functions of an ULA with seven sensors are reported, irradiated with a front wave parallel to the axis of the sensors and with unitary weights.

In general, as shown in (9.81), the radiation diagram represents the normalized diagram with respect to the direction of maximum gain $R_d(e^{j\omega},\theta)$, evaluated for a specific frequency and, in general, displayed on a polar plot, with values expressed in decibels or in natural values as, for example, shown in Fig. 9.18 (relative to the Cartesian beampattern of Fig. 9.17).

In Fig. 9.19, the normalized beampattern $R_d^{\mathrm{dB}}(\omega,\theta)$ is reported and in logarithmic scale for an ULA of 5 microphones away from each other 4.3 cm, operating in audio frequencies with sampling frequency equal to $f_c=8$ kHz.

From the previous examples it can be observed that, for an ULA, the width of beam is wider at low frequencies.

9.4.1.2 DSBF Gains

For a wave with DOA $=\theta_0$ the delay between the sensors is zero for which, indicating with $\mathbf{1}\triangleq\begin{bmatrix}1 & \cdots & 1\end{bmatrix}_{P\times1}^{T}$, the vector of P unitary elements is $\mathbf{a}(\omega,\theta_0)=\mathbf{1}$. Note, as shown in Fig. 9.17, that for a DSBF with unit weights, in the direction of maximum gain θ_0, the response $R(\omega,\theta_0)$ is precisely equal to the number

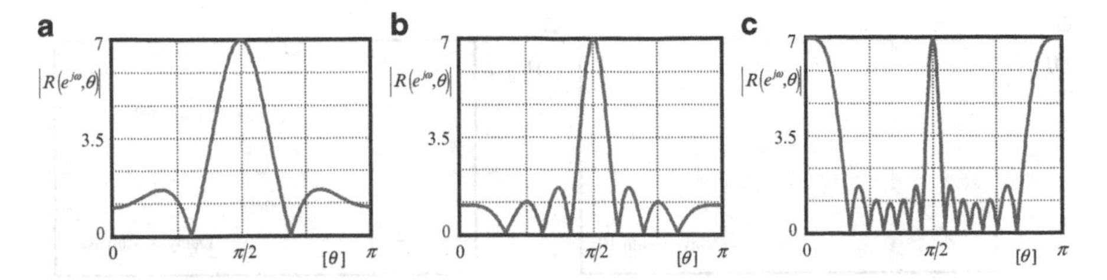

Fig. 9.17 Narrowband array beampattern for an incident wave $\theta = [-90°, 90°]$, for $P = 7$, and unitary weights. Distance between the sensors: (**a**) $d = \lambda/4$; (**b**) $d = \lambda/2$; (**c**) $d = \lambda$

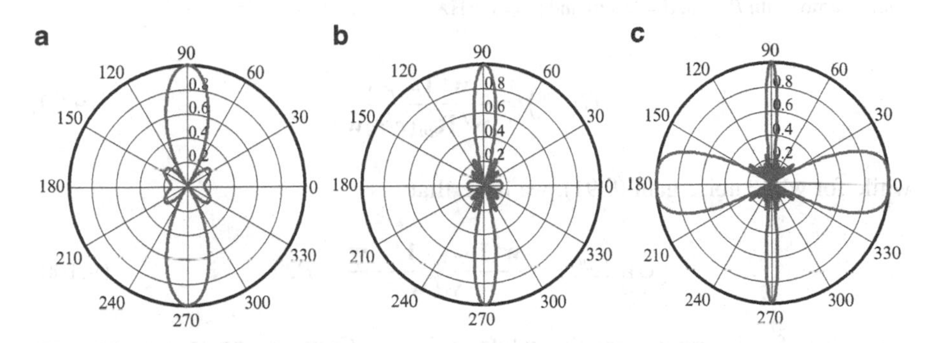

Fig. 9.18 Directivity diagrams in normalized polar form for narrowband ULA described in examples (**a**), (**b**), and (**c**) of Fig. 9.17

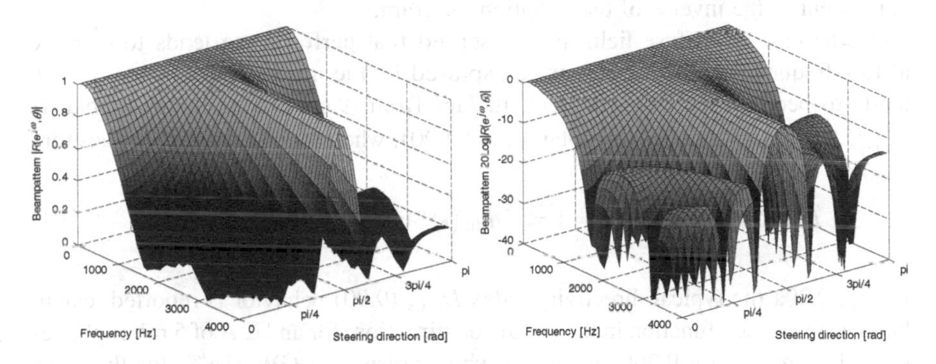

Fig. 9.19 3D Normalized beampattern in natural values and in dB, as a function of angle of arrival and frequency, for $P = 5$, $d = 4.3$ cm, and $f_c = 8$ kHz

of sensors, i.e., $\mathbf{w}^T\mathbf{a}(\omega,\theta_0) = P$. However, it is usual to consider a unity gain at θ_0 for which is $\mathbf{w}^T\mathbf{a}(\omega,\theta_0) = 1$. This is equivalent to impose the weights equal to

$$\mathbf{w} = \mathbf{1}/P. \tag{9.98}$$

Therefore, for any value of constant weights, called $\mathbf{1}_{P\times P} = \mathbf{1}_{P\times 1} \cdot \mathbf{1}_{P\times 1}^T$ the matrix of unitary elements, for the definition (9.84), the array gain is

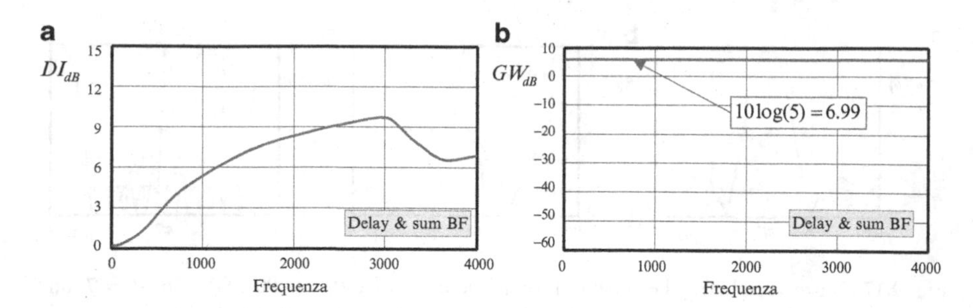

Fig. 9.20 Performance indices: (**a**) directivity; (**b**) white noise gain. For an ULA-DSBF microphones array with $P = 5$, $d = 5$ cm and $f_c = 8$ kHz

$$G(e^{j\omega}) = \frac{\mathbf{w}^T \mathbf{1}_{P\times P}\mathbf{w}}{\mathbf{w}^T \hat{\mathbf{R}}_{nn}(e^{j\omega})\mathbf{w}} \tag{9.99}$$

while for white noise gain (9.91), we have that

$$GW(e^{j\omega}) = \frac{\mathbf{w}^T \mathbf{1}_{P\times 1}\cdot \mathbf{1}_{P\times 1}^T \mathbf{w}}{\mathbf{w}^T \mathbf{w}} = P. \tag{9.100}$$

Note that, for isotropic spatially white noise or Gaussian noise coming from all directions, the DSBF maximizes the white noise gain $GW(e^{j\omega})$. In addition, for incoherent noise field by (9.84), the achievable noise reduction is in practice equivalent to the inverse of the radiation diagram.

In the case of diffuse field, it is observed that performance tends to degrade at low frequencies. In fact, the noise captured by the microphones, when $d \ll \lambda$, tends to become spatially coherent. In fact from (9.68), the columns of matrix $\boldsymbol{\Gamma}_{nn}^{\text{diffuse}}(e^{j\omega})$ tends to become unitary. For (9.90), when $\mathbf{a}(\omega,\theta_0) = \mathbf{1}$, it is therefore

$$DI(e^{j\omega}) = \frac{\mathbf{w}^T \mathbf{1}_{P\times P}\mathbf{w}}{\mathbf{w}^T \mathbf{1}_{P\times P}\mathbf{w}} = 1 \Rightarrow DI_{\text{dB}}(e^{j\omega}) = 0, \quad \text{for } \omega \to 0. \tag{9.101}$$

In Fig. 9.20a the typical directivity index DI_{dB} (9.90) behavior is reported, calculated as frequency function in the broadside direction, for an ULA of 5 microphones spaced 5 cm. Figure 9.20b shows the white noise gain $\text{GW}_{\text{dB}}(e^{j\omega})$ for the same array. From the physical point of view, as already illustrated in Fig. 9.19, at low frequencies, for the scarce spatial diversity, the ULA tends to lose directivity for which it acquires "in phase" both for the SOI and the noise coming from all directions.

Remark The DSBF is very sensible to noise especially at low frequencies and for arrays with few elements; moreover, the DSBF is very sensible to the sensors characteristics dispersion (gain, phase, position, etc.). To decrease the coherence at low frequencies, it is convenient to increase, as much as possible, the distance

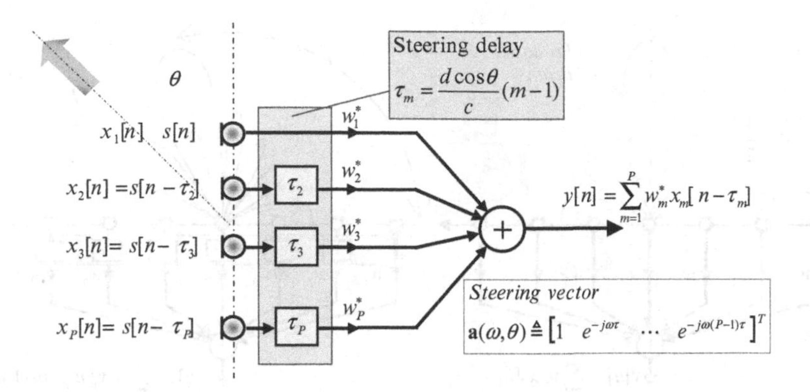

Fig. 9.21 Incident wave on an ULA with an angle $\theta \neq 90°$

between the sensors. This suggests the use of different array topology, appropriately spaced for each frequency range, as for the harmonic distribution ULA described in Sect. 9.2.3.3.

9.4.1.3 Radiation Pattern Orientation: Steering Delay

As seen in (9.97) the simple sum produces a sinc(x)-like radiation function (see also Fig. 9.17) with a main lobe in the front direction of the array. To change the orientation of the radiation (or steer) and produce such a lobe at any angle, in addition to the trivial solution of physically orienting the array toward the SOI, you can insert an artificial delay, called *steering delay*, in order to put in phase the response with a certain angle $\theta \neq 90°$, as shown in Fig. 9.21.

For a single radiating source, the ULA steering vector, as already defined in (9.37), is defined as the vector whose elements are a function of the phase delay, relative to each receiver, associated to the incident plane wave with an angle θ.

For example, in Fig. 9.22 a narrowband BF is illustrated, wherein the beam orientation is achieved through steering time delay inserted at the BF's input, downstream of the sensors.

For an incident plane wave with zero-phase reference sensor, for which $x_1[n] \equiv s[n]$, with the narrowband signal defined directly in DT as $s[n] = e^{j\omega n}$ (with the appropriate assumptions and simplifications), the output of the array is

$$y[n] = \sum_{m=1}^{P} w_m^* x_m[n] = e^{j\omega n} \sum_{m=1}^{P} w_m^* e^{-j\omega(m-1)\tau} \tag{9.102}$$

that, for the sinusoidal signal and ULA, is equivalent to an FIR filter, defined in the spatial domain, with delay elements equal to $z^{-\tau}$. The BF's radiation pattern is expressed as a steering delay τ and the BF's weight function and calculated as $R(\omega,\theta) = \mathbf{w}^H \mathbf{a}(\omega,\theta)$ and defined by expression (9.97).

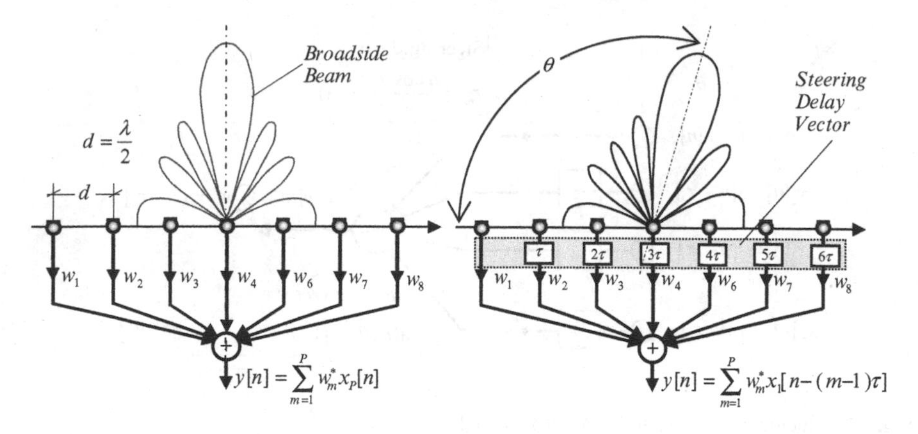

Fig. 9.22 Beampattern orientation of a delay and sum beamformer with the insertion of steering time delay

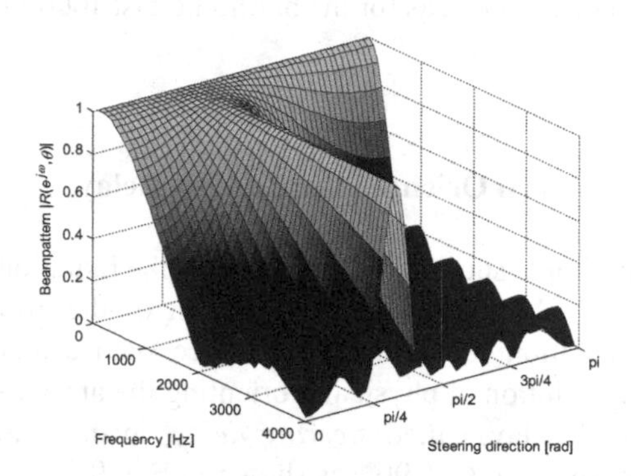

Fig. 9.23 Radiation diagram in natural values, for an ULA of $P = 8$ microphones spaced 4.3 cm, with a sampling frequency equal to CF $= 8$ kHz and a steering angle equal to $\pi/3$

Figure 9.23 shows the 3D plot of the radiation pattern in normalized natural values, for an ULA microphones array with $P = 8$, $d = 4.3$ cm, working for audio frequency with $f_c = 8$ kHz and a steering angle equal to 60°.

9.4.2 Differential Sensors Array

The conventional BFs have the sensors spaced at a distance $d \approx \lambda/2$ (related to the maximum frequency to be acquired), with a directivity proportional to the number of sensors P [(9.99) and (9.100)]. Another example of data-independent beamforming consists of a linear array (also not uniform) with distance between the sensors $d \ll \lambda$, i.e., sensors *almost coincident* and fixed look-direction in the *endfire direction*, and with a theoretical maximum gain equal to P^2. This type of array is also called superdirective BF (SDBF).

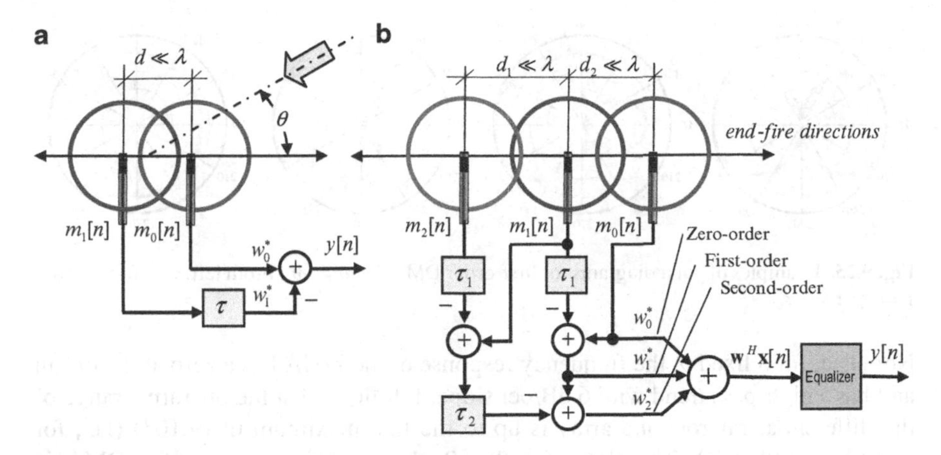

Fig. 9.24 Examples of differential microphones: (**a**) first order; (**b**) second order

In the case of acoustic sensors, the system is referred to as differential microphones array (DMA) or *gradient array* of microphones [14] (but it can be applied also to loudspeakers) implemented with the structure of Fig. 9.24.

The conventional arrays behave as a low-pass FIR filters defined in the spatial domain, for which the directional gain depends on its physical size. Conversely, the differentials arrays, having high-pass characteristics, are defined with different theoretical assumptions with respect to the standard delay-and-sum BF and with mandatory endfire direction, i.e., $\theta = 0$, of the desired signal. Moreover, delay-and-sum beamformer uses delay elements in order to steer the beam direction, whereas DMA may, in certain situations, steers the null direction.

Indeed, the differential microphones array can be considered as a finite-time-difference approximation of spatiotemporal derivatives of the scalar acoustic pressure field [15].

The DMA is built with an array of P omnidirectional capsules placed at a distance as small as possible, compatibly with the size of the mechanical structure and the low frequency noise. The order of the microphone is equal to $P - 1$.

9.4.2.1 DMA Radiation Diagram

Refer to Fig. 9.24a for a wave coming from $\theta = 0$; the delay between sensors is equal to $\tau_d = d/c$ and for (9.6) we have that $kd = \omega\tau_d$. For $P = 2$, $d \ll \lambda$, and inserting a steering delay $0 \le \tau \le \tau_d$ on one of the microphones, such that $\omega\tau \ll \lambda$, the expression of radiation diagram $R(e^{j\omega},\theta)$ can be written as

$$R\left(e^{j\omega},\theta\right) = 1 - e^{-j\omega(\tau+\tau_d\cos\theta)}. \tag{9.103}$$

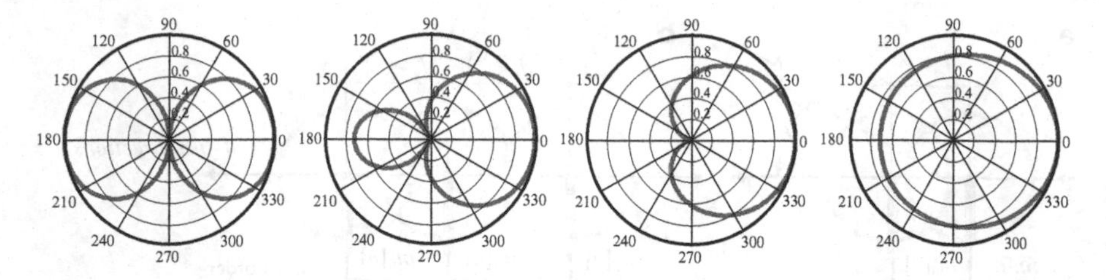

Fig. 9.25 Examples of polar diagrams for first-order DMA for $\omega = \omega_c$. From left, $\tau = 0, \tau = \tau_d/3$, $\tau = \tau_d, \tau = 3\tau_d$

Note that, for a fixed θ, the frequency response of the DMA has a zero at the origin and has a high-pass trend with 6 dB/oct slope. It follows that the operative range of the differential microphone array is up to the first maximum of (9.103) (i.e., for $\omega(\tau+\tau_d \cos\theta) = \pi$). Therefore, for $\theta = 0$, the *cut-off-frequency* of a DMA is $\omega_c = \pi/(\tau+\tau_d)$.

DMA Polar Diagram

For $\omega \leq \omega_c$, analyzing the expressions (9.103) with the approximation $\sin(\alpha) \sim \alpha$ for $\alpha \to 0$, $R(e^{j\omega},\theta)$ can be approximated as $R(e^{j\omega},\theta) \approx \omega(\tau+\tau_d \cos\theta)$. The radiation diagram in the θ domain, for fixed ω, can be written as $\tau+\tau_d \cos\theta$ which is not dependent on frequency.

As illustrated in Fig. 9.25, with $\omega = \omega_c$, for $\tau = 0$ and the normalized polar diagram is a *dipole* or *figure eight*, for $\tau = \tau_d$ is a *cardioid*, while for $\tau < \tau_d$ the diagram is of *hypercardioid* type.

DMA Frequency Response

Considering the expression (9.103) for the power diagram, we have that

$$\left|R\left(e^{j\omega},\theta\right)\right|^2 = 2 - 2\cos\left(\omega(\tau + \tau_d\cos\theta)\right). \tag{9.104}$$

Figure 9.26a shows the frequency response of a DMA with $P = 2, \tau = \tau_d$ (i.e., with a cardioid polar plot) and $d = 2.5$ cm, (i.e., with the cut-off frequency $f_c = 3.44$ kHz). Figure 9.26b shows the frequency response for a cardioid DMA considering difference distance between microphones. Figure 9.26c shows the frequency response for different radiation patterns with fixed distance $d = 2.5$ cm.

Due to the high-pass characteristic of $R(e^{j\omega},\theta)$, the MDA is very susceptible to disturbance. For this reason, the distance d is chosen as a compromise between the hypothesis $kd \ll 1$ and d should not be too small, to be not sensitive to noise.

Usually, the DMA requires the insertion of an equalizer to compensate for the high-pass trend of (9.104). For low frequencies, the equalization takes very

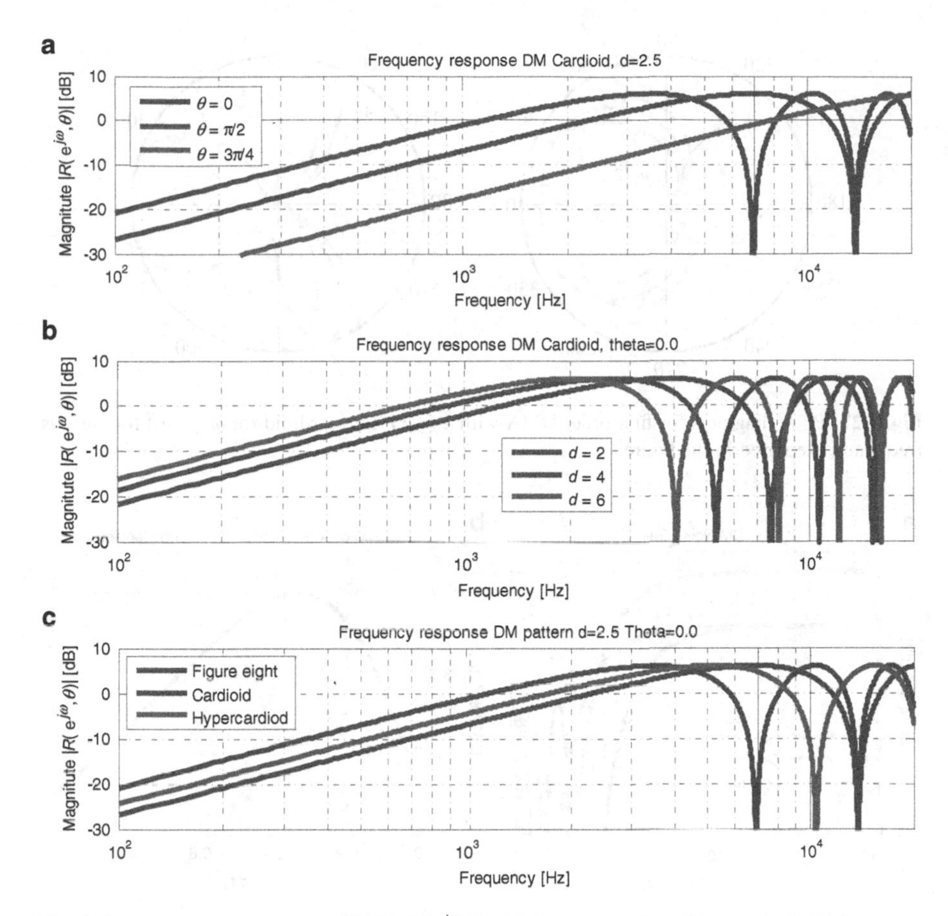

Fig. 9.26 Frequency response of DMA $R(e^{j\omega},\theta)$: (**a**) for some value of the angle θ; (**b**) for some value of the distance; (**c**) for different pattern

high gain. This means that any disturbance is strongly amplified. A lower limit for signal disturbance is represented by sensor noise. It determines the minimum limit for the frequency range that is reasonable for the operation of a differential array. Again, microphones mismatch puts the lower limit at higher frequencies.

In Fig. 9.27 the polar diagrams are shown for $\tau = \tau_d$ and for different frequencies. Note that for $\omega > \omega_c$ the polar plot is not a *cardioid.*

9.4.2.2 DMA Array Gain for Spherically Symmetric Isotropic-Noise

The array gain for spherical isotropic noise field can be computed by the expression (9.85). Considering $N(e^{j\omega},\theta) = 1$ and combining with (9.104)

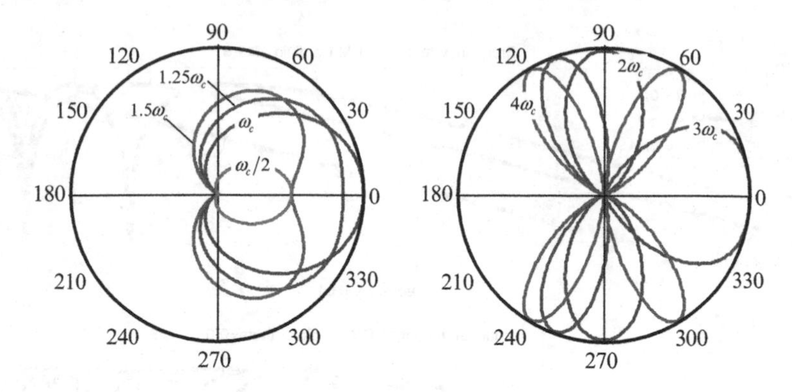

Fig. 9.27 Polar diagrams for first-order DMA with $\tau = \tau_d$ (i.e., cardioid for $\omega \leq \omega_c$) for various frequencies reported in the figure

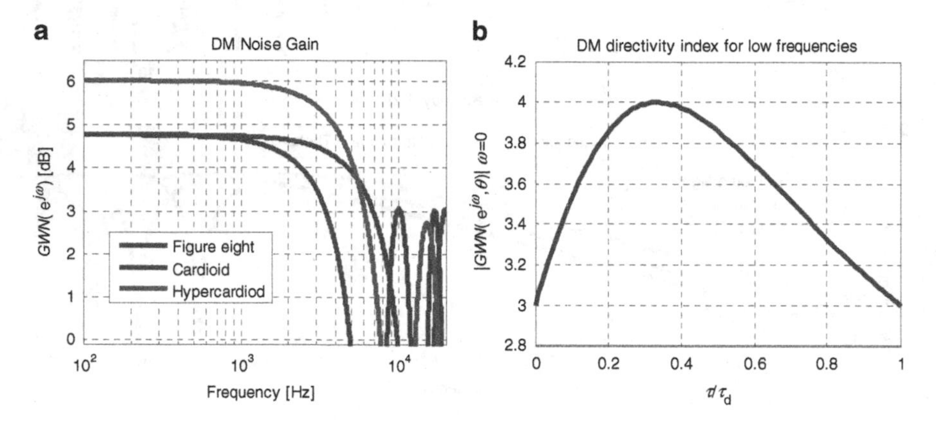

Fig. 9.28 DMA *Array Gain*: (**a**) directivity index $kd \ll 1$, for $\tau = \tau_d$, $\tau = \tau_d/3$, and $\tau = 0$; (**b**) gain at low frequency

$$G\left(e^{j\omega}, \theta_0\right) = \frac{2 - 2\cos\left(\omega(\tau + \tau_d \cos\theta)\right)}{\frac{1}{4\pi}\int_0^{2\pi}\int_0^{\pi}\left[2 - 2\cos\left(\omega(\tau + \tau_d\cos\theta)\right)\right]\sin\theta \cdot d\theta d\phi}. \tag{9.105}$$

Solving, the array gain assumes the form (see Fig. 9.28a) [16]:

$$G\left(e^{j\omega}\right) = \frac{2\sin^2\left(\frac{\omega}{2}(\tau_d + \tau)\right)}{1 - \cos\left(\omega\tau\right)\cdot\left(\sin\left(\omega\tau_d\right)/\omega\tau_d\right)} \tag{9.106}$$

for which, let $r = \tau/\tau_d$, the gain at low frequency is

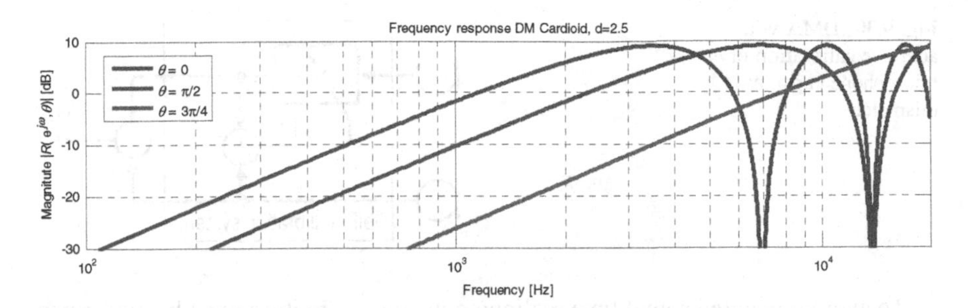

Fig. 9.29 Frequency response of a second-order differential microphone array, shown in Fig. 9.24b, for $d_1 = d_2$ and $\tau_1 = \tau_2$. Note that the maximum gain is equal to P^2, i.e., 9.5 dB

$$\lim_{\omega \to 0} G\left(e^{j\omega}\right) = \frac{3(1+r)^2}{1+3r^2}. \tag{9.107}$$

The low frequency gain has a maximum equal to P^2 for $r = \frac{1}{3}$, i.e., for hypercardioid configuration (see Fig. 9.28b).

Other considerations on the array gain performance of an endfire line array will be made later in Sect. 9.5.1.3.

For $P > 2$, the expression (9.103) with $\tau_{d_i} = d_i/c$ can be generalized as

$$R\left(e^{j\omega}, \theta\right) = \prod_{i=1}^{P} \left[1 - e^{-j\omega\left(\tau_i + (d_i/c)\cos\theta\right)}\right] \tag{9.108}$$

and for $\omega \leq \omega_c$, $R(e^{j\omega}, \theta) \approx \omega^P \prod_{i=1}^{P} (\tau + \tau_{d_i} \cos\theta)$. The latter can be written as a power series of the type

$$R\left(e^{j\omega}, \theta\right) \approx A\omega^P (a_0 + a_1 \cos\theta + \ldots + a_P \cos\theta), \quad \text{with} \quad \sum\nolimits_i a_i = 1. \tag{9.109}$$

Figure 9.29 shows the frequency response for a second-order DMA.

By inserting complex weights in addition to delays, you can get the BF with beampattern approximating specific masks. The design criteria are very similar to that of digital filters. Consequently, we can get response curves of the type max-flat, equiripple, min L_2 norm, etc.

9.4.2.3 DMA with Adaptive Calibration Filter

In the case of higher orders ($P > 2$), equalizers with high gains (>60 dB) at low frequencies are required (see Fig. 9.29). Therefore, microphone mismatch and noise can cause severe degradation of performance in the low frequency range. A simple expedient to overcome this limitation can be made with an *adaptive calibration filter* as shown in Fig. 9.30 [17].

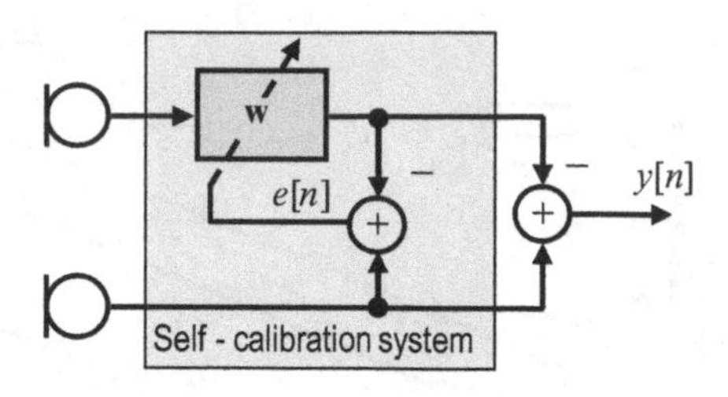

Fig. 9.30 DMA with adaptive calibration of microphone capsules mismatch

To avoid unwanted signal time realignments, the calibration must be performed a priori, e.g., considering a plane wave coming from the broadside direction.

9.4.3 Broadband Beamformer with Spectral Decomposition

The narrowband processing is conceptually simpler than the broadband one because the temporal frequency is not considered. This situation suggests a simple way for the realization of a broadband beamformer through the input signal domain transformation, typically made via a sliding DTF, DTC, etc. transform (see Sect. 7.5.1), so as to obtain a sum of narrowband processes. As illustrated in Fig. 9.31, the set of narrowband contributions of identical frequency, called *frequency bins*, are processed in many narrowband-independent BF units related to each frequency [8, 18].

The BF is operating in the transformed domain and can be considered as a MISO TDAF (see Sect. 7.5). We denote with $\boldsymbol{X} \in \mathbb{C}^{P \times N_f}$ the matrix containing the N_f frequency bins of each of the P receivers (calculated with sliding transform of N_f length), and with $\boldsymbol{W} \in \mathbb{C}^{P \times N_f}$ the matrix containing in each column the BF's weights relative to each frequency. Considering the DFT transform implemented by FFT, the BF output is calculated as

$$\mathbf{y} = \mathrm{FFT}^{-1}\left(\boldsymbol{W}^H \boldsymbol{X}\right).$$

The output of the receivers is transformed into the frequency domain, and signals relating to the same frequency (frequency bin) are combined with simple delay and sum BF.

A second decomposition mode consists in dividing the signals into time-space subbands. The division into spatial subbands is performed with a suitable array distribution, for example, the harmonic linear arrays described in Sect. 9.2.3.3, while the temporal processing is performed by a filters bank as described in Sect. 7.6.

The subbands are determined by the selection of a subset of sensors. Each subband subset is considered as a BF that can be implemented in the time or frequency domain. Each subband BF processes a narrower-band signal compared to that of the input signal $s[n]$ and, in the case of a high number of spatial subbands, the subband processing can be executed with a simple DSBF.

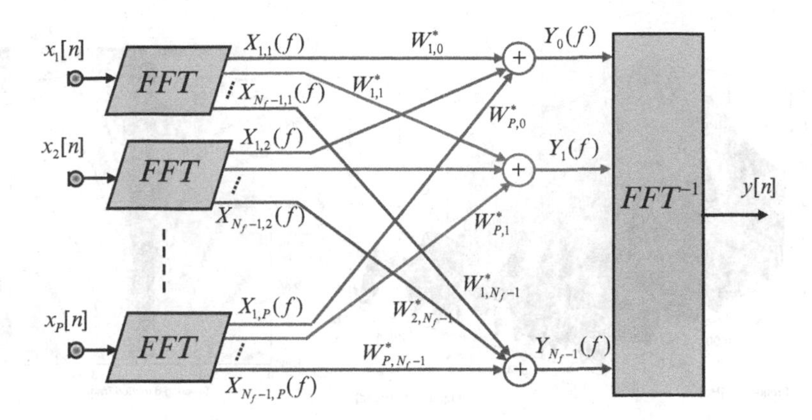

Fig. 9.31 Principle diagram of a broadband frequency domain beamformer with narrow-band decomposition

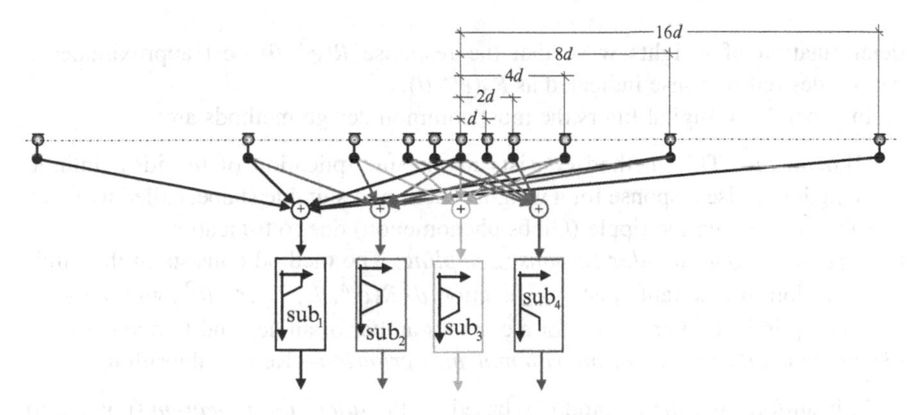

Fig. 9.32 Example of an 11 microphones beamformer with nested structure with 4 subbands using DSBF sub-array. For each subband only 5 microphones are selected

By way of example, in Fig. 9.32, a four subbands nested structure with 11 microphones is shown. Figure 9.33 shows the beampattern of the BF of Fig. 9.32 for a distance between the sensors $d = 3.5$ cm and $f_c = 16$ kHz.

Note that even if a FSBF structure is used, the subdivision into subbands still allows the use of much shorter filters compared to the full band case.

9.4.4 Spatial Response Direct Synthesis with Approximate Methods

The methods described below can be viewed as a generalization of the approximation techniques used in the digital filtering design, in which the specifications are given both in the frequency and in space domain. The BF design consists in the

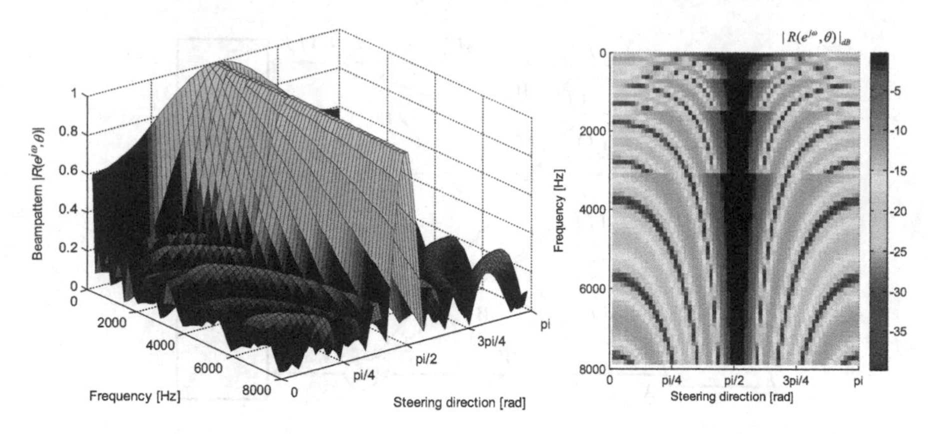

Fig. 9.33 Radiation diagram in natural values for the nested structure BF of Fig. 9.32, with $d = 3.5$ cm and $f_c = 16$ kHz. The subbands are defined as: $\text{sub}_1 = (0, 800]$, $\text{sub}_2 = (800, 1600]$, $\text{sub}_3 = (1600, 3200]$, and $\text{sub}_4 = (3200, 8000]$ [Hz]

determination of weights $\mathbf{w}$ so that the response $R(e^{j\omega},\theta)$ best approximates a certain desired response indicated as $R_d(e^{j\omega},\theta)$.

In general, for digital filters the most common design methods are:

(1) *Windowing*: This method consists in the multiplication of the ideal infinite length impulse response for a weight function of suitable shape, called *window*, able to mitigate the ripple (Gibbs phenomenon) due to truncation.
(2) *Frequency and angular response sampling*: The method consists in the minimization of a suitable distance function $d\{R(e^{j\omega},\theta), R_d(e^{j\omega},\theta)\}$, with a specified optimization criterion, for a certain number of angles and frequencies.
(3) *Polynomial approximation with min–max criterion*—Remez algorithm.

In beamforming, the method (3), based on the *alternation theorem* (relative to the techniques of polynomial approximation), is applicable only in the case of linear array with uniform distribution.

9.4.4.1 Windowing Method

The analogy of narrow-band arrays with FIR filters expressed by (9.102) implies also common design methodologies. For unit weights BF, the array behaves as an MA FIR filter (see Sect. 1.6.3) for which increasing the length of the filter (in this case the number of sensors) decreases the width of the lobe but not the level of the secondary lobes. To decrease the level of secondary lobes is necessary to determine appropriate weighting schemes similar to those of the windowing method of linear phase FIR filters.

The choice of the window allows to determine the acceptable level of the secondary lobes while the number of sensors determines the width of the beampattern or the array spatial resolution.

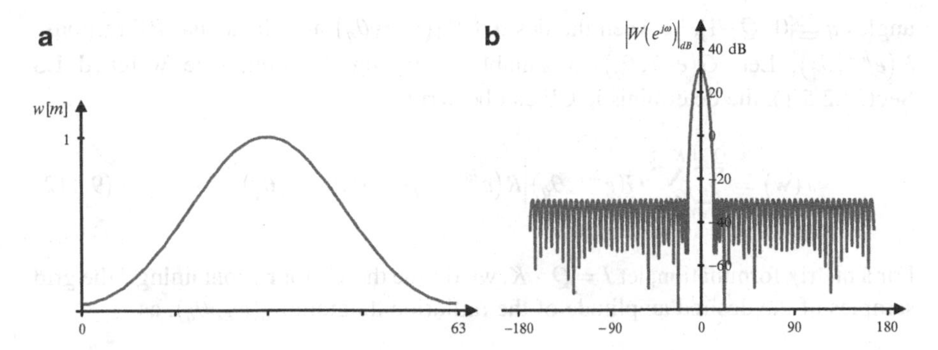

Fig. 9.34 Dolph–Chebyshev window white 20 log $_{10}(10^{\alpha}) = 60$ dB: (**a**) time domain response; (**b**) frequency domain response

A very common choice made in the antenna array is the Dolph–Chebyshev window that has the property of having the secondary lobes all at the same level (almost equiripple characteristic) and a rather narrow spatial band.

Calling $W(m)$, $m \in [-P+1, P-1]$, the DFT of the weights filter, the Dolph–Chebyshev window is computed as (for details see [19])

$$W(m) = (-1)^m \frac{\cos\left[P\cos^{-1}\left[\beta\cos\left(\pi\frac{m}{P}\right)\right]\right]}{\cosh\left[P\cosh^{-1}(\beta)\right]}; \quad 0 \leq |m| \leq P-1 \tag{9.110}$$

in which the term β, is defined by $\beta = \cosh\left[\frac{1}{N}\cosh^{-1}(10^{\alpha})\right]$ and

$$\cos^{-1}(X) = \begin{cases} \frac{\pi}{2} - \tan^{-1}\left[X/\sqrt{1.0 - X^2}\right], & |X| < 1.0 \\ \ln\left[X + \sqrt{X^2 - 1.0}\right], & |X| \geq 1.0 \end{cases}. \tag{9.111}$$

The parameter α is proportional to the desired secondary lobes attenuation in dB, which is equal to 20 log $_{10}(10^{\alpha})$. To obtain the weights w_m^*, it is sufficient to perform the inverse DFT of the samples $W(m)$ in (9.110).

Figure 9.34 shows the plot of the weights and the spatial response for an array of $P = 64$ elements with weights $w[m]$ calculated using the Dolph–Chebyshev window. Other types of windows are described in [19].

9.4.4.2 Spatial Response Synthesis with Frequency-Angular Sampling

The frequency-angular response sampling method, coincides with the analogous frequency sampling method of the digital filters. In practice, it minimizes the LS distance in a finite number of points of frequencies $k \in [0, K-1]$ and angles $q \in [0, Q-1]$, between the desired $R_{\mathrm{d}}(e^{j\omega_k}, \theta_q)$ and the actual BF response $R(e^{j\omega_k}, \theta_q)$. Let $G(e^{j\omega_k}, \theta_q)$ a suitable weighing function, (see weighed LS Sect. 4.2.5.1), the deterministic CF can be written as

$$J(\mathbf{w}) = \sum_{q=0}^{Q-1} \sum_{k=0}^{K-1} G(e^{j\omega_k}, \theta_q) \left| R(e^{j\omega_k}, \theta_q) - R_{\mathrm{d}}(e^{j\omega_k}, \theta_q) \right|^2. \tag{9.112}$$

For a matrix formulation, let $J = Q \cdot K$, we define the vector $\mathbf{r}_d$ containing J the grid samples of the desired amplitude of the radiation diagram $R_{\mathrm{d}}(\omega_k, \theta_q)$, as

$$\mathbf{r}_d \in \mathbb{R}^{J \times 1} \triangleq \left[R_d(\omega_0, \theta_0) \cdots R_d(\omega_k, \theta_0) \cdots R_d(\omega_k, \theta_{Q-1}) \cdots R_d(\omega_{K-1}, \theta_{Q-1}) \right]^T \tag{9.113}$$

and similarly the vector $\mathbf{r}$ is defined as the J samples of the actual BF responses

$$\mathbf{r} \in \mathbb{R}^{J \times 1} \triangleq \left[R(\omega_0, \theta_0) \cdots R(\omega_k, \theta_0) \cdots R(\omega_k, \theta_{Q-1}) \cdots R(\omega_{K-1}, \theta_{Q-1}) \right]^T \tag{9.114}$$

The steering matrix $\mathbf{A} \in \mathbb{C}^{PM \times J}$ is defined as containing the steering vectors in the $Q \times K$ sampling points of the response. Moreover, the steering matrix can be decomposed into a real and an imaginary parts, $\mathbf{A} = \mathbf{A}_R + j\mathbf{A}_I$. Since $\mathbf{A}_I$ is anti-symmetric (Sect. 9.3.3.2), considering only the real part, the beampattern can be written as $|R(\omega, \theta)|^2 = \mathbf{w}^T \mathbf{A}_R \mathbf{w}$. Formally

$$\mathbf{A}_R \in \mathbb{R}^{PM \times J} \triangleq \mathrm{Re}\left[\mathbf{a}(\omega_0, \theta_0) \quad \mathbf{a}(\omega_0, \theta_1) \quad \cdots \quad \mathbf{a}(\omega_{K-1}, \theta_{Q-1}) \right]_{1 \times J} \tag{9.115}$$

and considering the weighing function matrix defined as

$$\mathbf{G} \triangleq \mathrm{diag}\left[g_k \in \mathbb{R}^+, \quad k = 0, \ldots, J-1 \right] \tag{9.116}$$

the weighed LS problem (9.113) can be formulated in a canonical way with normal equations, of the type

$$\mathbf{w}_{opt} \therefore \min_{\mathbf{w}} \left\| \mathbf{r}_d - \mathbf{A}_R^T \mathbf{w} \right\|_{\mathbf{G}}^2 \tag{9.117}$$

Therefore, minimizing with respect to the parameter vector $\mathbf{w}$, we obtain an $PM \times J$ linear equations system with optimal (regularized) solution of the type

$$\mathbf{w}_{opt} \in \mathbb{R}^{PM \times 1} = \left(\mathbf{A}_R \mathbf{G} \mathbf{A}_R^T + \delta \mathbf{I} \right)^{-1} \mathbf{A}_R \mathbf{G} \mathbf{r}_d \tag{9.118}$$

Figure 9.35 shows the broadband ULA beampattern, with sixteen sensors ($P = 16$) and sixteen taps FIR filters ($M = 16$), with coefficients evaluated by the (9.118). The desired response has unity gain and linear phase for $f \in [2, 4]$ kHz, and

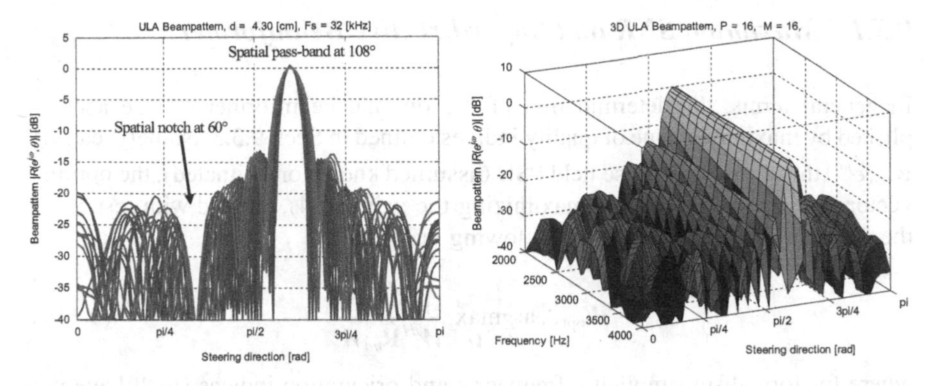

Fig. 9.35 Radiation function of a data-independent ULA with $P = 16$, $M = 16$, distance between the sensors $d = 4.3$ cm, with $F_s = 32$ kHz; evaluated for $K = 16$ frequencies and for $K = 120$ angles, in the range $f \in [2, 4]$ kHz and $\theta \in [0, 180°]$. The desired response is unity gain and linear phase for $f \in [2, 4]$ kHz, for an angle of $\theta = 108°$ and a spatial notch for $\theta = 60°$

for an angle equal to $\theta = 108°$. Furthermore, by an appropriate choice of the weighting function $\mathbf{G}$ is considered a null response to an angle equal to $\theta = 60°$ (spatial notch filter).

Remark Unlike the adaptive filtering case, where the normal equations are determined considering the estimated second order input signal statistics, in this case, the matrix $\mathbf{A}^{\#}$ is entirely deterministic because it specifies the BF desired response.

The LS method for beamforming problems, can be easily extended by considering particular constraints on the desired space-frequency response as, for example, null-response (or zeros) in certain directions.

Remark In case that the actual size J of the space spanned by the vectors $\mathbf{a}(\omega_k, \theta_q)$, for $k = 0, 1, ..., K-1, q = 0, 1, ..., Q-1$, is less than the PM, the matrix $\mathbf{A}$ is ill-conditioned. This situation may occur when only one direction of arrival is sampled. In this case, it is proved that the image space of $\mathbf{A}$, $\mathcal{R}(\mathbf{A})$, is approximately equal to the TBWP for that direction [8].

From (9.118), for ill-conditioned $\mathbf{A}$, the vector $\mathbf{w}$ norm tends to become too high, resulting in poor white noise gain performance (9.91). In these situations, in order to not excessively increasing the norm of $\mathbf{w}$, for the (9.118) calculation, in addition or alternatively to a regularized solution, is convenient to use an reduced rank approximation of $\mathbf{A}$, using, for example, the SVD decomposition [8, 12, 20, 21].

9.5 Data-Dependent Beamforming

In this section, we extend the method to the least squares described for the deterministic case (See Sect. 9.4.4.2), to the case where the CF depends from the SOI or from interference that you want to suppress (or both).

9.5.1 *Maximum SNR and Superdirective Beamformer*

In general terms, the determination of the optimal beamformer can be accomplished by maximizing one of quality indices defined in Sect. 9.3.3. Namely, calling $\hat{\mathbf{R}}_{nn}(e^{j\omega})$ the normalized noise field PSD (assumed known or estimated), the optimal vector $\boldsymbol{W}_{\text{opt}}$ can be calculated, maximizing the gain (9.84), defined with respect to the considered noise field, by the following criterion

$$\boldsymbol{W}_{\text{opt}} \therefore \underset{\boldsymbol{W}\in\Omega}{\operatorname{argmax}} \frac{\left|\boldsymbol{W}^H\mathbf{a}\right|^2}{\boldsymbol{W}^H\hat{\mathbf{R}}_{nn}\boldsymbol{W}} \tag{9.119}$$

where for formalism simplicity, frequency and orientation indices (ω,θ) have not been reported. The solution of the latter can be determined using the Lagrange multipliers method or, more simply, considering its gradient with respect to $\boldsymbol{W}^H$.

9.5.1.1 Standard Capon Beamforming

A simple solution (9.119), proposed in [22, 23], and known as the *standard Capon beamforming*, is directly obtained by imposing unity gain along the LD θ_0. In this case, the CF (9.119) is equivalent to the minimization of its denominator and imposing the unitary gain constraint. Therefore, for this type of problem, the CF can be defined as

$$\boldsymbol{W}_{\text{opt}} \therefore \underset{\boldsymbol{W}\in\Omega}{\operatorname{argmin}} \left(\boldsymbol{W}^H\hat{\mathbf{R}}_{nn}\boldsymbol{W}\right) \quad \text{s.t.} \quad \boldsymbol{W}^H\mathbf{a} = 1 \tag{9.120}$$

The solution of the optimization problem (9.119), and reformulated in (9.120), may be performed by applying the method of Lagrange multiplier as in Sect. 4.2.5.5 (see also Sect. B.3.2). Therefore, we can write

$$L(\mathbf{w},\boldsymbol{\lambda}) = \frac{1}{2}\boldsymbol{W}^H\hat{\mathbf{R}}_{nn}\boldsymbol{W} + \boldsymbol{\lambda}^H\left(\boldsymbol{W}^H\mathbf{a} - 1\right) \tag{9.121}$$

where $L(\mathbf{w},\boldsymbol{\lambda})$ is the Lagrangian and the term $\frac{1}{2}$ is added for later simplifications. The gradient of (9.121) with respect to $\mathbf{w}$ is $\nabla_{\mathbf{w}}L(\mathbf{w},\boldsymbol{\lambda}) = \hat{\mathbf{R}}_{nn}\boldsymbol{W} + \mathbf{a}^H\boldsymbol{\lambda}$, and to determine the optimal solution we set it equal to zero. The optimal solution in terms of Lagrange multipliers is $\boldsymbol{W}_{\text{opt}} = -\hat{\mathbf{R}}_{nn}^{-1}\mathbf{a}^H\boldsymbol{\lambda}$. Since $\boldsymbol{W}_{\text{opt}}$ must also satisfy the constraint of the CF, it follows that $\mathbf{a}^H\boldsymbol{W}_{\text{opt}} = \mathbf{a}^H\left(-\hat{\mathbf{R}}_{nn}^{-1}\mathbf{a}^H\boldsymbol{\lambda}\right) = 1$, i.e., $\boldsymbol{\lambda} = -\left[\mathbf{a}^H\hat{\mathbf{R}}_{nn}^{-1}\mathbf{a}\right]^{-1}$. Then we get

$$\boldsymbol{W}_{\text{opt}} = \frac{\hat{\mathbf{R}}_{nn}^{-1}\mathbf{a}}{\mathbf{a}^H\hat{\mathbf{R}}_{nn}^{-1}\mathbf{a}}. \tag{9.122}$$

Note that spherical isotropic noise with Gaussian distribution is $\hat{\mathbf{R}}_{nn} = \mathbf{I}$. The optimal solution, in this case, results to be the conventional DSBF.

$$W_{\text{opt}} = \frac{\mathbf{a}}{\mathbf{a}^H\mathbf{a}} = \frac{1}{P}. \tag{9.123}$$

9.5.1.2 Cox's Regularized Solutions with Robustness Constraints

Another possibility to improve the expression (9.120) consists in defining a CF in which the gain $G(e^{j\omega})$ is maximized and, in addition, imposing a certain white noise gain $GW(e^{j\omega})$ less than the maximum possible. Formally, the CF becomes

$$W_{\text{opt}} \therefore \underset{W\in\Omega}{\operatorname{argmax}}\, G\left(e^{j\omega}\right) \quad \text{s.t.} \quad GW\left(e^{j\omega}\right) = \beta^2 \leq P. \tag{9.124}$$

Equivalently, to have more design flexibility and get a regularized solution, as proposed by Cox et al. in [13, 24], instead of (9.124), it is possible to minimize the expression

$$W_{\text{opt}} \therefore \underset{W\in\Omega}{\operatorname{argmin}} \left(\frac{1}{G(e^{j\omega})} + \delta \frac{1}{GW(e^{j\omega})} \right)$$

where δ is interpreted as a Lagrange multiplier. Substituting the expressions of the gains (9.84) and (9.91), the CF can be defined as

$$W_{\text{opt}} \therefore \underset{W\in\Omega}{\operatorname{argmin}} \left(\frac{W^H\hat{\mathbf{R}}_{nn}W}{\left|W^H\mathbf{a}\right|^2} + \delta \frac{W^H W}{\left|W^H\mathbf{a}\right|^2} \right) = \underset{W\in\Omega}{\operatorname{argmin}} \left(\frac{W^H\left(\hat{\mathbf{R}}_{nn}+\delta\mathbf{I}\right)W}{\left|W^H\mathbf{a}\right|^2} \right). \tag{9.125}$$

The solution of the previous is similar to (9.119) in which the matrix $\hat{\mathbf{R}}_{nn}$ was replaced by its regularized form $\hat{\mathbf{R}}_{nn} \to \left(\hat{\mathbf{R}}_{nn}+\delta\mathbf{I}\right)$.

Therefore, by imposing unity gain along the LD, as in (9.122), we get

$$W_{\text{opt}} = \frac{\left(\hat{\mathbf{R}}_{nn}+\delta\mathbf{I}\right)^{-1}\mathbf{a}}{\mathbf{a}^H\left(\hat{\mathbf{R}}_{nn}+\delta\mathbf{I}\right)^{-1}\mathbf{a}}. \tag{9.126}$$

Modulating the regularization terms δ, it is possible to obtain optimal solutions depending on the noise field characteristic. For example, for $\delta \to \infty$, we obtain the conventional DSBF (see Fig. 9.37).

Remark The possibility of knowing the noise or signal characteristics is limited to a few typical applications: for example, in radar, in active sonar, where the characteristic of the transmitted signal is a priori known, or in the seismic, in which the noise can be estimated before the wave arrival. Only in these, and a few other situations, it is possible to estimate the noise or signal characteristic in the absence of the signal or noise.

More likely, in *passive* cases it is possible to estimate the PSD of the entire signal received from the sensors $\mathbf{R}_{xx}(e^{j\omega})$ that is coming from all directions and also contains the noise component. In this case in (9.120) and then in (9.126), it is sufficient to replace $\hat{\mathbf{R}}_{nn}(e^{j\omega}) \rightarrow \mathbf{R}_{xx}(e^{j\omega})$.

Note that in the array gain maximization, considering also the white noise gain equality constraint, the following three quadratic forms are alternatively considered

$$\left|\boldsymbol{W}^H\mathbf{a}\right|^2, \quad \boldsymbol{W}^H\mathbf{R}_{nn}\boldsymbol{W} \quad \text{and} \quad \boldsymbol{W}^H\boldsymbol{W}. \tag{9.127}$$

Since in the output power, in array gain, in white noise gain, and in generalized supergain ratio (see Sect. 9.3.3.3), only two of the quadratic forms in (9.127) are considered; we can define some equivalent forms of the optimization problem. Following this philosophy, in Cox [13, 24], the problem of the optimal constrained array determination can be formalized in the following ways.

Problem A Maximizing the array gain (9.84), as in (9.119), with constraints on the white noise gain and on the unitary gain along the LD, the CF can be written as

$$\boldsymbol{W}_{\text{opt}} \therefore \underset{\boldsymbol{W}\in\Omega}{\operatorname{argmax}} \frac{\left|\boldsymbol{W}^H\mathbf{a}\right|^2}{\boldsymbol{W}^H\bar{\mathbf{R}}\boldsymbol{W}} \quad \text{s.t.} \quad \frac{\left|\boldsymbol{W}^H\mathbf{a}\right|^2}{\boldsymbol{W}^H\boldsymbol{W}} = \delta^2, \quad \boldsymbol{W}^H\mathbf{a} = 1. \tag{9.128}$$

Problem B Maximizing the array gain (9.84), with constraints on the $\boldsymbol{W}$ norm and on the unitary gain along the LD, i.e.,

$$\boldsymbol{W}_{\text{opt}} \therefore \underset{\boldsymbol{W}\in\Omega}{\operatorname{argmax}} \frac{\left|\boldsymbol{W}^H\mathbf{a}\right|^2}{\boldsymbol{W}^H\bar{\mathbf{R}}\boldsymbol{W}} \quad \text{s.t.} \quad \boldsymbol{W}^H\boldsymbol{W} = \delta^{-2}, \quad \boldsymbol{W}^H\mathbf{a} = 1. \tag{9.129}$$

the matrix $\bar{\mathbf{R}}$, depending on the a priori knowledge of the specific problem, can be replaced with the noise or signal matrix: $\bar{\mathbf{R}} \rightarrow \hat{\mathbf{R}}_{nn}(e^{j\omega})$ or $\bar{\mathbf{R}} \rightarrow \hat{\mathbf{R}}_{xx}(e^{j\omega})$.

In other words for $\bar{\mathbf{R}} \rightarrow \hat{\mathbf{R}}_{xx}(e^{j\omega})$, from a physical point of view, only the signal not coming from the LD θ_0 that, mainly, should contain the noise is attenuated. As said above, a general solution of the problems (A) and (B), considering a solution of (9.126), is

$$\boldsymbol{W}_{\text{opt}} = \frac{(\mathbf{R}_{xx}+\delta\mathbf{I})^{-1}\mathbf{a}}{\mathbf{a}^H(\mathbf{R}_{xx}+\delta\mathbf{I})^{-1}\mathbf{a}}. \tag{9.130}$$

Finally, note that in the presence of multiple constraints, the formalization of the problem appears to be of the type

$$\boldsymbol{W}_{\text{opt}} \therefore \underset{\boldsymbol{W}\in\Omega}{\operatorname{argmin}}\ \boldsymbol{W}^H\mathbf{R}_{xx}\boldsymbol{W} \quad \text{s.t.} \quad \boldsymbol{C}^H\boldsymbol{W} = \boldsymbol{F}, \tag{9.131}$$

where $\boldsymbol{C}$ represents a suitable matrix of constraint and $\boldsymbol{F}$ the gain (typically $\boldsymbol{F} = 1$).

In this case the solution calculated with the Lagrange multipliers method (see Sect. 4.2.5.5) has the form

$$\boldsymbol{W}_{\text{opt}} = \mathbf{R}_{xx}^{-1}\boldsymbol{C}\left[\boldsymbol{C}^H\mathbf{R}_{xx}^{-1}\boldsymbol{C}\right]^{-1}\boldsymbol{F}. \tag{9.132}$$

This solution, derived from Cox, coincides with the Frost BF discussed in more detail and depth, in Sect. 9.5.3.

As a corollary of the above, it is observed that the BF weight vector, $\boldsymbol{W}$, can be decomposed into two orthogonal components

$$\boldsymbol{W} = \boldsymbol{G} + \boldsymbol{V}. \tag{9.133}$$

By defining the projection operators (see Sect. A.6.5) relating to the $\boldsymbol{C}$ as

$$\tilde{\mathbf{P}} = \boldsymbol{C}\left(\boldsymbol{C}^H\boldsymbol{C}\right)^{-1}\boldsymbol{C}^H, \quad \textit{projection on} \quad \Psi \in \mathcal{R}(\boldsymbol{C}) \tag{9.134}$$

$$\mathbf{P} = \mathbf{I} - \tilde{\mathbf{P}}, \qquad \textit{projection on} \quad \Sigma \in \mathcal{N}\left(\boldsymbol{C}^H\right) \tag{9.135}$$

such that

$$\boldsymbol{G} = \tilde{\mathbf{P}}\boldsymbol{W} \tag{9.136}$$

$$\boldsymbol{V} = \mathbf{P}\boldsymbol{W} \tag{9.137}$$

projecting the optimal solution (9.132), the image space of $\boldsymbol{C}$ is

$$\boldsymbol{G} = \boldsymbol{C}\left[\boldsymbol{C}^H\boldsymbol{C}\right]^{-1}\boldsymbol{F} \tag{9.138}$$

that does not depend on $\mathbf{R}_{xx}^{-1}$.

Insights and adaptive solutions of (9.132) and of the forms (9.136), (9.137), and (9.138) are presented and discussed in the following paragraphs.

9.5.1.3 Line-Array Superdirective Beamformer

The conventional beamformer for $d \approx \lambda/2$ has a directivity in the broadside direction approximately equal to the number of sensors P. In the case of ULA for $d \to 0$, as for the differential microphones (see Sect. 9.4.2), the gain of the array is, depending on the noise field characteristics, higher than that of conventional BF. In particular, in [12, 14–17, 25, 26], it is shown that for $d \ll \lambda/2$, in the endfire direction, for diffuse field with spherical symmetry, the array has a directivity index tending asymptotically to P^2 (see, for example, the Fig. 9.28, for $P = 2$). While in the case of cylindrical symmetry, the gain tends asymptotically to $2P$. However, as illustrated in Fig. 9.36, this relationship tends to be exactly verified only for low order array, $P = 2$ and $P = 3$.

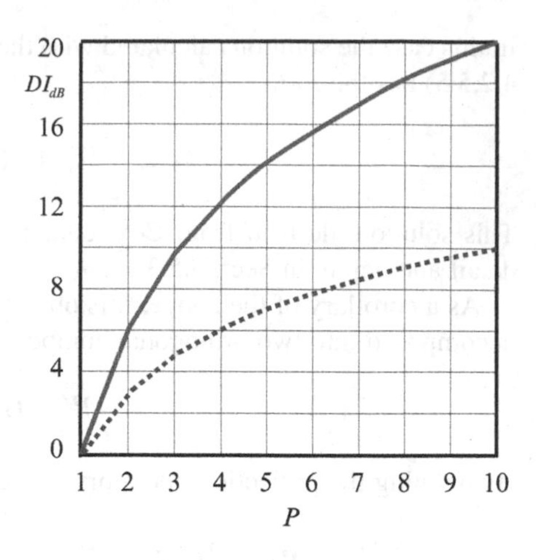

Fig. 9.36 Directivity index for P coincident omnidirectional microphones. Case of isotropic spherical (*continuous line*) and cylindrical noise (*dotted line*) (modified from [14])

This type of array, for $d \ll \lambda/2$, i.e., $d \to 0$, is said to be *superdirective BF* (SDBF) and in the case of filter-and-sum array, the filters weights can be determined using the same optimization criteria defined in the previous paragraph. In particular, the SDBF can be defined with the following specificity

(i) Endfire array
(ii) Distance between the sensors $d \ll \frac{\lambda}{2}$ and $d \to 0$
(iii) A priori known isotropic noise characteristics
(iv) Optimal weights determined by appropriate constraints

For the study and SDBF synthesis, we consider the regularized solution with robustness constraints, expressions (9.124), (9.125), and (9.126), when the noise is diffuse, with cylindrical or spherical symmetry. In this situation, the optimal solution is determined for $\hat{\mathbf{R}}_{nn} \to \mathbf{\Gamma}_{nn}^{\text{diffuse}}$. The CF is then

$$W_{\text{opt}} = \frac{\left(\mathbf{\Gamma}_{nn}^{\text{diffuse}} + \delta\mathbf{I}\right)^{-1}\mathbf{a}}{\mathbf{a}^H\left(\mathbf{\Gamma}_{nn}^{\text{diffuse}} + \delta\mathbf{I}\right)^{-1}\mathbf{a}}. \tag{9.139}$$

The correlation between the regularization parameter δ and the constraint on the white noise gain β^2 [see (9.124)] is rather complex and depends on the nature of the noise. However, for $\delta \to 0$, in (9.139) the noise statistics yielding a BF with optimal directivity and low white noise gain. On the contrary, for $\delta \to \infty$ the diagonal matrix $\delta\mathbf{I}$ prevails and we get the conventional DSBF characterized by a optimal white noise gain $GW(e^{j\omega}) \sim P$.

Figure 9.37 shows, by way of example, the curves with the relationship between G_{dB} and GW_{dB}, to vary the regularization parameter $(0 \le \delta < \infty)$, for an ULA with $P = 8$, with sensors spaced from $d = 0.1\lambda$ to $d = 0.4\lambda$, for cylindrically and spherically, isotropic noise. From the figure it can be observed that for $\delta \to \infty$, the gain tends to become that of the conventional BF, while for $\delta \to 0$, and small d, it tends to exceed that value and become proportional to $2P$ or P^2, respectively, for

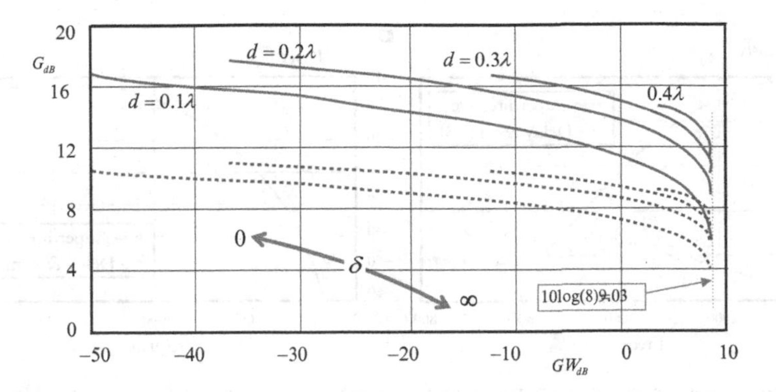

Fig. 9.37 Gain array performance of an endfire line array as a function of the white noise gain (G_{dB} vs. $\mathrm{GW}_{\mathrm{dB}}$), where the regularizing parameter δ is the variable in the case of spherical (*solid line*) and cylindrical (*dotted line*) isotropic noise. Case of ULA with $P = 8$, $\theta = \theta_{\mathrm{endfire}}$, for d shown in the figure (modified from [24])

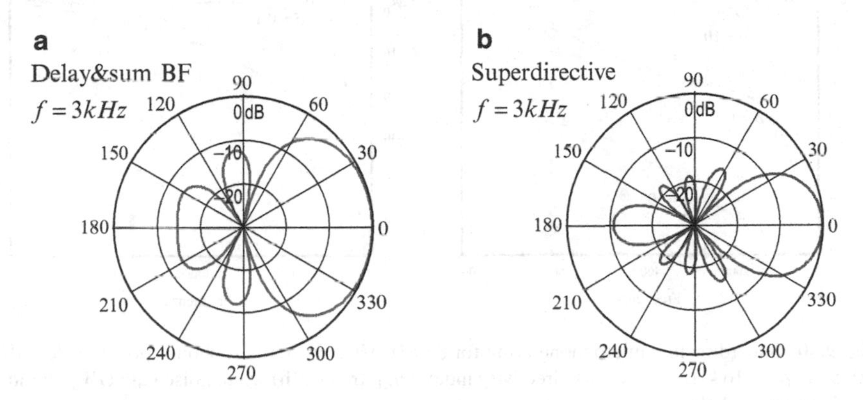

Fig. 9.38 Radiation patterns at 3 kHz, for a microphones array with $P = 5$, $d = 3$ cm, $f_c = 16$ kHz: (**a**) delay and sum BF; (**b**) filter and sum BF with optimum design

cylindrical and spherical symmetry noise. So, for $d \to 0$, and weight calculated with $\delta \to 0$, the line array is said to be *supergain array*.

In Fig. 9.38 is reported a radiation pattern comparison of a conventional ULA and superdirective BF, with weights determined with (9.139), while in Fig. 9.39 is the comparison of the directivity index *DI* and of the white noise gain, for the same BF.

Figure 9.40 presents the directivity index *DI* and the white noise gain GW_{dB} performance, of an array with $P = 3$ omnidirective microphones. The BF weights were determined by the minimization of the CF (9.124) with the constraints $\mathbf{W}^H\mathbf{a}(\omega,\theta_0) = 1$ and $\mathbf{W}^H\mathbf{W} \leq \beta$, with solution (9.139).

Note that for $\delta = 0$, the beamformer tends to be superdirective with *DI* tending to maximum theoretical value ($DI_{\mathrm{dB}} = 10\log_{10}(P^2) = 9.54$ dB) but with low *GW*

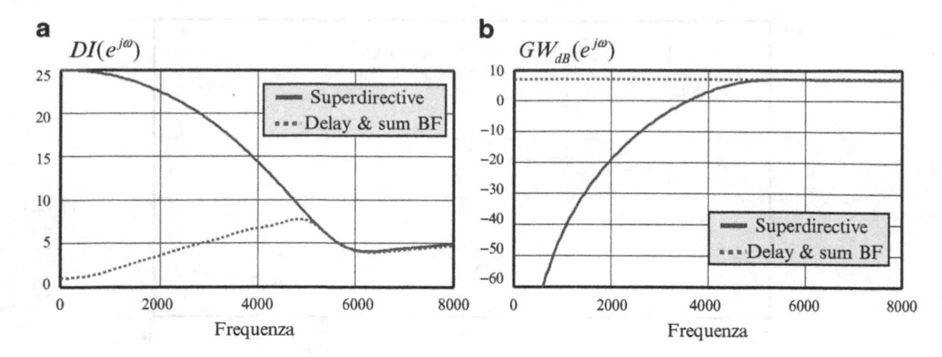

Fig. 9.39 Trends of directivity index "DI," in natural values, and of the white noise gain GW_{dB}, for arrays with radiation patterns of Fig. 9.38

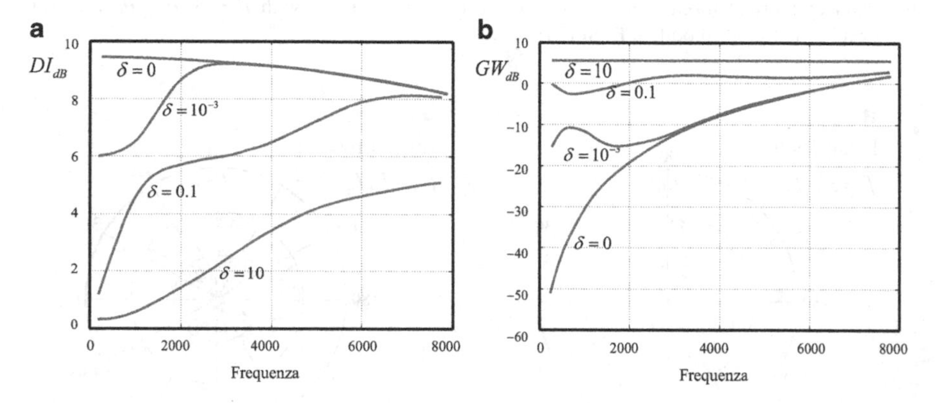

Fig. 9.40 Superdirective microphone array for $\delta = 0$, 10, $P = 3$, microphones positions [0 0.01 0.025] m, $f_c = 16$ kHz, θ_{endfire}: (**a**) directivity index DI_{dB} trends; (**b**) white noise gain GW_{dB} trends (modified from [12])

especially at low frequencies. On the contrary, for $\delta = 10$, the beamformer tends to be a DSBF ($GW_{\text{dB}} = 10\log_{10}(P) = 4.77$ dB), but with low directivity.

Remark In the case of loudspeakers cluster, the superdirective beamformers are often appealed simply as *line arrays.*

9.5.2 *Post-filtering Beamformer*

For microphone arrays operating in high reverberant environments, the diffuse field, coming from all directions, is not entirely eliminated even through a superdirective radiation diagrams synthesis. Furthermore, the noise component is also present in the LD. In these cases, to improve performance an adaptive Wiener filter can be inserted, downstream of the BF. The method, called *post-filtering* and proposed by Zelinski in [27], calculates the AF coefficients, using the cross-spectral density between the array channels. In other words, as shown in Fig. 9.41, the use of

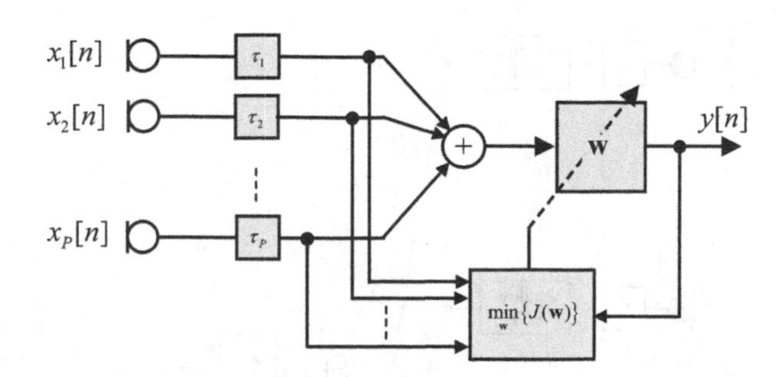

Fig. 9.41 Post-filtering beamforming

the post-filter together with a conventional beamformer is used to add to the filter operating in the spatial domain a noise canceller operating in the frequency domain.

The signal model for the post-filter adaptation is derived from (9.32) $(\mathbf{x}[n] = \mathbf{a}(\omega,\theta)s[n]+\mathbf{n}[n])$ with $\mathbf{n}[n]$ white spatially uncorrelated noise and independent from the signal $s[n]$. The CF for minimizing the SNR, in the LS sense, is

$$J(\mathbf{w}) \triangleq E\left\{\left(y[n] - s[n]\right)^2\right\}. \tag{9.140}$$

The optimal vector $\mathbf{w}_{\text{opt}}$ (Wiener filter, see Chap. 3) is calculated as

$$W_{\text{opt}} = \frac{\mathbf{R}_{ss}(e^{j\omega})}{\mathbf{R}_{xx}(e^{j\omega})} = \frac{\mathbf{R}_{ss}(e^{j\omega})}{\mathbf{R}_{ss}(e^{j\omega}) + \mathbf{R}_{nn}(e^{j\omega})}. \tag{9.141}$$

For the estimation of spectra $\mathbf{R}_{ss}(e^{j\omega})$ and $\mathbf{R}_{nn}(e^{j\omega})$, we observe that the cross-correlation, not considering the steering, can be written as

$$\begin{aligned} E\{x_i[n]x_j[n+m]\} &= E\left\{\left(s[n] + n_i[n]\right) + \left(s[n+m] + s_j[n+m]\right)\right\} \\ &= E\{s[n]s[n+m]\} + E\{s[n]n_j[n+m]\} \\ &+ E\{n_i[n]s[n+m]\} + E\{n_i[n]n_j[n+m]\} \end{aligned} \tag{9.142}$$

where the last three terms of the above, if the noise is not correlated, are null. For which from the (9.142), it is possible to estimate the PSD of the signal. In fact, for $i \neq j$, we get

$$\begin{aligned} \mathbf{R}_{ss}(e^{j\omega}) &= \text{DTFT}\left(E\{x_i[n]x_j[n+m]\}\right) \\ &\approx \text{DTFT}\left(E\{s[n]s[n+m]\}\right) \quad i \neq j. \end{aligned} \tag{9.143}$$

The adaptation formula is

$$W_{\text{opt}} = \frac{\text{DTFT}\left(E\left\{\left[x_i[n]x_j[n+m]\right]\right\}\right) \quad i \neq j}{\text{DTFT}\left(E\left\{\left[x_i[n]x_j[n+m]\right]\right\}\right)}. \tag{9.144}$$

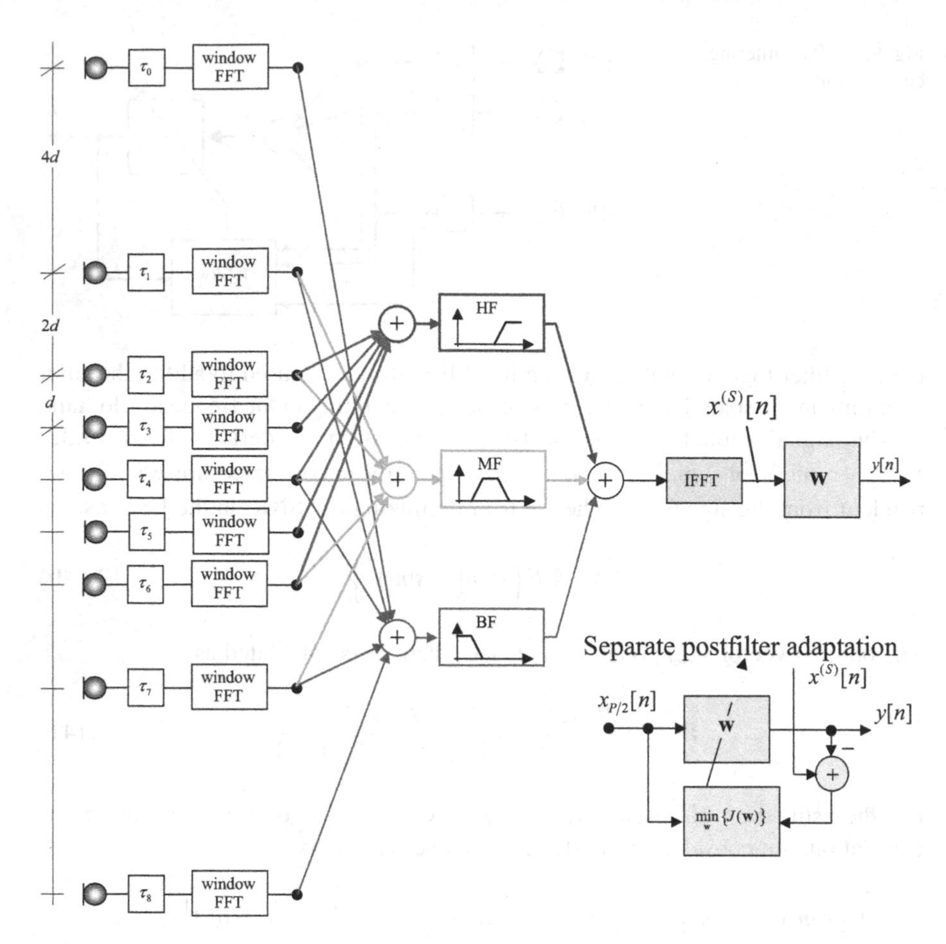

Fig. 9.42 Example of linear harmonic array with nested sub-array with $d = 5$ cm and a possible scheme for separate adaptation of the post-filter **w**

Remark To ensure uncorrelated spatial noise, namely a null coherence function $\gamma(\mathbf{r}_{i,j},\omega) \approx 0$, microphones must be far between each other. However, large distance between microphones may produce spatial aliasing (i.e., lower bandwidth) and poor performance for coherent noise case. Moreover, high interelement distance results in very narrow beamwidth at higher frequencies and, consequently, high sensitive to steering misadjustment.

In literature, there are numerous variants of the post-filtering beamformer as, for example, in [28, 29], in which the authors suggest the use of linear harmonic array with nested sub-array as shown in Fig. 9.42.

9.5.2.1 Separate Post-filter Adaptation

A simple alternative way for adapting the weights, $\mathbf{w}$, can be determined considering, as input of the adaptive post-filter, the signal coming from the central sensors, e.g., $x_{P/2}[n]$, and as desired signal $d[n]$, the output of the DSBF, i.e., considering $d[n] \equiv x^{(S)}[n]$.

9.5.3 *Minimum Variance Broadband Beamformer: Frost Algorithm*

The approach described in this paragraph, proposed by Frost [30], reformulates beamforming as a constrained LS problem in which the desired signal, by definition, unknown, is replaced by the suitable constraints imposed on the array frequency response. In other words, the *Frost algorithm* can be seen as a generalization, with a different interpretation, of the LS method for maximum SNR BFs, described in Sect. 9.5.1. The adaptation is then a linearly constrained optimization algorithm (see Sect. 4.2.5.5). The AP algorithm, described in [8], is indicated as *linearly constrained minimum variance* (LCMV) broadband beamforming.

We proceed defining the desired spatial response toward the LD, simultaneously minimizing the noise power from all other directions, through a simple relationship between the LD, the desired frequency response, and the array weights.

The model illustrated in Fig. 9.43 is a FSBF of P sensors with M-length FIR filters downstream of each sensor. The input signal is defined by (9.32), for which the output, considering the composite MISO model, is equal to

$$y[n] = \mathbf{w}^H \mathbf{x} \tag{9.145}$$

with $\mathbf{w}, \mathbf{x} \in (\mathbb{R},\mathbb{C})^{P(M)\times 1}$. The input noise snap-shot $\mathbf{n}[n] = \begin{bmatrix} n_1[n] & \cdots & n_P[n] \end{bmatrix}^T$, by assumption with spatial zero mean, consists precisely in the signal coming from all different directions with respect to the LD.

For the theoretical development, we consider the SOI as a single plane wave, incident on the array, with parallel front with respect to the sensors line, or with broadside direction $\theta = 90°$. Obviously, the SOI snap-shot $\mathbf{s}[n]$ is the same (in phase) on all the sensors (and in the filters delay line), while signals coming from directions $\theta \neq 90°$ are not in phase. To produce the output, the signal and the noise, by (9.49), are filtered and summed.

Regarding the signal $\mathbf{s}[n]$, for plane wave hypothesis, it is assumed identical on all the sensors. Therefore, due to the system linearity, its processing is equivalent to convolution with a single FIR filter. The impulse response of such filter, indicated as $\mathbf{f} \in (\mathbb{R},\mathbb{C})^{M\times 1} = \begin{bmatrix} f[0] & \cdots & f[M-1] \end{bmatrix}^T$, is the sum by columns, of the FIR filters coefficients placed downstream of individual sensors.

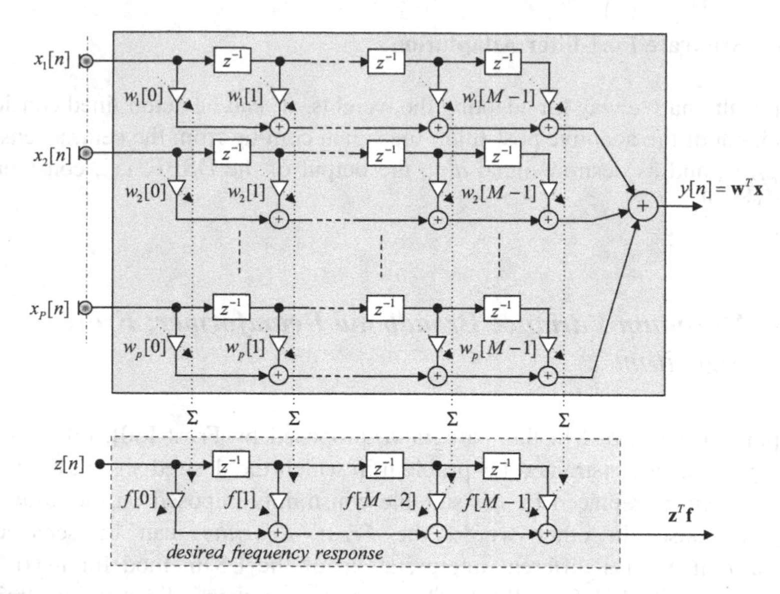

Fig. 9.43 Linearly constrained minimum variance (LCMV) broadband beamformer and equivalent process imposed as a constraint (modified from [30])

Formally, as shown below in Fig. 9.43, called $\mathbf{w}[k] = \left[w_1[k] \cdots w_P[k]\right]^T$, for $0 \leq k \leq M{-}1$, the vectors containing the coefficients of the FIR filters related to the k-th delay line elements, we have that

$$f[k] = \sum_{j=1}^{P} w_j[k], \quad \text{for } 0 \leq k \leq M-1. \tag{9.146}$$

In other words, the filter **f** is determined considering the desired frequency response along the LD. For example, **f** could be a bandpass FIR filter designed using the windowing method or it may be a simple delay. Since the signal coming from all the other directions is supposed to be noise with zero mean, in practice, the filter **f** is relative only to the incident "in phase" signal on the sensors or coming just from the LD.

Overall, the Frost BF consists of $P \cdot M$ free parameters and from (9.146), for the frequency response determination along the LD, it is necessary to choose a priori the M coefficients of the filter **f** which, therefore, may represent the constraint of the optimization problem. It follows that the Frost BF has $PM - M$ degrees of freedom that can be used to minimize the total output power of the array. Given that the frequency response along the LD is imposed by the filter **f** constraint, this corresponds to the power minimization along all directions different from the LD.

In the case of LD not perpendicular to the sensors line $\theta_0 \neq 90°$, the array can be oriented by inserting a suitable steering delay vector as previously illustrated.

9.5.3.1 Linearly Constrained LS

The method, also called *constrained power minimization*, sets through the vector **f**, a certain frequency response of the filter along a desired direction. The array weights are chosen according to minimization of the variance (energy) along the other directions.

We define the error function $e[n]$ as

$$e[n] = d_\theta[n] - y[n] = d_\theta[n] - \mathbf{w}^H\mathbf{x} \tag{9.147}$$

where the desired output $d_\theta[n]$ can be considered zero or different from zero depending on the angle of observation of the array. In practice, we want to minimize the energy in directions different from that of observation and, *vice versa*, maximize it in the LD. In general, we will have

$$d_\theta[n] = \begin{cases} 0 & \text{for } \theta \neq \text{LD} \\ \max & \text{for } \theta = \text{LD} \end{cases}. \tag{9.148}$$

Minimizing the (9.147) with LS criterion, for $d_{\theta \neq \text{LD}}[n] = 0$, we get a CF identical to those obtained by the maximization of the quality indices (see Sect. 9.3.3), which is

$$J(\mathbf{w}) = E\left\{\left|e[n]\right|^2\right\} = E\left\{\left|y[n]\right|^2\right\} = \mathbf{w}^H E\left\{\mathbf{x}\mathbf{x}^H\right\}\mathbf{w} = \mathbf{w}^H\mathbf{R}_{xx}\mathbf{w}. \tag{9.149}$$

The minimum of (9.149) coincides with the minimization of the output power of the array. The nontrivial solution, $\mathbf{w} \neq 0$, can be determined by imposing some gain along the LD or, more generally, a constraint on the desired frequency response for the SOI. This constraint, derived from the reasoning previously done, practically coincides with the expression (9.146), for which the constrained optimization problem can be formulated as

$$\mathbf{w}_{\text{opt}} \therefore \underset{\mathbf{w}}{\operatorname{argmin}} \left\{\mathbf{w}^H\mathbf{R}_{xx}\mathbf{w}\right\} \quad \text{s.t.} \quad \mathbf{C}^H\mathbf{w} = \mathbf{f} \tag{9.150}$$

with linear constraints, expressed by the filter weights **f**, are due to the BF frequency response along the LD. Note that (9.150) is similar to (9.120), for $\hat{\mathbf{R}}_{nn}(e^{j\omega}) \rightarrow \mathbf{R}_{xx}(e^{j\omega})$.

The objective of linearly constrained minimization is to determine the coefficients **w** that satisfy the constraint in (9.150) and simultaneously reduce the mean square value of the noise output components. Note that the above expression can be interpreted as a generalization of (9.120).

9.5.3.2 Matrix Constrain Determination

The matrix $\mathbf{C}^H \in \mathbb{R}^{M \times P(M)}$ is defined in such a way the constraint of (9.150) coincides with (9.146). Then this depends on the type of representation of the MISO beamformer. For better understanding, as an example, we evaluate the matrix $\mathbf{C}$ for an array with three sensors ($P = 3$) and four delays ($M = 4$). In agreement with (9.146), explicitly writing the constraint, we get

$$\begin{aligned} w_1[0] + w_2[0] + w_3[0] &= f[0] \quad 1^\circ\text{snap-shot} \\ w_1[1] + w_2[1] + w_3[1] &= f[1] \quad 2^\circ\text{snap-shot} \\ w_1[2] + w_2[2] + w_3[2] &= f[2] \quad 3^\circ\text{snap-shot} \\ w_1[3] + w_2[3] + w_3[3] &= f[3] \quad 4^\circ\text{snap-shot}. \end{aligned} \tag{9.151}$$

From the definition of weights vector as $\mathbf{w} \in (\mathbb{R},\mathbb{C})^{P(M)\times 1}$ (see (9.47)), the previous can be expressed in matrix terms $\mathbf{C}^H\mathbf{w} = \mathbf{f}$, as

$$\begin{bmatrix} 1 & 0 & 0 & 0 & 1 & 0 & 0 & 0 & 1 & 0 & 0 & 0 \\ 0 & 1 & 0 & 0 & 0 & 1 & 0 & 0 & 0 & 1 & 0 & 0 \\ 0 & 0 & 1 & 0 & 0 & 0 & 1 & 0 & 0 & 0 & 1 & 0 \\ 0 & 0 & 0 & 1 & 0 & 0 & 0 & 1 & 0 & 0 & 0 & 1 \end{bmatrix} \begin{bmatrix} w_1[0] \\ w_1[1] \\ w_1[2] \\ w_1[3] \\ w_2[0] \\ w_2[1] \\ w_2[2] \\ w_2[3] \\ w_3[0] \\ w_3[1] \\ w_3[2] \\ w_3[3] \end{bmatrix} = \begin{bmatrix} f[0] \\ f[1] \\ f[2] \\ f[3] \end{bmatrix} \tag{9.152}$$

for which $\mathbf{C}$ is a *circulating sparse matrix*, constructed with P blocks of unitary matrices $\mathbf{I}_{M\times M}$,

$$\mathbf{C}^H \in \mathbb{R}^{M\times M(P)} = \begin{bmatrix} \mathbf{I}_{M\times M} & \mathbf{I}_{M\times M} & \cdots & \mathbf{I}_{M\times M} \end{bmatrix}_{1\times P}. \tag{9.153}$$

Note that $\mathbf{C}^H\mathbf{C} = P\mathbf{I}$ and $\det(\mathbf{C}) = P^M$. In theory, you can choose any matrix $\mathbf{C}$ as long as whatever it is worth the constraint $\mathbf{C}^H\mathbf{w} = \mathbf{f}$.

The expression (9.150) is a linearly constrained optimization problem, referred to the covariance matrix $\mathbf{R}_{xx}$, for which it is minimized to the total power of the BF output. Therefore, it is appropriate to define this method as a *linearly constrained minimum power* [1]. To minimize the noise power, as discussed in superdirective beamforming (see Sect. 9.5.1), from a formal point of view, it is more appropriate to refer to the generic noise covariance matrix $\mathbf{Q}_{nn}$, for which the function to be minimized is $\mathbf{w}^H\mathbf{Q}_{nn}\mathbf{w}$. The appellative *linearly constrained minimum variance* (LCMV) is more properly referred to this case. It is common, however, to use the term LCMV for both situations.

9.5.3.3 Lagrange Multipliers Solution for Constrained LS

The LS solution of the problem (9.150) can be performed by applying the Lagrange multiplier method as developed in Sect. 4.2.5.5. Therefore, we can write

$$L(\mathbf{w},\boldsymbol{\lambda}) = \frac{1}{2}\mathbf{w}^T\mathbf{R}_{xx}\mathbf{w} + \boldsymbol{\lambda}^T\left(\mathbf{C}^H\mathbf{w} - \mathbf{f}\right). \tag{9.154}$$

The trivial LS solution is $\mathbf{w}_{\text{LS}} = 0$, and the nontrivial solution correspond to the Cox solution for multiple constraints (9.132), i.e.,

$$\mathbf{w}_{\text{opt}} = \frac{\mathbf{R}_{xx}^{-1}\mathbf{C}}{\mathbf{C}^H\mathbf{R}_{xx}^{-1}\mathbf{C}}\mathbf{f}. \tag{9.155}$$

The previous, in robust mode, can be written as [see (9.126)]

$$\mathbf{w}_{\text{opt}} = \frac{(\mathbf{R}_{xx} + \delta\mathbf{I})^{-1}\mathbf{C}}{\mathbf{C}^H(\mathbf{R}_{xx} + \delta\mathbf{I})^{-1}\mathbf{C}}\mathbf{f} \tag{9.156}$$

where the parameter $0 \le \delta < \infty$ represents the regularization term. By varying δ it is possible to obtain optimal solutions depending on the noise field type.

Remark The described LS method, in the case where the LD frequency response in distortionless condition is flat and linear phase, is such that the filter output coincides with the ideal *maximum likelihood estimation* of a stationary process immersed in Gaussian noise. For this reason, at times, this method is defined as maximum likelihood distortionless estimator (MLDE) or least squares unbiased estimator (LSUB).

9.5.3.4 Constrained Stochastic Gradient LMS Recursive Solution

The recursive procedure of the Frost's algorithm can be determined proceeding as in Sect. 5.3.4.1. In this case the recursive procedure is written as

$$\mathbf{w}_n = \mathbf{w}_{n-1} - \mu\nabla_{\mathbf{w}}L(\mathbf{w},\boldsymbol{\lambda}) = \mathbf{w}_{n-1} - \mu[\mathbf{R}_{xx}\mathbf{w}_{n-1} + \mathbf{C}\boldsymbol{\lambda}] \tag{9.157}$$

that with the constraint on the weights $\mathbf{C}^H\mathbf{w}_n = \mathbf{f}$ is

$$\mathbf{w}_n = \mathbf{P}[\mathbf{w}_{n-1} - \mu\mathbf{R}_{xx}\mathbf{w}_{n-1}] + \mathbf{g}. \tag{9.158}$$

The *projection operators* $\mathbf{P}$ (see (9.135)) and the *quiescent vector* $\mathbf{g}$ (see (9.138) and Sect. 5.3.4.2) are defined as

$$\begin{aligned}\tilde{\mathbf{P}} &\in (\mathbb{R},\mathbb{C})^{PM\times PM} \triangleq \mathbf{C}(\mathbf{C}^H\mathbf{C})^{-1}\mathbf{C}\\ \mathbf{P} &\in (\mathbb{R},\mathbb{C})^{PM\times PM} \triangleq \left[\mathbf{I}-\tilde{\mathbf{P}}\right]\\ \mathbf{g} &\in (\mathbb{R},\mathbb{C})^{PM\times 1} \triangleq \mathbf{C}(\mathbf{C}^H\mathbf{C})^{-1}\mathbf{f}.\end{aligned} \tag{9.159}$$

In practice, considering the instantaneous SDA approximation $\mathbf{R}_{xx} \approx \mathbf{x}_n\mathbf{x}_n^H$ and $y[n] = \mathbf{x}_n^H\mathbf{w}_{n-1}$, the formulation with gradient projection LCLMS (GP-LCLMS) (see (5.112) for $d[n] = 0$) assumes the form

$$\mathbf{w}_n = \mathbf{P}\left[\mathbf{w}_{n-1} - \mu y^*[n]\mathbf{x}_n\right] + \mathbf{g} \tag{9.160}$$

where $y[n]$ represents the array output and the weight vector is initialized as $\mathbf{w}_0 = \mathbf{g}$. The adaptation step μ that controls the convergence speed and the steady-state noise is, in general, normalized as in the NLMS. For which it is

$$\mu = \frac{\mu_0}{\mu_1 + \sum_{j=1}^{P}\sum_{k=0}^{M-1} x_j^2[n-k]} \tag{9.161}$$

with μ_0 and μ_1 appropriate scalar value (see Chaps. 3 and 4).

Remark The reader can easily verify that for $\mathbf{C}^H$ defined as in (9.153), the projection matrix, for $M = 3$ and $P = 4$, is equal to

$$\tilde{\mathbf{P}} \in \mathbb{R}^{P(M)\times P(M)} = \begin{bmatrix} \frac{1}{P}\mathbf{I}_{M\times M} & \frac{1}{P}\mathbf{I}_{M\times M} & \frac{1}{P}\mathbf{I}_{M\times M}\\ \frac{1}{P}\mathbf{I}_{M\times M} & \frac{1}{P}\mathbf{I}_{M\times M} & \frac{1}{P}\mathbf{I}_{M\times M}\\ \frac{1}{P}\mathbf{I}_{M\times M} & \frac{1}{P}\mathbf{I}_{M\times M} & \frac{1}{P}\mathbf{I}_{M\times M}\end{bmatrix}_{P\times P} \tag{9.162}$$

for which, for large array P ($P > 20$), results $\mathbf{P} = \left(\mathbf{I} - \tilde{\mathbf{P}}\right) \approx \mathbf{I}$. Therefore, the update formula (9.160) can be simplified as $\mathbf{w}_n = \mathbf{w}_{n-1} - \mu y^*[n]\mathbf{x}_n + \mathbf{g}$.

9.5.3.5 Geometric Interpretation

For the GP-LCLMS algorithm in (9.160) a geometric interpretation can be given, useful for the error correction properties [30].

We define *constraint subspace plane* as the nullspace of the $\mathbf{C}$ matrix, indicated as $\mathcal{N}(\mathbf{C}^H)$ and which includes the null vector, the resulting plane from homogeneous form of the constraint equation (see Sect. A.6.2). For this it is $\mathcal{N}(\mathbf{C}^H) \triangleq \{\mathbf{w} ∴ \mathbf{C}^H\mathbf{w} = \mathbf{0}\}$.

For negligible error, the weight vector, $\mathbf{w} \in (\mathbb{R},\mathbb{C})^{P(M)\times 1}$, satisfies the constraint equation in (9.150) and therefore terminates in the *hyperplane constraint* Λ, of size $(PM - M)$ and parallel to $\mathcal{N}(\mathbf{C}^H)$, defined as the space $\Lambda = \{\mathbf{w} ∵ \mathbf{C}^H\mathbf{w} = \mathbf{f}\}$,

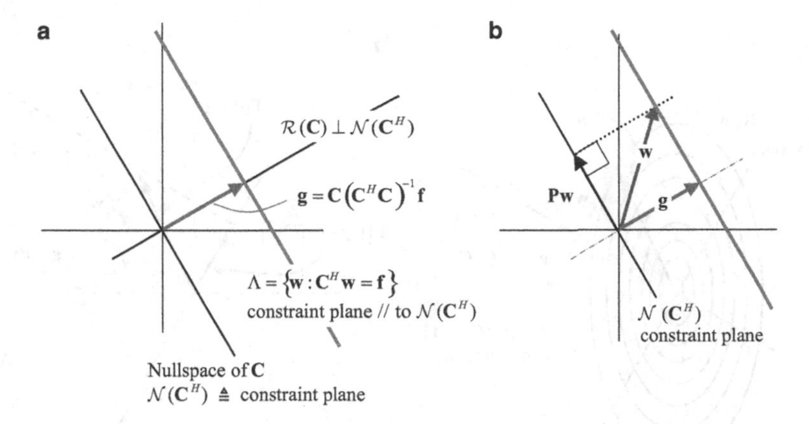

Fig. 9.44 Geometric interpretation: (**a**) the constraint plane and the subspace defined by the constraint; (**b**) projection **P** of **w** in the constraint subspace (modified from [30])

schematically illustrated in Fig. 9.44a. From linear algebra, it is known that vectors oriented in the normal direction to the constraint plane are linear combinations of the columns of the constraint matrix itself and that, in this case, is equal to **C**. It follows that the vector **g**, defined in the last of (9.159) as appearing in the adaptation formula (9.160) and used for algorithm initialization, points to the normal direction of the constraint plane. From algebra (see Sect. A.6), we have that the column space (or image) of the **C** matrix is orthogonal to its nullspace i.e., $\mathcal{R}(\mathbf{C}) \perp \mathcal{N}(\mathbf{C}^H)$, thus by definition it appears that $\mathbf{g} \in \mathcal{R}(\mathbf{C})$. As illustrated in Fig. 9.44a, **g** ends just in the constraint plane and is perpendicular to it $\mathbf{g} \perp \Lambda$. Since, by definition $\mathbf{C}^H\mathbf{g} = \mathbf{f}$; therefore, **g** is the shorter vector that ends in the constraint plane.

Note, also, that the matrix **P**, which appears in the definition (9.159), is just a projection operator (see (9.64)). So, pre-multiplying any vector **w** by **P**, as shown in Fig. 9.44b, this is projected in the $\Sigma \in \mathcal{N}(\mathbf{C}^H)$ plane.

Figure 9.45b shows the geometric representation of the CLMS, implemented with (9.160). Note how the solution at (k+1)th instant is given by the sum of the vector **g**, perpendicular to the constraint plane Λ and the vector $\mathbf{P}[\mathbf{w}_k - \mu y^*[k]\mathbf{x}_k]$ lying on the constraint subspace $\mathcal{N}(\mathbf{C}^H)$.

Remark Note that in expression (9.158), the last term on the right, just for the constraint definition (9.150), can be neglected, $[\mathbf{f} - \mathbf{C}^H\mathbf{w}_n] \to 0$. With this approximation, the expression (9.160) can be written in simplified form as

$$\mathbf{w}_n = \mathbf{P}\big[\mathbf{w}_{n-1} - \mu y[n]\mathbf{x}_n\big] \approx \mathbf{w}_{n-1} - \mu \mathbf{P} y[n]\mathbf{x}_n. \tag{9.163}$$

The geometric interpretation of this expression, known as the *gradient projection algorithm*, can be easily obtained by the reader.

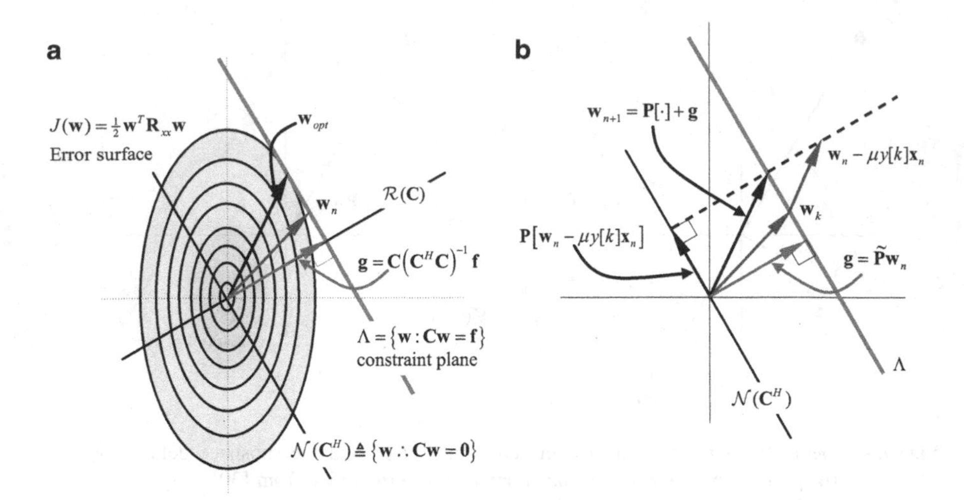

Fig. 9.45 2D representation of CLMS. (**a**) The optimal solution vector $\mathbf{w}_{\text{opt}}$, has its vertex at tangency point between the constraint plane and the isolevel contour of the error surface; (**b**) vector representation of the CLMS (modified from [30])

9.5.3.6 LCMV Constraints Determination

The Frost BF can be easily modified considering multiple constraints from different directions and at different frequencies. It is possible, for example, to impose a maximum in one or more directions and/or attempt to cancel some interferences which come from known directions.

Beamformer *Minimum Variance Distortionless Response*

Suppose we have a space-frequency constraint, which requires a certain frequency response for a certain direction θ_k. In this case, (9.150) takes the form $\mathbf{w}^H\mathbf{a}(\omega,\theta_k) = g_k$, where $g_k \in \mathbb{C}$ is a complex constant that indicates the desired BF gain for signals of frequency ω and direction θ_k. With this formalism the expression of constrained LS (9.155) takes the simplified form

$$\mathbf{w}_{\text{opt}} = \frac{\mathbf{R}_{xx}^{-1}\mathbf{a}(\omega,\theta_k)}{\mathbf{a}^H(\omega,\theta_k)\mathbf{R}_{xx}^{-1}\mathbf{a}(\omega,\theta_k)} g_k. \tag{9.164}$$

In the case in which $\theta_k = \text{LD}$ and $g_k = 1$, the BF (9.164) is called the minimum variance distortionless response (MVDR). In practice, the MVDR is a simple form of LCMV with a single constraint that requires a unitary response along the LD.

Remark The expression (9.164) can be extended by considering the presence of multiple constraints. Suppose, for example, that we want a gain g_0 along the direction θ_0, a zero ($g_1 = 0$) along the direction θ_1, a gain g_2 for the direction θ_2. The constraint relationship, then, can be written as a vector

$$\begin{bmatrix} \mathbf{a}^H(\omega,\theta_0)\mathbf{w} \\ \mathbf{a}^H(\omega,\theta_1)\mathbf{w} \\ \mathbf{a}^H(\omega,\theta_2)\mathbf{w} \end{bmatrix} = \begin{bmatrix} g_0^* \\ 0 \\ g_2^* \end{bmatrix}. \tag{9.165}$$

For $J < M$ linear constraints in $\mathbf{w}$, it is always possible to write them in the form $\mathbf{C}^H\mathbf{w} = \mathbf{f}$. In this case, the constraints are linearly independent if $\mathbf{C}$ has rank equal to J, i.e., $\mathcal{R}(\mathbf{C}) = J$.

Multiple Amplitude-Frequency Derivative Constraints

An important aspect of Frost's LCMV is that the beam orientation, with an appropriate steering vector applied to the input, cannot be inserted without affecting the performance of the beamformer itself.

A simple variant to overcome this problem is to modify the linear constraints structure by means of appropriate weighing space. In practice, as suggested in [31], the matrix $\mathbf{C}^H \in \mathbb{R}^{M\times(M)P}$ can be changed as

$$\mathbf{C}^H = \begin{bmatrix} \operatorname{diag}(c_{1,0}, c_{1,1}, \ldots, c_{1,M-1}) & \cdots & \operatorname{diag}(c_{P,0}, c_{P,1}, \ldots, c_{P,M-1}) \end{bmatrix}_{1\times P} \tag{9.166}$$

where, unlike (9.153), the $\mathbf{C}^H$ matrix blocks are diagonal but are no longer unitary. Then, the constraint matrix can be redefined on the basis of different philosophies such as the presence of multiple constraints, and constraints on the directional beam derivative.

In [32], for example, an optimal weighing method has been defined for the insertion of $J < M$ gain and directional-derivative constraints. Although the adaptation algorithm is formally identical, the inclusion of more constraints leads to define a $\mathbf{C}^H$ matrix of size ($JM\times PM$) rather than ($M\times PM$), i.e.,

$$\mathbf{C}^H \in \mathbb{R}^{J(M)\times 1(PM)} = \begin{bmatrix} \tilde{\mathbf{C}}_0 & \tilde{\mathbf{C}}_1 & \cdots & \tilde{\mathbf{C}}_{J-1} \end{bmatrix}_{J\times 1}^H \tag{9.167}$$

where

$$\tilde{\mathbf{C}}_j^H \in \mathbb{R}^{M\times(M)P} = \begin{bmatrix} \tilde{\mathbf{c}}_1 & \cdots & \tilde{\mathbf{c}}_P \end{bmatrix}_{1\times P}$$

with $\tilde{\mathbf{c}}_p \in \mathbb{R}^{M\times M} = \operatorname{diag}(c_{p,0}, c_{p,1}, \ldots, c_{p,M-1})$, while the vector $\mathbf{f}$, which appears in the constraint, is redefined as

$$\mathbf{f} = \begin{bmatrix} \tilde{\mathbf{f}}_0 & \tilde{\mathbf{f}}_2 & \cdots & \tilde{\mathbf{f}}_{J-1} \end{bmatrix}^T. \tag{9.168}$$

Each constraint vector $\tilde{\mathbf{c}}_p$ with the corresponding scalar $f_{j,p}$ places a constraint on the weight vector $\mathbf{w}_p$. The coefficients $\tilde{\mathbf{c}}_p$ describe the radiation pattern, in the LD (with amplitude and first derivative constraints). To zero forcing the constraints of higher derivative order, the vector in (9.168) must be such that $\tilde{\mathbf{f}}_j = \mathbf{0}_M$ for $j = 1, 2, ..., J-1$.

In practice, the derivative constraints are used to influence the response, on a specific region, forcing the beampattern derivative to assume null value in certain frequency-direction points. These constraints are used in addition to those in space. An example where the derivative constraints are useful, is the one in which the direction of arrival is approximately known. If the signal comes close to the constraint point, the derivative constraint prevents the possibility of having a null response in the desired direction [8].

Eigenvector Constraints

These constraints are based on the LS approximation of the desired beampattern and used to control the beamformer response.

Consider a set of constraints, which allow the space-frequency beampattern control, toward a source of direction θ_0 in the frequency band $[\omega_a, \omega_b]$. The size of the steering vector $\mathbf{a}(\omega,\theta_0)$ span, on that frequency band, is approximately given by the product TBWP (previously discussed). Choosing the number of constraint points J, significantly larger related to TBWP, the subspace constraints derived from the normal equations (9.117), as an approximation of rank M of the steering matrix $\mathbf{A}$, can be defined by its SVD

$$\mathbf{A}_M = \mathbf{V}\mathbf{\Sigma}_M\mathbf{U}^H \tag{9.169}$$

where $\mathbf{\Sigma}_M$ is a diagonal matrix containing the singular values of $\mathbf{A}$, while the M columns of $\mathbf{V}$ and $\mathbf{U}$ are, respectively, the left and right singular vectors of $\mathbf{A}$ corresponding to those singular values. With the decomposition (9.169), equation (9.117) can be reformulated as

$$\mathbf{V}^H\mathbf{w} = \mathbf{\Sigma}_M^{-1}\mathbf{U}^H\mathbf{r}_d. \tag{9.170}$$

Note that the latter has the same form of the constraint equation $\mathbf{C}^H\mathbf{w} = \mathbf{f}$, in which the constraint matrix, in this case, is equal to $\mathbf{V}$ that contains the eigenvectors of $\mathbf{A}\mathbf{A}^H$ (from which the name *eigenvector constraints*).

9.6 Adaptive Beamforming with Sidelobe Canceller

In this section, introduce some adaptive methods for BFs operating on-line in time-varying conditions. The adaptive algorithms are, by definition, data dependent; then the parameters update can be performed considering the noise field and/or SOI statistical characteristics. MISO algorithms are presented, implemented in the time or frequency domain, and based on the first- and second-order statistics.

9.6.1 Introduction to Adaptive Beamforming: The Multiple Adaptive Noise Canceller

A first example of adaptive AP, previously discussed in Chap. 2, consists in the multiple adaptive interference/noise canceller (AIC), for the acoustic case, illustrated in Fig. 9.46. The structure consists of a primary sensor that captures prevalently the SOI and superimposed noise and a secondary array that captures mostly the noise sources. The noise signal, coming from the secondary array, after a suitable processing, is subtracted from that provided by the primary sensor. In the context of beamforming, this architecture is indicated as *multiple sidelobe canceller* (MSC).

The determination of the optimal weights, easily derivable from the LS method, is briefly reported below.

With reference to Fig. 9.46, called $y_p[n]$, the signal of the primary source, $\mathbf{x}_a$, is the vector of signals coming from the secondary array, $y_a[n]$ the FIR filters bank output, $\mathbf{w}$ the weight vector of the whole bank [see (9.47)], and $y[n] = y_p[n] - \mathbf{w}^H\mathbf{x}_a$ the MSC output. Furthermore, defining $\mathbf{r}_{pa} = E\{\mathbf{x}_a y_p^*[n]\}$ the vector containing the ccf between the primary input and the auxiliary inputs $\mathbf{R}_{aa} = E\{\mathbf{x}_a\mathbf{x}_a^H\}$ the acf matrix relative to the auxiliary inputs, the optimal filter weights can be calculated, according with Wiener's theory, as $\mathbf{w}_{\text{opt}} = \mathbf{R}_{aa}^{-1}\mathbf{r}_{pa}$.

Remark As stated previously on AIC (see Sect. 3.4.5), this method is much more consistent as far as the noise signal is absent from the primary input. The adaptive solution can be easily implemented with consideration already developed in Chap. 5.

9.6.2 Generalized Sidelobe Canceller

An adaptive approach, more general than the Frost's LCMV structure, proposed by Griffiths and Jim [13], is the generalized sidelobe canceller (GSC). The GSC is an alternative form to implement the LCMV, in which the unconstrained components are separated from the constrained ones. The fixed components are constituted

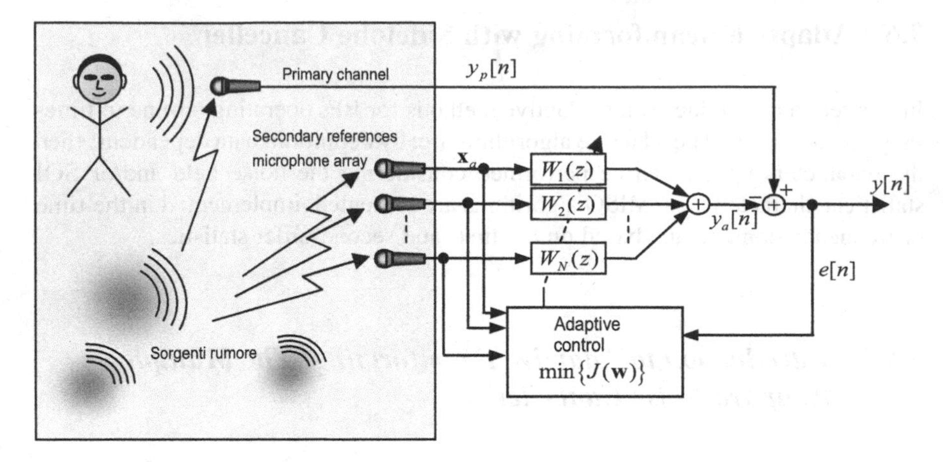

Fig. 9.46 Multiple sidelobe canceller (MSC) with microphone array

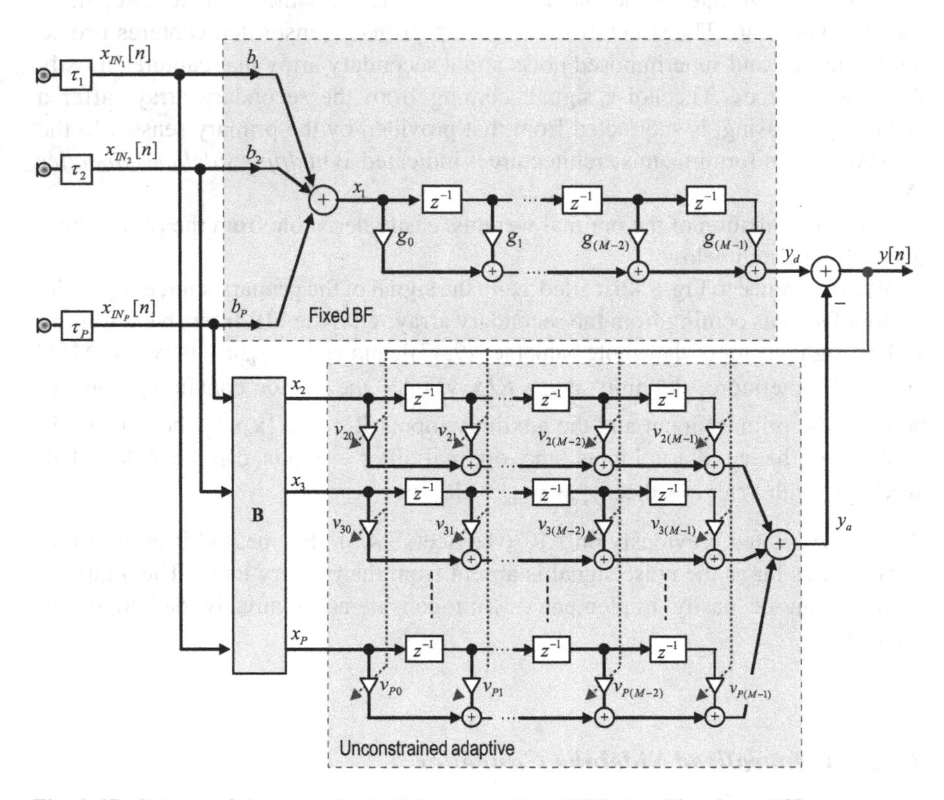

Fig. 9.47 Scheme of the generalized sidelobe canceller (GSC) (modified from [13])

by a data-independent BF designed to satisfy the CF constraints, whereas, the unconstrained components cancel interference in an adaptive way. In GSC, with the general scheme shown in Fig. 9.47, the BF is derived as the superposition of two effects, namely the Frost's BF is divided into two distinct sublinear processes (1) a conventional fixed beamformer and (2) an adaptive unconstrained BF which can be interpreted as an MISO Widrow's AIC.

As for the LCMV, the desired signal is downstream to the steering time-delay τ_k. In such a way each sensor output is an *in-phase* replica of the signal coming from the LD, while the signals coming from other directions are cancelled because of destructive interference.

The fixed sub-process (upper part in the figure) is a conventional beamformer with output

$$x_1[n] = \mathbf{b}^H \mathbf{x}_{\text{IN}} \tag{9.171}$$

where the coefficients $\mathbf{b}$, all different from zero, are chosen according to the spatial specifications on the main lobe width and the secondary lobes attenuation. Furthermore, the gain constraint is defined with the FIR filter $\mathbf{g}$, which acts on the signal $x_1[n]$, determining the prescribed frequency and phase response, for which

$$y_d[n] = \sum_{k=0}^{M-1} g_k x_1[n-k]. \tag{9.172}$$

Usually, it also requires the normalization

$$\sum_{k=0}^{M-1} g_k = 1. \tag{9.173}$$

The adaptive subprocess (lower part in the figure), also called *interference canceller*, is an adaptive beamformer that acts on disturbances that will be subtracted from the main process. The interference canceller is formed by a transformation matrix $\mathbf{B} \in (\mathbb{R},\mathbb{C})^{(P-1)\times P}$, called *block matrix*, followed by a bank of M-length adaptive FIR filters, whose coefficients, in composite notation, are referred to as $\mathbf{v} \in (\mathbb{R},\mathbb{C})^{(P-1)M\times 1}$. With reference to Fig. 9.47, called $\mathbf{x}[n] \in (\mathbb{R},\mathbb{C})^{(P-1)\times 1}$, the n-th time instant snap-shot at the block matrix output, we have that

$$\mathbf{x}[n] = \mathbf{B}\mathbf{x}_{\text{IN}}[n]. \tag{9.174}$$

Saying $\mathbf{s}[n] = \mathbf{a}(\omega,\theta)s[n]$, the SOI, the signal model to the receivers is defined as $\mathbf{x}_{\text{IN}}[n] = \mathbf{s}[n] + \mathbf{n}[n]$. In the GSC, the transformation with the block matrix $\mathbf{B}$ has the task of eliminating the SOI (i.e., in-phase) component of the signal $\mathbf{x}_{\text{IN}}[n] \in (\mathbb{R},\mathbb{C})^{P\times 1}$ (i.e., $\mathbf{s}[n]$), from the input to the filter bank $\mathbf{v}$. In this way, the input to the adaptive process presents only the interference $\mathbf{n}[n]$, which will be subtracted from the fixed process, by imposing the minimization of the output power.

9.6.2.1 Block Matrix Determination

The signal $\mathbf{s}[n]$ is, by definition, incident on the sensors with identical phase. The signals coming from different directions, noise, interference, and reverberation have a different phase on each sensor. It follows that in order to obtain the cancellation of $\mathbf{s}[n]$, it is sufficient that for the block matrix, the sum of the elements of each row is zero, i.e., the matrix $\mathbf{B} = [b_{ij}]$ for $i = 2, ..., P$ and $j = 1, ..., P$, such that

$$b_{ij} \therefore \sum_{j=1}^{P} b_{ij} = 0, \quad 2 \le i \le P. \tag{9.175}$$

In fact, with the previous condition, (9.174) makes the cancellation of the in-phase component of each snap-shot. For better understanding, we consider a case with four sensors with a choice matrix $\mathbf{B}$, as shown in (9.175), so that the sum of the elements of each row is zero. Indicating with $x_{IN_k}[n] = s[n] + n_k[n]$, the signal at the k-th sensor, writing explicitly (9.174), omitting for simplicity writing the index $[n]$, we have that

$$\begin{aligned} x_2 &= (b_{21} + b_{22} + b_{23} + b_{24})s + b_{21}n_1 + b_{22}n_2 + b_{23}n_3 + b_{24}n_4 \\ x_3 &= (b_{31} + b_{32} + b_{33} + b_{34})s + b_{31}n_1 + b_{32}n_2 + b_{33}n_3 + b_{34}n_4 \\ x_4 &= (b_{41} + b_{42} + b_{43} + b_{44})s + b_{41}n_1 + b_{42}n_2 + b_{43}n_3 + b_{44}n_4 \end{aligned} \tag{9.176}$$

in which, it is clear that the component $s[n]$, which is identical for all the sensors, is eliminated by each equation by imposing the constraint $b_{k1}+b_{k2}+b_{k3}+b_{k4} = 0$ for $k = 2, 3, 4$. It follows that the signal $x_k[n]$ is a linear combination of only the interfering signals.

The constraint (9.175) indicates that $\mathbf{B}$ is characterized by $P-1$ linearly independent rows with zero sum. Among all the block matrices that satisfy (9.175), for $P = 4$, some possible choices for $\mathbf{B}$ are, for example

$$\begin{bmatrix} 0.9 & -0.3 & -0.3 & -0.3 \\ 0 & 0.8 & -0.4 & -0.4 \\ 0 & 0 & 0.7 & -0.7 \end{bmatrix}; \quad \begin{bmatrix} 1 & -1 & 0 & 0 \\ 0 & 1 & -1 & 0 \\ 0 & 0 & 1 & -1 \end{bmatrix}; \quad \begin{bmatrix} 1 & -1 & 1 & -1 \\ 1 & 1 & -1 & -1 \\ 1 & -1 & -1 & 1 \end{bmatrix}. \tag{9.177}$$

For the choice of $\mathbf{B}$, some authors suggest to determine the coefficients so that the transformation can be carried out with only sum-difference operations.

Remark In the presence of reverberation, the cancellation carried out by the block matrix concerns only the direct component of $\mathbf{s}[n]$. The reflected components, arriving from *all* directions, are no longer in-phase on the sensors and are not blocked by $\mathbf{B}$. It follows that the GSC attenuates, in addition to the not-in-phase disturbance, also the reverberated components of the SOI.

9.6.3 GSC Adaptation

The output of the adaptive beamformer section in composite notation is equal to

$$y_a[n] = \mathbf{v}^H\mathbf{x} \tag{9.178}$$

in which, similarly to (9.48), $\mathbf{x} = \begin{bmatrix}\mathbf{x}_2^T & \mathbf{x}_3^T & \cdots & \mathbf{x}_P^T\end{bmatrix}^T$, $\mathbf{v} = \begin{bmatrix}\mathbf{v}_2^T & \mathbf{v}_3^T & \cdots & \mathbf{v}_P^T\end{bmatrix}^T$, and the element $\mathbf{x}_k$ contains the delay line values of the k-th filter, namely $\mathbf{x}_k^T = \begin{bmatrix}x_k[n] \ x_k[n-1] \cdots x_k[n-M+1]\end{bmatrix}$.

9.6.3.1 GSC with On-line Algorithms

The total GSC output is equal to

$$y[n] = y_d[n] - y_a[n] \tag{9.179}$$

as for AIC, coinciding with the error signal of the adaptation process, which then can be done without any constraint. In case of using the simple LMS algorithm, the expression of adaptation is equal to

$$\mathbf{v}_{n+1} = \mathbf{v}_n - \mu y[n]\mathbf{x}_n. \tag{9.180}$$

For which, the LMS-like adaptation coincides with that of an ordinary multichannel MISO adaptive filter (see Sect. 5.3.5). Moreover, the implementation of more efficient algorithm, like APA, RLS, etc., appears to be trivial, while the frequency domain implementation is discussed later in Sect. 9.6.5.

Remark The GSC tends to reduce the output noise contribution, and in order to avoid SOI distortions, the filters should be adapted when at the input only the noise (i.e., in the absence of the SOI itself) is present. For example, in the case of speech enhancement beamforming, it is therefore necessary to add a further block processing, called *voice activity detector* (VAD) [33, 34], which allows the definition of the presence or less of the SOI and, accordingly, adjust the learning rate.

9.6.3.2 GSC with Block Algorithms

Consider the simplified scheme in Fig. 9.48 where, with reference to Fig. 9.47, for simplicity, the coefficients $b_i = 1$, $i = 1, \ldots, P$. The GSC output is

$$y[n] = y_d[n] - y_a[n] = \mathbf{x}_{\mathrm{IN}}^H\mathbf{g} - \mathbf{x}_{\mathrm{IN}}^H\mathbf{B}\mathbf{v} \tag{9.181}$$

for which the structure of (9.181) reveals the similarity with the LCMV previously discussed.

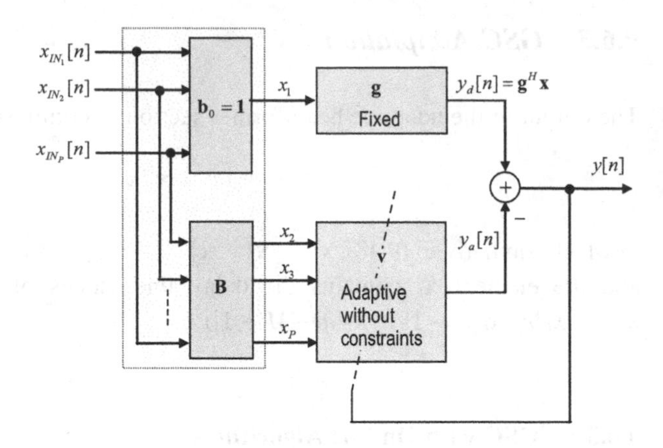

Fig. 9.48 Block structure of GSC beamforming

Defining $\mathbf{w}$, the vector of all the GSC parameters such that $y[n] = \mathbf{x}_{\text{IN}}^H\mathbf{w}$, namely,

$$\mathbf{w} \triangleq \mathbf{g} - \mathbf{B}\mathbf{v} \tag{9.182}$$

the Frost's block adaptation formula (9.150) can be rewritten as

$$\min_{\mathbf{v}} \left\{ (\mathbf{g} - \mathbf{B}\mathbf{v})^H \mathbf{R}_{x_{\text{IN}}x_{\text{IN}}} (\mathbf{g} - \mathbf{B}\mathbf{v}) \right\}. \tag{9.183}$$

In fact, by definition of $\mathbf{w}$, it is noted that the above equation includes also the gain constraint along the LD.

Moreover, the solution of (9.183) with respect to $\mathbf{v}$ can be expressed as

$$\mathbf{v}_{\text{opt}} = \left(\mathbf{B}^H \mathbf{R}_{x_{\text{IN}}x_{\text{IN}}} \mathbf{B}\right)^{-1} \mathbf{B}^H \mathbf{R}_{x_{\text{IN}}x_{\text{IN}}} \mathbf{g}. \tag{9.184}$$

For simplicity, by defining the covariance matrix of $\mathbf{x}$ as $\mathbf{R}_{xx} = \mathbf{B}^H\mathbf{R}_{x_{\text{IN}}x_{\text{IN}}}\mathbf{B}$, and the cross-correlation vector between $\mathbf{x}$ and $\mathbf{y}_d$ as $\mathbf{p}_{xy_d} = \mathbf{B}^H\mathbf{R}_{x_{\text{IN}}x_{\text{IN}}}\mathbf{g}$, the optimal solution for the adaptive GSC section can be rewritten in a compact Wiener's form as

$$\mathbf{v}_{\text{opt}} = \mathbf{R}_{xx}^{-1}\mathbf{p}_{xy_d}. \tag{9.185}$$

Remark The formulation (9.184) is suitable to an interesting connection with the LCMV method. By considering the linear constraint of the Frost's beamformer, $\mathbf{C}^H\mathbf{w} = \mathbf{f}$ and for the (9.182), we have that

$$\mathbf{C}^H(\mathbf{g} - \mathbf{B}\mathbf{v}) = \mathbf{f} \tag{9.186}$$

for which, wanting to determine $\mathbf{B}$ so that the GSC coincides with the LCMV, it is sufficient to impose the optimal solution (9.155) rewritten as

$$\mathbf{w}_{\text{opt}} = \mathbf{R}_{x_{\text{IN}}x_{\text{IN}}}^{-1}\mathbf{C}\left[\mathbf{C}^H\mathbf{R}_{x_{\text{IN}}x_{\text{IN}}}^{-1}\mathbf{C}\right]^{-1}\mathbf{f} \tag{9.187}$$

that, by replacing $\mathbf{w}_{\text{opt}} = \mathbf{g}-\mathbf{B}\mathbf{v}_{\text{opt}}$, and for (9.184), can be written as

$$\left[\mathbf{I} - \mathbf{B}\left(\mathbf{B}^H\mathbf{R}_{x_{\text{IN}}x_{\text{IN}}}\mathbf{B}\right)^{-1}\mathbf{B}^H\mathbf{R}_{xx}\right]\mathbf{g} = \mathbf{R}_{x_{\text{IN}}x_{\text{IN}}}^{-1}\mathbf{C}\left[\mathbf{C}^H\mathbf{R}_{x_{\text{IN}}x_{\text{IN}}}^{-1}\mathbf{C}\right]^{-1}\mathbf{f}. \tag{9.188}$$

Multiplying both sides by $\mathbf{B}^H\mathbf{R}_{xx}$, we get

$$\mathbf{B}^H\mathbf{R}_{x_{\text{IN}}x_{\text{IN}}}\underbrace{\left[\mathbf{I} - \mathbf{B}^H\mathbf{R}_{x_{\text{IN}}x_{\text{IN}}}\mathbf{B}\left(\mathbf{B}^H\mathbf{R}_{x_{\text{IN}}x_{\text{IN}}}\mathbf{B}\right)^{-1}\right]}_{=0}\mathbf{g} = \mathbf{B}^H\mathbf{C}\left[\mathbf{C}^H\mathbf{R}_{x_{\text{IN}}x_{\text{IN}}}^{-1}\mathbf{C}\right]^{-1}\mathbf{f} \tag{9.189}$$

where the left part is equal to zero. Simplifying is easy to verify that, being by definition $\left[\mathbf{C}^H\mathbf{R}_{x_{\text{IN}}x_{\text{IN}}}^{-1}\mathbf{C}\right]^{-1}\mathbf{f} \neq \mathbf{0}$, that is, necessarily, the condition

$$\mathbf{B}^H\mathbf{C} = \mathbf{0}. \tag{9.190}$$

Furthermore, in agreement with what has been developed in Sect. 9.5.3.4, we have that

$$\mathbf{g} = \mathbf{C}\left(\mathbf{C}^H\mathbf{C}\right)^{-1}\mathbf{C}^H\mathbf{w}. \tag{9.191}$$

For (9.182), this implies that each column of the block matrix $\mathbf{B}$ must be orthogonal to the weights vector $\mathbf{g}$, namely $\mathbf{B}\mathbf{g} = \mathbf{0}$. Moreover, since the $\mathbf{C}^H$ matrix is rectangular, with a number of columns equal to the number of linear constraints $(J < M)$, the size of the nullspace of $\mathbf{B}$ is exactly equal to the number of constraints $J = \dim(\mathbf{f})$ and the blocking properties of $\mathbf{B}$, derived precisely from this nullspace. In fact, it appears that the $\mathbf{B}$ dimensions are $\left[(P{-}J)\times P\right]$, and the matrix $\mathbf{B}$ has linearly independent columns that satisfy the condition $\mathbf{B}^H\mathbf{C} = \mathbf{0}$.

Note that, (9.191) does not depend from $\mathbf{R}_{x_{\text{IN}}x_{\text{IN}}}$ and that the matrix $\mathbf{C}[\mathbf{C}^H\mathbf{C}]^{-1}$ is the $\mathbf{C}$ pseudoinverse. The equation (9.191) provides the solution to minimum energy (minimum norm) of the optimization problem with constraints $\mathbf{C}^H\mathbf{w} = \mathbf{f}$. Moreover, it is interesting to observe that

$$\mathbf{g}^H\mathbf{g} = \mathbf{f}^H\left[\mathbf{C}^H\mathbf{C}\right]^{-1}\mathbf{f}. \tag{9.192}$$

9.6.3.3 Geometric Interpretation of GSC

Consider the LCMV solution $\mathbf{w}_{\text{opt}}$ (9.187), and consider the optimal vector decomposition as the sum of two orthogonal vectors, namely

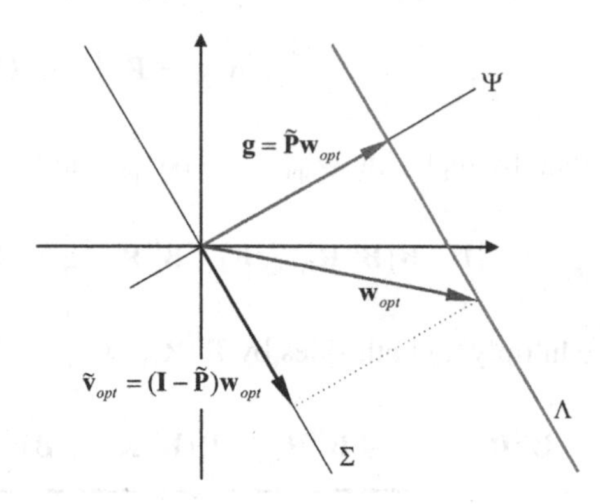

Fig. 9.49 Geometric GSC interpretation $\mathbf{w}_{\text{opt}} = \mathbf{g} + \tilde{\mathbf{v}}_{\text{opt}}$ such that $\mathbf{g} \perp \tilde{\mathbf{v}}_{\text{opt}}$

$$\mathbf{w}_{\text{opt}} = \mathbf{g} - \mathbf{B}\mathbf{v}_{\text{opt}} = \mathbf{g} + \tilde{\mathbf{v}}_{\text{opt}}, \qquad \text{where} \qquad \mathbf{g} \perp \tilde{\mathbf{v}}_{\text{opt}}. \tag{9.193}$$

For (9.191) evaluated for $\mathbf{w}_{\text{opt}}$, the vector $\mathbf{g}$ is a projection of $\mathbf{w}_{\text{opt}}$ on the column space of $\mathbf{C}$, $\Psi \in \mathcal{R}(\mathbf{C})$. Likewise, the vector $\tilde{\mathbf{v}}_{\text{opt}}$ is a projection of $\mathbf{w}_{\text{opt}}$ on the nullspace of $\mathbf{C}^H$, $\Sigma \in \mathcal{N}(\mathbf{C}^H)$. Therefore, similarly to what was discussed previously in (9.134) and (9.135) and for the Frost's BF, for which the projection matrix defined in (9.159), it holds that

$$\begin{aligned} \mathbf{g} &= \tilde{\mathbf{P}}\mathbf{w}_{\text{opt}} \\ \tilde{\mathbf{v}}_{\text{opt}} &= \mathbf{P}\mathbf{w}_{\text{opt}} = \left(\mathbf{I} - \tilde{\mathbf{P}}\right)\mathbf{w}_{\text{opt}} \end{aligned} \tag{9.194}$$

with the graphical representation shown in Fig. 9.49.

Remark The expression (9.193) coincides with the structure of Fig. 9.48. From the above considerations, it follows that $\mathbf{g}$ represents a data-independent deterministic beamformer, which gives a response in the subspace Ψ, that minimizes the white noise power. In the GSC lower path, the matrix $\mathbf{B}$ blocks the $\mathbf{x}$ elements in the subspace Ψ. The vector $\mathbf{v}$ combines the block matrix $\mathbf{B}$ output, so as to minimize the all output power outside to the subspace Ψ. In practice, the GSC constraints are implemented in fixed mode, while the $\mathbf{v}$ filters optimization consists of a simple unconstrained adaptive MISO process.

9.6.4 Composite-Notation GSC with J *constraints*

If there are a number of constraints $J > 1$, it is convenient to refer to the more general structure shown in Fig. 9.50, where the BF structure is defined on a dual path, and GSC is redrawn in a simplified style as a single space-temporal filter.

Without loss of generality, we consider the filters of the fixed and adaptive sections of the same length M. For which we can define the vectors

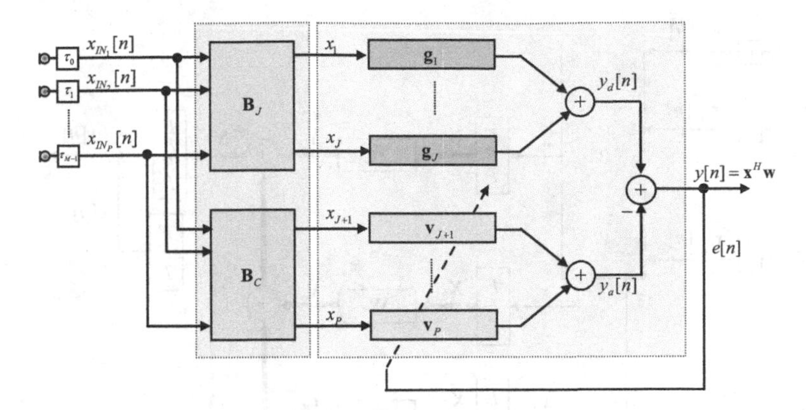

Fig. 9.50 GSC in composite notation

$$\mathbf{g} \in (\mathbb{R},\mathbb{C})^{J(M)\times 1} = \begin{bmatrix} \mathbf{g}_1^T & \cdots & \mathbf{g}_J^T \end{bmatrix}_{J\times 1}^T \tag{9.195}$$

$$\mathbf{v} \in (\mathbb{R},\mathbb{C})^{(P-J)(M)\times 1} = \begin{bmatrix} \mathbf{v}_{J+1}^T & \cdots & \mathbf{v}_P^T \end{bmatrix}_{(P-J)\times 1}^T. \tag{9.196}$$

In addition, consider the composite input and weights vectors, according to the following partitions

$$\mathbf{x} \in (\mathbb{R},\mathbb{C})^{P(M)\times 1} = \begin{bmatrix} [\mathbf{x}_1^T & \cdots & \mathbf{x}_J^T] & [\mathbf{x}_{J+1}^T & \cdots & \mathbf{x}_P^T] \end{bmatrix}_{2\times 1}^T \tag{9.197}$$

$$\mathbf{w} \in (\mathbb{R},\mathbb{C})^{P(M)\times 1} = \begin{bmatrix} \mathbf{g}^T & -\mathbf{v}^T \end{bmatrix}_{2\times 1}^T \tag{9.198}$$

such that the GSC output can be rewritten in compact form as

$$\begin{aligned} y[n] &= \begin{bmatrix} [\mathbf{x}_1^T & \cdots & \mathbf{x}_J^T] & [\mathbf{x}_{J+1}^T & \cdots & \mathbf{x}_P^T] \end{bmatrix} \begin{bmatrix} \mathbf{g} \\ -\mathbf{v} \end{bmatrix} \\ &= \mathbf{x}^H \mathbf{w} \end{aligned} \tag{9.199}$$

where $\mathbf{g}$ is the fixed part and $\mathbf{v}$ the variable part of the global MISO $\mathbf{w}$ filter (Fig. 9.51).

Furthermore, the matrix $\mathbf{B}$ can be partitioned with a constraint matrix $\mathbf{B}_J$ and a matrix of block $\mathbf{B}_C$, such that

$$\mathbf{B} \triangleq \begin{bmatrix} \mathbf{B}_J^{J\times P} \\ \mathbf{B}_C^{(P-J)\times P} \end{bmatrix}. \tag{9.200}$$

Note that in the GSC block structure in Fig. 9.48, it is supposed a matrix $\mathbf{B}_J$ constituted with a single vector. In fact, for simplicity very often, it refers to the case where there is only one constraint ($J = 1$). In this case, the matrix $\mathbf{B}_J$ is formed by the row vector whose all elements are different from zero, $b_{0,i} \neq 0$ for $i = 1, \ldots, P$. For example, with a number of sensors $P = 4$, a choice of the matrix $\mathbf{B}$ may be the following

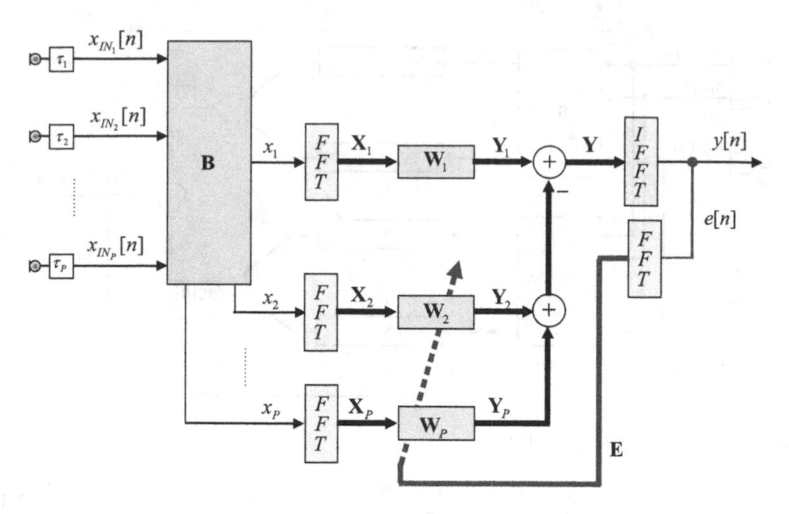

Fig. 9.51 Frequency domain GSC in compact notation for $J = 1$

$$\begin{bmatrix} b_{0,1} & b_{0,2} & b_{0,3} & b_{0,4} \\ 1 & -1 & 0 & 0 \\ 0 & 1 & -1 & 0 \\ 0 & 0 & 1 & -1 \end{bmatrix}. \tag{9.201}$$

9.6.5 Frequency Domain GSC

The algorithms described in the previous sections can easily be reformulated as transformed domain algorithms (see Chap. 7). Let's see, by way of example, the reformulation of the GSC in the frequency domain [35, 36].

We define the vector $\mathbf{x}_{m,k}$ as a input signal block of length L, of the m-th sensor, relative to the k-th instant. Whereby

$$\mathbf{x}_{m,k} = \begin{bmatrix} x_m[kL] & x_m[kL-1] & \cdots & x_m[kL-L+1] \end{bmatrix}^T. \tag{9.202}$$

Calling M the GSC filter length and for the overlap-and-save method implementation (see Sect. 7.3.3) N the DFT length (calculated with the FFT), it is necessary that $N \geq L+M-1$ in such a way that the L samples of the output block can be properly calculated. Moreover, for simplicity it is assumed that the filter length is an entire multiple of the block length ($M = S \cdot L$) and, again for simplicity, we impose that the FFT length is equal to $N = L+M = L(S+1)$. Then, to obtain the N-points FFT, indicated with $\boldsymbol{X}_{m,k}$, the last $(S+1)$ blocks of the input vectors are needed. For which

$$\boldsymbol{X}_{m,k} = \mathbf{F} \cdot \begin{bmatrix} \mathbf{x}_{m,k-S} & \mathbf{x}_{m,k-S+1} & \cdots & \mathbf{x}_{m.k} \end{bmatrix}^T \tag{9.203}$$

wherein (see Sect. 1.3.2 for details), by defining the phasor, $F_N = e^{-j2\pi/N}$; the DFT matrix $\mathbf{F}$, such that $\mathbf{F}^{-1} = \mathbf{F}^H/N$, is defined as $\left(\mathbf{F} \therefore f_{kn} = F_N^{kn}\ k, n \in [0, N-1]\right)$. Indicating with $\odot$, the Hadamard operator, the output of the m-th adaptive filter channel, can be written as

$$\boldsymbol{Y}_{m,k} = \boldsymbol{W}_{m,k} \odot \boldsymbol{X}_{m,k} \tag{9.204}$$

for which the frequency domain output of the whole beamformer is

$$\boldsymbol{Y}_k = \boldsymbol{Y}_{1,k} - \sum_{m=2}^{P} \boldsymbol{Y}_{m,k}. \tag{9.205}$$

With the overlap-save method, the time domain output samples

$$\mathbf{y}_k = \begin{bmatrix} y[kL] & y[kL+1] & \cdots & y[kL+L-1] \end{bmatrix}^T \tag{9.206}$$

are determined by selecting only the last L samples of the N-length output vector $\mathbf{F}^{-1}\boldsymbol{Y}_k$. Therefore, the output block, expressed as the inverse transformation, is

$$\mathbf{y}_k = \mathbf{g}_{0,L}\mathbf{F}^{-1}\boldsymbol{Y}_k \tag{9.207}$$

where for $N = L+M$, $\mathbf{g}_{0,L} \in \mathbb{R}^{N \times N}$ is a weighing matrix, also called the *projection output matrix*, defined as

$$\mathbf{g}_{0,L} \in \mathbb{R}^{(M+L)\times(M+L)} = \begin{bmatrix} \mathbf{0}_{M,M} & \mathbf{0}_{M,L} \\ \mathbf{0}_{L,M} & \mathbf{I}_{L,L} \end{bmatrix}. \tag{9.208}$$

In practice, the multiplication by $\mathbf{g}_{0,L}$ forces to zero the first M samples, leaving unchanged the last L.

The BF is a MISO system, whereby for the adaptation can proceed by adapting the individual channels of bank using one of the FDAF procedures described in Sect. 7.3.

However, due to the block processing approach, in some typical BF applications, the systematic delay between input and output is not allowed. Consider, for example, a microphone array used as a hearing aid for people with hearing problems. In these cases, the frequency domain approach, as it was illustrated in the previous paragraph, cannot be used.

A possible remedy, in cases of very long filters, is possible by partitioning the impulse response in the various section. The algorithm already presented in Sect. 7.4, for the case of single channel adaptive filtering, is the *partitioned frequency domain adaptive filter*. Given the linearity of the system, the total output of the filter can be calculated as the sum of the outputs relating to impulse response partitions.

The block diagram of the so-called *partitioned frequency domain adaptive BF* (PFDABF) is shown in Fig. 9.52.

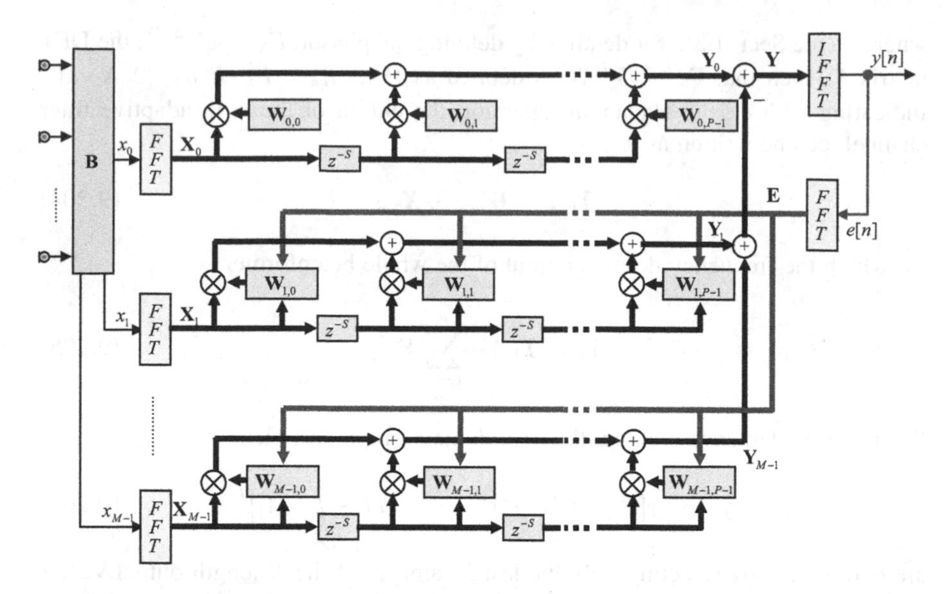

Fig. 9.52 Block diagram of the *partitioned frequency domain adaptive beamformer* (modified from [35])

9.6.6 Robust GSC Beamforming

As already indicated for maximum SNR BF (see Sect. 9.5.1), to increase the LCMV robustness or GSC beamformer, in addition to single or multiple linear constraints, it is possible to place constraints on the sensitivity function. Considering noise and signal uncorrelated, it appears that the BF sensitivity is inversely proportional to the white noise gain. For which, a simple expedient to increase the BF robustness is to limit the white noise gain GW. As suggested by Cox (see Sect. 9.5.1.2), to increase the robustness to random perturbations on the sensors, it is usual to add a supplementary quadratic inequality constraint. For which the CF is of the type

$$\min_{\mathbf{w}} \left\{ \mathbf{w}^H \mathbf{R}_{xx} \mathbf{w} \right\} \quad \text{s.t.} \quad \mathbf{C}^H \mathbf{w} = \mathbf{f}; \quad \mathbf{w}^H \mathbf{w} \le GW_{\max}. \tag{9.209}$$

with LS solution which takes the form (9.156)

$$\mathbf{w}_{\text{opt}} = \frac{(\mathbf{R}_{xx} + \delta \mathbf{I})^{-1} \mathbf{C}}{\mathbf{C}^H (\mathbf{R}_{xx} + \delta \mathbf{I})^{-1} \mathbf{C}} \mathbf{f} \tag{9.210}$$

where the regularization parameter $0 \le \delta < \infty$ is chosen based on the noise field characteristics.

For the GSC, $\mathbf{w} \in (\mathbb{R}, \mathbb{C})^{(P(M) \times 1)} = \begin{bmatrix} \mathbf{g}^T & -\mathbf{v}^T \end{bmatrix}^T$ represents the vector of all the BF free parameters, and the optimal solution $\mathbf{w}_{\text{opt}} = \mathbf{g} - \mathbf{B}\mathbf{v}_{\text{opt}}$, as described above, can be decomposed into two orthogonal components. So, the white noise gain constraint in (9.209) can be expressed as $\mathbf{g}^H\mathbf{g}+\mathbf{v}^H\mathbf{B}^H\mathbf{B}\mathbf{v} \leq GW_{\max}$ that determines a form of the type

$$\mathbf{v}^H\mathbf{v} \leq \beta^2 = \mathrm{GW}_{\max} - \mathbf{g}^H\mathbf{g}. \tag{9.211}$$

In this case, the solution $\mathbf{v}_{\text{opt}}$ is an extension of (9.185) and is equal to

$$\mathbf{v}_{\text{opt}} = \left(\mathbf{R}_{xx} + \delta\mathbf{I}\right)^{-1}\mathbf{p}_{xy_d} \tag{9.212}$$

that can be expressed as

$$\mathbf{v}_{\text{opt}} = \left(\mathbf{I} + \delta\mathbf{R}_{xx}\right)^{-1}\mathbf{R}_{xx}^{-1}\mathbf{p}_{xy_d} = \left(\mathbf{I} + \delta\mathbf{R}_{xx}\right)^{-1}\widehat{\mathbf{v}}_{\text{opt}} \tag{9.213}$$

where with $\widehat{\mathbf{v}}_{\text{opt}}$ is indicated as the optimal solution, without quadratic constraints, defined by (9.185).

9.6.7 *Beamforming in High Reverberant Environment*

In some BF applications, as in the capture of speech signals in high reverberant environments, the noise in which we operate may have coherent and/or incoherent nature. As previously shown, the presence of reverberation generates a hardly predictable diffuse noise field. In these cases, a too rigid BF architecture may not be optimal in all working conditions.

In this paragraph, some BF variants able to operate in the reverberant field are illustrated and discussed.

9.6.7.1 Linearly Constrained Adaptive Beamformer with Adaptive Constrain

An LCMV variant, which allows it to operate in environments with coherent and incoherent noise field, is presented. The method proposed in [37] allows to reduce the disturbance coming outside the LD and, simultaneously, to adapt a post-filter to suppress the diffused uncorrelated noise coming, by definition, also from LD. The method, called *linearly constrained adaptive beamforming with adaptive constraint*, in practice coincides with the LCMV described above, in which the constraint filter is not a priori fixed but is also adaptive and implemented as a Wiener filter. In practice, the optimization criterion maximizes the BF power along the LD in the presence of an adaptive constraint. Equation (9.150) is redefined as

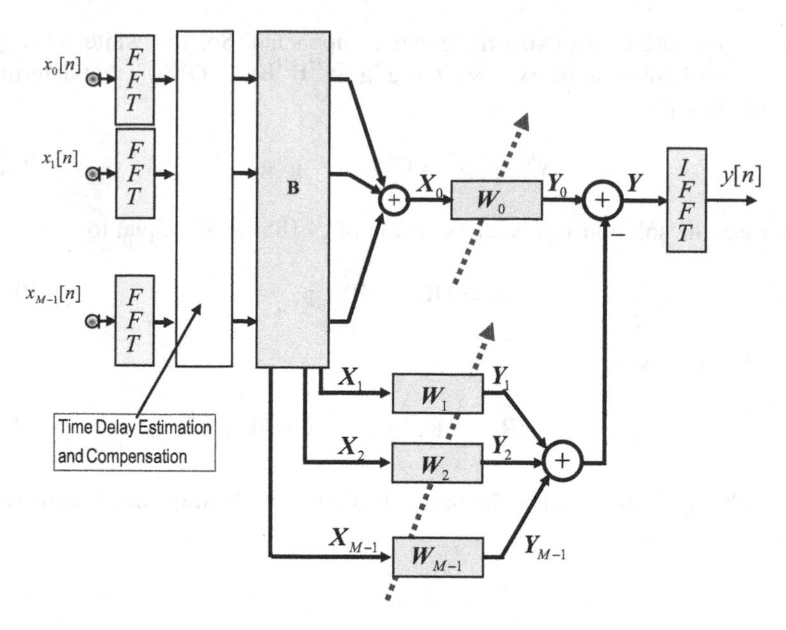

Fig. 9.53 Schematic of GSC with adaptive constraint (modified from [37])

$$\min_{\mathbf{w}} \left\{ \mathbf{w}^H \mathbf{R}_{xx} \mathbf{w} \right\} \quad \text{s.t.} \quad \mathbf{C}^H \mathbf{w} = \mathbf{R}_{ff}^{-1} \mathbf{r}_{vv} \tag{9.214}$$

where $\mathbf{R}_{ff}$ represents the autocorrelation matrix of the signal coming only from the LD, while $\mathbf{r}_{vv}$ is the autocorrelation of the SOI (usually speech) estimated, for example, with a spatial cross-correlation. For (9.155), the optimal regularized solution is

$$\mathbf{w}_{\text{opt}} = \frac{(\mathbf{R}_{xx} - \delta \mathbf{I})^{-1} \mathbf{C}}{\mathbf{C}^H (\mathbf{R}_{xx} - \delta \mathbf{I})^{-1} \mathbf{C}} \mathbf{R}_{ff}^{-1} \mathbf{r}_{vv}. \tag{9.215}$$

The estimate of the matrix $\mathbf{R}_{ff}$ and the vector $\mathbf{r}_{vv}$ can be made directly from the input data in the time or frequency domain with the method described in Sect. 9.5.2, as indicated by Zelinski in [27].

The described method, which can be interpreted as a combination of the techniques of post-filtering beamformer, described in Sect. 9.5.2, and of the standard LCMV, is easily extensible to the GSC (Fig. 9.53).

Remark The filtering techniques, in the most general term, allow the extraction of information immersed in noise characterized by a certain statistic. The determination of the filter, adaptive or static, may be driven from a priori knowledge of (1) the noise characteristics in which it operates, or, if this is unknown and (2) the characteristics of the signal to be extracted.

In the case of beamforming, the methodologies LCMV or GSC mainly operate according to the paradigm (1) and, by definition, are therefore optimal for coherent noise, for example, coming from specific directions and in the absence of diffuse noise as reverberation or multipath. The performance of such beamformer in reverberant environments, in which there is a high diffuse field without a specific direction of arrival, as also reported in the literature (see, for example [33, 37]) is not too different from the simple DSBF model described above.

On the contrary, the post-filtering methodology described in Sect. 9.5.2, being based on desired signal autocorrelation estimate, operates with the paradigm of the a priori knowledge of the signal statistics and is more suitable in the presence of diffuse field and with high reverberation time.

The LCMV with adaptive constraint methodology try, somehow, to merge the (1) and (2) paradigms, by considering two distinct adaptive processes: the first as in the GSC, operating so as to cancel the spatially coherent noise, i.e., from specific directions, and the second adaptive, following the post-filtering philosophy, is based on the estimation of the desired signal acf, allowing the diffuse noise cancellation.

9.7 Direction of Arrival and Time Delay Estimation

In the sensors array, the DOA estimation or its dual problem, the time delay estimation (TDE), is of central interest in many AP applications.

In the DOA estimation, we can distinguish the narrow- and broadband cases, and the nonparametric and parametric spectral methods. Similarly, for the TDE, it is necessary to distinguish the cases in which the propagation model is anechoic or reverberant.

9.7.1 Narrowband DOA

The narrowband DOA estimation is one of the most characteristic and oldest AP issues. The DOA applications include radar, telecommunications, underwater acoustics, GPS, sound localization, etc. The first proposed techniques are the standard beamforming, with resolution limited by the array geometry, and the classical methods of spectral estimation. For waves with close arrival angles and low SNR, the parametric approaches and/or based on the maximum likelihood (ML) estimation (see Sect. C.3.2.2) have the higher resolution.

In the stochastic ML methods, the signals are assumed Gaussian while for deterministic ML they are regarded as arbitrary. The noise is considered stochastic in both methods. In ideal stochastic ML conditions, it is possible to reach the optimal statistical solution or the so-called Cramér-Rao bound (CRB) at the expense, however, of a high computational complexity required to solve a complex multidimensional nonlinear optimization problem which, moreover, does not guarantee global convergence [38–40].

However, the so-called super-resolution approaches, i.e., based on the signal and noise subspaces decomposition of the input correlation matrix $\mathbf{R}_{xx}$, guarantee best performance and a high computational efficiency than the ML methods.

9.7.1.1 Narrowband DOA Signal Model

For the analytical development, consider the signal model in the presence of multiple narrowband sources. For an array of P elements irradiated by $N_s < P$ sources, considering only the dependence of the angle θ, the model (9.23) is

$$\begin{aligned}\mathbf{x}[n] &= \sum_{k=1}^{N_S} \mathbf{a}_k(\theta)s_k[n] + \mathbf{n}[n] \\ &= \mathbf{A}(\theta)\mathbf{s}[n] + \mathbf{n}[n]\end{aligned} \tag{9.216}$$

where $\mathbf{A}(\theta) \in \mathbb{C}^{(P\times N_S)} = [\mathbf{a}_1(\theta) \quad \mathbf{a}_2(\theta) \quad \cdots \quad \mathbf{a}_{N_S}(\theta)]$ is the steering matrix (see Sect. 9.2.2).

9.7.1.2 DOA with Conventional Beamformer: Steered Response Power Method

For narrowband signals the DOA is generally done through a scan of the field-of-view (FOV) $\Theta \equiv [\theta_{\min},\theta_{\max}]$, related to the array geometry (see Sect. 9.2.1). In practical terms, we proceed to the array output power evaluation for steering of various angles, for which the method is indicated as steered response power (SRP). From scanning of the FOV, the estimated direction is relative to the angles in which there is the maximum BF output power.

BF output is $y[n] = \mathbf{w}^H\mathbf{x}$, for which the output power as a function of the angle is defined as $P(\theta) = E\left\{|y[n]|^2\right\}_{\mathbf{w}=\mathbf{w}_{\mathrm{opt}}}$, for $\theta \in \Theta$; i.e.

$$\begin{aligned}P(\theta) &= E\left\{|y[n]|^2\right\}_{\mathbf{w}=\mathbf{w}_{\mathrm{opt}}} \qquad \theta \in [\theta_{\min}, \theta_{\max}] \\ &= \mathbf{w}_{\mathrm{opt}}^H \mathbf{R}_{xx} \mathbf{w}_{\mathrm{opt}}.\end{aligned} \tag{9.217}$$

The previous quantity, calculated with suitable angular resolution, can be regarded as a spectrum in which, instead of the frequency, the DOA angle is considered. In practice, (9.217) is estimated for θ variable within the FOV and its maximum determines the directions of arrival.

For conventional DSBF with isotropic Gaussian noise, the optimal beamformer (see Sect. 9.5.1) can be computed as

$$\mathbf{w}_{\text{opt}} = \frac{\mathbf{a}(\theta)}{\mathbf{a}^H(\theta)\mathbf{a}(\theta)} \tag{9.218}$$

for which $\alpha = \left(\mathbf{a}^H(\theta)\mathbf{a}(\theta)\right)^{-2}$; by substituting in (9.217) we get

$$P_{\text{DSBF}}(\theta) = \alpha\mathbf{a}^H(\theta)\mathbf{R}_{xx}\mathbf{a}(\theta), \quad \theta \in \Theta \tag{9.219}$$

which represents a *spatial spectrum*.

9.7.1.3 DOA with Capon's Beamformer

In the standard Capon method (see Sect. 9.5.1.1), the optimal BF vector is

$$\mathbf{w}_{\text{opt}} = \frac{\mathbf{R}_{xx}^{-1}\mathbf{a}(\theta)}{\mathbf{a}^H(\theta)\mathbf{R}_{xx}^{-1}\mathbf{a}(\theta)}$$

Substituting the latter in (9.217), the DOA can be done by defining the following quantity

$$\begin{aligned} P_{\text{CAPON}}(\theta) &= \frac{\mathbf{a}^H(\theta)\mathbf{R}_{xx}^{-1}\mathbf{R}_{xx}\mathbf{R}_{xx}^{-1}\mathbf{a}(\theta)}{\left[\mathbf{a}^H(\theta)\mathbf{R}_{xx}^{-1}\mathbf{a}(\theta)\right]^2} \qquad \theta \in \Theta \\ &= \frac{1}{\mathbf{a}^H\mathbf{R}_{xx}^{-1}\mathbf{a}}. \end{aligned} \tag{9.220}$$

The DOA estimation with (9.220) has a resolution that is not able to resolve more signals coming from rather close angles. The peaks of (9.220), in fact, represent the power of the incident signal only in an approximate way. The method has a robustness degree of the typical nonparametric methods of spectral analysis and does not require any rank reduced signal modeling.

9.7.1.4 DOA with Subspace Analysis

The DOA can be determined by the subspace properties of the input signal covariance matrix (see 9.3.1.2). In fact, for a consistent estimate of the components of signal and noise, it is possible to perform the eigen analysis of the spatial covariance matrix defined in (9.55). In the reduced rank methods only the signal subspace is considered while the contribution due to noise, assumed Gaussian and uncorrelated, is discarded.

Multiple Signal Classification Algorithm

For a P elements array irradiated by N_s sources, with $N_s < P$, the spatial correlation (9.55) can be written as

$$\mathbf{R}_{xx} = \mathbf{A}\mathbf{R}_{ss}\mathbf{A}^H + \mathbf{R}_{nn} \tag{9.221}$$

where for white Gaussian noise $\mathbf{R}_{nn} = \sigma_n^2\mathbf{I}$. Place $\mathbf{\Lambda}_n \equiv \mathbf{R}_n = \sigma_n^2\mathbf{I}$, proceeding to the spatial covariance matrix spectral factorization (see Sect. 9.3.1.2), from (9.62), we get

$$\mathbf{R}_{xx} = \mathbf{U}_s\mathbf{\Lambda}_s\mathbf{U}_s^H + \sigma_n^2\mathbf{U}_n\mathbf{U}_n^H. \tag{9.222}$$

where $\mathbf{U}_s$ and $\mathbf{U}_n$ are unitary matrices and $\mathbf{\Lambda} = \text{diag}[\lambda_1, \lambda_2, ..., \lambda_P]$ the diagonal matrix $\mathbf{R}_{xx}$ with real eigenvalues ordered as $\lambda_1 \geq \lambda_2 \geq \cdots \geq \lambda_P > 0$. Assuming the noise variance σ_n^2 a priori known (or in some way estimate), you can partition the eigenvectors and eigenvalues belonging to the signal and noise in the following way

$$\lambda_1, \lambda_2, ..., \lambda_{N_s} \geq \sigma_n^2, \qquad \textit{signal space}$$

and

$$\lambda_{N_s+1}, \lambda_{N_s+2}, ..., \lambda_P = \sigma_n^2, \quad \textit{noise space}.$$

Assuming Gaussian noise and independent between the sensors, it appears that the noise subspace eigenvectors are orthogonal to the column space of the steering matrix $\mathcal{R}(\mathbf{A}^H) \perp \mathbf{U}_n$. In other words we can write

$$\mathbf{U}_n^H\mathbf{a}(\theta_i) = 0, \quad \text{for} \quad \theta_i = [\theta_1, ..., \theta_{N_s}] \tag{9.223}$$

From the above properties, the estimation algorithm called Multiple SIgnal Classification (MUSIC) [38–42], can be derived by defining the so-called *MUSIC spatial spectrum* $P_{\text{MUSIC}}(\theta)$, i.e., the following quantity

$$P_{\text{MUSIC}}(\theta) = \frac{1}{\mathbf{a}^H(\theta)\mathbf{U}_n\mathbf{U}_n^H\mathbf{a}(\theta)} \tag{9.224}$$

where the number of sources N_s must be a priori known or estimated. Note that $P_{\text{MUSIC}}(\theta)$ is not a real power; in fact (9.224) represents the distance between two subspaces and is therefore defined as *pseudo spectrum*. The quantity $P_{\text{MUSIC}}(\theta)$ represents an estimate of the input signal $\mathbf{x}[n]$ pseudo spectrum, calculated by an estimate of the eigenvectors of the correlation matrix $\mathbf{R}_{xx}$.

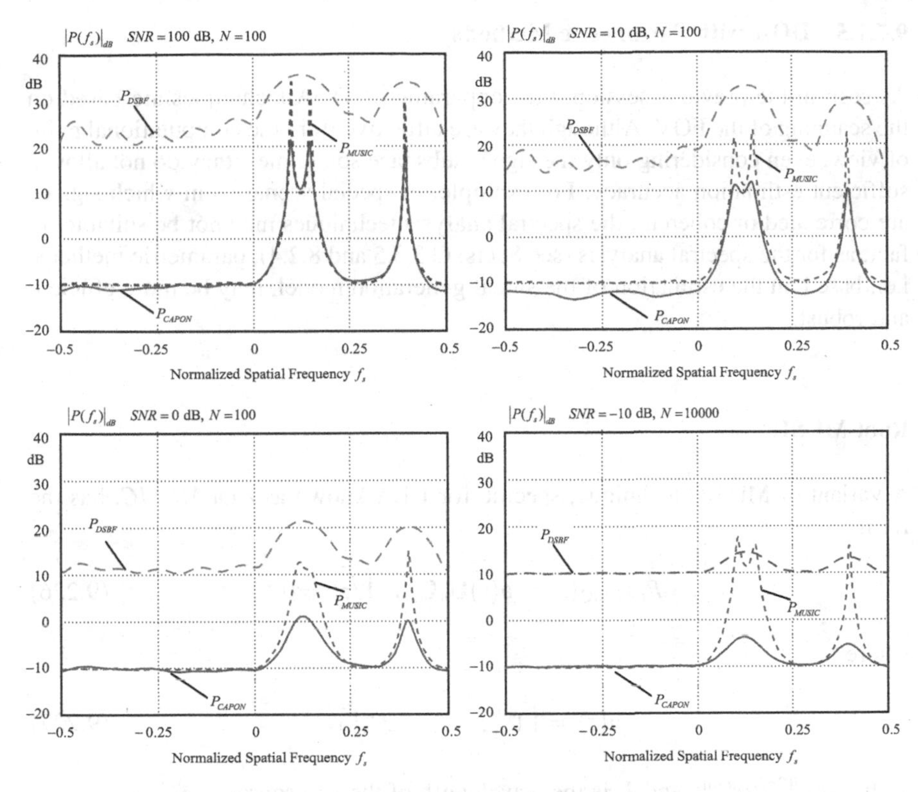

Fig. 9.54 Narrowband DOA estimation with the conventional DSBF compared with the Capon and MUSIC methods, for an ULA with $P = 10$ isotropic sensors, interspaced by $d = \lambda_s/2$. The lengths of the sequences of the signal and the SNR are shown directly in figure

The computational cost of the MUSIC may be high for a fine-grained scan of the FOV. In addition, the MUSIC has been and is a research topic very fertile, and in the literature there are numerous MUSIC algorithm variations and specializations.

Example For a more clear perception of achievable performances, consider an ULA with $P = 10$ isotropic sensors spaced with interdistance $d = \lambda_s/2$. Consider the presence of three radiating sources of the same power, with spatial frequency $f_s = 0.1, 0.15$, and 0.4, defined as

$$f_s = \frac{d}{\lambda_s}\cos(\theta). \tag{9.225}$$

The analysis window length is equal to $N = 100$, 10,000 samples, in the presence additive Gaussian noise (complex) with unit variance, with SNR = 100, 10, 0, and −10 [dB]. With reference to Fig. 9.54, the DOA estimation is performed by comparing the results obtained with the conventional technique, referred to as P_{DSBF}, (9.219), the standard method of Capon, (9.220), and the MUSIC method (9.224). The SNR data and the sequence length are shown directly in the figure.

9.7.1.5 DOA with Parametric Methods

As seen in the previous session, the nonparametric DOA techniques are based on the scanning of the FOV. Although they are attractive from the computational point of view, even considering only the signal subspace sometimes, they do not allow a sufficient estimation accuracy. For example, in special scenarios in which signals are correlated or coherent, the spectral analysis techniques may not be suitable. In fact, as for the spectral analysis (see Sects. C.3.3.5 and 8.2.4), parametric methods, i.e., based on the estimation of the signal generation model, may be more efficient and robust.

Root MUSIC

A variant of MUSIC technique, specific for ULA known as *root MUSIC*, has the form

$$P_{\mathrm{rMUSIC}}(z) = \mathbf{a}(z)\mathbf{U}_n\mathbf{U}_n^H\mathbf{a}(1/z) = 0 \tag{9.226}$$

where

$$\mathbf{a}(z) = \begin{bmatrix} 1 & z & \cdots & z^{P-1} \end{bmatrix}^T \tag{9.227}$$

with $z = e^{j(2\pi/\lambda_s)d\sin\theta}$ and λ_s is the wavelength of the s-th source.

The DOA is estimated by the roots of the polynomial (9.227), available in complex conjugate pairs, closest to the unit circle. In the contrary, the roots more internal, related to the noise. For low SNR the root MUSIC presents better performance than MUSIC.

ESPRIT Algorithm

The estimation of signal parameters via rotational invariance technique (ESPRIT) represents one of the most efficient and robust narrowband DOA methods. Proposed by Paulraj, Roy, and Kailath [43, 44], the ESPRIT is an algebraic method that does not require any scanning procedure. The basic idea of the method is to exploit the properties of the underlying rotational invariance of the signal subspace, through the invariance to the translation (rotation in this case) of the array.

Consider an ULA illuminated by N_s sources, with a steering matrix $\mathbf{A}$, defined as

$$\mathbf{A} = \begin{bmatrix} 1 & 1 & 1 & 1 \\ e^{j\Omega_1} & e^{j\Omega_2} & \cdots & e^{j\Omega_{N_s}} \\ \vdots & \vdots & \ddots & \vdots \\ e^{j\Omega_1(P-1)} & e^{j\Omega_2(P-1)} & \cdots & e^{j\Omega_{N_s}(P-1)} \end{bmatrix} \tag{9.228}$$

with $\Omega_i = (2\pi/\lambda)d\cos\theta_i$. The algorithm uses the steering matrix structure in different way than the other methods. First we observe that the matrix $\mathbf{A} \in \mathbb{C}^{P\times N_s}$ defined in (9.228) has a cyclic structure. Second we define two matrices $\mathbf{A}_1, \mathbf{A}_2 \in \mathbb{C}^{(P-1)\times N_s}$, $\mathbf{A}_1 = \lceil \mathbf{A} \rceil$ and $\mathbf{A}_2 = \lfloor \mathbf{A} \rfloor$, by erasing the first and the last row of $\mathbf{A}$, such that

$$\mathbf{A} = \begin{bmatrix} \mathbf{A}_1 \\ \text{last row} \end{bmatrix} = \begin{bmatrix} \text{first row} \\ \mathbf{A}_2 \end{bmatrix} \tag{9.229}$$

and, in addition, the following relationship holds

$$\mathbf{A}_2 = \mathbf{A}_1\mathbf{\Phi} \tag{9.230}$$

where $\mathbf{\Phi}$ is a diagonal matrix defined as $\mathbf{\Phi} = \text{diag}\begin{bmatrix} e^{j\Omega_1} & e^{j\Omega_2} & \cdots & e^{j\Omega_{N_s}} \end{bmatrix}_{N_s\times N_s}$. Similarly, we define two matrices formed with the eigenvectors of the signal subspace matrix $\mathbf{U}_s$, such that $\mathbf{U}_1 = \lceil \mathbf{U}_s \rceil \in \mathbb{C}^{(P-1)\times P}$ and $\mathbf{U}_2 = \lfloor \mathbf{U}_s \rfloor \in \mathbb{C}^{(P-1)\times P}$. Recalling that $\mathbf{U}_s$ and $\mathbf{A}$ are related to the same column space span $\left(\mathcal{R}(A^H)\right)$, there is a full rank matrix $\mathbf{T}$ such that

$$\mathbf{U}_s = \mathbf{AT} \tag{9.231}$$

So, it is also true that

$$\begin{aligned} \mathbf{U}_1 &= \mathbf{A}_1\mathbf{T} \\ \mathbf{U}_2 &= \mathbf{A}_2\mathbf{T} = \mathbf{A}_1\mathbf{\Phi}\mathbf{T}. \end{aligned} \tag{9.232}$$

Combining with (9.230) we obtain

$$\begin{aligned} \mathbf{U}_2 &= \mathbf{A}_1\mathbf{\Phi}\mathbf{T} \\ \mathbf{U}_1 &= \mathbf{T}^{-1}\mathbf{\Phi}\mathbf{T} \end{aligned} \tag{9.233}$$

for which, by defining the matrix $\mathbf{\Psi} = \mathbf{T}^{-1}\mathbf{\Phi}\mathbf{T}$, we can write

$$\mathbf{U}_2 = \mathbf{U}_1\mathbf{\Psi} \tag{9.234}$$

where, it is noted that, both the matrices $\mathbf{T}$ and $\mathbf{\Phi}$ are unknowns.

From (9.234) the $\mathbf{\Psi}$ matrix can be determined using the LS approach; therefore we can write

$$\mathbf{\Psi} = \left(\mathbf{U}_2^H\mathbf{U}_2\right)^{-1}\mathbf{U}_2^H\mathbf{U}_1 \quad \text{or} \quad \mathbf{\Psi} = \mathbf{U}_1^H\left(\mathbf{U}_1^H\mathbf{U}_1\right)^{-1}\mathbf{U}_2 \tag{9.235}$$

Moreover, from (9.234), it appears that the diagonal elements $\mathbf{\Phi}$ coincide with the eigenvalues of $\mathbf{\Psi}$, i.e., they shall be treated as a *similarity transformations* characterized by the same eigenvalues.

The ESPRIT algorithm can be formalized by the following steps:

1. Decomposition of the covariance matrix $\mathbf{R}_{xx}$ and determining the signal subspace of $\mathbf{U}_s$.
2. Determination of the matrices $\mathbf{U}_1$ and $\mathbf{U}_2$.
3. Computation of $\mathbf{\Psi} = \mathbf{U}_1^H\left(\mathbf{U}_1^H\mathbf{U}_1\right)^{-1}\mathbf{U}_2$.
4. Eigenvector of $\mathbf{\Psi}$, ψ_n, $n = 1, 2, \ldots, N_s$ and determination of the DOA estimate from the angles ψ_n.

Remark The ESPRIT computational cost is lower than other parametric techniques such as root MUSIC and, like the latter, does not provide any scanning or research in the array FOV. Moreover, note that the matrix $\mathbf{\Psi}$ can be determined with one of the methods described in Chap. 4, such as TLS or other techniques.

9.7.2 Broadband DOA

In the broadband case, each source no longer corresponds to a full-rank covariance matrix and parametric subspace methods require a reduced rank analysis. Then, the extension of the narrowband methods (parametric or not) for broadband signals may be done, as usual in BF described in the preceding paragraphs, by replacing the complex weights (phase shift and sum) with filters. As previously developed, for the FSBF output, we have $y[n] = \mathbf{w}^H\mathbf{x}$, for which the DOA can be estimated by the maximum output power as a function of the angle. For example, in the frequency domain, indicating with $W_p(e^{j\omega})$, the BF filters TFs downstream of the p-th sensors, considering a ULA-FSBF with P inputs and one steering, the following relation holds:

$$Y\left(\theta, e^{j\omega}\right) = \sum_{m=1}^{P} W_m\left(e^{j\omega}\right)X_m\left(e^{j\omega}\right)e^{-jk(m-1)d\cos\theta} \tag{9.236}$$

where $\tau_s = \left((d\cos\theta)/c\right)\big|_{\theta=\theta_s}$ (see Sect. 9.4.1.1) is the appropriate steering delay to focus the array in the spatial direction of the source θ_s. So, considering the DTFT, the output power $P(\theta)$ is

$$\begin{aligned} P(\theta) &= \frac{1}{2\pi}\int_{-\pi}^{\pi} \left|Y(\theta, e^{j\omega})\right|^2 d\omega \\ &= \text{DTFT}^{-1}\left\{\left|Y(\theta, e^{j\omega})\right|^2\right\} \end{aligned} \tag{9.237}$$

so, with the steered response power (SRP) method, the DOA estimated θ_s equal to

$$\hat{\theta}_s = \underset{\theta \in \Theta}{\text{argmax}}\left\{P(\theta)\right\}. \tag{9.238}$$

The DOA estimation is obtained by (9.237) evaluated for $\theta \in [\theta_{\min} + k\Delta\theta]$, for $k = 0, 1, \ldots, K$, where $\Delta\theta \triangleq (\theta_{\max} - \theta_{\min})/K$ and K an integer of appropriate value, related to the minimum angle for the desired spatial resolution.

9.7.3 Time Delay Estimation Methods

The time delay estimation (TDE) consists in estimating the propagation time of a wave impinging on two or more receivers [45]. The TDE is a problem intimately related to the DOA estimated. Considering, for example, an ULA (see Fig. 9.7), with a priori known spatial sensors coordinates, through the TDE it is possible to calculate the DOA. In fact, from (9.36) with known τ , c and d, it is $\theta = \cos^{-1}(\tau c/d)$.

However, the TDE appears to be rather complicated in the presence of low SNR and/or in complex multipath propagation environments, as in the case of acoustic reverberation.

9.7.3.1 Method of Cross-Correlation

As already explained in Chap. 3 (see Sect. 3.4.3), the TDE can be traced to a simple identification problem.

Considering a source $s[n]$ impinging on two sensors with interdistance d, the received signal is

$$\begin{aligned} x_1[n] &= s[n] + n_1[n] \\ x_2[n] &= \alpha s[n + D] + n_2[n] \end{aligned} \tag{9.239}$$

with D representing the time delay of arrival, α an attenuation, and $n_1[n]$, $n_2[n]$ the measurement noise assumed uncorrelated with $s[n]$. A simple way to estimate the delay D is the analysis of the cross-correlation function (ccf)

$$r_{x_1x_2}[k] = E\{x_1[n]x_2[n-k]\}. \tag{9.240}$$

In this case, saying $\hat{r}_{x_1x_2}[n]$ the estimate ccf calculated on a N-length window time-average, the estimate of the delay D is equal to

$$\hat{D} = \underset{n\in[0,N-1]}{\text{argmax}} \{\hat{r}_{x_1x_2}[n]\}. \tag{9.241}$$

Remark Whereas in the signal model (9.239), and in ideal Gaussian noise conditions, the cross-correlation between the inputs can be written as

$$r_{x_1x_2}[n] = \alpha r_{ss}[n-D] + r_{n_1n_2}[n] \tag{9.242}$$

It follows that in the frequency domain considering the DTFT operator, the CPSD $R_{x_1x_2}(e^{j\omega})$ is defined as

$$R_{x_1x_2}(e^{j\omega}) = \alpha R_{ss}(e^{j\omega})e^{-j\omega D} + R_{n_1n_2}(e^{j\omega}). \tag{9.243}$$

9.7.3.2 Knapp–Carter's Generalized Cross-Correlation Method

The TDE can be improved by inserting the filters with TF $W_1(e^{j\omega})$ and $W_2(e^{j\omega})$ suitably determined on the input sensors as illustrated in Fig. 9.55. For development, indicating with $R_{y_1y_2}(e^{j\omega}) = \text{DTFT}\{r_{y_1y_2}[n]\}$, the CPSD between the outputs of such filters is

$$\begin{aligned} r_{y_1y_2}[n] &= \frac{1}{2\pi}\int_{-\pi}^{\pi} R_{y_1y_2}(e^{j\omega})e^{j\omega n}d\omega \\ &= \text{DTFT}^{-1}\{R_{y_1y_2}(e^{j\omega})\} \end{aligned} \tag{9.244}$$

where the CPSD $R_{y_1y_2}(e^{j\omega})$ can be expressed as

$$R_{y_1y_2}(e^{j\omega}) = W_1(e^{j\omega})W_2^*(e^{j\omega})R_{x_1x_2}(e^{j\omega}). \tag{9.245}$$

The Knapp–Carter method [46], also called generalized cross correlation (GCC) method, with reference to Fig. 9.54 is based on the sensors signals pre-filtering with TFs subject to the constraint

$$F_g(e^{j\omega}) = W_1(e^{j\omega})W_2^*(e^{j\omega}) \tag{9.246}$$

With $F_g(e^{j\omega})$ defined as *window* or real *weighing function*. With this position, the ccf (9.244) is defined as

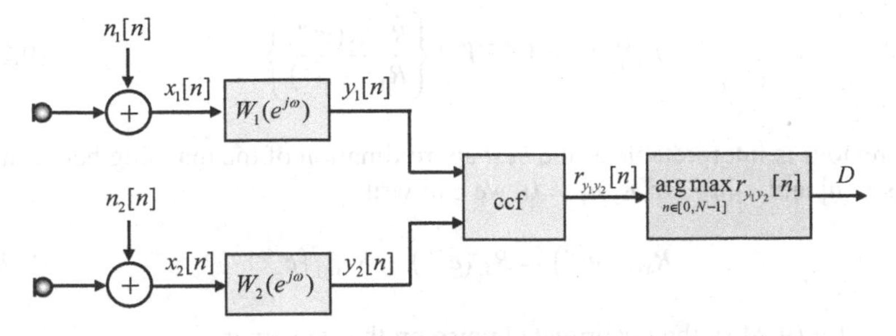

Fig. 9.55 Knapp–Carter's generalized cross-correlation (GCC) for TDE estimation

$$r_{y_1y_2}^{(g)}[n] = \mathrm{DTFT}^{-1}\left\{F_g\left(e^{j\omega}\right)R_{x_1x_2}\left(e^{j\omega}\right)\right\}. \tag{9.247}$$

In fact, for $F_g(e^{j\omega})$ real, $W_1(e^{j\omega})$ and $W_2(e^{j\omega})$ *necessarily* have identical phase for which they do not affect the ccf peak location.

From weighting function $F(e^{j\omega})$ choice, we can define various algorithms. For example, for $F(e^{j\omega}) = 1$ (9.247) coincides with the simple cross-correlation method (9.241).

In general, the CPSD is not a priori known but is estimated with a temporal average and indicated as $\hat{R}_{x_1x_2}(e^{j\omega})$. In this case, considering a generic real weighing function $F(e^{j\omega})$, (9.247) is written as

$$\hat{r}_{y_1y_2}^{(g)}[n] = \mathrm{DTFT}^{-1}\left\{F\left(e^{j\omega}\right)\hat{R}_{x_1x_2}\left(e^{j\omega}\right)\right\}. \tag{9.248}$$

Remark (9.247) is also interpreted as a frequency windowing of the CPSD, before making the DTFT^{-1}, in order to reduce the delay estimation error (9.241). In the literature [38, 45, 46], various methods for the determination of such optimal window have been proposed. Some based on different paradigms including ML estimation are listed below.

Roth's Weighing Method

In the Roth method [45], the weighing function, indicated as $F_R(e^{j\omega})$, is defined as

$$F_R\left(e^{j\omega}\right) = \frac{1}{R_{x_1x_1}\left(e^{j\omega}\right)}. \tag{9.249}$$

With this position, the ccf (9.248) is equivalent to the impulse response of the optimal Wiener filter (see Sect. 3.3.1), defined as

$$\hat{r}_{x_1x_2}^{(R)}[n] = \text{DTFT}^{-1}\left\{\frac{\hat{R}_{x_1x_2}(e^{j\omega})}{R_{x_1x_1}(e^{j\omega})}\right\}. \tag{9.250}$$

The previous is interpretable as the *best* approximation of the mapping between the inputs $x_1[n]$ and $x_2[n]$. For $n_1[n] \neq 0$, we can write

$$R_{x_1x_1}(e^{j\omega}) = R_{ss}(e^{j\omega}) + R_{n_1n_1}(e^{j\omega}) \tag{9.251}$$

whereby for (9.243), the uncorrelated noise on the sensors is

$$r_{x_1x_2}^{(R)}[n] = \delta[n-D] * \text{DTFT}^{-1}\left\{\frac{\alpha R_{ss}(e^{j\omega})}{R_{ss}(e^{j\omega}) + R_{n_1n_1}(e^{j\omega})}\right\}. \tag{9.252}$$

Remark As in the original Knapp–Carter paper [48], for the theoretical development, the weighing functions are determined based on the *true* PSD. In general, however, these are not available in practice and replaced with their time-average estimates. In this sense, the Roth weighing has the effect of suppressing the frequency region in which $R_{n_1n_2}(e^{j\omega})$ is large and where the CPSD estimate $\hat{R}_{x_1x_1}(e^{j\omega})$ can be affected by large errors.

Smoothed Coherence Transform Method

In the method called smoothed coherence transform (SCOT) [38, 45, 46], the weighing function, indicated as $F_S(e^{j\omega})$, is defined as

$$F_S(e^{j\omega}) = \frac{1}{\sqrt{R_{x_1x_1}(e^{j\omega})R_{x_2x_2}(e^{j\omega})}} \tag{9.253}$$

for which (9.248) takes the form

$$\begin{aligned}\hat{r}_{y_1y_2}^{(S)}[n] &= \text{DTFT}^{-1}\left\{F_S(e^{j\omega})\hat{R}_{x_1x_2}(e^{j\omega})\right\} \\ &= \text{DTFT}^{-1}\left\{\frac{\hat{R}_{x_1x_2}(e^{j\omega})}{\sqrt{R_{x_1x_1}(e^{j\omega})R_{x_2x_2}(e^{j\omega})}}\right\} \\ &= \text{DTFT}^{-1}\left\{\hat{\gamma}_{x_1x_2}(e^{j\omega})\right\}.\end{aligned} \tag{9.254}$$

Note in fact that the estimated coherence function [see (9.67)] is defined as

$$\hat{\gamma}_{x_1x_2}(e^{j\omega}) = \frac{\hat{R}_{x_1x_2}(e^{j\omega})}{\sqrt{R_{x_1x_1}(e^{j\omega})R_{x_2x_2}(e^{j\omega})}}. \tag{9.255}$$

Setting $W_1(e^{j\omega}) = \left[R_{x_1x_1}(e^{j\omega})\right]^{-1/2}$ and $W_2(e^{j\omega}) = \left[R_{x_2x_2}(e^{j\omega})\right]^{-1/2}$, the SCOT method can be interpreted as a pre-whitening filtering performed before the cross-correlation computation.

Note that in (9.255), the PSD on the k-th sensor is assumed known, and, under certain conditions, this assumption may be reasonable. In the case where these conditions are not satisfied, even in this case, we can consider the estimated PSD and CPSD $\hat{R}_{x_kx_k}(e^{j\omega}) \sim R_{x_kx_k}(e^{j\omega})$, for $k = 1, 2$.

TDE as ML Estimation

In the ML estimation case, it is demonstrated that (see [38] for details), called $C_{x_1x_2}(e^{j\omega}) = \left|\gamma_{x_1x_2}(e^{j\omega})\right|^2$ the magnitude squared coherence (MSC), the weighting function can be defined as

$$F_{\mathrm{ML}}\left(e^{j\omega}\right) = \frac{C_{x_1x_2}(e^{j\omega})}{\left|R_{x_1x_2}(e^{j\omega})\right|\left[1 - C_{x_1x_2}(e^{j\omega})\right]}. \tag{9.256}$$

Therefore, substituting in (9.244), we get

$$r_{y_1y_2}^{(\mathrm{ML})}[n] = \mathrm{DTFT}^{-1}\left\{\frac{C_{x_1x_2}(e^{j\omega})}{\left[1 - C_{x_1x_2}(e^{j\omega})\right]} \cdot e^{j\hat{\phi}\left(e^{j\omega}\right)}\right\} \tag{9.257}$$

where

$$e^{j\hat{\phi}\left(e^{j\omega}\right)} = \frac{\hat{R}_{x_1x_2}(e^{j\omega})}{\left|R_{x_1x_2}(e^{j\omega})\right|}. \tag{9.258}$$

and in the case of additive uncorrelated noise is

$$e^{j\hat{\phi}\left(e^{j\omega}\right)} \sim e^{-j\omega\hat{D}} \tag{9.259}$$

for which the phase $\hat{\phi}\left(e^{j\omega}\right)$ represents a measure of time and is $\hat{D} = \hat{\phi}\left(e^{j\omega}\right)$.

The ML method assigns a large weight to the phase, especially in the frequency region in which the MSC is relatively large. Furthermore, the correlation (9.257) has its maximum value for $n = D$, i.e., when $e^{j\phi(e^{j\omega})}e^{j\omega D} = 1$.

Note that, in the estimated PSD and CPSD case, the method is said to be approximate maximum likelihood (AML).

The Phase Transform Method

In this method, the weighing function is defined as

$$F_P(e^{j\omega}) = \frac{1}{|R_{x_1x_2}(e^{j\omega})|} \tag{9.260}$$

and is indicated as phase transform (PHAT) for which (9.248) can be expressed as

$$\hat{r}^{(P)}_{y_1y_2}[n] = \text{DTFT}^{-1}\left\{\frac{\hat{R}_{x_1x_2}(e^{j\omega})}{|R_{x_1x_2}(e^{j\omega})|}\right\} \tag{9.261}$$

i.e., for (9.258) is $\hat{r}^{(P)}_{y_1y_2}[n] = \text{DTFT}^{-1}\left\{e^{j\hat{\phi}(e^{j\omega})}\right\}$. Then, it follows that for the signal model (9.239) and uncorrelated noise (i.e., $R_{n_1n_2}(e^{j\omega}) = 0$) we have

$$\left|R_{x_1x_2}(e^{j\omega})\right| = \alpha R_{ss}(e^{j\omega}) \tag{9.262}$$

and in this case, always for (9.258), we have $r^{(P)}_{y_1y_2}[n] = \delta\left[n - \hat{D}\right]$. In other words, the correlation provides a direct estimate of the delay.

Remark The PATH technique, for the signal model (9.239) and uncorrelated noise, ideally, does not suffer from spread as other methods. In practice, however, if $\hat{R}_{x_1x_2}(e^{j\omega}) \neq R_{x_1x_2}(e^{j\omega})$ then $e^{j\hat{\phi}(e^{j\omega})} \neq e^{-j\omega D}$, and the estimate of the correlation $\hat{r}^{(P)}_{y_1y_2}[n]$ is not a delta function. Other problems can arise when the energy of the input signal is small. In the event that, for some frequencies $R_{x_1x_2}(e^{j\omega}) = 0$, the phase $\phi(e^{j\omega})$ is undefined and the TDE is impossible. This suggests that for the function $F_P(e^{j\omega})$, a further weighting should be considered to compensate for the cases of absence of the input signal.

9.7.3.3 Steered Response Power PHAT Method

The steered response power PHAT (SRP-PHAT) method is the combination between the SRP and the GCC PHAT weighing method [47]. From a simple visual inspection of Fig. 9.55, we can see that this corresponds exactly to a two-channel conventional FSBF.

With reference to Fig. 9.56, generalizing the GCC PHAT weighing function to the P-channel case, from (9.246) it results as $F_{kp}(e^{j\omega}) = W_k(e^{j\omega})W_p^*(e^{j\omega})$. For which, for the FSBF's output power, we have that

$$P(\theta) = \sum_{k=1}^{P}\sum_{p=1}^{P} \frac{1}{2\pi}\int_{-\pi}^{\pi} F_{kp}(e^{j\omega})X_k(e^{j\omega})X_p^*(e^{j\omega})e^{j\omega\left(n_{\tau_k} - n_{\tau_p}\right)}d\omega \tag{9.263}$$

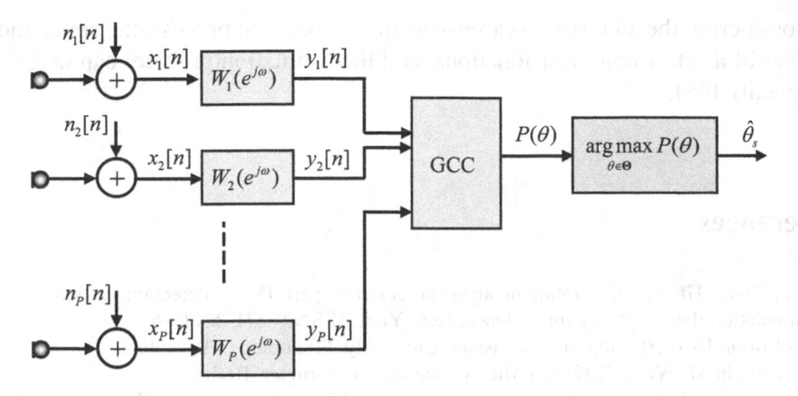

Fig. 9.56 SRP-PATH or P-channel GCC method

Note that for more than two channels, the PHAT weighing takes the form

$$F_{kp}\left(e^{j\omega}\right) = \frac{1}{\left|X_k(e^{j\omega})X_p^*(e^{j\omega})\right|} \tag{9.264}$$

that, in practice, for the FSBF corresponds to individual channel filters defined as

$$W_p\left(e^{j\omega}\right) = \frac{1}{\left|X_p(e^{j\omega})\right|}, \quad \text{for } p = 1,\ 2,\ \ldots,\ P \tag{9.265}$$

referred to as SRP-PHAT filters. Note that, by indicating with $r_{kp}[n_\tau]$ the GCC-PHAT between the $y_k[n]$ and $y_p[n]$ output, in the discrete time domain, (9.263) can be expressed as

$$P(\theta) = \sum_{k=1}^{P}\sum_{p=1}^{P} r_{kp}\left[n_{\tau_p} - n_{\tau_k}\right]. \tag{9.266}$$

The previous is the sum of the GCC, calculated between all possible permutations between inputs pairs, shifted by the difference between the steering delay.

Remark Knowing the location and the estimate delays between pairs of sensors, the determination of the position of the source can be formalized as an estimation problem. In fact, given the measured TDE τ_{ij} and known the coordinate of microphone position $m_p = \left[x_p \quad y_p \quad z_p\right]$ for $p = 1, \ldots, P$, the problem of localization can be formalized as the estimation of the radiating source coordinate $s_n = \left[x_n \quad y_n \quad z_n\right]$ for $n = 1, \ldots, Ns$.

Let $h_p(m_n) = \|D_p - D_n\| - \|D_n\|$ a *range difference function*, that by definition is nonlinear, and defining ε as *measurement errors*, we can write the relation as $d_{n1} = h_p(m_n) + \varepsilon_p$, for $p = 2, \ldots, P$.

Therefore, it is possible to define a set of $P - 1$ nonlinear equation system that in vector form can be written as $\mathbf{d} = \mathbf{h}(m_n) + \boldsymbol{\varepsilon}$.

Considering the additive $\boldsymbol{\varepsilon}$ as a zero mean-independent process, the signal model consists of a set of nonlinear functions, and the estimation problem can take some complexity [48].

References

1. Van Trees HL (2002) Optimum array processing: part IV of detection, estimation, and modulation theory. Wiley Interscience, New York, ISBN 0-471-22110-4
2. McCowan IA (2001) report. Queensland University Technology, Australia
3. Brandstein M, Wards D (2001) Microphone arrays. Springer, Berlin
4. Applebaum SP (1966) Adaptive arrays. Syracuse University Research Corp, Report SURC SPL TR 66-001 (reprinted in IEEE Trans on AP, vol AP-24, pp 585–598, Sept 1976)
5. Johnson D, Dudgeon D (1993) Array signal processing: concepts and techniques. Prentice-Hall, Englewood Cliffs, NJ
6. Haykin S, Ray Liu KJ (2009) Handbook on array processing and sensor networks. Wiley, New York. ISBN 978-0-470-37176-3
7. Seto WW (1971) Acoustics. McGraw-Hill, New York
8. Van Veen B, Buckley KM (1988) Beamforming a versatile approach to spatial filtering. IEEE Signal Process Mag 5(2):4–24
9. Fisher S, Kammeyer KD, Simmer KU (1996) Adaptive microphone arrays for speech enhancement in coherent and incoherent noise fields. In: 3rd Joint meeting of the acoustical society of America and the acoustical society of Japan, Honolulu, Hawaii, 2–6 Dec 1996
10. Krim H, Viberg M (1996) Two decades of array signal processing research—the parametric approach. IEEE Signal Process Mag 13(4):67–94
11. Elko GW (2001) Spatial coherence functions for differential microphones in isotropic fields. In: Brandstein M, Ward D (eds) Microphone arrays. Springer, Heidelberg, pp 61–85. ISBN 3-540-41953-5
12. Doclo S, Moonen M (2007) Superdirective beamforming robust against microphone mismatch. IEEE Trans Audio Speech Lang Process 15(2):617–631
13. Cox H, Zeskind RM, Kooij T (1986) Practical supergain. IEEE Trans ASSP ASSP-34(3): 393–398
14. Elko GW (2004) Differential microphones array. In: Huang Y, Benesty J (eds) Audio signal processing for next generation multimedia communication systems. Kluwer Academic, Dordrecht. ISBN 1-4020-7768-8
15. Kolundžija M, Faller C, Vetterli M (2011) Spatiotemporal gradient analysis of differential microphone arrays. J Audio Eng Soc 59(1/2):20–28
16. Buck M, Rößler M (2001) First order differential microphone arrays for automotive applications. In: International workshop on acoustic echo and noise control, Darmstadt, Germany, pp 19–22
17. Buck M (2002) Aspects of first-order differential microphone arrays in the presence of sensor imperfections. Eur Trans Telecommun 13(2):115–122
18. Benesty J, Chen J, Huang Y, Dmochowski J (2007) On microphone-array beamforming from a mimo acoustic signal processing perspective. IEEE Trans Audio Speech Lang Process 15(3): 1053–1065
19. Harris FJ (1978) On the use of windows for harmonic analysis with the discrete Fourier transform. Proc IEEE 66(1):51–84
20. Doclo S, Moonen M (2003) Design of broadband beamformers robust against gain and phase errors in the microphone array characteristics. IEEE Trans Signal Process 51(10):2511–2526

21. Chen H, Ser W (2009) Design of robust broadband beamformers with passband shaping characteristics using Tikhonov regularization. IEEE Trans Audio Speech Lang Proc 17(4):665–681
22. Capon J, Greenfield RJ, Kolker RJ (1967) Multidimensional maximum-likelihood processing of a large aperture seismic array. Proc IEEE 55:192–211
23. Capon J (1969) High resolution frequency-wavenumber spectrum analysis. Proc IEEE 57:1408–1418
24. Cox H, Zeskind RM, Owen MM (1987) Robust adaptive beamforming. IEEE Trans ASSP ASSP-35:1365–1375
25. Bitzer J, Kammeyer KD, Simmer KU (1999) An alternative implementation of the superdirective beamformer. In: Proceedings of 1999 I.E. workshop on applications of signal processing to audio and acoustics, New Paltz, New York
26. Trucco A, Traverso F, Crocco M (2013) Robust superdirective end-fire arrays. MTS/IEEE Oceans. doi: 10.1109/OCEANS-Bergen.2013.6607994
27. Zelinski R (1988) A microphone array with adaptive post-filtering for noise reduction in reverberant rooms. In: Proceedings of IEEE international conference on acoustics, speech, and signal processing, ICASSP-88, vol 5, pp 2578–2581
28. Marro C, Mahieux Y, Simmer K (1996) Performance of adaptive dereverberation techniques using directivity controlled arrays. In: Proceedings of European signal processing conference EUSIPCO96, Trieste, Italy, pp 1127–1130
29. Fisher S, Kammeyer KD (1997) Broadband beamforming with adaptive postfiltering for speech acquisition in noisy environments. In: Proceedings of IEEE international conference on acoustics, speech, and signal processing, ICASSP-97, vol 1, pp 359–362, April 1997
30. Frost OL III (1972) An algorithm for linearly constrained adaptive array processing. Proc IEEE 60(8):926–935
31. Buckley KM, Griffiths LJ (1986) An adaptive generalized sidelobe canceller with derivative constrains. IEEE Trans Antenn Propag AP34(3):311–319
32. Er MH, Cantoni A (1983) Derivative constraints for broad-band element space antenna array processors. IEEE Trans Antenn Propag AP31:1378–1393
33. Karray L, Martin A (2003) Toward improving speech detection robustness for speech recognition in adverse environments. Speech Commun 3:261–276
34. Mousazadeh S, Cohen I (2013) Voice activity detection in presence of transient noise using spectral clustering. IEEE Trans Audio Speech Lang Process 21(6):1261–1271
35. Joho M, Moschytz GS (1997) Adaptive beamforming with partitioned frequency-domain filters. In: IEEE workshop on applications of signal processing to audio and acoustics, New Palz, NY, USA, 19–22 Oct 1997
36. Bitzer J, Kammeyer KD, Simmer KU (1999) An alternative implementation of the superdirective beamformer. In: Proceedings of 1999 I.E. workshop on applications of signal processing to audio and acoustics, New Paltz, New York, 17–20 Oct 1999
37. Fischer S, Simmer KU (1996) Beamforming microphone arrays for speech acquisition in noisy environments. Speech Commun 20(3–4):215–227 (special issue on acoustic echo control and speech enhancement techniques)
38. Scarbrough K, Ahmed N, Carter GC (1981) On the simulation of a class of time delay estimation algorithms. IEEE Trans Acoust Speech Signal Process ASSP-29(3):534–540
39. Schmidt RO (1981) A signal subspace approach to multiple emitter location and spectral estimation. PhD dissertation, Stanford University, Stanford, CA
40. Li T, Nehorai A (2011) Maximum likelihood direction finding in spatially colored noise fields using sparse sensor arrays. IEEE Trans Signal Process 59:1048–1062
41. Schmidt RO (1986) Multiple emitter location and signal parameter estimation. IEEE Trans Antenn Propag 34(3):276–280
42. Stoica P, Nehorai A (1989) MUSIC, maximum likelihood and Cramér-Rao bound. IEEE Trans Acoust Speech Signal Process 37:720–741

43. Paulraj A, Roy R, Kailath T (1985) Estimation of signal parameters via rotational invariance techniques-ESPRIT. In: Proceedings of 19th Asilomar conference on signals systems, and computers. Asilomar, Pacific Grove, CA
44. Roy R, Kailath T (1989) ESPRIT—estimation of signal parameters via rotational invariance techniques. IEEE Trans Acoust Speech Signal Process 37(7):984–995
45. Special Issue (1981) Time delay estimation. IEEE Trans Acoust Speech Signal Process ASSP-29(3)
46. Knapp CH, Carter GC (1976) The generalized correlation method for estimation of time delay. IEEE Trans Acoust Speech Signal Process ASSP-24(4):320–327
47. Di Biase JH, Silverman HF, Brandstein MS (2001) Robust localization in reverberant rooms. In: Brandstein M, Ward D (eds) Microphone arrays: signal processing techniques and applications. Springer, Berlin. ISBN 3-540-42953-5
48. Huang Y, Benesty J, Chen J (2008) Time delay estimation and source localization. In: Springer handbook of speech processing. Springer, Berlin, ISBN: 978-3-540-49125-5

Appendix A: Linear Algebra Basics

A.1 Matrices and Vectors

A matrix **A** [1, 24, 25, 27], here indicated in bold capital letter, consists of a set of ordered elements arranged in rows and columns. A matrix with N rows and M columns is indicated with the following notations:

$$\mathbf{A} = \mathbf{A}_{N \times M} = \begin{bmatrix} a_{11} & a_{12} & \cdots & a_{1M} \\ a_{21} & a_{11} & \cdots & a_{2M} \\ \vdots & \vdots & \ddots & \vdots \\ a_{N1} & a_{N1} & \cdots & a_{NM} \end{bmatrix} \tag{A.1}$$

or

$$\mathbf{A} = \left[a_{ij}\right] \quad i = 1, 2, ..., N; \; j = 1, 2, ..., M, \tag{A.2}$$

where i and j are, respectively, row and column indices. The elements a_{ij} may be real or complex variables. An N rows and M columns $(N \times M)$ real matrix can be indicated as $\mathbf{A} \in \mathbb{R}^{N \times M}$ while for the complex case as $\mathbf{A} \in \mathbb{C}^{N \times M}$. When property holds both in the real and complex case, the matrix can be indicated as $\mathbf{A} \in (\mathbb{R}, \mathbb{C})^{N \times M}$ or as $\mathbf{A}\ (N \times M)$ or as simply $\mathbf{A}_{N \times M}$.

A.2 Notation, Preliminary Definitions, and Properties

A.2.1 Transpose and Hermitian Matrix

Given a matrix $\mathbf{A} \in \mathbb{R}^{N \times M}$ the *transpose matrix*, indicated as $\mathbf{A}^T \in \mathbb{R}^{M \times N}$, is obtained by interchanging the rows and columns of **A**, for which

A. Uncini, *Fundamentals of Adaptive Signal Processing*, Signals and Communication Technology, DOI 10.1007/978-3-319-02807-1,

$$\mathbf{A}^T = \begin{bmatrix} a_{11} & a_{21} & \cdots & a_{N1} \\ a_{12} & a_{11} & \cdots & a_{N2} \\ \vdots & \vdots & \ddots & \vdots \\ a_{1M} & a_{2M} & \cdots & a_{NM} \end{bmatrix} \tag{A.3}$$

or

$$\mathbf{A}^T = \left[a_{ji}\right] \qquad i = 1, 2, ..., N; \;\; j = 1, 2, ..., M. \tag{A.4}$$

It is therefore $(\mathbf{A}^{\mathrm{T}})^{\mathrm{T}} = \mathbf{A}$.

In the case of complex matrix $\mathbf{A} \in \mathbb{C}^{N \times M}$ we define Hermitian matrix the matrix transpose and complex conjugate

$$\mathbf{A}^H = \left[a_{ji}^*\right] \qquad i = 1, 2, ..., N; \;\; j = 1, 2, ..., M. \tag{A.5}$$

If the matrix is indicated as $\mathbf{A}(N \times M)$, the symbol $(^H)$ can be used to indicate both the transpose of the real case and the Hermitian of the complex case.

A.2.2 Row and Column Vectors of a Matrix

Given a matrix $\mathbf{A} \in (\mathbb{R},\mathbb{C})^{N \times M}$, its ith row vector is indicated as

$$\mathbf{a}_{i:} \in (\mathbb{R}, \mathbb{C})^{M \times 1} = \begin{bmatrix} a_{i1} & a_{i2} & \cdots & a_{iM} \end{bmatrix}^H \tag{A.6}$$

while the jth column vector as

$$\mathbf{a}_{:j} \in (\mathbb{R}, \mathbb{C})^{M \times 1} = \begin{bmatrix} a_{1j} & a_{2j} & \cdots & a_{Nj} \end{bmatrix}^H \tag{A.7}$$

A matrix $\mathbf{A} \in (\mathbb{R},\mathbb{C})^{N \times M}$ can be represented by its N row vectors as

$$\mathbf{A} = \begin{bmatrix} \mathbf{a}_{1:}^H \\ \mathbf{a}_{2:}^H \\ \vdots \\ \mathbf{a}_{N:}^H \end{bmatrix} = \begin{bmatrix} \mathbf{a}_{1:} & \mathbf{a}_{2:} & \cdots & \mathbf{a}_{N:} \end{bmatrix}^H \tag{A.8}$$

or by its M column vectors as

$$\mathbf{A} = \begin{bmatrix} \mathbf{a}_{:1} & \mathbf{a}_{:2} & \cdots & \mathbf{a}_{:M} \end{bmatrix} = \begin{bmatrix} \mathbf{a}_{:1}^H \\ \mathbf{a}_{:2}^H \\ \vdots \\ \mathbf{a}_{:M}^H \end{bmatrix}^H \tag{A.9}$$

Given a matrix $\mathbf{A} \in (\mathbb{R},\mathbb{C})^{N\times M}$ you can associate a vector $\text{vec}(\mathbf{A}) \in (\mathbb{R},\mathbb{C})^{NM\times 1}$ containing, stacked, all column vectors of $\mathbf{A}$

$$\begin{aligned} \text{vec}(\mathbf{A}) &= \begin{bmatrix} \mathbf{a}_{:1}^H & \mathbf{a}_{:2}^H & \cdots & \mathbf{a}_{:M}^H \end{bmatrix}_{NM\times 1}^H \\ &= \begin{bmatrix} a_{11},\ldots,a_{N1}, & a_{12},\ldots,a_{N2},\cdots\cdots,a_{1M},\ldots,a_{NM} \end{bmatrix}_{NM\times 1}^H. \end{aligned} \tag{A.10}$$

Remark Note that in Matlab you can extract entire columns or rows of a matrix with the following instructions:

`A(i,:)`, extracts the entire ith row in a row vector of dimension M;
`A(:,j)`, extracts the entire jth column in column vector of size N;
`A(:)`, extracts the entire matrix into a column vector of dimension $N \times M$.

A.2.3 Partitioned Matrices

Sometimes it can be useful to represent a matrix $\mathbf{A}_{(N+M)\times(P+Q)}$ in partitioned form of the type

$$\mathbf{A} = \begin{bmatrix} \mathbf{A}_{11} & \mathbf{A}_{12} \\ \mathbf{A}_{21} & \mathbf{A}_{22} \end{bmatrix}_{(M+N)\times(P+Q)} \tag{A.11}$$

in which the elements $\mathbf{A}_{ij}$ are in turn matrices defined as

$$\begin{aligned} &\mathbf{A}_{11} \in (\mathbb{R},\mathbb{C})^{M\times P}, \mathbf{A}_{12} \in (\mathbb{R},\mathbb{C})^{M\times Q} \\ &\mathbf{A}_{21} \in (\mathbb{R},\mathbb{C})^{N\times P}, \mathbf{A}_{22} \in (\mathbb{R},\mathbb{C})^{N\times Q} \end{aligned} \tag{A.12}$$

The partitioned product follows the same rules as the product of matrices. For example applies

$$\begin{bmatrix} \mathbf{A}_{11} & \mathbf{A}_{12} \\ \mathbf{A}_{21} & \mathbf{A}_{22} \end{bmatrix} \begin{bmatrix} \mathbf{B}_1 \\ \mathbf{B}_2 \end{bmatrix} = \begin{bmatrix} \mathbf{A}_{11}\mathbf{B}_1 + \mathbf{A}_{12}\mathbf{B}_2 \\ \mathbf{A}_{21}\mathbf{B}_1 + \mathbf{A}_{22}\mathbf{B}_2 \end{bmatrix} \tag{A.13}$$

Obviously, the dimensions of the partition matrices must be compatible.

A.2.4 Diagonal, Symmetric, Toeplitz, and Hankel Matrices

A given matrix $\mathbf{A} \in (\mathbb{R},\mathbb{C})^{N\times N}$ is called *diagonal* if $a_{ji} = 0$ for $i \neq j$. Is called *symmetric* if $a_{ji} = a_{ij}$ or $a_{ji} = a_{ij}^*$ if the complex case, whereby $\mathbf{A}^T = \mathbf{A}$ for real case and $\mathbf{A}^H = \mathbf{A}$ for complex case.

A matrix $\mathbf{A} \in (\mathbb{R},\mathbb{C})^{N\times N} = [a_{ij}]$ such that $[a_{i,j}] = [a_{i+1,j+1}] = [a_{i-j}]$ is *Toeplitz*, i.e., each descending diagonal from left to right is constant.

Moreover, a matrix $\mathbf{A} \in (\mathbb{R},\mathbb{C})^{N\times N} = [a_{ij}]$ such that $[a_{i,j}] = [a_{i-1,j+1}] = [a_{i+j}]$ is *Hankel*, i.e., each ascending diagonal from left to right is constant.

For example, the following $\mathbf{A}_T$, $\mathbf{A}_H$ matrices:

$$\mathbf{A}_T = \begin{bmatrix} a_i & a_{i-1} & a_{i-2} & a_{i-3} & \cdots \\ a_{i+1} & a_i & a_{i-1} & a_{i-2} & \ddots \\ a_{i+2} & a_{i+1} & a_i & a_{i-1} & \ddots \\ a_{i+3} & a_{i+2} & a_{i+1} & a_i & \ddots \\ \vdots & \ddots & \ddots & \ddots & \ddots \end{bmatrix}, \quad \mathbf{A}_H = \begin{bmatrix} a_{i-3} & a_{i-2} & a_{i-1} & a_i & \cdots \\ a_{i-2} & a_{i-1} & a_i & a_{i+1} & \iddots \\ a_{i-1} & a_i & a_{i+1} & a_{i+2} & \iddots \\ a_i & a_{i+1} & a_{i+2} & a_{i+3} & \iddots \\ \vdots & \iddots & \iddots & \iddots & \ddots \end{bmatrix} \tag{A.14}$$

are Toeplitz and Hankel matrices.

Given a vector $\mathbf{x} = \begin{bmatrix} x(0) & \cdots & x(M-1) \end{bmatrix}^T$, a special kind of Toeplitz/Hankel matrix, called *circulant matrix* obtained rotating the elements of $\mathbf{x}$ for each column (or row) as

$$\mathbf{A}_T = \begin{bmatrix} x(0) & x(3) & x(2) & x(1) \\ x(1) & x(0) & x(3) & x(2) \\ x(2) & x(1) & x(0) & x(3) \\ x(3) & x(2) & x(1) & x(0) \end{bmatrix}, \quad \mathbf{A}_H = \begin{bmatrix} x(0) & x(1) & x(2) & x(3) \\ x(1) & x(2) & x(3) & x(0) \\ x(2) & x(3) & x(0) & x(1) \\ x(3) & x(0) & x(1) & x(2) \end{bmatrix} \tag{A.15}$$

Remark The circulant matrices are important in DSP because they are diagonalized [see (A.9)] by a discrete Fourier transform, using a simple FFT algorithm.

A.2.5 Some Basic Properties

The following fundamental properties are valid:

$$\begin{aligned} (\mathbf{ABC}\cdots)^{-1} &= \mathbf{C}^{-1}\mathbf{B}^{-1}\mathbf{A}^{-1}\cdots \\ (\mathbf{A}^H)^{-1} &= (\mathbf{A}^{-1})^H \\ (\mathbf{A}+\mathbf{B})^H &= \mathbf{A}^H + \mathbf{B}^H \\ (\mathbf{AB})^H &= \mathbf{B}^H\mathbf{A}^H \\ (\mathbf{ABC}\cdots)^H &= \cdots\mathbf{C}^H\mathbf{B}^H\mathbf{A}^H. \end{aligned} \tag{A.16}$$

A.3 Inverse, Pseudoinverse, and Determinant of a Matrix

A.3.1 Inverse Matrix

A square matrix $\mathbf{A} \in (\mathbb{R},\mathbb{C})^{N\times N}$ is called *invertible* or *nonsingular* if there exists a matrix $\mathbf{B} \in (\mathbb{R},\mathbb{C})^{N\times N}$ such that $\mathbf{BA} = \mathbf{I}$, where $\mathbf{I}_{N\times N}$ is the so-called *identity matrix* or *unit matrix* defined as $\mathbf{I} = \text{diag}(1,1, \ldots,1)$. In such case the matrix $\mathbf{B}$ is uniquely determined from $\mathbf{A}$ and is defined as the *inverse* of $\mathbf{A}$, also indicated as $\mathbf{A}^{-1}$ (or $\mathbf{A}^{-1}\mathbf{A} = \mathbf{I}$).

Note that if $\mathbf{A}$ is nonsingular the system equation

$$\mathbf{Ax} = \mathbf{b} \tag{A.17}$$

has a unique solution, given by $\mathbf{x} = \mathbf{A}^{-1}\mathbf{b}$.

A.3.2 Generalized Inverse or Pseudoinverse of a Matrix

The *generalized inverse* or *Moore–Penrose pseudoinverse* of a matrix represents a general way to the determination of the solution of a linear real or complex system equations of the type (A.17), in the case of $\mathbf{A} \in (\mathbb{R},\mathbb{C})^{N\times M}$, $\mathbf{x} \in (\mathbb{R},\mathbb{C})^{M\times 1}$, $\mathbf{b} \in (\mathbb{R},\mathbb{C})^{N\times 1}$. In general terms, considering a generic matrix $\mathbf{A}_{N\times M}$ we can define its pseudoinverse $\mathbf{A}^{\#}_{M\times N}$ a matrix such that the following four properties are true:

$$\begin{aligned} \mathbf{AA}^{\#}\mathbf{A} &= \mathbf{A} \\ \mathbf{A}^{\#}\mathbf{AA}^{\#} &= \mathbf{A}^{\#} \end{aligned} \tag{A.18}$$

and

$$\begin{aligned} \mathbf{AA}^{\#} &= \left(\mathbf{AA}^{\#}\right)^{H} \\ \mathbf{A}^{\#}\mathbf{A} &= \left(\mathbf{A}^{\#}\mathbf{A}\right)^{H}. \end{aligned} \tag{A.19}$$

Given a linear system (A.17) for its solution we can distinguish the following three cases:

$$\mathbf{A}^{\#} = \begin{cases} \mathbf{A}^{-1} & N = M, \quad \textit{square matrix} \\ \mathbf{A}^{H}\left(\mathbf{AA}^{H}\right)^{-1} & N < M, \quad \textit{“fat” matrix} \\ \left(\mathbf{A}^{H}\mathbf{A}\right)^{-1}\mathbf{A}^{H} & N > M, \quad \textit{“tall” matrix} \end{cases} \tag{A.20}$$

where by the solution of the system (A.17) may always be expressed as

$$\mathbf{x} = \mathbf{A}^{\#}\mathbf{b}. \tag{A.21}$$

The proof of (A.20) for the case of a square and fat matrix is immediate. The case of tall matrix can be easily demonstrated after the introduction of SVD decomposition presented below. Different method for calculating the pseudoinverse refers to possible decompositions of the matrix **A**.

A.3.3 Determinant

Given square matrix $\mathbf{A}_{N\times N}$ the *determinant*, indicated as $\det(\mathbf{A})$ or $\Delta_{\mathbf{A}}$, is a scalar value associated with the matrix itself, which summarizes some of its fundamental properties, calculated by the following rule.

If $\mathbf{A} = a \in \mathbb{R}^{1\times 1}$, by definition the determinant is $\det(\mathbf{A}) = a$. The determinant of a square matrix $\mathbf{A} \in \mathbb{R}^{N\times N}$ is defined in terms of the determinant of order $N-1$ with the following recursive expression:

$$\det(\mathbf{A}) = \sum_{j=1}^{N} a_{ij}\left[(-1)^{j+i}\det\left(\mathbf{A}_{ij}\right)\right], \tag{A.22}$$

where $\mathbf{A}_{ij} \in \mathbb{R}^{(N-1)\times(N-1)}$ is a matrix obtained by eliminating the ith row and the jth column of **A**.

Moreover, it should be noted that the value $\det(\mathbf{A}_{ij})$ is called *complementary minor* of a_{ij}, and the product $(-1)^{j+i}\det(\mathbf{A}_{ij})$ is called *algebraic complement* of the element a_{ij}.

Property Given the matrices $\mathbf{A}_{N\times N}$ and $\mathbf{B}_{N\times N}$ the following properties are valid:

$$\begin{aligned}
\det(\mathbf{A}) &= \prod\nolimits_i \lambda_i, \quad \lambda_i = \mathrm{eig}(\mathbf{A}) \\
\det(\mathbf{AB}) &= \det(\mathbf{A})\det(\mathbf{B}) \\
\det\left(\mathbf{A}^H\right) &= \det(\mathbf{A})^* \\
\det\left(\mathbf{A}^{-1}\right) &= 1/\det(\mathbf{A}) \\
\det(c\mathbf{A}) &= c^N \det(\mathbf{A}) \\
\det\left(\mathbf{I} + \mathbf{ab}^H\right) &= \left(1 + \mathbf{a}^H\mathbf{b}\right)* \\
\det(\mathbf{I} + \delta\mathbf{A}) &\cong 1 + \det(\mathbf{A}) + \delta\mathrm{Tr}(\mathbf{A}) \\
&\quad + \tfrac{1}{2}\delta^2\mathrm{Tr}(\mathbf{A})^2 - \tfrac{1}{2}\delta^2\mathrm{Tr}\left(\mathbf{A}^2\right) \quad \text{for small } \delta.
\end{aligned} \tag{A.23}$$

A matrix $\mathbf{A}_{N\times N}$ with $\det(\mathbf{A}) \neq 0$ is called *nonsingular* and is always invertible. Note that the determinant of a diagonal or triangular matrix is the product of the values on the diagonal.

A.3.4 *Matrix Inversion Lemma*

Very useful in the development of adaptive algorithms, the matrix inversion lemma (MIL) (also known as the Sherman–Morrison–Woodbury formula [1, 2]) states that: if $\mathbf{A}^{-1}$ and $\mathbf{C}^{-1}$ exist, the following equation algebraically verifiable is true[1]:

$$[\mathbf{A}+\mathbf{BCD}]^{-1} = \mathbf{A}^{-1} - \mathbf{A}^{-1}\mathbf{B}\left[\mathbf{C}^{-1}+\mathbf{DA}^{-1}\mathbf{B}\right]^{-1}\mathbf{DA}^{-1}, \tag{A.24}$$

where $\mathbf{A} \in \mathbb{C}^{M\times M}$, $\mathbf{B} \in \mathbb{C}^{M\times N}$, $\mathbf{C} \in \mathbb{C}^{N\times N}$, and $\mathbf{D} \in \mathbb{C}^{N\times M}$. Note that (A.24) has numerous variants the first of which, for simplifying, is that for $\mathbf{D} = \mathbf{B}^H$

$$\left[\mathbf{A}+\mathbf{BCB}^H\right]^{-1} = \mathbf{A}^{-1} - \mathbf{A}^{-1}\mathbf{B}\left[\mathbf{C}^{-1}+\mathbf{B}^H\mathbf{A}^{-1}\mathbf{B}\right]^{-1}\mathbf{B}^H\mathbf{A}^{-1} \tag{A.25}$$

The Kailath's variant is defined for $\mathbf{D} = \mathbf{I}$, in which (A.24) takes the form

$$[\mathbf{A}+\mathbf{BC}]^{-1} = \mathbf{A}^{-1} - \mathbf{A}^{-1}\mathbf{B}\left[\mathbf{I}+\mathbf{CA}^{-1}\mathbf{B}\right]^{-1}\mathbf{CA}^{-1} \tag{A.26}$$

A variant of the previous one is when the matrices $\mathbf{B}$ and $\mathbf{D}$ are vectors, or for $\mathbf{B} \rightarrow \mathbf{b}, \in \mathbb{C}^{M\times 1}$, $\mathbf{D} \rightarrow \mathbf{d}^H \in \mathbb{C}^{1\times M}$, and $\mathbf{C} = \mathbf{I}$, for which (A.24) becomes

$$\left[\mathbf{A}+\mathbf{bd}^H\right]^{-1} = \mathbf{A}^{-1} - \frac{\mathbf{A}^{-1}\mathbf{bd}^H\mathbf{A}^{-1}}{1+\mathbf{d}^H\mathbf{A}^{-1}\mathbf{b}}. \tag{A.27}$$

A case of particular interest in adaptive filtering is when in the above we have $\mathbf{d} = \mathbf{b}^H$.

In all variants the inverse of the sum $\mathbf{A} + \mathbf{BCD}$ is a function of the inverse of the matrix $\mathbf{A}$. It should be noted, in fact, that the term that appears in the denominator of (A.27) is a scalar value.

A.4 Inner and Outer Product of Vectors

Given two vectors $\mathbf{x} \in (\mathbb{R},\mathbb{C})^{N\times 1}$ and $\mathbf{w} \in (\mathbb{R},\mathbb{C})^{N\times 1}$ we define *inner product* (or *scalar product* or sometime *dot product*) indicated as $\langle \mathbf{x},\mathbf{w}\rangle \in (\mathbb{R},\mathbb{C})$; the product is defined as

---

[1] The algebraic verification can be done developing the following expression:

$$\begin{aligned}
&[\mathbf{A}+\mathbf{BCD}]\left[\mathbf{A}^{-1} - \mathbf{A}^{-1}\mathbf{B}\left[\mathbf{C}^{-1}+\mathbf{DA}^{-1}\mathbf{B}\right]^{-1}\mathbf{DA}^{-1}\right] \\
&\quad = \mathbf{I}+\mathbf{BCDA}^{-1} - \mathbf{B}\left[\mathbf{C}^{-1}+\mathbf{DA}^{-1}\mathbf{B}\right]^{-1}\mathbf{DA}^{-1} - \mathbf{BCDA}^{-1}\mathbf{B}\left[\mathbf{C}^{-1}+\mathbf{DA}^{-1}\mathbf{B}\right]^{-1}\mathbf{DA}^{-1} \\
&\quad = \ldots \\
&\quad = \mathbf{I}.
\end{aligned}$$

$$\langle \mathbf{x}, \mathbf{w} \rangle = \mathbf{x}^H \mathbf{w} = \sum_{i=1}^{N} x_i^* w_i. \tag{A.28}$$

The *outer product* between two vectors $\mathbf{x} \in (\mathbb{R},\mathbb{C})^{M\times 1}$ and $\mathbf{w} \in (\mathbb{R},\mathbb{C})^{N\times 1}$, denoted as $\rangle\mathbf{x}, \mathbf{w}\langle \in (\mathbb{R},\mathbb{C})^{M\times N}$, is a matrix defined by the product

$$\rangle\mathbf{x}, \mathbf{w}\langle = \mathbf{x}\mathbf{w}^H = \begin{bmatrix} x_1 w_1^* & \cdots & x_1 w_N^* \\ \vdots & \ddots & \vdots \\ x_M w_1^* & \cdots & x_M w_N^* \end{bmatrix}_{M\times N}. \tag{A.29}$$

Given two matrices $\mathbf{A}_{N\times M}$ and $\mathbf{B}_{P\times M}$, represented by the respective column vectors

$$\begin{aligned} \mathbf{A} &= \begin{bmatrix} \mathbf{a}_{:1} & \mathbf{a}_{:2} & \cdots & \mathbf{a}_{:M} \end{bmatrix}_{1(N)\times M} \\ \mathbf{B} &= \begin{bmatrix} \mathbf{b}_{:1} & \mathbf{b}_{:2} & \cdots & \mathbf{b}_{:M} \end{bmatrix}_{1(P)\times M} \end{aligned} \tag{A.30}$$

with $\mathbf{a}_{:j} = \begin{bmatrix} a_{1j} & a_{2j} & \cdots & a_{Nj} \end{bmatrix}^T$ and $\mathbf{b}_{:j} = \begin{bmatrix} b_{1j} & b_{2j} & \cdots & b_{Pj} \end{bmatrix}^T$, we define the *matrix outer product* as

$$\mathbf{A}\mathbf{B}^H \in (\mathbb{R},\mathbb{C})^{N\times P} = \sum_{i=1}^{M} \mathbf{a}_{i:} \mathbf{b}_{i:}^H \tag{A.31}$$

Note that the above expression indicates the sum of the outer product of the column vectors of the respective matrices.

A.4.1 Geometric Interpretation

The inner product of a vector for itself $\mathbf{x}^H\mathbf{x}$ is often referred to as

$$\|\mathbf{x}\|_2^2 \triangleq \langle \mathbf{x}, \mathbf{x} \rangle = \mathbf{x}^H \mathbf{x} \tag{A.32}$$

that, as better specified below, corresponds to the square of its length in a Euclidean space. Moreover, in Euclidean geometry, the inner product of vectors expressed in an orthonormal basis is related to their length and angle.

Let $\|\mathbf{x}\| \triangleq \sqrt{\|\mathbf{x}\|_2^2}$ the length of $\mathbf{x}$, if $\mathbf{w}$ is another vector, such that θ is the angle between $\mathbf{x}$ and $\mathbf{w}$ we have

$$\mathbf{x}^H \mathbf{w} = \|\mathbf{x}\| \cdot \|\mathbf{w}\| \cos\theta. \tag{A.33}$$

A.5 Linearly Independent Vectors

Given a set of vectors in $(\mathbb{R},\mathbb{C})^P$, $\{\mathbf{a}_i\}$, $\{\mathbf{a}_i \in (\mathbb{R},\mathbb{C})^P, \forall\, i,\, i = 1, \ldots, N\}$ and a set of scalars $c_1, c_2, \ldots, c_N$, we define the vector $\mathbf{b} \in (\mathbb{R},\mathbb{C})^P$ as a linear combination of the vectors $\{\mathbf{a}_i\}$ as

$$\mathbf{b} = \sum_{i=1}^{N} c_i \mathbf{a}_i. \tag{A.34}$$

The vectors $\{\mathbf{a}_i\}$ are defined as *linearly independent* if, and only if, (A.34) is zero only in the case that all scalars c_i are zero.

Equivalently, the vectors are called *linearly dependent* if, given a set of scalars $c_1, c_2, \ldots, c_N$, not all zero,

$$\sum_{i=1}^{N} c_i \mathbf{a}_i = \mathbf{0}. \tag{A.35}$$

Note that the columns of the matrix $\mathbf{A}$ are linearly independent if, and only if, the matrix $(\mathbf{A}^H\mathbf{A})$ is nonsingular or, as explained in the next section, is a *full rank matrix*. Similarly, the rows of the matrix $\mathbf{A}$ are linearly independent if, and only if, $(\mathbf{A}\mathbf{A}^H)$ is nonsingular.

A.6 Rank and Subspaces Associated with a Matrix

Given $\mathbf{A}_{N\times M}$, the *rank of the matrix* $\mathbf{A}$, indicated as $r = \text{rank}(\mathbf{A})$, is defined as the scalar indicating the maximum number of its linearly independent columns. Note that $\text{rank}(\mathbf{A}) = \text{rank}(\mathbf{A}^H)$; it follows that a matrix is called *reduced rank matrix* when $\text{rank}(\mathbf{A}) < \min(N,M)$ and is *full rank matrix* when $\text{rank}(\mathbf{A}) = \min(N,M)$. It is also

$$\text{rank}(\mathbf{A}) = \text{rank}\left(\mathbf{A}^H\right) = \text{rank}\left(\mathbf{A}^H\mathbf{A}\right) = \text{rank}\left(\mathbf{A}\mathbf{A}^H\right). \tag{A.36}$$

A.6.1 Range or Column Space of a Matrix

We define *column space* of a matrix $\mathbf{A}_{N\times M}$ (also called *range* or *image*), indicated as $\mathcal{R}(\mathbf{A})$ o $\text{Im}(\mathbf{A})$, the subspace obtained from the set of all possible linear combinations of its linearly independent column vectors. So, called $\mathbf{A} = [\,\mathbf{a}_1 \ \cdots \ \mathbf{a}_M\,]$ the columns partition of the matrix, $\mathcal{R}(\mathbf{A})$ represents the *linear span*[2] (also called the *linear hull*) of the column vectors set in a vector space

[2] The term span $(v_1, v_2, \ldots, v_n)$ is the set of all vectors, or the space, that can be represented as the linear combination of $v_1, v_2, \ldots, v_n$.

$$\begin{aligned}\mathcal{R}(\mathbf{A}) &\triangleq \text{span}\begin{bmatrix}\mathbf{a}_1 & \mathbf{a}_2 & \cdots & \mathbf{a}_M\end{bmatrix} \\ &= \left\{\mathbf{y}\in\mathbb{R}^N \therefore \ \mathbf{y}=\mathbf{A}\mathbf{x}, \quad \text{for} \quad \text{some} \ \ \mathbf{x}\in\mathbb{R}^N\right\}.\end{aligned} \tag{A.37}$$

Moreover, calling $\mathbf{A} = [\,\mathbf{b}_1 \quad \cdots \quad \mathbf{b}_N\,]$ the row matrix partition, the dual definition is

$$\begin{aligned}\mathcal{R}\left(\mathbf{A}^H\right) &\triangleq \text{span}\begin{bmatrix}\mathbf{b}_1 & \mathbf{b}_2 & \cdots & \mathbf{b}_N\end{bmatrix} \\ &= \left\{\mathbf{x}\in\mathbb{R}^N \therefore \mathbf{x}=\mathbf{A}\mathbf{y}, \quad \text{for} \quad \text{some} \ \ \mathbf{y}\in\mathbb{R}^M\right\}.\end{aligned} \tag{A.38}$$

It appears, for the previous definition, that the rank of $\mathbf{A}$ is equal to the size of its column space

$$\text{rank}(\mathbf{A}) = \dim\left[\mathcal{R}(\mathbf{A})\right]. \tag{A.39}$$

A.6.2 Kernel or Nullspace of a Matrix

The *kernel* or *nullspace* of matrix $\mathbf{A}_{N\times M}$, indicated as $\mathcal{N}(\mathbf{A})$ or $\text{Ker}(\mathbf{A})$, is the set of all vector $\mathbf{x}$ for which $\mathbf{A}\mathbf{x} = \mathbf{0}$. More formally

$$\mathcal{N}(\mathbf{A}) \triangleq \left\{\mathbf{x}\in(\mathbb{R},\mathbb{C})^M \therefore \ \mathbf{A}\mathbf{x}=\mathbf{0}\right\}. \tag{A.40}$$

Similarly, the dual definition of *left nullspace* is

$$\mathcal{N}\left(\mathbf{A}^H\right) \triangleq \left\{\mathbf{y}\in(\mathbb{R},\mathbb{C})^N \therefore \mathbf{A}^H\mathbf{y}=\mathbf{0}\right\}. \tag{A.41}$$

The size of the kernel is called *nullity of the matrix*

$$\text{null}(\mathbf{A}) = \dim\left[\mathcal{N}(\mathbf{A})\right]. \tag{A.42}$$

In fact, the expression $\mathbf{A}\mathbf{x} = \mathbf{0}$ is equivalent to a homogeneous linear equations system and is equivalent to the span of the solutions of that system. Whereby calling $\mathbf{A} = [\,\mathbf{a}_1 \quad \cdots \quad \mathbf{a}_N\,]^H$ the rows partition of $\mathbf{A}$, the product $\mathbf{A}\mathbf{x} = \mathbf{0}$ can be expressed as

$$\mathbf{A}\mathbf{x} = \mathbf{0} \Leftrightarrow \begin{bmatrix}\mathbf{a}_1^H\mathbf{x} \\ \mathbf{a}_2^H\mathbf{x} \\ \vdots \\ \mathbf{a}_N^H\mathbf{x}\end{bmatrix} = \mathbf{0} \tag{A.43}$$

It follows that $\mathbf{x}\in\mathcal{N}(\mathbf{A})$ if, and only if, $\mathbf{x}$ is orthogonal to the space described by the row vectors of $\mathbf{A}$, or

$$\mathbf{x}\perp\text{span}\begin{bmatrix}\mathbf{a}_1 & \mathbf{a}_2 & \cdots & \mathbf{a}_N\end{bmatrix}$$

Namely, a vector $\mathbf{x}$ is located in the nullspace of $\mathbf{A}$ iff it is perpendicular to every vector in the space of row $\mathbf{A}$. In other words, the column space of the matrix $\mathbf{A}$ is orthogonal to its nullspace $\mathcal{R}(\mathbf{A}) \perp \mathcal{N}(\mathbf{A})$.

A.6.3 Rank–Nullity Theorem

For any matrix $\mathbf{A}_{N\times M}$,

$$\dim\left[\mathcal{N}(\mathbf{A})\right] + \dim\left[\mathcal{R}(\mathbf{A})\right] = \text{null}(\mathbf{A}) + \text{rank}(\mathbf{A}) = M. \tag{A.44}$$

The above equation is known as *rank–nullity theorem.*

A.6.4 The Four Fundamental Matrix Subsapces

When the matrix $\mathbf{A}_{N\times M}$ is full rank, i.e., $r = \text{rank}(\mathbf{A}) = \min(N,M)$, the matrix always admits a left-inverse $\mathbf{B}$ or an right-inverse $\mathbf{C}$ or, in the case of $N = M$, admits the inverse $\mathbf{A}^{-1}$.

As a corollary, it is appropriate to recall the fundamental concepts related to the subspaces definable for a matrix $\mathbf{A}_{N\times M}$

1. *Column space* of $\mathbf{A}$: indicted as $\mathcal{R}(\mathbf{A})$, is defined by the $\mathbf{A}$ columns span.
2. *Nullspace* of $\mathbf{A}$: indicted as $\mathcal{N}(\mathbf{A})$, contains all vectors $\mathbf{x}$ such that $\mathbf{Ax} = \mathbf{0}$.
3. *Row space* of $\mathbf{A}$: equivalent to the column space of $\mathbf{A}^H$, indicated as $\mathcal{R}(\mathbf{A}^H)$, is defined by the span of the rows of $\mathbf{A}$.
4. *Left nullspace* of $\mathbf{A}$: equivalent to the nullspace of $\mathbf{A}^H$, indicated as $\mathcal{N}(\mathbf{A}^H)$, contains all vectors x such that $\mathbf{A}^H\mathbf{x} = \mathbf{0}$.

Indicating with $\mathcal{R}^\perp(\mathbf{A})$ and $\mathcal{N}^\perp(\mathbf{A})$ the orthogonal complements, respectively, of $\mathcal{R}(\mathbf{A})$ and $\mathcal{N}(\mathbf{A})$, the following relations are valid (Fig. A.1):

$$\begin{aligned} \mathcal{R}(\mathbf{A}) &= \mathcal{N}^\perp(\mathbf{A}^H) \\ \mathcal{N}(\mathbf{A}) &= \mathcal{R}^\perp(\mathbf{A}^H) \end{aligned} \tag{A.45}$$

and the dual

$$\begin{aligned} \mathcal{R}^\perp(\mathbf{A}) &= \mathcal{N}(\mathbf{A}^H) \\ \mathcal{N}^\perp(\mathbf{A}) &= \mathcal{R}(\mathbf{A}^H). \end{aligned} \tag{A.46}$$

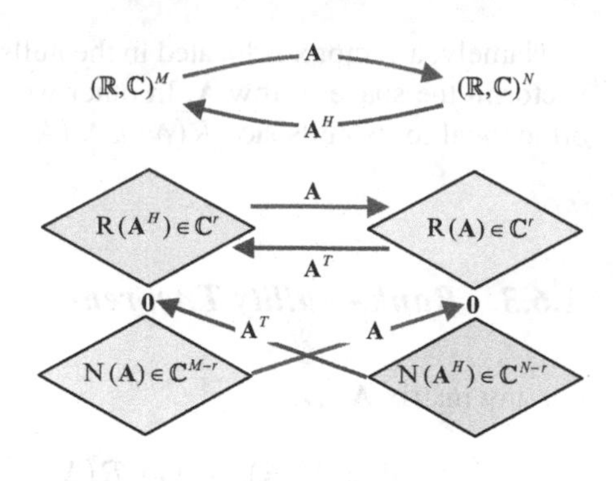

Fig. A.1 The four subspaces associated with the matrix $\mathbf{A} \in (\mathbb{R},\mathbb{C})^{N \times M}$. These subspaces determine an orthogonal decomposition of the space, into the column space $\mathcal{R}(\mathbf{A})$, and the left nullspace $\mathcal{N}(\mathbf{A}^H)$. Similarly an orthogonal decomposition of $(\mathbb{R},\mathbb{C})^N$ into the row space $\mathcal{R}(\mathbf{A}^H)$ and the nullspace $\mathcal{N}(\mathbf{A})$

A.6.5 Projection Matrix

A square matrix $\mathbf{P} \in (\mathbb{R},\mathbb{C})^{N \times N}$ is defined *projection operator* iff $\mathbf{P}^2 = \mathbf{P}$, i.e., is *idempotent*. If $\mathbf{P}$ is symmetric, then the projection is orthogonal. Furthermore, if $\mathbf{P}$ is a projection matrix, it is also $(\mathbf{I}\text{–}\mathbf{P})$.

Examples of orthogonal projection matrices are matrices associated with the pseudoinverse $\mathbf{A}^\#$ in the over- and under-determined cases.

In the case of overdetermined case $N > M$ and $\mathbf{A}^\# = (\mathbf{A}^H\mathbf{A})^{-1}\mathbf{A}^H$, we have that

$$\mathbf{P} = \mathbf{A}\left(\mathbf{A}^H\mathbf{A}\right)^{-1}\mathbf{A}^H, \qquad \textit{projection operator} \tag{A.47}$$

$$\mathbf{P}^\perp = \mathbf{I} - \mathbf{A}\left(\mathbf{A}^H\mathbf{A}\right)^{-1}\mathbf{A}^H \qquad \textit{orthogonal complement projection oper.} \tag{A.48}$$

such that $\mathbf{P} + \mathbf{P}^\perp = \mathbf{I}$, i.e., $\mathbf{P}$ projects a vector on the subspace $\Psi = \mathcal{R}(\mathbf{A})$, while $\mathbf{P}^\perp$ on its orthogonal complement $\Psi^\perp = \mathcal{R}^\perp(\mathbf{A})$ or $\sum = \mathcal{N}\left(\mathbf{A}^H\right)$. Indeed, calling $\mathbf{x} \in (\mathbb{R},\mathbb{C})^{M \times 1}$ and $\mathbf{y} \in (\mathbb{R},\mathbb{C})^{N \times 1}$, such that $\mathbf{A}\mathbf{x} = \mathbf{y}$, we have that $\mathbf{P}\mathbf{y} = \mathbf{u}$ and $\mathbf{P}^\perp\mathbf{y} = \mathbf{v}$ such that $\mathbf{u} \in \mathcal{R}(\mathbf{A})$ and $\mathbf{v} \in \mathcal{N}\left(\mathbf{A}^H\right)$ (see Fig. A.2).

In the underdetermined, case where $N < M$ and $\mathbf{A}^\# = \mathbf{A}^H(\mathbf{A}\mathbf{A}^H)^{-1}$, we have

$$\mathbf{P} = \mathbf{A}^H\left(\mathbf{A}\mathbf{A}^H\right)^{-1}\mathbf{A} \tag{A.49}$$

$$\mathbf{P}^\perp = \mathbf{I} - \mathbf{A}^H\left(\mathbf{A}\mathbf{A}^H\right)^{-1}\mathbf{A}. \tag{A.50}$$

A.7 Orthogonality and Unitary Matrices

In DSP, the conditions of orthogonality, orthonormality, and bi-orthogonality, represent a tool of primary importance. Here are some basic definitions.

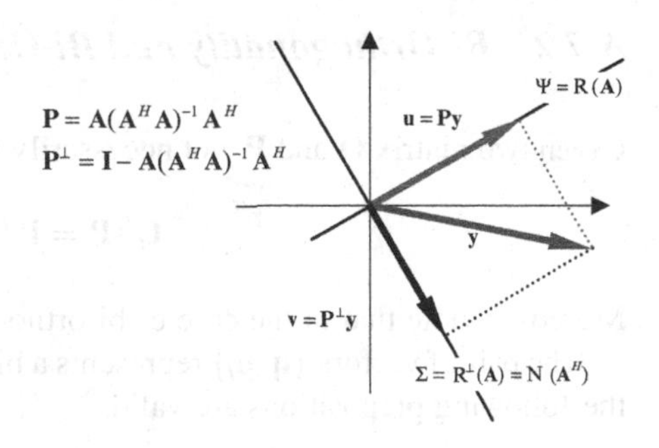

Fig. A.2 Representation of the orthogonal projection operator

A.7.1 Orthogonality and Unitary Matrices

Two vectors **x** and **w** $\{\mathbf{x},\ \mathbf{w} \in (\mathbb{R},\mathbb{C})^N\}$ are orthogonal if their inner product is zero $\langle \mathbf{x},\mathbf{w}\rangle = 0$. This situation is sometimes referred to as $\mathbf{x} \perp \mathbf{w}$.

A set of vectors $\{\mathbf{q}_i\}$, $\left(\mathbf{q}_i \in (\mathbb{R},\mathbb{C})^N,\ \forall\ i,\ i = 1, \ldots, N\right)$ is called *orthogonal* if

$$\mathbf{q}_i^H \mathbf{q}_j = 0 \quad \text{for} \ \ i \neq j \tag{A.51}$$

A set of vectors $\{\mathbf{q}_i\}$ is called *orthonormal* if

$$\mathbf{q}_i^H \mathbf{q}_j = \delta_{ij} = \delta[i - j], \tag{A.52}$$

where δ_{ij} is the Kronecker symbol defined as $\delta_{ij} = 1$ for $i = j$; $\delta_{ij} = 0$ for $i \neq j$.

A matrix $\mathbf{Q}_{N \times N}$ is orthonormal if its columns are an orthonormal set of vectors. Formally

$$\mathbf{Q}^H \mathbf{Q} = \mathbf{Q}\mathbf{Q}^H = \mathbf{I}. \tag{A.53}$$

Note that in the case of orthonormality is $\mathbf{Q}^{-1} = \mathbf{Q}^H$. Moreover, a matrix for which $\mathbf{Q}^H \mathbf{Q} = \mathbf{Q}\mathbf{Q}^H$ is defined as *normal matrix*.

An important property of orthonormality is that it has no effect on inner product, which is

$$\langle \mathbf{Q}\mathbf{x}, \mathbf{Q}\mathbf{y}\rangle = (\mathbf{Q}\mathbf{x})^H \mathbf{Q}\mathbf{y} = \mathbf{x}^H \mathbf{Q}^H \mathbf{Q}\mathbf{y} = \langle \mathbf{x}, \mathbf{y}\rangle. \tag{A.54}$$

Furthermore, if multiplied to a vector does not change its length

$$\|\mathbf{Q}\mathbf{x}\|_2^2 = (\mathbf{Q}\mathbf{x})^H \mathbf{Q}\mathbf{x} = \mathbf{x}^H \mathbf{Q}^H \mathbf{Q}\mathbf{x} = \|\mathbf{x}\|_2^2. \tag{A.55}$$

A.7.2 Bi-Orthogonality and Bi-Orthogonal Bases

Given two matrix $\mathbf{Q}$ and $\mathbf{P}$, not necessarily square, these are called bi-orthogonal if

$$\mathbf{Q}^H\mathbf{P} = \mathbf{P}^H\mathbf{Q} = \mathbf{I}. \tag{A.56}$$

Moreover, note that in the case of bi-orthonormality $\mathbf{Q}^H = \mathbf{P}^{-1}$ and $\mathbf{P}^H = \mathbf{Q}^{-1}$.

The pair of vectors $\{\mathbf{q}_i,\mathbf{p}_j\}$ represents a bi-orthogonal basis if, and only if, both of the following prepositions are valid:

1. For each $i,\, j \in \mathbb{Z}$

$$\left\langle \mathbf{q}_i \mathbf{p}_j \right\rangle = \delta[i-j] \tag{A.57}$$

2. There are $A, B, \tilde{A}, \tilde{B} \in \mathbb{R}^+$ such that $\forall\ \mathbf{x} \in E$; the following inequalities are valid:

$$A\|\mathbf{x}\|^2 \le \sum\nolimits_k \left|\langle \mathbf{q}_k, \mathbf{x}\rangle\right|^2 \le B\|\mathbf{x}\|^2 \tag{A.58}$$

$$\tilde{A}\|\mathbf{x}\|^2 \le \sum\nolimits_k \left|\langle \mathbf{p}_k, \mathbf{x}\rangle\right|^2 \le \tilde{B}\|\mathbf{x}\|^2. \tag{A.59}$$

The pair of vectors that satisfy (1.) and inequality (2.) are called Riesz bases [2]. For which the following expansion formulas apply:

$$\mathbf{x} = \sum\nolimits_k \langle \mathbf{q}_k, \mathbf{x}\rangle \mathbf{p}_k = \sum\nolimits_k \langle \mathbf{p}_k, \mathbf{x}\rangle \mathbf{q}_k. \tag{A.60}$$

Comparing the previous inequalities with (A.52), we observe that the term bi-orthogonal is used as the non-orthogonal basis $\{\mathbf{q}_i\}$ and is associated with a dual basis $\{\mathbf{p}_j\}$ that satisfies the condition (A.57). If $\{\mathbf{p}_i\}$ was the orthogonal expansion (A.60) would be the usual orthogonal expansion.

A.7.3 Paraunitary Matrix

A matrix $\mathbf{Q} \in (\mathbb{R},\mathbb{C})^{N\times M}$ is called *paraunitary matrix* if

$$\mathbf{Q} = \mathbf{Q}^H \tag{A.61}$$

In the case of the square matrix then

$$\mathbf{Q}^H\mathbf{Q} = c\mathbf{I} \tag{A.62}$$

A.8 Eigenvalues and Eigenvectors

The *eigenvalues* of a square matrix $\mathbf{A}_{N\times N}$ are the solutions of the characteristic polynomial $p(\lambda)$, of order N, defined as

$$p(\lambda) \triangleq \det(\mathbf{A} - \lambda\mathbf{I}) = 0 \tag{A.63}$$

for which the eigenvalues $\{\lambda_1, \lambda_2, \ldots, \lambda_N\}$ of the matrix $\mathbf{A}$, denoted as $\lambda(\mathbf{A})$ or $\mathrm{eig}(\mathbf{A})$, are the roots of the characteristic polynomial $p(\lambda)$.

For each eigenvalue λ is associated with an *eigenvector* $\mathbf{q}$ defined by the equation

$$(\mathbf{A} - \lambda\mathbf{I})\mathbf{q} = \mathbf{0} \qquad \text{or} \qquad \mathbf{A}\mathbf{q} = \lambda\mathbf{q}. \tag{A.64}$$

Consider a simple example of a real matrix $\mathbf{A}_{2\times 2}$ defined as

$$\mathbf{A} = \begin{bmatrix} 2 & 1 \\ 1 & 2 \end{bmatrix}. \tag{A.65}$$

For (A.63) the characteristic polynomial is

$$\det(\mathbf{A} - \lambda\mathbf{I}) = \det\begin{bmatrix} 2-\lambda & 1 \\ 1 & 2-\lambda \end{bmatrix} = \lambda^2 - 4\lambda + 3 = 0 \tag{A.66}$$

with two distinct and real roots: $\lambda_1 = 1$ and $\lambda_2 = 3$, for which $\lambda_i(\mathbf{A}) = (1{,}3)$.

The eigenvector related to $\lambda_1 = 1$ is

$$\begin{bmatrix} 2 & 1 \\ 1 & 2 \end{bmatrix}\begin{bmatrix} q_1 \\ q_2 \end{bmatrix} = \begin{bmatrix} q_1 \\ q_2 \end{bmatrix} \quad \Rightarrow \quad \mathbf{q}_1 = \begin{bmatrix} 1 \\ -1 \end{bmatrix} \tag{A.67}$$

while the eigenvector related to $\lambda_2 = 3$ is

$$\begin{bmatrix} 2 & 1 \\ 1 & 2 \end{bmatrix}\begin{bmatrix} q_1 \\ q_2 \end{bmatrix} = 3\begin{bmatrix} q_1 \\ q_2 \end{bmatrix} \quad \Rightarrow \quad \mathbf{q}_2 = \begin{bmatrix} 1 \\ 1 \end{bmatrix}. \tag{A.68}$$

The eigenvectors of a matrix $\mathbf{A}_{N\times N}$ are sometimes referred to as $\mathrm{eigenvect}(\mathbf{A})$.

A.8.1 Trace of Matrix

The *trace of matrix* $\mathbf{A}_{N\times N}$ is defined as the sum of its elements in the main diagonal and, equivalently, and is equal to the sum of its (complex) eigenvalues

$$\text{tr}[\mathbf{A}] = \sum_{i=1}^{N} a_{ii} = \sum_{i=1}^{N} \lambda_i. \tag{A.69}$$

Moreover we have that

$$\begin{aligned} &\text{tr}[\mathbf{A}+\mathbf{B}] = \text{tr}[\mathbf{A}] + \text{tr}[\mathbf{B}] \\ &\text{tr}[\mathbf{A}] = \text{tr}\left[\mathbf{A}^H\right] \\ &\text{tr}[c\mathbf{A}] = c \cdot \text{tr}[\mathbf{A}] \\ &\text{tr}[\mathbf{ABC}] = \text{tr}[\mathbf{BCA}] = \text{tr}[\mathbf{CAB}] \\ &\mathbf{a}^H\mathbf{a} = \text{tr}[\mathbf{a}^H\mathbf{a}]. \end{aligned} \tag{A.70}$$

Matrices have the *Frobenius inner product*, which is analogous to the vector inner product. It is defined as the sum of the products of the corresponding components of two matrices **A** and **B** having the same size:

$$\langle \mathbf{A}, \mathbf{B} \rangle = \sum_i \sum_j a_{ij} b_{ij} = \text{tr}\left(\mathbf{A}^H\mathbf{B}\right) = \text{tr}\left(\mathbf{A}\mathbf{B}^H\right).$$

A.9 Matrix Diagonalization

A matrix $\mathbf{A}_{N \times N}$ is called *diagonalizable matrix* if there is an invertible matrix **Q** such that there exists a decomposition

$$\mathbf{A} = \mathbf{Q}\mathbf{\Lambda}\mathbf{Q}^{-1} \tag{A.71}$$

or, equivalently,

$$\mathbf{\Lambda} = \mathbf{Q}^{-1}\mathbf{A}\mathbf{Q}. \tag{A.72}$$

This is possible if, and only if, the matrix **A** has N linearly independent eigenvectors and the matrix **Q**, partitioned as column vectors $\mathbf{Q} = [\mathbf{q}_1 \;\; \mathbf{q}_2 \;\; \cdots \;\; \mathbf{q}_N]$, is built with independent eigenvectors of **A**. In this case, $\mathbf{\Lambda}$ is a diagonal matrix built with the eigenvalues of **A**, i.e., $\mathbf{\Lambda} = \text{diag}(\lambda_1, \lambda_2, \ldots, \lambda_N)$.

A.9.1 Diagonalization of a Normal Matrix

The matrix $\mathbf{A}_{N \times N}$ is said *normal matrix* if $\mathbf{A}^H\mathbf{A} = \mathbf{A}\mathbf{A}^H$. A matrix **A** is normal iff it can be factorized as

$$\mathbf{A} = \mathbf{Q}\mathbf{\Lambda}\mathbf{Q}^H \tag{A.73}$$

where $\mathbf{Q}^H\mathbf{Q} = \mathbf{Q}\mathbf{Q}^H = \mathbf{I}$, $\mathbf{Q} = [\mathbf{q}_1 \;\; \mathbf{q}_2 \;\; \cdots \;\; \mathbf{q}_N]$, $\mathbf{\Lambda} = \text{diag}(\lambda_1, \lambda_2, \ldots, \lambda_N)$, and $\mathbf{\Lambda} = \mathbf{Q}^H\mathbf{A}\mathbf{Q}$.

The set of all eigenvectors of **A** is defined as the *spectrum of the matrix*. The radius of the spectrum or *spectral radius* is defined as the eigenvalue of maximum modulus

$$\rho(\mathbf{A}) = \max_i \left(|\mathrm{eig}(\mathbf{A})|\right). \tag{A.74}$$

Property If the matrix $\mathbf{A}_{N\times N}$ is nonsingular, then all the eigenvalues are nonzero and the eigenvalues of the inverse matrix $\mathbf{A}^{-1}$ are the reciprocal of $\mathrm{eig}(\mathbf{A})$.

Property If the matrix $\mathbf{A}_{N\times N}$ is symmetric and semi-definite positive then all eigenvalues are real and positive. So we have that

1. The eigenvalues λ_i of **A** are real and nonnegative:

$$\mathbf{q}_i^H\mathbf{A}\mathbf{q}_i = \lambda_i\mathbf{q}_i^T\mathbf{q}_i \quad \Rightarrow \quad \lambda_i = \frac{\mathbf{q}_i^H\mathbf{A}\mathbf{q}_i}{\mathbf{q}_i^H\mathbf{q}_i} \quad (\textit{Rayleigh quotient}) \tag{A.75}$$

2. The eigenvectors of **A** are orthogonal for distinct λ_i

$$\mathbf{q}_i^H\mathbf{q}_j = 0, \quad \text{for } i \neq j \tag{A.76}$$

3. The matrix **A** can be diagonalized as

$$\mathbf{A} = \mathbf{Q}\boldsymbol{\Lambda}\mathbf{Q}^H \tag{A.77}$$

where $\mathbf{Q} = [\,\mathbf{q}_1 \;\; \mathbf{q}_2 \;\; \cdots \;\; \mathbf{q}_N\,]$, $\boldsymbol{\Lambda} = \mathrm{diag}(\lambda_1,\lambda_2, \ldots,\lambda_N)$, and **Q** is a unitary matrix or $\mathbf{Q}^T\mathbf{Q} = \mathbf{I}$

4. An alternative representation for **A** is then

$$\mathbf{A} = \sum_{i=1}^{N}\lambda_i\mathbf{q}_i\mathbf{q}_i^H = \sum_{i=1}^{N}\lambda_i\mathbf{P}_i \tag{A.78}$$

where the term $\mathbf{P}_i = \mathbf{q}_i\mathbf{q}_i^H$ is defined as *spectral projection*.

A.10 Norms of Vectors and Matrices

A.10.1 Norm of Vectors

Given a vector $\mathbf{x} \in (\mathbb{R},\mathbb{C})^N$, its *norm* refers to its length relative to a vector space. In the case of a space of order p, called L_p *space*, the norm is indicated as $\|\mathbf{x}\|^{L_p}$ or $\|\mathbf{x}\|_p$ and is defined as

$$\|\mathbf{x}\|_p \triangleq \left[\sum_{i=1}^{N} |x_i|^p\right]^{1/p}, \quad \text{for} \quad p \geq 1. \tag{A.79}$$

L_0 *norm* The expression (A.79) is valid even when $0 < p < 1$; however, the result is not exactly a norm. For $p = 0$, (A.79) becomes

$$\|\mathbf{x}\|_0 \triangleq \lim_{p\to 0} \|\mathbf{x}\|_p = \sum_{i=1}^{N} |x_i|^0. \tag{A.80}$$

Note that (A.80) is equal to the number of nonzero entries of the vector $\mathbf{x}$.

L_1 *norm*

$$\|\mathbf{x}\|_1 \triangleq \sum_{i=1}^{N} |x_i|, \quad L_1\,\textit{norm}. \tag{A.81}$$

The previous expression represents the sum of modules of the elements of the vector $\mathbf{x}$.

L_{inf} *norm* For $p \to \infty$ the (A.79) becomes

$$\|\mathbf{x}\|_\infty \triangleq \max_{i=1,N}\{|x_i|\} \tag{A.82}$$

called *uniform norm* or *norm of the maximum* and denoted as L_{inf}.

Euclidean or L_2 *norm* The Euclidean norm is defined for $p = 2$ and expresses the standard length of the vector.

$$\|\mathbf{x}\|_2 \triangleq \sqrt{\sum_{i=1}^{N} |x|_i^2} = \sqrt{\mathbf{x}^H\mathbf{x}}, \quad \textit{Euclidean or } L_2 \textit{ norm} \tag{A.83}$$

$$\|\mathbf{x}\|_2^2 \triangleq \mathbf{x}^H\mathbf{x}, \quad \textit{quadratic Euclidean norm} \tag{A.84}$$

$$\|\mathbf{x}\|_{\mathbf{G}}^2 \triangleq \|\mathbf{x}^H\mathbf{G}\mathbf{x}\|, \quad \textit{quadratic weighted Euclidean norm,} \tag{A.85}$$

where $\mathbf{G}$ is a diagonal weighing matrix.

Frobenius norm Similar to the L_1 norm, it is defined as

$$\|\mathbf{x}\|_F \triangleq \sqrt{\sum_{i=1}^{N} |x_i|^2}, \quad \textit{Frobenius norm} \tag{A.86}$$

Property For each norm we have the following property:

1. $\|\mathbf{x}\| \geq 0$, the equality holds only for $\mathbf{x} = \mathbf{0}$

2. $\|\alpha\mathbf{x}\| = \alpha\|\mathbf{x}\|, \quad \forall\alpha$
3. $\|\mathbf{x}+\mathbf{y}\| \leq \|\mathbf{x}\| + \|\mathbf{y}\|$ triangle inequality.

The distance between two vectors $\mathbf{x}$ and $\mathbf{y}$ is defined as

$$\|\mathbf{x}-\mathbf{y}\|_p \triangleq \left[\sum_{i=1}^{N}|x_i - y_i|^p\right]^{1/p}, \qquad \text{for} \qquad p > 0. \tag{A.87}$$

It is called distance or similarity measure in the Minkowsky metric [1].

A.10.2 *Norm of Matrices*

With regard to the norm of a matrix, similar to the vectors norms, these may be defined in the following mode. Given an $\mathbf{A}_{N\times N}$ matrix

L_1 *norm*

$$\|\mathbf{A}\|_1 \triangleq \max_j \sum_{i=1}^{N}|a_{ij}|, \quad L_1 norm \tag{A.88}$$

represents the column of $\mathbf{A}$ with largest sum of absolute values

Euclidean or L_2 *norm* The Euclidean norm is defined for the space $p = 2$ and expresses the standard length of the vector

$$\|\mathbf{A}\|_2 \triangleq \sqrt{\lambda_{\max}} \Rightarrow \max_{\lambda_i} \text{eig}\left(\mathbf{A}^H\mathbf{A}\right) \quad \text{o} \quad \max_{\lambda_i} \text{eig}\left(\mathbf{A}\mathbf{A}^H\right) \tag{A.89}$$

L_{inf} *norm*

$$\|\mathbf{A}\|_\infty \triangleq \max_i \sum_{j=1}^{N}|a_{ij}|, \quad L_{\text{inf}}\ norm \tag{A.90}$$

that represents the row with greater sum of the absolute values.

Frobenius norm

$$\|\mathbf{A}\|_F \triangleq \sqrt{\sum_{i=1}^{N}\sum_{j=1}^{M}|a_{ij}|^2}, \quad Frobenius\ norm \tag{A.91}$$

A.11 Singular Value Decomposition Theorem

Given a matrix $\mathbf{X} \in (\mathbb{R},\mathbb{C})^{N\times M}$ with $K = \min(N,M)$, of rank $r \leq K$, there are two orthonormal matrices $\mathbf{U} \in (\mathbb{R},\mathbb{C})^{N\times N}$ and $\mathbf{V} \in (\mathbb{R},\mathbb{C})^{M\times M}$ containing for columns, respectively, the eigenvectors of $\mathbf{X}\mathbf{X}^H$ and the eigenvectors of $\mathbf{X}^H\mathbf{X}$, namely,

$$\mathbf{U}_{N\times N} = \text{eigenvect}\left(\mathbf{X}\mathbf{X}^H\right) = \begin{bmatrix}\mathbf{u}_0 & \mathbf{u}_1 & \cdots & \mathbf{u}_{N-1}\end{bmatrix} \tag{A.92}$$

$$\mathbf{V}_{M\times M} = \text{eigenvect}\left(\mathbf{X}^H\mathbf{X}\right) = \begin{bmatrix}\mathbf{v}_0 & \mathbf{v}_1 & \cdots & \mathbf{v}_{M-1}\end{bmatrix} \tag{A.93}$$

such that the following equality is valid:

$$\mathbf{U}^H\mathbf{X}\mathbf{V} = \boldsymbol{\Sigma}, \tag{A.94}$$

equivalently,

$$\mathbf{X} = \mathbf{U}\boldsymbol{\Sigma}\mathbf{V}^H \tag{A.95}$$

or

$$\mathbf{X}^H = \mathbf{V}\boldsymbol{\Sigma}\mathbf{U}^H. \tag{A.96}$$

The expressions (A.94)–(A.96) represent the *SVD decomposition of the matrix* $\mathbf{A}$, shown graphically in Fig. A.3

The matrix $\boldsymbol{\Sigma} \in \mathbb{R}^{N\times M}$ is characterized by the following structure:

$$\begin{aligned} K &= \min(M,N) \quad & \boldsymbol{\Sigma} &= \begin{bmatrix}\boldsymbol{\Sigma}_K & \mathbf{0}\\ \mathbf{0} & \mathbf{0}\end{bmatrix}, \\ K &= N = M & \boldsymbol{\Sigma} &= \boldsymbol{\Sigma}_K \end{aligned} \tag{A.97}$$

where $\boldsymbol{\Sigma}_K \in \mathbb{R}^{K\times K}$ is a diagonal matrix containing the positive square root of the eigenvalues of the matrix $\mathbf{X}^H\mathbf{X}$ (or $\mathbf{X}\mathbf{X}^H$) defined as *singular values*.[3] In formal terms

$$\boldsymbol{\Sigma}_K = \text{diag}(\sigma_0, \sigma_1, \ldots, \sigma_{K-1}) \triangleq \sqrt{\text{diag}\left(\text{eig}\left(\mathbf{X}^H\mathbf{X}\right)\right)} \equiv \sqrt{\text{diag}\left(\text{eig}\left(\mathbf{X}\mathbf{X}^H\right)\right)}, \tag{A.98}$$

where

$$\sigma_0 \geq \sigma_1 \geq \ldots \geq \sigma_{K-1} > 0 \qquad \text{and} \qquad \sigma_K = \cdots = \sigma_{N-1} = 0. \tag{A.99}$$

Note that the *singular values* σ_i of $\mathbf{X}$ are in descending order. Moreover, the column vectors $\mathbf{u}_i$ and $\mathbf{v}_i$ are defined, respectively, as *left singular vectors* and *right singular vectors* of $\mathbf{X}$. Since $\mathbf{U}$ and $\mathbf{V}$ are orthogonal, it is easy to see that the matrix $\mathbf{X}$ can be written as a product

[3] Remember that the nonzero eigenvalues of the matrices $\mathbf{X}^H\mathbf{X}$ and $\mathbf{X}\mathbf{X}^H$ are identical.

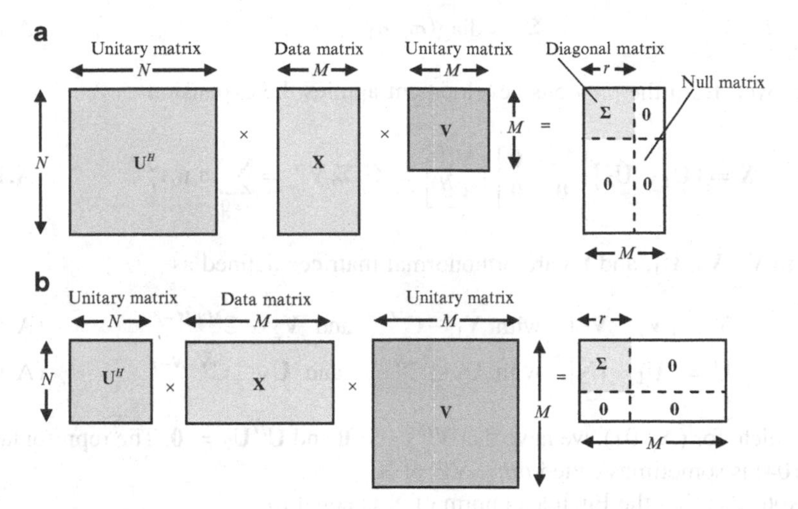

Fig. A.3 Schematic of the SVD decomposition in the cases (**a**) overdetermined (matrix **X** is *tall*); (**b**) underdetermined (matrix **X** is *fat*)

$$\mathbf{X} = \mathbf{U}\boldsymbol{\Sigma}\mathbf{V}^H = \sum_{i=0}^{K-1} \sigma_i \mathbf{u}_i \mathbf{v}_i^H. \tag{A.100}$$

A.11.1 Subspaces of Matrix X and SVD

The SVD reveals important property of the matrix **X**. In fact, for $r < K$ we have $r = \text{rank}(\mathbf{X})$, for which the first r columns of **U** form an orthonormal basis of the column space $\mathcal{R}(\mathbf{X})$, while the first r columns of **V** form an orthonormal basis for the nullspace (or kernel) $\mathcal{N}(\mathbf{X}^H)$ of **X**, i.e.,

$$\begin{aligned} r &= \text{rank}(\mathbf{X}) \\ \mathcal{R}(\mathbf{X}) &= \text{span}(\mathbf{u}_0, \mathbf{u}_1, \ldots, \mathbf{u}_{r-1}) \\ \mathcal{N}(\mathbf{X}^H) &= \text{span}(\mathbf{v}_r, \mathbf{v}_{r+1}, \ldots, \mathbf{v}_{N-1}). \end{aligned} \tag{A.101}$$

In the case that $r < K$, also, for (A.99) is

$$\sigma_0 \geq \sigma_1 \geq \ldots \geq \sigma_{r-1} > 0 \quad \text{and} \quad \sigma_r = \ldots = \sigma_{N-1} = 0. \tag{A.102}$$

It follows that (A.97), for the cases over/under-determined, becomes

$$\boldsymbol{\Sigma} = \begin{bmatrix} \boldsymbol{\Sigma}_r & \mathbf{0} \\ \mathbf{0} & \mathbf{0} \end{bmatrix},$$

where

$$\mathbf{\Sigma}_r = \text{diag}(\sigma_0, \sigma_1, ..., \sigma_{r-1}). \tag{A.103}$$

Moreover, from the previous development applies the expansion

$$\mathbf{X} = [\mathbf{U}_1 \quad \mathbf{U}_2] \begin{bmatrix} \mathbf{\Sigma}_r & \mathbf{0} \\ \mathbf{0} & \mathbf{0} \end{bmatrix} \begin{bmatrix} \mathbf{V}_1^H \\ \mathbf{V}_2^H \end{bmatrix} = \mathbf{U}_1 \mathbf{\Sigma}_r \mathbf{V}_1^H = \sum_{i=0}^{r-1} \sigma_i \mathbf{u}_i \mathbf{v}_i^H, \tag{A.104}$$

where $\mathbf{V}_1$, $\mathbf{V}_2$, $\mathbf{U}_1$, and $\mathbf{U}_2$ are orthonormal matrices defined as

$$\mathbf{V} = [\mathbf{V}_1 \quad \mathbf{V}_2] \quad \text{with } \mathbf{V}_1 \in \mathbb{C}^{M \times r} \text{ and } \mathbf{V}_2 \in \mathbb{C}^{M \times M-r} \tag{A.105}$$

$$\mathbf{U} = [\mathbf{U}_1 \quad \mathbf{U}_2] \quad \text{with } \mathbf{U}_1 \in \mathbb{C}^{N \times r} \text{ and } \mathbf{U}_2 \in \mathbb{C}^{N \times N-r} \tag{A.106}$$

for which, for (A.101), we have that $\mathbf{V}_1^H \mathbf{V}_2 = \mathbf{0}$ and $\mathbf{U}_1^H \mathbf{U}_2 = \mathbf{0}$. The representation (A.104) is sometimes called *thin* SVD of $\mathbf{X}$.

Note also that the Euclidean norm of $\mathbf{X}$ is equal to

$$\|\mathbf{X}\|_2 = \sigma_0 \tag{A.107}$$

while its Frobenius norm is equal to

$$\|\mathbf{X}\|_F \triangleq \sqrt{\sum_{i=0}^{N-1} \sum_{j=0}^{M-1} |x_{ij}|^2} = \sqrt{\sigma_0^2 + \sigma_1^2 + \cdots + \sigma_r^2}. \tag{A.108}$$

Remark A special important case of SVD decomposition occurs when the matrix $\mathbf{X}$ is symmetric and nonnegative. In this case it is

$$\mathbf{\Sigma} = \text{diag}(\lambda_0, \lambda_1, ..., \lambda_{r-1}), \tag{A.109}$$

where $\lambda_0 \geq \lambda_1 \geq \ldots \geq \lambda_{r-1} \geq 0$ are the real eigenvalues of $\mathbf{X}$ corresponding to the eigenvectors $\mathbf{v}_i$.

A.11.2 Pseudoinverse Matrix and SVD

The Moore–Penrose pseudoinverse of the overdetermined case is defined as $\mathbf{X}^\# = (\mathbf{X}^H\mathbf{X})^{-1}\mathbf{X}^H$, while for the underdetermined case is $\mathbf{X}^\# = \mathbf{X}^H(\mathbf{X}\mathbf{X}^H)^{-1}$. It should be noted that in expression (A.95), $\mathbf{X}^\#$ always results in the following forms:

$$\mathbf{X}^{\#} = \left(\mathbf{X}^H\mathbf{X}\right)^{-1}\mathbf{X}^H = \mathbf{V}\begin{bmatrix} \boldsymbol{\Sigma}_K^{-1} & \mathbf{0} \\ \mathbf{0} & \mathbf{0} \end{bmatrix}\mathbf{U}^H \quad N > M$$
$$\mathbf{X}^{\#} = \mathbf{X}^H\left(\mathbf{X}\mathbf{X}^H\right)^{-1} = \mathbf{V}\begin{bmatrix} \boldsymbol{\Sigma}_K^{-1} & \mathbf{0} \\ \mathbf{0} & \mathbf{0} \end{bmatrix}\mathbf{U}^H \quad N < M, \tag{A.110}$$

where for $K = \min(N,M)$, $\boldsymbol{\Sigma}_K^{-1} = \mathrm{diag}(\sigma_0^{-1},\sigma_1^{-1}, \ldots, \sigma_{K-1}^{-1})$, and for $r \leq K$,

$$\mathbf{X}^{\#} = \mathbf{V}_1\boldsymbol{\Sigma}_r^{-1}\mathbf{U}_1^H. \tag{A.111}$$

For both over and under-determined, by means of (A.95), the partitions (A.105) and (A.106) are demonstrable.

Remark Remember that the right singular vectors $\mathbf{v}_0, \mathbf{v}_1,\ldots,\mathbf{v}_{M-1}$, of the data matrix $\mathbf{X}$, are equal to the eigenvectors of the oversized matrix $\mathbf{X}^H\mathbf{X}$, while the left singular vectors $\mathbf{u}_0, \mathbf{u}_1,\ldots,\mathbf{u}_{N-1}$ are equal to the eigenvectors of the undersized matrix $\mathbf{X}\mathbf{X}^H$. It is, also, true that $r = \mathrm{rank}(\mathbf{X})$, i.e., the number of positive singular values is equivalent to the rank of the data matrix $\mathbf{X}$. Therefore, the SVD decomposition provides a practical tool for determining the rank of a matrix and its pseudoinverse.

Corollary For the calculation of the pseudoinverse it is also possible to use other types of decomposition such as that shown below.

Given a matrix $\mathbf{X} \in (\mathbb{R},\mathbb{C})^{N\times M}$ with $\mathrm{rank}(\mathbf{X}) = r < \min(N,M)$, there are two matrices $\mathbf{C}_{M\times r}$ and $\mathbf{D}_{r\times N}$ such that $\mathbf{X} = \mathbf{CD}$. Using these matrices it is easy to verify that

$$\mathbf{X}^{\#} = \mathbf{D}^H\left(\mathbf{D}\mathbf{D}^H\right)^{-1}\left(\mathbf{C}^H\mathbf{C}\right)^{-1}\mathbf{C}^H. \tag{A.112}$$

A.12 Condition Number of a Matrix

In numerical analysis the *condition number*, indicated as $\chi(\cdot)$, associated with a problem is the degree of *numerical tractability* of the problem himself. A matrix $\mathbf{A}$ is called *ill-conditioned* if $\chi(\mathbf{A})$ takes large values. In this case, some methods of matrix inversion can present a high numerical nature error.

Given a matrix $\mathbf{A} \in (\mathbb{R},\mathbb{C})^{N\times M}$, the condition number is defined as

$$\chi(\mathbf{A}) \triangleq \|\mathbf{A}\|_p\|\mathbf{A}^{\#}\|_p \quad 1 \leq \chi(\mathbf{A}) \leq \infty, \tag{A.113}$$

where $p = 1, 2, \ldots, \infty$, $\|\cdot\|_p$ may be the Frobenius norm and $\mathbf{A}^{\#}$ the pseudoinverse of $\mathbf{A}$. The number $\chi(\mathbf{A})$ depends on the type of chosen norm. In particular, in the case of L_2 norm it is possible to prove that

$$\chi(\mathbf{A}) = ||\mathbf{A}||_2||\mathbf{A}^{\#}||_2 = \frac{\sigma_{\max}}{\sigma_{\min}}, \tag{A.114}$$

where $\sigma_{\max} = \sigma_1$ and $\sigma_{\min}(=\sigma_M \text{ o } \sigma_N)$ are, respectively, the maximum and minimum singular values of $\mathbf{A}$. In the case of a square matrix

$$\chi(\mathbf{A}) = \frac{\lambda_{\max}}{\lambda_{\min}}, \tag{A.115}$$

where $\lambda_{\max}$ and $\lambda_{\min}$ are the maximum and minimum eigenvalues of $\mathbf{A}$.

A.13 Kroneker Product

The Kronecker product between two matrices $\mathbf{A} \in (\mathbb{R},\mathbb{C})^{P\times Q}$ and $\mathbf{B} \in (\mathbb{R},\mathbb{C})^{N\times M}$, usually indicated as $\mathbf{A} \otimes \mathbf{B}$, is defined as

$$\mathbf{A} \otimes \mathbf{B} = \begin{bmatrix} a_{11}\mathbf{B} & \cdots & a_{1Q}\mathbf{B} \\ \vdots & \ddots & \vdots \\ a_{P1}\mathbf{B} & \cdots & a_{PQ}\mathbf{B} \end{bmatrix} \in (\mathbb{R},\mathbb{C})^{PN\times QM}. \tag{A.116}$$

The Kronecker product can be convenient to represent linear systems equations and some linear transformations.

Given a matrix $\mathbf{A} \in (\mathbb{R},\mathbb{C})^{N\times M}$, you can associate with it a vector, $\text{vec}(\mathbf{A}) \in (\mathbb{R},\mathbb{C})^{NM\times 1}$, containing all its column vectors [see (A.10)].

For example, given the matrices $\mathbf{A}_{N\times M}$ and $\mathbf{X}_{M\times P}$, it is possible to represent their product as

$$\mathbf{AX} = \mathbf{B}, \tag{A.117}$$

where $\mathbf{B}_{N\times P}$; using the definition (A.10) and the Kronecker product, we have that

$$(\mathbf{I} \otimes \mathbf{A})\text{vec}(\mathbf{X}) = \text{vec}(\mathbf{B}) \tag{A.118}$$

that represents a system of linear equations of NP equations and MP unknowns.

Similarly, given the matrices, $\mathbf{A}_{N\times M}$, $\mathbf{X}_{M\times P}$, and $\mathbf{B}_{P\times Q}$ it is possible to represent their product

$$\mathbf{AXB} = \mathbf{C} \tag{A.119}$$

in a equivalent manner as a QN linear system equation in MP unknowns or as

$$(\mathbf{B}^T \otimes \mathbf{A})\text{vec}(\mathbf{X}) = \text{vec}(\mathbf{C}). \tag{A.120}$$

Appendix B: Elements of Nonlinear Programming

B.1 Unconstrained Optimization

The term nonlinear programming (NLP) indicates the process of solving linear or nonlinear systems of equations, rather than a closed mathematical–algebraic approach with a methodology that minimizes or maximizes some cost function associated with the problem.

This Appendix briefly introduces the basic concepts of NLP. In particular, it presents some fundamental concepts of the unconstrained and the constrained optimization methods [3–15].

B.1.1 Numerical Methods for Unconstrained Optimization

The problem of unconstrained optimization can be formulated as follows: *find a vector* $\mathbf{w} \in \Omega \equiv \mathbb{R}^{M}$[4] *that minimizes (maximizes) a scalar function* $J(\mathbf{w})$. Formally

$$\mathbf{w}^* = \min_{\mathbf{w}\in\Omega} J(\mathbf{w}). \tag{B.1}$$

The real function $J(\mathbf{w})$, $J : \mathbb{R}^M \rightarrow \mathbb{R}$, is called cost function (CF), or loss function or objective function or energy function, $\mathbf{w}$ is an M-dimensional vector of variables that could have any values, positive or negative, and Ω is the variables or search space. Minimizing a function is equivalent to maximizing the negative of the function itself. Therefore, without loss of generalities, minimizing or maximizing a function are equivalent problems.

A point $\mathbf{w}^*$ is a global minimum for function $J(\mathbf{w})$ if

[4]For uniformity of writing, we denote by Ω the *search space*, which in the absence of constraints coincides with the whole space or $\Omega \equiv \mathbb{R}^M$. As we will see later, in the presence of constraints, there is a reduced search space, i.e., $\Omega \subset \mathbb{R}^M$.

A. Uncini, *Fundamentals of Adaptive Signal Processing*, Signals and Communication Technology, DOI 10.1007/978-3-319-02807-1,

$$J(\mathbf{w}^*) \leq J(\mathbf{w}), \quad \forall \mathbf{w} \in \mathbb{R}^M \tag{B.2}$$

and $\mathbf{w}^*$ is a strict local minimizer if (B.2) holds for a ε-radius ball centered in $\mathbf{w}^*$ indicated as $B(\mathbf{w}^*,\varepsilon)$.

B.1.2 Existence and Characterization of the Minimum

The admissible solutions of a problem can be characterized in terms of some sufficient and necessary conditions

First-order necessary condition (FONC) (for minimization or maximization) is that

$$\nabla J(\mathbf{w}) = 0, \tag{B.3}$$

where the operator $\nabla J(\mathbf{w}) \in \mathbb{R}^M$ is a vector indicating the gradient of function $J(\mathbf{w})$ defined as

$$\nabla J(\mathbf{w}) \triangleq \frac{\partial J(\mathbf{w})}{\partial \mathbf{w}} = \begin{bmatrix} \frac{\partial J(\mathbf{w})}{\partial w_1} & \frac{\partial J(\mathbf{w})}{\partial w_2} & \cdots & \frac{\partial J(\mathbf{w})}{\partial w_M} \end{bmatrix}^T. \tag{B.4}$$

Second-order necessary condition (SONC) is that the Hessian matrix $\nabla^2 J(\mathbf{w}) \in \mathbb{R}^{M \times M}$, defined as

$$\nabla^2 J(\mathbf{w}) \triangleq \frac{\partial}{\partial \mathbf{w}} \left[\frac{\partial J(\mathbf{w})}{\partial \mathbf{w}} \right]^T = \frac{\partial}{\partial \mathbf{w}} [\nabla J]^T$$

$$= \begin{bmatrix} \frac{\partial}{\partial w_1} \left[\frac{\partial J(\mathbf{w})}{\partial \mathbf{w}} \right]^T \\ \frac{\partial}{\partial w_2} \left[\frac{\partial J(\mathbf{w})}{\partial \mathbf{w}} \right]^T \\ \vdots \\ \frac{\partial J}{\partial w_M} \left[\frac{\partial J(\mathbf{w})}{\partial \mathbf{w}} \right]^T \end{bmatrix} = \begin{bmatrix} \frac{\partial^2 J(\mathbf{w})}{\partial w_1^2} & \frac{\partial^2 J(\mathbf{w})}{\partial w_1 \partial w_2} & \cdots & \frac{\partial^2 J(\mathbf{w})}{\partial w_1 \partial w_M} \\ \frac{\partial^2 J(\mathbf{w})}{\partial w_2 \partial w_1} & \frac{\partial^2 J(\mathbf{w})}{\partial w_2^2} & \cdots & \frac{\partial^2 J(\mathbf{w})}{\partial w_2 \partial w_M} \\ \vdots & \vdots & \ddots & \vdots \\ \frac{\partial^2 J(\mathbf{w})}{\partial w_M \partial w_1} & \frac{\partial^2 J(\mathbf{w})}{\partial w_M \partial w_2} & \cdots & \frac{\partial^2 J(\mathbf{w})}{\partial w_M^2} \end{bmatrix}, \tag{B.5}$$

is positive semi-definite (PSD) or

$$\mathbf{w}^T \cdot \nabla^2 J(\mathbf{w}) \cdot \mathbf{w} \geq 0, \quad \text{for all } \mathbf{w}. \tag{B.6}$$

Second-order sufficient condition (SONC) is that: given FONC satisfied, the Hessian matrix $\nabla^2 J(\mathbf{w})$ is definite positive that is $\mathbf{w}^T \cdot \nabla^2 J(\mathbf{w}) \cdot \mathbf{w} > 0$ for all $\mathbf{w}$.

A *necessary and sufficient condition* for which $\mathbf{w}^*$ is a strict local minimizer of $J(\mathbf{w})$ can be formalized by the following theorem:

Theorem The point $\mathbf{w}^*$ is a strict local minimizer of $J(\mathbf{w})$ iff:

Given nonsingular $\nabla^2 J(\mathbf{w})$ evaluated at the point $\mathbf{w}^*$, then $J(\mathbf{w}^*) < J(\mathbf{w})\ \forall$ $\varepsilon > 0, \forall\ \mathbf{w}$ such that $0 < \|\mathbf{w} - \mathbf{w}^*\| < \varepsilon$, if $\nabla J(\mathbf{w}) = 0$ and $\nabla^2 J(\mathbf{w})$ is symmetric and positive defined.

B.2 Algorithms for Unconstrained Optimization

In the field of unconstrained optimization, it is known that some general principles can be used to study most of the algorithms. This section describes some of these fundamental principles.

B.2.1 Basic Principles

Our problem is to determine (or better *estimate*) the vector $\mathbf{w}^*$, called *optimal* solution, which minimizes the CF $J(\mathbf{w})$. If the CF is smooth and its gradient is available, the optimal solution can be computed (estimated) by an iterative procedure that minimizes the CF, i.e., starting from some initial condition (IC) $\mathbf{w}_{-1}$, a suitable solution is available only after a certain number of adaptation steps: $\mathbf{w}_{-1} \to \mathbf{w}_0 \to \mathbf{w}_1 \ldots \mathbf{w}_k \ldots \to \mathbf{w}^*$. The recursive estimator has a form of the type

$$\mathbf{w}_{k+1} = \mathbf{w}_k + \mu_k \mathbf{d}_k \tag{B.7}$$

or as

$$\mathbf{w}_k = \mathbf{w}_{k-1} + \mu_k \mathbf{d}_k, \tag{B.8}$$

where k is the adaptation index. The vector $\mathbf{d}_k$ represents the adaptation direction and the parameter μ_k is the step size also called adaptation rate, step length, learning rate, etc., that can be obtained by means of a one-dimensional search.

An important aspect of recursive procedure (B.7) concerns the algorithm order. In the first-order algorithms, the adaptation is carried out using only knowledge about the CF gradient, evaluated with respect to the free parameters $\mathbf{w}$. In the second-order algorithms, to reduce the number of iterations needed for convergence, information about the $J(\mathbf{w})$ curvature, i.e., the CF Hessian, is also used.

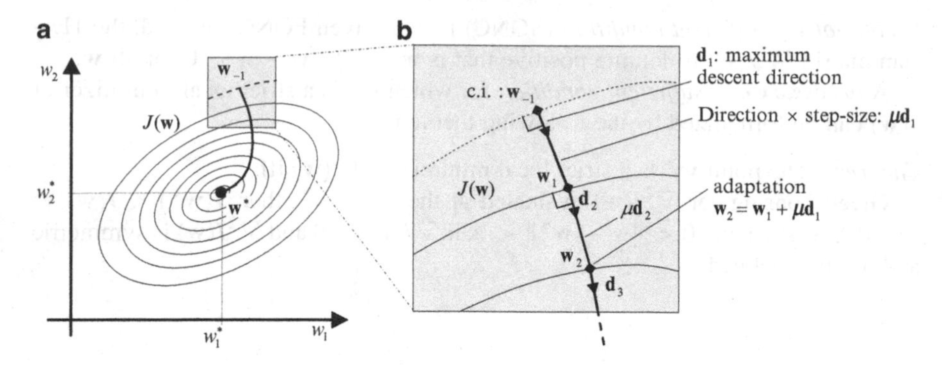

Fig. B.1 Qualitative evolution of the trajectory of the weights w_k, during the optimization process towards the optimal solution w^*, for a generic two-dimensional objective function (**a**) qualitative trend of steepest descent along the negative gradient of the surface $J(\mathbf{w})$; (**b**) particularly concerning the direction and the step size

Figure B.1 shows a qualitative evolution of the recursive optimization algorithm.

B.2.2 First- and Second-Order Algorithms

Let $J(\mathbf{w})$ be the CF to be minimized, if the CF gradient is available at learning step k, indicated as $\nabla J(\mathbf{w}_k)$, it is possible to define a family of iterative methods for the optimum solution computation. These methods are referred to as search methods or searching the performance surface, and the best-known algorithm of this class is the *steepest descent algorithm* (SDA) (Cauchy 1847). Note that, given the popularity of the SDA, this class of search methods is often identified with the name SDA algorithms.

Considering the general adaptation formula (B.7), indicating for simplicity the gradient as $\mathbf{g}_k = \nabla J(\mathbf{w}_k)$, the direction vector $\mathbf{d}_k$ is defined as follows:

$$\mathbf{d}_k = -\mathbf{g}_k, \qquad \textit{SDA algorithms.} \tag{B.9}$$

The SDA are first-order algorithms because adaptation is determined by knowledge of the gradient, i.e., only the first derivative of the CF. Starting from a given IC $\mathbf{w}_{-1}$, they proceed by updating the solution (B.7) along the opposite direction to the CF gradient with a step length μ.

The learning algorithms performances can be improved by using second-order derivative. In the case that the Hessian matrix is known, the method, called exact Newton, has a form of the type

$$\mathbf{d}_k = -\left[\nabla^2 J(\mathbf{w}_k)\right]^{-1}\mathbf{g}_k, \qquad \textit{exact Newton.} \tag{B.10}$$

In the case the Hessian is unknown the method, called quasi-Newton (Broyden 1965; [3] and [4] for other details), has a form of the type

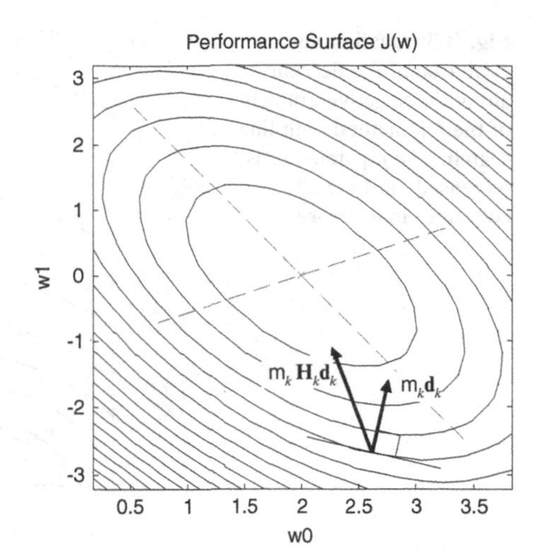

Fig. B.2 In the second-order algorithms, the matrix $\mathbf{H}_k$ determines a transformation in terms of rotation and gain, of the vector $\mathbf{d}_k$ in the direction of the minimum of the surface $J(\mathbf{w})$

$$\mathbf{d}_k = -\mathbf{H}_k\mathbf{g}_k, \quad \textit{quasi-Newton}, \tag{B.11}$$

where the matrix $\mathbf{H}_k$ is an approximation of the inverse of the Hessian matrix

$$\mathbf{H}_k \approx \left[\nabla^2 J(\mathbf{w}_k)\right]^{-1}. \tag{B.12}$$

The matrix $\mathbf{H}_k$ is a weighing matrix that can be estimated in various ways. As Fig. B.2 shows, the product $\mu_k\mathbf{H}_k$ can be interpreted as an optimum choice in direction and step-size length, calculated so as to follow the surface-gradient descent in very few steps. As the lower limit, as in the exact Newton's method, only with one step.

B.2.3 Line Search and Wolfe Condition

The step size μ of the unconstrained minimization procedure can be chosen *a priori* (according to certain rules) and kept fixed during the entire process or may be variable, and mentioned as μ_k. In this case, the step size can be optimized according to some criterion, e.g., the *line search* method defined as

$$\mu_k = \min_{\mu_{\min}<\mu<\mu_{\max}} J(\mathbf{w}_k + \mu\mathbf{d}_k). \tag{B.13}$$

With this technique, the parameter μ_k is (locally) increased, using a certain step, until the CF continues to decrease. The length of the learning rate is variable and usually with smaller size approaching to the optimal solution.

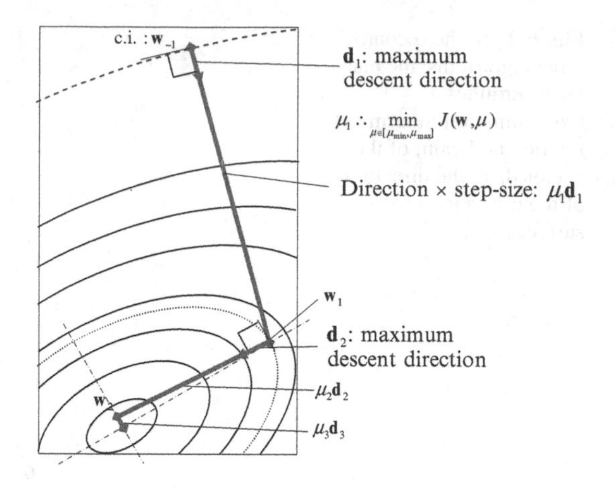

Fig. B.3 Qualitative evolution of the descent along the negative gradient of the CF method with line search. The μ parameter is increased until the CF continues to decrease

A typical qualitative evolution of line search during descent along the gradient of the CF is shown in Fig. B.3.

As illustrated in Fig. B.3, in certain situations, the number of iterations to reach the optimal point can be drastically reduced, however, with a considerable increase in computational cost due to the calculation of the expression (B.13).

For noisy or rippled CF the expression (B.13) can be computed with some difficulties. So algorithms for determination of optimal step size should be used with some cautions.

The Wolfe conditions are a set of inequalities for performing inexact line search, especially in second-order methods, in order to determine an optimal step size. Then inexact line searches provide an efficient way of computing an acceptable step size μ that reduces the objective function "sufficiently," rather than minimizing the objective function over $\mu \in \mathbb{R}^+$ exactly. A line search algorithm can use Wolfe conditions as a requirement for any guessed μ, before finding a new search direction $\mathbf{d}_k$. A step length μ_k is said to satisfy the Wolfe conditions if the following two inequalities hold:

$$\begin{aligned} J(\mathbf{w}_k + \mu_k \mathbf{d}_k) - J(\mathbf{w}_k) &\leq \sigma_1 \mu_k \mathbf{d}_k^T \mathbf{g}_k \\ \mathbf{d}_k^T \nabla J(\mathbf{w}_k + \mu_k \mathbf{d}_k) &\geq \sigma_2 \mathbf{d}_k^T \mathbf{g}_k, \end{aligned} \tag{B.14}$$

where $0 < \sigma_1 < \sigma_2 < 1$. The first inequality ensures that the CF J_k is reduced sufficiently. The second, called curvature condition, ensures that the slope has been reduced sufficiently. It is easy to show that if $\mathbf{d}_k$ is a descent direction, if J_k is continuously differentiable and if J_k is bounded below along the ray $\{\mathbf{w}_k + \mu \mathbf{d}_k | \mu > 0\}$ then there always exist step size satisfying (B.14). Algorithms that are guaranteed to find, in a finite number of iterations, a point satisfying the Wolfe conditions have been developed by several authors (see [4] for details).

If we modify the curvature condition

$$\left|\mathbf{d}_k^T \nabla J(\mathbf{w}_k + \mu_k \mathbf{d}_k)\right| \leq \left|\sigma_2 \mathbf{d}_k^T \mathbf{g}_k\right| \tag{B.15}$$

known as strong Wolfe condition, this can result in a value for the step size that is close to a minimizer of $J(\mathbf{w}_k + \mu_k \mathbf{d}_k)$.

B.2.3.1 Line Search Condition for Quadratic Form

Let $\mathbf{A} = \mathbb{R}^{M\times M}$ be a symmetric and positive definite matrix, for a quadratic CF defined as

$$J(\mathbf{w}) = c - \mathbf{w}^T\mathbf{b} + \frac{1}{2}\mathbf{w}^T\mathbf{A}\mathbf{w} \tag{B.16}$$

the optimal step size is

$$\mu = \frac{\mathbf{d}_{k-1}^T \mathbf{d}_{k-1}}{\mathbf{d}_{k-1}^T \mathbf{A} \mathbf{d}_{k-1}}. \tag{B.17}$$

Proof The line search is a procedure to find a *best* step size along steepest direction which minimizes the derivative $\frac{\partial}{\partial\mu}\{J(\mathbf{w})\} \to 0$. Using the chain rule, we can write

$$\frac{\partial J(\mathbf{w}_k)}{\partial \mu} = \left[\frac{J(\mathbf{w}_k)}{\partial \mathbf{w}_k}\right]^T \frac{\partial \mathbf{w}_k}{\partial \mu} = \left[\nabla J(\mathbf{w}_k)\right]^T \mathbf{d}_{k-1}.$$

Intuitively, from the current point reached by the line search procedure, the next direction is orthogonal to the previous direction that is $\mathbf{d}_k \perp \mathbf{d}_{k-1}$ (see Fig. B.3). For the determination of the optimal step size μ, we see that $\nabla J(\mathbf{w}_k) = -\mathbf{d}_k$. It follows

$$\begin{aligned} \mathbf{d}_k^T \mathbf{d}_{k-1} &= 0 \\ [\nabla J(\mathbf{w}_k)]^T \mathbf{d}_{k-1} &= 0. \end{aligned} \tag{B.18}$$

For a CF of the type (B.16), at the kth iteration, the negative gradient (search direction) is $\nabla J(\mathbf{w}_k) = \mathbf{b} - \mathbf{A}\mathbf{w}_k$. Let weight's correction equal to $\mathbf{w}_k = \mathbf{w}_{k-1} + \mu \mathbf{d}_{k-1}$, the expression (B.18) can be written as

$$\begin{aligned} [\mathbf{b} - \mathbf{A}\mathbf{w}_k]^T \mathbf{d}_{k-1} &= 0 \\ [\mathbf{b} - \mathbf{A}(\mathbf{w}_{k-1} + \mu \mathbf{d}_{k-1})]^T \mathbf{d}_{k-1} &= 0 \end{aligned}$$

by the latter, with the position $\mathbf{d}_{k-1} = -\mathbf{b} + \mathbf{A}\mathbf{w}_{k-1}$,

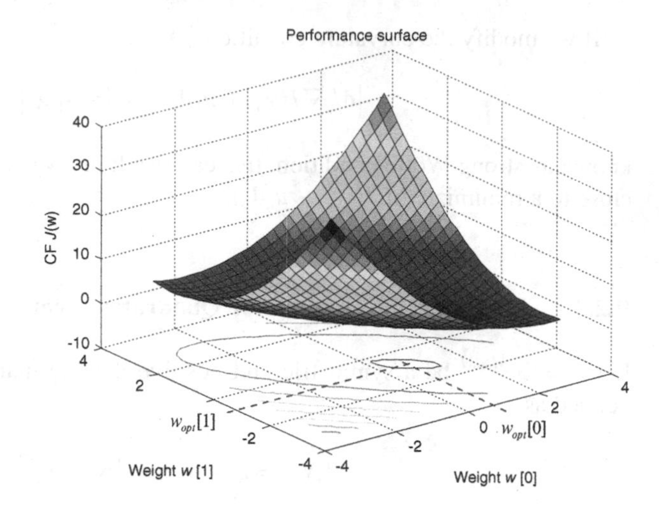

Fig. B.4 Trend of the cost function considered in the example

$$
\begin{gathered}
\left[\mathbf{d}_{k-1} - \mu \mathbf{A}\mathbf{d}_{k-1})\right]^T \mathbf{d}_{k-1} = 0 \\
\mathbf{d}_{k-1}^T \mathbf{d}_{k-1} - \mu \mathbf{d}_{k-1}^T \mathbf{A}\mathbf{d}_{k-1} = 0
\end{gathered}
$$

Finally solving for μ we have

$$
\mu = \frac{\mathbf{d}_{k-1}^T \mathbf{d}_{k-1}}{\mathbf{d}_{k-1}^T \mathbf{A}\mathbf{d}_{k-1}}.
$$

Q.E.D.

Example Consider a quadratic CF (B.16) with $\mathbf{A} = \begin{bmatrix} 1 & 0.8 \\ 0.8 & 1 \end{bmatrix}$, $\mathbf{b} = [0.1 \ \ -0.2]^T$ and $c = 0.1$, the plot of the performance surface is reported in Fig. B.4.

Problem Find the optimal solution, using a Matlab procedure, with tolerance $Tol = 1\mathrm{e}^{-6}$ starting with IC $\mathbf{w}_{-1} = [0 \ \ 3]^T$.

In Fig. B.5 the weights trajectories, plotted over the isolevel performance surface, are reported for the standard SDA and SDA plus the Wolfe condition.

Computed optimum solution

```
w[0] = 0.72222
w[1] = -0.77778
```

SDA Computed optimum solution with $\mu = 0.1$

```
n. Iter = 1233
w[0] = 0.72222
w[1] = -0.77778
```

SDA2_Wolfe optimal solution mu computed with Eq. (B.17)

```
n. Iter = 30
w[0] = 0.72222
w[1] = -0.77778
```

Matlab Functions

```
% ------------------------------------------------------------------------
% Standard Steepest Descent Algorithm
%
%   Copyright 2013 - A. Uncini
%   DIET Dpt - University of Rome 'La Sapienza' - Italy
%   $Revision: 1.0$  $Date: 2013/03/09$
%-------------------------------------------------------------------------
function [w, k] = SDA(w, g, R, c, mu, tol, MaxIter)
% Steepest descent ------------------------------------------------------
    for k = 1 : MaxIter
        gradJ   = grad_CF(w, g, R); % Gradient computation
        w       = w - mu*gradJ;     % up-date solution
        if ( norm(gradJ) < tol ), break, end % end criteria
    end
end

% ------------------------------------------------------------------------
% Standard Steepest Descent Algorithm and Wolf condition
% for quadratic CF
%    J(w) = c - w'b + (1/2)w'Aw;
%
%   Copyright 2013 - A. Uncini
%   DIET Dpt - University of Rome 'La Sapienza' - Italy
%   $Revision: 1.0$  $Date: 2013/03/09$
%-------------------------------------------------------------------------
function [w, k] = SDA2(w, g, R, c, mu, tol, MaxIter)
    for k=1:MaxIter
        gradJ = (R*w-g);  % Gradient computation or grad_CF(w, g, R);
        mu    = gradJ'*gradJ/(gradJ'*R*gradJ); % Opt. step-size eqn. (B.17)
        w     = w - mu*gradJ;   % up-date solution
        if ( norm(gradJ) < tol ), break, end % end criteria
    end
end

% ------------------------------------------------------------------------
% Standard quadratic cost function and gradient computation
%
%   Copyright 2013 - A. Uncini
%   DIET Dpt - University of Rome 'La Sapienza' - Italy
%   $Revision: 1.0$  $Date: 2013/03/09$
%-------------------------------------------------------------------------
function [Jw] = CF(w, c, b, A)
    Jw = c - 2*w'*b + w'*A*w;
end
%-------------------------------------------------------------------------
function [gradJ] = grad_CF(w, b, A)
     gradJ = (A*w-b);
end
```

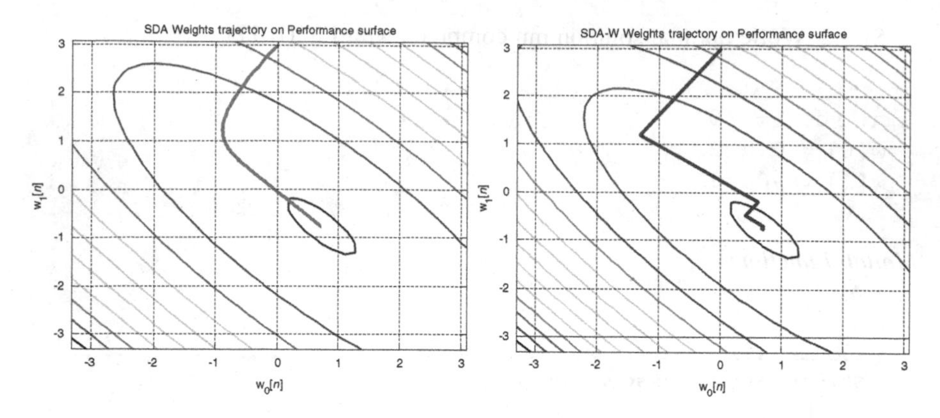

Fig. B.5 Trajectories of the weights on the isolevel CF curves for steepest descent algorithm (SDA) and Wolfe SDA

B.2.4 The Standard Newton's Method

The Newton methods are based on the exact minimum computation of a quadratic local approximation of the CF. In other words, rather than directly determine the approximate minimum of the true CF, the minimum of a locally quadratic approximation of the CF is exactly computed.

The method can be formalized by considering the truncated second-order Taylor series expansion of the CF $J(\mathbf{w})$ around a point $\mathbf{w}_k$ defined as

$$J(\mathbf{w}) \cong J(\mathbf{w}_k) + [\mathbf{w} - \mathbf{w}_k]^T \nabla J(\mathbf{w}_k) + \frac{1}{2}[\mathbf{w} - \mathbf{w}_k]^T \nabla^2 J(\mathbf{w}_k)[\mathbf{w} - \mathbf{w}_k]. \quad \text{(B.19)}$$

The minimum of the (B.19) is determined by imposing $\nabla J(\mathbf{w}_k) \rightarrow \mathbf{0}$, so, the $\mathbf{w}_{k+1}$ point (that minimizes the CF) necessarily satisfies the relationship[5]

$$\nabla J(\mathbf{w}_k) + \frac{1}{2}\nabla^2 J(\mathbf{w}_k)[\mathbf{w}_{k+1} - \mathbf{w}_k] = \mathbf{0}. \quad \text{(B.20)}$$

If the inverse of the Hessian matrix exists, the previous expression can be written in the following form of finite difference equation (FDE):

$$\mathbf{w}_{k+1} = \mathbf{w}_k - \mu_k \cdot \left[\nabla^2 J(\mathbf{w}_k)\right]^{-1} \nabla J(\mathbf{w}_k) \qquad \text{for} \qquad \nabla^2 J(\mathbf{w}_k) \neq \mathbf{0}, \quad \text{(B.21)}$$

where $\mu_k > 0$ is a suitable constant. The expression (B.21) represents the standard form of discrete Newton's method.

[5] In the optimum point $\mathbf{w}_{k+1}$, by definition, is $J(\mathbf{w}_{k+1}) \cong J(\mathbf{w}_k)$. It follows that the (B.19) can be written as $\mathbf{0} = [\mathbf{w}_{k+1} - \mathbf{w}_k]^T \nabla J(\mathbf{w}_k) + \frac{1}{2}[\mathbf{w}_{k+1} - \mathbf{w}_k]^T \nabla^2 J(\mathbf{w}_k)[\mathbf{w}_{k+1} - \mathbf{w}_k]$. So, simplifying the term $[\mathbf{w}_{k+1} - \mathbf{w}_k]^T$ gives (B.20).

Remark The CF approximation with a quadratic form is significant because $J(\mathbf{w})$ is usually an energy function. As explained by the Lyapunov method [5], you can think of that function as the energy associated with a continuous-time dynamical system described by a system of differential equations of the form

$$\frac{d\mathbf{w}}{dt} = -\mu_0 \left[\nabla^2 J(\mathbf{w}_k)\right]^{-1} \nabla J(\mathbf{w}_k) \tag{B.22}$$

such that for $\mu_0 > 0$, (B.21) corresponds to its numeric approximation. In this case, the convergence properties of Newton's method can be studied in the context of a quadratic programming problem of the type

$$\mathbf{w}_* = \underset{\mathbf{w} \in \Omega}{\arg\min} \; J(\mathbf{w}) \tag{B.23}$$

when the CF has a quadratic form of type (B.16). Note that for $\mathbf{A}$ positive definite function $J(\mathbf{w})$ is strictly convex and admits an absolute minimum $\mathbf{w}_*$ that satisfies

$$\mathbf{A}\mathbf{w}_* = \mathbf{b} \qquad \Rightarrow \qquad \mathbf{w}_* = \mathbf{A}^{-1}\mathbf{b}. \tag{B.24}$$

Also, observe that the gradient and the Hessian of the expression (B.16) are calculated explicitly as $\nabla J(\mathbf{w}) = \mathbf{A}\mathbf{w} - \mathbf{b}$ and $\nabla^2 J(\mathbf{w}) = \mathbf{A}$, and replacing these values in the form (B.21) for $\mu_k = 1$, the recurrence becomes

$$\mathbf{w}_{k+1} = \mathbf{w}_k - \mathbf{A}^{-1}(\mathbf{A}\mathbf{w}_k - \mathbf{b}) = \mathbf{A}^{-1}\mathbf{b}. \tag{B.25}$$

The above expression indicates that the Newton method converges theoretically at the minimum point, in only one iteration. In practice, however, the gradient calculation and the Hessian inverse pose many difficulties. In fact, the Hessian matrix is usually ill-conditioned and its inversion represents *an ill-posed problem.* Furthermore, the IC $\mathbf{w}_{-1}$ can be quite far from the minimum point and the Hessian at that point, it may not be positive definite, leading the algorithm to diverge. In practice, a way to overcome these drawbacks is to slow the adaptation speed by including a step-size parameter μ_k on the recurrence. It follows that in *causal form*, (B.25) can be written as

$$\mathbf{w}_k = \mathbf{w}_{k-1} - \mu_k \mathbf{A}^{-1}(\mathbf{A}\mathbf{w}_{k-1} - \mathbf{b}). \tag{B.26}$$

As mentioned above, in the simplest form of Newton method, the weighting of the equations (B.10) is made with the inverse Hessian matrix, or by its estimation. We then have

$$\mathbf{H}_k = \left[\nabla^2 J_{k-1}\right]^{-1}, \quad \textit{Exact Newton algorithms} \tag{B.27}$$

$$\mathbf{H}_k \approx \left[\nabla^2 J_{k-1}\right]^{-1}, \quad \textit{Quasi-Newton algorithms} \tag{B.28}$$

thereby forcing both direction and the step size to the minimum of the gradient function. Parameter learning can be constant ($\mu_k < 1$) or also estimated with a suitable optimization procedure.

B.2.5 The Levenberg–Marquardt's Variant

A simple method to overcome the problem of ill-conditioning of the Hessian matrix, called *Levenberg–Marquardt variant* [6, 7], consists in the definition of an adaptation rule of the type

$$\mathbf{w}_k = \mathbf{w}_{k-1} = \mu_k \left[\delta \mathbf{I} + \nabla^2 J_{k-1}\right]^{-1} \mathbf{g}_k \tag{B.29}$$

where the constant $\delta > 0$ must be chosen considering two contradictory requirements: small to increase the convergence speed and sufficiently large as to make the Hessian matrix always positive definite.

Levenberg–Marquardt method is an approximation of the Newton algorithm. It has, also, quadratic convergence characteristics. Furthermore, convergence is guaranteed even when the estimate of initial conditions is far from minimum point.

Note that the sum of the term $\delta\mathbf{I}$, in addition to ensure the positivity of the Hessian matrix, is strictly related to the Tikhonov regularization theory. In the presence of noisy CF, the term $\delta\mathbf{I}$ can be viewed as a Tikhonov regularizing term which determines the optimal solution of a smooth version of CF [8].

B.2.6 Quasi-Newton Methods or Variable Metric Methods

In many optimization problems, the Hessian matrix is not explicitly available. In the *quasi-Newton*, also known as *variable metric methods*, the inverse Hessian matrix is determined iteratively and in an approximate way. The Hessian is updated by analyzing successive gradient vectors. For example, in the so-called *sequential quasi-Newton methods*, the estimate of the inverse Hessian matrix is evaluated by considering two successive values of the CF gradient.

Consider the second-order CF approximation and let $\Delta\mathbf{w} = [\mathbf{w} - \mathbf{w}_k]$, $\mathbf{g}_k = \nabla J(\mathbf{w}_k)$, and $\mathbf{B}_k$ an approximation of the Hessian matrix $\mathbf{B}_k \approx \nabla^2 J(\mathbf{w}_k)$; from Eq. (B.19) we can write

$$J(\mathbf{w} + \Delta\mathbf{w}) \approx J(\mathbf{w}_k) + \Delta\mathbf{w}^T \mathbf{g}_k + \tfrac{1}{2}\Delta\mathbf{w}^T \mathbf{B}_k \Delta\mathbf{w}. \tag{B.30}$$

The gradient of this approximation (with respect to $\Delta\mathbf{w}$) can be written as

$$\nabla J(\mathbf{w}_k + \Delta\mathbf{w}_k) \approx \mathbf{g}_k + \mathbf{B}_k\Delta\mathbf{w}_k \tag{B.31}$$

called *secant equation*. The Hessian approximation can be chosen in order to *exactly* satisfy Eq. (B.31); so, $\Delta\mathbf{w}_k \rightarrow \mathbf{d}_k$ and setting this gradient to zero provides the Quasi-Newton adaptations

$$\Delta\mathbf{w}_k \rightarrow \mathbf{d}_k \tag{B.32}$$

In particular, in the method of Broyden–Fletcher–Goldfarb–Shanno (BFGS) [3, 9–11], the adaptation takes the form

$$\begin{aligned} \mathbf{d}_k &= -\mathbf{B}_k^{-1}\mathbf{g}_k \\ \mathbf{w}_{k+1} &= \mathbf{w}_k + \mu_k\mathbf{d}_k \\ \mathbf{B}_{k+1} &= \mathbf{B}_k - \frac{\mathbf{B}_k\mathbf{s}_k\mathbf{s}_k^T\mathbf{B}_k^T}{\mathbf{s}_k^T\mathbf{B}_k\mathbf{s}_k} + \frac{\mathbf{u}_k\mathbf{u}_k^T}{\mathbf{u}_k^T\mathbf{s}_k} \\ \mathbf{s}_k &= \mathbf{w}_{k+1} - \mathbf{w}_k \\ \mathbf{u}_k &= \mathbf{g}_{k+1} - \mathbf{g}_k, \end{aligned} \tag{B.33}$$

where the step size μ_k satisfies the above Wolfe conditions (B.14). It has been found that for the optimal performance a very loose line search with suggested values of the parameters in (B.14), equal to $\sigma_1 = 10^{-4}$ and $\sigma_2 = 0.9$, is sufficient.

A method that can be considered as a serious contender of the BFGS [4] is the so-called *symmetric rank-one* (SR1) method where the update is given by

$$\mathbf{B}_{k+1} = \mathbf{B}_k + \frac{(\mathbf{d}_k - \mathbf{B}_k\mathbf{s}_k)(\mathbf{d}_k - \mathbf{B}_k\mathbf{s}_k)^T}{\mathbf{s}_k^T(\mathbf{d}_k - \mathbf{B}_k\mathbf{s}_k)}. \tag{B.34}$$

It was first discovered by Davidon (1959), in his seminal paper on quasi-Newton methods, and rediscovered by several authors. The SR1 method can be derived by posing the following simple problem. Given a symmetric matrix $\mathbf{B}_k$ and the vectors $\mathbf{s}_k$ and $\mathbf{d}_k$, find a new symmetric matrix $\mathbf{B}_{k+1}$ such that $(\mathbf{B}_{k+1}–\mathbf{B}_k)$ has rank one, and such that

$$\mathbf{B}_k\mathbf{s}_k = \mathbf{d}_k. \tag{B.35}$$

Note that, to prevent the method from failing, one can simply set $\mathbf{B}_{k+1} = \mathbf{B}_k$ when the denominator in (B.34) is close to zero, though this could slow down the convergence speed.

Remark In order to avoid the computation of inverse matrix $\mathbf{B}_k$, denoting $\mathbf{H}_k$ as an approximation of the inverse Hessian matrix $\left(\mathbf{H}_k \approx [\nabla^2 J(\mathbf{w}_k)]^{-1}\right)$, and approximating $(\mathbf{d}_k \approx \Delta\mathbf{w}_k)$ the recursion (B.33) can be rewritten as

$$\begin{aligned} \mathbf{w}_{k+1} &= \mathbf{w}_k + \mu_k \mathbf{d}_k \\ \mathbf{d}_k &\simeq \mathbf{w}_{k+1} - \mathbf{w}_k = -\mathbf{H}_k \mathbf{g}_k \\ \mathbf{u}_k &= \mathbf{g}_{k+1} - \mathbf{g}_k \\ \mathbf{H}_{k+1} &= \left[\mathbf{I} - \frac{\mathbf{d}_k \mathbf{u}_k^T}{\mathbf{d}_k^T \mathbf{u}_k}\right] \mathbf{H}_k \left[\mathbf{I} - \frac{\mathbf{u}_k \mathbf{d}_k^T}{\mathbf{d}_k^T \mathbf{u}_k}\right] + \frac{\mathbf{d}_k \mathbf{d}_k^T}{\mathbf{d}_k^T \mathbf{u}_k}, \end{aligned} \tag{B.36}$$

where usually, the step size μ_k is optimized by a one-dimensional line search procedure (B.13) that takes the form

$$\mu_k \therefore \min_{\mu \in \mathbb{R}^+} J[\mathbf{w}_k - \mu \mathbf{H}_k \nabla J_k]. \tag{B.37}$$

The procedure is initialized with arbitrary IC $\mathbf{w}_{-1}$ and with the matrix $\mathbf{H}_{-1} = \mathbf{I}$. Alternatively, in the last of (B.36) $\mathbf{H}_k$ can be calculated with the Barnes–Rosen formula (see for [3] details)

$$\mathbf{H}_{k+1} = \mathbf{H}_k + \frac{(\mathbf{d}_k - \mathbf{H}_k \mathbf{u}_k)(\mathbf{d}_k - \mathbf{H}_k \mathbf{u}_k)^T}{(\mathbf{d}_k - \mathbf{H}_k \mathbf{u}_k)^T \mathbf{u}_k}. \tag{B.38}$$

The variable metric method is computationally more efficient than that of Newton. In particular, good line search implementations of BFGS method are given in the *IMSL* and *NAG* scientific software library. The BFGS method is fast and robust and is currently being used to solve a myriad of optimization problems [4].

B.2.7 Conjugate Gradient Method

Introduced by Hestenes–Stiefel [12] the *conjugate gradient algorithm* (CGA) marks the beginning of the field of large-scale nonlinear optimization. The CGA, while representing a simple change compared to SDA and the quasi-Newton method, has the advantage of a significant increase in the convergence speed and requires storage of only a few vectors.

Although there are many recent developments of limited memory and discrete Newton, CGA is still the one of the best choice for solving very large problems with relatively inexpensive objective functions. CGA, in fact, has remained one of the most useful techniques for solving problems large enough to make matrix storage impractical.

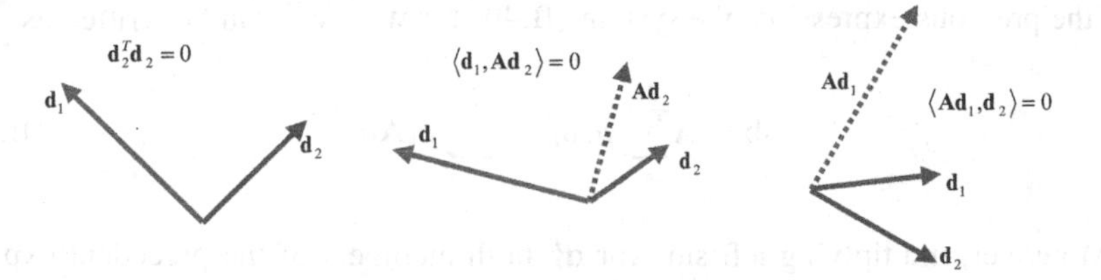

Fig. B.6 Example of orthogonal and **A**-conjugate directions

B.2.7.1 Conjugate Direction

Two vectors $(\mathbf{d}_1,\mathbf{d}_2) \in \mathbb{R}^{M\times 1}$ are defined orthogonal if $\mathbf{d}_1{}^T\mathbf{d}_2 = 0$ or $\langle \mathbf{d}_1,\mathbf{d}_2 \rangle = 0$. Given a symmetric and positive defined matrix $\mathbf{A} \in \mathbb{R}^{M\times M}$ the vectors are defined as **A**-*orthogonal* or **A**-*conjugate*, indicated as $\langle \mathbf{d}_1,\mathbf{d}_2 \rangle|_\mathbf{A} = 0$, if $\mathbf{d}_1^T\mathbf{A}\mathbf{d}_2 = 0$. Result in terms of scalar product is $\langle \mathbf{A}\mathbf{d}_1, \mathbf{d}_2 \rangle = \langle \mathbf{A}^T\mathbf{d}_1, \mathbf{d}_2 \rangle = \langle \mathbf{d}_1, \mathbf{A}^T\mathbf{d}_2 \rangle = \langle \mathbf{d}_1, \mathbf{A}\mathbf{d}_2 \rangle = 0$.

Preposition The conjugation implies the linear independence and for $\mathbf{A} \in \mathbb{R}^{M\times M}$ symmetric and positive definite, the set of **A**-conjugate vectors, $\langle \mathbf{d}_{k-1},\mathbf{d}_k \rangle|_\mathbf{A} = 0$, for $k = 0,\ldots,M-1$, indicated as $[\mathbf{d}_k]_{k=0}^{M-1}$, are linearly independent (Fig. B.6).

B.2.7.2 Conjugate Direction Optimization Algorithm

Given the standard optimization problem (B.1) with the hypothesis that the CF is a quadratic form of the type (B.16), the following theorem holds.

Theorem Given a set of nonzero **A**-conjugate directions, $[\mathbf{d}_k]_{k=0}^{M-1}$ for each IC $\mathbf{w}_{-1} \in \mathbb{R}^{M\times 1}$, the sequence $\mathbf{w}_k \in \mathbb{R}^{M\times 1}$ generated as

$$\mathbf{w}_{k+1} = \mathbf{w}_k + \mu_k\mathbf{d}_k \qquad \text{for} \qquad k = 0, 1, \ldots \tag{B.39}$$

with μ_k determined as *line search* criterion (B.17), converges in M steps to the unique optimum solution $\mathbf{w}^*$.

Proof The Proof is performed in two steps (1) computation of the step size μ_k; (2) Proof of the *subspace optimality Theorem*.

1. *Computation of the step size* μ_k

Consider the standard quadratic CF minimization problem for which

$$\nabla J(\mathbf{w}) \to 0 \quad \Rightarrow \quad \mathbf{A}\mathbf{w} = \mathbf{b} \tag{B.40}$$

with optimal solution $\mathbf{w}^* = \mathbf{A}^{-1}\mathbf{b}$. A set of nonzero **A**-conjugate directions $[\mathbf{d}_k]_{k=0}^{M-1}$ forming a base over $\mathbb{R}^M$ such that the solution can be expressed as

$$\mathbf{w}^* = \sum_{k=0}^{M-1} \mu_k\mathbf{d}_k. \tag{B.41}$$

For the previous expression, the system (B.40) for $\mathbf{w} = \mathbf{w}^*$ can be written as

$$\mathbf{b} = \mathbf{A}\sum_{k=1}^{M} \mu_k \mathbf{d}_k = \sum_{k=1}^{M} \mu_k \mathbf{A}\mathbf{d}_k \tag{B.42}$$

Moreover, multiplying left side for $\mathbf{d}_i^T$ both members of the precedent expression, and being by definition $\langle \mathbf{d}_i^T\mathbf{A},\, \mathbf{d}_j\rangle = 0$ for $i \neq j$, we can write

$$\mathbf{d}_i^T\mathbf{b} = \mu_k \mathbf{d}_i^T\mathbf{A}\mathbf{d}_k \tag{B.43}$$

which allows the calculation of the coefficients of the base (B.41) μ_k as

$$\mu_k = \frac{\mathbf{d}_k^T\mathbf{b}}{\mathbf{d}_k^T\mathbf{A}\mathbf{d}_k}. \tag{B.44}$$

For the definition of the CGA method, we consider a recursive solution for CF minimization, in which in the (k–1)th iteration we consider negative gradient around $\mathbf{w}_k$, called in this context, *residue*. Indicating the negative direction of the gradient as $\mathbf{g}_{k-1} = -\nabla J(\mathbf{w}_{k-1})$, we have

$$\mathbf{g}_{k-1} = \mathbf{b} - \mathbf{A}\mathbf{w}_{k-1} \tag{B.45}$$

The expression (B.44) can be rewritten as

$$\mu_k = \frac{\mathbf{d}_k^T(\mathbf{g}_{k-1} + \mathbf{A}\mathbf{w}_{k-1})}{\mathbf{d}_k^T\mathbf{A}\mathbf{d}_k}. \tag{B.46}$$

From definition of $\mathbf{A}$-conjugate directions $\mathbf{d}_k^T\mathbf{A}\mathbf{w}_{k-1} = 0$ we have

$$\mu_k = \frac{\mathbf{d}_k^T\mathbf{g}_{k-1}}{\mathbf{d}_k^T\mathbf{A}\mathbf{d}_k}. \tag{B.47}$$

Remark Expression (B.47) represents an alternative formulation for the optimal step-size computation (B.17).

2. *Subspace optimality Theorem*

Given a quadratic CF $J(\mathbf{w}) = \frac{1}{2}\mathbf{w}^T\mathbf{A}\mathbf{w} - \mathbf{w}^T\mathbf{b}$, and a set of nonzero $\mathbf{A}$-conjugate directions, $[\mathbf{d}_k]_{k=0}^{M-1}$ for any IC $\mathbf{w}_{-1} \in \mathbb{R}^{M\times 1}$ the sequence $\mathbf{w}_k \in \mathbb{R}^{M\times 1}$ generated as

$$\mathbf{w}_{k+1} = \mathbf{w}_k + \mu_k\mathbf{d}_k, \qquad \text{for} \qquad k \geq 0 \tag{B.48}$$

with

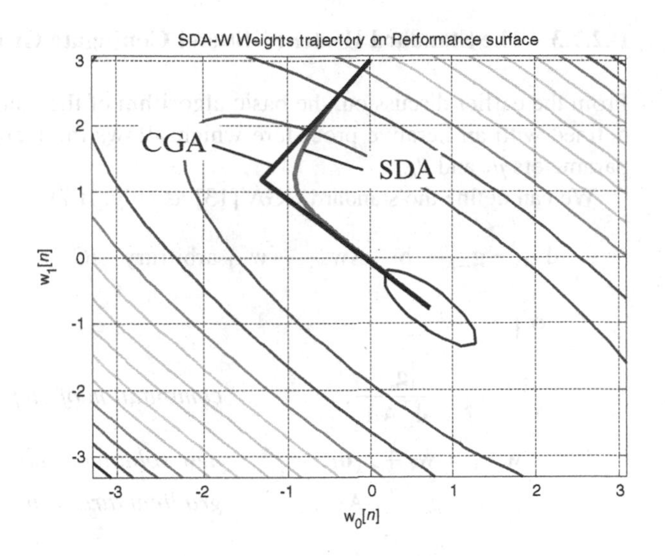

Fig. B.7 Trajectories of the weights on the isolevel CF curves for steepest descent algorithm (SDA) and the standard Hestenes–Stiefel conjugate gradient algorithm

$$\mu_k = \frac{\mathbf{d}_k^T \mathbf{g}_{k-1}}{\mathbf{d}_k^T \mathbf{A} \mathbf{d}_k} \tag{B.49}$$

reaches its minimum $\mathbf{w}_{k+1} \to \mathbf{w}^*$ value in the set $\left[\mathbf{w}_{-1} + \text{span}\{\, \mathbf{d}_0 \quad \cdots \quad \mathbf{d}_k \,\}\right]$. Equivalently, considering the general solution $\overline{\mathbf{w}}$, we have that $\left[\nabla J(\overline{\mathbf{w}})\right]^T \mathbf{d}_k = 0$. Then there is, necessarily, a parameter $\beta_i \in \mathbb{R}$ such that

$$\overline{\mathbf{w}} = \mathbf{w}_{-1} + \beta_0 \mathbf{d}_0 + \cdots + \beta_k \mathbf{d}_k \tag{B.50}$$

Then

$$\begin{aligned} 0 &= \left[\nabla J(\overline{\mathbf{w}})\right]^T \mathbf{d}_i \\ &= \left[\mathbf{A}\left[\mathbf{w}_{-1} + \beta_0 \mathbf{d}_0 + \cdots + \beta_k \mathbf{d}_{k-1}\right] + \mathbf{b}\right]^T \mathbf{d}_i \\ &= \left[\mathbf{A}\mathbf{w}_{-1} + \mathbf{b}\right]^T + \beta_0 \mathbf{d}_0^T \mathbf{A} \mathbf{d}_i + \cdots + \beta_k \mathbf{d}_k^T \mathbf{A} \mathbf{d}_i \\ &= \left[\nabla J(\mathbf{w}_{-1})\right]^T \mathbf{d}_i + \beta_i \mathbf{d}_i^T \mathbf{A} \mathbf{d}_i, \end{aligned} \tag{B.51}$$

whereby we can calculate the parameter β_i as

$$\beta_i = \frac{-\left[\nabla J(\mathbf{w})\right]^T \mathbf{d}_k}{\mathbf{d}_k^T \mathbf{A} \mathbf{d}_k} = \frac{\mathbf{g}_{k+1}^T \mathbf{A} \mathbf{d}_k}{\mathbf{d}_k^T \mathbf{A} \mathbf{d}_k} \tag{B.52}$$

Q.E.D.

B.2.7.3 The Standard Hestenes–Stiefel Conjugate Gradient Algorithm

From the earlier discussion, the basic algorithm of the conjugate directions can be defined with an iterative procedure which allows the recursive calculation of the parameters μ_k and β_k.

We can define the standard CGA [13] as (Fig. B.7)

$$\mathbf{d}_{-1} = \mathbf{g}_{-1} = \mathbf{b} - \mathbf{A}\mathbf{w}_{-1} \quad (\mathbf{w}_{-1} \text{ arbitrary}) \; IC \tag{B.53}$$

do {

$$\mu_k = \frac{|\mathbf{g}_k|^2}{\mathbf{d}_k^T \mathbf{A} \mathbf{d}_k}, \qquad \textit{computation of step size} \tag{B.54}$$

$$\mathbf{w}_{k+1} = \mathbf{w}_k + \mu_k \mathbf{d}_k, \qquad \textit{new solution or adaptation} \tag{B.55}$$

$$\mathbf{g}_{k+1} = \mathbf{g}_k - \mu_k \mathbf{A} \mathbf{d}_k, \qquad \textit{gradient direction update}$$

$$\beta_k = \frac{|\mathbf{g}_{k+1}|^2}{|\mathbf{g}_k|^2}, \qquad \textit{computation of "beta" parameter} \tag{B.56}$$

$$\mathbf{d}_{k+1} = \mathbf{g}_{k+1} + \beta_k \mathbf{d}_k, \qquad \textit{search direction} \tag{B.57}$$

} while $\left(\|\mathbf{g}_k\| > \varepsilon\right)$ *end criterion* : *output for* $\|\mathbf{g}_k\| < \varepsilon$.

```
%---------------------------------------------------------------- ----
% The type 1 Hestenes - Stiefel Conjugate Gradient Algorithm
%  for CF: J(w) = c - w'b + (1/2)w'Aw;
%
%   Copyright 2013 - A. Uncini
%   DIET Dpt - University of Rome 'La Sapienza' - Italy
%   $Revision: 1.0$  $Date: 2013/03/09$
%--------------------------------------------------------------------
function [w, k] = CGA1(w, b, A, c, mu, tol, MaxIter)
    d = b - A*w;
    g = d;
    g1 = g'*g;
    for k=1:MaxIter
        Ad = A*d;
        mu = g1/(d'*Ad);    % Optimal step-size (B.54)
        w  = w + mu*d;      % up-date solution (B.55)
        g  = g - mu*Ad;     % up-date gradient or residual (B.55)
        g2 = g'*g;
        be = g2/g1;         % 'beta' parameter (B.56)
        d  = g + be*d;;     % up-date direction (B.57)
        g1 = g2;
        if ( g2 <= tol ), break, end % end criteria
    end
end
% Hestenes - Stiefel Conjugate Gradient Algorithm type 1 -----------
```

Remark In place of the formulas (B.54) and (B.56) one may use

$$\mu_k = \frac{\mathbf{d}_k^T \mathbf{g}_k}{\mathbf{d}_k^T \mathbf{A}\mathbf{d}_k} \tag{B.58}$$

$$\beta_k = -\frac{\mathbf{g}_{k+1}^T \mathbf{A}\mathbf{d}_k}{\mathbf{d}_k^T \mathbf{A}\mathbf{d}_k}. \tag{B.59}$$

These formulas, although more complicated than (B.54) and (B.56), have μ and β parameters more easily changed during the iterations.

Moreover note that the direction of estimated gradients (or residual) $\mathbf{g}_k$ is mutually orthogonal $\langle \mathbf{g}_{k+1}, \mathbf{g}_k \rangle = 0$, while the direction of vectors $\mathbf{d}_k$ is mutually $\mathbf{A}$-conjugate $\langle \mathbf{d}_{k+1}, \mathbf{A}\mathbf{d}_k \rangle = 0$.

```
%-----------------------------------------------------------------------
% The type 2 Hestenes - Stiefel Conjugate Gradient Algorithm
%    J(w) = c - w'b + (1/2)w'Aw;
%
%   Copyright 2013 - A. Uncini
%   DIET Dpt - University of Rome 'La Sapienza' - Italy
%   $Revision: 1.0$  $Date: 2013/03/09$
%-----------------------------------------------------------------------
function [w,,k] = CGA2(w, b, A, c, mu, tol, MaxIter)
    d = b - A*w;
    g = d;
    for k = 1 : MaxIter
        Ad  = A*d;
        dAd = d'*Ad;
        mu  = (d'*g)/dAd;    % Optimal step-size (B.58)
        w   = w + mu*d;      % up-date solution
        g   = g - mu*Ad;     % up-date direction
        be  = -(g'*Ad)/dAd;  % 'beta' param (B.59)
        d   = g + be*d;
        if ( norm(g) <= tol ), break, end % end criteria
    end
end
% Hestenes - Stiefel Conjugate Gradient Algorithm type 2 -----------
```

B.2.7.4 Gradient Algorithm for Generic CF

The method of conjugate gradient can be generalized to find a minimum of a generic CF. In this case the search method is sometimes called nonlinear CGA [14]. In this case the gradient cannot explicitly be computed but only estimated in various ways. In particular the residual cannot be directly found but, let $\nabla J(\mathbf{w}_k)$ an estimation of the CF's gradient at the kth iteration, we set residual as $\mathbf{g}_k = -\nabla J(\mathbf{w}_k)$.

The line search procedure cannot be computed as in the Hestenes–Stiefel CGA previously described and could be substituted by minimizing the expression

$$\left[\nabla J(\mathbf{w}_k + \mu_k \mathbf{d}_k)\right]^T \mathbf{d}_k. \tag{B.60}$$

Moreover, the estimated Hessian of CF $\nabla^2 J(\mathbf{w}_k)$ plays the role of matrix $\mathbf{A}$.

A simple modified CGA method form is defined by the following recurrence. Starting from IC $\mathbf{w}_{-1}$ and $\beta_0^{\text{XY}} = 0$

$$\mathbf{w}_{-1} \quad (\mathbf{w}_{-1} \text{ arbitrary}) \text{ IC} \tag{B.61}$$

$$\mathbf{d}_{-1} = \mathbf{g}_{-1} = -\nabla J(\mathbf{w}_{-1}) \text{ IC} \tag{B.62}$$

`do{`

$\quad$ `determine` μ_k, *Wolfe conditions*

$$\mathbf{w}_{k+1} = \mathbf{w}_k + \mu_k \mathbf{d}_k, \quad \textit{Adaptation} \tag{B.63}$$

$\quad$ $\mathbf{g}_k = -\nabla J(\mathbf{w}_k)$, *gradient estimation*

$$\texttt{compute } \beta_k = \beta_k^{\text{XY}}, \quad \textit{"beta"parameter} \tag{B.64}$$

$$\mathbf{d}_{k+1} = \mathbf{g}_{k+1} + \beta_k \mathbf{d}_k, \quad \textit{Compute the search direction} \tag{B.65}$$

$$\texttt{if} \left|\mathbf{g}_{k+1}^T \mathbf{g}_k\right| > 0.2 \left\|\mathbf{g}_{k+1}\right\|^2 \texttt{then} \, \mathbf{d}_{k+1} = \mathbf{g}_{k+1}, \quad \textit{Restartcondition} \tag{B.66}$$

`} while` $(\|\mathbf{g}_k\| > \varepsilon)$ *end criterion. Exit when* $\|\mathbf{g}_k\| < \varepsilon$

The parameter β_k^{XY}, which plays a central role for nonlinear CGA, can be determined through various philosophies of calculation. Below are shown the most common methods for the calculation of the *beta* parameter (see for details [15])

$$\beta_k^{\text{HS}} = \frac{\mathbf{d}_k^T \mathbf{g}_{k+1}}{\mathbf{d}_k^T (\mathbf{w}_{k+1} - \mathbf{w}_k)}, \quad \textit{Hestenes} - \textit{Stiefel}\,(\textit{HS}) \tag{B.67}$$

$$\beta_k^{\text{PR}} = \frac{\mathbf{d}_{k+1}^T \mathbf{g}_{k+1}}{\mathbf{g}_k^T \mathbf{g}_k}, \quad \textit{Polak} - \textit{Ribi\`{O}re} - \textit{Polyak}\,(\textit{PRP}) \tag{B.68}$$

$$\beta_k^{\text{HS}} = \frac{\mathbf{d}_k^T \mathbf{g}_{k+1}}{\mathbf{d}_k^T (\mathbf{w}_{k+1} - \mathbf{w}_k)}, \quad \textit{Liu} - \textit{Storey}\,(\textit{LS}) \tag{B.69}$$

$$\beta_k^{\text{FR}} = \frac{\mathbf{g}_{k+1}^T \mathbf{g}_{k+1}}{\mathbf{g}_k^T \mathbf{g}_k}, \quad \textit{Fletcher} - \textit{Reevs}\,(\textit{FR}) \tag{B.70}$$

$$\beta_k^{\text{CD}} = -\frac{\mathbf{g}_{k+1}^T \mathbf{g}_{k+1}}{\mathbf{g}_k^T \mathbf{d}_k}, \quad \textit{Conjugate Descent} - \textit{Fletcher}\,(\textit{CD}) \tag{B.71}$$

$$\beta_k^{\text{DY}} = \frac{\mathbf{g}_{k+1}^T \mathbf{g}_{k+1}}{\mathbf{d}_k^T (\mathbf{w}_{k+1} - \mathbf{w}_k)}, \quad \textit{Dai} - \textit{Yuan}\,(\textit{DY}). \tag{B.72}$$

Note that, in the specialized literature, there are many other variants (see, for example [4]). For strictly quadratic CF this method reduces to the linear search provided μ_k is the exact minimizer [3]. Other choices of the parameter β_k^{XY} in (B.65) also possess this property and give rise to distinct algorithms for nonlinear problems.

In the CGA, the increase in convergence speed is obtained from information on the search direction that depends on the previous iteration $\mathbf{d}_{k-1}$, moreover for a quadratic CF, it is conjugated to the gradient direction. Theoretically, the algorithm, for $\mathbf{w} \in \mathbb{R}^M$, converges in M or less iterations.

To avoid numerical inaccuracy in the direction search calculation or for the non-quadratic CF nature, the method requires a periodic reinitialization. Indeed, over certain conditions, (B.67)–(B.72) may assume negative value. So a more appropriate choice is

$$\beta_k = \max\{\beta_k^{\mathrm{XY}}, 0\}. \tag{B.73}$$

Thus if a negative value of β_k^{PR} occurs, this strategy will restart the iteration along the correct steepest descent direction.

The CGA can be considered as an intermediate approach between the SDA and the quasi-Newton method. Unlike other algorithms, the CGA main advantage is derived from the fact of not needing to explicitly estimate the Hessian matrix which is, in practice, replaced by the β_k parameter.

B.3 Constrained Optimization Problem

The problem of *constrained optimization* can be formulated as: find a vector $\mathbf{w} \in \Omega \subset \mathbb{R}^M$ that minimizes (maximizes) a scalar function

$$\min_{\mathbf{w} \in \Omega} J(\mathbf{w}) \tag{B.74}$$

subject to (s.t.) the constraints

$$g_i(\mathbf{w}) \geq 0, \quad \text{for } i = 1, 2..., M. \tag{B.75}$$

Methods for solving constrained optimization problems are often characterized by two conflicting needs:

- Finding admissible solutions,
- Finding the algorithm to minimize the objective function.

In general, there are two basic approaches:

- Transform the problems into simpler constrained problems,
- Transform the problems into a sequence (in the limit a single) of unconstrained problems.

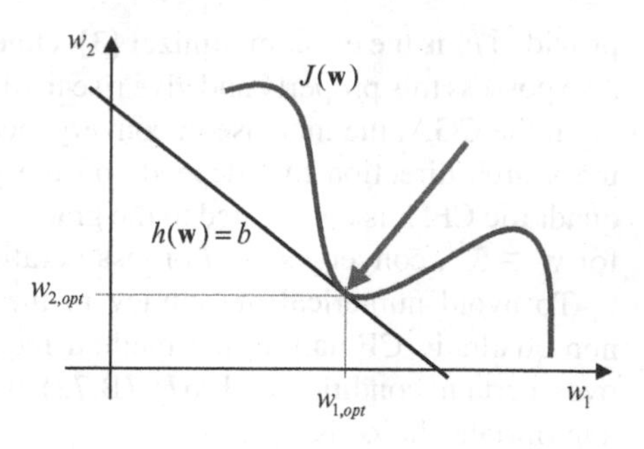

Fig. B.8 In the optimal point curve $J(\mathbf{w})$ and $h(\mathbf{w})$ are necessarily tangent

B.3.1 Single Equality Constraint: Existence and Characterization of the Minimum

As in unconstrained optimization problems (see Sect. B.1.2), to have admissible solution some sufficient and sufficient and necessary conditions must be satisfied.

For example, in the case of single equality constraint the problem can be formulated as

$$\min_{\mathbf{w}\in\Omega} J(\mathbf{w}) \quad \text{s.t.} \quad h(\mathbf{w}) = b. \tag{B.76}$$

First-order necessary condition (FONC) for minimum (or maximum) is that the functions $J(\mathbf{w})$ and $h(\mathbf{w})$ have continuous first-order partial derivative and that there exists some free parameter scalar λ such that

$$\nabla J(\mathbf{w}) + \lambda \nabla h(\mathbf{w}) = 0 \tag{B.77}$$

or, as illustrated in Fig. B.8, the two surface must be tangent. Note that $h(\mathbf{w}) = b$ or $-h(\mathbf{w}) = -b$ are the same and that there is non-restriction on λ.

B.3.2 Constrained Optimization: Methods of Lagrange Multipliers

The *method of Lagrange multipliers* (MLM) is the fundamental tool for analyzing and solving nonlinear constrained optimization problems. Lagrange multipliers can be used to find the extreme of a multivariate function $J(\mathbf{w})$ subject to the constraint function $h(\mathbf{w}) = b$, where J and h are functions with continuous first partial derivatives on the open set, containing the curve $h(\mathbf{w}) - b = 0$, and $\nabla h(\mathbf{w}) \neq 0$ at any point on the curve.

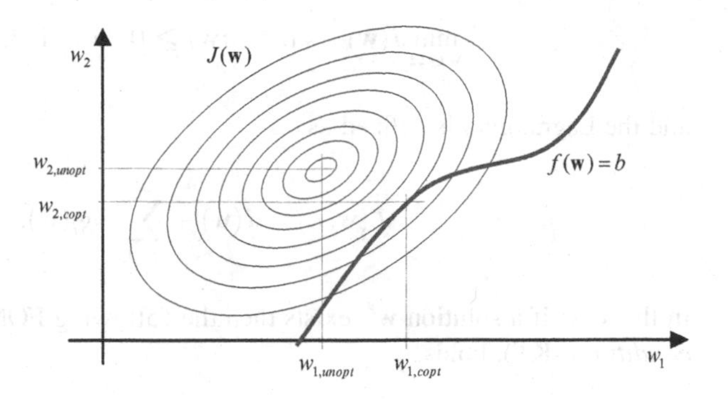

Fig. B.9 Example of a constrained optimization problem for $M = 2$. The constrained optimum value is the closest point to the unconstrained optimum, belonging to the constraint curve $f(\mathbf{w}) = b$

B.3.2.1 Optimization with Single Constraint

In the case of a single equality constrained optimization problem (B.76), we define the *Lagrangian* or *Lagrange function* as

$$L(\mathbf{w}, \lambda) = J(\mathbf{w}) + \lambda\left[h(\mathbf{w}) - b\right] \tag{B.78}$$

such that, in the case that the existence condition is verified, the solution can be found solving the following unconstrained optimization problem associated with (B.76):

$$\min_{\mathbf{w}\in\Omega} L(\mathbf{w}, \lambda) \tag{B.79}$$

$$\min_{\lambda\in L} L(\mathbf{w}, \lambda) \tag{B.80}$$

That is, $\nabla L(\mathbf{w}, \lambda) = 0$, or

$$\nabla_{\mathbf{w}} L(\mathbf{w}, \lambda) = \nabla J(\mathbf{w}) + \lambda \nabla h(\mathbf{w}) = 0 \tag{B.81}$$

$$\nabla_{\lambda} L(\mathbf{w}, \lambda) = h(\mathbf{w}) - b = 0. \tag{B.82}$$

If (B.81) and (B.82) hold then $(\mathbf{w}, \lambda)$ is a stationary point for the Lagrange function.

In other words, the Lagrange multiplier method represents a *necessary condition* for the existence of optimal solution in such constrained optimization problems.

Fig. B.9 shows an example of a constrained optimization problem for $M = 2$.

B.3.2.2 Optimization Problem with Multiple Inequality Constraints: Kuhn–Tucker Conditions

The generalization for multiple constraints can be formulated as

$$\min_{\mathbf{w}\in\Omega} J(\mathbf{w}) \quad \text{s.t.} \quad g_i(\mathbf{w}) \geq 0 \quad i = 1, 2, ..., K \tag{B.83}$$

and the Lagrangian is defined as

$$L(\mathbf{w}, \boldsymbol{\lambda}) = J(\mathbf{w}) + \sum_{i=1}^{K} \lambda_i g_i(\mathbf{w}). \tag{B.84}$$

In this case if a solution $\mathbf{w}^*$ exists then the following FONC, called *Kuhn–Tucker conditions* (KT), holds:

$$\begin{aligned} \nabla J(\mathbf{w}^*) + \sum_{i=1}^{K} \lambda_i^* \nabla g_i(\mathbf{w}^*) &= 0 \\ g_i(\mathbf{w}^*) &\geq 0 \\ \lambda_i^* &\leq 0, \qquad \text{for} \qquad i = 1, 2, ..., K \\ \lambda_i^* g_i(\mathbf{w}^*) &= 0. \end{aligned} \tag{B.85}$$

A feasible point $\mathbf{w}^*$ for the minimization problem (B.83) is *regular point* if the set of vectors $\nabla g_i(\mathbf{w}^*)$ is linearly independent over a set of indices corresponding to the equality constraints at optimal point $\mathbf{w}^*$, formally

$$\nabla g_i(\mathbf{w}^*) \ i \in I_0, \qquad \text{for} \qquad I_0 \triangleq \{ i \in [1, K] \therefore g_i(\mathbf{w}^*) = 0 \} \tag{B.86}$$

In eqns. (B.85) we have assumed that the first derivatives $\nabla J(\mathbf{w})$ and $\nabla g(\mathbf{w})$ exist and that $\mathbf{w}^*$ is a *regular point* or that the constraints satisfy the *regularity conditions*. Moreover, a point $\mathbf{w} \in \Omega \subset \mathbb{R}^M$ is called *feasible point*, and the optimization problem is called consistent, if the set of feasible points is nonempty. A feasible point $\mathbf{w}^*$ is a local minimizer if $f(\mathbf{w}^*)$ is a minimum on the set of feasible points.

A point $(\mathbf{w}^*, \boldsymbol{\lambda}^*)$ at which KT conditions hold is called a *saddle point* for the Lagrangian function if $J(\mathbf{x})$ is convex and all $g_i(\mathbf{w})$ are concave. At the saddle point the Lagrangian satisfies the inequalities

$$L(\mathbf{w}^*, \boldsymbol{\lambda}) \leq L(\mathbf{w}^*, \boldsymbol{\lambda}^*) \leq L(\mathbf{w}, \boldsymbol{\lambda}^*). \tag{B.87}$$

So, for the Lagrange function a minimum exists with respect to $\mathbf{x}$ and a maximum with respect to $\boldsymbol{\lambda}$.

Note also that the last of (B.85) conditions, that is, $\lambda_i^* g_i(\mathbf{x}^*) = 0 \quad i = 1, 2, ..., K$, is called *complementary slackness condition.*

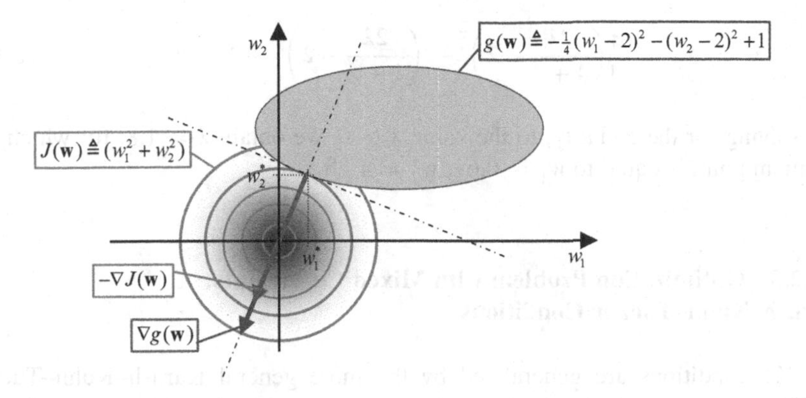

Fig. B.10 In the optimum point, the surface of the CF $J(\mathbf{w})$ is tangent to the curve of the constraint $g(\mathbf{w})$

Example Consider the problem

$$\min_{\mathbf{w}\in\Omega}\left(w_1^2+w_2^2\right) \quad \text{s.t.} \quad -\tfrac{1}{4}(w_1-2)^2-\left(w_2-2\right)^2+1\geq 0\,. \tag{B.88}$$

The KT is defined as

$$\begin{gathered}\begin{bmatrix}2w_1\\2w_2\end{bmatrix}-\lambda\cdot\begin{bmatrix}\frac{1}{2}(2-w_1)\\2(2-w_2)\end{bmatrix}=0\\ -\tfrac{1}{4}(w_1-2)^2-\left(w_2-2\right)^2+1\geq 0\\ \lambda\geq 0\\ \lambda\left[1-\tfrac{1}{4}(w_1-2)^2-(w_2-2)^2\right]=0.\end{gathered} \tag{B.89}$$

Geometrically illustrated in Fig. B.10.

Calculation of the solution with the KT

$$\begin{bmatrix}2w_1\\2w_2\end{bmatrix}-\lambda\cdot\begin{bmatrix}\frac{1}{2}(2-w_1)\\2(2-w_2)\end{bmatrix}=0 \tag{B.90}$$

for which

$$w_1=\frac{2\lambda}{4+\lambda};\quad w_2=\frac{2\lambda}{1+\lambda}. \tag{B.91}$$

For $\lambda=0$, one has $w_1=0$ and $w_2=0$, which, however, is not a feasible solution as the constraint conditions (B.88) are not met. It follows that λ must necessarily be positive. Substituting the values (B.91) in the constraint

$$-\frac{1}{4}\left(\frac{2\lambda}{4+\lambda}-2\right)^2-\left(\frac{2\lambda}{1+\lambda}-2\right)^2+1\geq 0 \tag{B.92}$$

and solving for the equality, to the value $\lambda > 0$, we obtain $\lambda = 1.8$, for which the optimum point is equal to $w_1^* = 0.61$, $w_2^* = 1.28$.

B.3.2.3 Optimization Problem with Mixed Constraints: Karush–Kuhn–Tucker Conditions

The KT conditions are generalized by the more general Karush–Kuhn–Tucker (KKT) conditions, which take into account equality and inequality constraints of the most general form $h_i(\mathbf{x}) = 0$, $g_i(\mathbf{x}) \geq 0$, and $f_i(\mathbf{x}) \leq 0$. The KKT conditions are necessary for a solution in nonlinear programming to be optimal, provided some regularity conditions are satisfied.

In the presence of equality and inequality constraints the nonlinear optimization problem can be written as

$$\min J(\mathbf{w}) \quad \text{s.t.} \quad \begin{cases} \sum_{i=1}^{K_l}\kappa_i l_i(\mathbf{w}) \leq b_i \\ \sum_{i=1}^{K_g}\sigma_i g_i(\mathbf{w}) \geq b_i \\ \sum_{i=1}^{K_e}\upsilon_i h_i(\mathbf{w}) = b_i, \end{cases} \tag{B.93}$$

where $J(\mathbf{w})$, $l_i(\mathbf{w})$, $g_i(\mathbf{w})$, and $h_i(\mathbf{w})$, for all i, have continuous first-order partial derivative on some subset $\Omega \subset \mathbb{R}^M$. Let

$$\boldsymbol{\lambda} \in \mathbb{R}^K = \begin{bmatrix} \kappa_1 & \cdots & \kappa_{K_l} & \sigma_1 & \cdots & \sigma_{K_g} & \upsilon_1 & \cdots & \upsilon_{K_e} \end{bmatrix}^T \tag{B.94}$$

with $K = K_l + K_g + K_e$, the vector containing all the Lagrange multipliers, and

$$\mathbf{f}(\mathbf{w}) = \begin{bmatrix} \mathbf{l}(\mathbf{w}) & \mathbf{g}(\mathbf{w}) & \mathbf{h}(\mathbf{w}) \end{bmatrix}^T \tag{B.95}$$

a vector of functions containing all the inequalities and equalities constraints, for the problem (B.93) the Lagrangian assumes the forms

$$\begin{aligned} L(\mathbf{w},\boldsymbol{\lambda}) &= J(\mathbf{w}) + \sum_{i=1}^{K} \lambda_i \left(f_i(\mathbf{w}) - b_i \right) \\ &= J(\mathbf{w}) + [\boldsymbol{\kappa} \;\; \boldsymbol{\sigma} \;\; \upsilon]^T \left[\mathbf{l}(\mathbf{w}) \;\; \mathbf{g}(\mathbf{w}) \;\; \mathbf{h}(\mathbf{w}) \right] \\ &= J(\mathbf{w}) + \boldsymbol{\lambda} \mathbf{f}(\mathbf{w}), \end{aligned} \tag{B.96}$$

where vectors $\boldsymbol{\kappa}$, $\boldsymbol{\sigma}$, and υ are called dual variables. Further, suppose that $\mathbf{w}^*$ is a regular point for the problem. If $\mathbf{w}^*$ is a local minimum that satisfies some regularity conditions, then there exist constants vector $\boldsymbol{\lambda}^*$ such that (KKT conditions)

$$\nabla J(\mathbf{w}^*) + \sum_{i=1}^{K} \lambda_i^* \nabla f_i(\mathbf{w}^*) = 0 \tag{B.97}$$

and

$$\begin{array}{ll} \kappa_i^* \geq 0 & i = 1, 2, \ldots, K_l \\ \sigma_i^* \leq 0 & i = 1, 2, \ldots, K_g \\ \upsilon_i^* = \pm 1 & i = 1, 2, \ldots, K_e \;\; \text{(arbitray sign)} \\ \lambda_i^* = 0 & i \in I_0, \;\; \text{for } I_0 \triangleq \left\{ i \in \left[1, K_l + K_g\right] \therefore f_i(\mathbf{w}^*) = 0 \right\} \\ \lambda_i^* \left[f(\mathbf{w}^*) - b_i \right] = 0 & i = 1, 2, \ldots, K_l + K_g, \end{array} \tag{B.98}$$

where I_0 means the set of indices i from $i \in [1, K_l + K_g]$ for which the inequalities are satisfied at $\mathbf{w}^*$ as strict inequalities.

In the case that the functions $J(\mathbf{w})$ and $f_i(\mathbf{w})$ are convex, then $\lambda_i^* > 0$, and concave, then $\lambda_i^* < 0$, for $i = 1, 2, \ldots, K$, then the point $(\mathbf{w}^*, \boldsymbol{\lambda}^*)$ is a *saddle point* of the Lagrangian function (B.96), and $\mathbf{w}^*$ is a global minimizer of the problem (B.93).

Observe that in the case only a equality constraint is present, $h_i(\mathbf{w}) = b_i$, $i = 1, 2, \ldots, K$ the above condition simplifies as

$$\nabla J(\mathbf{w}^*) + \sum_{i=1}^{K} \upsilon_i^* \nabla h_i(\mathbf{w}^*) = 0 \tag{B.99}$$

and the conditions (B.98) are vacuous.

Remarks The KKT conditions provide that the intersection of the set of feasible directions with the set of descent directions coincides with the intersection of the set of feasible directions for linearized constraints with the set of descent directions.

To ensure that the necessary KKT conditions allow to identify local minimum point, assumption of regularity of constraints must be satisfied. In general, it may require the regularity of all admissible solutions, but, in practice, it is sufficient that the regularity conditions are satisfied only for such point.

In some cases, the necessary conditions are also sufficient for optimality. This is the case when the objective function J and the inequality constraints l_i, g_i are continuously differentiable convex functions and the equality constraints h_j are affine functions. Moreover, the broader class of functions in which KKT conditions guarantees global optimality are the so-called *invex functions*.

The invex functions, which represent a generalization of convex functions, are defined as differentiable vector functions $\mathbf{r}(\mathbf{w})$, for which there exists a vector valued function $\mathbf{q}(\mathbf{w}, \mathbf{u})$, such that

$$\mathbf{r}(\mathbf{w}) - \mathbf{r}(\mathbf{u}) \geq \mathbf{q}(\mathbf{w}, \mathbf{u}) \cdot \nabla \mathbf{r}(\mathbf{u}) \quad \forall \mathbf{w}, \mathbf{u}. \tag{B.100}$$

In other words, a function $\mathbf{r}(\mathbf{w})$ is an *invex function* iff each stationary point (a point of a function where the derivative is zero) is a global minimum point.

So, if equality constraints are affine functions and inequality constraints and the objective function are continuously differentiable invex functions, then KKT conditions are sufficient for global optimality.

B.3.3 Dual Problem Formulation

Consider the previously treated optimization problem (Sect. B.3.2.2), with multiple inequality constraints (B.83) and Lagrangian (B.84), with a convex objective function $J(\mathbf{w})$ and concave constraint functions $g_i(\mathbf{w})$, here called *primal inequality-constrained problem.*

For this problem, at the *saddle point* the Lagrangian satisfies the inequalities (B.87), that is, $L(\mathbf{w}^*, \boldsymbol{\lambda}) \leq L(\mathbf{w}^*, \boldsymbol{\lambda}^*) \leq L(\mathbf{w}, \boldsymbol{\lambda}^*)$, and the followings properties hold:

$$\begin{aligned} \nabla_{\mathbf{w}} L(\mathbf{w}^*, \boldsymbol{\lambda}^*) &= \mathbf{0} \\ \nabla_{\boldsymbol{\lambda}} L(\mathbf{w}^*, \boldsymbol{\lambda}^*) &= \mathbf{0} \\ \nabla_{\mathbf{w}}\ L(\mathbf{w}, \boldsymbol{\lambda}^*) &\geq \mathbf{0} \\ \nabla_{\boldsymbol{\lambda}} L(\mathbf{w}^*, \boldsymbol{\lambda}) &\leq \mathbf{0}. \end{aligned} \tag{B.101}$$

Note that, since the Lagrangian exhibits a minimum with respect to $\mathbf{w}$ and a maximum with respect to $\boldsymbol{\lambda}$, we can reformulate the primal inequality-constrained problem (B.83, B.84) as the min–max problem of finding a vector $\mathbf{w}^*$ which solves

$$\min_{\mathbf{w} \in \Omega} \max_{\lambda_i \leq 0} L(\mathbf{w}, \boldsymbol{\lambda}) = \min_{\mathbf{w} \in \Omega} \max_{\lambda_i \leq 0} \left\{ J(\mathbf{w}) + \sum_{i=1}^{K} \lambda_i g_i(\mathbf{w}) \right\}. \tag{B.102}$$

The above expression allows us to transform the primal min–max problem (B.102) in an equivalent *dual max–min problem* defined as

$$\max_{\mathbf{w}\in\Omega} L(\mathbf{w},\boldsymbol{\lambda}) \quad \text{s.t.} \quad \nabla J(\mathbf{w}) + \sum_{i=1}^{K} \lambda_i \nabla g_i(\mathbf{w}) = 0, \ \ (\lambda_i \leq 0). \tag{B.103}$$

Assuming that there is a unique minimum $(\mathbf{w}^*, \boldsymbol{\lambda}^*)$ to the problem: $\min_{\mathbf{w}\in\Omega} L(\mathbf{w},\boldsymbol{\lambda})$, then for each fixed vector $\boldsymbol{\lambda} \leq 0$ we can define a Lagrange function in terms of the alone Lagrange multipliers $\boldsymbol{\lambda}$ as

$$L(\boldsymbol{\lambda}) \triangleq \min_{\mathbf{w}\in\Omega} L(\mathbf{w},\boldsymbol{\lambda}). \tag{B.104}$$

The optimization problem can be now defined, in a more simple and elegant dual form, as

$$\max L(\boldsymbol{\lambda}) \quad \text{s.t.} \quad \lambda_i \leq 0, \quad i = 1, 2, ..., K, \tag{B.105}$$

where the Lagrange multipliers $\boldsymbol{\lambda}$ are called *dual variables* and $L(\boldsymbol{\lambda})$ is called dual *objective function*. So, let $\mathbf{g}(\mathbf{w}) = \begin{bmatrix} g_1(\mathbf{w}) & \cdots & g_K(\mathbf{w}) \end{bmatrix}^T$ the vector containing the constraint functions we obtain a simple relation

$$\nabla_{\boldsymbol{\lambda}} L(\boldsymbol{\lambda}) = \mathbf{g}\big(\mathbf{w}(\boldsymbol{\lambda})\big). \tag{B.106}$$

The dual form may or may not be simpler than the original (primal) optimization. For some particular case, when the problem presents some special structure, dual problem can be easier to solve. For example, the dual problem can show some advantage for separable and partial separable problems.

Appendix C: Elements of Random Variables, Stochastic Processes, and Estimation Theory

C.1 Random Variables

A *random variable* (RV) (or *stochastic variable*) is a variable that can assume different values depending on some random phenomenon [16–23].

Definition of RV (Papoulis [16]) An RV is a number $x(\zeta) \in (\mathbb{R}, \mathbb{C})$ *assigned to every* $\zeta \in S$ *outcome of an experiment. This number can be the gain in a game of chance, the voltage of a random source,..., or any numerical quantity that is of interest in the performance of the experiment.*

An RV is indicated as $x(\zeta)$, $y(\zeta)$, $z(\zeta)$, ... or $x_1(\zeta)$, $x_2(\zeta)$, ...; and can be defined with discrete or continuous values. For example, we consider a poll of students at a certain University. The set of all students is denoted by $S = (\zeta_1, \zeta_2, \ldots, \zeta_N)$ while, as shown in Fig. C.1, the discrete RVs $x_1(\zeta)$ and $x_2(\zeta)$ represent, respectively, the age (in years) and the number of passed exams, while the continuous RVs $x_4(\zeta)$ and $x_5(\zeta)$ represent, respectively, the height and the weight of students.

In other words, the RV $x(\zeta) \in \mathbb{R}$ represents a function with domain (or range) S, defined as *abstract probability space* (or *universal set of experimental results*), of possible infinite dimension, (e.g., the 52-cards deck, the six faces of a die, the value of a voltage generator, the temperature of an oven, etc.), which assign for each $\zeta_k \in S$ a number, i.e., $x : S \to \mathbb{R}$.

More formally, the result of the experiment ζ_k is defined as a *stochastic event* or *occurrence* $\zeta_k \in F \subseteq S$, where the subset F called *events* is a σ*-field*, which represents a subset collection of S, with closure property.[6]

Remark The value related to a specific event or occurrence of an RV is denoted as $x(\zeta_k) = x$ (e.g., if the kth student is 22 years old, $x_1(\zeta_k) = 22$). Instead, the

[6] A σ-field or σ-algebra or *Borel field* is a collection of sets, where a given measure is defined. This concept is important in probability theory, where it is interpreted as a collection of events to which can be attributed probabilities.

A. Uncini, *Fundamentals of Adaptive Signal Processing*, Signals and Communication Technology, DOI 10.1007/978-3-319-02807-1,

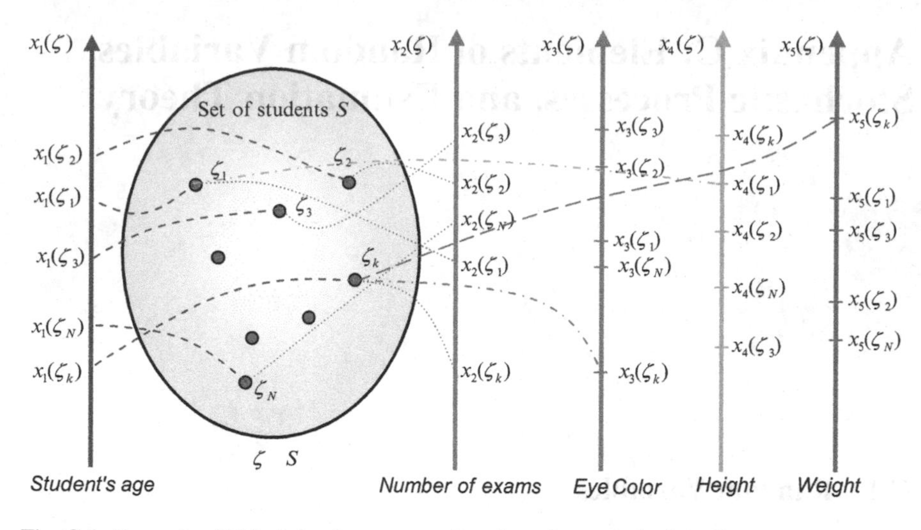

Fig. C.1 Example of RVs defined over a set of students for a scholastic poll

notation $x(\zeta) = x$ is interpreted as an event defined by all occurrences of ζ such that $x(\zeta) = x$. For example, $x_2(\zeta) = 15$ denotes all the students who have passed 15 exams. Moreover, in the case of continuous RVs, the notation $x(\zeta) \leq x$ or $a \leq x(\zeta) \leq b$ is interpreted as an interval. For example, $1.72 \leq x_4(\zeta) \leq 1.82$ denotes all the students with a height between 1.75 and 1.85 [m]. Indeed, for continuous RVs, a fixed value is a *non-sense* and should always be considered a range of values (e.g., $x_4(\zeta) = 1.8221312567125367$ is, obviously, a non-sense).

In the study of RVs an important question concerns to the *probability*[7] related to an event $\zeta_k \in S$, which can be defined by a nonnegative quantity denoted as $p(\zeta_k)$, $k = 1,2,\ldots$. However, it should be noted that the abstract probability space may be not a *metric space*. So, rather than referring to the elements $\zeta_k \in S$, we consider the RVs $x(\zeta) \in \mathbb{R}$ associated with the events that, by definition, are defined on a metric space. For example, what is the probability that $x_1(\zeta) \leq 24$ or that $x_2(\zeta) = 20$? Or that $x_4(\zeta) \leq 1.85$ or $71.3 \leq x_5(\zeta) \leq 90.2$? For this reason, the predictability of the events $x(\zeta_k) = x$; or considering the continuous case, that $x(\zeta) \leq x$, or $a \leq x(\zeta) \leq b,\ldots$, is manipulated through a *probability function* $p(\cdot)$, characterized by the following axiomatic properties:

$$
\begin{aligned}
p\left[x(\zeta) = +\infty\right] = 0 \\
p\left[x(\zeta) = -\infty\right] = 0
\end{aligned}
\tag{C.1}
$$

From the above definitions the random phenomena can be characterized by (1) the definition of an *abstract probability space* described by the triple (S, F, p) and (2) the *axiomatic definition of probability* of an RV.

[7] From the Latin *probare*, test, try, and *ilis*, be able to.

Remark In the context of RVs, care must be taken in the notation used. Sometime RVs are indicated as $X(\zeta)$ or as $\mathbf{x}(\zeta)$ (as in [16]). In these notes we prefer using the italic font $x(\zeta)$ for RV, bold font $\mathbf{x}(\zeta)$ for RV vectors, and the form $\mathbf{x}(t,\zeta)$ or $x(t,\zeta)$ $\left(\mathbf{x}[n,\zeta]\right.$ or $x[n,\zeta]$ in DT$\left.\right)$ for stochastic processes. Moreover, a complex RV $z(\zeta)\in\mathbb{C}$ is defined as the sum: $z(\zeta)=x(\zeta)+j\cdot y(\zeta)$, where $x(\zeta), y(\zeta)\in\mathbb{R}$.

C.1.1 Distributions and Probability Density Function

The elements of an event $x(\zeta)\leq x$ change depending on the number x; it follows that the probability of this event, indicated as $p\big(x(\zeta)\leq x\big)$, is a function of x itself.

Let $x(\zeta)$ be an RV, we define the *probability density function* (pdf), denoted as $f_x(x)$, that is a nonnegative integrable function, such that

$$p\big(a\leq x(\zeta)\leq b\big)=\int_a^b f_x(x)dx,\quad \textit{probability density function.}\tag{C.2}$$

Therefore, from the basic axioms (C.1) it is possible to demonstrate that the *probability of sure event* can be written as

$$\int_{-\infty}^{+\infty} f_x(x)dx=1.\tag{C.3}$$

Moreover, the event $x(\zeta)\leq x$ is characterized by the *cumulative density function* (cdf) defined as

$$F_x(x)=p(x(\zeta)\leq x),\quad \text{for}\ -\infty<x<\infty\tag{C.4}$$

or, from (C.2)

$$F_x(x)=\int_{-\infty}^{x} f_x(v)dv,\quad \textit{cumulative density function.}\tag{C.5}$$

In fact, we have that $f_x(x)=\mathrm{d}\,F_x(x)/\mathrm{d}\,x$, and the value of cdf represents a *measure* of probability $p(x(\zeta)\leq x)$.

For the cdf the following properties apply:

$$\begin{aligned}&0\leq F_x(x)\leq 1;\ \ F_x(-\infty)=0;\ \ F_x(+\infty)=1\\&F_x(x_1)\leq F_x(x_2)\ \ \text{if}\ \ x_1<x_2.\end{aligned}$$

It follows that the cdf is a nondecreasing monotone function.

Note that $f_x(x)$ is not a probability measure. To obtain the probability of the event $x<x(\zeta)\leq x+\Delta x$, we must multiply the pdf for the interval Δx. That is,

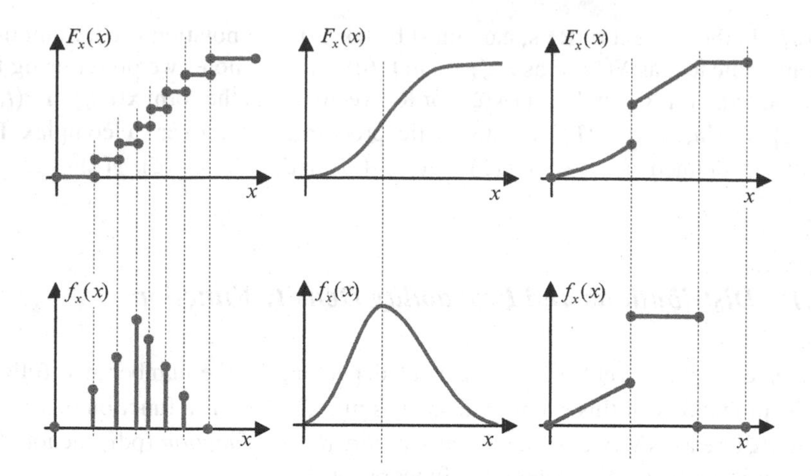

Fig. C.2 Example of trends of the *cumulative distribution functions* (*top figure*) and of the *probability density function* (*lower*) figure for discrete RV (*left*), continuous RV (*middle*), and mixed discrete–continuous RV (*right*)

$$f_x(x)\Delta x \approx \Delta F_x(x) \triangleq F_x(x+\Delta x) - F_x(x) = p\big(x < x(\zeta) \le x + \Delta x\big). \qquad \text{(C.6)}$$

Some example of continuous, discrete, and mixed pdf and cdf are reported in Fig. C.2.

C.1.2 Statistical Averages

The pdf completely characterizes an RV. However, in many situations it is convenient or necessary to represent more concisely the RV through a few specific parameters that describe its average behavior. These numbers, defined as *statistical averages* or *moments*, are determined by the *mathematical expectation.* Note that even if for the determination of statistical averages, formally, the pdf knowledge is necessary, somehow, those averages can be estimated without explicit knowledge of the pdf.

C.1.2.1 Expectation Operator

The *mathematical expectation*, usually indicated as $E\{x(\zeta)\}$, is a number defined by the following integral:

$$E\{x(\zeta)\} = \int_{-\infty}^{\infty} x f_x(x) dx, \qquad \text{(C.7)}$$

where the function $E\{\cdot\}$ indicates the *expected value* or the *average value* or *mean value*. The expected value is also indicated as $\mu = E\{x(\zeta)\}$.

C.1.2.2 Moments and Central Moments

Considering a function of RV denoted as $g(x(\zeta))$, the expected value becomes

$$E\{g[x(\zeta)]\} = \int_{-\infty}^{\infty} g(x)f_x(x)dx \tag{C.8}$$

in the case that $g[x(\zeta)] = x^m(\zeta)$ (elevation to the mth power) the previous expression is defined as *moment of order m*

$$E\{x^m(\zeta)\} = \int_{-\infty}^{\infty} x^m f_x(x)dx. \tag{C.9}$$

The calculation of the moment is of particular significance when from the RV is removed its expected value μ, i.e., considering the RV $(x(\zeta) - \mu)$. In this case the statistical function, called *central moment*, is defined as

$$E\{[x(\zeta) - \mu]^m\} = \int_{-\infty}^{\infty} (x - \mu)^m f_x(x)dx. \tag{C.10}$$

C.1.3 Statistical Quantities Associated with Moments of Order m

The moments computed with the previous expressions are of particular significance for certain orders. For example, the first-order moment $m = 1$ is just the *expected value* μ defined by (C.7). Generalizing, *moments* and *central moments* of any order can be written as

$$\begin{aligned} r_x^{(m)} &= E\{x^m(\zeta)\} \\ c_x^{(m)} &= E\{[x(\zeta) - \mu]^m\}. \end{aligned} \tag{C.11}$$

In particular, note that $c_x^{(0)} = 1$ and $c_x^{(1)} = 0$; moreover, it is obvious that for zero-mean processes the *central moment* is identical to the *moment*.

C.1.3.1 Variance and Standard Deviation

We define the *variance*, indicated as σ_x^2, as the value of the second-order central moment

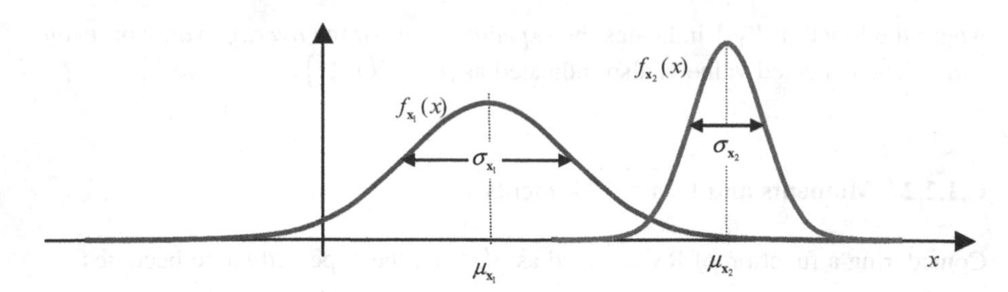

Fig. C.3 Typical trends of *Gaussian* or *normal* pdf with the indication of the *expected value* and *standard deviation*

$$\sigma_x^2 = c_x^{(2)} = E\left\{[x(\zeta) - \mu]^2\right\} = \int_{-\infty}^{\infty} (x - \mu)^2 f_x(x)dx, \tag{C.12}$$

where the positive constant $\sigma_x = \sqrt{\sigma_x^2}$ is defined as *standard deviation* of **x**.

Figure C.3 shows the pdf of two overlapped Gaussian (or normal) processes with representations of the expected value and standard deviation. (The expression of the normal distribution pdf is given in Sect. C.1.5.2.)

C.1.3.2 The Third- and Fourth-Order Moments: Skewness and Kurtosis

The *skewness* is defined as the statistic quantity associated with the third-order central moment, defined by the following relation:

$$k_x^{(3)} \triangleq E\left\{\left[\frac{x(\zeta) - \mu}{\sigma_x}\right]^3\right\} = \frac{1}{\sigma_x^3} c_x^{(3)}. \tag{C.13}$$

The skewness, as illustrated in Fig. C.4a for $k_x^{(3)} > 0$ and $k_x^{(3)} < 0$, represents the degree of asymmetry of a generic pdf. In fact, in the case where the pdf is symmetric the skewness size is zero.

The kurtosis is a statistical quantity related to the fourth-order moment defined as

$$k_x^{(4)} \triangleq E\left\{\left[\frac{x(\zeta) - \mu}{\sigma_x}\right]^4\right\} - 3 = \frac{1}{\sigma_x^4} c_x^{(4)} - 3. \tag{C.14}$$

Note that the term -3, as we shall see later, provides a zero kurtosis in the case of Gaussian distribution processes. As illustrated in Fig. C.4b, for $k_x^{(4)} > 0$, there is a "narrow" distribution trend that is called *super-Gaussian.* If $k_x^{(4)} < 0$, the trend of the pdf is more "broad" and is called *sub-Gaussian.*

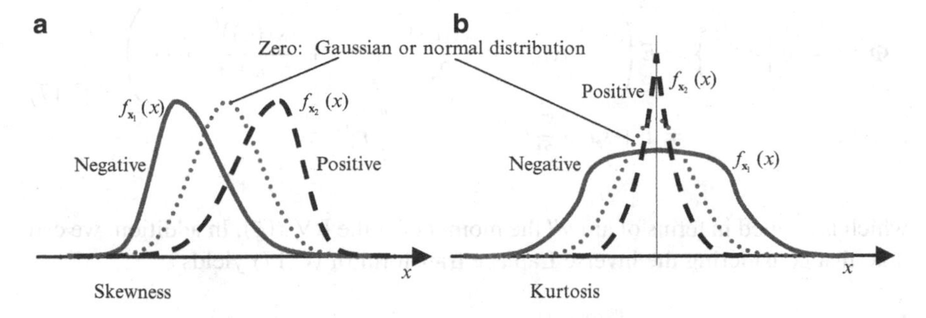

Fig. C.4 Typical trends of distribution with positive and negative (**a**) *skewness*; (**b**) *kurtosis*

C.1.3.3 Chebyshev's Inequality

Given an RV $x(\zeta)$ with the mean value μ and standard deviation σ_x, for any real number $k > 0$, the following inequality is true:

$$p\left\{|x(\zeta) - \mu| \geq k\sigma_x\right\} \leq \frac{1}{k^2} \quad k > 0. \tag{C.15}$$

An RV deviates k times from its average value with probability less than or equal to $1/k^2$. The Chebyshev's inequality (C.15) is a useful result for a generic distribution $f_{\mathbf{x}}(x)$ regardless of its form.

C.1.3.4 Characteristic Function and Cumulants

Consider the sign reversal Laplace (or Fourier) transform of the pdf $f_x(x)$ that, in the context of statistics, is called *characteristic function*, defined as

$$\Phi_x(s) = \int_{-\infty}^{\infty} f_x(x) e^{sx} dx, \tag{C.16}$$

where s is the complex Laplace variable.[8]

Equation (C.16) can be interpreted as the *moment-generating function*. In fact, the development in Taylor series of (C.16) for $s = 0$ yields

[8] The complex Laplace variable can be written $s = \alpha + j\xi$. Note that the complex part $j\xi$ should not be interpreted as a frequency.

$$\begin{aligned}\Phi_x(s) &\triangleq E\left\{e^{sx(\zeta)}\right\} = E\left(1 + sx(\zeta) + \frac{(sx(\zeta))^2}{2!} + \cdots + \frac{(sx(\zeta))^m}{m!} + \cdots\right) \\ &= 1 + s\mu + \frac{s^2}{2!}r_x^{(2)} + \cdots + \frac{s^m}{m!}r_x^{(m)} + \cdots\end{aligned} \tag{C.17}$$

which is defined in terms of are *all* the moments of the RV $x(\zeta)$. In addition, we can note that considering the inverse Laplace transform of (C.17) yields

$$r_x^{(m)} = \left.\frac{d^m\Phi_x(s)}{ds^m}\right|_{s=0}, \quad \text{for} \quad m = 1, 2, ... \tag{C.18}$$

The *cumulants* are statistical descriptors, similar to the moments, which allow having "more information" in the case of high-order statistics.

The *cumulant-generating function* is defined as the logarithm of the moment-generating function

$$\Psi_x(s) \triangleq \ln\Phi_x(s). \tag{C.19}$$

Hence, we define the *m-order cumulant* as the expression

$$\kappa_x^{(m)} \triangleq \left.\frac{d^m\Psi_x(s)}{ds^m}\right|_{s=0}, \quad \text{for } m = 1, 2, ... \tag{C.20}$$

from the above definition we can see that for a zero-mean RV, the first five cumulants are

$$\begin{aligned}\kappa_x^{(1)} &= r_x^{(1)} = \mu = 0 \\ \kappa_x^{(2)} &= r_x^{(2)} = \sigma_x^2 \\ \kappa_x^{(3)} &= c_x^{(3)} \\ \kappa_x^{(4)} &= c_x^{(4)} - 3\sigma_x^4 \\ \kappa_x^{(5)} &= c_x^{(5)} - 10c_x^{(3)}\sigma_x^2.\end{aligned} \tag{C.21}$$

Note that the first two are identical to central moments.

C.1.4 Dependent RVs: The Joint and Conditional Probability Distribution

If there is some dependence between two (or more) RVs, you need to study how the probability of one affects the other and vice versa.

For example, considering the experiment described in Fig. C.1 where the RVs x_4 and x_5, representing, respectively, the height and weight of students, are statistically dependent, as well as the age x_1 and the number of exams x_2. In probabilistic terms, this means that tall students are probably heavier, or considering the random variables x_1 and x_2, that younger students are likely to have sustained less exams.

In terms of pdf, given two RVs $x(\zeta)$ and $y(\zeta)$, we define the *joint pdf*, denoted as $f_{xy}(x,y)$, the pdf of the event obtained by the intersection between the sets $p(a \leq x(\zeta) \leq b)$ and $p(c \leq y(\zeta) \leq d)$, i.e., the distribution probability of occurrence of the two events. Therefore, extending the definition (C.2), the *joint pdf*, denoted as $f_{xy}(x,y)$, can be defined by the following integral:

$$p(a \leq x(\zeta) \leq b, c \leq y(\zeta) \leq d) = \int_c^d \int_a^b f_{xy}(x,y)dxdy, \quad \textit{joint pdf} \tag{C.22}$$

namely, the probability that $x(\zeta)$ and $y(\zeta)$ assume value inside the interval $[a, b]$ and $[c, d]$, respectively. Let us define, also, $f_{x|y}(x|y)$ the *conditional pdf* of $x(\zeta)$ given $y(\zeta)$, such that it is possible to evaluate the probability of the events $p(a \leq x(\zeta) \leq b, y(\zeta) = c)$ as

$$p(a \leq x(\zeta) \leq b, y(\zeta) = c) = \int_a^b f_{x|y}(x\,|\,y)dx, \quad \textit{conditional pdf} \tag{C.23}$$

i.e., the probability that $x(\zeta)$ assumes value inside the interval $[a, b]$ given that $y(\zeta) = c$.

Let $f_y(y)$ be the pdf of $y(\zeta)$, called in the context *marginal pdf*, from the previous expressions the joint pdf, in the case that the $x(\zeta)$ is conditioned by $y(\zeta)$, can be written as $f_{xy}(x,y) = f_{x|y}(x|y)f_y(y)$. This expression indicates how the probability of event $x(\zeta)$ is conditioned by the probability of $y(\zeta)$. Moreover, let $f_x(x)$ be the marginal pdf of $x(\zeta)$, for simple symmetry it follows that the joint pdf is also $f_{xy}(x,y) = f_{y|x}(y|x)f_x(x)$; so, now we can relate the joint and conditional pdfs by a *Bayes' rule*, which states that

$$f_{xy}(x,y) \equiv f_{x|y}(x\,|\,y)f_y(y) = f_{y|x}(y\,|\,x)f_x(x), \quad \textit{Bayes rule} \tag{C.24}$$

Moreover, we have

$$\int_x \int_y f_{xy}(x,y)dydx = 1. \tag{C.25}$$

Definition Two (or more) RVs are *independent* iff

$$f_{x|y}(x\,|\,y) = f_x(x) \quad \text{and} \quad f_{y|x}(y\,|\,x) = f_y(y) \tag{C.26}$$

or, considering (C.24), iff

$$f_{xy}(x,y) = f_x(x)f_y(y). \tag{C.27}$$

Property If two RVs are independent they are necessarily uncorrelated.

The covariance and the correlation of joint RV are respectively defined as

$$c_{xy}^{(2)} = E\{x(\zeta)y(\zeta)\} - E\{y(\zeta)\} \cdot E\{x(\zeta)\} \tag{C.28}$$

$$r_{xy}^{(2)} = c_{xy}^{(2)} \big/ (\sigma_x \sigma_y). \tag{C.29}$$

Two RVs $x(\zeta)y(\zeta)$ are uncorrelated, iff their cross-correlation (covariance) is zero. Consequently, if (C.27) holds, then $E\{x(\zeta)y(\zeta)\} = E\{y(\zeta)\} \cdot E\{x(\zeta)\}$, and for (C.28) their cross-correlation is zero.

Finally note that, if two RV are uncorrelated, they are not necessarily independent.

C.1.5 Typical RV Distributions

C.1.5.1 Uniform Distribution

The *uniform distribution* is appropriate for the description of an RV with equi-probable events in the interval $[a, b]$. The pdf of the uniform distribution is defined as.

$$f_x(x) = \begin{cases} \dfrac{1}{b-a} & a \le x \le b \\ 0 & \text{elsewhere} \end{cases} \tag{C.30}$$

The corresponding cdf is

$$F_x(x) = \int_{-\infty}^{x} f_x(v)dv = \begin{cases} 0 & x < a \\ \dfrac{x-a}{b-a} & a \le x \le b \\ 1 & x > b \end{cases} \tag{C.31}$$

Its characteristic function is

$$\Phi_x(s) = \frac{e^{sb} - e^{sa}}{s(b-a)}. \tag{C.32}$$

Finally, the mean value and the variance are

$$\mu = \frac{a+b}{2} \quad \text{and} \quad \sigma_x^2 = \frac{(b-a)^2}{12}. \tag{C.33}$$

C.1.5.2 Normal Distribution

The *normal distribution*, also called *Gaussian distribution*, is one of the most useful and appropriate description of many statistical phenomena.

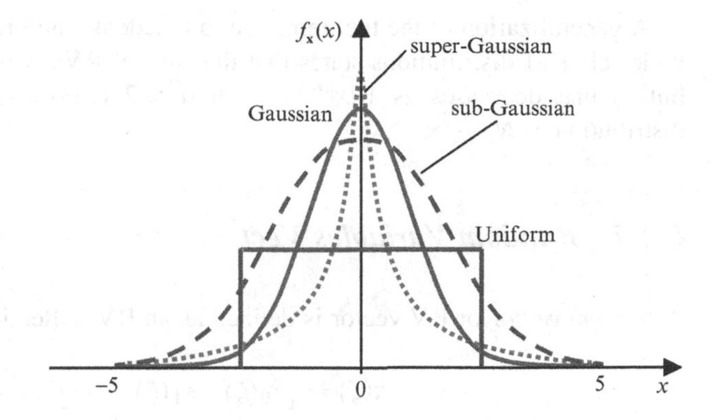

Fig. C.5 Qualitative behavior of some typical distributions

The normal distribution pdf, already illustrated in Fig. C.3, with mean value μ and standard deviation σ_x, is

$$f_x(x) = \frac{1}{\sqrt{2\pi\sigma_x^2}} e^{-\frac{1}{2\sigma_x^2}(x-\mu)^2} \tag{C.34}$$

with a CF

$$\Phi_x(s) = e^{\mu s - \frac{1}{2}\sigma_x^2 s^2}. \tag{C.35}$$

From previous equations an RV with normal pdf, often referred to as $N(\mu,\sigma_x^2)$, is defined by its mean value μ and its variance σ_x^2. Note also that the moments of higher order can be determined in terms of only the first two moments. In fact, we have (Fig. C.5)

$$c_x^{(m)} = E\left\{\left|x(\zeta) - \mu\right|^m\right\} = \begin{cases} 1\cdot 3\cdot 5\cdot\cdots\cdot(m-1)\sigma_x^m & \text{for } m \text{ even} \\ 0 & \text{for } m \text{ odd.} \end{cases} \tag{C.36}$$

In particular, the fourth-order moments are $c_x^{(4)} = 3\sigma_x^4$ and for the Gaussian distribution the kurtosis is zero.

Remark From (C.36) we observe that an RV with Gaussian distribution is *fully* characterized only by the mean value and variance and that the moments of higher order do not contain any useful information.

C.1.6 The Central Limit Theorem

An important theorem is the statistical *central limit theorem* whose statement says that the sum of N independent RVs with the same distribution, i.e., iid with finite variance, tends to the normal distribution as $N \to \infty$.

A generalization of the theorem, due to Gnedenko and Kolmogorov, valid for a wider class of distributions states that the sum of RVs with low power-tail distribution that decreases as $1/|x|^{\alpha+1}$ with $\alpha \leq 2$ tends to the Lévy alpha-stable distribution as $N \to \infty$.

C.1.7 Random Variables Vectors

A *random vector* or RV vector is defined as an RV collection of the type

$$\mathbf{x}(\zeta) = \begin{bmatrix} x_0(\zeta) & x_1(\zeta) & \cdots \end{bmatrix}^T.$$

By a generalization of the definition (C.7), the expectation of random vector is also a vector that, omitting the writing of event (ζ), is defined as

$$\boldsymbol{\mu} = E\{\mathbf{x}\} = \begin{bmatrix} E\{x_0\} & E\{x_1\} & \cdots \end{bmatrix}^T = [\mu_0 \quad \mu_1 \quad \cdots]^T. \tag{C.37}$$

C.1.8 Covariance and Correlation Matrix

In the case of random vector, the second-order statistic is a matrix. Therefore, the *covariance matrix* is defined as

$$\mathbf{C}_x = E\left\{(\mathbf{x} - \boldsymbol{\mu})(\mathbf{x} - \boldsymbol{\mu})^T\right\}, \quad \textit{Covariance matrix.} \tag{C.38}$$

For example, given a two-dimensional random vector $\mathbf{x} = [x_0 \quad x_1]^T$ the covariance is defined as

$$\begin{aligned} \mathbf{C}_x &= E\left\{\begin{bmatrix} x_0 - \mu_0 \\ x_1 - \mu_1 \end{bmatrix} \begin{bmatrix} (x_0 - \mu_0) & (x_1 - \mu_1) \end{bmatrix}\right\} \\ &= \begin{bmatrix} E|x_0 - \mu_{x_0}|^2 & E\left\{(x_0 - \mu_{x_0})(x_1 - \mu_{x_1})\right\} \\ E\left\{(x_1 - \mu_{x_1})(x_0 - \mu_{x_0})\right\} & E|x_1 - \mu_{x_1}|^2 \end{bmatrix} \end{aligned} \tag{C.39}$$

so, the autocovariance matrix is symmetric

$$\mathbf{C}_x = \mathbf{C}_x^T, \tag{C.40}$$

where the superscript "T" indicates the matrix transposition. Moreover, the *autocorrelation matrix* is defined as

$$\mathbf{R}_x = E\left\{\mathbf{x}\mathbf{x}^T\right\}, \quad \textit{Autocorrelation matrix.} \tag{C.41}$$

For the two-dimensional RV previously defined it is then

$$\mathbf{R}_x = \begin{bmatrix} E|x_0|^2 & E\{x_0 x_1\} \\ E\{x_1 x_0\} & E|x_1|^2 \end{bmatrix} \tag{C.42}$$

and

$$\mathbf{R}_x = \mathbf{R}_x^T. \tag{C.43}$$

Property The autocorrelation matrix of an RV vector $\mathbf{x}$ is always defined nonnegative, i.e., for each vector $\mathbf{w} = [w_0 \quad w_1 \quad \cdots \quad w_{M-1}]^T$ the quadratic form $\mathbf{w}^T\mathbf{R}_x\mathbf{w}$ is positive semi-definite or nonnegative

$$\mathbf{w}^T\mathbf{R}_x\mathbf{w} \geq 0. \tag{C.44}$$

Proof Consider the inner product between $\mathbf{x}$ and $\mathbf{w}$

$$\alpha = \mathbf{w}^T\mathbf{x} = \mathbf{x}^T\mathbf{w} = \sum_{k=0}^{M-1} w_k x_k. \tag{C.45}$$

The RV mean squared value of α is defined as

$$E\{\alpha^2\} = E\{\mathbf{w}^T\mathbf{x}\mathbf{x}^T\mathbf{w}\} = \mathbf{w}^T E\{\mathbf{x}\mathbf{x}^T\}\mathbf{w} = \mathbf{w}^T\mathbf{R}_x\mathbf{w}. \tag{C.46}$$

since, by definition, $\alpha^2 \geq 0$, it is $\mathbf{w}^T\mathbf{R}_x\mathbf{w} \geq 0$.

Q.E.D.

C.1.8.1 Eigenvalues and Eigenvectors of the Autocorrelation Matrix

From geometry (see Sect. A.8), the eigenvalues can be computed by solving the characteristic polynomial $p(\lambda)$, defined as $p(\lambda) \triangleq \det(\mathbf{R} - \lambda\mathbf{I}) = 0$.

A real or complex autocorrelation matrix $\mathbf{R} \in \mathbb{R}^{M \times M}$ is symmetric and positive semi-definite. We know that for this type of matrix the following properties listed below are valid.

1. The eigenvalues λ_i of $\mathbf{R}$ are real and nonnegative. In fact, for (A.61) we have that $\mathbf{R}\mathbf{q} = \lambda\mathbf{q}$, and by left multiplying for $\mathbf{q}_i^T$, we get

$$\mathbf{q}_i^T\mathbf{R}\mathbf{q}_i = \lambda_i\mathbf{q}_i^T\mathbf{q}_i \Rightarrow \lambda_i = \frac{\mathbf{q}_i^T\mathbf{R}\mathbf{q}_i}{\mathbf{q}_i^T\mathbf{q}_i} \geq 0, \quad \textit{Rayleigh quotient}. \tag{C.47}$$

2. The eigenvectors $\mathbf{q}_i$ $i = 0, 1,\ldots,M - 1$, of $\mathbf{R}$ are orthogonal for distinct values of λ_i

$$\mathbf{q}_i^T \mathbf{q}_j = 0, \text{ for } \quad i \neq j. \tag{C.48}$$

3. The matrix $\mathbf{R}$ can always be diagonalized as

$$\mathbf{R} = \mathbf{Q}\boldsymbol{\Lambda}\mathbf{Q}^T, \tag{C.49}$$

where $\mathbf{Q} = [\mathbf{q}_0 \quad \mathbf{q}_1 \quad \cdots \quad \mathbf{q}_{M-1}]$,

$$\boldsymbol{\Lambda} = \mathrm{diag}(\lambda_0, \lambda_1, \ldots, \lambda_{M-1}) \tag{C.50}$$

and $\mathbf{Q}$ is a unitary matrix, i.e., $\mathbf{Q}^T\mathbf{Q} = \mathbf{I}$.
4. An alternative representation for $\mathbf{R}$ is

$$\mathbf{R} = \sum_{i=0}^{M-1} \lambda_i \mathbf{q}_i \mathbf{q}_i^T = \sum_{i=0}^{M-1} \lambda_i \mathbf{P}_i, \tag{C.51}$$

where the term $\mathbf{P}_i = \mathbf{q}_i \mathbf{q}_i^T$ is defined *spectral projection*.
5. The trace of the matrix $\mathbf{R}$ is

$$\mathrm{tr}[\mathbf{R}] = \sum_{i=0}^{M-1} \lambda_i \Rightarrow \frac{1}{M} \sum_{i=0}^{M-1} \lambda_i = r_{xx}[0] = \sigma_x^2. \tag{C.52}$$

C.2 Stochastic Processes

Generalizing the concept of RV, a stochastic process (SP) is a rule to assign each result ζ to a function $\mathbf{x}(t, \zeta)$. Hence, SP is a family of two-dimensional functions, of the variables t and ζ, where the domain is defined over the set of all the experimental results $\zeta \in S$, while the time variable t represents the set of real numbers $t \in \mathbb{R}$. If $\mathbb{R}$ represents the real axis of time, then $\mathbf{x}(t, \zeta)$ is a *continuous-time stochastic process*. In the case that $\mathbb{R}$ represents a set of integers, then we have a *discrete-time stochastic process*, and the time index is denoted by $n \in \mathbb{Z}$.

In general terms, a discrete-time SP is a time-series $\mathbf{x}[n, \zeta]$, consisting of all possible sequences of the process. Each individual sequence, corresponding to a specific result $\zeta = \zeta_k$, indicated as $\mathbf{x}[n, \zeta_k]$, represents an RV sequence (indexed by n) that is called *realization* or *sample sequence* of the process.

Since the SP is a two-variable function, then there are four possible interpretations

i) $\mathbf{x}[n, \zeta]$ is an SP $\Rightarrow$ n variable, ζ variable;
ii) $\mathbf{x}[n, \zeta_k]$ is an RV sequence $\Rightarrow$ n variable, ζ fixed;
iii) $\mathbf{x}[n_k, \zeta]$ is an RV $\Rightarrow$ n fixed, ζ variable;
iv) $\mathbf{x}[n_k, \zeta_k]$ is a number $\Rightarrow$ n fixed, ζ fixed.

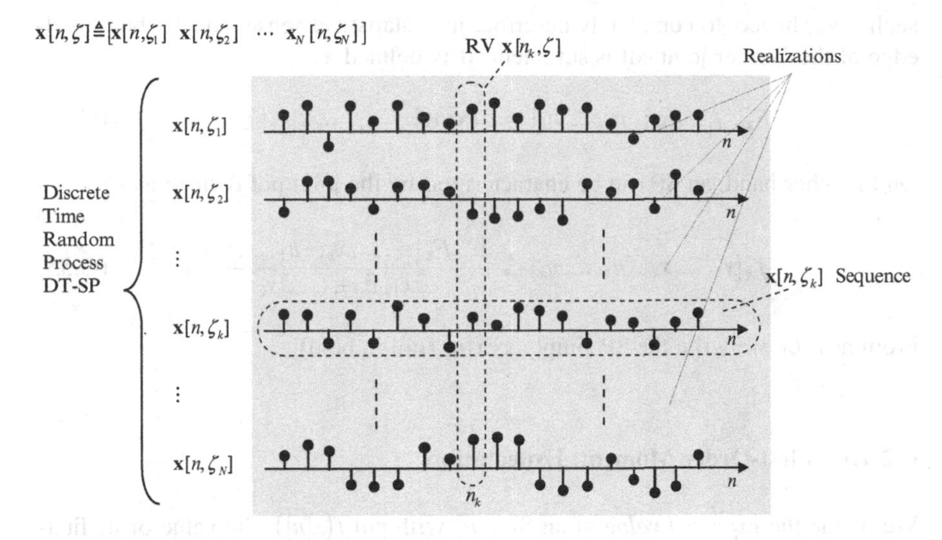

Fig. C.6 Representation of the stochastic process $\mathbf{x}[n,\zeta]$. As usual in context of DSP, the process sample is simply indicated as $\mathbf{x}[n]$

For clarity of presentation, as usual in many scientific contexts (signal processing, neural networks, etc.), writing ζ parameter is omitted and, later in the text, the SP $\mathbf{x}[n,\zeta]$ is indicated only with $\mathbf{x}[n]$ or $x[n]$ (sometimes bold is omitted) and the sample process sequence $\mathbf{x}[n,\zeta_k]$ is often simply referred to as $x_k[n]$.

Definition We define *discrete-time stochastic process* (DT-SP), denoted as $\mathbf{x}[n] \in \mathbb{R}^N$, an RV vector, defined as

$$\mathbf{x}[n] = \left[x_1[n], x_2[n], \ldots, x_N[n]\right], \tag{C.53}$$

where the integer $n \in \mathbb{Z}$ represents the time index. Note, as illustrated in Fig. C.6, that in (C.53) each realization $x_k[n]$ represents an RV sequence of the same process.

C.2.1 Statistical Averages of an SP

The determination of the statistical averages of SPs can be performed exactly as for the RVs. In fact, note that for a given fixed temporal index, see property iii), the process consists in a simple RV so that it is possible to evaluate all the statistical functions proceeding as in Sect. C.1.2. Similarly, setting the parameter ζ and considering two different temporal indexes n_1 and n_2 we are in the presence of joint RVs so that it is possible to characterize the process by the joint cdf $F_{\mathbf{x}}[x_1,\ x_2;\ n_1,\ n_2]$. However, in general an SP contains an infinite number of such RVs; hence, to completely describe, in a statistical sense, an SP, the knowledge of the k-order joint cdf is sufficient. It is defined as

$$F_{\mathbf{x}}[x_1, ..., x_k;\ n_1, ..., n_k] = p\big(\mathbf{x}[n_1] \le x_1, ..., \mathbf{x}[n_k] \le x_k\big). \tag{C.54}$$

On the other hand, an SP can be characterized by the joint pdf defined as

$$f_{\mathbf{x}}[x_1, ..., x_k;\ n_1, ..., n_k] \triangleq \frac{\partial^{2k} F_{\mathbf{x}}[x_1, ..., x_k;\ n_1, ..., n_k]}{\partial x_1, \partial x_2, ..., \partial x_k}. \tag{C.55}$$

From now on we write the SP simply as $x[n]$ (not in bold).

C.2.1.1 First-Order Moment: Expectation

We define the *expected value* of an SP $x[n]$ with pdf $f(x[n])$, the value of its first-order moment at a given time index n. According with Eq. (C.7), the *expected value* is defined as

$$\mu_n = E\{x[n]\}. \tag{C.56}$$

Referring to Fig. C.6, and considering the notation $\mathbf{x}[n,\zeta]$, the *expectation operator* $E\{\cdot\}$ represents the *ensemble average* of the RV $\mu_{n_k} = E\{\mathbf{x}[n_k, \zeta]\}$.

Equation (C.56) can be also interpreted in terms of *relative frequency* by the following expression:

$$\mu_{n_k} = \lim_{N\to\infty}\left[\frac{1}{N}\sum_{j=1}^{N} x_j[n_k]\right]. \tag{C.57}$$

In other words (see Fig. C.6), the expectation represents the mean value of the set of RV $x[n_k]$ at a fixed time instant.

If the process is not stationary, i.e., its statistics changes in time, its mean value is variable during time. So, in general, we have

$$\mu_n \ne \mu_m, \quad \text{for } n \ne m. \tag{C.58}$$

C.2.1.2 Second-Order Moment: Autocorrelation and Autocovariance

We define *autocorrelation*, or second-order *moment*, the sequence

$$r[n, m] = E\{x[n]x[m]\}. \tag{C.59}$$

In terms of relative frequency Eq. (C.59) can be written as

$$r[n,m] = \lim_{N\to\infty}\left[\frac{1}{N}\sum_{k=1}^{N} x_k[n]x_k[m]\right]. \tag{C.60}$$

The autocorrelation is a measure that indicates the association degree or dependency between the process at time n and at time m.

Moreover, we have that

$$r[n,n] = E\{x^2[n]\}, \qquad \textit{average power of the sequence.}$$

We define autocovariance, or second-order central moment, the sequence

$$c[n,m] = E\Big\{\big(x[n]-\mu_n\big)\big(x[m]-\mu_m\big)\Big\} = r[n,m]-\mu_n\mu_m. \tag{C.61}$$

C.2.1.3 Variance and Standard Deviation

Similarly for the definition in Sect. C.1.3.1, the *variance* of an SP is a value related to the central second-order moment defined as

$$\sigma_{xn}^2 = E\Big\{\big(x[n]-\mu_n\big)^2\Big\} = E\{x^2[n]\}-\mu_n^2. \tag{C.62}$$

The quantity σ_{xn} is defined as *standard deviation*, which represents a measure of the observation dispersion $x[n]$ around its mean value μ_n.

Remark For zero-mean processes, the central moment coincides with moment. It follows then $\sigma_{xn}^2 = r[n,n] = E\{x^2[n]\}$; in other words, the variance coincides with the signal power.

C.2.1.4 Cross-correlation and Cross-covariance

The statistical relationships between two *jointly distributed* SP $x[n]$ and $y[n]$ (i.e., defined over the same space results S) can be described by their joint second-order moments (the *cross-correlation* and *cross-covariance*) defined, respectively, as

$$r_{xy}[n,m] = E\{x[n]y[m]\} \tag{C.63}$$

$$c_{xy}[n,m] = E\Big\{\big(y[n]-\mu_{yn}\big)\big(x[m]-\mu_{ym}\big)\Big\} = r_{xy}[n,m]-\mu_{xn}\mu_{ym}. \tag{C.64}$$

Moreover, the normalized cross-correlation is defined as

$$r_{xy}[n,m] = \frac{c_{xy}[n,m]}{\sigma_{xn}\sigma_{xm}}. \tag{C.65}$$

C.2.2 High-Order Moments

In linear systems the high-order moments are rarely used with respect to the first- and second-order ones. The interest in higher order moments, in fact, is increasing in nonlinear systems.

C.2.2.1 Moments of Order *m*

Generalizing the foregoing for first- and second-order statistics, moments and central moments of any order can be written as

$$\begin{aligned} r^{(m)}[n_1, ..., n_m] &= E\{x[n_1] \cdot x[n_2] \cdots x[n_m]\} \\ c^{(m)}[n_1, ..., n_m] &= E\{(x[n_1] - \mu_{n_1})(x[n_2] - \mu_{n_2}) \cdots (x[n_m] - \mu_{n_m})\}. \end{aligned}$$

For a particular index *n* the previous expressions are simplified as

$$\begin{aligned} r_x^{(m)} &= E\{(x[n])^m\} \\ c_x^{(m)} &= E\{(x[n] - \mu_x)^m\}. \end{aligned}$$

Note, also, that $c_x^{(0)} = 1$ and $c_x^{(1)} = 0$. It is obvious that, for zero-mean processes, the central moment is identical to the moment.

C.2.2.2 Moments of Third Order

The third-order moments are defined as

$$\begin{aligned} r^{(3)}[k, m, n] &= E\{x[k] \cdot x[m] \cdot x[n]\} \\ c^{(3)}[k, m, n] &= E\{(x[k] - \mu_k)(x[m] - \mu_m)(x[n] - \mu_n)\}. \end{aligned}$$

C.2.3 Property of Stochastic Processes

C.2.3.1 Independent SP

An SP is called *independent* iff

$$f_{\mathbf{x}}[x_1, ..., x_k;\ n_1, ..., n_k] = f_1[x_1;\ n_1] \cdot f_2[x_2;\ n_2] \cdot ... \cdot f_k[x_k;\ n_k] \qquad \text{(C.66)}$$

$\forall\ k,\ n_i\ i = 1, ..., k$; or else, $x[n]$ is an SP formed with independent RV $x_1[n]$, $x_2[n]$,....

For two, or more, independent sequences $x[n]$ and $y[n]$ we also have that

$$E\{x[n]\cdot y[n]\} = E\{x[n]\}\cdot E\{y[n]\}. \tag{C.67}$$

C.2.3.2 Independent Identically Distributed SP

If all the SP sequences are independent and with equal pdf, i.e., $f_1[x_1;\ n_1] = \ldots = f_k[x_k;\ n_k]$, then the SP is defined as iid.

C.2.3.3 Uncorrelated SP

An SP is called *uncorrelated* if

$$c[n,m] = E\Big\{\big(x[n]-\mu_n\big)\big(x[m]-\mu_m\big)\Big\} = \sigma_{xn}^2\delta[n-m]. \tag{C.68}$$

Two processes $x[n]$ and $y[n]$ are uncorrelated if

$$c_{xy}[n,m] = E\Big\{\big(x[n]-\mu_{xn}\big)\big(y[m]-\mu_{xm}\big)\Big\} = 0 \tag{C.69}$$

and if

$$r_{xy}[n,m] = \mu_{xn}\mu_{xm}. \tag{C.70}$$

Remark If the SP $x[n]$ and $y[n]$ are independent they are, also, necessarily *uncorrelated* while the contrary is not always true, i.e., the assumption of independency is stronger than the uncorrelation.

C.2.3.4 Orthogonal SP

Two processes $x[n]$ and $y[n]$ are defined as *orthogonal* iff

$$r_{xy}[n,m] = 0. \tag{C.71}$$

C.2.4 Stationary Stochastic Processes

An SP is defined *stationary* or *time invariant* if the statistic of $x[n]$ is identical to the translated process $x[n-k]$ statistics. Very often in real situations we consider the processes as stationary. This is due to the simplifications of the correlation functions associated with them.

In particular, a sequence is called *strict sense stationary* (SSS) or *stationary of order N* if we have

$$f_{\mathbf{x}}[x_1, \ldots, x_N;\ n_1, \ldots, n_N] = f_{\mathbf{x}}[x_1, \ldots, x_N; n_1 - k, \ldots, n_N - k] \quad \forall k \tag{C.72}$$

An SP is *wide sense stationary* (WSS) if its first-order statistics do not change over time

$$E\{x[n]\} = E\{x[n+k]\} = \mu \quad \forall n, k. \tag{C.73}$$

As a corollary, consider also the following definitions. An SP is defined *wide sense periodic* (WSP) if

$$E\{x[n]\} = E\{x[n+N]\} = \mu. \quad \forall n \tag{C.74}$$

An SP is *wise sense cyclostationary* (WSC) if the following relations are true:

$$\begin{aligned} E\{x[n]\} &= E\{x[n+, N]\} \\ r[m,n] &= r[m+N, n+N] \quad \forall m, n. \end{aligned} \tag{C.75}$$

Let us define $k = n - m$ as *correlation lag* or *correlation delay*, the correlation is usually written as

$$r[k] = E\{x[n]x[n-k]\} = E\{x[n+k]x[n]\}. \tag{C.76}$$

The latter is often referred to as autocorrelation function (acf).

Similarly, considering two joint WSS processes, the autocovariance (C.61) is defined as

$$c[k] = E\Big\{(x[n+k] - \mu)\,(x[n] - \mu)\Big\} = r[k] - \mu^2. \tag{C.77}$$

Property The acf of WSS processes has the following properties:

1. The autocorrelation sequence $r[k]$ is symmetric with respect to delay

$$r[-k] = r[k] \tag{C.78}$$

2. The correlation sequence is defined nonnegative. So, for any $M > 0$ and $\mathbf{w} \in \mathbb{R}^M$ we have that

$$\sum_{k=1}^{M}\sum_{m=1}^{M} w[k]r[k-m]w[m] \geq 0 \tag{C.79}$$

Such property represents a necessary and sufficient condition so that $r[k]$ is an acf.

3. The zero time delay term is that of maximum amplitude

$$E\{x^2[n]\} = r[0] \geq |r[k]| \quad \forall n, k. \tag{C.80}$$

Given two joint WSS processes $x[n]$ and $y[n]$, the cross-correlation function (ccf) is defined as

$$r_{xy}[k] = E\{x[n]y[n-k]\} = E\{x[n+k]y[n]\}. \tag{C.81}$$

Finally, the cross-covariance sequence is defined as

$$c_{xy}[k] = E\Big\{\big(x[n+k]-\mu_x\big)\big(y[n]-\mu_y\big)\Big\} = r_{xy}[k] - \mu_x\mu_y. \tag{C.82}$$

C.2.5 *Ergodic Processes*

An SP is called *ergodic* if the ensemble averages coincide with the time averages. The consequence of this definition is that an ergodic process must, necessarily, also be strict sense stationary.

C.2.5.1 Statistics Averages of Ergodic Processes

For the determination of the statistics of an ergodic processes it is necessary to define the *time-average* mathematical operation. For a discrete-time random signal $x[n]$ the mathematical operator of time average, indicated as $\langle x[n]\rangle$, is defined as

$$\begin{aligned}\langle x[n]\rangle &= \lim_{N\to\infty}\frac{1}{N}\sum_{n=0}^{N-1}x[n]\\ \langle x[n+k]x[n]\rangle &= \lim_{N\to\infty}\frac{1}{N}\sum_{n=0}^{N-1}x[n+k]x[n].\end{aligned} \tag{C.83}$$

It is possible to define all the statistical quantities and functions by replacing the ensemble-average operator $E(\cdot)$ with the time-average operator $\langle\ \cdot\ \rangle$ also indicated as $\hat{E}\{\cdot\}$. In other words, if $x[n]$ is an ergodic process, we have that

$$\mu = \langle x[n]\rangle = E\{x[n]\}. \tag{C.84}$$

If $x[n]$ is an ergodic process for the correlation we have

$$\langle x[n+k]x[n]\rangle = E\{x[n+k]x[n]\}. \tag{C.85}$$

If a process is ergodic then it is WSS, i.e., only stationary processes can be ergodic. On the contrary, a WSS process cannot be ergodic.

Considering the sequence $x[n]$, we have that

$$\langle x[n] \rangle, \qquad \textit{Mean Value} \tag{C.86}$$

$$\langle x^2[n] \rangle, \qquad \textit{Mean Square Value} \tag{C.87}$$

$$\langle (x[n]-\mu)^2 \rangle, \qquad \textit{Variance} \tag{C.88}$$

$$\langle x[n+k]x[n] \rangle, \qquad \textit{Autocorrelation} \tag{C.89}$$

$$\langle (x[n+k]-\mu)(x[n]-\mu) \rangle, \qquad \textit{Autocovariance} \tag{C.90}$$

$$\langle x[n+k]y[n] \rangle, \qquad \textit{Cross-correlation} \tag{C.91}$$

$$\langle (x[n+k]-\mu_x)(y[n]-\mu_y) \rangle, \qquad \textit{Cross-covariance} \tag{C.92}$$

For deterministic *power signals*, it is important to mention the similarities among the correlation sequences, calculated by the temporal average (C.89), and determined by the definition (C.76). Although this is a formal similarity due to the fact that random sequences are power signals, the time averages are (for the closure property) RVs, and the corresponding quantities for deterministic power signals are numbers or deterministic sequences.

Two individually ergodic SPs $x[n]$ and $y[n]$ have the property of *joint ergodicity* if the cross-correlation is identical to Eq. (C.91), i.e.,

$$E\{x[n+k]y[n]\} = \langle x[n+k]y[n] \rangle. \tag{C.93}$$

Remark The ergodic processes are very important in applications as very often only one realization of the process is available: in many practical situations, however, the processes are stationary ergodic. Therefore, the assumption of ergodicity allows the *estimation* of statistical functions starting from the time averages available only for the single realization of the process. Moreover, in the case of ergodic sequences of finite duration, the expression (C.83) is calculated as

$$r[k] = \begin{cases} \dfrac{1}{N}\displaystyle\sum_{n=0}^{N-1-k} x[n+k]x[n] & k \geq 0 \\ r[-k] & k < 0 \end{cases}. \tag{C.94}$$

C.2.6 Correlation Matrix of Random Sequences

A stochastic process can be represented as an RV vector and, as defined in Sect. C.1.7, its second-order statistics are defined by the mean values vectors and by the *correlation matrix*. Considering a random vector $\mathbf{x}_n$ from the SP $x[n]$ as follows:

$$\mathbf{x}_n \triangleq \begin{bmatrix} x[n] & x[n-1] & \cdots & x[n-M+1] \end{bmatrix}^T \tag{C.95}$$

for the definition (C.37), its mean value is defined as

$$\boldsymbol{\mu}_{x_n} = \begin{bmatrix} \mu_{x_n} & \mu_{x_{n-1}} & \cdots & \mu_{x_{n-M+1}} \end{bmatrix}^T \tag{C.96}$$

and for (C.41) and (C.63), the autocorrelation matrix is defined as

$$\mathbf{R}_{x_n} = E(\mathbf{x}_n\mathbf{x}_n^T) = \begin{bmatrix} r_x[n,n] & \cdots & r_x[n,n-M+1] \\ \vdots & \ddots & \vdots \\ r_x[n-M+1,n] & \cdots & r_x[n-M+1,n-M+1] \end{bmatrix}. \tag{C.97}$$

since $r_x[n-i,\ n-j] = r_x[n-j,\ n-i]$ for $0 \leq (i,j) \leq M-1$, $\mathbf{R}_{x_n}$ is symmetric (or Hermitian for complex processes).

In the case of stationary process the acf is independent from index n and, by defining the correlation lag as $k = j - i$, we obtain

$$r_x[n-i, n-j] = r_x[j-i] = r_x[k]. \tag{C.98}$$

Then the autocorrelation matrix is a symmetric *Toeplitz matrix* of the form

$$\mathbf{R}_x = E\{\mathbf{x}\mathbf{x}^T\} = \begin{bmatrix} r[0] & r[1] & \cdots & r[M-1] \\ r[1] & r[0] & \cdots & r[M-2] \\ \vdots & \vdots & \ddots & \vdots \\ r[M-1] & r[M-2] & \cdots & r[0] \end{bmatrix}. \tag{C.99}$$

The autocorrelation matrix of stationary process is always Toeplitz (see Sect. A.2.4) and, for (C.44), nonnegative.

C.2.7 Stationary Random Sequences and TD LTI Systems

For random sequences processed by TD LTI systems, it is necessary to study the relationship between the input and output pdfs. For simplicity, consider a stable circuit TD LTI characterized by the impulse response $h[n]$, where the input $x[n]$ is a random, real or complex, stationary sequence WSS. The output $y[n]$ is computed by the DT convolution defined as

$$y[n] = \sum_{l=-\infty}^{\infty} h[l]x[n-l]. \tag{C.100}$$

C.2.7.1 Input–Output Cross-correlation Sequence

Consider the expression (C.100), and pre-multiplying both sides by $x[n+k]$, and performing the expectation we get

$$E\{x[n+k]y[n]\} = \sum_{l=-\infty}^{\infty} h[l]E\{x[n+k]x[n-l]\}, \tag{C.101}$$

i.e.,

$$r_{xy}[k] = \sum_{l=-\infty}^{\infty} h[l]r_{xx}[k+l] = \sum_{m=-\infty}^{\infty} h[-m]r_{xx}[k-m]. \tag{C.102}$$

In other words, the following relations are valid:

$$r_{xy}[k] = h[-k] * r_{xx}[k] \tag{C.103}$$

and similarly

$$r_{yx}[k] = h[k] * r_{xx}[k]. \tag{C.104}$$

From the previous we also have that

$$r_{xy}[k] = r_{yx}[-k]. \tag{C.105}$$

C.2.7.2 Output Autocorrelation Sequence

Multiplying both sides of (C.100) for $y[n-k]$ and computing the expectation we get

$$E\{y[n]y[n-k]\} = \sum_{l=-\infty}^{\infty} h[l]E\{x[n-l]y[n-k]\} \tag{C.106}$$

or

$$r_{yy}[k] = \sum_{l=-\infty}^{\infty} h[l]r_{xy}[k-l] = h[k] * r_{xy}[k]. \tag{C.107}$$

In other words, we can write

$$r_{yy}[k] = h[k] * h[-k] * r_{xx}[k]. \tag{C.108}$$

By defining the term $r_{hh}[k]$ as

$$r_{hh}[k] \triangleq h[k] * h[-k] = \sum_{l=-\infty}^{\infty} h[l]h[l-k]. \tag{C.109}$$

(C.108) can be written as

$$r_{yy}[k] = r_{hh}[k] * r_{xx}[k]. \tag{C.110}$$

Therefore, in the case of a stationary signal, $x[n]$ is filtered with a circuit of the impulse response $h[n]$, and the output autocorrelation is equivalent to the input autocorrelation filtered with an impulse response equal to $r_{hh}[n] = h[k] * h[-k]$.

C.2.7.3 Output Pdf

The output pdf determination of a DT-LTI system is usually a difficult task. However, for Gaussian input process, also the output is always a Gaussian process with a correlation (C.110). In the case of multiple iid inputs, the output is determined by the weighted sum of the independent input SPs. Therefore, the output pdf is equal to the convolution of the pdf of each SP.

C.2.7.4 Stationary Random Sequences Spectral Representation

Given a stationary zero-mean discrete-time signal $x[n]$ for $-\infty < n < \infty$, this has not, in general, finite energy for which the DTFT, and more generally the z-transform, does not converge. The autocorrelation sequence $r_{xx}[n]$, computed by (C.76) or in terms of relative frequency, however, is "almost always" with finite energy, and when this is true, its envelope decays (goes to zero) when the delay increases. In these cases the sequence of autocorrelation always results absolutely summable and its z-transform, defined as

$$R_{xx}(z) = \sum_{k=-\infty}^{\infty} r_{xx}[k] z^{-k},$$

admits some convergence region on the z-plane. Note, also, that for the symmetry properties of (C.78), we have that $R_{xx}(z^{-1}) = R_{xx}(z)$.

C.2.7.5 Power Spectral Density

We define the *power spectral density* (PSD) as the DTFT of the autocorrelation

$$R_{xx}(e^{j\omega}) = \sum_{k=-\infty}^{\infty} r_{xx}[k] e^{-j\omega k}. \tag{C.111}$$

The PSD is a nonnegative real function that does not preserve the phase information. The $R_{xx}(e^{j\omega})$ provides a distribution measure of the average power of a random process, in function of the frequency.

We define *cross-spectrum* or *cross-PSD* (CPSD) the DTFT of the sequence of cross-correlation

$$R_{xy}\left(e^{j\omega}\right) = \sum_{k=-\infty}^{\infty} r_{xy}[k]e^{-j\omega k}. \tag{C.112}$$

The CPSD is a complex function. Its amplitude describes the frequencies of the SP $x[n]$ associated, with a large or small amplitude, with those of the SP $y[n]$. The phase $\measuredangle R_{xy}(e^{j\omega})$ indicates the *phase delay* of $y[n]$ with respect to $x[n]$ for each frequency.

From equation (C.105), the following property holds:

$$R_{xy}\left(e^{j\omega}\right) = R_{yx}^{*}\left(e^{j\omega}\right) \tag{C.113}$$

so, $R_{xy}(e^{j\omega})$ and $R_{yx}^{*}(e^{j\omega})$ have the same module but opposite phase.

C.2.7.6 Spectral Representation of Stationary SP and TD LTI systems

For an impulse response $h[n]$, with z-transform $H(z) = Z\{h[n]\}$, we have the following property:

$$Z\{h[n]\} = H(z) \quad \Leftrightarrow \quad Z\{h^{*}[-n]\} = H^{*}(1/z^{*}). \tag{C.114}$$

From the above and for (C.103)–(C.106), then

$$R_{xy}(z) = H^{*}(1/z^{*})R_{xx}(z) \tag{C.115}$$

$$R_{yx}(z) = H(z)R_{xx}(z) \tag{C.116}$$

$$R_{yy}(z) = H(z)H^{*}(1/z^{*})R_{xx}(z). \tag{C.117}$$

For $z = e^{j\omega}$, we can write

$$R_{xy}\left(e^{j\omega}\right) = H^{*}\left(e^{j\omega}\right)R_{xx}\left(e^{j\omega}\right) \tag{C.118}$$

$$R_{yx}\left(e^{j\omega}\right) = H\left(e^{j\omega}\right)R_{xx}\left(e^{j\omega}\right) \tag{C.119}$$

$$R_{yy}\left(e^{j\omega}\right) = H\left(e^{j\omega}\right)H^{*}\left(e^{j\omega}\right)R_{xx}\left(e^{j\omega}\right) = \left|H\left(e^{j\omega}\right)\right|^{2}R_{xx}\left(e^{j\omega}\right). \tag{C.120}$$

Moreover, for (C.118) and (C.119) we have that

$$R_{yx}\left(e^{j\omega}\right) = R_{xy}^{*}\left(e^{j\omega}\right). \tag{C.121}$$

Example Consider the sum of two SPs $w[n] = x[n] + y[n]$, and evaluate the $r_{ww}[k]$.

By applying the definition (C.76) we have that

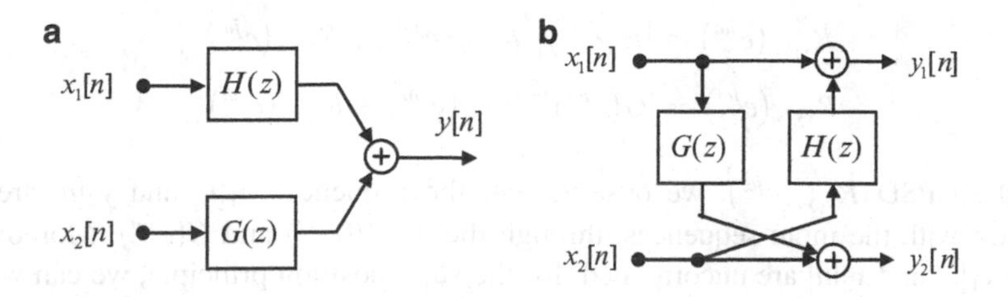

Fig. C.7 Block diagrams of TD LTI systems illustrated in the examples

$$\begin{aligned} r_{ww}[k] &= E\{w[n]w[n-k]\} = E\Big\{(x[n]+y[n])\cdot(x[n-k]+y[n-k])\Big\} \\ &= E\Big\{(x[n]x[n-k]\Big\} + E\{x[n]y[n-k]\} + E\{y[n]x[n-k]\} + E\{y[n]y[n-k]\} \\ &= r_{xx}[k] + r_{xy}[k] + r_{yx}[k] + r_{yy}[k]. \end{aligned}$$

For uncorrelated sequences the cross contributions are zero [see (C.67)]. Hence, we obtain that $r_{ww}[k] = r_{xx}[k] + r_{yy}[k]$; therefore, for the PSD we have

$$R_{ww}\left(e^{j\omega}\right) = R_{xx}\left(e^{j\omega}\right) + R_{yy}\left(e^{j\omega}\right).$$

Example Evaluate the output PSD $R_{yy}(e^{j\omega})$, for the TD LTI system illustrated in Fig. C.7a), with random uncorrelated input sequences $x_1[n]$ and $x_2[n]$.

The inputs $x_1[n]$ and $x_2[n]$ are mutually uncorrelated and, since the system is linear, can be considered separately with the *superposition principle*. The output PSD is calculated as the sum of the single contributions when the other is null. So we have

$$R_{yy}\left(e^{j\omega}\right) = R_{yy}^{x_1}\left(e^{j\omega}\right) + R_{yy}^{x_2}\left(e^{j\omega}\right).$$

For the (C.120), we get

$$\begin{aligned} R_{yy}^{x_1}\left(e^{j\omega}\right) &\triangleq R_{yy}\left(e^{j\omega}\right)\big|_{x_2[n]=0} = \left|H\left(e^{j\omega}\right)\right|^2 R_{x_1x_1}\left(e^{j\omega}\right) \\ R_{yy}^{x_2}\left(e^{j\omega}\right) &\triangleq R_{yy}\left(e^{j\omega}\right)\big|_{x_1[n]=0} = \left|G\left(e^{j\omega}\right)\right|^2 R_{x_2x_2}\left(e^{j\omega}\right). \end{aligned}$$

Finally, we have that

$$R_{yy}\left(e^{j\omega}\right) = \left|H\left(e^{j\omega}\right)\right|^2 R_{x_1x_1}\left(e^{j\omega}\right) + \left|G\left(e^{j\omega}\right)\right|^2 R_{x_2x_2}\left(e^{j\omega}\right).$$

Example Evaluate the PSDs $R_{y_1y_2}(e^{j\omega}), R_{y_2y_1}(e^{j\omega}), R_{y_1y_1}(e^{j\omega})$, and $R_{y_2y_2}(e^{j\omega})$, for the TD LTI system illustrated in Fig. C.7b), with random uncorrelated input sequences $x_1[n]$ and $x_2[n]$.

For (C.118)–(C.120), the output PSD we obtain is

$$R_{y_1y_1}(e^{j\omega}) = \left|H(e^{j\omega})\right|^2 R_{x_2x_2}(e^{j\omega}) + R_{x_1x_1}(e^{j\omega})$$
$$R_{y_2y_2}(e^{j\omega}) = \left|G(e^{j\omega})\right|^2 R_{x_1x_1}(e^{j\omega}) + R_{x_2x_2}(e^{j\omega}).$$

For the CPSD $R_{y_1y_2}(e^{j\omega})$, we observe that the sequences $y_1[n]$ and $y_2[n]$ are in relation with the input sequences, through the TF $H(e^{j\omega})$ and $G(e^{j\omega})$. Moreover, since $x_1[n]$ and $x_2[n]$ are uncorrelated, for the superposition principle, we can write

$$R_{y_1y_2}(e^{j\omega}) = R_{y_1y_2}^{x_1}(e^{j\omega}) + R_{y_1y_2}^{x_2}(e^{j\omega}).$$

Note that for $x_2[n] = 0$ is $y_1[n] \equiv x_1[n]$ for which, for (C.119), we obtain

$$R_{y_1y_2}^{x_1}(e^{j\omega}) = R_{y_1y_2}(e^{j\omega})\big|_{x_2[n]=0} = H(e^{j\omega})R_{x_1x_1}(e^{j\omega}).$$

Similarly, for the other input when $x_1[n] = 0$, for (C.118), we obtain

$$R_{y_1y_2}^{x_2}(e^{j\omega}) = R_{y_1y_2}(e^{j\omega})\big|_{x_1[n]=0} = G^*(e^{j\omega})R_{x_2x_2}(e^{j\omega}).$$

The CPSD $R_{y_1y_2}(e^{j\omega})$ is then

$$R_{y_1y_2}(e^{j\omega}) = H(e^{j\omega})R_{x_1x_1}(e^{j\omega}) + G^*(e^{j\omega})R_{x_1x_1}(e^{j\omega}).$$

Similarly, for the CPSD $R_{y_2y_1}(e^{j\omega})$, we get

$$R_{y_2y_1}(e^{j\omega}) = H^*(e^{j\omega})R_{x_1x_1}(e^{j\omega}) + G(e^{j\omega})R_{x_1x_1}(e^{j\omega}).$$

C.3 Basic Concepts of Estimation Theory

In many real applications the distribution functions are not *a priori* known and should be determined by appropriate experiments carried out using a finite set of measured data. The estimation of such statistics can be performed by the use of methodologies defined in the context of the *Estimation Theory*[9] (ET) [16–22].

---

[9] The *Estimation Theory* is a very ancient discipline and famous scientists as Lagrange, Gauss, Legendre, etc., have used it in the past, and in the last century, attention to it has considerably increased. In fact, many were scientists who have worked in this field (Wold, Fisher, Kolmogorov, Wiener, Kalman, etc.). Among these N. Wiener, between 1930 and 1940, was among those who most emphasized the importance that not only the noise but also signals should be considered as stochastic processes.

C.3.1 Preliminary Definitions and Notation

Let Θ be defined as the *parameters space*, the general problem of *parameters estimation* is the determination of a parameter $\theta \in \Theta$ or, more generally, of a vector of unknown parameters $\boldsymbol{\theta} \in \Theta \triangleq [\theta[n]]_0^{L-1}$, starting from a series of observations or measurements $\mathbf{x} \triangleq [x[n]]_0^{N-1}$, by means of *estimation function* $h(\cdot)$, called *estimator*, i.e., such that the *estimate* is $\hat{\boldsymbol{\theta}} = h(\mathbf{x})$.

Before proceeding to further developments, let us introduce some preliminary formal definitions.

$\boldsymbol{\theta} \in \Theta$ In general, $\boldsymbol{\theta}$ indicates the parameters vector to be estimated. Depending on the *estimation paradigm* adopted, as better illustrated in the following, $\boldsymbol{\theta}$ can be considered as n RV, characterized by a certain *a priori* known supposed (or hypothesized) distribution, or simply considered as a deterministic unknown.

$h(\mathbf{x})$ This function, that is itself an RV, indicates the *estimator*, namely, the law which would determine the value of the parameters to be estimated starting from the observations $\mathbf{x}$.

$\hat{\boldsymbol{\theta}}$ This symbol indicates the result, i.e., $\hat{\boldsymbol{\theta}} = h(\mathbf{x})$. Note that the *estimated value* is *always* an RV characterized by a certain pdf and/or values of its moments.

C.3.1.1 Sampling Distribution

The above definitions show that the estimator relative to the ζ_kth event, denoted by $h\left(\{\mathbf{x}[n,\zeta_k]\}_0^{N-1}\right)$, is defined in an N dimensional space, whose distribution can be obtained from the joint distribution of the RVs $\{\mathbf{x}[n,\zeta]\}_0^{N-1}$ and $\boldsymbol{\theta}$. This distribution, in the case of a single deterministic parameter estimation, is shown as $f_{\mathbf{x};\theta}(\mathbf{x};\theta)$ and is defined as *sampling distribution*.

Note that *sampling distribution* represents one of the fundamental concepts in the estimation theory because it contains *all* the information needed to define the *estimator quality* characteristics. In fact, it is intuitive to think that the sampling distribution of a "*good*" estimator may be the most concentrate as possible. Thus it has a *small variance* around the true value of the parameter to be estimated.

C.3.1.2 Estimation Theory: Classical and Bayesian Approaches

In classical estimation theory $\boldsymbol{\theta}$ represents an unknown *deterministic* vector of parameters. Therefore, the formalism $f_{\mathbf{x};\boldsymbol{\theta}}(\mathbf{x};\boldsymbol{\theta})$ indicates a *parametric dependency* of the pdf related to the measures $\mathbf{x}$, from the parameters $\boldsymbol{\theta}$. For example, consider the simple case where $N = 1$ where the parameter θ represents a certain (mean)

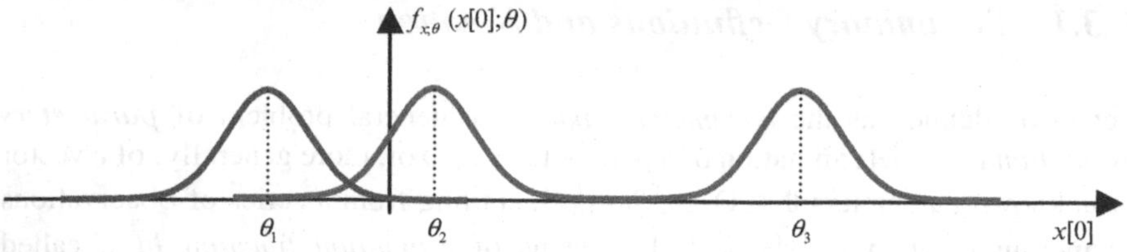

Fig. C.8 Dependency of pdf $f_{x;\theta}\big(x[0];\,\theta\big)$ form the unknown parameter θ

value and the pdf $f_{x;\theta}\big(x[0];\,\theta\big)$ is *normally distributed* around this value $x[0] \sim N(\theta,\sigma_{x[0]}^2)$ so that

$$f_{x[0];\theta}\big(x[0];\theta\big) = \frac{1}{\sqrt{2\pi}\sigma_{x[0]}} e^{-\frac{1}{2\sigma_{x[0]}^2}(x[0]-\theta)^2} \tag{C.122}$$

illustrated, by way of example, in Fig. C.8, for some value of the parameter θ. In other words, the parameter θ is *not* an RV and $f_{x;\theta}\big(x[0];\,\theta\big)$ indicates a parametric pdf that depends on a deterministic value θ .

On the contrary, in *Bayesian estimation theory* θ is an RV characterized by its pdf $f_\Theta(\theta)$, *a priori* pdf, which contains all the *a priori* known information (or believed). The quantity to be estimated is then interpreted as a realization of the RV θ. Subsequently, the estimation process is described by the *joint pdf* through the Bayes rule, as [see Sect. C.1.4, Eq. (C.24)]

$$f_{\mathbf{x},\theta}(\mathbf{x},\theta) = f_{\mathbf{x},\theta}(\mathbf{x}\,|\,\theta) f_\theta(\theta) = f_{\mathbf{x},\theta}(\theta\,|\,\mathbf{x}) f_{\mathbf{x}}(\mathbf{x}), \tag{C.123}$$

where $f_{\mathbf{x}|\theta}(\mathbf{x}|\theta)$ is the conditional pdf that represents the knowledge carried from the data $\mathbf{x}$ conditioned by knowledge of distribution $f_\theta(\theta)$.[10]

From the definition of the estimator quality, it is not always possible to know the sampling distribution $f_{\mathbf{x};\theta}(\mathbf{x};\theta)$. In practice, however, it is possible to use the low-order moments as the expectation $E\big(\hat{\theta}\big)$, the variance, denoted as $\mathrm{var}\big(\hat{\theta}\big)$ or $\sigma_{\hat{\theta}}^2$, and the mean squares error (MSE) denoted as $\mathrm{mse}\big(\hat{\theta}\big)$.

C.3.1.3 Estimator, Expectation, and Bias

An estimator is called *unbiased*, if the expectation of the estimated value tends to the true value of the parameter to be estimated. In other words,

---

[10] The notation $f_{\mathbf{x};\theta}(\mathbf{x};\theta)$ indicates a parametric pdf family where θ is the free parameter. Moreover, remember that the notation $f_{\mathbf{x},\theta}(\mathbf{x},\theta)$ indicates the *joint pdf*, while $f_{\mathbf{x}|\theta}(\mathbf{x}|\theta)$ indicates the *conditional pdf*.

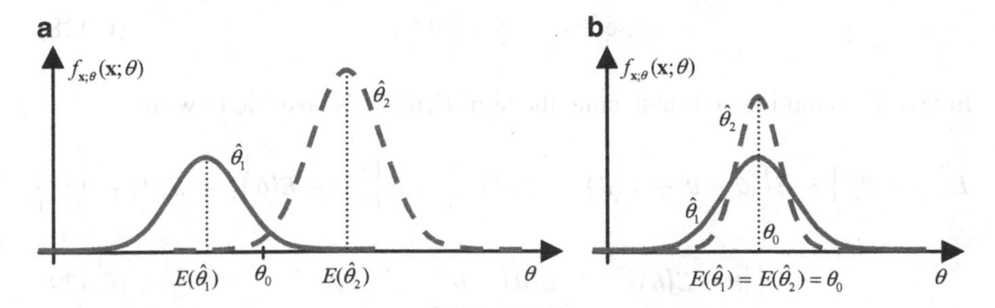

Fig. C.9 Estimator bias and variance (**a**) biased estimator; (**b**) *unbiased* estimator

$$E(\hat{\theta}) = \theta. \tag{C.124}$$

If $E(\hat{\theta}) \neq \theta$, it is possible to define a quantity called *deviation* or *bias* as

$$b(\hat{\theta}) \triangleq E(\hat{\theta}) - \theta. \tag{C.125}$$

Remark The presence of a *bias* term, probably, indicates the presence of a *systematic error*, i.e., due to the measure process (or due to estimation algorithm). Note that an unbiased estimator not necessarily is a "good" estimator. In fact, the only guarantee is that, in *average*, it tends to the true value.

C.3.1.4 Estimator Variance

For better characterizing the estimation quality we define the *estimator variance* as

$$\operatorname{var}(\hat{\theta}) = \sigma_{\hat{\theta}}^2 \triangleq E\left\{|\hat{\theta} - E(\hat{\theta})|^2\right\} \tag{C.126}$$

that represents a *dispersion measure* of the pdf of $\hat{\theta}$ around its expected value (Fig. C.9).

C.3.1.5 Estimator's Mean Square Error and Bias-Vs.-Variance Trade-off

Given the *true* value θ and its estimated value $\hat{\theta}$, the MSE of the related estimator $\hat{\theta} = h(\mathbf{x})$ can be defined as

$$\operatorname{mse}(\hat{\theta}) = E\left\{|\hat{\theta} - \theta|^2\right\}. \tag{C.127}$$

So the $\operatorname{mse}(\cdot)$ is a measure of the average quadratic deviation of the estimated value with respect to the true value. Note that, considering the definitions (C.125) and (C.126), the $\operatorname{mse}(\hat{\theta})$ can be written as

$$\operatorname{mse}(\hat{\theta}) = \sigma_{\hat{\theta}}^2 + |b(\hat{\theta})|^2. \tag{C.128}$$

In fact, by summing and subtracting the term $E(\hat{\theta})$, it is possible to write

$$\begin{aligned} E\{|\hat{\theta} - \theta|^2\} &= E\{|\hat{\theta} - \theta + E(\hat{\theta}) - E(\hat{\theta})|^2\} = E\{|[\hat{\theta} - E(\hat{\theta})] + [E(\hat{\theta}) - \theta]|^2\} \\ &= \underbrace{E\{|\hat{\theta} - E(\hat{\theta})|^2\}}_{\sigma_{\hat{\theta}}^2} + \underbrace{|E(\hat{\theta}) - \theta|^2}_{|b(\hat{\theta})|^2}. \end{aligned} \tag{C.129}$$

The expression (C.128) shows that the MSE is formed by the sum of two contributes: one due to the estimation *variance*, while the other due to its *bias*.

C.3.1.6 Example: Estimate the Current Gain of a White Gaussian Sequence

As example we consider the estimation of discrete sequence $x[n]$ consisting of N independent samples defined as

$$x[n] = \theta + w[n], \tag{C.130}$$

where θ represents the constant component (by analogy with the constant electrical *direct current* (DC)) and $w[n]$ is *additive white Gaussian noise* (AWGN) with zero mean and indicated as $w[n] \sim N(0,\sigma_w^2)$.

Intuitively reasoning, we can define different algorithms for the estimation of θ. For example, two very commonly used estimators are defined as

$$\hat{\theta}_1 = h_1(\mathbf{x}) \triangleq x[0] \tag{C.131}$$

$$\hat{\theta}_2 = h_2(\mathbf{x}) \triangleq \frac{1}{N}\sum_{n=0}^{N-1} x[n]. \tag{C.132}$$

To assess the quality of the estimators $h_1(\mathbf{x})$ and $h_2(\mathbf{x})$, we calculate the respective expected values and variances. For the expected values we have

$$E(\hat{\theta}_1) = E(x[0]) = \theta \tag{C.133}$$

$$E(\hat{\theta}_2) = E\left(\frac{1}{N}\sum_{n=0}^{N-1} x[n]\right) = \frac{1}{N}\sum_{n=0}^{N-1} E(x[n]) = \frac{1}{N}[N\theta] = \theta. \tag{C.134}$$

Therefore, both estimators converge to the same expected value that coincides with the true value of θ parameter to estimate. By reasoning in a similar way, for the variance we have

$$\mathrm{var}\left(\hat{\theta}_1\right) = \mathrm{var}\left(x[0]\right) = \sigma_w^2 \tag{C.135}$$

and

$$\mathrm{var}\left(\hat{\theta}_2\right) = \mathrm{var}\left(\frac{1}{N}\sum_{n=0}^{N-1} x[n]\right). \tag{C.136}$$

The latter, for the hypothesis of independency, can be rewritten as

$$\mathrm{var}\left(\hat{\theta}_2\right) = \frac{1}{N^2}\sum_{n=0}^{N-1} \mathrm{var}\left(x[n]\right) = \frac{1}{N^2}\left[N\sigma_w^2\right] = \frac{\sigma_w^2}{N}. \tag{C.137}$$

Then, it follows that the variance of the estimator $\mathrm{var}\left[h_2(\mathbf{x})\right] < \mathrm{var}\left[h_1(\mathbf{x})\right]$ and for $N \to \infty$, $\mathrm{var}\left(\hat{\theta}_2\right) \to 0$. For this reason, the estimator $h_2(\mathbf{x})$ turns out to be better than $h_1(\mathbf{x})$. In fact, according to certain paradigms, as we shall see later, $h_2(\mathbf{x})$ is the *best possible estimator.*

C.3.1.7 Minimum Variance Unbiased (MVU) Estimator

Ideally a good estimator should have the MSE which tends to zero. Unfortunately, the adoption of this criterion produces, in most cases, a not "feasible" estimator. In fact, the expression of the MSE (C.128) is formed by the contribution of the variance added to that of bias. For better understanding consider the example of the average value estimator (C.132), redefined using the following expression:

$$\hat{\theta} = h(\mathbf{x}) \triangleq a\frac{1}{N}\sum_{n=0}^{N-1} x[n], \tag{C.138}$$

where a is a suitable constant. The problem, now, consists in determining the value of the constant a such that the MSE of the estimator is minimal.

Since by definition, $E\left(\hat{\theta}\right) = a\theta$, $\mathrm{var}\left(\hat{\theta}\right) = a^2\sigma\hat{\theta}^2/N$ and for Eq. (C.128) we have

$$\mathrm{mse}\left(\hat{\theta}\right) = \frac{a^2\sigma_{\hat{\theta}}^2}{N} + (a-1)^2\theta^2. \tag{C.139}$$

Hence, differentiating the MSE with respect to a, we obtain

$$\frac{d\left(\mathrm{mse}\left(\hat{\theta}\right)\right)}{da} = \frac{2a\sigma_{\hat{\theta}}^2}{N} + 2(a-1)\theta^2. \tag{C.140}$$

The optimum value a_{opt} is obtained by setting these equations to zero and solving with respect to A. It follows:

$$a_{opt} = \frac{\theta^2}{\theta^2 + \sigma_{\hat{\theta}}^2/N}. \tag{C.141}$$

The previous expression shows that the value a_{opt} depends on θ, i.e., the estimator goodness depends on the parameter θ, which should be determined by the estimator itself. Such paradox indicates the *non-computability* of a_{opt} parameter, i.e., the non-feasibility of the estimator. Generally, with certain exceptions, any criteria that depends on the bias determines a not feasible estimator.

On the other hand, the optimal estimator is not the one with minimum MSE but is what constrains the bias to zero and minimizes the estimated variance. For such reason, this estimator is called *minimum variance unbiased* (MVU) estimator. For one MVU estimator, from definition (C.128),

$$\text{mse}\left(\hat{\theta}\right) = \sigma_{\hat{\theta}}^2, \quad \textit{MVU estimator}. \tag{C.142}$$

C.3.1.8 Bias Vs. Variance Trade-off

From what was said, a "*good*" estimator should be unbiased and with minimum variance. Often in practical situations, the two features are mutually contradictory, i.e., when reducing the variance the bias increases. This situation reflects a kind of *indeterminacy* between bias and variance often referred to as *bias–variance trade-off*.

The MVU estimator does not always exist and this is generally true when the variance of the estimator depends on the value of the parameter to be estimated. Note also that the existence of the MVU estimator does not imply its determination. In other words, although theoretically it exists, it is not guaranteed that we can determine it.

C.3.1.9 Consistent Estimator

An estimator is said to be *weakly consistent*, if it converges in probability to the true parameter value, for a sample length N which tends to infinity

$$\lim_{N\to\infty} p\left\{\left|h(\mathbf{x}) - \theta\right| < \varepsilon\right\} \quad \forall \varepsilon > 0. \tag{C.143}$$

An estimator is called *strong sense consistent*, if it converges with probability one to parameter value, for a sample length N which tends to infinity

$$\lim_{N\to\infty} p\{h(\mathbf{x}) = \theta\} = 1. \tag{C.144}$$

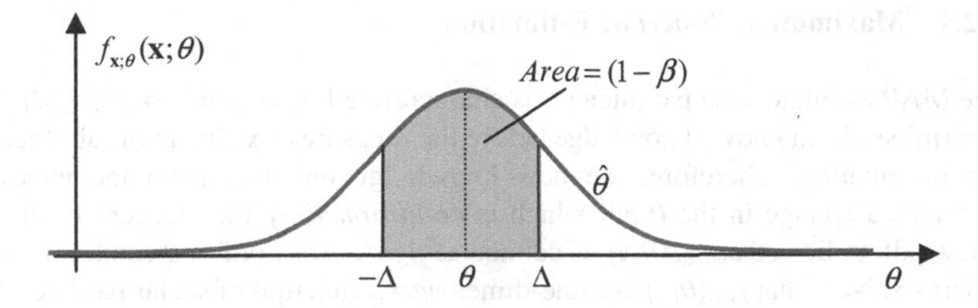

Fig. C.10 Confidence interval around the true θ value

Sufficient conditions for a *weak sense consistent estimator* are that the variance and the bias tend to zero, for sample length N tending to infinity, i.e.,

$$\begin{aligned} \lim_{N\to\infty} E\{h(\mathbf{x})\} &= \theta \\ \lim_{N\to\infty} \operatorname{var}\{h(\mathbf{x})\} &= 0. \end{aligned} \tag{C.145}$$

In this case, the sampling distribution tends to become an impulse centered on the value to be estimated.

C.3.1.10 Confidence Interval

Increasing the sample length N, under sufficiently general conditions, the estimate tends to the true value $\left(\hat{\theta} \xrightarrow[N\to\infty]{} \theta\right)$. Moreover, for the central limit theorem, if N increases, the pdf of $\hat{\theta}$ is well approximated by the normal distribution.

Knowing the sampling distribution of an estimator, it is possible to calculate a certain interval $(-\Delta, \Delta)$, which defines a specified probability. Such interval, called *confidence interval*, indicates that the event $\hat{\theta}$ is in the range $(-\Delta, \Delta)$, around θ, with probability $(1-\beta)$ or confidence $(1-\beta)\cdot 100\%$ (see Fig. C.10).

C.3.2 Classical and Bayesian Estimation

In the classical ET, as previously indicated, the problem is addressed considering the parameter to be estimated as deterministic, while in Bayesian ET, the estimate parameter is considered stochastic. If the parameter is an RV, it is characterized by a certain pdf that reflects *a priori* knowledge on the parameter itself.

Both theories have found several applications in signal processing and, in particular, the three main estimation paradigms used are the following:

i) the *maximum a posteriori estimation* (MAP);
ii) the *maximum likelihood* estimation (ML);
iii) the *minimum mean squares error* estimation (MMSE).

C.3.2.1 Maximum *a Posteriori* Estimation

In the MAP estimator, the parameter θ is characterized by an *a priori* pdf $f_\theta(\theta)$ that is determined from known knowledge before the measure of $\mathbf{x}$ data in the absence of other information. Therefore, the new knowledge obtained from the measure determines a change in the θ pdf which is *conditioned* by the measure itself. So the new pdf, indicated as $f_{\mathbf{x}|\theta}(\theta|\mathbf{x})$, is defined as *a posteriori* pdf of θ conditioned by measures $\mathbf{x}$. Note that $f_{\mathbf{x}|\theta}(\theta|\mathbf{x})$ is a one-dimensional function of scalar parameter θ, but it is also subject to conditioning due to the measures.

Therefore, the MAP estimate consists in determining the maximum *a posteriori* pdf. Indeed, this can be obtained by differentiating $f_{\mathbf{x}|\theta}(\theta|\mathbf{x})$ with respect to the parameter θ, and equating the result to be

$$\theta_{\text{MAP}} \triangleq \left\{\theta \therefore \frac{\partial f_{\mathbf{x}|\theta}(\theta|\mathbf{x})}{\partial\theta} = 0\right\}. \tag{C.146}$$

Sometimes, instead of the maximum of $f_{\mathbf{x}|\theta}(\theta|\mathbf{x})$, we consider its natural logarithm. So θ_{MAP} can be found from the maximum of the function $\ln f_{\mathbf{x}|\theta}(\theta|\mathbf{x})$, for which

$$\theta_{\text{MAP}} \triangleq \left\{\theta \therefore \frac{\partial \ln f_{\mathbf{x}|\theta}(\theta|\mathbf{x})}{\partial\theta} = 0\right\}. \tag{C.147}$$

Since the logarithm is a monotonically increasing function, the value found is the same as that in (C.146). However, the determination of the $f_{\mathbf{x}|\theta}(\theta|\mathbf{x})$ or $\ln f_{\mathbf{x}|\theta}(\theta|\mathbf{x})$ is often problematic, and using the rule derived from the Bayes theorem, for (C.123) it is possible to write the conditioned pdf as

$$f_{\mathbf{x}|\theta}(\theta|\mathbf{x}) = \frac{f_{\mathbf{x}|\theta}(\mathbf{x}|\theta) f_\theta(\theta)}{f_{\mathbf{x}}(\mathbf{x})}. \tag{C.148}$$

Considering the logarithm of both sides of the previous, we can write

$$\ln f_{\mathbf{x}|\theta}(\theta|\mathbf{x}) = \ln f_{\mathbf{x}|\theta}(\mathbf{x}|\theta) + \ln f_\theta(\theta) - \ln f_{\mathbf{x}}(\mathbf{x}).$$

Thus, the procedure for the MAP estimate is

$$\frac{\partial}{\partial\theta}\left(\ln f_{\mathbf{x}|\theta}(\mathbf{x}|\theta) + \ln f_\theta(\theta) - \ln f_{\mathbf{x}}(\mathbf{x})\right) = 0$$

and, since $\ln f_{\mathbf{x}}(\mathbf{x})$ does not depend on θ , we can write

$$\theta_{\text{MAP}} \triangleq \left\{ \theta \therefore \frac{\partial}{\partial \theta} \left(\ln f_{\mathbf{x}|\theta}(\mathbf{x}|\theta) + \ln f_{\theta}(\theta) \right) = 0 \right\}. \tag{C.149}$$

Finally, note that it is possible to determine the MAP solution equivalently through (C.146), (C.147) or, (C.149).

C.3.2.2 Maximum-Likelihood Estimation

In the *maximum-likelihood* (ML) estimation, the parameter θ to be estimated is considered as a simple deterministic unknown. Therefore, in the ML estimation the determination of θ_{ML} is carried out through the maximization of the function $f_{\mathbf{x};\theta}(\mathbf{x};\theta)$ defined as a parametric pdf family, where θ is the deterministic parameter. In this respect, the function $f_{\mathbf{x};\theta}(\mathbf{x};\theta)$ is sometimes referred to as the *likelihood function* L_θ . Note that if $f_{\mathbf{x};\theta}(\mathbf{x};\theta_1) > f_{\mathbf{x};\theta}(\mathbf{x};\theta_2)$, then the value of θ_1 is "more plausible" of the value θ_2, so that the ML paradigm indicates that the estimated value θ_{ML} is the most likely according to the observations $\mathbf{x}$. As for the MAP method, also for ML estimator it is often considered the natural logarithm function $\ln f_{\mathbf{x};\theta}(\mathbf{x};\theta)$. Note that, although θ is a deterministic parameter, the likelihood function L_θ (or $\ln L_\theta$) has stochastic nature and is considered as an RV. In this case, if the estimates solution exists, it can be found as the only solution of the equation that maximize the *likelihood equation* defined as

$$\theta_{\text{ML}} \triangleq \left\{ \theta \therefore \frac{\partial \ln f_{\mathbf{x};\theta}(\mathbf{x};\theta)}{\partial \theta} = 0 \right\}. \tag{C.150}$$

Such solution is defined as *maximum-likelihood estimate* (MLE).

In other words, the ML methods search for the *most likely* value of θ, namely, research within the space Θ of all possible θ values, the value of the parameter that maximizes the probability that θ_{ML} is the most plausible sample. From a mathematical point of view, calling $L_\theta = f_{\mathbf{x};\theta}(\mathbf{x};\theta)$ the *likelihood function*, we have

$$\theta_{\text{ML}} = \max_{\theta \in \Theta} \{L_\theta\}. \tag{C.151}$$

The MLE also has the following properties:

- *Sufficient*—if there is a *sufficient statistic*[11] for θ then the MLE is also a sufficient statistic;
- *Efficient*—an estimator is called efficient if there is a lower limit of the variance obtained from an unbiased estimator. An estimator that reaches this limit is

[11] A *sufficient statistic* is a statistic such that "no other statistic which can be calculated from the same sample provides any additional information as to the value of the parameter" [18]. In other words, a statistic is sufficient for a pdf family if the sample from which it is calculated gives no additional information than does the statistic.

called *fully efficient estimator*. Although, for a finite set of observations N, the fully efficient estimator does not exist, in many practical cases, the ML estimator turns out to be asymptotically fully efficient.

- *Gaussianity*—the MLE turns out to be asymptotically Gaussian.

In case the efficient estimator does not exist, then the lower limit of the MLE cannot be achieved and, in general, it is difficult to measure the distance from this limit.

Remark By comparing the ML and MAP estimators it should be noted that in the latter the estimate is derived using a combination of *a priori* and *a posteriori* known information on θ, where such knowledge is formulated in terms of the pdf $f_\theta(\theta)$. However, the ML estimation results potentially more feasible in practical problems because it does not require any *a priori* knowledge. Both procedures require knowledge of the joint *a posteriori* pdf of the observations.

Note also that the ML estimator can be derived starting from the MAP and considering the parameter θ as an RV with uniformly distributed pdf between $[-\infty, +\infty]$.

C.3.2.3 Example: Noisy Measure of a Parameter with a Single Observation

As a simple example to illustrate the methodology MAP and ML, consider a single measure x consisting of the sum of a parameter θ and a normal distributed zero-mean RV w (AWGN) $w \sim N(0,\sigma_w^2)$. Then, the process is defined as

$$x = \theta + w. \tag{C.152}$$

It appears that (1) in ML estimating the parameter θ is a deterministic unknown constant, while, (2) in the MAP estimate θ is an RV with an *a priori* pdf of the normal type $N(\theta,\sigma_\theta^2)$.

ML Estimation

In ML method, the likelihood function $L_\theta = f_{x,\theta}(x;\theta)$ appears to be a scalar function of a single variable. From equation (C.152) x is, by definition, a Gaussian signal with mean value θ and variance equal to σ_w^2. It follows that the likelihood function L_θ reflects this dependence and appears to be defined as

$$L_\theta = f_x(x;\theta) = \frac{1}{\sqrt{2\pi\sigma_w^2}} e^{-\frac{1}{2\sigma_w^2}(x-\theta)^2}. \tag{C.153}$$

Its logarithm is

$$\ln L_\theta = \ln f_x(x;\theta) = -\frac{1}{2}\ln\left(2\pi\sigma_w^2\right) - \frac{1}{2\sigma_w^2}(x-\theta)^2. \tag{C.154}$$

To determine the maximum, we differentiate with respect to θ, and we equate to zero

$$\theta_{\mathrm{ML}} \triangleq \left\{\theta \therefore \frac{1}{\sigma_w^2}(x-\theta) = 0\right\},$$

that is,

$$\theta_{\mathrm{ML}} = x. \tag{C.155}$$

It follows, then, that the best estimate in the ML sense is just the x value of the measure. This is an intuitive result since, in the absence of other information, it is not in any way possible to refine the estimate of the parameter θ.

The variance associated with the estimated value appears to be

$$\mathrm{var}(\theta_{\mathrm{ML}}) = E\left(\theta_{\mathrm{ML}}^2\right) - E^2(\theta_{\mathrm{ML}}) = E\left(x^2\right) - E^2(x)$$

that, for $x = \theta + w$, is

$$\mathrm{var}(\theta_{\mathrm{ML}}) = \theta^2 + \sigma_w^2 - \theta^2 = \sigma_w^2$$

which obviously coincides with the variance of the superimposed noise w.

MAP Estimation

In MAP method we have $x = \theta + w$ with $w \sim N(0,\sigma_w^2)$ and we suppose the *a priori* known pdf $f(\theta)$ that is normal distributed: $N(\theta_0,\sigma_\theta^2)$. The MAP estimation is that obtained from Eq. (C.149) as

$$\theta_{\mathrm{MAP}} \triangleq \left\{\theta \therefore \frac{\partial}{\partial\theta}\left(\ln f_{x|\theta}(x|\theta) + \ln f_\theta(\theta)\right) = 0\right\}. \tag{C.156}$$

Given the θ value, the pdf of x is Gaussian with mean value θ and variance σ_w^2. It follows that the logarithm of the density is

$$\ln f_{x,\theta}(x|\theta) = -\frac{1}{2}\ln\left(2\pi\sigma_w^2\right) - \frac{1}{2\sigma_w^2}(x-\theta)^2. \tag{C.157}$$

while the *a priori* known density $f(\theta)$ is equal to

$$f_\theta(\theta) = \frac{1}{\sqrt{2\pi\sigma_\theta^2}} e^{-\frac{1}{2\sigma_\theta^2}(\theta-\theta_0)^2} \tag{C.158}$$

with logarithm

$$\ln f_\theta(\theta) = -\frac{1}{2}\ln\left(2\pi\sigma_\theta^2\right) - \frac{1}{2\sigma_\theta^2}(\theta - \theta_0)^2. \tag{C.159}$$

By substituting (C.157) and (C.159) in (C.156) we obtain

$$\theta_{\text{MAP}} \triangleq \left\{\theta \therefore \frac{\partial}{\partial\theta}\left(-\frac{1}{2}\ln\left(2\pi\sigma_w^2\right) - \frac{1}{2\sigma_w^2}(x-\theta)^2 - \frac{1}{2}\ln\left(2\pi\sigma_\theta^2\right) - \frac{1}{2\sigma_\theta^2}(\theta-\theta_0)^2\right) = 0\right\}.$$

Differentiating we obtain

$$\frac{(x - \theta_{\text{MAP}})}{\sigma_w^2} - \frac{(\theta_{\text{MAP}} - \theta_0)}{\sigma_\theta^2} = 0, \tag{C.160}$$

that is,

$$\theta_{\text{MAP}} = \frac{x\sigma_\theta^2 - \theta_0\sigma_w^2}{\sigma_w^2 + \sigma_\theta^2} = \frac{x + \theta_0\left(\sigma_w^2/\sigma_\theta^2\right)}{1 + \left(\sigma_w^2/\sigma_\theta^2\right)}. \tag{C.161}$$

Comparing the latter with the ML estimate (C.155), we observe that the MAP estimate can be viewed as a weighted sum of the ML estimate x and of the *a priori* mean value θ_0. In (C.161), the ratio of the variances $(\sigma_w^2/\sigma_\theta^2)$ can be seen as a *measure of confidence* of the value θ_0. The lower the value of σ_θ^2, the greater the ratio $(\sigma_w^2/\sigma_\theta^2)$, and the greater the confidence in θ_0, less is the weight of the observation x.

In the limit case where $(\sigma_w^2/\sigma_\theta^2) \to \infty$, the MAP estimate is simply given by the value of the *a priori* mean θ_0. At the opposite extreme, if σ_θ^2 increases, then the MAP estimate coincides with the ML estimate $\theta_{MAP} \to x$.

C.3.2.4 Example: Noisy Measure of a Parameter by *N* Observations

Let's consider, now, the previous example where N measurements are available

$$x[n] = \theta + w[n], \quad n = 0, 1, ..., N - 1, \tag{C.162}$$

where samples $w[n]$ are iid, zero-mean Gaussian distributed $N(0,\sigma_w^2)$.

ML Estimation

In the MLE, the likelihood function $L_\theta = f_{\mathbf{x};\theta}(\mathbf{x};\theta)$ is an N-dimensional multivariate Gaussian defined as

$$L_\theta = f_{\mathbf{x};\theta}(\mathbf{x};\theta) = \frac{1}{\left(2\pi\sigma_w^2\right)^{N/2}} e^{-\frac{1}{2\sigma_w^2}\sum_{n=0}^{N-1}\left(x[n]-\theta\right)^2}. \tag{C.163}$$

Its logarithm is

$$\ln L_\theta = \ln f_{\mathbf{x};\theta}(\mathbf{x};\theta) = -\frac{N}{2}\ln\left(2\pi\sigma_w^2\right) - \frac{1}{2\sigma_w^2}\sum_{n=0}^{N-1}\left(x[n]-\theta\right)^2.$$

Differentiating with respect to θ, and setting to zero

$$\frac{\partial \ln L_\theta}{\partial \theta} = \sum_{n=0}^{N-1}\left(x[n]-\theta_{\mathrm{ML}}\right) = 0$$

we obtain

$$\theta_{\mathrm{ML}} = \frac{1}{N}\sum_{n=0}^{N-1} x[n]. \tag{C.164}$$

It follows, then, that the *best estimate* in the ML sense coincides with the average value of the observed data. This represents an intuitive result, already previously reached, since, in the absence of other information, it is not possible to do better.

MAP Estimation

In MAP estimation we have $x[n] = \theta + w[n]$ where $w \sim N(0,\sigma_w^2)$, and we suppose that the *a priori* pdf is normally distributed $f(\theta)$, $N\left(,\hat{\theta},\sigma_\theta^2\right)$. The MAP estimation, proceeding as in the latter case, is obtained as

$$\frac{\sum_{n=0}^{N-1}\left(x[n]-\theta_{\mathrm{MAP}}\right)}{\sigma_w^2} - \frac{\left(\theta_{\mathrm{MAP}}-\hat{\theta}\right)}{\sigma_\theta^2} = 0, \tag{C.165}$$

that is,

$$\theta_{\text{MAP}} = \frac{\frac{1}{N}\sum_{n=0}^{N-1} x[n] + \hat{\theta} \cdot \left(\sigma_w^2/N\sigma_\theta^2\right)}{1 + \left(\sigma_w^2/N\sigma_\theta^2\right)}. \tag{C.166}$$

Again, comparing the latter with the ML estimation, we observe that the MAP estimate can be viewed as a weighted sum of the MLE and the *a priori* mean value. Comparing with the case of single observation [Eq. (C.161)], one can observe that the increase in the number of observations N is a reduced dependence of the *a priori* density by a factor N. This result is reasonable and intuitive: each new observation reduces the variance of the observations and reduces the dependence of the model *a priori*.

C.3.2.5 Example: Noisy Measure of L Parameters with N Observations

We consider now the general case where we have N measurements $\mathbf{x} \triangleq \left[x[n]\right]_0^{N-1}$, and we estimate a number of L parameters $\boldsymbol{\theta} \triangleq \left[\theta[n]\right]_0^{L-1}$, where samples of $w[n]$ are zero-mean Gaussian $N(0,\sigma_w^2)$, iid.

MAP Estimation

Proceed in this case prior to the MAP estimate. We seek to maximize the posterior density $f_{\mathbf{x}|\boldsymbol{\theta}}(\boldsymbol{\theta}|\mathbf{x})$ or, equivalently, the $\ln f_{\mathbf{x},\boldsymbol{\theta}}(\boldsymbol{\theta}|\mathbf{x})$, with respect to $\boldsymbol{\theta}$. This is achieved by differentiating with respect to each component of $\boldsymbol{\theta}$ and equating to zero. It is then

$$\frac{\partial \ln f_{\mathbf{x},\boldsymbol{\theta}}(\boldsymbol{\theta}|\mathbf{x})}{\partial \theta[n]} = 0, \quad n = 0, 1, ..., L-1. \tag{C.167}$$

By separating the derivatives we obtain L equations in L unknowns in the parameters $\theta[0], \theta[1], \ldots, \theta[L-1]$ that, changing the type of notation, can be expressed as

$$\nabla_{\boldsymbol{\theta}}\left\{f_{\mathbf{x},\boldsymbol{\theta}}(\boldsymbol{\theta}|\mathbf{x})\right\} = 0, \tag{C.168}$$

where the symbol $\nabla_{\boldsymbol{\theta}}$ indicates the differential operator, called "*gradient*," defined as

$$\nabla_{\boldsymbol{\theta}} \triangleq \left[\frac{\partial}{\partial\theta[0]}, \frac{\partial}{\partial\theta[1]}, \cdots, \frac{\partial}{\partial\theta[L-1]}\right]^T.$$

As in the case of a single parameter, the Bayes rule holds, so we have

$$f_{\mathbf{x}|\boldsymbol{\theta}}(\boldsymbol{\theta}|\mathbf{x}) = \frac{f_{\mathbf{x}|\boldsymbol{\theta}}(\mathbf{x}|\boldsymbol{\theta}) f_{\boldsymbol{\theta}}(\boldsymbol{\theta})}{f_{\mathbf{x}}(\mathbf{x})},$$

which, considering the logarithm, can be written as

$$\ln f_{\mathbf{x}|\boldsymbol{\theta}}(\boldsymbol{\theta}|\mathbf{x}) = \ln f_{\mathbf{x}|\boldsymbol{\theta}}(\mathbf{x}|\boldsymbol{\theta}) + \ln f_{\boldsymbol{\theta}}(\boldsymbol{\theta}) - f_{\mathbf{x}}(\mathbf{x}),$$

where $f_{\mathbf{x}}(\mathbf{x})$ do not depends from $\boldsymbol{\theta}$, so we can write

$$\boldsymbol{\theta}_{\mathrm{MAP}} \triangleq \left\{ \boldsymbol{\theta} \therefore \frac{\partial \left(\ln f_{\mathbf{x}|\boldsymbol{\theta}}(\mathbf{x}|\boldsymbol{\theta}) + \ln f_{\boldsymbol{\theta}}(\boldsymbol{\theta}) \right)}{\partial \theta[n]} = 0, \quad \text{for } n = 0, 1, \ldots, L-1 \right\}. \tag{C.169}$$

Finally, the solution of the above simultaneous equations consists in the MAP estimation.

ML Estimation

In ML estimation, the likelihood function is $L_\theta = f_{\mathbf{x};\boldsymbol{\theta}}(\mathbf{x};\boldsymbol{\theta})$ or, equivalently, its logarithm is $\ln L_\theta = \ln f_{\mathbf{x},\boldsymbol{\theta}}(\mathbf{x};\boldsymbol{\theta})$. Its maximum is defined as

$$\boldsymbol{\theta}_{\mathrm{ML}} \triangleq \left\{ \boldsymbol{\theta} \therefore \frac{\partial \ln f_{\mathbf{x},\boldsymbol{\theta}}(\mathbf{x};\boldsymbol{\theta})}{\partial \theta[n]} = 0, \quad \text{for } n = 0, 1, \ldots, L-1 \right\}. \tag{C.170}$$

C.3.2.6 Variance Lower Bound: Cramér–Rao Lower Bound

A very important issue of estimation theory concerns the existence of the *lower limit of variance* of the MVU estimator. This limit, known in the literature as the Cramér–Rao lower bound (CRLB) (also known as the *Cramér–Rao inequality* or *information inequality*), in honor of the mathematicians: Harald Cramér and Calyampudi Radhakrishna Rao, who first derived this limit [23], expresses the minimum value of variance that can be achieved in the estimation of a vector of deterministic parameters $\boldsymbol{\theta}$.

For the determination of the limit we consider a classical estimator and a vector of RVs $\mathbf{x}(\zeta) = \begin{bmatrix} x_0(\zeta) & x_1(\zeta) & \cdots & x_{N-1}(\zeta) \end{bmatrix}^T$, and a unbiased estimator $\hat{\boldsymbol{\theta}} = h(\mathbf{x})$, such that, by definition $E\left(\boldsymbol{\theta} - \hat{\boldsymbol{\theta}}\right) = 0$, also characterized by the covariance matrix $\mathbf{C}_{\boldsymbol{\theta}}$ $(L \times L)$ defined as [see (C.38)]

$$\mathbf{C}_{\boldsymbol{\theta}} = \operatorname{cov}(\hat{\boldsymbol{\theta}}) = E\left[(\boldsymbol{\theta} - \hat{\boldsymbol{\theta}})(\boldsymbol{\theta} - \hat{\boldsymbol{\theta}})^T\right]. \tag{C.171}$$

Moreover, define the *Fisher information* matrix $\mathbf{J} \in \mathbb{R}^{(L \times L)}$, whose elements are[12]

$$J(i,j) = -E\left[\frac{\partial^2 \ln f_{\mathbf{x},\boldsymbol{\theta}}(\mathbf{x};\boldsymbol{\theta})}{\partial\theta[i]\partial\theta[j]}\right], \quad \text{for } i,j = 0, 1, \ldots, L-1. \tag{C.172}$$

The CRLB is defined by the inequality

$$\mathbf{C}_{\boldsymbol{\theta}} \geq \mathbf{J}^{-1}. \tag{C.173}$$

The above indicates that *the variance of the estimator cannot exceed the inverse of the amount of information contained in the random vector* **x**. In other words, inequality (C.173) expresses the lower limit of variance obtained from an unbiased estimator for a vector of parameters **x**.

As defined in Sect. C.3.1.6, an estimator with this property, in the sense of equality (C.173), is a *Minimum Variance Unbiased* (MVU) estimator. Note that (C.173) can be interpreted as $[\mathbf{C}_{\boldsymbol{\theta}} - \mathbf{J}^{-1}] \geq 0$ (positive semi-definite). An estimator which has the property (C.173), in the sense of equality, is fully efficient.

Equation (C.173) expresses a general condition for the limit of the covariance matrix of the parameters. Sometimes, it is useful to limit the individual parameters variances of the estimate: this corresponds to the diagonal elements of the matrix $[\mathbf{C}_{\boldsymbol{\theta}} - \mathbf{J}^{-1}]$. It follows that the diagonal elements of the matrix are nonnegative, i.e.,

$$\operatorname{var}\left(\theta[i]\right) \geq \frac{1}{J(i,i)}, \quad \text{for } i = 0, 1, \ldots, L-1 \tag{C.174}$$

from which we have that

$$\operatorname{var}\left(\hat{\theta}\right) \geq \frac{1}{-E\left[\frac{\partial^2 \left[\ln f_{\mathbf{x},\theta}(\mathbf{x};\theta)\right]}{\partial\theta^2}\right]} \tag{C.175}$$

or

[12] The *Fisher information* is defined as variance of the derivative associated with the likelihood function logarithmic. The Fisher information can be interpreted as the amount of information carried by an observable RV **x**, related to a nonobservable parameter θ, upon which the *likelihood function* of θ, $L_\theta = f_{\mathbf{x};\theta}(\mathbf{x};\theta)$, depends.

$$\operatorname{var}(\hat{\theta}) \geq \frac{1}{E\left[\left(\frac{\partial \ln f_{\mathbf{x},\theta}(\mathbf{x};\theta)}{\partial \theta}\right)^2\right]} \tag{C.176}$$

which represents an equivalent form of the CRLB.

Proof We have that

$$\begin{aligned}\frac{\partial^2 \ln f_{\mathbf{x},\theta}(\mathbf{x};\theta)}{\partial \theta^2} &= \frac{\frac{\partial^2}{\partial \theta^2}\ln f_{\mathbf{x},\theta}(\mathbf{x};\theta)}{f_{\mathbf{x},\theta}(\mathbf{x};\theta)} - \left(\frac{\frac{\partial}{\partial \theta}\ln f_{\mathbf{x},\theta}(\mathbf{x};\theta)}{f_{\mathbf{x},\theta}(\mathbf{x};\theta)}\right)^2 \\ &= \frac{\frac{\partial^2}{\partial \theta^2}\ln f_{\mathbf{x},\theta}(\mathbf{x};\theta)}{f_{\mathbf{x},\theta}(\mathbf{x};\theta)} - \left(\frac{\partial \ln f_{\mathbf{x},\theta}(\mathbf{x};\theta)}{\partial \theta}\right)^2\end{aligned}$$

since $\int \frac{\partial \ln f(x;\theta)}{\partial \theta} f(x;\theta)dx = \int \frac{\partial f(x;\theta)}{\partial \theta}dx = \frac{\partial}{\partial \theta}\int f(x;\theta)dx = \frac{\partial}{\partial \theta}1 = 0$, we get

$$E\left(\frac{\frac{\partial^2}{\partial \theta^2}\ln f_{\mathbf{x},\theta}(\mathbf{x};\theta)}{f_{\mathbf{x},\theta}(\mathbf{x};\theta)}\right) = \ldots = \frac{\partial^2}{\partial \theta^2}\cdot \int f(x;\theta)dx = \frac{\partial^2}{\partial \theta^2}\cdot 1 = 0.$$

Therefore

$$E\left[\left(\frac{\partial \ln f_{\mathbf{x},\theta}(\mathbf{x};\theta)}{\partial \theta}\right)^2\right] = -E\left[\frac{\partial^2 \ln f_{\mathbf{x},\theta}(\mathbf{x};\theta)}{\partial \theta^2}\right]$$

Q.E.D.

Remark The CRLB expresses the minimum error variance of the estimator $h(\mathbf{x})$ of θ in terms of the pdf $f_{\mathbf{x};\theta}(\mathbf{x};\theta)$ of the observations $\mathbf{x}$. So any unbiased estimator has an error variance greater than the CRLB.

Example As an example, consider the ML estimator for a single observation already studied in Sect. C.3.2.3, where we have [see (C.154)]

$$\ln L_\theta = \ln f_{x;\theta}(x;\theta) = -\frac{1}{2}\ln\left(2\pi\sigma_w^2\right) - \frac{1}{2\sigma_w^2}(x-\theta)^2.$$

From (C.176), the CRLB is

$$\operatorname{var}(\hat{\theta}) \geq \frac{1}{-E\left[\frac{\partial^2\left[\ln f_{\mathbf{x},\theta}(\mathbf{x};\theta)\right]}{\partial \theta^2}\right]} = \frac{1}{E\left[\frac{\partial^2}{\partial \theta^2}\left(\frac{1}{2\sigma_w^2}(x-\theta)^2\right)\right]}. \tag{C.177}$$

Simplifying it is noted that the CRLB is given by the simple relationship

$$\mathrm{var}(\hat{\theta}) \geq \sigma_w^2. \tag{C.178}$$

The lower limit coincides with the ML estimator variance and, in this case, one can conclude that the ML estimator reaches the CRLB on a finite set of N observations.

C.3.2.7 Minimum Mean Squares Error Estimator

Suppose we want to estimate the parameter θ using a single measure x, such that the mean squares error defined in (C.127) is minimized. Let $\hat{\theta} = h(x)$, it appears that $\mathrm{mse}(\hat{\theta}) = E\left\{|\hat{\theta} - \theta|^2\right\}$; so, we have

$$\mathrm{mse}(\hat{\theta}) = E\left\{|h(x) - \theta|^2\right\}. \tag{C.179}$$

The expected value of the latter can be rewritten as

$$\mathrm{mse}(\hat{\theta}) = \int_{-\infty}^{\infty}\int_{-\infty}^{\infty} |h(x) - \theta|^2 f_{x,\theta}(x,\theta)d\theta dx. \tag{C.180}$$

Remember that the joint pdf $f_{x,\theta}(x,\theta)$ can be expanded as

$$f_{x,\theta}(x,\theta) = f_{x,\theta}(\theta|x)f_x(x). \tag{C.181}$$

Then, we obtain

$$\mathrm{mse}(\hat{\theta}) = \int_{-\infty}^{\infty} f_x(x)\left[\int_{-\infty}^{\infty} |h(x) - \theta|^2 f_{x,\theta}(\theta|x)d\theta\right]dx. \tag{C.182}$$

In the previous expression, both integrals are positive everywhere (by pdf definition). Moreover, the external integral is fully independent from the function $h(x)$. It follows that the minimization of the (C.182) is equivalent to the minimization of the internal integral

$$\int_{-\infty}^{\infty} |h(x) - \theta|^2 f_{x,\theta}(\theta|x)d\theta. \tag{C.183}$$

Differentiating with respect to $h(x)$ and setting to zero

$$2h'(x)\int_{-\infty}^{\infty} |h(x) - \theta| f_{x,\theta}(\theta|x)d\theta = 0$$

or

$$h(x)\int_{-\infty}^{\infty} f_{x,\theta}(\theta|\mathrm{x})d\theta = \int_{-\infty}^{\infty} \theta f_{x,\theta}(\theta|x)d\theta.$$

by definition we have that $\int_{-\infty}^{\infty} f_{x,\theta}(\theta|x)d\theta = 1$ which is

$$\theta_{\mathrm{MMSE}} = h(x) \triangleq \int_{-\infty}^{\infty} \theta f_{x,\theta}(\theta|x)d\theta = E(\theta|x). \tag{C.184}$$

The MMSE estimator is obtained when the function $h(x)$ is equal to the expectation of θ conditioned to the data x. Moreover, note that differently from MAP and ML, the MMSE estimator requires knowledge of the conditioned expected value of the *a posteriori* pdf but does not require its explicit knowledge. The $\theta_{\mathrm{MMSE}} = E(\theta|x)$ is, in general, a nonlinear function of the data. An important exception is when the *a posteriori* pdf is Gaussian. In this case, in fact, θ_{MMSE} became a linear function of x.

It is interesting to compare the MAP estimator described above and the MMSE. The two estimators consider the parameter to estimate θ an RV for which both can be considered Bayesian. Both also produce estimates based on *a posteriori* pdf of θ and the distinction between the two is the optimization criteria. The MAP takes the maximum (peak) of the function while on the MMSE criterion considers the expected value. Moreover, note that for symmetrical density, the peak and the expected value (and thus the MAP and MMSE) coincide, and note also that this class includes the most common class of Gaussian *a posteriori* density.

Comparing classical and Bayesian estimators we observe that in the former case, quality is defined in terms of bias, consistency, and efficiency, etc. In Bayesian estimation of the θ RV implies the non-appropriateness of these indicators: the performance is evaluated in terms of *cost function* such as in (C.182). Note that the MMSE cost function is not the only possible choice. In terms of principle, you can choose other features such as, for example, the *minimum absolute value* or *Minimum Absolute Error* (MAE)

$$\mathrm{mae}(\hat{\theta}) = E\big(|h(x) - \theta|\big). \tag{C.185}$$

Indeed, the MAP estimator can be derived from different forms of cost function. The optimal estimator in the sense MAE coincides with the median of the *a posteriori* density. For symmetric density, the MAE coincides with the MMSE and the MAP. In the case of unimodal symmetric density, optimal solution can be obtained with a wide class of cost functions that, moreover, coincides with the solution θ_{MMSE}.

Finally, note that in the case of multivariate density, expression (C.184) can be generalized as

$$\boldsymbol{\theta}_{\mathrm{MMSE}} = E(\boldsymbol{\theta}|\mathbf{x}). \tag{C.186}$$

C.3.2.8 Linear MMSE Estimator

The expression of the MMSE estimator (C.184) or (C.186), as noted in the previous paragraph, is generally nonlinear. Suppose, now, to impose the form of the MMSE estimator, the constraint of linearity with respect to the observed data **x**. With this constraint, the estimator consists of a simple linear combination of measures. It, therefore, assumes the form

$$\theta^*_{\text{MMSE}} = h(\mathbf{x}) \triangleq \sum_{i=0}^{N-1} h_i \cdot x[i] = \mathbf{h}^T\mathbf{x}, \tag{C.187}$$

where the coefficients **h** are the weights that can be determined by the minimization of the mean squares error, defined as

$$\mathbf{h}_{\text{opt}} \triangleq \left(\mathbf{h} \therefore \frac{\partial}{\partial \mathbf{h}}\left\{E\left|\theta - \mathbf{h}^T\mathbf{x}\right|^2\right\} = 0\right). \tag{C.188}$$

For the derivative computation it is convenient to define the quantity "*error*" as

$$e = \theta - \theta^*_{\text{MMSE}} = \theta - \mathbf{h}^T\mathbf{x} \tag{C.189}$$

and, using previous definition, it is possible express the mean squares error as a function of the estimator parameters **h** as

$$J(\mathbf{h}) \triangleq E\{e^2\} = E\left\{\left|\theta - \mathbf{h}^T\mathbf{x}\right|^2\right\}. \tag{C.190}$$

With previous positions, the derivative of (C.188) is

$$\frac{\partial J(\mathbf{h})}{\partial \mathbf{h}} = \frac{\partial E\left\{|e|^2\right\}}{\partial \mathbf{h}} = 2e\frac{\partial E\{\theta - \mathbf{h}^T\mathbf{x}\}}{\partial \mathbf{h}} = -2e\mathbf{x}. \tag{C.191}$$

The optimal solution can be computed for $\left(\partial J(\mathbf{h})/\partial \mathbf{h}\right) = 0$, which is

$$E\{e \cdot \mathbf{x}\} = 0. \tag{C.192}$$

The above expression indicates that at best solution point, there is the orthogonality between the error e and the vector of data **x** (measures). In other words, (C.192) expresses the *principle of orthogonality* that represents a fundamental property of the linear MMSE estimation approach.

C.3.2.9 Example: Signal Estimation

We extend, now, the concepts presented in the preceding paragraphs to the estimation of signals defined as time sequences.

With this assumption the vector of measured data is represented by the sequence $\mathbf{x} = \begin{bmatrix} x[n] & x[n-1] & \cdots & x[n-N+1] \end{bmatrix}^T$, while the vector of parameters to be estimated is another sequence, in this context called *desired signal*, indicated as $\mathbf{d} = \begin{bmatrix} d[n] & d[n-1] & \cdots & d[n-L+1] \end{bmatrix}^T$. In this situation, the estimator is defined by the operator

$$\hat{\mathbf{d}} = T\{\mathbf{x}\}. \tag{C.193}$$

In other words, $T\{\cdot\}$ *maps* the sequence $\mathbf{x}$ to another sequence $\hat{\mathbf{d}}$.

For such problem the estimators MAP, ML, MMSE, and linear MMSE are defined as follows:

1. MAP

$$\arg\max f_{\mathbf{x}|\mathbf{d}}\big(d[n]|x[n]\big), \tag{C.194}$$

2. ML

$$\arg\ max f_{\mathbf{x};\mathbf{d}}\big(d[n];x[n]\big), \tag{C.195}$$

3. MMSE

$$\hat{d}\,[n] = E\big\{d[n]|x[n]\big\}, \tag{C.196}$$

4. Linear MMSE

$$\hat{d}\,[n] = \mathbf{h}^T\mathbf{x}. \tag{C.197}$$

Comparing the four procedures we can say that the linear MMSE estimator, while it is the less general, has the simplest implementative form. In fact, the methods 1.–3. require the explicit knowledge of the density of signals (and parameters to estimate) or, at least, conditional expectations. The linear MMSE, however, can be obtained only by knowledge of the second-order moments (acf, ccf) of the data and parameters and, even if they are not known, these could easily be estimated directly from data. As another strong point of the linear MMSE method, note that the structure of the operator $T\{\cdot\}$ has the form of a convolution (inner or dot product) and it takes the form of an FIR filter; so we have

$$\hat{d}[n] = \sum_{k=0}^{M-1} w[k]x[n-k] = \mathbf{w}^T\mathbf{x} \tag{C.198}$$

for which the parameters **h** in (C.197) are replaced with the coefficients of the linear FIR filter **w**. This solution, which happens to be one of the best and most widely used in *adaptive signal processing*, is, also, extended to many *artificial neural networks* architectures.

C.3.3 Stochastic Models

An extremely powerful paradigm, useful for statistic characterization of many types of time series, is to consider a stochastic sequence as the output of a linear time-invariant filter whose input is white noise sequence. This type of random sequence is defined as *linear stochastic process*. For stationary sequences this model is general and the following theorem holds.

C.3.3.1 Wold Theorem

A stationary random sequence $x[n]$ that can be represented as an output of a causal, stable, time-invariant filter, characterized by the impulse response $h[n]$, for white noise input $\eta[n]$,

$$x[n] = \sum_{k=0}^{\infty} h[k]\eta[n-k], \tag{C.199}$$

is defined as *linear stochastic process*.

Moreover, let $H(e^{j\omega})$ be the frequency response of the $h[n]$ [see (C.120)], the *Power Spectral Density* (PSD) of $x[n]$ is defined as

$$R_{xx}\left(e^{j\omega}\right) = \left|H\left(e^{j\omega}\right)\right|^2\sigma_\eta^2, \tag{C.200}$$

where σ_η^2 represents the variance (the power) of the white noise $\eta[n]$.

C.3.3.2 Autoregressive Model

The *autoregressive* (AR) time-series model is characterized by the following difference equation:

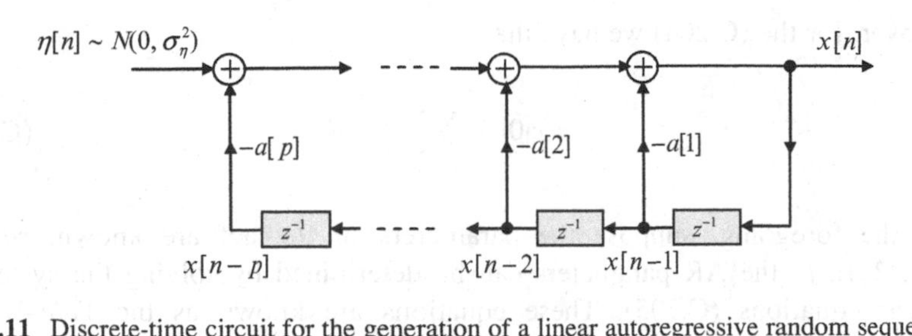

Fig. C.11 Discrete-time circuit for the generation of a linear autoregressive random sequence

$$x[n] = -\sum_{k=1}^{p} a[k]x[n-k] + \eta[n], \tag{C.201}$$

which defines the pth-order autoregressive model that is indicated as AR(p). The filter coefficients $\mathbf{a} = \begin{bmatrix} a_1 & a_2 & \cdots & a_p \end{bmatrix}^T$ are called autoregressive parameters.

The frequency response of the AR filter is

$$H(e^{j\omega}) = \frac{1}{1 + \sum_{k=1}^{p} a[k]e^{-j\omega k}} \tag{C.202}$$

so it is an *all-pole* filter. Therefore, the PSD of the process is (Fig. C.11)

$$R_{xx}(e^{j\omega}) = \frac{\sigma_\eta^2}{\left| 1 + \sum_{k-1}^{p} a[k]e^{-j\omega k} \right|^2}. \tag{C.203}$$

Moreover, it is easy to show that the acf of an AR(p) model satisfies the following difference equation:

$$r[k] = \begin{cases} -\sum_{l=1}^{p} a[l]r[k-l] & k \geq l \\ -\sum_{l=1}^{p} a[l]r[l] + \sigma_\eta^2 & k = 0. \end{cases} \tag{C.204}$$

Note that the latter can be written in matrix form as

$$\begin{bmatrix} r[0] & r[1] & \cdots & r[p-1] \\ r[1] & r[0] & \cdots & r[p-2] \\ \vdots & \vdots & \ddots & \vdots \\ r[p-1] & r[p-2] & \cdots & r[0] \end{bmatrix} \begin{bmatrix} a[1] \\ a[2] \\ \vdots \\ a[p] \end{bmatrix} = - \begin{bmatrix} r[1] \\ r[2] \\ \vdots \\ r[p] \end{bmatrix}. \tag{C.205}$$

Moreover, for the (C.204) we have that

$$\sigma_\eta^2 = r[0] + \sum_{k=1}^{p} a[k]r[k]. \tag{C.206}$$

From the foregoing, suppose the parameters of the acf are known, $r[k]$ for $k = 1, 2,\ldots, p$, the AR parameters can be determined by solving the system of p linear equations (C.205). These equations are known as the *Yule–Walker equations*.

Example: First-Order AR process: Markov Process Consider a first-order AR process in which, for simplicity of exposition, it is assumed $a = -a[1]$, we have that

$$x[n] = ax[n-1] + \eta[n] \quad n \geq 0,\ x[-1] = 0. \tag{C.207}$$

The TF has a single pole $H(z) = 1/(1 - az^{-1})$. For the (C.204)

$$r[k] = \begin{cases} ar[k-1] & k \geq 1 \\ ar[1] + \sigma_\eta^2 & k = 0 \end{cases}, \tag{C.208}$$

which can be solved as

$$r[k] = r[0]a^k \quad k > 0. \tag{C.209}$$

Hence from (C.206) we have that

$$\sigma_\eta^2 = r[0] - ar[1]. \tag{C.210}$$

It is possible to derive the acf in function of the parameter a as

$$r[k] = \frac{\sigma_\eta^2}{1-a^2}a^k. \tag{C.211}$$

The process generated with the (C.207) is typically defined as *first-order Markov stochastic process* (Markov-I model). In this case, the AR filter has an impulse response that decreases geometrically with a rate a determined by the position of the pole on the z-plane.

Narrowband First-Order Markov Process with Unitary Variance

Usually, the measurement of the performance of adaptive algorithms is made with narrowband unit variance SP. Very often, these SPs are generated with Eq. (C.207) for values of a very close to 1, i.e., $0 \ll a < 1$.

In addition, from (C.211), to have a $x[n]$ process with unit variance, it is sufficient that the input GWN has a variance equal to $1 - a^2$. In other words for $\eta[n] = N(0,1)$, it is sufficient to have a TF $H(z) = \sqrt{1-a^2}/(1 - az^{-1})$ which corresponds to a difference equation

$$x[n] = ax[n-1] + \sqrt{1-a^2}\eta[n] \quad n \geq 0,\ x[-1] = 0. \tag{C.212}$$

In this case the acf is $r[k] = \sigma_\eta^2 a^k$ for $k = 0, 1, \ldots, M$, so the autocorrelation matrix is

$$\mathbf{R}_{xx} = \sigma_\eta^2 \begin{bmatrix} 1 & a & a^2 & \cdots & a^{M-1} \\ a & 1 & a & \cdots & a^{M-2} \\ a^2 & a & 1 & \cdots & \vdots \\ \vdots & \vdots & \vdots & \ddots & a \\ a^{M-1} & a^{M-2} & \cdots & a & 1 \end{bmatrix}. \tag{C.213}$$

For example, in case $M = 2$, the condition number of the $\mathbf{R}_{xx}$, given by the ratio between maximum and minimum eigenvalue, is equal to[13]

$$\chi(\mathbf{R}_{xx}) = \frac{1+a}{1-a} \tag{C.214}$$

for which, in order to test the algorithms under extreme conditions, it is possible to generate a process with predetermined value of the condition number. In fact solving the latter for a, we get

$$a = \frac{\chi(\mathbf{R}_{xx}) - 1}{\chi(\mathbf{R}_{xx}) + 1}. \tag{C.215}$$

C.3.3.3 Moving Average Model

The *moving average* (MA) time-series model is characterized by the following difference equation:

$$x[n] = \sum_{k=0}^{q} b[k]\eta[n-k] \tag{C.216}$$

which defines the order q moving average model, indicated as MA(q). The coefficients of the filter $\mathbf{b} = \begin{bmatrix} b_0 & b_1 & \cdots & b_q \end{bmatrix}^T$ are called moving average parameters. The scheme of the moving average circuit model is illustrated in Fig. C.12.

---

[13] $p(\lambda) = \det\begin{bmatrix} 1-\lambda & a \\ a & 1-\lambda \end{bmatrix} = \lambda^2 - 2\lambda + (1 - a^2)$. For which $\lambda_{1,2} = 1 \pm a$.

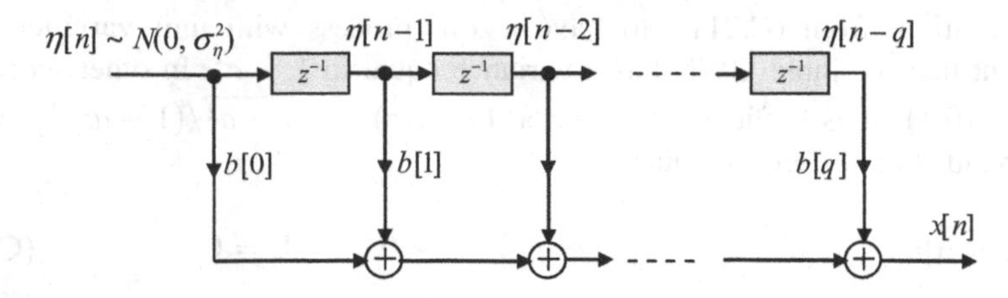

Fig. C.12 Discrete-time circuit for the generation of a linear moving average random sequence

The frequency response of the filter is

$$H\left(e^{j\omega}\right) = \sum_{k=0}^{q} b[k]e^{-j\omega k}. \tag{C.217}$$

The filter has a multiple pole in the origin and is characterized only by zeros. The PSD of the process is

$$R_{xx}\left(e^{j\omega}\right) = \sigma_{\eta}^{2}\left|\sum_{k=0}^{q} b[k]e^{-j\omega k}\right|^{2}. \tag{C.218}$$

The acf of the MA(q) model is

$$r[k] = \begin{cases} \sigma_{\eta}^{2}\sum_{l=0}^{q-|k|} b[l]b[l+|k|] & |k| \le q \\ 0 & k > q. \end{cases} \tag{C.219}$$

C.3.3.4 Spectral Estimation with Autoregressive Moving Average Model

If the generation filter has poles and zeros the model is an autoregressive moving average (ARMA). Denoted by q and p, respectively, the degree of the polynomial at numerator and at the denominator of the transfer function $H(z)$, the model is indicated as ARMA(p, q). The model is then characterized by the following difference equation:

$$x[n] = -\sum_{k=1}^{p} a[k]x[n-k] + \sum_{k=0}^{q} b[k]\eta[n-k]. \tag{C.220}$$

For the PSD we have then

$$\begin{aligned} R_{xx}(e^{j\omega}) &= \sigma_{\eta}^{2}\left|H(e^{j\omega})\right|^{2} \\ &= \sigma_{\eta}^{2}\frac{\left|b_0 + b_1e^{-j\omega} + b_2e^{-j2\omega} + \cdots + b_Me^{-jq\omega}\right|^{2}}{\left|1 + a_1e^{-j\omega} + a_2e^{-j2\omega} + \cdots + a_Ne^{-jp\omega}\right|^{2}}. \end{aligned} \tag{C.221}$$

Remark The models AR, MA, or ARMA are widely used in digital signal processing applications such as in many contexts: the analysis and synthesis of signals, signals compression, signals classification, quality enhancement, etc.

The expression (C.221) defines a power spectral density, which represents an estimate of the spectrum of the signal $x[n]$. In other words, (C.221) allows the estimation of the PSD through the estimation of the parameters a and b of the model generation stochastic ARMA signal. In techniques of signal analysis such methods are referred to as parametric methods of spectral estimation [17].

References

1. Golub GH, Van Loan CF (1989) Matrix computation. John Hopkins University press, Baltimore, MD. ISBN 0-80183772-3
2. Sherman J, Morrison WJ (1950) Adjustment of an inverse matrix corresponding to a change in one element of a given matrix. Ann Math Stat 21(1):124–127
3. Fletcher R (1986) Practical methods of optimization. Wiley, New York. ISBN 0471278289
4. Nocedal J (1992) Theory of algorithms for unconstrained optimization. Acta Numerica (199):242
5. Lyapunov AM (1966) Stability of motion. Academic, New York
6. Levenberg K (1944) A method for the solution of certain problems in least squares. Quart Appl Math 2:164–168
7. Marquardt D (1963) An algorithm for least squares estimation on nonlinear parameters. SIAM J Appl Math 11:431–441
8. Tychonoff AN, Arsenin VY (1977) Solution of Ill-posed problems. Winston & Sons, Washington, DC. ISBN 0-470-99124-0
9. Broyden CG (1970) The convergence of a class of double-rank minimization algorithms. J Inst Math Appl 6:76–90
10. Goldfarb D (1970) A family of variable metric updates derived by variational means. Math Comput 24:23–26
11. Shanno DF (1970) Conditioning of quasi-Newton methods for function minimization. Mathe Comput 24:647–656
12. Magnus MR, Stiefel E (1952) Methods of conjugate gradients for solving linear systems. J Res Natl Bur Stand 49:409–436
13. Hestenes MR, Stiefel E (1952) Methods of conjugate gradients for solving linear systems. J Res Natl Bur Stand 49(6):409–436, available on-line http://nvlpubs.nist.gov/nistpubs/jres/049/6/V49.N06.A08.pdf
14. Shewchuk JR (1994) An introduction to the conjugate gradient method without the agonizing pain. School of Computer Science, Carnegie Mellon University, Pittsburgh, PA
15. Andrei N (2008) Conjugate gradient methods for large-scale unconstrained optimization scaled conjugate gradient algorithms for unconstrained optimization. Ovidius University, Constantza, on-line available on http://www.ici.ro/camo/neculai/cg.ppt
16. Papoulis A (1991) Probability, random variables, and stochastic processes, 3rd edn. McGraw-Hill, New York
17. Kay SM (1998) Fundamentals of statistical signal processing detection theory. Prentice Hall, Upper Saddle River, NJ
18. Fisher RA (1922) On the mathematical foundations of theoretical statistics. Philos Trans R Soc A 222:309–368

A. Uncini, *Fundamentals of Adaptive Signal Processing*, Signals and Communication Technology, DOI 10.1007/978-3-319-02807-1,

19. Manolakis DG, Ingle VK, Kogon SM (2005) Statistical and adaptive signal processing. Artech House, Norwood, MA
20. Widrow B, Stearns SD (1985) Adaptive signal processing. Prentice Hall ed, Englewood Cliffs, NJ
21. Sayed AH (2003) Fundamentals of adaptive filtering. IEEE Wiley Interscience, Hoboken, NJ
22. Wiener N (1949) Extrapolation, interpolation and smoothing of stationary time series, with engineering applications. Wiley, New York
23. Rao C (1994) Selected papers of C.R. Rao. In: Das Gupta S (ed), Wiley. ISBN:978-0470220917
24. Strang G (1988) Linear algebra and its applications, 3rd edn. Thomas Learning, Lakewood, CO. ISBN 0-15-551005-3
25. Petersen KB, Pedersen MS (2012) The matrix cookbook, Ver. November 15
26. Daubechies I (1988) Orthonormal bases of compactly supported wavelets. Commun Pure Appl Math 41:909–996
27. Wikipedia: http://en.wikipedia.org/wiki/Matrix_theory

Index

A. Uncini, *Fundamentals of Adaptive Signal Processing*, Signals and Communication Technology, DOI 10.1007/978-3-319-02807-1,

B

Printed in the United States
By Bookmasters